Advanced Vibrational Spectroscopy for Biomedical Applications

University of Cambridge, United Kingdom

21–23 March 2016

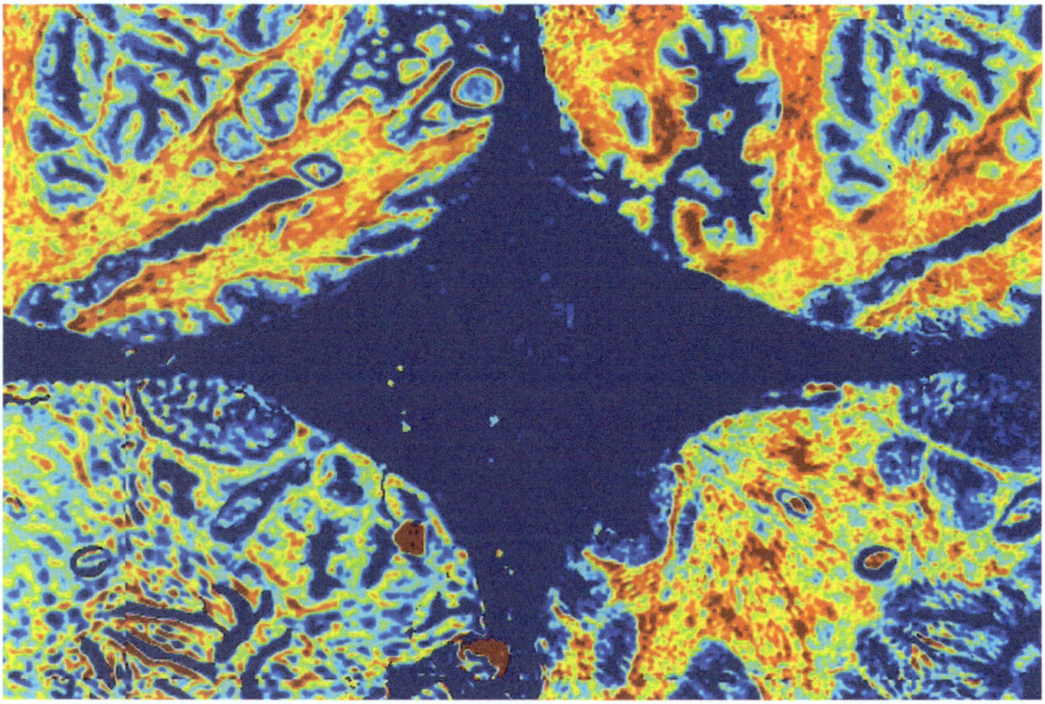

FARADAY DISCUSSIONS

Volume 187, 2016

ROYAL SOCIETY OF CHEMISTRY

The Faraday Division of the Royal Society of Chemistry, previously the Faraday Society, founded in 1903 to promote the study of sciences lying between Chemistry, Physics and Biology.

Faraday Discussions (Print ISSN 1359-6640, Electronic ISSN 1364-5498) is published 10 times a year by the Royal Society of Chemistry, Thomas Graham House, Science Park, Milton Road, Cambridge, UK CB4 0WF.

Volume 187 ISBN-13: 978-1-78262-482-0

2016 annual subscription price: print+electronic £929 US $1635; electronic only £883, US $1553. Customers in Canada will be subject to a surcharge to cover GST. Customers in the EU subscribing to the electronic version only will be charged VAT. All orders, with cheques made payable to the Royal Society of Chemistry, should be sent to RSC Order Department, Royal Society of Chemistry, Thomas Graham House, Science Park, Milton Road, Cambridge, CB4 0WF, UK.

Tel +44 (0) 1223 432398; E-mail www.orders@rsc.org

If you take an institutional subscription to any RSC journal you are entitled to free, site-wide web access to that journal. You can arrange access via Internet Protocol (IP) address at www.rsc.org/ip. Customers should make payments by cheque in sterling payable on a UK clearing bank or in US dollars payable on a US clearing bank.

Printed in the UK

MIX Paper from responsible sources FSC www.fsc.org FSC® C013604

Faraday Discussions are unique international discussion meetings that focus on rapidly developing areas of chemistry and its interfaces with other scientific disciplines.

Advanced Vibrational Spectroscopy for Biomedical Applications

Faraday Discussions

www.rsc.org/faraday_d

A General Discussion on Advanced Vibrational Spectroscopy for Biomedical Applications was held in Cambridge, UK on the 21st, 22nd and 23rd of March 2016.

CONTENTS

ISSN 1359-6640; ISBN: 978-1-78262-482-0

Cover
Front cover image courtesy of Mike Pilling and Peter Gardner, University of Manchester.

A false colour representation of infrared QCL discrete frequency chemical images of prostate biopsy tissue.

INTRODUCTORY LECTURE

PAPERS AND DISCUSSIONS

SPONSORS

Agilent Technologies

AstraZeneca

ThermoFisher SCIENTIFIC

EXHIBITORS

Princeton Instruments

see the nanoworld neaspec

LOT QuantumDesign

ANASYS INSTRUMENTS The nanoscale analysis company

Faraday Discussions

PAPER

Infrared micro-spectroscopy of human tissue: principles and future promises

Max Diem,*[ab] Ayşegül Ergin,[b] Stan Remiszewski,[b] Xinying Mu,[bc] Ali Akalin[d] and Dan Raz[e]

Received 23rd February 2016, Accepted 14th March 2016
DOI: 10.1039/c6fd00023a

This article summarizes the methods employed, and the progress achieved over the past two decades in applying vibrational (Raman and IR) micro-spectroscopy to problems of medical diagnostics and cellular biology. During this time, several research groups have verified the enormous information contained in vibrational spectra; in fact, information on protein, lipid and metabolic composition of cells and tissues can be deduced by decoding the observed vibrational spectra. This decoding process is aided by the availability of computer workstations and advanced algorithms for data analysis. Furthermore, commercial instrumentation for the fast collection of both Raman and infrared micro-spectral data has enabled the collection of images of cells and tissues based solely on vibrational spectroscopic data. The progress in the field has been manifested by a steady increase in the number and quality of publications submitted by established and new research groups in vibrational spectroscopy in the biological and biomedical arenas.

1. Introduction

The authors of this paper wish to thank the organizers of the conference "Advanced Vibrational Spectroscopy for Biomedical Applications: Faraday Discussion", held 21–23 March 2016, in Cambridge, UK, for the opportunity to present this summary of the field as it has evolved over the past 20 or so years. The main aim of this paper, aside from introducing the concepts and methodology, is to point out the enormous potential of this endeavor in the general area of medical diagnostics, and in histopathology in particular. Recent results suggest that techniques of vibrational (infrared and Raman) spectroscopy have molecular sensitivity toward detecting biochemical changes in tissue that rivals that of

[a]*Laboratory for Spectral Diagnosis (LSpD), Department of Chemistry and Chemical Biology, Northeastern University, 316 Hurtig Hall, 360 Huntington Ave, Boston, MA, USA. E-mail: m.diem@neu.edu*
[b]*Cireca Theranostics, LLC, 19 Blackstone St, Cambridge, MA, USA*
[c]*Department of Mathematics and Statistics and Program in Bioinformatics, Boston University, Boston, MA, USA*
[d]*Department of Pathology, University of Massachusetts Medical School, Worcester, MA, USA*
[e]*Division of Thoracic Surgery, City of Hope Medical Center, Duarte, CA, USA*

immuno-histo-chemistry (IHC), and require less effort, less material, and are faster and cheaper. Vibrational spectroscopic techniques detect changes in the proteome and metabolome, and even some changes in the nucleic acid composition in cancerous cells and tissues in one experiment, albeit not at the same specificity as other techniques. However, the spectral methods introduced here do maintain the *spatial* information content that is offered by classical histopathology and immunohistochemistry, but which is lost in genomic and most proteomic methods.

The motivation for using spectral methods to aid classical histopathology arises from the fact that morphological classification of certain tumors is exceedingly difficult. Some diseases, such as uterine cervical cancers, are mostly due to one particular cancer type, squamous cell carcinoma (SqCC), and their diagnosis by cytology and histopathology is well established. Furthermore, cervical cancer generally progresses slowly, has high rates of cure, and the intervention methods are well established. Other cancers, however, present a direr picture. For lung cancer, the subject of this review paper, there are three major cancer types: small cell lung cancer (SCLC), with an incidence of about 18% of all lung cancers, and adeno- (ADC) and squamous cell carcinomas, each with about 30% incidence,[1] with the rest split between large cell carcinomas and mixed types (the distribution of lung cancer types reported by different authors is somewhat variable[2]). The treatment options depend on the outcome of histopathologic diagnosis, in particular for ADC and SqCC. According to ref. 1, there is "...*relative* consensus among pathologists on the diagnosis of SCLC", which determines the choice of treatment. However, the morphological subtyping of non-small cell lung cancer (ADC and SqCC) is less reliable and may require revision in the diagnosis between biopsy and resection specimens on the same tumor. Chemotherapeutic agents that are effective against ADC, such as the bevacizumab, may have fatal pulmonary hemorrhage side effects in a SqCC patient;[3] thus, the accuracy of histopathologic diagnosis is of prime importance for patient benefit. In the past, the role of the pathologist was mostly to distinguish small cell from non-small cell carcinomas, whereas with the new treatment options available, pathology is required to differentiate ADC and SqCC as well, often on small biopsies.

Due to these new requirements, the variability in morphological diagnoses, and the need for more accurate classifications, new and more objective methods of cancer classification have been generated.[4] Among these are novel immunohistochemical approaches, such as the ones summarized in ref. 3, or modern genomic methods that sample hundreds or thousands of gene expression markers for these cancers.[1] Immuno-histochemical methods have the advantage that the information is available at a microscopic scale, and spatial information on tumor heterogeneity is available. In gene-based techniques, a larger sample volume is needed, and the microscopic information on tissue architecture or tumor structure is lost.

The techniques introduced in this paper, infrared spectral histopathology (SHP) and infrared spectral cytopathology (SCP),[5] offer the advantages of microscopic methods that require less material than even immunohistochemistry, can easily be incorporated into the workflow of standard histopathology, and render a classification that is based on biochemical composition and compositional changes, rather than on morphological criteria. These label-free techniques use inherent molecular fingerprint signatures that are well understood, and do not require any chemical agents or elaborate sample preparation schemes. SHP and SCP have been advanced by a number of research groups to a level where medical applications are forthcoming.

This review article is organized as follows. A short historical view will be offered that presents the earliest attempts to use spectral results for aiding medical diagnoses. This introduction is followed by a review of the method of action of SHP in terms of the information content, and the extraction of relevant medical information from the spectral datasets. There is still some discussion about the details of optimizing the methodology for SHP, but most research groups in the field have settled on quite similar procedures.

The success of SHP applied to a cohort of *ca.* 380 lung cancer patients will be reviewed after introducing the methodology. This study is referred to as the LC70x study according to the source of the samples (see Section 3a), and was published in detail.[6,7] At the end of this review, the continuation of this work will be introduced briefly. The aim of this continuation study, the accurate distinction between ADC and SqCC of the lung, and the sub-classification of ADC, requires hundreds of samples and data acquisition has been ongoing. This study will make extensive use of IHC for the ADC/SqCC differentiation, and the detection of certain cancer markers.

Next, the paper will present a different approach to the analysis of the LC70x data. The previously published results[6,7] used unsupervised multivariate methods to segment data as a first step of analysis, followed by classification of unknown samples by algorithms based on artificial neural networks (ANNs) or support vector machines (SVMs). These studies produced excellent average prediction accuracy (for normal/necrosis/SCLC and ADC/SqCC classification) of over 90% in spite of the heterogeneity of the input data. In this approach, the "spectroscopist's view" of data analysis was lost to some degree since datasets are analyzed mathematically and no correlation between spectral and biological changes is obtained. The next section of this paper addresses this point, and sheds light into the mode of action of SHP. This approach, for the first time, analyzes the source of spectral heterogeneity, in particular, in the case of ADC, from a biochemical viewpoint. The data heterogeneity may have imposed an upper limit on the diagnostic accuracy achieved in previous SHP studies, but the enhanced understanding of it will improve future accuracy of SHP and may shed light on the opportunity to enhance cancer diagnostics by including infrared signatures of disease as a complementary tool to overcome existing limitations in classical histopathology.

Finally, the paper will briefly summarize results from the companion methodology to SHP, namely the study of individual cells by spectral methods. This technique, referred to as spectral cytopathology (SCP), has been reviewed recently in several papers.[8–10]

2. Background and method of action of SHP

The idea of using vibrational spectroscopic methods as adjunct medical diagnostic tools dates back over 65 years to a time when infrared spectroscopy itself was in its infancy[11,12] and infrared instrumentation incorporated grating or prism-based monochromators, manual wavelength scanning and manual plotting of visually read galvanometer signals. Acquisition of a single spectrum took several hours. The paper on infrared spectral data of tissue by Blout and Mellors[11] concludes with the comment "Whether infrared techniques will allow an exact differentiation and identification of tissues must await further work". This sentence was written before protein secondary structural motifs, the DNA double-helical structure, the process of transcription and translation of the genetic code (the dogma of modern molecular biology) and the molecular origin of disease were understood or even anticipated.

Yet even then, forward-looking scientists thought of the possibility of using the biochemical information contained in vibrational spectra for medical classification and diagnosis, and to supplement the then state-of-the-art morphological information available from classical cyto- and histopathology. However, it was not until the first decade of the 21st century that the promise for spectral histopathology and spectral cytopathology became practical. Notwithstanding a flurry of review articles in the 1990s which proclaimed spectral diagnostic successes, it took over a dozen years of intense effort by several research groups to understand even the basic effects that confound the infrared spectroscopy of cells and tissues,[13,14] to develop the computational methods to detect the often minute changes in the spectra of cells and tissues with disease,[15] and to develop medically and statistically acceptable methods for the comparison of spectral and classical diagnostic results. Advances in the spectroscopic efforts were enormously aided by concomitant improvement in measurement technology in the early 2000s, and an explosive growth of computational power available to spectroscopists. Interestingly, in the eyes of the author, the increased computational power, along with the development of some fundamental theoretical underpinnings, were the most important developments to propel SCP and SHP toward widespread acceptance.

At the root of this last statement lies the fact that changes in the raw spectral data due to different states of disease, tissue types or cell biological event are generally small, and often are superimposed on much larger spectral variations that are mostly due to confounding features caused by morphological changes within the sample, and due to the heterogeneity of tissue spectra. This heterogeneity in both Raman and infrared SHP is due to the fact that the volume interrogated by the highly focused light beam is smaller than that of a cell; thus, within a distance of a few micrometers, cancerous areas, normal tissue with various degrees of inflammatory cells, blood vessels, nerve tissue and others may be found.[13] Even within a solid tumor, areas of abundant necrotic cells will cause a change in the spectral patterns, and are detectable by SHP.

The discussion of the past paragraphs thus suggests an approach to spectral diagnostic methods that relies heavily on imaging and image analysis techniques, and may be summarized as follows. Spectral data need to be collected microscopically,† either as absorption spectra in infrared-based techniques, or as scattering spectra in Raman or CARS spectroscopy, from pixel elements the size of which is determined by the diffraction limit. Hundreds of thousands or millions of pixel spectra are combined into raw "hyperspectral datasets" that conserve spatial information, and contain the entire vibrational spectrum for each pixel. The pixel spectra, along with the inherent spatial information, are subsequently

† An exception to this statement is the analysis of biofilms, or films of dried body fluids which can be analyzed macroscopically.

analyzed, most commonly at two levels. The first level of analysis is carried out by comparing an image of the conventionally stained sample to the spectral image obtained through analysis of the hyperspectral data set. This step is depicted in Fig. 1. At the second level, a comparison between the visual image (such as the hematoxylin–eosin- (H&E) or IHC-based images) with the spectral image results in the correlation between spectral features and tissue attributes such as disease state, tissue structure, presence of cancer marker, *etc.*, that allows the training of suitable algorithms to recognize spectral features and lead to a classification that is based on spectral patterns.

An example of the first level of analysis is shown in Fig. 1. Panel (a) depicts a photomicrograph of a cancer core sample, about 2 mm in diameter, that was stained by H&E and imaged using visible microscopy (see Section 3b). Prior to H&E staining, the same sample tissue, on the same substrate, was imaged in the infrared, and about 100 000 pixel spectra, each originating from a pixel about 6 μm on edge, were collected. The hyperspectral dataset was converted, after suitable spectral pre-processing, to a pseudo-color image shown in Panel (b), Fig. 1. This step was carried out in a completely un-supervised manner by a method referred to as hierarchical cluster analysis (HCA) without operator input about tissue types, disease states or sample origin, but based on spectral similarity (correlation coefficients). The spectral image agrees very well with the photomicrograph taken from the H&E stained tissue sections, and regions of connective tissue, connective tissue with inflammatory response, necrosis and small cell lung cancer are readily distinguished in the HCA-based spectral image. The labels attached to these regions were established by correlation to a pathologic diagnosis.

Thus, the infrared spectral image in Fig. 1(b) is based on multivariate analysis of an infrared hyperspectral dataset collected from the *unstained* tissue section.

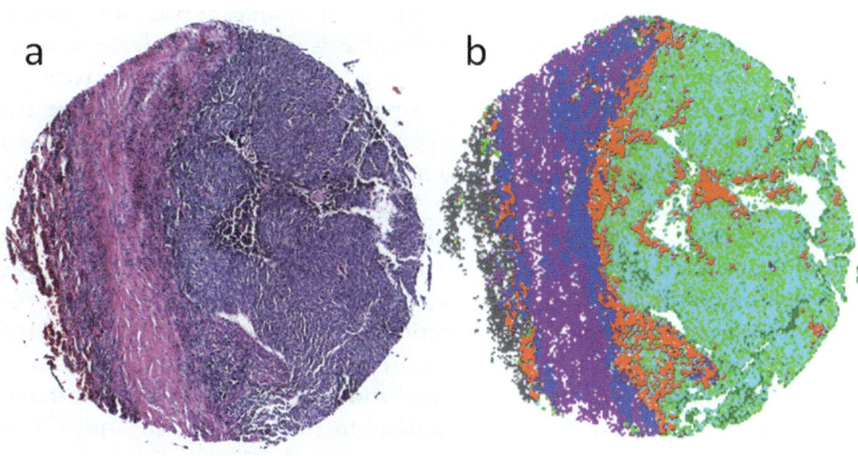

Fig. 1 Photomicrograph (a) of a H&E stained tissue section from a patient with SCLC (tissue spot A2 on TMA LC703). Normal tissue (pink region) with and without inflammatory response is visible. (b) 7 cluster infrared spectral image, based on unsupervised hierarchical cluster analysis (HCA), of the same tissue spot. Regions of connective tissue (purple), connective tissue with various inflammatory response (red and dark blue), necrosis (gray) and small cell lung cancer (light blue, light and dark green) are readily distinguished.

Each infrared spectrum covered the entire mid-infrared spectral range (800–4000 cm^{-1}) and contains the entire fingerprint of biochemical composition of the pixel from which the spectrum was collected. Thus, in contrast to classical histopathology which uses morphological descriptors of the tissue, SHP uses biochemical changes between tissue pixels to arrive at an image. Fig. 1 further demonstrates that spectra of the cancerous regions show small variations, as indicated by the different cluster colors; similarly, cancer-adjacent normal (connective) tissue may exhibit a natural variance, due to a number of factors such as inflammatory response or tissue architecture. Thus, any valid changes in the spectral characteristics of tissue must be correlated to high resolution visual images in which a pathologist "annotates", or assigns, regions of spectral characteristics to readily identifiable morphological features. An example of the second level analysis by a trained classifier is shown in Fig. 5, and will be discussed in Section 4.

Data-sets containing several hundred patients, comparable in size to the dataset reported here, have been, or are being acquired at other laboratories for lung tissue[16,17] and for different organs[18,19] or diseases. The accuracies obtained in these studies compare very favorably with the accuracy obtained in the studies presented here, and may be limited by the accuracy of the techniques against which SHP is correlated. The principles of data collection, analysis and correlation to medical and meta-data introduced below is similar in these studied; furthermore, they can equally be applied to infrared SCP, as well as data collected *via* complementary techniques, such as Raman micro-spectroscopy or CARS microscopy.

Finally, it is appropriate at this point to compare SHP and SCP to other areas of modern bio-spectroscopy. Spectroscopy of cells and tissues differs substantially from spectroscopy of biological molecules in homogeneous phases, as carried out in standard biophysical measurements (solution phase NMR, CD, infrared, Raman, *etc.*): first, the spatial heterogeneity of the observed spectra needs to be taken into account. Second, there is a significant patient-to-patient variation in the data of cancerous samples that is rarely encountered in spectroscopy of normal biological samples. Third, the spectra of tissue pixels are more complicated than the spectra of a single compound since they represent superpositions of component spectra of unknown abundance. Finally, the pixel spectra are somewhat dependent on sample morphology. The effects of sample morphology on the observed spectra is pronounced if the sample is not a homogeneous film, but consists of discrete particles, such as cells or cellular nuclei. Since the size of the nuclei or small cells is approximately the same as the wavelength of the infrared light, scattering effects will confound the observed infrared spectra.[14,20] The resulting scattering effect has been referred to as "Resonance Mie scattering" (R-Mie) by the group that first described it in detail.[14] R-Mie scattering may mix the real and imaginary parts of the infrared spectrum, thereby causing large distortions in the observed absorptive line shapes. Furthermore, reflections from the sample/air interface can have a similar effect,[20,21] and artifacts from the sample substrate may further distort the observed spectra.[22] Thus, the (infrared) spectra of cells and tissues are inherently subject to confounding effects that need to be corrected before they can be used for medical diagnostics.

3. Methods

In this section, an approach followed for the analysis of lung tissue, and published previously[6,7,23] will be presented. This approach has resulted in a dataset and algorithms that have high accuracy and high statistical significance. The methods developed by other groups in infrared SHP, in general, are quite similar, although there may be small differences in the experimental procedures. However, the similarity in classification accuracy achieved by other groups[24] suggests that the methods are stable, and not dependent on irreproducible classifier performance that have plagued very early efforts in machine learning algorithms.[25]

3a. Sample selection

The main part of this paper deals with the analysis, by different methods, of a dataset that is referred to herein as the LC70x data set. The samples were selected to provide the necessary sampling statistics for the envisioned goals. To this end, special lung cancer tissue microarrays (TMAs) were assembled at Biomax (Rockville, MD) for the lung cancer classification study.[6,7] These TMAs included about 380 patient samples on LC701 (normal tissue), LC702 (necrotic tissue), LC703 (SCLC), LC704 (SqCC), and LC705 and LC706 (ADC) (for abbreviations used, please see the list of abbreviations). In addition, non-cancerous lung lesion samples (not in TMA format) were available from the Department of Pathology, University of Massachusetts School of Medicine, Worcester, MA.

All samples were formalin-fixed and paraffin-embedded (FFPE). The use of FFPE samples for these studies has been criticized in the past by reviewers. However, the FFPE sampling and archiving procedure is by far the most prevalent method in longitudinal histopathological studies, and is widely accepted in pathology. Although formalin-fixation causes some cross-linking of proteins, FFPE tissues are widely used in immuno-histopathology which relies on the recognition of protein structure by highly structurally sensitive antibodies. Thus, one may conclude that protein structure is reasonably well preserved in spite of the fixation. In addition, since all samples used underwent the same fixation/preservation steps, comparison between these samples is a valid endeavor. Of course, FFPE and otherwise fixed tissue sections should never be compared in a SHP study. Finally, spectral differences upon fixation were quantified in recent studies,[26,27] but were found to be quite small if the sample morphology is not altered.

Tissue sections were cut, using a microtome, to a thickness of 5 μm from TMA blocks, mounted on "low-e" slides (see below) and de-paraffinized using standard protocols. After infrared data acquisition, the tissue sections were stained with hematoxylin–eosin (H&E) to permit correlation of the visual and infrared spectral images. Adjacent parallel sections, mounted on glass slides, were used for immuno-histochemical analysis.

3b. Data acquisition

A number of recent publications have dealt with the methods to collect high quality infrared spectral datasets for tissue sections.[28,29] The commercial instrument platforms available at this time were liquid nitrogen cooled focal plane array detector equipped Fourier transform (FT) infrared imaging systems using reflective (Cassegrain) objectives, or quantum cascade laser (QCL) based imaging systems using room temperature micro bolometer array detectors and refractive objectives. While the future, most likely, belongs to the latter type instruments, all data presented here were collected using the former instrumental approach. Data for the LC70x lung cancer study were collected using Perkin-Elmer Corporation's (Shelton, CT) Spectrum One/Spotlight 400 instruments at the LSpD. These instruments used linear, 8 × 2 element, photoconductive HgCdTe (MCT) detectors. Details of these studies have been published.[23] More recent data were acquired using the Agilent Corporation (Santa Clara, CA) model 620/670 FT systems, incorporating 128 × 128 pixel detectors in rapid scanning mode. The Agilent instruments provide a field of view of 704 μm × 704 μm per exposure (tile); each exposure taking about 12 minutes. In recent work, the compatibility of data collected using different instrument platforms, and instruments of the same type in different laboratories, has been demonstrated.[30]

All infrared spectra of tissue presented here were collected in transflection (reflection–absorption) mode from samples mounted on "low-e" (also known as MirrIR) slides (Kevley Technologies, Chesterland, OH, USA) at a spectral resolution of 4 cm^{-1}. The shortcomings of using reflective slides have been reported in the literature;[22] yet, for the limited spectral range employed in this study, and by using the pre-processing methods discussed below, the artifacts introduced by the transflection measurement can be minimized.[29]

Subsequent to infrared data acquisition, the tissue sections on the low-e slides were stained with hematoxylin/eosin (H&E) using standard procedures, and imaged at 20× magnification using an Olympus BX51 microscope equipped with a high resolution digital camera (Qimaging, Burnaby, BC, Canada, Model QICAM Fast 1394), a Prior Scientific, Inc. (Rockland, MA) digitally controlled microscope stage with 0.1 μm resolution, and operated under the *Surveyor* Software with Turboscan (Objective Imaging, Ltd, Cambridge, UK). A typical 2 mm diameter tissue spot yielded 72 individual tiles that were automatically stitched to yield a 2 × 2 mm image which was stored as a 100 MByte.bmp file. When imported into the annotation software (see Section 3d), the individual high resolution tiles and their coordinates were read to allow expansion to the full optical resolution of the image. Annotation of the sample, *i.e.*, the correlation between spectral features and tissue morphological features, should be carried out on the same section of tissue, which required that the tissue is first scanned *via* the infrared microscope, subsequently stained and imaged using the visible microscope. Only under these conditions can microscopic variations in tissue structure and spectral changes be properly correlated.

3c. Data preprocessing

All data manipulation and analysis was carried out, using software developed in house using the MATLAB (Mathworks, Natick, MA) environment. Although not all researchers involved in SHP utilize the same approach, the methods described in the next paragraph have been thoroughly tested and yield high signal-to-noise (S/N) spectra devoid of water vapor residuals and un-confounded by resonance Mie scattering.

Data pre-processing starts by importing the data in raw instrument data format. Each dataset consisted of between 25 000 to 30 000 pixel spectra per mm^2, depending on the pixel size of the instrument. Thus, datasets may contain from hundreds of thousands to millions of individual spectra. For the original LC70x study, 4 pixel spectra were binned to reduce the dataset size. Each spectrum covered the entire fingerprint and C–H, N–H and O–H stretching spectral regions up to 4000 cm^{-1}, yet the analysis presented below utilized spectral vectors between 800 and 1800 cm^{-1} only. The data pre-processing was performed with a program known as CIRECA_SPP (spectral pre-processor), formerly described under the name of ViChe (Vibrational Chemometrics[29]). Pre-processing included quality tests, the reduction of residual water vapor spectral features following a method reported by Bruun, *et al.*,[31] noise reduction by noise-adjusted principal component analysis (NA-PCA),[32,33] conversion to second derivative spectra[34] and correction of resonance Mie (R-Mie)-caused band shape distortion.[20] The last two steps will be discussed next in more detail.

Arguments for using second derivative spectra, rather than absorption spectra, have been presented before.[15] Since the process of taking second derivatives reduces the half width of spectral peaks, it increases the sensitivity toward detecting shoulders or second peaks that may not be apparent in the raw spectra. This is shown in Fig. 2. When computing statistical parameters, such as the correlation coefficient between two spectra, the second derivatives can distinguish between the

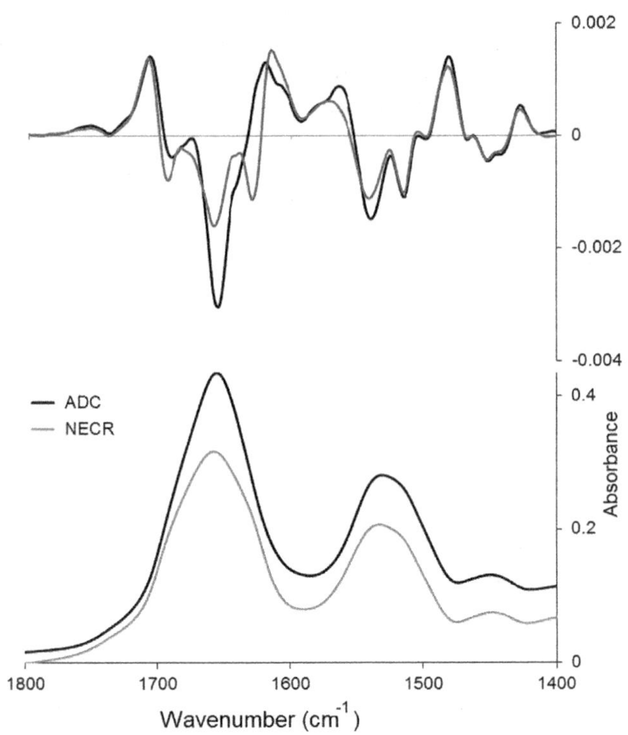

Fig. 2 Comparison between spectra from regions of ADC (black traces) and ADC with necrosis (gray traces) in the amide I/II spectral region. Top: second derivative spectra; bottom: absorption spectra. Notice the characteristic necrosis marker at *ca.* 1630 cm^{-1} which is clearly resolved in the second derivative spectrum, but only appears as a broad shoulder in the absorbance spectrum.

two spectral types shown in Fig. 2 more easily than the absorbance spectra. However, the computation of second derivatives is accompanied by a loss in S/N; thus, the noise reduction step provided by NA-PCA is essential.

The use of a phase correction methodology to un-mix the real (refractive index) and imaginary (absorption) spectral contributions that may be mixed by Mie scattering, has been described before.[20] This approach parallels the phase correction carried out within the control software of any commercial FTIR spectrometer to correct for "chirped" interferograms[20,35] and corrects all but the most distorted spectra. However, it is valid for a relatively narrow wavenumber range (*e.g.*, 800–1800 cm^{-1}), as pointed out before.[20] Incidentally, the absorption spectra shown in Fig. 2 were from phase-corrected second derivative after dual integration.

Each pre-processed pixel spectrum contains a snapshot of the biochemical composition of a cell; in addition, differences in protein secondary structure can be detected in these pixel spectra. This information is invaluable for the detection and diagnosis of diseases that result from protein mis-folding or aggregation.

The final step in the spectral pre-processing routine is the construction of pseudo-color spectral images like the one shown in Fig. 1(b). This is accomplished by a totally unsupervised method, hierarchical cluster analysis (HCA). HCA segments the pixel spectra in one dataset into clusters according to their similarity without any input from histology. The resulting pseudo-color images generally reproduce the boundaries between tissue types visible on the H&E stained images (see Fig. 1) very well, and are completely reproducible and objective. In this way, images are obtained that can be used by a pathologist for correlating morphological features and staining patterns with objective spectral features. For each dataset, 14 cluster images (2 to 15 clusters) were computed and stored as bitmap (.bmp) files.

In the original LC70x studies,[6,7,23] 4 pixels of the raw data sets were binned into one pixel. This process increased the average S/N ratio of the datasets, but reduced the spatial resolution. This was done mostly for computational reasons, since hierarchical clustering could not be carried out on un-binned datasets. More recently, all original datasets were re-processed without binning, using workstations with 24 or 48 core Intel processors and up to 256 GByte RAM. Execution time for an un-binned, 100 000 pixel spectra data set was about 10 min.

3d. Sample annotation

One of the most important steps in understanding spectral patterns that change with disease, and for the training of algorithms for the automatic classification or diagnosis of tissue sections, is the correlation process between spectral data and the corresponding histopathological assessment. This process was carried out using software known as CIRECA_Annotate, written explicitly for this task.

CIRECA_Annotate automatically registers the H&E image, and the HCA-based spectral images to permit the precise selection of regions with particular morphology in the H&E image. These regions are subsequently located in the spectral pseudo-color image. This is shown in Fig. 3. Panel A shows the H&E-stained image with the cluster demarcations indicated in red. The pathologist then draws freeform regions, ideally within the cluster boundaries, as shown by the orange lines in Fig. 3(a) and (b). Pixel spectra from the selected regions were extracted, checked for consistency with spectral clusters by a majority criterion,

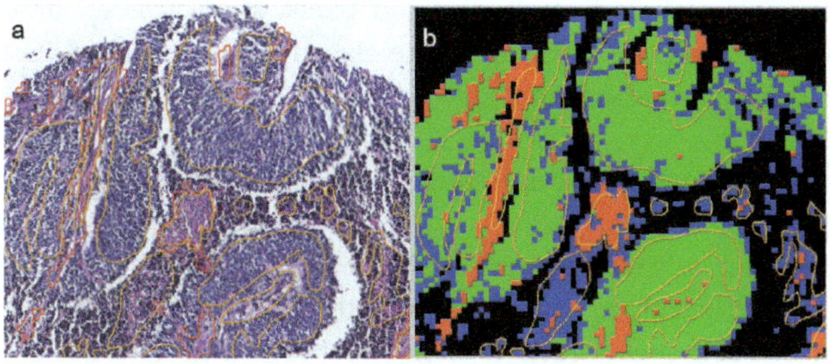

Fig. 3 (a) Image of H&E stained tissue with the cluster boundaries displayed in red. (b) Annotation regions (within orange lines) selected by the pathologist.

and subsequently associated with a disease or tissue type code. An example of the codes used in this work for adenocarcinomas is shown in Table 1; the overall list of tissue types contained over 160 different descriptors.

The pixel-by-pixel correlation afforded by this annotation method enabled complete pixel traceability, since each annotated pixel spectrum has an associated disease/tissue type code.

For tissue spots belonging to the training set (see Section 3e below), the pixel spectra were used to train the classifier to recognize specific features in the training data. For tissue spots belonging to the test (validation) set, the pixel spectra, along with their tissue codes were not used in the training, but were used to test the trained classifier to ascertain that the annotated pixel spectra were properly classified. The "vaulted" dataset was used to determine the overall accuracy of the SHP procedure.

3e. Training and testing of supervised classifiers

Pixel spectra with the same disease/tissue type code from different patients were incorporated into two spectral databases containing annotated spectra and associated descriptors. Random selection, by patient, was used for this split. One of these databases, referred to as the training set, was used for all efforts to train

Table 1 Partial list of ADC tissue subtypes and modifiers used in the annotation process

Main type	Subtype	Modifier	Code
ADC	Lepidic	Non-mucinous	31
ADC	Lepidic	Mucinous	32
ADC	Acinar	Non-mucinous	34
ADC	Acinar	Mucinous	35
ADC	Papillary	Non-mucinous	37
ADC	Papillary	Mucinous	38
ADC	Micro-papillary	Non-mucinous	40
ADC	Micro-papillary	Mucinous	41
ADC	Solid	Non-mucinous	43
ADC	Solid	Mucinous	44
ADC	Colloid		46
ADC	Enteric		49
ADC	Fetal (low grade)		52
ADC	Fetal (high grade)		55

the classifiers. The other dataset, referred to as the "vaulted" or test set, was "locked" away until all the training of the classifier was finished. Only then was the vaulted dataset opened and used to assess the accuracy of the trained classifier.

The training set itself was split into a training subset and a validation subset in order to optimize the classifier performance. This optimization included the choice of classifier, either Support Vector Machines (SVMs), Random Forests (RFs) or Artificial Neural Nets (ANNs). Once the choice was made for SVMs, they were optimized using different kernel functions and operating parameters. Details of the statistical methods have previously been reported.[7]

In order to assess the accuracy of SHP, the results of the classifier were compared to the gold standard, classical histopathology. To this end, the known diagnostic outcome of the validation subset was considered the "true" outcome, and "true positive (TP)", "false negative (FN)", "false positive (FP)", and "true negative (TN)" predictions were obtained. The sensitivity and specificity of SHP then were computed using the equations

$$\text{Sensitivity} = (\text{TP})/[(\text{TP}) + (\text{FN})]$$

and

$$\text{Specificity} = (\text{TN})/[(\text{TN}) + (\text{FP})]$$

The accuracy is defined as the mean of the sensitivity and specificity.

The resulting classifiers were applied, after testing and refinement, to the vaulted test set on a pixel-by-pixel basis, and the overall sensitivity and specificity were reported. For this result, only annotated pixel spectra in the vaulted dataset were used. The accuracy of SHP was also evaluated on a patient-to-patient basis, and the outcome was very similar to that of the pixel-to-pixel basis, and attested the stability of the spectral method.

The method described so far has been developed and tested in the author's laboratory on a variety of samples, from lung,[6,7,23] cervix[36] and sentinel breast lymph nodes,[37,38] and has produced excellent results that can be obtained fully automatically, offers complete traceability of each pixel spectrum, and is scalable between different cancer types. The approaches developed by other groups using SHP, in general, are similar, but may differ in some of the pre-processing steps, and may use different classification algorithms. Large datasets of hundreds of patients have been collected by the groups of K. Gerwert for lung[17] and colon[16] cancer, by R. Bhargava's group for prostate[39] and breast[19] cancer, and P. Gardner's group for prostate[18] cancer.

Efforts in infrared spectral cytopathology have paralleled those for tissue, and will be reviewed at the end of this paper. However, the methodologies and instrumentation used by different groups are much more variable, and the reader is referred to recent reviews of this field.[5,8,40,41] Efforts in applications of Raman spectroscopy have been very successful as well, but have been more in the area of *in vivo* detection of cancerous lesions,[42] and cytology in aqueous media.[43–45] The results in these arenas have been very impressive, but use quite different methods and will not be reviewed here.

4. SHP results for lung tissue

In this section, results from the authors' efforts for the diagnosis and prognosis of lung disease by infrared spectral methods will be presented. First, results from previously published studies will be summarized (Sections 4a and 4b) to provide an indication of the classification accuracy that can be achieved with SHP, and to indicate where improvements of the methodology were required, namely in the classification of cancer adjacent normal tissue, and the differentiation of non-small cell lung cancers. These sections also will summarize briefly the goals of the ongoing study that is being carried out in collaboration with a National Cancer Center in the US, City of Hope Medical Center, Duarte, CA.

Sections 4a and 4b are based on a strictly statistical approach to the analysis of the datasets, and ignore the "spectroscopist's view" of data analysis. This particular aspect will be presented in Section 4c, and involves a discussion of the spectral heterogeneity within the dataset, the prototypical spectral patterns of normal and diseased tissue types, and aspects of interpreting the large variations in the protein spectra of cancerous cells. This discussion should be seen as an approach to aid the multivariate data analysis, since it defines how the inherent tissue heterogeneity manifests itself as spectral heterogeneity for certain cancer classes, and helps to understand the biochemical causes of the spectral variance.

Section 4d will introduce results from efforts to pre-process data into a more manageable grouping of spectral data. These efforts were carried out by extending the principles of HCA over a large number of patients ("cross-patient" cluster analysis). This cpHCA approach revealed that there are less than *ca.* 20 types of tissue spectra in the lung dataset that are reproduced over different patients. This approach also revealed the limits of unsupervised methods, because ADC and SqCC were grouped into several clusters that could not be further classified using HCA.

4a. Review of the LC70x study – pixel based

The overall goals of the LC70x study were the distinction by SHP between normal (NOR), necrotic (NECR) and cancerous tissue at a first level, and between SCLC, SqCC and ADC at a second level. This classification was carried out in several ways that were described in detail in the original literature,[7] and are summarized below. The original paper also presented results on the dependence of the confidence interval on the number of patients required in the training dataset, see Fig. 4; thus, it laid the foundation for the presently ongoing work in terms of the required number of patients for the sub-classification of ADC. In Fig. 4, the experimental determination of the confidence interval was obtained by running 10 independent SVM training and test runs on 10 000 training spectra from five classes (NOR, NECR, SCLC, SqCC and ADC) with the number of patients contributing to the 10 000 spectra varying between 30 and 135. The overall accuracy increased, as expected, and the uncertainty decreased by a factor of 5. The observed confidence intervals agreed well with those from analytical equations.[7]

The results of the binary, hierarchical pixel-based test and the patient-based leave-one-out-cross validation (LOOCV) are shown in Table 2. The binary, hierarchical pixel-based test was carried out in the following sequence of steps: cancer

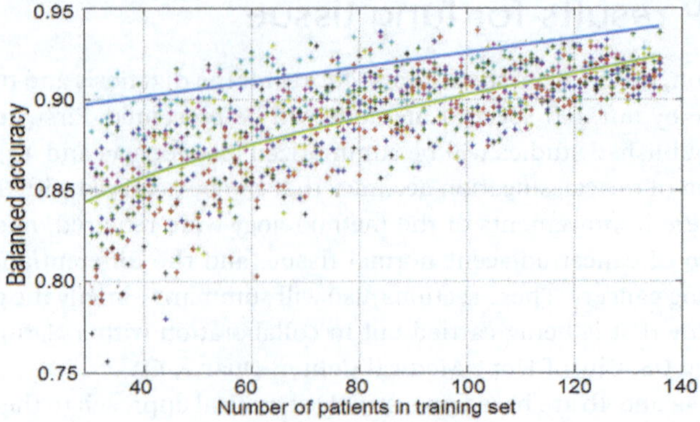

Fig. 4 Experimental simulation (dots) and theoretical balanced accuracy (blue line) and confidence interval (green line) as a function of patient number. Each of the experimental dots represents a simulation run with a fixed number of patients.

vs. normal, SCLC *vs.* not SCLC, NECR *vs.* not NECR, to SqCC *vs.* ADC. The discrimination between SqCC and ADC was achieved with somewhat lower accuracy than all other classifications steps, *ca.* 87%. This classification is also the most difficult to carry out in classical histopathology, in particular in the case of poorly differentiated carcinomas. Thus, the SHP classification accuracy may be reduced because of uncertainties in the annotation. Furthermore, these two cancer types can occur as mixed adeno-squamous carcinomas which will aggravate both the annotation process as well as the SHP classification.

Patient-based LOOCV was carried out on a subset of 170 patients, since the computations are very time consuming. LOOCV yielded classification accuracies, also shown in Table 2, that were nearly identical to those reported for the pixel-based hierarchical SVM. LOOCV is frequently used in medical statistics when the size of the dataset is small since the training of the classifier is being carried out for nearly the entire dataset. The fact that the LOOCV results here are nearly identical to those of the pixel-based entire dataset indicates that the patient number was sufficient to ascertain a statistically significant result. The classification accuracy in these studies was excellent, with an overall average accuracy better than 90%.

SVM classifiers also were trained to perform further ADC differentiation into the five major ADC subclasses (lepidic, acinar, solid, papillary and micropapillary). Due to limitations in the number of patients in these classes, this effort

Table 2 Results of the SVM-based classification of lung cancer tissue

Classification	Pixel-based hierarchical SVM Average accuracy	Patient-based LOOCV Average accuracy
Cancer *vs.* normal	98.3%	98.6%
SCLC *vs.* not SCLC	92.4%	95.5%
Necrosis *vs.* not	94.7%	94.2%
SqCC *vs.* ADC	87.5%	86.4%

has to be regarded as a pilot study; nevertheless, the overall accuracy obtained (>90%) rivals the best methods presently available for this classification task.

Next, the trained classifiers were applied to the entire tissue spots. The results here were very positive for the cancerous regions, whether they were annotated as such, or cancerous regions in close proximity to the cancerous regions but not annotated, and high patient-based accuracies on the prediction of the entire tissue spot were obtained as well.[4] However, the accuracy of classifying cancer-adjacent normal tissue regions was found to be reduced, and some cancer-adjacent normal tissue areas were predicted as cancerous tissue, in particular in the presence of inflammatory response. This is shown in Fig. 5.

The reason for this was that all non-cancerous regions, distal or proximal to cancer, had been combined into a "normal" tissue class to facilitate training of the classifiers. However, valuable information was lost in this step. The raw HCA images often contained the distinction between pure connective tissue and

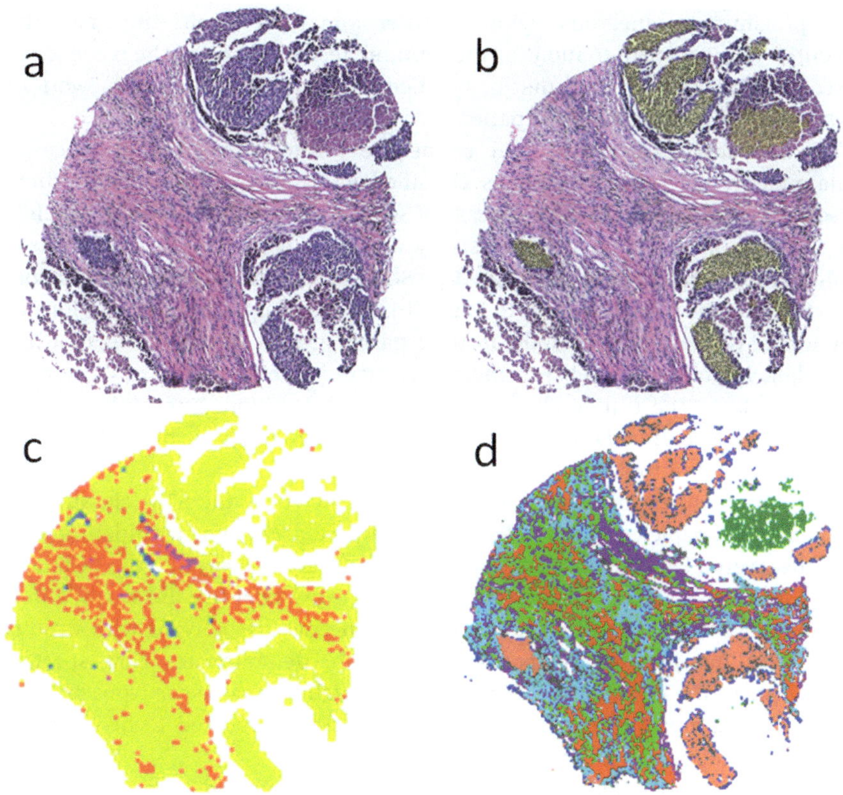

Fig. 5 Results of the trained classifier analysis (level 2 analysis) of sample C3 of TMA LC703 (SCLC). (a) Photomicrograph of H&E-stained tissue spot. (b) Superposition of annotated areas (yellow) on the H&E image. (c) Prediction image of trained classifier that misclassified some of the cancer adjacent normal tissue as cancer (see text for detail). (d) HCA image of same spot demonstrating the discrimination between SCLC (salmon), necrosis (dark green), loose connective tissue (purple), connective tissue with abundant inflammatory response (light blue), connective tissue with fibroblasts (red), and dense connective tissue (green).

connective tissue with inflammatory response (see Fig. 5). These aspects will be discussed in the next section.

4b. Review of the LC70x study – whole spot analysis

In a small pilot study, the SHP-based distinction between blood vessel walls (endothelium) in cancer adjacent (proximal) and cancer distal tissue was found to be possible with an accuracy of 90%, and the distinction of all classes of connective tissue between cancer distal and cancer proximal regions with an accuracy of 88%.[6] The sensitivity of SHP to detect such differences had been reported before,[46–48] and it appears that changes in proximal normal tissue occur due to the presence of cancerous cells. For connective tissue, the spectral differences were not surprising, given the amount of inflammatory cells and fibroblasts found in both dense and loose connective tissue surrounding cancerous tissue. These results suggested the necessity of including cancer adjacent and cancer proximal tissue types as separate classes in the training of classifiers, since spectral differences existed between tissue types depending on their proximity to cancerous regions. Furthermore, these results emphasize the potential for using SHP to monitor the immune response toward the presence of disease, an area of research that has gained attention within the framework of immunotherapy to treat cancer patients.[49]

As indicated earlier, proximal connective tissue with an abundance of inflammatory cells was sometimes classified as cancerous by the algorithms described above, particularly in the case of SCLC. This is shown in Fig. 5, which is an example of level 2 image analysis by a trained classifier. Panel (a) shows a photomicrograph of spot LC703 C3. The SCLC areas selected by the pathologist are shown as the yellow overlays in Panel (b). In this sample, the large regions of connective tissue differ markedly in their pathology: the center of the spot has abundant inflammatory cells; furthermore, there is a broad band of these cells extending from the center toward the 11 o'clock position. The pink areas of connective tissue directly next to the cancerous tissue (dark purple in Panel (a)) have embedded fibroblast.

The pure connective tissue was correctly predicted by the SVM classifier as normal tissue, depicted in red in Panel (c). However, areas of heavy inflammatory response were classified as cancerous tissue, and are shown, like the SCLC, in yellow. However, the raw cluster analysis, shown in Panel (d), did easily differentiate between SCLC, necrosis, and three connective tissue types with various degrees of inflammatory response. This figure demonstrates that spectral methods have the inherent specificity to distinguish normal tissue types by their degree of inflammatory response, but that the classifiers need to be explicitly trained to recognize these "normal" tissue classes.

These issues were addressed by alternate and improved methods of data analysis and algorithms that will be introduced in Section 4c. One of the major improvements in the data analysis was the ability to carry out the spectral pre-processing on un-binned data sets which improved the clarity of the spectral images. Furthermore, spectral homogeneity for the training of SVMs was improved by carrying out cluster analysis over different patients, and using different decision sequences for hierarchical SVMs. These improvements will be discussed next.

4c. Cluster analysis and cross-patient cluster analysis of high spatial resolution data

As expected, the cluster images of the un-binned data exhibited significantly better spatial agreement with the corresponding H&E images. The concomitant increase in computation time for cluster analysis was about ten-fold, from *ca.* 40 s for a binned, 25 000 spectra dataset to *ca.* 10 min for an un-binned, 100 000 spectra dataset. Such computations, however, require large amounts of RAM and parallel processing capabilities.

The improvement in the agreement between the HCA-based infrared images and the high resolution H&E-stained images can be gauged from Fig. 6 that depicts a 1 mm × 1 mm section of an ADC tissue sample in un-binned format. Inspection of Panel (b) reveals that the un-binned (6.25 μm pixel size) image was able to resolve individual cell nuclei in the HCA image, see arrows. This was observed for some of the non-small cell lung cancer (NSCLC) samples, where it was possible at the 5 or 6 cluster level to detect spectral differences between areas of the nuclei and cytoplasm. Such spectral differences are shown in Fig. 7 that demonstrates shifts in the nucleic acid spectral regions (*ca.* 1235 and 1085 cm^{-1}), attributed to the different distribution of DNA and RNA between the nucleus and cytoplasm. In the cytoplasm, only RNA (mostly in the form of r-RNA) is found, whereas the nucleus contains all the DNA in a cell, and small amounts of m-RNA. DNA and RNA differ in the frequency of the antisymmetric phosphate (PO_2^-) vibration which occurs in RNA at about 5 cm^{-1} higher wavenumber. Similar findings about the improvement in image quality with decreasing pixel size (below the actual diffraction limit) had been reported by Nasse, *et al.*[50]

The higher resolution, un-binned data also were used for refined unsupervised analyses of the dataset. In order to better understand the heterogeneity of tissue spectra of the same disease class between patients, a "cross-patient cluster analysis" (cpHCA) was carried out to define which spectral types were distinguishable by un-supervised methods, and thereby delineate the variance in the data. Since the entire annotated dataset encompassed millions of spectra, HCA could not be carried out due to the matrix size and computational time restrictions. *K*-means clustering was deemed to be too dependent on the seed conditions to produce reproducible results. Thus, an approach was used in which the cluster centroids and a weighting factor depending on the size of the cluster, from single patient HCA were used as inputs for cpHCA. This analysis resulted in hierarchical

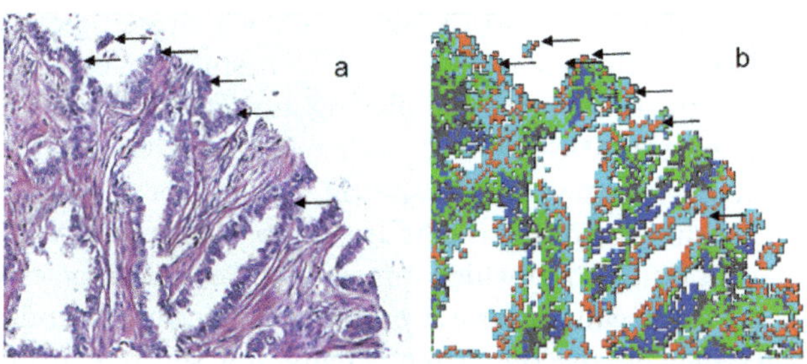

Fig. 6 (a) Photomicrograph of a H&E stained ADC tissue section (LC706 A4) and (b) HCA-based spectral images at 6.25 μm pixel size, showing cell nuclei (red clusters, see arrows).

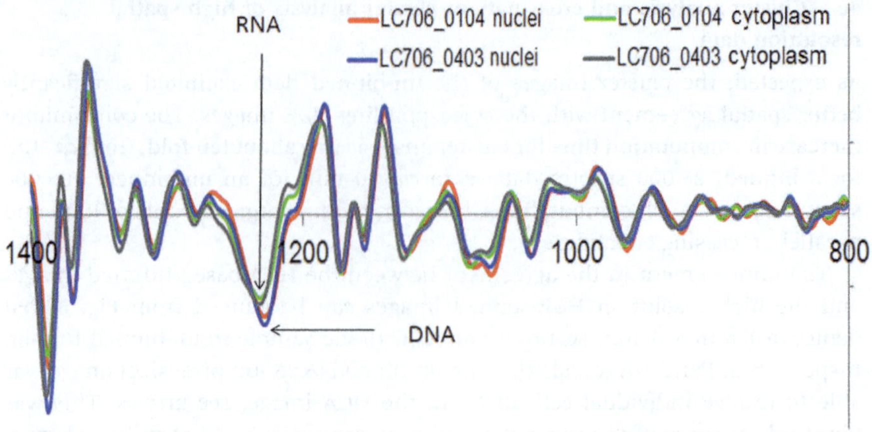

Fig. 7 Cluster mean second derivative spectra of nuclear and cytoplasmic regions in ADC samples.

clustering of cluster centroids, and provided new mean cluster spectra that were reproducible between patients. cpHCA was carried out for a subset of 49 tissue spots including the SCLC, SqCC and ADC cancer types. This analysis revealed that there are only about 20 different spectral classes that can be distinguished by unsupervised methods. Distinct classes were found to exist for normal connective tissue, loose and dense connective tissue with strong inflammatory response or fibroblast abundance, necrosis, blood components, keratin and SCLC. However, ADC and grade 3 SqCC could not be separated into unique spectral classes, but rather, appeared distributed in three clusters more or less randomly. This observation led to the conclusion that the non-small cell lung cancers, ADC and SqCC, have spectra that are either so similar, or exhibit such large heterogeneity, that they cannot be distinguished by unsupervised methods.

The difficulty in distinguishing ADC and SqCC by unsupervised methods was traced to the large heterogeneity caused by necrosis in all cancer types, and to the presence of connective tissue features in the ADC spectra (to be discussed in Section 4d). This resulted in revisions in the order (decision tree structure) of the sequential SVM classifiers. This work is presently ongoing. Another factor influencing the spectral ADC/SqCC differentiation is the fact that these cancers may occur as mixed types, such as an adeno-squamous subtypes; furthermore, there is some uncertainties in the histological diagnosis for these cancers, particularly in the case of poorly differentiated ADCs and SqCCs[1] which are notoriously difficult to diagnose by histopathology. The continuation study reported in Section 4e partially addresses the last point by employing immunohistochemistry for a reliable distinction between ADC and SqCC, whereas the next section addresses the patient-to-patient heterogeneity of the spectra patterns for all three cancer types. It is expected that the supervised methods for ADC/SqCC classification, and the ADC sub-classification (both of which already had good accuracies in the previously described studies) will improve substantially with a better understanding of the factors affecting the spectral heterogeneity and spectral mixing that were revealed by the cpHCA analysis.

4d. Mean spectral patterns for different lung cancer tissue types

As pointed out in the above section, unsupervised methods reached their limits in distinguishing spectral features from ADC and SqCC. The spectral heterogeneity of the datasets collected for these disease classes had been suspected to be the cause of the difficulties in distinguishing them, and was first pointed out in ref. 6. This heterogeneity seems to be particularly pronounced in the ADC classes, as shown in Fig. 8, which depicts principal component analysis (PCA) 'scores' plots of one particular ADC subtype, and grade 3 SqCC. Panels (a) and (c) show the clustering of all spectra, each represented by a (+) symbol, whereas Panels (b) and (d) show patient-based 95% contours and cluster centroids. Although the SqCC data contains many more spectra (over 12 000), their distribution is much more homogeneous than that of the ADC spectral data. For the ten annotated ADC subtypes, the heterogeneity within the data set was much larger.

In order to further investigate the heterogeneity of the cancer spectra, mean cluster spectra and the standard deviations between patients were calculated for the three major cancer classes. To this end, all cluster spectra were extracted that contained at least one cancer annotation region by pathology. If a patient sample contained more than one subtype (in the case of ADC) or grade (in the case of SqCC), mean spectra for all clusters were included. The spectral heterogeneity was assessed separately in the "proteome" region (the term "proteome region" implies

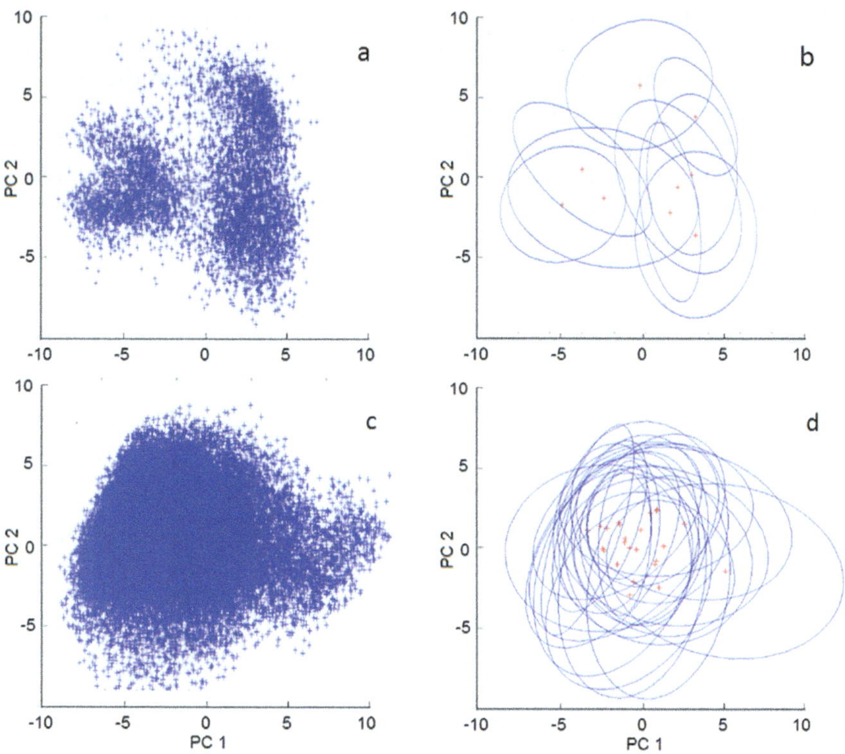

Fig. 8 PCA scores plots of all annotated spectra for lepidic, mucinous ADC and Grade 3 SqCC. See text for details.

the part of the spectrum that contains the protein amide I and amide II vibrations (1500–1750 cm^{-1})), and a "low wavenumber" region that mostly contains the phosphodiester modes of DNA and RNA.

The proteome spectral region was found to be so variable in all three cancer types that it may be of restricted value for cancer classification, but invaluable in the detection of necrosis. It appears that infrared spectral methods are enormously sensitive toward the presence of β-sheet signatures in the protein region. Since these signals are a hallmark of necrosis or apoptosis,[40,51] it appears that SHP is more sensitive than classical pathology in detecting these conditions. The presence of β-sheet protein signals, and signals of precipitated proteins (*ca.* 1690 and 1630 cm^{-1}, see Fig. 2) generally reduces the main protein peak intensity at 1655 cm^{-1}; therefore, large changes in the overall protein intensity and spectral band envelopes are observed. This is shown in Fig. 9 that separately depicts the mean patient cluster spectra for SCLC, SqCC and ADC patients. Each spectrum represents one patient. The heterogeneity of these spectra is manifested in significant variations in the amide I peak intensity at *ca.* 1655 cm^{-1} by more than a factor of 2, and the presence of a shoulder at 1630 cm^{-1} (the necrosis marker). Shifts in the main amide I peak wavenumber at 1655 cm^{-1} can furthermore be attributed to the presence of various contributions of collagen, to be discussed below. In collagen the amide I band occurs at 1648 cm^{-1}; therefore, the second derivative spectra often exhibit two resolved amide I peaks at 1655 and 1648 cm^{-1}.

The intensity variations of the main protein peak at 1655 cm^{-1} were generally correlated with the presence of prominent β-sheet features at *ca.* 1630 cm^{-1}: most patient mean spectra with low 1655 cm^{-1} intensity exhibited large 1630 cm^{-1} contribution which suggested that the total protein concentration was fairly constant, but that the protein intensity was distributed between the 1655 and 1630 cm^{-1} regions. This further suggests that SHP monitors the transition from functional proteins in their native conformation toward the misfolded (and possibly precipitated) states in apoptosis and necrosis. Thus, this spectral region is highly sensitive toward the detection of necrosis, but is less useful for the classification of the three cancer types, and may be the cause of the inability of cpHCA to distinguish ADC and SqCC. Trained classifiers mostly ignored this spectral region for distinguishing the cancer types.[23]

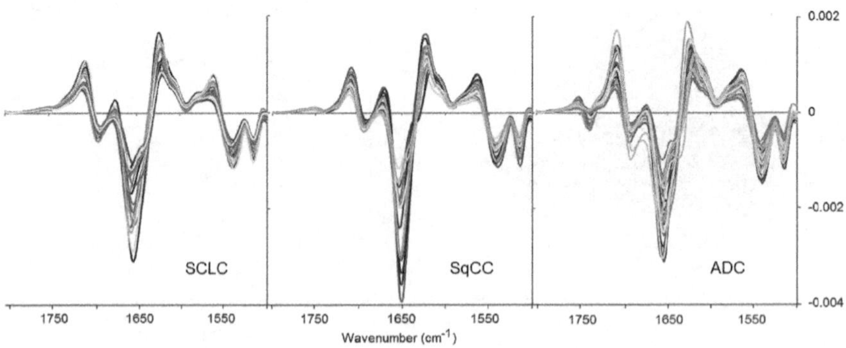

Fig. 9 Patient-to-patient spectral variations in the amide I/II region for the three main lung cancer types. See text for details.

The "low wavenumber" spectral region (800 to 1400 cm^{-1}) shows more consistent spectral patterns and less variation in intensity within each cancer type. Of the three cancer types analyzed here, SqCC showed the smallest standard deviation between patients, shown as the gray areas around the mean spectra in Fig. 10, trace (a), even though it included several different grades. Inspection of Fig. 10(a) also reveals that the mean SqCC and SCLC spectra are quite different, particularly in the "nucleic acid" spectral regions, at 1235 and 1085 cm^{-1}.

However, the mean and standard deviation spectra shown in Fig. 10(b) show much larger heterogeneity in ADC, in particular at *ca.* 1235 and 1085 cm^{-1}. This region is commonly associated with the antisymmetric and symmetric P–O stretching vibrations of the phosphodiester vibration of DNA and RNA (see Section 4c).[5] Yet a careful analysis of patient mean ADC spectra revealed that the heterogeneity in these two spectral regions could be traced to the contributions of collagen. This was an unexpected finding since collagen spectral signatures are generally confined to the region of connective tissue and the stroma, not the cancerous regions. However, in many ADC regions, the collagen contributions could easily be detected in the second derivative spectra.

The common protein secondary structural motifs, α-helical and β-sheet conformations, show little structure-specific spectral information below 1500 cm^{-1}. However, collagen, the major constituent of connective tissue and the extracellular matrix, has distinct spectral signatures in the low wavenumber region. A triad of peaks at *ca.* 1202, 1226 and 1278 cm^{-1} and a triad at 1030, 1060, and 1082 cm^{-1} are observed for all different collagen types[52] and have been assigned to the extended amide III vibration,[5] and glycosylation of the triple helix, respectively. In addition, as already mentioned, the amide I peak occurs at 1648 to 1650 cm^{-1} for collagen, sufficiently shifted from the α-helical amide I frequency (1655 cm^{-1}) that it frequently is resolved and appears as a doublet in the second

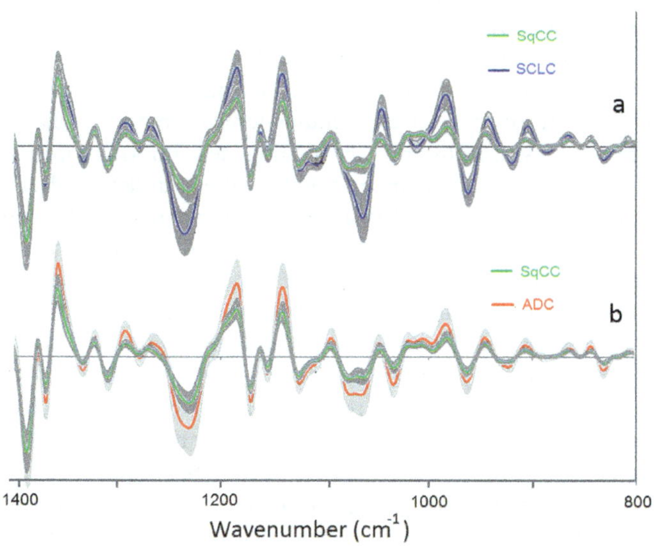

Fig. 10 Mean spectra (solid lines) and between-patient standard deviation (gray shaded areas) of SqCC and SCLC (a) and SqCC and ADC (b).

derivative spectra. Analogous results for the microenvironment of breast ADC, and the influence of collagen on the extracellular matrix, have been reported by E. Goormaghtight's group.[48]

These collagen spectral signatures contribute strongly to the heterogeneity of spectral traces for ADC, as shown by the very large standard deviation in the ADC spectra in Fig. 10. These deviations are largest at *ca.* 1240 and 1080 cm^{-1} and occur at very similar wavenumber values as the DNA/RNA bands. Many of the ADC patient mean spectra, in fact, distinctly show the collagen triad of peaks at 1200, 1226 and 1278 cm^{-1} in the second derivatives, with the strongest of them at 1226 appearing as a low frequency shoulder of the phosphodiester antisymmetric stretching vibration at 1235 cm^{-1}. The presence of collagen in lung ADC is well known in the pathology literature,[53,54] and certain subtypes of ADC were previously referred to as "scar carcinomas" due to the high abundance of scar-like connective tissue, and the abundance of fibroblasts. Although the heterogeneity created by different collagen contributions may aggravate the classification of lung ADC, the possibility of detecting collagen spectral signatures in the extracellular matrix is of high diagnostic interest since it may allow the diagnosis of fibrotic processes indicating the host response to cancer.

Thus, the heterogeneity of the ADC, SqCC and SCLC spectra in the "proteome region" was traced to the sensitivity of the spectral methods to detect necrosis, whereas the heterogeneity of the ADC spectra in the low wavenumber region was attributed to the presence of collagen within the cancerous regions. In this case, the other cause of spectral variance is most likely due to different ADC subtypes. Since the supervised SVM algorithm referred to earlier was able to distinguish the different ADC subtypes quite well, there are, apparently, spectral signatures outside the necrosis and connective tissue regions that allow for this distinction. At this time, visual inspection of the spectral patterns has not allowed pinpointing subtype-specific spectral signatures, nor any correlation between subtype and collagen abundance.

In summary, the analysis of mean patient spectra, and the variance between them, has established the following novel insights into the sensitivity of infrared SHP toward several aspects of lung disease. First, these studies solidified the understanding of the mode of action of SHP. Second, the combination of spatial and biochemical information at the cellular level permits the classification of disease stages microscopically, thereby adding new information not available from cornerstone platforms that are on the frontline to understand cancer and to design and target drug therapies. Third, SHP may prove to be an IHC-compatible, efficient alternative to complement routine proteomic and genomic methods. Fourth, SHP delivers new objective, quantitative and spatially resolved data to advance discovery programs and characterizes similarities and differences within clinical cohorts and individual samples. Fifth, the heterogeneity of the observed spectral patterns, first thought to be an obstacle to diagnostic classification, was found to have two major advantages: the possibility of detecting necrotic spectral patterns before necrosis manifests itself as a morphological change, and the sensitivity to detect the presence of collagen signatures in ADC (but not in SqCC and SCLC). The presence of collagen spectral patterns may present an additional method for the distinction of the non-small cell lung cancers that is not easily available from the inspection of H&E stained tissue. Finally, the presence of different collagen types in lung tissue may be invaluable for the diagnosis of diseases other than lung cancer, such as fibrotic diseases. Concomitant efforts in algorithm development have determined that accuracies higher than those reported earlier[7] can be achieved when the supervised algorithms are trained to account for the spectral heterogeneity due to necrosis.

4e. Lung cancer continuation study

The LC70x studies that were previously reported[6,7,23] and were discussed in Sections 4a–c, are being continued at the laboratory of Cireca, LLC, in collaboration with a National Cancer Center. The goals of this continuation study follow directly out of the discussion above: a highly accurate distinction of ADC and SqCC, and sub-classification of ADC and grading of SqCC. The need for accurate SqCC–ADC diagnosis has been discussed above and has been reviewed in the literature,[1,3,4] since the treatment of SqCC patients with ADC specific drugs can have fatal effects.

Due to the importance of accurate ADC and SqCC classification, and due to the relatively low classification accuracy in standard histopathology, in particular for poorly differentiated tumors, more accurate tests have been sought over the past decade. Although there are some distinguishing features, such as gland formation (specific for ADC) or keratinization or intercellular bridges (specific for SqCC), these criteria are subjective and non-quantitative. The use of immunohistochemical staining has increased the overall accuracy of classification to about 80% sensitivity and 90% specificity, using mainly Thyroid Transcription Factor-1 (TTF-1) and Napsin A antibodies as markers for ADC, and p40 for SqCC.[3] TTF-1 is a small protein that regulates transcription of genes specific for the lung and a few other organs. Lung ADCs are usually TTF-1 positive, while SqCCs are rarely positive. These IHC methods are presently considered to be the gold standard for distinguishing these two cancer classes, and will be used in the continuation study to confirm the pathology report, and to exploit the possibility of detecting spectral signals that may be surrogates for IHC staining.

Next, some results for the first part of the continuation study are presented. Since this study encompasses over 400 additional patients and incorporates analysis of H&E and IHC-stained slides, and the spectral datasets, its conclusion is not expected until later in 2016. However, these ongoing efforts have revealed results that point to the high specificity of SHP toward the detection of specific cancer markers. Previous results by Kallenbach-Thieltges, *et al.*[16] have demonstrated, using immunofluorescent markers for Ki-67 and p53 in colon cancer samples, that there are spectral patterns in the SHP results that co-localize with the IHC result. Fig. 11 shows high resolution IHC and SHP images that demonstrate that such co-localization can be achieved for IHC stains, and that there is a direct correlation between the two imaging modalities. The IHC and SHP images were collected from adjacent slides cut from the same tissue block. One sample was stained with a TTF-1 antibody that was subsequently visualized using standard streptavidin–biotin based methods, yielding a dark brown positive response. SHP images were collected from samples mounted on 'low-e' slides as discussed in Section 3.

Panel (a) of Fig. 11 shows the TTF-1 stained section, and panel (b) a semi-transparent overlay of the TTF-1 image and a four-cluster HCA image. It is obvious that the gray cluster in the HCA images co-localized exceedingly well with the

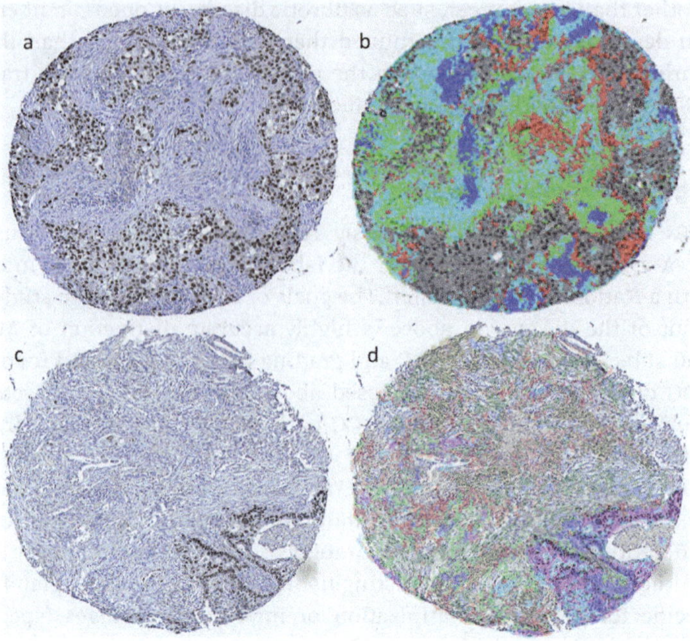

Fig. 11 Photomicrographs (a and c) of TTF-1 stained tissue samples. Semitransparent overlays (b and d) of TTF-1 and HCA images, demonstrating the co-localization of TTF-1 signatures with corresponding spectral regions.

TTF-1 positive regions. However, this image does not necessarily demonstrate that HCA detects the cancer marker TTF-1, but rather, that at the four-cluster level, HCA distinguished mainly between normal tissue, inflammation, and cancer, and that the cancerous regions interacted heavily with the TTF-1 stain. In many other tissue samples, co-localization of the TTF-1 positive regions with cluster images occurred at a much a higher cluster level. In (d), for example, the purple region that co-localized with TTF-1 positive areas only appeared at the eight-cluster level, whereas the cancerous regions were detected at the three-cluster level. If co-localization occurs at a cluster level much higher than the cancer *vs.* normal differentiation, it is likely that HCA detects some changes in the tissue composition that are a direct effect of the TTF-1 presence. In the case of TTF-1, which is a small, mostly helical protein with a few disordered sections, it is unlikely that SHP detects the overexpression of TTF-1 *per se* and may detect different protein expression caused by TTF-1. In other IHC stains used in this study, the overexpressed protein belonged to a class of immunoglobulins with very different secondary structures. In this case, the ability of infrared spectral methods to detect regions of high protein expression comes as less of a surprise.

The second goal of this study is the improved sub-classification of ADC. Based on previous estimates of sampling size and confidence intervals,[6] data of hundreds of patients need to be acquired for the distinction of five subtypes (lepidic, acinar, solid, colloid, and papillary/micro-papillary) with several descriptors (*e.g.*, mucinous *vs.* non-mucinous). The motivation of these efforts is to establish SHP as a prognostic tool, since the different subtypes affect the disease outcome, with lepidic, acinar and papillary ADC showing intermediate response to intervention, whereas colloid, solid and micro-papillary subtypes show a much lower response and poor survival prognosis.[55]

5. Spectral cytopathology

The work in SCP can be divided into two main areas, the analysis of exfoliated cells and the analysis of cultured cells. Although both efforts deal with spectroscopy of individual cells, they are conceptually quite different: cultured cells display a much higher cell-to-cell homogeneity, and any spectral differences can be attributed to effects such as the cell cycle,[56] stem-cell like characteristics,[57] cell differentiation,[58] and even the degree of aggressiveness.[59] A review of many of these efforts has been published recently.[41]

Efforts on exfoliated cells have been reviewed recently as well.[5,9,10] In order not to duplicate the information presented in these reviews, the present summary deals with SCP of exfoliated cells only. In exfoliated cells, the heterogeneity of observed spectra is even larger than that discussed in the previous discussion. This is mainly due to the fact that an exfoliate contains many cell types, from immature cells, mature cells, inflammatory cells, bacteria, cellular debris and non-cellular materials. Interestingly, it was the field of analysis of exfoliated cells that first attracted infrared spectroscopists, but the first reliable results on larger datasets were not published until 1998 by Cohenford & Rigas.[60]

The methodology that works best for SCP is analogous to the methods introduced for tissue samples in Section 3, except that the dataset is not presorted by HCA, but rather, cell spectra are computed from individual pixels in an approach that has been referred to as the PapMap method.[61]

This section of the summary will briefly review the efforts in the University laboratory of the author, the Laboratory for Spectral Diagnostics, to collect and interpret large datasets for exfoliated cells from the oral cavity, cervix and esophagus. This discussion can be summarized as follows:

• SCP detects disease earlier than standard cytology. In fact, most of the SCP results reported in this review were carried out on "morphologically normal" cells; that is, from cells that were harvested from patients with disease, but that still exhibited normal morphology. These cells already show spectral abnormalities that are similar to the spectra of cells that were diagnosed as cancerous or otherwise diseased cells. Thus, SCP was shown to present spectral patterns that can aide in the earliest detection of cellular abnormalities.

• The correlations between SCP and standard cytology are even more difficult than the correlation between SHP and histopathology since the number of truly abnormal cells in an exfoliate is very small.

5a. SCP of the oral mucosa

In this section, the efforts carried out at the LSpD toward establishing SCP as a viable method for detecting the onset of disease in exfoliated cells of the oral cavity will be summarized. This study, reported in 2010,[62] was the first study ever in which many thousands of individual cell spectra were available. This initial study revealed three major findings:

1. The cellular spectra of different collection sites within the oral cavity (the tongue, the cheeks, and the mouth floor) were slightly different; thus, for a reliable classification, abnormalities should be compared only if the collection site is the same.

2. The spectra of normal, dysplastic and cancerous cells exhibited a gradual change that permitted reliable detection of oral cancer.

3. The findings of Cohenford and Rigas[60] were confirmed in that most cells from patients diagnosed with disease showed abnormal spectral patterns, even if they still exhibited normal morphology.

In a much larger study including over 90 normal and nearly 30 clinical samples (*i.e.*, samples with confirmed disease), the earlier results[62] on the differences between cells from the cheeks, the mouth floor and the tongue were further validated. The main motivation of this study was the development of a screening tool for oral disease. The resulting dataset contained many thousand spectra of individual cells. Analysis of this dataset was carried out by unsupervised and supervised methods. PCA, the unsupervised method used, generally showed good discrimination of the spectra into normal and disease classes (data not shown). Analysis by trained methods (ANNs) revealed excellent sensitivity and specificity, 96 and 94.3%, respectively. However, the overall number of cells analyzed was relatively small (300 each of normal and diseased cells in the training and test sets) due to the small number of cells that could be positively identified as diseased by classical cytopathology. The analysis of the oral mucosa also included cells that were infected by the herpes simplex virus. These cells showed significant differences in the protein spectral region, indicating the change in protein expression upon viral infection.

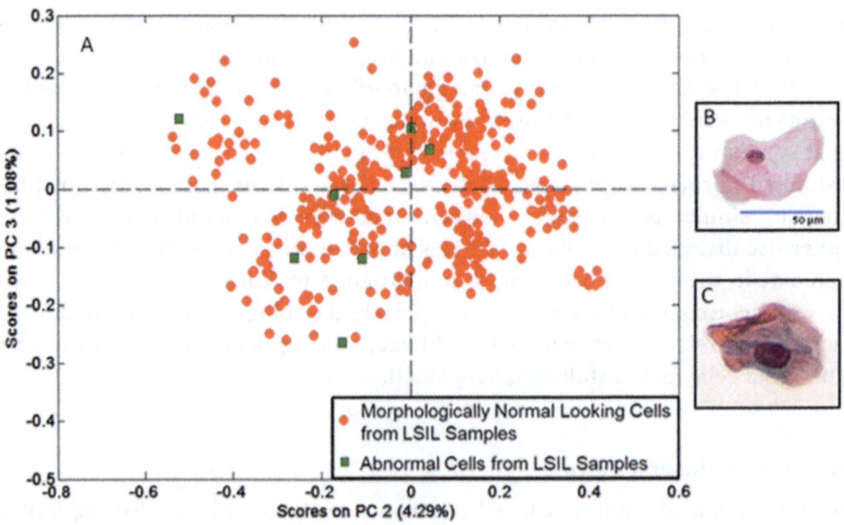

Fig. 12 (A) PCA scores plot of a few cells with abnormal morphology (green squares) within the PCA cluster of morphologically normal cells (red dots) from a sample with confirmed disease. (B) Morphologically normal cell in a sample diagnosed disease. (C) Cell with abnormal morphology in a sample with disease.

5b. SCP of the cervical mucosa

In cervical cytology, which was the original goal of many research efforts in SCP, completely analogous results were observed, namely that the majority of cells from abnormal samples showed spectral abnormality, although they exhibited normal morphology. These results mirror the earlier findings by Cohenford,[60,63] but take into account additional confounding factors, such as hormone-mediated maturation processes that include, for example, the accumulation of glycogen in the final stage of maturation. Since the glycogen absorption bands masked a large part of the low frequency (1000–1200 cm^{-1}) spectrum, valuable information in the nucleic acid phosphate stretching region (*ca.* 1085 and 1235 cm^{-1}) is inaccessible. Thus, the studies reported below were from subjects using oral contraceptives which prevent complete maturation of the squamous cervical epithelium, and therewith, reduce the glycogen abundance. The changes in the maturation patterns of cervical cells in response to menopause could be demonstrated nicely using SCP.[64]

To demonstrate the validity of the assumption that the "abnormal" spectra observed for morphologically normal cells was not an artifact due to hitherto unknown confounding effects, a few morphologically abnormal cells were identified in the samples from patients diagnosed with cervical pre-cancer. These few cells are the ones on which an "abnormal" cytology diagnosis is based. The spectra of these cells were found to occur within the cluster of abnormal cells with normal morphology in the PCA scores plot (see Fig. 12). This observation demonstrated that in a sample of cervical disease, the cells with normal morphology already exhibited the same spectral patterns as the cells that are morphologically abnormal. The implications of this observation are quite far reaching in that the detection of abnormality is not restricted to the few cells in a cervical exfoliate that exhibit abnormal morphology. Rather, SCP detects abnormal spectral signatures that are exhibited by most cells, even if they still have normal morphology. This study also included tests to detect infection by the high risk human papilloma virus (hrHPV) *via* SCP. Here, the Digene Hybrid Capture test (Qiagen, Valencia, CA) was used as a confirming test.

5c. SCP of esophageal cells

The work at the LSpD to detect and diagnose esophageal disease followed similar efforts by Bakker Schut *et al.*[65] and Stone *et al.*[66,67] who used *in vivo* and *in vitro* Raman spectral methods to investigate esophageal tissue. The SCP results reported here differ from the two Raman approaches in that exfoliated cells were used.[10,40] Cytological methods have the advantage over point-by-point spectral methods that a much larger and more representative area of the organ's surface can be sampled, since the exfoliation methods collect cells from quite large areas.

In the SCP results presented above, the classical cytology diagnosis was based on detecting a few cells with abnormal morphology and inferring from these few diagnostic cells that the patient was suffering from a given disease. In the esophageal SCP study described below, the spectra of hundreds of cells that were morphologically described by a physician as normal, Barrett's esophagus and dysplasia, were analyzed and produced data-sets with unambiguous annotation.

There are two major histological sub-types of esophageal cancer, squamous cell carcinoma and adenocarcinoma; the latter of which has seen an alarming increase in incidence in Western countries over the past several decades.[68] Early diagnosis of esophageal adenocarcinoma is of the utmost importance to allow intervention, as prognosis for patients with symptomatic and advanced adenocarcinoma is very poor. Esophageal adenocarcinoma typically arises in the distal one-third portion of the esophagus and its primary risk factor is Barrett's esophagus; a metaplastic transformation of the esophageal mucosa that is thought to arise as a protective response to chronic gastro-esophageal reflux disease (GERD).[69] Barrett's esophagus is recognized by the transition from stratified squamous epithelial cells to specialized columnar cells and is present in about 15% of patients undergoing an endoscopy for symptoms of GERD.[70] Although these columnar cells appear to be more resistant to acid reflux, they are predisposed to carcinogenesis.

Esophageal cytology is being considered as an early screening method for esophageal disease, and collection devices such as the Cytosponge™, which have shown acceptability in a primary care setting,[71] would have the advantage of sampling a large part of the esophageal surface. If applied in clinical practice, esophageal cytology would require cytological expertise and consume significant cytopathology resources. This aspect, coupled with the low sensitivity of cytology, has precluded this methodology from wide acceptance. The promise of SCP, therefore, lies in its ability to diagnose exfoliated cells automatically with good sensitivity and specificity.

Here, the results of a SCP study are reported for an annotated dataset of 299 squamous, 145 Barrett's and 14 dysplastic individual cell spectra, a total of 458 diagnosed cells from 10 patients. In addition to the annotated cells, thousands of un-annotated cells were analyzed in this study. The overall results were reported previously.[9,40] The most abundant cell types found in each sample were squamous cells of the esophagus, which are more easily exfoliated than the columnar cells.

These studies had vastly lower patient numbers than the tissue study results reported in Section 4; thus, the confidence interval of these results is much broader. The analysis of this dataset was carried out, as before, by unsupervised (PCA) and supervised methodologies (ANNs). The unsupervised method revealed differences in the mean class spectra that could be interpreted in terms of biochemical differences between the disease states: the spectra of dysplastic cells showed an increase of the nucleic acid bands (1085 and 1235 cm^{-1}) and a concomitant decrease of the intensity at 1020 cm^{-1}. This latter change is most likely due to variations in carbohydrate (glycogen) concentration. A decrease in the glycogen signal of diseased cells has also been reported previously in the analysis of cervical cells[72] and has been reported for Barrett's associated changes by Stone, *et al.*[73,74] Furthermore, a prominent low wavenumber shoulder of the amide I band at *ca.* 1635 cm^{-1} was detected in the squamous cells of patients with esophageal disease, indicating apoptosis or necrosis in these cells.

Unsupervised analysis of the spectral data demonstrates that there are interpretable and reproducible changes that occur between the spectra of normal and diseased samples. In order to achieve a supervised classification, ANNs were trained to establish the sensitivity and specificity for this method when screening for esophageal disease.

The training and test phases of this ANN followed standard procedures. The results of this analysis are summarized in Table 3. The results from this ANN are

Table 3 Binary classification sensitivity, specificity and overall accuracy of ANNs trained for supervised analysis. BE: Barrett's esophagus, SQ: squamous cells, DYS: dysplasia

	Sensitivity	Specificity	Accuracy
BE *vs.* SQ	95.5%	94.7%	95.1%
BE *vs.* DYS	88.7%	91.4%	90.0%
SQ *vs.* DYS	93.4%	90.9%	92.1%

a clear indication of the sensitivity and specificity that can be achieved for the classification of cells exfoliated from the esophagus and analyzed by SCP.

6. Conclusions

This review summarized efforts in infrared SHP and SCP, mainly from the authors' view of creating spectral methods that are applicable in medical diagnostics. To this end, a set of procedures have been developed, tested and documented as work instructions that range from the testing of 'low-e' slides for reflectance homogeneity and flatness, sample de-paraffination, spectral data acquisition, pre-processing, annotation and algorithm training. The methods are being developed to be instrument platform independent, and have been utilized for paraffin-embedded samples in TMA format, as well as for larger tissue biopsies. Normal tissue, various cancers from different organs, as well as benign lesions have been studied. Using similar protocols, fresh-frozen tissue data banks have been constructed by other groups.[17] Medical diagnostics on exfoliated cells is equally possible, but correlation to classical methods of cytology is more difficult due to the scarcity of abnormal, diagnosable cells in cell exfoliates.

In contrast to an introductory lecture/paper presented on the same theme at an earlier Faraday discussion, this summary is presented with a very optimistic outlook on the subject. It appears that the last decade has produced enormous progress in the field, due to the collaboration and communication between the research groups involved, and the authors of this review wish to thank all the research groups in spectral diagnostics for their efforts, openness and collegiality that have made the field a particularly exciting research area. With the technical problems largely solved, the research now can concentrate on issues such as addressing real problems in diagnostic need, and commercialization of the technology.

Abbreviations

ADC	Adenocarcinoma
ANN	Artificial neural network
cpHCA	Cross-patient hierarchical cluster analysis
FT	Fourier transform
FFPE	Formalin-fixed, paraffin-embedded
FTIR	Fourier transform infrared spectroscopy
IHC	Immuno-histochemistry
IR	Infrared

H&E	Hematoxylin–eosin
HCA	Hierarchical cluster analysis
hrHPV	High risk human papilloma virus
LOOCV	Leave-one-out cross validation
MCT	Mercury cadmium telluride
NA-PCA	Noise-adjusted principal component analysis
NECR	Necrosis
NOR	Normal
NSCLC	Non-small cell lung carcinoma
PCA	Principal component analysis
QCL	Quantum cascade laser
RF	Random forest
S/N	Signal-to-noise ratio
SCLC	Small cell lung carcinoma
SCP	Spectral cytopathology
SHP	Spectral histopathology
SqCC	Squamous cell carcinoma
SVM	Support vector machine
TMA	Tissue microarray
TTF-1	Thyroid transcription factor

References

1 M. E. Garber, *et al.*, Diversity of gene expression in adenocarcinoma of the lung, *Proc. Natl. Acad. Sci. U. S. A.*, 2001, **98**, 13784–13789.

2 W. D. Travis, Pathology of Lung Cancer, *Clin. Chest Med.*, 2011, **32**, 669–692.

3 S. Mukhopadhyay and A.-L. A. Katzenstein, Subclassification of Non-Small Cell Lung Carcinomas Lacking Morphological Differentiation on Biopsy specimens: Utility of an Immunohistochemical Panel Containing TTF-1, Napsin A, p63 and CK5/6, *Am. J. Surg. Pathol.*, 2011, **35**(1), 15–25.

4 G. Rossi, *et al.*, Subtyping non-small cell lung cancer: Relevant issues and operative recommendations for the best pathology practice, *Int. J. Surg. Pathol.*, 2013, **21**(4), 326–336.

5 M. Diem, *Modern Vibrational Spectroscopy and Micro-Spectroscopy: Theory, Instrumentation and Biomedical Applications*, Wiley, Chichester, UK, 2015.

6 A. Akalin, *et al.*, Classification of Malignant and Benign Tumors of the Lung by Infrared Spectral Histopathology (SHP), *Laboratory Investigations*, 2015, **95**(4), 406–421.

7 X. Mu, *et al.*, Statistical Analysis of a Lung Cancer Spectral Histopathology (SHP) Data Set, *Analyst*, 2015, **140**, 2449–2464.

8 M. J. Baker, *et al.*, Using Fourier transform IR spectroscopy to analyze biological materials, *Nat. Protoc.*, 2014, **9**(8), 1771–1791.

9 M. Diem, *et al.*, Cancer Screening *via* Infrared Spectral Cytopathology (SCP): Results for the Upper Respiratory and Digestive Tracts, *Analyst*, 2016, **141**, 416–428.

10 M. Diem, *et al.*, Medical Applications of Infrared Spectral Imaging of Individual Cells, in *Infrared and Raman Spectroscopic Imaging*, ed. R. Salzer and H. W. Siesler, Wiley-VCH Publishing, Weinheim, Germany, 2014, pp. 181–223.

11 E. R. Blout and R. C. Mellors, Infrared Spectra of Tissues, *Science*, 1949, **110**, 137–138.

12 D. L. Woernley, Infrared Absorption Curves for Normal and Neoplastic Tissues and Related Biological Substances, *Cancer Res.*, 1952, **12**, 516–523.

13 B. Mohlenhoff, M. J. Romeo, M. Diem and B. R. Wood, Mie-type scattering and non-Beer–Lambert absorption behavior of human cells in infrared microspectroscopy, *Biophys. J.*, 2005, **88**(5), 3635–3640.

14 P. Bassan, *et al.*, Resonant Mie scattering in infrared spectroscopy of biological materials – understanding the 'dispersion artefact', *Analyst*, 2009, **134**, 1586–1593.

15 M. J. Romeo, *et al.*, Infrared and Raman Microspectroscopic Studies of Individual Human Cells, in *Vibrational Spectroscopy for Medical Diagnosis*, ed. M. Diem, P. R. Griffiths and J. M. Chalmers, John Wiley & Sons, Chichester, UK, 2008, pp. 27–70.

16 A. Kallenbach-Thieltges, *et al.*, Immunohistochemistry, histopathology and infrared spectral histopathology of colon cancer tissue sections, *J. Biophotonics*, 2013, **6**(1), 88–100.

17 F. Großerüschkamp, *et al.*, Marker-free automated histopathological annotation of lung tumour subtypes by FTIR imaging, *Analyst*, 2015, **140**, 2114–2120.

18 P. Bassan, *et al.*, Automated high-throughput assessment of prostate biopsy tissue using infrared spectroscopic chemical imaging, *Proc. SPIE*, 2014, **12**(4), 1–10.

19 F. N. Pounder and R. Bhargava, Toward Automated breast Histopathology Using Mid-IR Spectroscopic Imaging, in *Vibrational Spectroscopic Imaging for Biomedical Applications*, ed. G. Srinivasa, McGraw Hill, New York, 2010, pp. 1–26.

20 M. Miljković, B. Bird and M. Diem, Dispersive line shape effects in infrared spectroscopy, *Analyst*, 2012, **137**, 3954–3964.

21 P. Bassan, *et al.*, Reflection contributions to the dispersion artefact in FTIR spectra of single biological cells, *Analyst*, 2009, **134**, 1171–1175.

22 P. Bassan, *et al.*, The inherent problem of transflection-mode infrared spectroscopic microscopy and the ramifications for biomedical single point and imaging applications, *Analyst*, 2013, **138**, 144–157.

23 B. Bird, *et al.*, Infrared Spectral Histopathology (SHP): A Novel Diagnostic Tool for the Accurate Classification of Lung Cancers, *Laboratory Investigations*, 2012, **92**, 1358–1373.

24 N. Wald and E. Goormaghtigh, Infrared imaging of primary melanomas reveals hints of regional and distant metastases, *Analyst*, 2015, **140**, 2144–2155.

25 H. L. Dreyfus and S. E. Dreyfus, What Artificial Experts Can and Cannot Do, *Artif. Intell.*, 1992, **6**, 18–26.

26 A. I. Mazur, *et al.*, Detection of B-Lymphocyte Activation by various Primary Tumors, *J. Biophotonics*, 2013, **6**(1), 101–109.

27 A. I. Mazur, *et al.*, Evaluating Different Fixation Protocols for Spectral Cytopathology, Part 1, *Anal. Chem.*, 2012, **84**, 1259–1266.

28 M. Diem, *et al.*, Molecular pathology *via* Infrared and Raman spectral imaging, in *Ex vivo and in vivo Optical Pathology*, ed. M. Schmitt and J. Popp, Wiley-VCH, 2013, pp. 45–103.

29 M. Miljković, *et al.*, Spectral cytopathology: new aspects of data collection, manipulation and confounding effects, *Analyst*, 2013, **138**, 3975–3982.

30 A. Ergin, *et al.*, A method for the comparison of multi-platform spectral histopathology (SHP) data sets, *Analyst*, 2015, **140**, 2465–2472.

31 S. W. Bruun, *et al.*, Correcting Attenuated Total Reflection-Fourier Transform Infrared Spectra for Water Vapor and Carbon Dioxide, *Appl. Spectros.*, 2006, **60**(9), 1029–1039.

32 R. K. Reddy and R. Bhargava, Accurate histopathology from low signal-to-noise ratio spectroscopic imaging data, *Analyst*, 2010, **135**, 2818–2825.

33 A. Green, *et al.*, *IEEE Trans. Geosci. Electron.*, 1988, **26**, 65–74.

34 A. Savitzky and M. J. E. Golay, Smoothing and differentiation of data by simplified least-squares procedures, *Anal. Chem.*, 1964, **36**(8), 1627–1639.

35 P. R. Griffiths and J. A. De Haseth, Chemical Analysis, *Fourier Transform Infrared Spectrometry*, ed. P. J. Elving and J. D. Winnefordner, John Wiley & Sons, New York, 1986, vol. 83.

36 M. Diem, *et al.*, Molecular pathology *via* infrared and Raman spectral imaging, *J. Biophotonics*, 2013, **6**(11–12), 855–886.

37 B. Bird, *et al.*, Spectral Detection of Micro-Metastases and Individual Metastatic Cells in Lymph Node Histology, *Technol. Cancer Res. Treat.*, 2011, **10**(2), 135–144.

38 B. Bird, *et al.*, Detection of Breast Micro-Metastases in Axillary Lymph Nodes by Infrared Micro-Spectral Imaging, *Analyst*, 2009, **134**, 1067–1076.

39 J. T. Kwak, *et al.*, Improving Prediction of Prostate Cancer Recurrence using Chemical Imaging, *Sci. Rep.*, 2015, **5**, 8758.

40 D. Townsend, *et al.*, Infrared micro-spectroscopy for cyto-pathological classification of esophageal cells, *Analyst*, 2015, **140**, 2215–2223.

41 G. Clemens, *et al.*, Vibrational spectroscopic methods for cytology and cellular research, *Analyst*, 2014, **139**, 4411–4444.

42 N. Stone, *et al.*, Raman spectroscopy for identification of epithelial cancers, *Faraday Discuss.*, 2004, **126**, 141–157.

43 S. Dochow, *et al.*, Tumour cell identification by means of Raman spectroscopy in combination with optical traps and microfluidic environments, *Lab Chip*, 2011, **11**(8), 1484–1490.

44 U. Neugebauer, *et al.*, Toward Detection and Identification of Circulating Tumor Cells Using Raman Spectroscopy, *Analyst*, 2010, **135**, 3178–3182.

45 U. Neugebauer, *et al.*, Identification and differentiation of single cells from peripheral blood by Raman spectroscopic imaging, *J. Biophotonics*, 2010, **3**(8–9), 579–587.

46 S. E. Holton, M. J. Walsh and R. Bhargava, Subcellular localization of early biochemical transformations in cancer-activated fibroblasts using infrared spectroscopic imaging, *Analyst*, 2011, **136**, 2953–2958.

47 S. Federman, L. M. Miller and I. Sagi, Following matrix metalloproteinases activity near the cell boundary by infrared micro-spectroscopy, *Matrix Biol.*, 2002, **21**(7), 567–577.

48 S. Kumar, *et al.*, Change in the microenvironment of breast cancer studied by FTIR imaging, *Analyst*, 2013, **138**, 4058–4065.

49 D. R. Carrizosa and K. A. Gold, New strategies in immunotherapy for non-small cell lung cancer, *Transl. Lung Cancer Res.*, 2015, **4**(5), 553–559.

50 M. J. Nasse, *et al.*, High-resolution Fourier-transform infrared chemical imaging with multiple synchrotron beams, *Nat. Methods*, 2011, **8**(5), 413–416.

51 N. Jamin, *et al.*, Chemical Heterogeneity in Cell Death: Combined Synchrotron IR and Fluorescence Microscopy Studies of Single Apoptotic and Necrotic Cells, *Biopolymers*, 2003, **72**, 366–373.

52 K. Belbachir, *et al.*, Collagen types analysis and differentiation by FTIR spectroscopy, *Anal. Bioanal. Chem.*, 2009, **395**(3), 829–837.

53 G. J. Laurent, Lung collagen: more than scaffolding, *Thorax*, 1986, **41**, 418–428.

54 R. K. Bobba, *et al.*, Scar Carcinoma of the Lung: A Historical Perspective, *Clin. Lung Canc.*, 2011, **12**(3), 148–154.

55 A. Yoshizawa, *et al.*, Impact of proposed IASLC/ATS/ERS classification of lung adenocarcinoma: prognostic subgroups and implications for further revision of staging based on analysis of 514 stage I cases, *Mod. Pathol.*, 2011, **24**, 653–664.

56 M. Jimenez-Hernandez, *et al.*, Exploring the spectroscopic differences of Caki-2 cells progressing through the cell cycle while proliferating *in vitro*, *Analyst*, 2013, **138**, 3957–3966.

57 C. Hughes, *et al.*, SR-FTIR spectroscopy of renal epithelial carcinoma side population cells displaying stem cell-like characteristics, *Analyst*, 2010, **135**, 3133–3141.

58 R. Gasper and E. Goormaghtigh, Effects of the confluence rate on the FTIR spectrum of PC-3 prostate cancer cells in culture, *Analyst*, 2010, **135**, 3048–3051.

59 T. J. Harvey, *et al.*, Factors influencing the discrimination and classification of prostate cancer cell lines by FTIR microspectroscopy, *Analyst*, 2009, **134**, 1083–1091.

60 M. Cohenford and B. Rigas, Cytologically Normal Cells From Neoplastic Cervical Samples Display Extensive Structural Abnormalities on IR Spectroscopy: Implications for Tumor Biology, *Proc. Natl. Acad. Sci. U. S. A.*, 1998, **95**, 15327–15332.

61 J. M. Schubert, *et al.*, Single Point *vs.* Mapping Approach for Spectral Cytopathology (SCP), *J. Biophotonics*, 2010, **3**(8–9), 588–596.

62 K. Papamarkakis, *et al.*, Cytopathology by Optical Methods: Spectral Cytopathology of the Oral Mucosa, *Laboratory Investigations*, 2010, **90**, 589–598.

63 M. Cohenford, *et al.*, Infrared Spectroscopy of Normal and Abnormal Cervical Smears: Evaluation by Principal Component Analysis, *Gynecol. Oncol.*, 1997, **66**, 59–65.

64 M. Diem, *et al.*, Infrared and Raman Spectroscopy and Spectral Imaging of Individual Cells, in *Infrared and Raman Spectroscopic Imaging*, ed. R. Salzer and H. W. Siesler, Wiley-VCH Publishing, Weinheim, Germany, 2009, pp. 173–202.

65 T. C. Bakker Schut, *et al.*, NIR Raman Spectroscopy of Healthy and Diseased Esophagus, in *Spectroscopy of Biological Molecules: Modern Trends*, ed. P. Carmona, R. Navarro and A. Hernanz, Springer, Heidelberg, Germany, 1997, pp. 455–456.

66 H. Barr, *et al.*, Endoscopic Screening and Surveillance for Barrett's Esophagus – Clinical Implications, *Medsc. Gen. Med.*, 2006, **8**(2), 88–96.

67 N. Stone, C. Kendall, and H. Barr, Raman Spectroscopy as a Potential Tool for Early Diagnosis of Malignancies in Esophageal and Bladder Tissues, *Vibrational Spectroscopy for Medical Diagnosis*, ed. M. Diem, P. R. Griffiths and J. M. Chalmers, John Wiley & Sons, Chichester, UK, 2008.

68 R. F. Souza and S. J. Spechler, Concepts in the Prevention of Adenocarcinoma of the Distal Esophagus and Proximal Stomach, *Ca-Cancer J. Clin.*, 2005, **55**(6), 334–351.

69 S. J. Spechler, *et al.*, American Gastroenterological Association Medical Position Statement on the Management of Barrett's Esophagus, *Gastroenterology*, 2011, **140**, 1084–1091.

70 L. B. Gerson, P. W. Groeneveld and G. Triadafilopoulos, Cost-effectiveness model of endoscopic screening and surveillance in patients with gastroesophageal reflux disease, *Clin. Gastroenterol. Hepatol.*, 2004, **2**(10), 868–879.

71 S. R. Kadri, *et al.*, Acceptability and accuracy of a non-endoscopic screening test for Barrett's oesophagus in primary care cohort study, *BMJ*, 2010, **341**, c4372.

72 L. Chiriboga, P. Xie, H. Yee, D. Zarou, D. Zakim and M. Diem, Infrared Spectroscopy of Human Tissue. IV. Detection of Dysplastic and Neoplastic Changes of Human Cervical Tissue *Via* Infrared Microspectroscopy, *Cell. Mol. Biol.*, 1998, **44**(1), 219–229.

73 C. Kendall, *et al.*, Evaluation of Raman probe for oesophageal cancer diagnostics, *Analyst*, 2010, **135**, 3038–3041.

74 N. Stone, *et al.*, Near-infrared Raman spectroscopy for the classification of epithelial pre-cancers and cancers, *J. Raman Spectrosc.*, 2002, **33**(7), 564–573.

Faraday Discussions

PAPER

Development of a practical spatial-spectral analysis protocol for breast histopathology using Fourier transform infrared spectroscopic imaging

F. Nell Pounder,[a] Rohith K. Reddy[a] and Rohit Bhargava*[ab]

Received 30th November 2015, Accepted 7th January 2016

DOI: 10.1039/c5fd00199d

Breast cancer screening provides sensitive tumor identification, but low specificity implies that a vast majority of biopsies are not ultimately diagnosed as cancer. Automated techniques to evaluate biopsies can prevent errors, reduce pathologist workload and provide objective analysis. Fourier transform infrared (FT-IR) spectroscopic imaging provides both molecular signatures and spatial information that may be applicable for pathology. Here, we utilize both the spectral and spatial information to develop a combined classifier that provides rapid tissue assessment. First, we evaluated the potential of IR imaging to provide a diagnosis using spectral data alone. While highly accurate histologic [epithelium, stroma] recognition could be achieved, the same was not possible for disease [cancer, no-cancer] due to the diversity of spectral signals. Hence, we employed spatial data, developing and evaluating increasingly complex models, to detect cancers. Sub-mm tumors could be very confidently predicted as indicated by the quantitative measurement of accuracy *via* receiver operating characteristic (ROC) curve analyses. The developed protocol was validated with a small set and statistical performance used to develop a model that predicts study design for a large scale, definitive validation. The results of evaluation on different instruments, at higher noise levels, under a coarser spectral resolution and two sampling modes [transmission and transflection], indicate that the protocol is highly accurate under a variety of conditions. The study paves the way to validating IR imaging for rapid breast tumor detection, its statistical validation and potential directions for optimization of the speed and sampling for clinical deployment.

Introduction

Breast cancer is the most prevalent non-skin malignancy in women in the United States, with more than 231 000 new diagnoses and over 40 000 deaths estimated

[a]*Department of Bioengineering and Beckman Institute for Advanced Science and Technology, University of Illinois at Urbana-Champaign, Urbana, IL, USA. E-mail: rxb@illinois.edu*

[b]*Departments of Chemical & Biomolecular Engineering, Electrical & Computer Engineering, Mechanical Science & Engineering and Chemistry, University of Illinois at Urbana-Champaign, Urbana, IL, USA*

in 2015.[1] Since mortality is reduced by early detection, widespread population screening for breast cancer by mammography is recommended[2] and leads to over 1.6 million breast biopsies each year.[3] Although 80% of these biopsies are not diagnosed as cancer,[4] each biopsy must be stained and manually evaluated by a trained pathologist.[5] Pathology examinations require extensive analysis of tissue morphology and structure, and add cost to the evaluation.[6] In addition, manual analysis is time consuming and patients often wait significant time periods to obtain a diagnosis.[7] Patient stress levels, as measured by biochemical indicators (cortisol, for example), are significantly elevated while waiting to learn results of a biopsy, regardless of the eventual diagnosis.[8] Therefore, efficient automated techniques for biopsy evaluation would provide a substantial benefit for preliminary biopsy analysis. However, at this time, no automated technology exists to provide initial biopsy screening or reduce pathologist workload.

Infrared spectroscopic imaging today[9] combines both measurement of morphological properties and extensive information about sample biochemistry, which may be applicable for high-throughput pathology.[10–12] Fourier transform infrared (FT-IR) imaging, in particular, has been widely used in a number of studies which have evaluated the spectral features of breast tissue, both related to clinical disease diagnosis[13–15] as well as to various properties of breast cancer-mimicking cell cultures,[16–18] lymph node involvement,[19] methods of measurement[20–22] and properties of tissues,[23–31] including the tumor microenvironment. These studies have identified spectral features that may be useful for cell type, receptor status or disease recognition, but have not provided a validation analysis to demonstrate diagnostic performance that may be used to inspire clinical translation. Two of the major drawbacks of older technology – slower speed and poor spatial definition – are being actively addressed by advances such as discrete frequency IR imaging,[32] especially using quantum cascade lasers,[33–37] and the development of high definition imaging.[38–40] In parallel, there is a need to develop practical protocols for application to breast cancer. One of the goals of this study is to explore the development of translatable protocols. We especially note that no study has examined the potential to combine the spectral and spatial information in IR tissue images to develop methods for automated breast biopsy screening or disease diagnosis. Hence, we focus on this aspect and seek to provide a combined spatial-spectral protocol that can later be validated extensively.

Methods

Tissue sampling

Seven paraffin-embedded breast tissue microarrays (TMAs) were obtained from U.S. Biomax and thin sections were mounted on barium fluoride (BaF_2) for transmission mode FT-IR imaging. An adjacent section of each TMA was obtained and stained with hematoxylin and eosin (H&E) for pathologist evaluation. The TMA employed for supervised classification calibration and optimization contained 30 invasive ductal carcinomas, 1 invasive lobular carcinoma, and 34 adjacent normal tissue sections. An additional adjacent section of this TMA was mounted on a reflective slide (Kevley Technologies) for transflection mode IR imaging. Preliminary validation was performed on a separate section of this TMA containing 35 invasive ductal carcinomas, 2 invasive lobular carcinomas, and 40 adjacent normal tissue sections. Paraffin was removed from the TMAs by immersion and stirring in hexane at 40 °C for 2–3 days. The disappearance of the paraffin peak at 1462 cm^{-1} was monitored to ensure the paraffin was completely removed prior to image acquisition.

FT-IR spectroscopic imaging data acquisition

TMA images were collected using a Perkin Elmer Spotlight 400. Six TMA datasets were acquired with a 4 cm^{-1} spectral resolution, 2.2 cm s^{-1} moving mirror speed, 6.25 μm nominal pixel size, and 2 scans per pixel. A further validation TMA with 182 cores was collected at a 16 cm^{-1} spectral resolution with all other acquisition parameters held constant. An adjacent section of the calibration TMA on a reflective slide was collected by light transflection with all other acquisition parameters the same as the original calibration TMA. A NB medium apodization and undersampling ratio (UDR) of 2 were used to convert interferograms to single beam images. An IR background was collected each day on a clean area of each BaF_2 slide with 120 scans. Any remaining air and water vapor contribution was removed using the atmospheric correction algorithm in the Spotlight software. Finally, all datasets were truncated to 750–4000 cm^{-1} for ease of handling and storage.

IR images for the 199 core validation TMA were also collected with a Varian 7000 FT-IR/600 UMA microspectrometer with a 128 × 128 focal plane array (FPA) detector. Images were acquired at 8 scans per pixel and a 16 cm^{-1} spectral resolution with a UDR of 2. Single beam spectra were computed using a triangular apodization and the spectral range was truncated to 900–4000 cm^{-1} due to the lower detector cut-off. An intensity ratio was computed to an IR background collected at 128 scans per pixel on a blank area of the BaF_2 substrate to remove spectral features not associated with tissue.

Image processing and classification

Individual TMA core images were compiled to build a single dataset for each TMA using Environment for Visualizing Images (ENVI) and software written in-house with Interactive Data Language (IDL). A supervised pattern recognition method based on a modified Bayes' Rule, described in detail elsewhere,[41] was used to segment image pixels as stroma or epithelium and segment epithelium pixels as cancer or normal. To increase computation speed, the spectral datasets were reduced by manual tissue spectrum examination to a set of 89 potentially useful metrics to test in algorithm development. These spectral metrics include the features of ratios of absorption peak heights, peak areas normalized to amide I (1652 cm^{-1}) and centers of gravity. All bands were analyzed after a simple two-point linear baseline correction was performed across the absorption feature. These metrics were first tested for cell-type classification. Spectral metric frequency distributions with 50 bins were computed for each metric from pixels manually selected as stroma (140 100) and epithelium (50 082) by comparison with H&E staining. These pdfs were used to estimate probability distribution functions for each metric.

These distributions were applied with the metric profile to estimate posterior probabilities for each pixel for each class, which were used to build a discriminant function. The metrics were arranged by minimum pairwise error and classifiers were built from the first metric, the first two metrics, the first three metrics, and so on, continuing until a classifier was built with all 89 metrics. Receiver operating characteristic (ROC) analysis was used to assess each classifier and the change in the area under the ROC curve (ΔAUC) was computed with the addition of each metric to test if the metric provided useful additional information for cell type classification. The metrics were reordered by ΔAUC until an optimal set of metrics was obtained to achieve a maximum AUC with a minimum number of features. Once the metrics for the classification model were finalized, optimal thresholds were selected to produce color-coded classified images. The model was then validated on independent datasets.

Epithelium pixels were extracted using this spectral histology classification model and the set of 89 spectral metrics were again used to evaluate for the discrimination of cancer and normal pixels with spectral information alone. In addition, two methods were evaluated for tumor identification by spatial polling. The first method involves segmenting each TMA core into boxes of a specified size ranging from 1 × 1 pixel (6.25 × 6.25 μm^2) to 12 × 12 pixels (75 × 75 μm^2). The epithelium fraction of each box is computed, and the percent of boxes above a specified epithelium threshold is calculated for each individual TMA core. The fractions for cancer and normal TMA cores are compared to select an epithelium fraction for cancer detection. Epithelium thresholds from 0.1 to 1.0 are considered, and a cutoff point for cancer detection is selected. The cutoff points for each threshold are used to build a least squares linear trendline, which becomes an operating line for tumor detection. An operating line for each TMA core is computed, and the corresponding y-intercept and slope values are plotted to evaluate separation of cancer and normal TMA cores.

A second method was developed to perform pixel-level tumor segmentation. In this algorithm, a separate classification model is built by the procedure described for stroma and epithelium segmentation to distinguish the cancer and normal epithelium pixels with *spatial* metrics. These metrics are developed for cancer classification by computing the mean and standard deviation of the epithelium fraction for all boxes that contain a given pixel. Boxes ranging in size from 4 × 4 pixels (25 × 25 μm^2) to 160 × 160 pixels (1 mm^2) are considered in order to evaluate tissue features of various sizes. Probability distributions for each metric are estimated from the cancer (1 030 376 pixels) and normal (181 350 pixels) epithelium labeled from the two class histology model. A separate classification model is built for each spatial metric and ROC analysis is performed to select an optimal box size for pixel-level cancer classification. Combinations of spatial metrics are evaluated by ROC analysis for pixel-level tumor detection potential. TMA core-level tumor identification for univariate and multivariate spatial classification models is evaluated by varying the fraction of pixels identified as cancer on a TMA core as a threshold for tumor discrimination. The AUC values for core level ROC curves are computed by the trapezoid rule,[42] which is known to provide a conservative estimate of the true AUC value.[43]

Statistical analysis

Confidence bands for core level ROC curves were calculated by evaluating the standard error $E(p)$ for sensitivity and specificity values (p) for individual operating points with the binomial approximation

$$E(p) = \sqrt{\frac{p(1-p)}{n}} \tag{1}$$

where n is the number of cancer samples when p is sensitivity and n is the number of normal samples when p is specificity.[44] The standard error for a TMA core level AUC value is evaluated as

$$\mathrm{SE(AUC)} = \sqrt{\frac{\mathrm{AUC}(1-\mathrm{AUC}) + (n_1-1)(Q_1-\mathrm{AUC}^2) + (n_0-1)(Q_2-\mathrm{AUC}^2)}{n_0 n_1}} \tag{2}$$

with n_0 as the number of normal samples and n_1 as the number of cancer samples and

$$Q_1 = \frac{\mathrm{AUC}}{2-\mathrm{AUC}} \quad Q_2 = \frac{2\mathrm{AUC}^2}{1+\mathrm{AUC}} \ \text{(ref. 43)}.$$

These methods for calculation of the AUC and corresponding standard error do not require any assumptions about the distribution of the underlying populations. A confidence band or AUC confidence interval half-width was computed by multiplying the standard error by a selected z-score that corresponds to the acceptable probability of type I error, α, for tumor classification. A value for α was selected to reflect the level of confidence required for a given question. The estimated probability of a confidence interval including the true AUC is equal to 1 − α. Therefore, a small α value is required for cancer classification. A α value of 0.05 and a corresponding z-score of 1.96 were used for 95% confidence interval calculation. Since the AUC is an estimate of the true population, the central limit theorem permits the assumption of normal distribution for large sampling.[45] The precision of an AUC value estimated from a set of samples is reflected by the width of the associated confidence interval. These equations indicate that the width of a confidence interval for sensitivity, specificity and AUC are determined by the respective values for these statistics, sample size and acceptable α value.

Results and discussion

Classification with spectral metrics

As seen in Fig. 1A, unstained fixed tissue does not have any inherent contrast and cannot be readily evaluated for disease diagnosis. Standard pathology practice involves the application of hematoxylin & eosin (H&E) dyes (Fig. 1B).[5] Hematoxylin dye is used visualize basic nucleic acid structures prevalent mainly, though not exclusively, in epithelial tissue lining breast ducts and lobules and eosin dye is used to visualize acidic protein structures mainly prevalent in connective stromal tissue. This lack of specificity and staining variability is often a barrier to the application of computational approaches.[46,47] As noted in Fig. 1C, FT-IR imaging can also provide some contrast between different types of tissue due to differences in relative IR spectral absorbance between different types of breast tissue. These spectral differences can be used for quantitative tissue classification to obtain color coded images (Fig. 1D) that provide high throughput histologic information using TMAs[48] and can even be used to simulate H&E images.[49] Identification of epithelium is an important first step in tissue analysis as over 99% of breast tumors arise in the epithelium,[50] and this tissue is the primary component of most breast malignancies.[51]

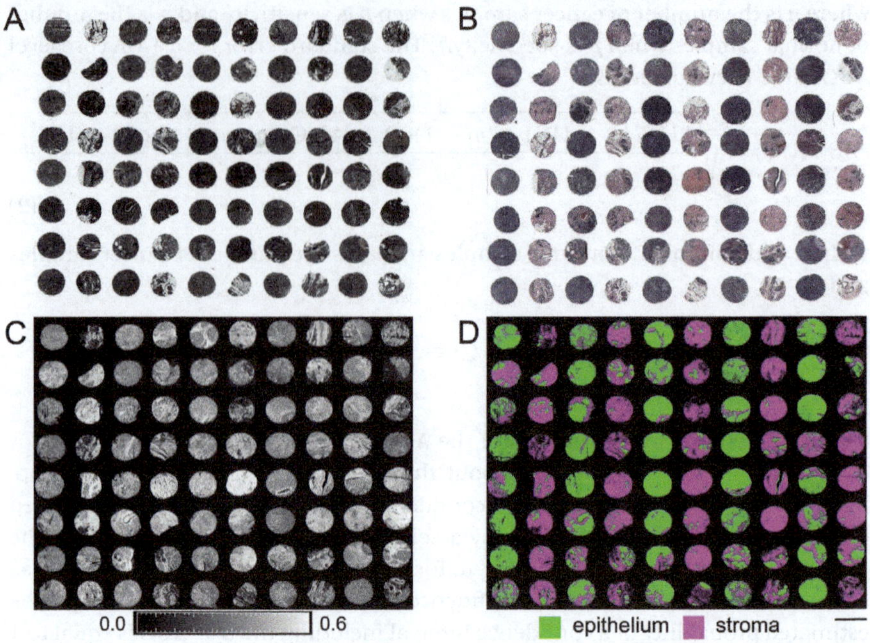

Fig. 1 Two-class breast histology. (A) Minimal tissue and tumor characteristics are visible on unstained tissue. (B) Stroma and epithelium are visible on tissue stained with hematoxylin and eosin (H&E) dyes. (C) Image of tissue absorbance, as per color bar scale below the image, of unstained tissue at 3294 cm^{-1} highlights differences in tissue, especially between stroma and epithelium. (D) Quantitative spectral data permits automated segmentation of stroma and epithelium, as noted by the color code below the image, without dyes or contrast agents. The scale bar represents the 1.5 mm diameter of a single core on this TMA.

Initial classification models are developed with spectral metrics to, first, segment epithelium and stroma and, second, segment the epithelium pixels as cancer or normal. As seen in Fig. 2A, spectral features for stroma and epithelium demonstrate substantial biochemical differences for these cell types. Average spectra are computed from manual labeling of epithelium (50 082 pixels) and stroma (140 100 pixels) on TMA cores in a calibration dataset with tissue from 40 different patients. A piecewise linear baseline is applied and a ratio is computed to the amide I absorbance at 1652 cm^{-1} to normalize for tissue sample thickness. While scattering from tissue is well-known to affect spectra,[52–55] an analysis of the variance in tissue[56] shows that there is a significant fraction of the spectrum that can be useful for analysis using a simple baseline correction. More sophisticated models for spectral corrections[57–59] and physics-based methods[60,61] are being developed that can potentially provide more information and will likely improve our results here. In the baselined spectra we note especial differences within the fingerprint region, which contains many overlapping spectral features prevalent in tissue.[62] The region includes symmetric PO_2 stretching and CO stretching vibrational modes at 1080 cm^{-1}, amide III protein modes within 1200–1338 cm^{-1}, CH_2 wagging at 1236 cm^{-1}, asymmetric PO_2 stretching at 1240 cm^{-1}, CH_2/CH_3 bending at 1400 cm^{-1}, asymmetric CH_3 bending at 1456 cm^{-1}, as well as the

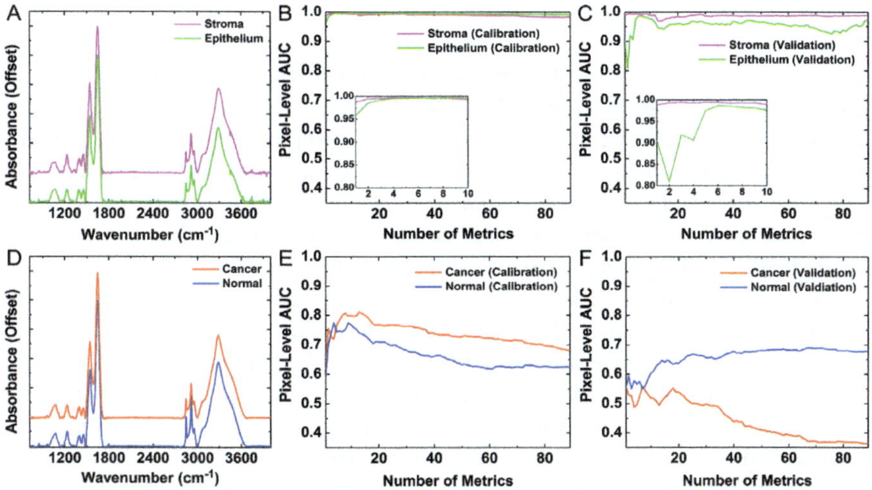

Fig. 2 Automated histology and pathology using only spectral metrics. (A) Average spectra for stroma and epithelium demonstrate clear biochemical differences between these cell types. (B) Spectral metrics provide accurate histologic segmentation of stroma and epithelium with AUC values of ~1 for each tissue class with only 6 metrics. (C) This classification is reproducible in validation on separate tissue samples. (D) Average spectra for cancer and normal epithelium indicate biochemical changes are less obvious in disease development. (E) ROC analysis indicates that spectral metrics demonstrate reduced discrimination in cancer and normal epithelium pixels with a maximum cancer pixel-level AUC of only 0.81. (F) Spectral metrics do not provide reproducible pathology discrimination, as demonstrated by a low cancer pixel-level AUC of 0.55 in validation samples.

amide II vibrational modes within 1542–1556 cm^{-1}.[63] A broad amide A vibrational feature at 3294 cm^{-1} (ref. 25) is also prominent, but we do not use the CH stretching region in our analysis due to additional strong variability arising from potentially residual paraffin. A comparison of the stroma and epithelium cell type spectra indicate that symmetric PO_2 vibrations in the DNA-rich tissue regions and CO stretching vibrations in secretory glycoproteins contribute a more significant relative absorbance in epithelium while asymmetric PO_2 stretching, amide III, and CH_2 wagging in the methyl side chains in collagen characterize the stroma IR spectrum. Clear spectral differences in these regions between epithelium and stroma indicate the potential for highly accurate classification.

As described in the methods section, a method for classification with spectral metrics provides accurate and reproducible differentiation of stroma and epithelium, as shown in Fig. 2B and C. This is demonstrated by the reproducible mean AUC value above 0.98 for both the calibration and validation TMA datasets with matched cancer and adjacent normal tissue samples from a set of 40 patients. Accurate histologic classification is accomplished with only 6 spectral metrics, as shown in the inset plots in Fig. 2B and C. Although each individual metric may not provide the same contribution to classification in the calibration and validation datasets, the AUC still reaches a maximum value in both datasets with the same six metrics. Notably, the classification contribution for individual metrics appears to vary more for epithelium than for stroma. This is reasonable, as collagen is a predominant component of breast stroma and has a distinct IR spectrum. Conversely, the IR spectrum for epithelium may vary more due to underlying physiologic conditions, for example between normal and different tumor regions, which would impact the performance of individual specific metrics in classification. It must be borne in mind that our model assumes two classes, but tissue is varied and the chemical diversity may not always fit the desired information model. Therefore, more than a few metrics are required to account for this diversity, compared to two-component polymeric systems for example.[64] A multi-feature classifier is advantageous to provide accurate and reproducible cell-type classification with spectral data.

Next, epithelium pixels were segmented into cancer and normal classes based on spectral metrics alone. The automated classification procedure was repeated with manually labeled epithelium from cancer (38 384) and normal (10 483) pixels as ground truth. From Fig. 2D, it can be seen that spectral differences are less obvious for cancer and normal epithelium. Although some small differences in absorbance are visible at 1400 cm^{-1} and 1456 cm^{-1}, these distinctions are less clear than those between different cell types and may not facilitate efficient classification in the manner of epithelium–stroma with only spectral metrics. Indeed, the classification potential for spectral metrics is significantly lower for discriminating cancer and normal pixels, as evidenced by the cancer pixel-level AUC of 0.81 with eight spectral metrics in the calibration data (Fig. 2E) and 0.55 with the same eight metric classification model on validation data from the same patients (Fig. 2F). Notably, the metric contributions appear to vary even more between the calibration and validation datasets for the cancer class than for the normal class. This may indicate that spectral variation is greater for cancer epithelium pixels than normal epithelium pixels. In addition, the AUC appears to fall below 0.5 in validation for the cancer class when more than 33 metrics are included in the model. This analysis indicates that metrics during calibration optimization may have completely different properties for cancerous epithelium between different datasets, even when both datasets contain tissue from the same set of patients. Therefore, techniques to improve data quality such as computational noise reduction[65] or other image enhancement techniques[66] would likely provide only limited benefit for this classification problem. Given the poor performance and reproducibility of pixel-level classification, sample level analysis will also provide a low level of accuracy. Inaccurate classification results will also have wide sensitivity and specificity confidence intervals and AUC error estimates,[43,44] and will not be useful for tumor discrimination. These data may also indicate the need for more complex and better spectral features, a more sophisticated disease model going beyond the two-class model here or it may not be possible to perform this segmentation with IR imaging. Instead of pursuing complex models, corrections or computational-heavy methods to explore the potential of IR imaging, our goal was to develop a rapid protocol. Complex models and time-consuming calculations are not conducive to this. Hence, instead of mining the spectrum with more complex methods or applying spectral corrections, we chose to pursue an alternate method.

Classification with spectral and spatial metrics

The classification protocol with additional information that we propose and examine in this manuscript involves a two-step procedure outlined in Fig. 3. In

Fig. 3A, the first step shows a spectral pixel-level segmentation of breast stroma and epithelium, which is highly accurate. In the second step, shown in Fig. 3B, a spatial information strategy is incorporated based on the epithelium–stromal segmentation. Computerized algorithms quantify epithelium content and arrangement with a technique termed spatial polling. Two algorithms for spatial polling are considered here. The first method involves TMA core-level spatial polling of a set of small tissue regions to obtain a diagnosis of cancer or normal for each individual core on a breast TMA. The second method involves pixel-level spatial polling of somewhat larger regions to obtain a diagnosis of cancer or normal for each individual pixel on a breast TMA. The advantages and challenges for each method are considered next prior to validation analysis. While there are

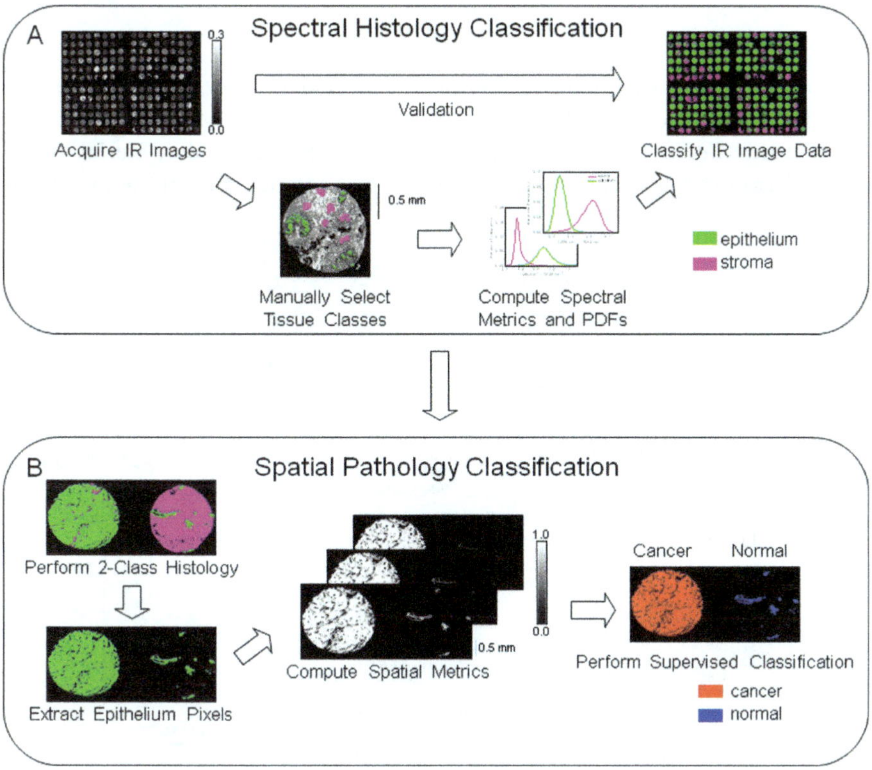

Fig. 3 Development of automated breast histopathology with spectral and spatial data. (A) Spectral histologic classification was performed with supervised pattern recognition by acquiring an IR imaging dataset of multiple samples from a TMA and comparing images at frequencies of known biological significance with H&E staining, the current gold standard in pathology. A large set of pixels were manually selected to represent the two tissue classes [stroma, epithelium]. Frequency distributions for the two classes for each important spectral feature were computed and used to classify individual pixels. After calibration of the two-class histology model additional validation TMA dataset images were automatically classified without any operator intervention. (B) Spatial information from resulting histologic images was used for pathologic classification by extraction of epithelium pixels and computational pixel-level spatial polling. Resulting spatial metrics were used as the input for the supervised classification procedure, used previously for histology analysis, to segment epithelium pixels into cancer and normal classes.

many other spatial analyses methods available, we sought to develop a fast and robust method. Doubtless, more complicated methods can be developed and the ideas proposed here can be extended to other methods.

First, TMA core-level spatial polling was conducted by dividing each TMA core into square boxes of specified dimensions (pixels). The percentage of boxes in each core with an epithelium fraction above a select threshold was computed. To minimize errors associated with inappropriate selection of a tissue region for tumor diagnosis, boxes containing no epithelium pixels were excluded from calculations. Square boxes sizes with pixel lengths of 1 × 1 to 12 × 12 were considered, and an 8 pixel (50 μm) box length was selected as an optimal size for tumor segmentation. As seen in Fig. 4A, cancer and normal cores are clearly separated at a wide range of epithelium fraction threshold values. A cutoff was selected for each epithelium threshold by the relationship

$$\frac{d_{\mathrm{C}}}{d_{\mathrm{N}}} = \frac{\sigma_{\mathrm{N}}}{\sigma_{\mathrm{C}}} \tag{3}$$

where d_{C} is the distance of the cutoff from the mean of the cancer TMA cores, d_{N} is the distance of the cutoff from the mean of the normal TMA cores, σ_{C} is the standard deviation for cancer TMA cores, and σ_{N} is the standard deviation for normal TMA cores. A least squares linear fit was computed from the individual

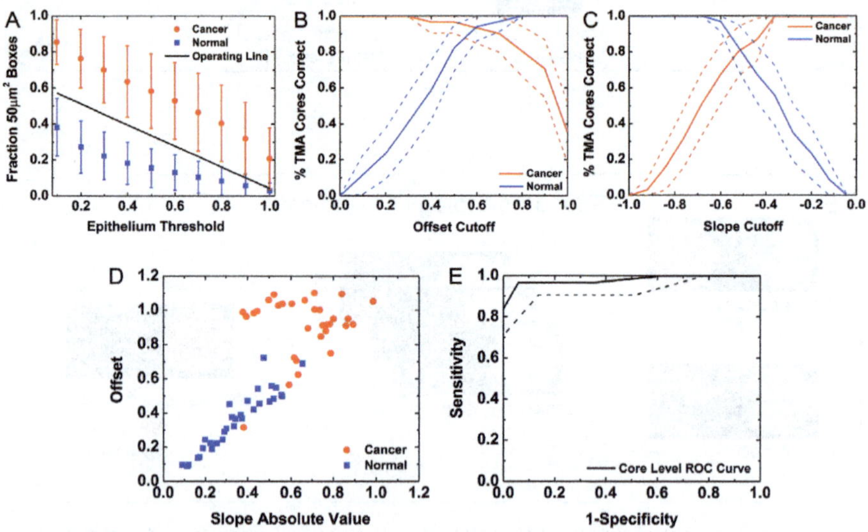

Fig. 4 Tumor TMA core segmentation by spatial polling. (A) A TMA core was divided into 8 × 8 pixel (50 μm^2) boxes and the fraction of boxes with epithelium content above a set of thresholds for each TMA core was computed. The mean value for cancer and normal classes was computed, and error bars represent standard deviation. An operating line was obtained for tumor TMA core classification. (B) A linear fit was calculated for each TMA core and the *y*-intercept offset cutoff for cancer detection was varied to assess classification sensitivity and specificity with this variable. (C) The slope cutoff for cancer detection was also varied to assess classification sensitivity and specificity with this variable. (D) A scatter plot of offset *vs.* slope absolute value for each TMA core demonstrates the contribution of each metric to tumor core identification. (E) The ROC curve indicates near-perfect tumor classification. The area between the dotted lines represents a 95% confidence region for the mean value.

cutoff points for each epithelium threshold to determine the operating line for tumor detection in Fig. 4A.

A least squares linear fit with offset and slope values was also computed from the plots for the fraction of 50 μm^2 boxes above a selected epithelium threshold for each individual TMA core. Cutoff values for cancer detection for *y*-intercept offset (Fig. 4B) and slope (Fig. 4C) were varied and the sensitivity and specificity of tumor detection was evaluated. The plots in Fig. 4B and C indicate that both of these metrics simultaneously provide high sensitivity and specificity for tumor identification. To assess the potential of both of these metrics in a single classifier, the slope and offset were then plotted together (Fig. 4D) and an operating line was moved to perform ROC analysis and evaluate classification potential. From this plot, cancer cores appear to have a greater *y*-intercept offset and slope absolute value than normal cores. Although most cases can be distinguished by the offset variable alone, the slope variable appears to add additional information that is useful to achieve the best possible classification. An AUC of 0.98 ± 0.04 (95% CI) is achieved on the calibration TMA with 65 cores (31 cancer and 34 normal) using this technique (Fig. 4E). This method is highly sensitive to the overall morphology of the tumor and intervening stromal scales. Though the technique seems to work well, the method will underestimate regions where the tumor may be close to the edge. This will result in designation of pixels in sparse tumor regions or edges as non-cancerous and in a smaller spatial extent of tumor than present. Therefore, other techniques of spatial polling were evaluated for cancer classification.

A second spatial polling technique was considered to accomplish pixel-level cancer and normal epithelium segmentation. Again, the method was developed for histology classification with spectral metrics with a second-level set of spatial metrics to evaluate epithelium content and distribution. In this method, we propose a computational selection of boxes ranging in size from 16×16 pixels to 160×160 pixels. This range was selected to evaluate regions varying in size from the approximate mean diameter of a normal breast duct[67] to an area approaching the size of a typical TMA core. A somewhat larger area is required for the pixel-level spatial polling than the core-level spatial polling from Fig. 4 because the metrics are computed separately for each individual pixel and are not averaged over an entire TMA core. The fraction of pixels in each box classified as epithelium was determined, and an average and standard deviation for the epithelium fraction was computed for each pixel from all boxes of a given size containing that pixel. These pixel-level computations were stored in an image metric vector with a format similar to the spectral metric vector used for histology classification. To evaluate the relative classification potential for spatial metrics, all pixels labeled as epithelium by histology classification considered as ground truth information with 1 030 376 cancer pixels and 181 350 normal pixels. Epithelium pixels were divided into cancer and normal classes by applying the same automated classification algorithm used to segment stroma and epithelium pixels but now with spatial metrics. Pixel-level ROC analysis was then performed for each class. Cancer and normal classification images were obtained at the operating point for each class where the difference between the fraction of pixels correctly and incorrectly classified was maximized. TMA core-level tumor identification was then accomplished by selecting an appropriate fraction of epithelium pixels on a TMA core labeled as cancer in the classification image as a threshold to diagnose the entire TMA core as cancer. This threshold was varied to produce an ROC curve to evaluate overall TMA core-level cancer classification potential.

To determine an appropriate region for pixel-level spatial polling, a single metric cancer and normal pixel-level classification model was developed for each box size metric and the area under the ROC curve (AUC) was plotted separately for each classifier. The plot for AUC *vs.* box size for 16 × 16 pixels to 160 × 160 pixels is displayed in Fig. 5A. The plot and images in this figure demonstrate that tumor identification is strongly influenced by the size of the spatial polling region. When a very small region of 16 × 16 pixels is considered, only a few cancer epithelium pixels are correctly classified and some normal pixels are misidentified as cancer (Fig. 5B). Clearly, this low sensitivity and specificity is not suitable for tumor detection. Therefore, a larger region must be considered for reasonable tumor identification. When a region greater than 48 × 48 pixels is considered the core-level classification AUC appears to plateau at a high value around 0.95 ± 0.06 (95% CI). A classified image for a spatial metric in this range with a box size of 80 × 80 pixels (Fig. 5C) indicates that a good separation of cancer and normal epithelium is achieved with some pixels labeled as cancer in 29 of 31 tumor TMA cores and normal pixels labeled as cancer in only 3 of 34 adjacent normal TMA cores. The core-level AUC begins to decline at a box size of 128 × 128 pixels, primarily due to a reduced sensitivity to small tumor regions in cancer TMA cores. A classified image for spatial polling with a box size of 160 × 160 pixels indicates that no pixels are labeled as cancer in 6 of 31 tumor TMA cores (Fig. 5D). These cores have small or more diffuse tumor regions that may not be detected when only a relatively large area is considered by spatial polling. This loss in classification accuracy with spatial polling at large regions indicates that both epithelium structure and content are important for TMA core-level tumor discrimination, as a classifier based only on epithelium content would likely produce asymptotic behavior after obtaining a maximum AUC value. Conversely, pixel-level classification accuracy increases at a relatively constant rate as the box size increases and levels off as the area considered begins to approach the size of a 1.5 mm diameter TMA core. The pixel-level classification AUC begins to reach a plateau at a box size of 120 × 120 pixels. For a pixel at the center of the TMA core with each 750 × 750 μm^2 box containing this pixel included a total region of approximately 2.25 mm^2 is actually considered by spatial polling, which encompasses the entire TMA core. Therefore, the pixel-level classification AUC approaches the core-level AUC when box sizes above 120 × 120 pixels are employed for spatial metric computation. Pixel-level and core-level classification do not follow the same trend for AUC *vs.* box size because different TMA cores have dramatically different numbers of epithelium pixels. Therefore, not all epithelium pixels are weighted equally in core-level ROC analysis.

While this simple approach yields encouraging results, multiple scales of spatial classification and morphologic diversity were evaluated by adding additional types of metrics and combining metrics from different box sizes. Metrics of the mean and standard deviation of the fraction of pixels classified as epithelium in all boxes of a selected size containing a given pixel were computed for box sizes ranging from 4 × 4 pixels (25 μm^2) to 160 × 160 pixels (1000 μm^2). This range was selected to evaluate areas that contain only a few cells to an entire TMA core. These metrics were combined to obtain an 80-metric image vector with fraction mean and standard deviation metrics for a square box length of pixels 4,8,16 and

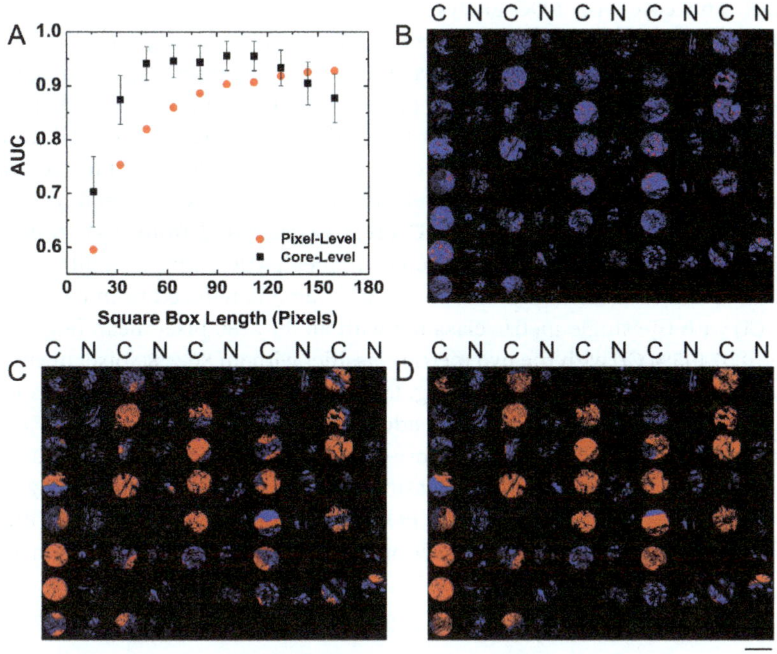

Fig. 5 Automated pathology with spatial metrics. (A) A plot of AUC *vs.* square box length in pixels indicates that accurate TMA core-level classification can be achieved at a range of box sizes and that pixel-level classification becomes more accurate as the box size increases. Error bars for the core-level AUC values represent standard error. (B) A classified image from a 16 × 16 pixel box size indicates low sensitivity when only a small spatial neighborhood is considered. Red represents pixels classified as cancer and blue represents pixels classified as normal. The notation [C, N] denotes [cancer, normal] samples as judged by pathologist review. (C) A classified image from an 80 × 80 pixel box size indicates increased sensitivity with high specificity when a larger spatial neighborhood is considered. (D) A classified image from a 160 × 160 pixel box size (1000 μm^2) indicates reduced sensitivity when a spatial area larger than the tumor area in some TMA cores is considered. The scale bar represents a 1.5 mm diameter of an individual core on this TMA.

so on up to 160. Cancer TMA cores are expected to have higher values for fraction mean metrics since a large mass of epithelium often signifies a tumor. Conversely, normal TMA cores are expected to have higher values for fraction standard deviation metrics since normal breast tissue contains ducts and lobules lined with a thin layer of epithelium. The metrics were sorted by the average estimated error from frequency distributions and used to build a classifier with 1 metric, 2 metrics, 3 metrics, and so on until all 80 metrics were included. A total of 80 classifiers were built, and ROC analysis was performed and an AUC value was computed for each classifier. The ΔAUC value was computed with the addition of each spatial metric and the metrics were re-sorted to select a set that provides optimal classification with a minimal number of spatial metrics. To evaluate the contribution of spatial metrics when only a smaller area of tissue is considered, this classification optimization was repeated with a smaller 40-metric vector containing fraction mean and standard deviation metrics for box sizes ranging from 4 × 4 pixels (25 μm^2) to 80 × 80 pixels (500 μm^2).

Pixel- and core-level ROC analysis for each classifier indicates that adding additional mean and standard deviation metrics has a varying impact on AUC. From the 40-metric vector with a maximum 80 × 80 pixel region an optimal classifier was obtained with 2 spatial metrics: an 80 × 80 pixel (500 μm^2) mean and a 16 × 16 pixel (100 μm^2) standard deviation. In Fig. 6A, cancer and normal pixel-level ROC curves are compared before and after the addition of the standard deviation spatial metric. The pixel-level cancer AUC value is increased from 0.87 to 0.9 and the pixel-level normal AUC value is decreased from 0.91 to 0.88 to provide a minimal overall increase of 0.001 in mean AUC with the addition of the standard deviation metric. The core-level AUC value is reduced from 0.94 ± 0.06 (95% CI) with the single metric classifier with an 80 × 80 pixel mean (Fig. 6B) to 0.92 ± 0.07 (95% CI) with the two metric classifier with an 80 × 80 pixel mean and a 16 × 16 pixel standard deviation (Fig. 6C). While this change is not statistically significant, it does indicate that these additional standard deviation metrics may not be independently beneficial for core-level tumor identification. The discrepancy between pixel and core level classification is explained by examining TMA classified images and the shape of the ROC curve with and without the additional spatial metrics. When the mean fraction metric is considered alone, on most

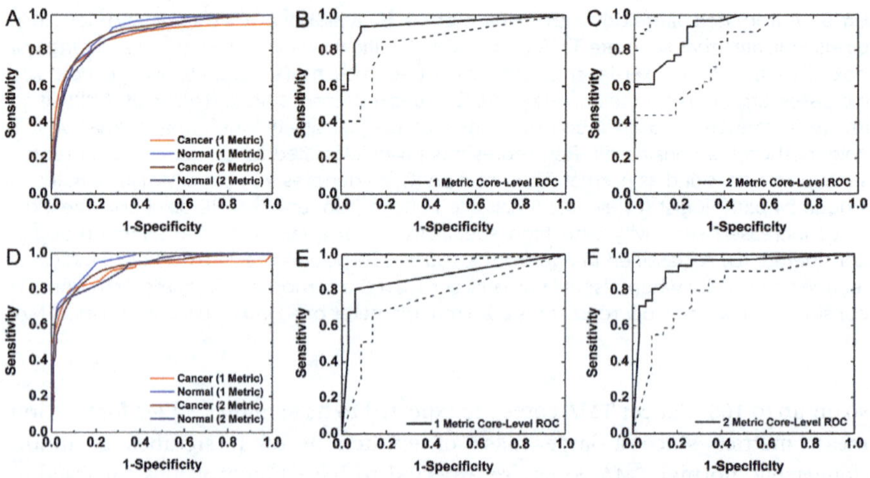

Fig. 6 Classification with multiple types of spatial metrics. (A) ROC analysis for multiple classification models for pixel-level cancer and normal segmentation using metrics computed from a box size range of 4 × 4 pixels to 80 × 80 pixels demonstrates some improvement in pixel-level cancer sensitivity with multivariate classification. (B) ROC analysis for core-level classification with a single metric of mean fraction for an 80 × 80 box indicates accurate overall tumor identification. (C) Core-level specificity is somewhat reduced with the addition of a standard deviation spatial metric for this 80 × 80 box size classifier. (D) ROC analysis for univariate and multivariate classification models for pixel-level cancer and normal segmentation using metrics computed from a box size range of 4 × 4 pixels to 160 × 160 pixels demonstrates some improvement with multivariate classification. (E) ROC analysis for core-level classification with a single mean fraction metric for a 160 × 160 box indicates reduced sensitivity in tumor detection. (F) Core-level sensitivity is improved with the addition of a standard deviation metric to this 160 × 160 box size spatial polling classifier. The areas between the dotted lines on the core-level ROC curves represent 95% confidence regions.

normal cores no pixels are classified as cancer. However, the addition of the standard deviation metric appears to increase the number of pixels classified as cancer on both tumor and normal TMA cores due to spatial heterogeneity. This small increase in non-specific pixels classified as cancer in normal TMA cores is responsible for the observed reduction in specificity on the core-level ROC curve and corresponding reduction in AUC with the addition of the standard deviation spatial metric.

Multi-level classification may provide advantages for core-level tumor detection when a larger tissue area is considered. When the entire 80-metric vector was employed to build an optimal classification model, 3 metrics were selected. These metrics include fraction mean metrics for box sizes of 160 × 160 pixels (1000 μm^2), 152 × 152 pixels (950 μm^2), and 148 × 148 pixels (925 μm^2). These additional metrics provide a minimal increase from 0.906 to 0.908 in pixel-level cancer AUC, from 0.952 to 0.953 in pixel-level normal AUC, and from 0.929 to 0.930 in pixel-level overall mean AUC. These increases in AUC are smaller than those observed when the smaller 80 × 80 pixel region was considered in spatial polling because the initial pixel-level AUC values with the single 160 × 160 mean metric classifier are greater, and may approach the maximum AUC value that can be attained with the information in these spatial metrics. The estimated core-level AUC was also increased by 0.01 from 0.88 ± 0.09 (95% CI) to 0.89 ± 0.08 (95% CI). The width of the confidence interval was reduced due to the increase in AUC,[43] even though the same set of 31 cancer and 34 normal samples were used for both analyses. However, the forward metric selection procedure employed in this classification model is not exhaustive and does not consider all potential metric combinations, and this selected model may not be the only useful model for classification with this spatial polling region.

To understand the influence of mean fraction metrics of different sizes, a classification model was built with the 160 × 160 pixel mean and 16 × 16 pixel standard deviation spatial metrics. With this classification model, the cancer pixel-level AUC was increased from 0.906 to 0.915, the normal pixel-level AUC was reduced from 0.95 to 0.92, and the mean pixel-level AUC was reduced from 0.93 to 0.92. The pixel-level ROC curves in Fig. 6D indicate that these changes in AUC with the multivariate classifier are not substantial for pixel-level cancer classification. Conversely, the core-level AUC was increased from 0.88 ± 0.09 (95% CI) with a single metric classifier with a 160 × 160 mean fraction (Fig. 6E) to 0.93 ± 0.07 (95% CI) with a 2 metric classifier with a 160 × 160 mean fraction and a 16 × 16 standard deviation (Fig. 6F). Although this increase is not statistically significant due to the limited number of samples included in the training analysis, the new classifier does appear to provide a substantial benefit for core-level tumor classification when large regions are considered. This change in AUC is reflected in the classified images and shape of the ROC curve. As noted earlier, classification with a single 160 × 160 fraction mean metric results in 6 missed tumors out of 31 total cancer TMA cores in the calibration dataset. This limitation in sensitivity is reflected in the TMA core-level ROC curve with this single metric classifier in Fig. 6E. When the optimal multivariate classifier with this metric is employed, pixels are labeled as cancer in 30 out of 31 cancer TMA cores. This increase in sensitivity is reflected in the multivariate core-level ROC curve in Fig. 6F and results in the increased observed AUC value. A previous study of 432 breast ducts with necrosis from 26 cancer patients and 520 ducts from 26 normal autopsy samples indicated that the ducts from advanced tumors with intraductal necrosis were rarely smaller than 240 μm in diameter while ducts from normal samples were rarely larger than 180 μm in diameter.[67] Therefore, this spatial metric may provide important information for pixel-level tumor discrimination at a single duct scale. As indicated in Fig. 5, a spatial metric computed from this spatial area alone does not provide accurate tumor discrimination. However, when it is combined with spatial polling over larger areas of tissue it may provide useful additional information for tumor classification.

After consideration of a broad range of spatial regions and multiple combinations of spatial metrics, a single spatial polling classifier was selected for extensive validation analysis. Although a maximum pixel-level AUC is only achieved by spatial polling over areas approaching the size of a TMA core, optimal core-level classification can be achieved with consideration of a much smaller area. Since high density validation TMAs contain cores of diameter 1 mm or smaller, TMA core-level validation analysis was performed with the smallest region to produce an estimated core-level AUC of at least 0.95 on the calibration dataset. Therefore, a spatial polling region of 52 × 52 pixels (325 μm^2) was selected for farther evaluation. This area is reasonable, as a study of 1285 breast ducts from 26 breast ductal carcinoma patients found a mean diameter of 349 μm.[67] Therefore, a region of this size with a high fraction of epithelium should encompass a tumor. Models with this metric were considered but rejected when the TMA core-level AUC was not increased.

Tumor classification validation and sample size analysis

The expected AUC and desired confidence interval half-width were considered in selecting the appropriate sample size for validation. For a binomial problem of unknown sample size with 2 classification options, *e.g.* cancer and normal, the AUC variance can be estimated as

$$\sigma^2(\theta) = \theta(1 - \theta) \tag{4}$$

where θ is the predicted AUC value.[68] This variance can be employed with an acceptable half-width for a confidence interval (L) and a given total sample size (n) to calculate a corresponding z-score by the equation

$$z_{\alpha/2} = L\sqrt{\frac{n}{\theta(1-\theta)}} \text{ (ref. 68).} \tag{5}$$

The p-value associated with the computed z-score is then obtained from tabulated z tables or software packages. In this manner, the z-score will increase linearly with L and the corresponding p-value will increase until it approaches a value of one. This p-value represents the probability that the true AUC is greater than or equal to the lower bound of the confidence interval. For example, for an AUC of 0.95 ± 0.02, a p-value of 0.95 would indicate that an AUC of at least 0.93 will be obtained in 95 out of 100 validation studies with the similar sampling. Therefore, a confidence of 0.95 can be assigned to the interval obtained in that study. The confidence assigned to an AUC value depends upon both the estimated AUC value θ and the sample size n. This is reflected in eqn (5), in which the $z_{\alpha/2}$ score and corresponding p-value increase with n and decrease with θ. This is expected, since populations with less overlap in metric distribution will be more easily separated and studies with a larger sample size will produce a smaller standard error. This trend is reflected in Fig. 7, where confidence increases with sample size for a given θ until it reaches a maximum value near 1. In this plot an interval half width L of 0.02 is assumed for all $z_{\alpha/2}$ and confidence calculations. If the expected AUC value is known from calibration studies, an appropriate sample size n can be selected where the p-value begins to approach 1. For an estimated θ of 0.95, the confidence levels off near a value of 1 with a sample size of $n = 700$. Therefore, our methods predict the number of samples required for definitive validation in addition to developing the protocol itself. The issue of sample size has received some attention as it is a critical aspect of the development of protocols. While at least one previous study examined the effect of sample size theoretically,[69] the integration of classification and statistical validation can be jointly accomplished in the manner above. The combination of spectral histology and spatial polling for tumor identification translates effectively to independent validation samples, as demonstrated by the consistent high AUC values for validation. Although the algorithm is trained exclusively on TMA cores of 1.5 mm in diameter, it translates directly to smaller cores of 1 mm diameter. We have further tested the samples on TMAs that contained cores of sizes as small as 1 mm as well as surgical resections. A detailed study of the validation and the limitations of this study is beyond the scope here and will be analyzed in a future report.

Applications for clinical translation

To implement this technique efficiently in a clinical setting, rapid data acquisition and analysis is necessary as classification of IR datasets can be performed in a matter of minutes. The "standard" data collection parameters of 4 cm^{-1} spectral resolution, 2 scans per pixel, and a 6.25 μm pixel size dictate an acquisition time of at least 1.5 hours to collect data for a 1.5 mm TMA core. These parameters, in our experience, provide an excellent trade-off between data quality and time of acquisition while providing accurate biomedical segmentation for other tissues. As breast tumors are normally evaluated in a clinical setting on much larger biopsy surgical resections, these data collection parameters are not reasonable for

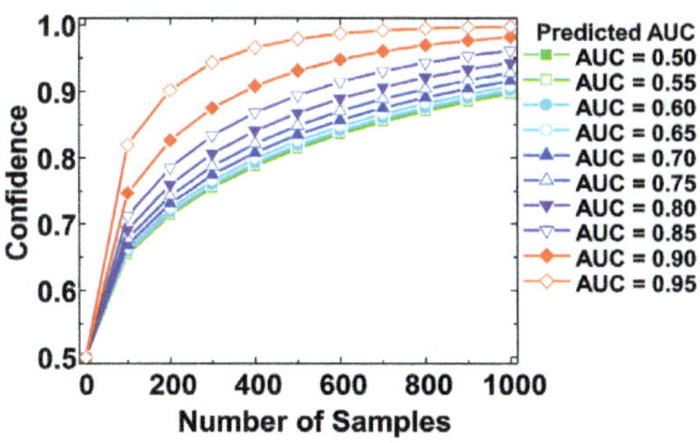

Fig. 7 Prediction of sample size for a potential validation study. Confidence in the AUC value shows a variation with the value of the AUC and the sample size for an interval of half-width 0.2 at a range of AUC values. Confidence increases with both AUC and sample size and the precise sample size needed can be read from the chart.

clinical implementation with conventional IR imaging instrumentation. One route to increasing the rate of data acquisition can be to reduce the required signal to noise ratio (SNR) or coarsen spectral resolution from these levels.[70] From the trading rules for IR spectroscopy, data collection time decreases linearly with spectral resolution and in quadrate with SNR,[71] trends that also hold for IR imaging.[72,73] In addition, a different hardware configuration using a FPA detector can be employed for rapid data acquisition.[74,75] The potential for each of these options is evaluated next in this manuscript to assess the impacts on single pixel spectra, spectral histology classification and spatial pathology classification. Qualitative image evaluation and quantitative ROC analysis were employed to determine the classification potential with reduced spectral quality and detail in IR image datasets.

In practice, a reduction in SNR can be accomplished by decreasing the number of scans per pixel required for data acquisition. To evaluate the potential of classification on high noise data, random Gaussian noise was added to one validation dataset and spectral and spatial classification for the original dataset and the dataset with noise added were compared to assess spectral histology and spatial pathology classification. The RMS noise for background pixels was estimated as 0.001 before adding noise and 0.016 after adding noise. When individual pixel spectra without and with added noise were compared, the added noise clearly obscured many important features within spectra, as shown in Fig. 8A. This is particularly apparent in the fingerprint region, which contains many of the key metrics used in the histology model to segment stroma and epithelium. This loss in spectral quality is reflected in the histology classification ROC analysis. In Fig. 8B, it can be seen that the AUC value for stroma is reduced from 0.99 to 0.95 and the AUC value for epithelium is reduced from 0.98 to 0.88 with added noise. Since a large number of pixels (30 140 stroma and 20 019 epithelium) are used in validation ROC analysis these changes are statistically significant. Notably, the epithelium AUC appears to be more adversely effected by noise than the stroma AUC. A comparison of classified histologic images before and after the addition of noise in Fig. 8E and F, respectively, indicates that randomly distributed epithelium pixels are misclassified as stroma. This reduction in epithelium sensitivity is the primary cause for the drop in the epithelium AUC value. This may be due to the epithelial spectra, consisting of a small set of broadly distributed peaks, leading to a higher apparent absorbance that starts to overlap with the stromal values. Conversely, the stroma spectrum appears more similar to collagen, with a larger set of more overlapping narrow peaks and an increase in values does not affect classification and a decrease in values folds into the tails of the lower noise distributions. By adding random noise, the epithelial spectra become more similar to stromal spectra, which would lead to random misclassifications of epithelium pixels as stroma. Since these misclassification events are random instead of systematic, both histology class images correspond reasonably well with H&E staining (Fig. 8D).

Conversely, the core-level tumor diagnosis AUC value only decreased from 0.97 $\pm$ 0.04 (95% CI) for the original dataset to 0.93 $\pm$ 0.06 (95% CI) for the dataset with added noise. As this change is not statistically significant, spatial polling can be performed on low SNR data without a statistically significant loss in classification potential. The ROC curve in Fig. 8C indicates that reasonable cancer detection sensitivity and specificity are achieved in the high noise dataset. A comparison of

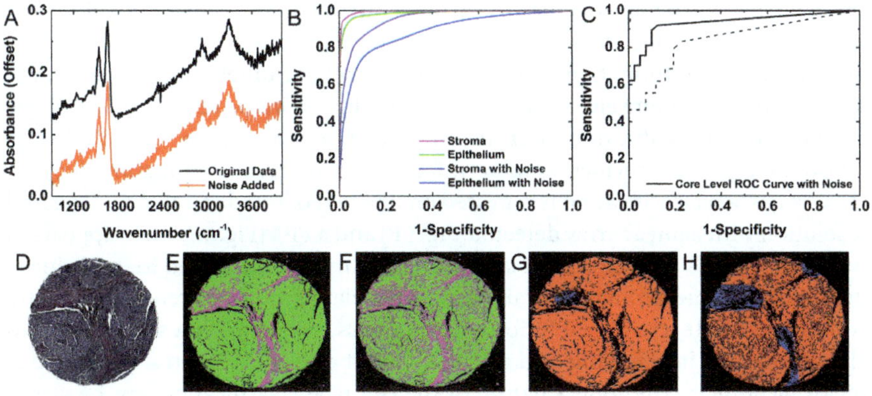

Fig. 8 Classification with reduced SNR. (A) A single pixel spectrum (epithelial) from the original dataset and the dataset with noise added demonstrates reduced visibility of spectral features, particularly in the fingerprint region. (B) Pixel-level segmentation accuracy of stroma and epithelium is decreased in high-noise data, but reasonable cell-type classification is still possible. (C) ROC analysis indicates that reasonable core-level tumor classification is possible with the addition of noise to the dataset. The area between the dotted lines represents bounds for a 95% confidence region. (D) H&E staining, (E) histology classification and (F) pathology classification by spatial polling on the original dataset indicate that this sample contains a dense invasive epithelial tumor. (G) This core classification from the dataset with noise added indicates some reduction in epithelium detection. (H) Tumor detection is also somewhat less sensitive in the core with noise added.

the classified images before and after the addition of noise indicates that the loss in AUC is due to a reduction in sensitivity, which may be a concern for cancer classification. However, most of this loss in sensitivity occurs near epithelium boundary pixels, as evidenced by the classification images to distinguish cancer (red) before and after adding noise in Fig. 8G and H, respectively. Therefore, large tumor regions can still be readily identified in high noise datasets.

Next, spectral resolution was evaluated for its effect on the spectral and spatial classification potential. Previous studies have demonstrated by simulation that a coarser spectral resolution of 16 cm^{-1} can be sufficient for histology classification.[70,76,77] This could potentially decrease data collection time by a factor of 64 due to the dual benefit of collecting fewer points and more light throughput per spectral element. Hence, we tested our approach for histology and pathology classification on two datasets acquired at a 16 cm^{-1} spectral resolution. The first dataset was acquired at 4 cm^{-1} and 16 cm^{-1} spectral resolutions on a 16 × 2 linear array detector and then at 16 cm^{-1} on a separate IR imaging instrument with a 128 × 128 FPA detector. The second dataset was acquired at only a 16 cm^{-1} spectral resolution on the linear array detector. This dataset was employed for validation of classification from rapid image acquisition. A spectrum from a pixel (stromal) was first examined from an image collected at 4 cm^{-1} with the linear array detector, at 16 cm^{-1} with the linear array detector, and at 16 cm^{-1} with the FPA detector in Fig. 9A. Since biomedical spectra are typically complex mixtures, many of the broad spectral features remain apparent at the coarser spectral resolution. However, some finer spectral features apparent in the 4 cm^{-1}

spectrum are less obvious in the 16 cm^{-1} resolution spectra. Notwithstanding, the 16 cm^{-1} spectra collected with the linear and focal plane array detectors appear to provide similar information. These observed spectral differences, however, are not important in themselves and need to be evaluated in the context of histologic classification. Classification of an IR image collected at a 4 cm^{-1} spectral resolution in Fig. 9D provides similar information to H&E staining in Fig. 9C. However, when classification is repeated on images collected at a 16 cm^{-1} spectral resolution with a linear array detector (Fig. 9E) and a FPA (Fig. 9F) there appears to be a bias towards epithelium classification. This may be attributed to the reduced definition of some stroma spectral features associated with collagen in the spectra collected at a 16 cm^{-1} resolution. Nevertheless, in many cores the observed classification of images collected at a 4 cm^{-1} and a 16 cm^{-1} spectral resolution is relatively similar. The robust nature of the cell type classification can be attributed to the types of metrics employed in classification. The peak heights and ratios are relatively consistent as long as the full width at half maximum (FWHM) value remains unchanged. The peak area and center of gravity metrics are also not highly impacted by small changes in peak shape or location.[70] Finally, many classified TMA cores appear similar when acquired at a 16 cm^{-1} spectral resolution with either the linear array or the FPA detector. Therefore, the spectral classification technique is broadly applicable across different spectral resolutions and IR imaging instruments.

Despite some differences in spectral histology, core-level pathology classification appears to be reproducible across different spectral resolutions and IR imaging instruments. Spectral histology images were classified by spatial polling to segment epithelium pixels as cancer or normal. Observed differences in pathology classification of the IR images collected at 4 cm^{-1} (Fig. 9G), at 16 cm^{-1} (Fig. 9H), and at 16 cm^{-1} with the FPA detector (Fig. 9I) follow the pattern of observed differences in histology classification, with the epithelial pixels in the section with more stroma from the 4 cm^{-1} IR image misclassified as normal due to the increased stromal content. In addition, minimal differences were observed in cancer identification between the 16 cm^{-1} images collected with different instruments. The square pixel size for the FPA is somewhat smaller, with a length 5.5 μm instead of 6.25 μm. Therefore the estimated box area on the FPA image is 286×286 μm^2, which is somewhat smaller than the 325×325 μm^2 area considered on all other images collected with the linear array detector. However, the core-level tumor detection ROC analysis in Fig. 9B on the image acquired with the FPA still appears to be highly accurate, with an AUC of 0.92 ± 0.04 (95% CI). Classification at a 16 cm^{-1} spectral resolution is confirmed on an additional validation TMA with 180 samples. Pixel-level ROC analysis for stroma and epithelium histology segmentation in Fig. 10A provides quantitative evidence of reproducible histology, with an AUC of 0.93 for epithelium and 0.96 for stroma. The small reduction in AUC for epithelium is due to a loss in specificity, which confirms the previous observation that some stromal pixels are mislabeled as epithelium. However, the histology AUC values for both classes are still acceptable. Next, core-level cancer detection ROC analysis was performed to assess tumor identification. From the plot in Fig. 10B, high sensitivity and specificity were achieved on this dataset with an AUC of 0.95 ± 0.03 (95% CI). These results indicate that data collection at a coarser 16 cm^{-1} spectral resolution is sufficient to achieve automated cell type classification and tumor detection in breast tissue.

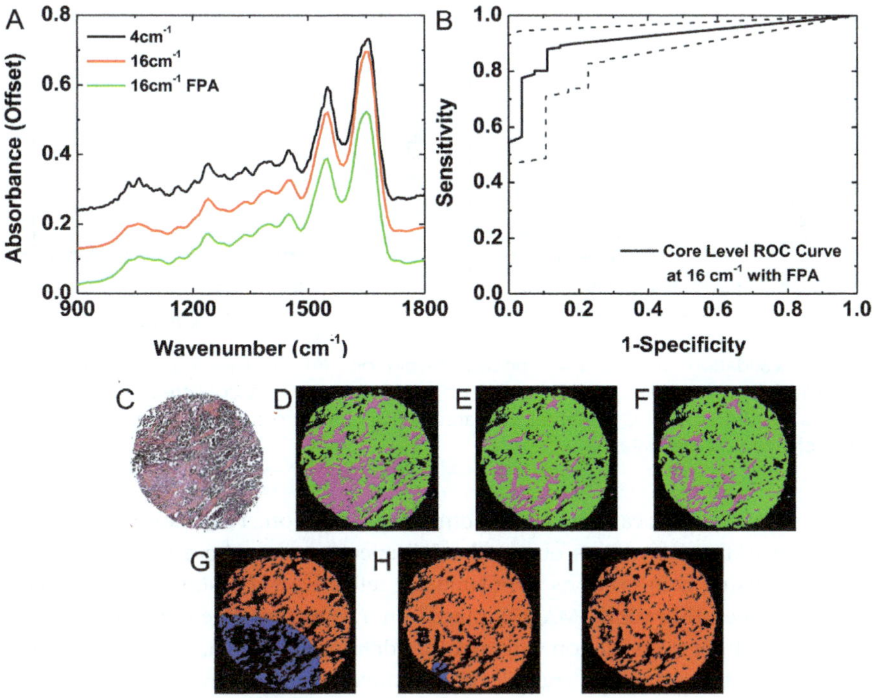

Fig. 9 Classification with a linear array and focal plane array detector with low resolution data. (A) A single-pixel stroma spectrum from images acquired at a 4 cm^{-1} spectral resolution with a linear array detector, 16 cm^{-1} spectral resolution with a linear array detector, and a 16 cm^{-1} spectral resolution with a FPA detector indicate some spectral changes associated with spectral resolution but minimal spectral variation associated with the detector. (B) ROC analysis indicates that accurate core-level classification is achieved at a 16 cm^{-1} spectral resolution with a FPA detector. (C) An epithelium tumor is detected by conventional H&E staining. (D) An IR image of a single TMA core collected at a 4 cm^{-1} spectral resolution with a linear array detector demonstrates epithelium (green) and stroma (magenta) segmentation that is consistent with H&E staining. (E) An IR image of this TMA core collected at a 16 cm^{-1} spectral resolution with a linear array detector demonstrates some additional pixels classified as epithelium. (F) An IR image of this TMA core collected at a 16 cm^{-1} spectral resolution with a FPA detector demonstrates similar classification to the image collected at 16 cm^{-1} with the linear array detector. (G) Pixel-level classification segments the epithelium pixels as cancer (red) or normal (blue) from the 4 cm^{-1} histology classified image. (H) Somewhat more epithelium pixels are identified as cancer from the 16 cm^{-1} image collected with the linear array. (I) The 16 cm^{-1} image collected with the FPA detector provides similar tumor identification as the image collected with the linear array detector.

To achieve cost-effective imaging in a clinical setting, sample preparation expenses must also be minimized. All evaluations to this point in this study have been performed on images collected by light transmission, which generally provides the best quality datasets. Other studies have reported using glass slides,[78] which involve a trade-off between practicality and reduction in spectral wavelength bandpass, or reflective glass slides, which require transflection model imaging. Another alternative is to use attenuated total reflection mode sampling,[79] which can offer higher spatial resolution[80] but requires good contact

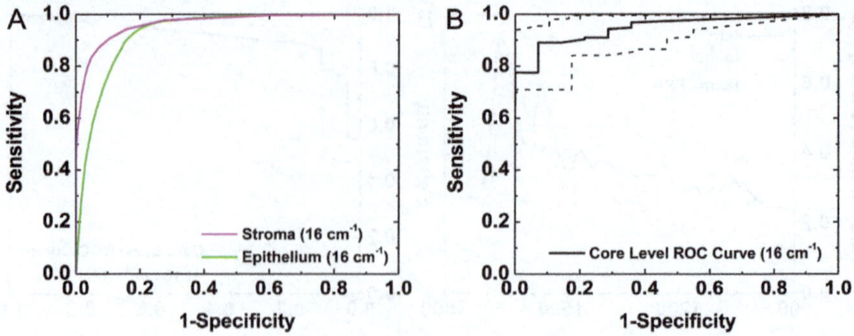

Fig. 10 Validation at a course spectral resolution. (A) Pixel-level segmentation of epithelium and stroma is demonstrated on a 180 patient TMA collected at a 16 cm^{-1} spectral resolution. (B) ROC analysis indicates that accurate core-level cancer detection is also achieved on this dataset.

with the sample that can be cumbersome for translation. To be compatible with clinical workflow as well as obtain the full spectrum, we chose to evaluate the approach using thin sections deposited on reflective glass slides. An adjacent section of the calibration TMA was placed on a reflective slide and images were acquired by light transflection with all other data collection parameters the same as transmission imaging. Transflection induced changes in spectra are well known,[81,82] but there is emerging evidence to demonstrate effective classifications despite these effects.[83] Spectral histologic classification was again performed to segment stroma and epithelium from this data set. Examination of classified images of a TMA core collected by transmission (Fig. 11D) and transflection (Fig. 11E) demonstrate similar classification in both datasets and reasonable correlation with H&E staining (Fig. 11C). Pixels were then labeled on the same cores as the calibration TMA, with 140 899 stroma pixels and 50 393 epithelium pixels manually annotated as a gold standard. Pixel-level histology ROC analysis was then performed to provide a quantitative assessment of the classification accuracy. An AUC of 0.99 was obtained for epithelium and 0.98 was obtained for stroma in transflection datasets. From the pixel-level AUC plots in Fig. 11A a minimal reduction in AUC is observed between the transmission and transflection images. These small differences can likely be attributed to optical effects and changes in SNR due to the double-pass nature of the measurement transflection imaging. Likewise, transmission and transflection images produce similar tumor identification. Pathology classified images for a single TMA core collected by transmission (Fig. 11F) and transflection (Fig. 11G) demonstrate that most epithelium pixels are correctly labeled as cancer in both images. This observation is confirmed by core-level ROC analysis on 75 samples in Fig. 11B with an AUC of 0.93 ± 0.06 (95% CI). While slightly lower than the transmission AUC of 0.95 ± 0.06 (95% CI), these results are not statistically different. Therefore, accurate tumor identification is possible by transflection image, spectral histology classification, and spatial pathology classification. Since our analysis is not based on complicated spectral analysis, rather on a simple spectral and spatial analysis, we anticipate that small changes in known confounding variables such as sample thickness will not have a significant impact. However, a comprehensive test needs to be conducted to quantify the effect, if any, of such changes.

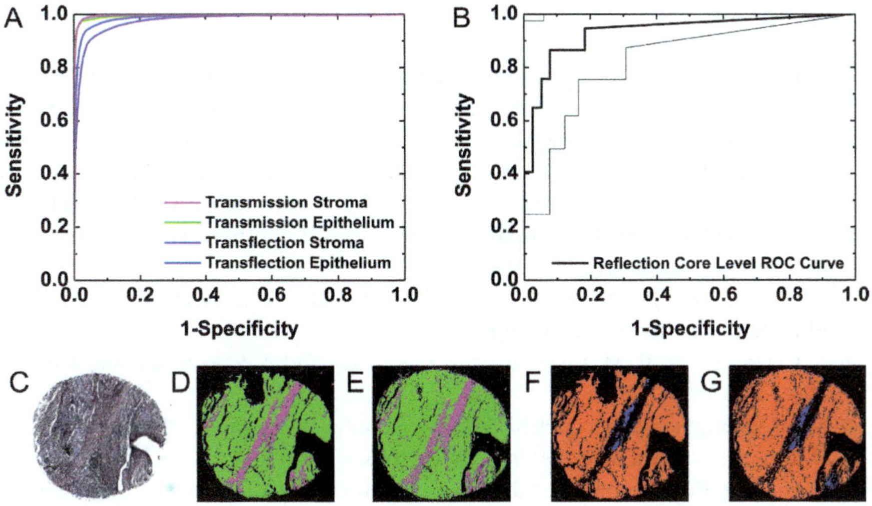

Fig. 11 Validation of transflection-mode imaging data. (A) Pixel-level ROC analysis indicates similar histologic classification accuracy is achieved in transmission and transflection images. (B) Core-level ROC analysis indicates that transflection images can be accurately classified to identify tumors. The area between the dotted lines represents a 95% confidence region. (C) H&E staining, (D) transmission histology, and (E) transflection histology images demonstrate similar cell-type segmentation. Green labels epithelium and magenta labels stroma as in previous images. (F) Transmission and (G) transflection images demonstrate similar patterns of tumor detection. Red labels cancer and blue labels normal epithelium pixels, as noted previously.

Conclusions

Combining FT-IR imaging, spectral histology classification, and spatial pathology classification is demonstrated to provide automated, accurate, and reproducible tumor identification. Each pixel is first labeled as stroma or epithelium using spectral recognition at the single pixel level; subsequently, epithelium pixels are labeled as cancer or normal by spatial polling based upon epithelium content and distribution. Robust classification is demonstrated in a definitive validation study. Options are considered for efficient clinical translation, including classification of data with increased noise or reduced spectral resolution. Tumor classification is also demonstrated on images collected with a FPA detector and on inexpensive reflective glass slides. The data demonstrate that a practical protocol for rapid breast cancer identification is possible and the various trade-offs in speeding up or reducing costs for clinical translation. Validation of this protocol and advances in instrumentation for rapid data acquisition can lead to a practical solution for breast cancer detection on biopsy samples.

References

1 *SEER Cancer Statistics Review, 1975–2012*, ed. N. Howlader, A. M. Noone, M. Krapcho, J. Garshell, D. Miller, S. F. Altekruse, C. L. Kosary, M. Yu, J. Ruhl, Z. Tatalovich, A. Mariotto, D. R. Lewis, H. S. Chen, E. J. Feuer and K. A. Cronin, National Cancer Institute, Bethesda, MD, http://seer.cancer.gov/csr/

1975_2012/, based on November 2014 SEER data submission, posted to the SEER web site, April 2015.

2 A. B. Chagpar and K. M. McMasters, *J. Surg. Res.*, 2007, **140**, 214–219.

3 *Thomson Reuters In-Patient and Out-Patient Views Market-Scan Database*, 2008.

4 S. H. Parker, F. Burbank, R. J. Jackman, C. J. Aucreman, G. Cardenosa, T. M. Cink, J. L. Coscia, G. W. Eklund, W. P. Evans, P. R. Garver, H. F. Gramm, D. K. Haas, K. M. Jacob, K. M. Kelly, L. K. Killebrew, M. C. Lechner, S. J. Perlman, A. P. Smid, L. Tabar, F. E. Taber and R. T. Wynn, *Radiology*, 1994, **193**, 359.

5 D. Carter, *Interpretation of Breast Biopsies*, Lippincott Williams & Wilkins, Philadelphia, 4th edn, 2004, pp. 37–50.

6 A. R. Hatmaker, R. M. J. Donahue, J. L. Tarpley and A. S. Pearson, *Am. J. Surg.*, 2006, **192**(5), e37.

7 M. Simunovic, A. Gagliardi, D. McCready, A. Coates, M. Levine and D. DePetrillo, *Can. Med. Assoc. J.*, 2001, **165**(4), 421.

8 E. V. Lang, K. S. Berbaum and S. K. Lutgendorf, *Radiology*, 2009, **250**, 631–637.

9 R. Bhargava, *Appl. Spectrosc.*, 2012, **66**, 1091–1120.

10 P. Lasch, L. Chiriboga, H. Yee and M. Diem, *Technol. Cancer Res. Treat.*, 2002, **1**, 1–7.

11 I. W. Levin and R. Bhargava, *Annu. Rev. Phys. Chem.*, 2005, **56**, 429–474.

12 G. Bellisola and C. Sorio, *Am. J. Cancer Res.*, 2012, **2**, 1–21.

13 H. Fabian, P. Lasch, M. Boese and W. Haensch, *J. Mol. Struct.*, 2003, **661**, 411–417.

14 H. Fabian, P. Lasch, M. Boese and W. Haensch, *Biopolymers*, 2002, **67**, 354–357.

15 T. Gao, J. Feng and Y. Ci, *Anal. Cell. Pathol.*, 1999, **18**, 87–93.

16 S. E. Holton, M. J. Walsh, A. Kajdacsy-Balla and R. Bhargava, *Biophys. J.*, 2011, **101**, 1513–1521.

17 S. E. Holton, A. Bergamaschi, B. S. Katzenellenbogen and R. Bhargava, *PLoS One*, 2014, **9**, e96878.

18 M. Smolina and E. Goormaghtigh, *Analyst*, 2015, **140**, 2336–2343.

19 B. Bird, M. Romeo, N. Laver and M. Diem, *J. Biophotonics*, 2009, **2**, 37–46.

20 J. Anastassopoulou, E. Boukaki, C. Conti, P. Ferraris, E. Giorgini, C. Rubini, S. Sabbatini, T. Theophanides and G. Tosi, *Vib. Spectrosc.*, 2009, **51**, 270–275.

21 R. Dukor, G. Story and C. Marcott, *Inst. Phys. Conf. Ser.*, 2000, **165**, 79–80.

22 S. Brady, M. N. Do and R. Bhargava, *16th IEEE International Conference on Image Processing(ICIP)*, 2009, pp. 829–832.

23 H. Fabian, M. Jackson, L. Murphy, P. H. Watson, I. Fichtner and H. H. Matsch, *Biospectroscopy*, 1995, **1**, 37–45.

24 M. Jackson, J. R. Mansfield, B. Dolenko, R. L. Somorjai, H. H. Mantsch and P. H. Watson, *Cancer Detect. Prev.*, 1999, **23**, 245–253.

25 H. Fabian, N. A. N. Thi, M. Eiden, P. Lasch, J. Schmitt and D. Naumann, *Biochim. Biophys. Acta*, 2006, **1758**, 874–882.

26 R. Eckel, H. Huo, W. W. Guan, X. Hu, W. D. Che and W. Huang, *Vib. Spectrosc.*, 2001, **27**, 165–173.

27 Y. Ci, T. Y. Gao, J. Feng and Z. Q. Guo, *Appl. Spectrosc.*, 1999, **53**, 312–315.

28 W. Y. Yang, X. L. Xiao, J. Tan and Q. Cai, *Vib. Spectrosc.*, 2009, **49**, 64–67.

29 C. Liu, Y. Zhang, X. Yan, C. Zhang, W. Li and D. Yang, *J. Lumin.*, 2006, **119–120**, 132–136.

30 T. Gao, J. Feng and Y. Ci, *Anal. Cell. Pathol.*, 1999, **18**, 87–93.

31 A. Benard, C. Desmedt, M. Smolina, P. Szternfeld, M. Verdonck, G. Rouas, N. Kheddoumi, F. Rothé, D. Larsimont, C. Sotiriou and E. Goormaghtigh, *Analyst*, 2014, **139**, 1044–1056.

32 A. K. Kodali, M. Schulmerich, J. Ip, G. Yen, B. T. Cunningham and R. Bhargava, *Anal. Chem.*, 2010, **82**, 5697–5706.

33 M. R. Kole, R. K. Reddy, M. V. Schulmerich, M. K. Gelber and R. Bhargava, *Anal. Chem.*, 2012, **84**, 10366–10372.

34 J. Rowlette, M. Weida, B. Bird, D. Arnone, M. Barre and T. Day, *BioOptics World*, 2014, **7**, 34–37.

35 N. Kröger, A. Egl, M. Engel, N. Gretz, K. Haase, I. Herpich, B. Kränzlin, S. Neudecker, A. Pucci, A. Schönhals, J. Vogt and W. Petrich, *J. Biomed. Opt.*, 2014, **19**, 111607.

36 N. Kröger-Lui, N. Gretz, K. Haase, B. Kränzlin, S. Neudecker, A. Pucci, A. Regenscheit, A. Schönhals and W. Petrich, *Analyst*, 2015, **140**, 2086–2092.

37 K. Yeh, S. Kenkel, J. N. Liu and R. Bhargava, *Anal. Chem.*, 2015, **87**, 485–493.

38 M. J. Nasse, M. J. Walsh, E. C. Mattson, R. Reninger, A. Kajdacsy-Balla, V. Macias, R. Bhargava and C. Hirschmugl, *Nat. Methods*, 2011, **8**, 413–416.

39 R. K. Reddy, M. J. Walsh, M. V. Schulmerich, P. S. Carney and R. Bhargava, *Appl. Spectrosc.*, 2013, **67**, 93–105.

40 H. Sreedhar, V. K. Varma, P. L. Nguyen, B. Davidson, S. Akkina, G. Guzman, S. Setty, A. Kajdacsy-Balla and M. J. Walsh, *J. Visualized Exp.*, 2015, **95**, 52332.

41 R. Bhargava, D. C. Fernandez, S. Hewitt and I. W. Levin, *Biochim. Biophys. Acta, Biomembr.*, 2006, **1758**, 830–845.

42 J. Padayachee, W. I. D. Rae and M. J. Alport, *IFMBE Proceedings*, 2007, **4**, 2476–2479.

43 J. A. Hanley and B. J. McNeil, *Radiology*, 1982, **143**, 29–36.

44 R. Harper and B. Reeves, *Br. Med. J.*, 1999, **318**, 1322–1323.

45 M. J. Wolfsegger and J. Thomas, *J. Pharmacokinet. Pharmacodyn.*, 2005, **32**, 757–766.

46 A. H. Beck, A. R. Sangoi, S. Leung, R. J. Marinelli, T. O. Nielsen, M. J. van de Vijver, R. B. West, M. van de Rijn and D. Koller, *Sci. Transl. Med.*, 2011, **3**, 108ra113.

47 J. T. Kwak, S. M. Hewitt, S. Sinha and R. Bhargava, *BMC Cancer*, 2011, **11**, 62, DOI: 10.1186/1471-2407-11-62.

48 D. C. Fernandez, R. Bhargava, S. M. Hewitt and I. W. Levin, *Nat. Biotechnol.*, 2005, **23**, 469–474.

49 D. Mayerich, M. J. Walsh, A. Kadjacsy-Balla, P. S. Ray, S. M. Hewitt and R. Bhargava, *Technology*, 2015, **3**, 27–31.

50 D. S. May and N. E. Stroup, *Plast. Reconstr. Surg.*, 1991, **87**, 193–194.

51 P. P. Rosen, *Rosen's Breast Pathology*, Lippincott, Williams, and Wilkins, Philadelphia, PA, 2nd edn, 2001, ch. 11, p. 249.

52 R. Bhargava, S.-Q. Wang and J. L. Koenig, *Appl. Spectrosc.*, 1998, **52**, 323–328.

53 M. Romeo and M. Diem, *Vib. Spectrosc.*, 2005, **38**, 129–132.

54 A. Kohler, J. Sulé-Suso, G. D. Sockalingum, M. Tobin, F. Bahrami, Y. Yang, J. Pijanka, P. Dumas, M. Cotte, D. G. van Pittius, G. Parkes and H. Martens, *Appl. Spectrosc.*, 2008, **62**, 259–266.

55 P. Bassan, H. J. Byrne, F. Bonnier, J. Lee, P. Dumas and P. Gardner, *Analyst*, 2009, **134**, 1586–1593.

56 J. T. Kwak, R. K. Reddy, S. Sinha and R. Bhargava, *Anal. Chem.*, 2012, **84**, 1063–1069.

57 B. Bird, M. Miljković and M. Diem, *J. Biophotonics*, 2010, **3**, 597–608.

58 P. Bassan, A. Sachdeva, A. Kohler, C. Hughes, A. Henderson, J. Boyle, J. H. Shanks, M. Brown, N. W. Clarke and P. Gardner, *Analyst*, 2012, **137**, 1370–1377.

59 K. R. Bambery, B. R. Wood and D. McNaughton, *Analyst*, 2012, **137**, 126–132.

60 T. van Dijk, D. Mayerich, P. S. Carney and R. Bhargava, *Appl. Spectrosc.*, 2013, **67**, 546–552.

61 B. J. Davis, P. S. Carney and R. Bhargava, *Anal. Chem.*, 2010, **83**, 525–532.

62 D. Naumann, *Appl. Spectrosc. Rev.*, 2001, **36**, 239–298.

63 M. Jackson, L. Choo, P. Watson, W. Halliday and H. H. Mantsch, *Biochim. Biophys. Acta*, 1995, **1270**, 1–6.

64 R. Bhargava, S.-Q. Wang and J. L. Koenig, *Adv. Polym. Sci.*, 2003, **163**, 137–191.

65 R. K. Reddy and R. Bhargava, *Analyst*, 2010, **135**, 2818–2825.

66 R. Bhargava, S.-Q. Wang and J. L. Koenig, *Appl. Spectrosc.*, 2000, **54**, 1690–1706.

67 N. A. Mayr, J. J. Staples, R. A. Robinson, J. E. Vanmetre and D. H. Hussey, *Cancer*, 1991, **67**, 2805–2812.

68 N. A. Obuchowski, *Stat. Meth. Med. Res.*, 1998, **7**, 371–392.

69 C. Beleites, U. Neugebauer, T. Bocklitz, C. Kraft and J. Popp, *Anal. Chim. Acta*, 2013, **760**, 25–33.

70 R. Bhargava, *Anal. Bioanal. Chem.*, 2007, **389**, 1155–1169.

71 P. R. Griffiths and J. A. de Haseth, *Fourier Transform Infrared Spectrometry*, John Wiley & Sons, Hoboken, NJ, 2nd edn, 2007, p. 254.

72 C. M. Snively and J. L. Koenig, *Appl. Spectrosc.*, 1999, **53**, 170–174.

73 R. Bhargava and I. W. Levin, *Anal. Chem.*, 2001, **73**, 5157–5167.

74 E. N. Lewis, P. J. Treado, R. C. Reeder, G. M. Story, A. E. Dowery, C. Marcott and I. W. Levin, *Anal. Chem.*, 1995, **67**, 3377–3381.

75 K. M. Dorling and M. J. Baker, *Trends Biotechnol.*, 2013, **31**, 437–438.

76 F. N. Keith, R. K. Reddy and R. Bhargava, *Proc. SPIE*, 2008, **6853**, 685306.

77 F. N. Pounder, R. K. Reddy, M. J. Walsh and R. Bhargava, *Proc. SPIE*, 2009, **7186**, 71860F.

78 P. Bassan, J. Mellor, J. Shapiro, K. Williams, M. Lisanti and P. Gardner, *Anal. Chem.*, 2014, **86**, 1648–1653.

79 R. Dukor, M. Liebman and B. Johnson, *Cell. Mol. Biol.*, 1998, **44**, 211–217.

80 M. J. Walsh, S. E. Holton, A. Kajdacsy-Balla and R. Bhargava, *Vib. Spectrosc.*, 2012, **60**, 23–28.

81 T. P. Wrobel, B. Wajnchold, H. J. Byrne and M. Baranska, *Vib. Spectrosc.*, 2013, **69**, 84–92.

82 A. Ergin, F. Großerüschkamp, O. Theisen, K. Gerwert, S. Remiszewski, C. M. Thompson and M. Diem, *Analyst*, 2015, **140**, 2465–2472.

83 D. Perez-Guaita, P. Heraud, K. M. Marzec, M. de la Guardia, M. Kiupel and B. R. Wood, *Analyst*, 2015, **140**, 2376–2382.

Faraday Discussions

PAPER

FTIR spectroscopic imaging and mapping with correcting lenses for studies of biological cells and tissues

James A. Kimber,[a] Liberty Foreman,[b] Benjamin Turner,[a] Peter Rich[b] and Sergei G. Kazarian*[a]

Received 10th November 2015, Accepted 4th December 2015
DOI: 10.1039/c5fd00158g

Histopathology of tissue samples is used to determine the progression of cancer usually by staining and visual analysis. It is recognised that disease progression from healthy tissue to cancerous is accompanied by spectral signature changes in the mid-infrared range. In this work, FTIR spectroscopic imaging in transmission mode using a focal plane array (96 × 96 pixels) has been applied to the characterisation of Barrett's oesophageal adenocarcinoma. To correct optical aberrations, infrared transparent lenses were used of the same material (CaF_2) as the slide on which biopsies were fixed. The lenses acted as an immersion objective, reducing scattering and improving spatial resolution. A novel mapping approach using a sliding lens is presented where spectral images obtained with added lenses are stitched together such that the dataset contained a representative section of the oesophageal tissue. Images were also acquired in transmission mode using high-magnification optics for enhanced spatial resolution, as well as with a germanium micro-ATR objective. The reduction of scattering was assessed using *k*-means clustering. The same tissue section map, which contained a region of high grade dysplasia, was analysed using hierarchical clustering analysis. A reduction of the trough at 1077 cm^{-1} in the second derivative spectra was identified as an indicator of high grade dysplasia. In addition, the spatial resolution obtained with the lens using high-magnification optics was assessed by measurements of a sharp interface of polymer laminate, which was also compared with that achieved with micro ATR-FTIR imaging. In transmission mode using the lens, it was determined to be 8.5 μm and using micro-ATR imaging, the resolution was 3 μm for the band at a wavelength of *ca.* 3 μm. The spatial resolution was also assessed with and without the added lens, in normal and high-magnification modes using a USAF target. Spectroscopic images of cells in transmission mode using two lenses are also presented, which are necessary for correcting chromatic aberration and refraction in both the condenser and objective. The use of lenses is shown to be necessary for obtaining high-quality spectroscopic images of cells

[a]*Department of Chemical Engineering, Imperial College London, London, SW7 2AZ, UK. E-mail: s.kazarian@imperial.ac.uk*

[b]*The Glynn Laboratory of Bioenergetics, Institute of Structural and Molecular Biology, University College London, London, WC1E 6BT, UK*

in transmission mode and proves the applicability of the pseudo hemisphere approach for this and other microfluidic systems.

Introduction

Fourier Transform Infrared (FTIR) mid-infrared micro-spectroscopy combined with a focal plane array detector,[1–4] typically containing 64 × 64 pixels, has been used to characterise a variety of materials and systems such as polymers,[5–8] crude oil deposits[9,10] and microfluidic devices.[11,12] The acquisition of spatial and chemical information simultaneously allows for the measurement of heterogeneous samples at a significantly faster rate than would be possible with point mapping. In particular, FTIR spectroscopic imaging is suitable for the analysis of cellular tissue and biological materials[13,14] due to its intrinsic label-free approach, requiring no modification of the tissue unlike the staining methods commonly used for visible assessment and can reveal information on protein conformation[15] and other cellular components.[16,17] This powerful approach, which relies on molecular vibrations, can be used to differentiate the disease states of cells[18] and thus is of great interest with respect to new diagnostic techniques as it can remove some degree of subjectivity that is present with visual histopathology.[19–21]

Measurement of tissue samples using FTIR spectroscopic imaging is typically done in transmission mode[22,23] or attenuated total reflection (ATR) mode.[14,16] Both sampling approaches have their advantages and disadvantages and the choice of one approach over the other depends on the sample being analysed.[3,16,24] ATR mode uses a high-refractive index prism such that infrared light undergoes total internal reflection. It probes samples using the evanescent wave, which penetrates into the sample between 1 μm and 10 μm depending on the wavelength of the light, angle of incidence and refractive indices of the sample and prism. As such, it is independent of overall sample thickness and offers higher spatial resolution due to the use of high-refractive index materials in the prism (*e.g.* germanium) which lowers the diffraction limit because of the increased numerical aperture.[3,14] With respect to cellular imaging, isolated cells can be studied in detail, with a projected pixel size of 1 μm and field of view of approximately 64 × 64 μm^2. However, ATR mode is a contact approach which has the possibility of deforming or damaging the sample such as a tissue, and introducing cross-contamination between measured regions, causing mapping to be particularly challenging.

Transmission mode allows for mapping with the use of an XY stage, and a wider field of view of 170 × 170 μm^2 with a projected pixel size of 2.67 μm.[25] Advances in microscope design, such as with the Agilent Cary 620 system, use 'high-magnification' (HM) optics with a projected pixel size of 1.1 μm – comparable to ATR.[26] However, the real spatial resolution of the system, R, *i.e.* the distinct resolution between two points (airy discs), is determined by the Rayleigh criterion (eqn (1)), which itself is a function of the numerical aperture of the condenser (NA_C) and objective (NA_O).[27] The numerical aperture is defined as $n \sin \theta$, λ is the wavelength of the light and R is the distance between two points.

$$R = \frac{1.22\lambda}{NA_O + NA_C} \quad (1)$$

Often the numerical aperture of the objective and condenser are the same so that, at 3000 cm^{-1}, the maximum spatial resolution for a microscope with a 0.62 NA objective and condenser is 3.3 μm. As the FPA is spatially discrete, it is necessary to oversample the imaged area to avoid aliasing effects and obtain true diffraction-limited images; hence the use of HM optics can reveal details at this limit. The Agilent high-magnification system produces a projected pixel size of 1.1 μm with a 15× objective by introducing additional 5× magnification optics that also include a 2.06× de-magnification unit resulting in an overall 36× magnification, so dividing the physical size of the pixel in the FPA detector by 36 produces the value of 1.1 μm. This projected pixel size has been cited by others as "spatial resolution" which is imprecise and thus it needs to be stated clearly that this value is a significant over-estimation of the achieved spatial resolution. Bhargava and co-workers[28] produced a recent report for practitioners of FTIR imaging about the effects of optical components on measured spectra as guidance for the optimisation of FTIR imaging systems. They highlighted that this is particularly important for biomedical analysis since scattering effects when measuring tissue samples affect the quality of the obtained spectra.[29–31] Our methodology presented here is one of the practical approaches for improving the FTIR spectra of tissues and cells by using additional optical components in imaging systems.

Transmission mode in FTIR spectroscopic imaging relies on infrared light passing through the whole sample and can have a higher signal to noise ratio compared with ATR mode. Its limitation is the requirement for thin samples to avoid saturation of spectral bands and also because the thickness of the samples should typically be less than the average size of domains as otherwise it could produce a spurious image where domains overlap. Another disadvantage of transmission mode is the potential for optical effects that degrade the image and spectral quality. One such effect is scattering which produces spectral anomalies and degrades the spatial resolution. For biological samples, scattering is predominantly Mie and resonant Mie scattering which is due to cells and their organelles being of similar dimensions to the wavelength of mid-infrared light (between 3 to 12 μm).[30,32,33] These distortions manifest themselves in the spectrum and interfere with subtle band shifts often purported to be due to changes in protein morphology. This is particularly problematic around the amide I band where a particularly strong dispersion artefact exists. Existing methods to correct spectra such as extended multiplicative scatter correction (EMSC) algorithms, and more recently, iterative resonant Mie scattering EMSC (RMieS-EMSC) have been used to correct scattering and the aforementioned dispersion artefact.[30,34]

Spectroscopic imaging in transmission mode through spectroscopic windows also introduces chromatic aberration, where different wavelengths are in focus at different distances from the objective. This is due to infrared light entering the window at an angle, causing degrees of refraction as a function of wavelength. This can be corrected using a lens of the same material, placed on top of the window to form a pseudo-hemisphere.[25] In this way, light enters normal to the lens surface and does not refract at the entry to the window material as the refractive indices of the lens and window are the same. This removes refraction and chromatic aberration, and acts as an immersion objective which increases the spatial resolution. Chan and Kazarian demonstrated this pseudo-hemisphere approach (Fig. 1) with calcium fluoride lenses, where the spatial resolution was increased by a factor of 1.4 for both a polymer interface and a 1951 USAF target.[25]

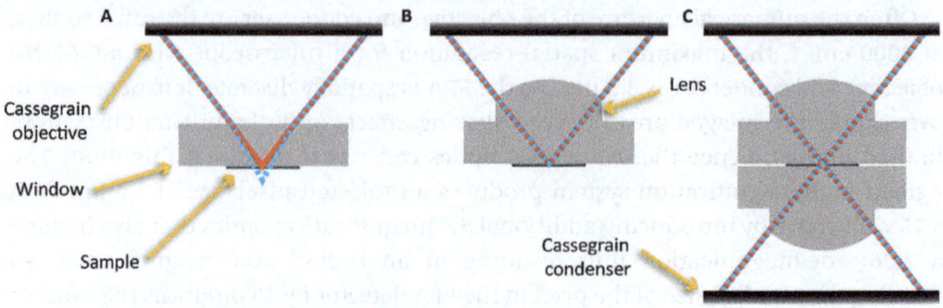

Fig. 1 Schematic of chromatic aberration through an infrared window (A) which is corrected with the use of a lens (B). The lens and window, both made of the same material, form a pseudo-hemisphere, acting as an immersion objective and increasing spatial resolution. For samples between two windows such as a microfluidic device, a bottom lens can be added for the condenser (C).

A potential issue with the use of a correcting lens is the reduction in the field of view. In cases without the lens, high resolution and large fields of view can be obtained by mapping the sample, obtaining spectral images which are tiled to form a large area. This is important as mapping large areas automatically can be obtained from thousands to millions of spectra for use with multivariate statistical analysis.[35] In this work, we demonstrate a method by which to map samples with the lens in place, increasing the spatial resolution and providing large area imaging capabilities. It was also reported that scattering is reduced using the lens,[25,36] the degree of which is also assessed in this work when imaging with the added lens is combined with the mapping approach. We also present the use of the lens with and without HM mode for the first time.

Methodology

A tissue sample similar to that used by Foreman *et al.* was used for these experiments.[37] The tissue sample in this work was obtained from a patient participating in a clinical research study in compliance with the relevant laws and institutional guidelines from the UK's National Research Ethics Service, REC no. 08/H0808/8. In brief, an oesophageal biopsy section of 8 μm thickness was mounted on a 2 mm thick calcium fluoride window. The sample was deparaffinised as per the procedure described by Foreman *et al.* and stored with a desiccant before use. For the spatial resolution tests, a 1951 USAF resolution target made from 1.5 mm float glass and vacuum deposited chromium supplied by Edmund Optics was used. As the spatial resolution using a USAF target is a highly idealised system, a *ca.* 20 μm thick sample containing a sharp interface between two spectrally dissimilar polymers was also imaged using the lenses in HM mode. The polymer interface, acting as a representative real sample, creates a test by which spectral contamination from one side of the interface to the other can be assessed, providing a measure of spatial resolution relevant to the spectroscopic imaging of tissues and as noted in previous work, it is actually 3 times worse than the resolution reported by the USAF target.[25]

Mapping experiments were conducted using a Bruker IRscope II infrared microscope with a 15 × 0.4 NA objective and 96 × 96 FPA detector, connected to a Bruker IFS 66s spectrometer. All other experiments were measured using an Agilent Cary 620 microscope with a 15 × 0.62 NA objective with HM optics where stated, connected to an Agilent 670 spectrometer. Tissue samples were measured at 4 cm^{-1} spectral resolution and with 192 scans. USAF target and polymer interface measurements were measured at 8 cm^{-1} with 128 scans.

The tissue samples were measured either with the tissue on the top-side of the calcium fluoride window for mapping without the lens *i.e.* tissue closest to the objective, or inverted for mapping with the lens, *i.e.* tissue closest to the condenser. In all cases, the tissue was in contact with the spectroscopic window on one side only and was not sandwiched between windows. The lens as used by Chan and Kazarian[25] was also applied in this work; it was originally designed for 4 mm thick calcium fluoride windows. In this case, an additional 2 mm thick window was placed between the mounted sample window and the lens to place the tissue at the lens focal point. To avoid off-centre distortions such as coma, the field aperture of the IRscope II was used to centre the lens and condenser with the objective. Placing the lens in a custom-designed holder ensured that the lens remained in alignment even if the window with the attached sample was moved by the XY stage. In this way, the window with the sample could slide underneath the lens, enabling mapping (Fig. 2). Custom OPUS® macros to control the stage and take measurements were used to obtain separate tiles. A blank region was marked for acquiring a background such that after every N scans, a background tile was also measured. As the background would change over time, each tile was ratioed against a linear interpolation of the two nearest (in time) background tiles to produce absorbance spectra. Tiles were then stitched together to produce a large array of spatially resolved absorbance spectra. The software used for imaging dataset ratioing and stitching was Matlab®.

For imaging with and without HM optics on the Cary 620 microscope, the tissue was imaged in a similar way but without the lens holder as mapping was not conducted for these experiments. The USAF target was measured with and without the lens, and in cases with the lens, an additional 4 mm calcium fluoride window was placed between the lens and the USAF target. For the polymer

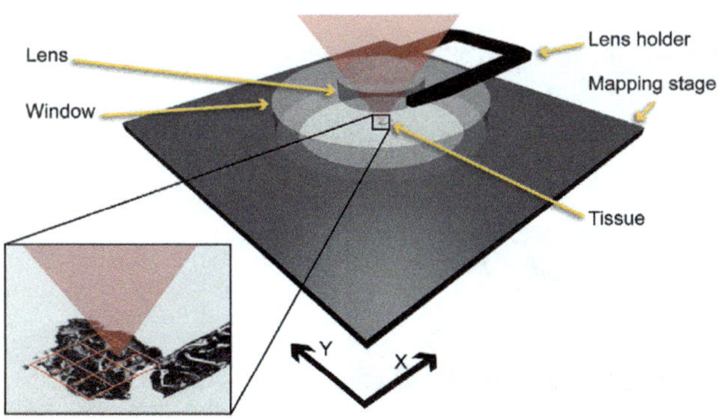

Fig. 2 Schematic showing imaging and mapping approach using the lens. The lens is kept in line with the objective while the window and sample are slid underneath using an XY stage, obtaining multiple tiles which are subsequently combined to form a large spectral image.

interface sample, it was sandwiched between two 4 mm calcium fluoride windows, and lenses were placed above and below the windows. This was also used for the imaging of live cells in a microfluidic device which is also presented in this work. The use of two lenses ensures the numerical aperture of the condenser and objective are equal, further increasing the spatial resolution more than is possible with only one lens. The polymer was imaged in HM mode.

In order to assess the degree of scattering from the tissue sample with each imaging set-up, spectra were processed using Cytospec software. Initially, spectra were passed through a quality test based on the integrated absorbance of the band corresponding to amide II (1485–1570 cm^{-1}), with pixels falling below an integrated absorbance value of 10 cm^{-1} being rejected from further analysis. Subsequently, spectra were vector-normalised using the region between 1200 cm^{-1} and 2000 cm^{-1} *i.e.* the fingerprint region. Although the FPA detector is sensitive to 900 cm^{-1}, the long path-length of calcium fluoride with the lens (8 mm) significantly reduced light throughput below 1200 cm^{-1} and as a result there was increased noise between 1000 cm^{-1} and 1200 cm^{-1}. Finally, *k*-means clustering to generate two clusters was performed on the spectra using the region between 1200 cm^{-1} and 2500 cm^{-1}, which would capture not only the fingerprint region but also the dispersion artefact. The spectral distance method used was Pearson's correlation coefficient (*D*-values). The *k*-means clustering would thus classify the spectra into clusters with and without the dispersion artefact, and the average spectra of the clusters from the original measured spectra are presented to demonstrate this along with the cluster maps. The integrated absorbance band corresponding to amide II (1485–1570 cm^{-1}) was also plotted as a function of all pixels to generate chemical images of the tissues.

Results and discussion

Mapping with and without the lens

Fig. 3 shows the results of mapping without the lens and Fig. 4 shows the results of mapping with the lens. The images are oriented differently due to the fact that for imaging without the lens, the tissue is mapped on the top of the window,

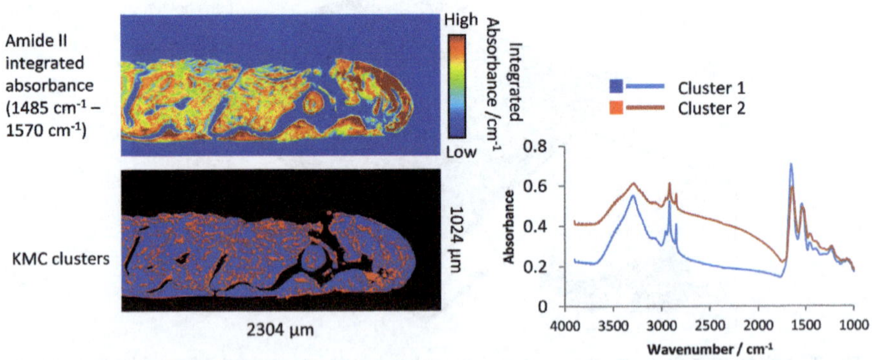

Fig. 3 FTIR images of oesophageal tissue without the lens obtained by mapping from multiple stitched images. Clusters from KMC are shown with average spectra from each cluster. Cluster 2 (red) indicates strong scattering at the edges of the tissue.

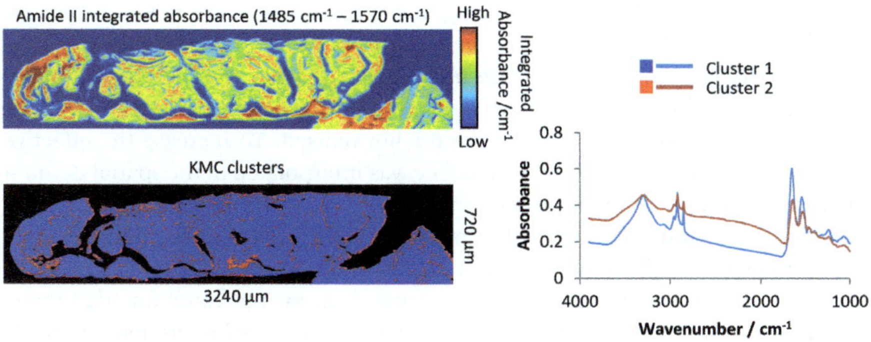

Fig. 4 FTIR images of oesophageal tissue with the lens obtained by mapping from multiple stitched images. Clusters from KMC are shown with average spectra from each cluster. Cluster 2 (red) indicates scattering at the edges of the tissue and the fraction of spectra classified into cluster 2 is reduced using the lens.

whereas with the lens, the window is flipped such that the tissue is on the underside. *k*-Means clustering was used to sort regions into those without the dispersion artefact and those with, as shown in the average spectra from each cluster (clusters 1 and 2 respectively). Although a large image area can be easily obtained, scattering throughout the image would hinder multivariate analysis without either correcting the spectra using existing algorithms, or using clustering to mask out those regions with the dispersion artefact from further analysis, which may exclude relevant parts of the tissue. Indeed, cancerous tissue with enlarged cell nuclei may be excluded should such a method be used. By contrast, Fig. 4 shows the same region of tissue with a markedly reduced number of pixels containing the scattering artefact, as determined by KMC. This demonstrates that it is possible to combine the lens-added approach with mapping, thus obtaining images with enhanced spatial resolution and reduced scattering while still imaging large areas of the sample.

Lastly, we performed hierarchical clustering analysis (HCA) to identify a region of high-grade dysplasia from the surrounding tissue. An adjacent section was

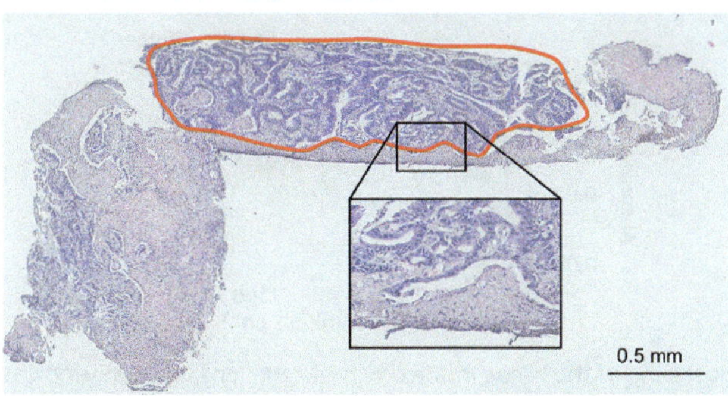

Fig. 5 H + E staining of an adjacent section of the tissue sample with a region marked as high-grade dysplasia in red, and a region of interest marked in black and enlarged which is similar to the area imaged in subsequent sections.

stained to highlight the region of high-grade dysplasia as shown in Fig. 5. Spectral bands within the 1300 cm^{-1} to 900 cm^{-1} region contain indicators for the presence of dysplasia and can therefore be used for classification.[38,39] Second derivative spectra (with 7 point smoothing) were obtained from the region between 1300 cm^{-1} to 1000 cm^{-1} and were vector normalised. To increase the effective signal to noise ratio, the spectral data cube was interpolated in the spatial domain by a reduction of 4× and subsequently, HCA was performed using Pearson's function and Ward's algorithm. The results of HCA applied to the tissue sample measured without the lens are shown in Fig. 6. The second derivative spectrum shows a decrease in the trough at 1077 cm^{-1} as an indicator for high-grade dysplasia as shown in cluster 3. For our configuration with the lens, there is approximately 9.4 mm of calcium fluoride. As such, it was not possible to obtain clustering results for the tissue imaged with the lens due to the decreased signal to noise ratio below 1200 cm^{-1}.

Assessment of spatial resolution for high-magnification FTIR imaging in transmission mode and FTIR imaging with the use of added lens using USAF resolution target

The USAF target was used to assess the spatial resolution and field of view change with the use of lenses and high magnification. Fig. 7 shows these results, with particular interest paid to the use of HM with and without the lens. It should be noted that without high magnification and without the lens, the projected pixel size of the Agilent Cary 620 is 5.5 μm, larger than that reported from the Bruker IRScope II, with a projected pixel size of 2.67 μm under the same conditions. This trade-off in spatial resolution produces a larger field of view and is clear when comparing similar results from our previous work.[25] Extracted transmission profiles from elements 4, 5 and 6 from group 7 in HM mode show that none of the elements are fully resolved, which is expected based on previous work[26] and the wavelength of infrared used (3.2 μm), even with a projected pixel size of 1.1 μm.

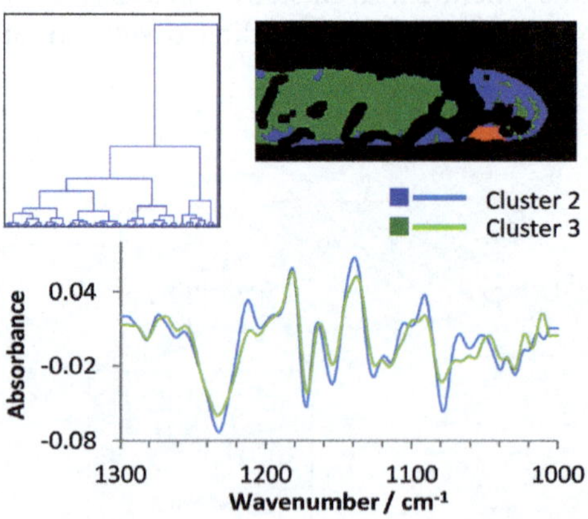

Fig. 6 HCA clustering of the tissue imaged without the lens, correctly indicating the high-grade dysplasia within the tissue. Average spectra from clusters 2 and 3 are plotted showing a decrease in the trough at 1077 cm^{-1} as one indicator of high-grade dysplasia. The average spectrum from the first cluster (red) is not plotted as it belongs to a corrupted data block, as also indicated in the rightmost branch of the dendrogram.

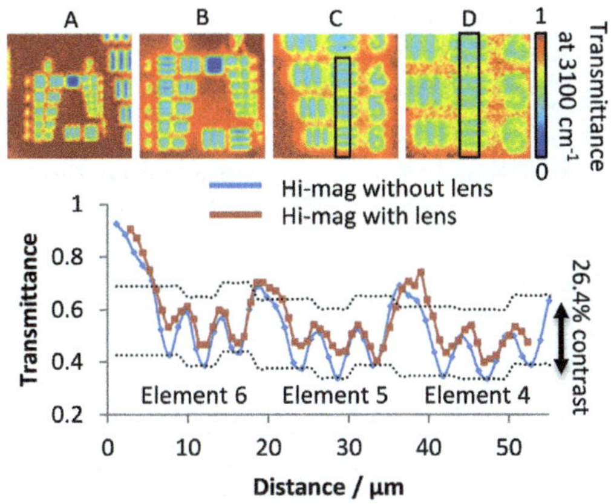

Fig. 7 Transmission FTIR images at 3100 cm^{-1} of the USAF target without HM, without the lens (A), without HM, with the lens (B), with HM, without the lens (C) and with HM, with the lens (D). The HM measurements were focused on elements 4, 5 and 6 of group 7. Regions of the HM images marked in black were extracted and averaged in the horizontal direction to produce transmittance profiles for the HM with and without the lens. Lines denoting the minimum contrast (26.4%) considered as fully resolved are plotted from the troughs of the transmittance profiles.

However, contrary to expectations, the contrast between the lines for each element imaged with the lens decreases compared to without the lens in HM mode. This may be due to optical phenomena within the system such as reflections at the interfaces between the lens, window and USAF target. The noise is also increased due to greater disparity between the condenser and objective lens numerical apertures resulting in a loss of throughput. These results show that for this particular configuration, using the lens does not increase the spatial resolution.

Assessment of spatial resolution for high-magnification FTIR imaging in transmission with the use of added lens with polymer interface sample and micro-ATR FTIR imaging

It is important to assess the spatial resolution obtained using the lens and HM optics with a sample that contains two spectrally distinct parts with a sharp interface. A polymer laminate sample was used (with no interdiffusion between two different polymers)[40] and Fig. 8 shows an absorbance image of the spectral band between 3251 cm^{-1} and 3448 cm^{-1} from using HM mode with top and bottom lenses. Also shown is an integrated absorbance profile along the line perpendicular to the interface between two polymers to assess the spatial resolution, generated from averaging within the vertical direction of the image. A sigmoid function of the form shown in eqn (2) was fitted to the integrated absorbance profile, and the coefficients (c_n) were optimised to give an R^2 value of 0.99.

$$f(x) = \frac{1}{c_1 + e^{(-c_2 x + c_3)}} + c_4 \quad (2)$$

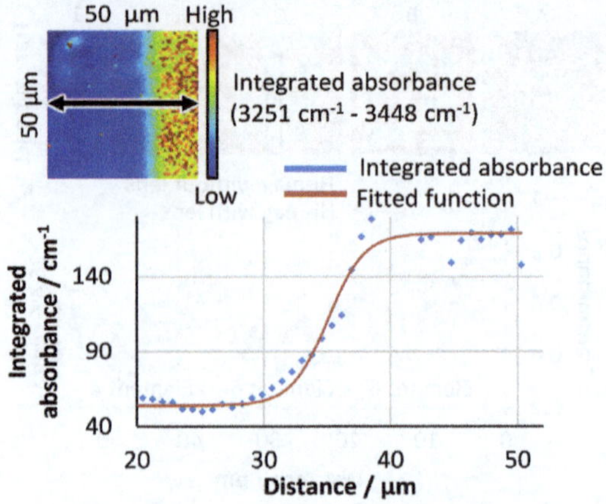

Fig. 8 FTIR image of polymer laminate obtained using transmission mode with HM and added lenses. The image is produced using the absorbance band between 3251 cm^{-1} and 3448 cm^{-1} and the integrated absorbance profile across the interface generated by averaging pixels in the vertical direction to produce a horizontal profile, as shown by the arrow. A sigmoid function is fitted and used to assess the spatial resolution.

Taking the limits of this function reveals the 'bulk' integrated absorbance of that spectral band within both halves of the sample. The apparent thickness of the interface was calculated as the distance between the points along the line perpendicular to interface at which the integrated absorbance values are between 5% and 95% of their maximum values. This experiment gave a measured value of 8.5 µm, an improvement over that which was obtained in our earlier work (11 µm).[25]

By contrast, micro-ATR imaging measurements of the same polymer interface produce a sharper profile and using the same analysis gives an apparent interface

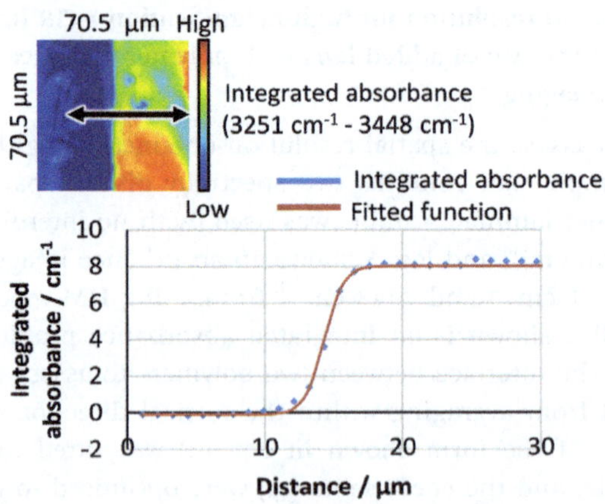

Fig. 9 FTIR image of polymer laminate obtained using micro-ATR mode with a germanium crystal. The image is produced using the absorbance band between 3251 cm^{-1} and 3448 cm^{-1} and the integrated absorbance profile across the interface is generated by averaging pixels in the vertical direction to produce a horizontal profile, as shown by the arrow. A sigmoid function is fitted and used to assess the spatial resolution.

thickness of 3 μm (Fig. 9). These results show that spatial resolution, when measured as a function of spectral contamination across a sharp interface, is often lower than that predicted by projected pixel size.

Tissue measurements with high magnification and the lens

To assess the effect of the lens with and without HM mode on the quality of the spectra and degree of scattering from biological samples, single spectral images of the tissue were measured. Fig. 10 shows a visible image of the tissue section used in this section and the previous mapping study, where the red and blue regions denote approximately where spectral images were obtained with and without HM. A similar region is marked in Fig. 5, showing the H&E image of an adjacent tissue section. Fig. 11 shows the spectroscopic imaging results of these measurements. It is clear from Fig. 11A and C, scattering as a fraction of the measured area is reduced with the lens, as the number of pixels classified as containing the scattering artefact is lower in C. However, in Fig. 11C, baseline undulations as a result of internal reflections between the windows are observed, though this type of interference is readily removable.[41] With HM mode as shown in Fig. 7B and D, the effect of the lens is less clear with respect to scattering. However, the signal to noise ratio is substantially worse with the lens, especially around 1000 cm^{-1}, which is due to the mis-match between the condenser and objective numerical aperture, and the thickness of the calcium fluoride attenuating infrared light at frequencies below approximately 1200 cm^{-1}. As concluded from the results above with the polymer interface sample, the use of HM mode and the lens does not yield any benefit, and decreases the signal to noise ratio considerably. It was noted by Foreman *et al.*[37] and others,[23] that the region between 1200 cm^{-1} and 900 cm^{-1} contains vital information on the disease state of the tissue, and that while light at these frequencies is attenuated, approaches using more powerful sources such as synchrotron light[23] or quantum cascade lasers can be used to increase throughput in this region.[42] Another solution to increase the light throughput of frequencies below 1200 cm^{-1} is to use alternative window materials instead of calcium fluoride. We attempted to use windows and lenses made from multispectral grade zinc sulphide (refractive index of 2.2) which

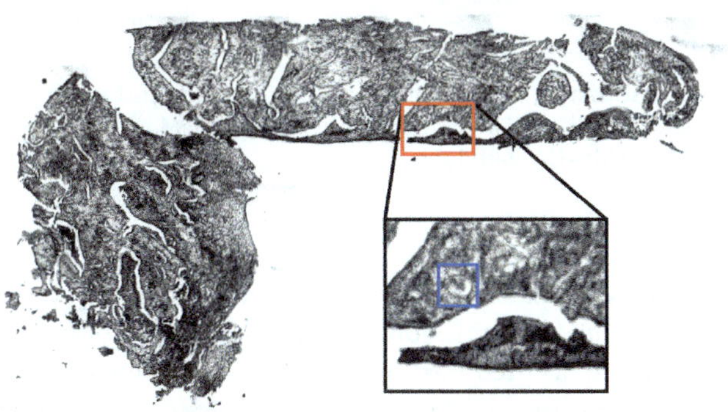

Fig. 10 Visible image of the tissue used for mapping and assessment of the lens and high magnification optics. The red area shows approximately the region used for imaging with and without the lens, the blue region shows approximately the region imaged in high magnification mode with and without the lens.

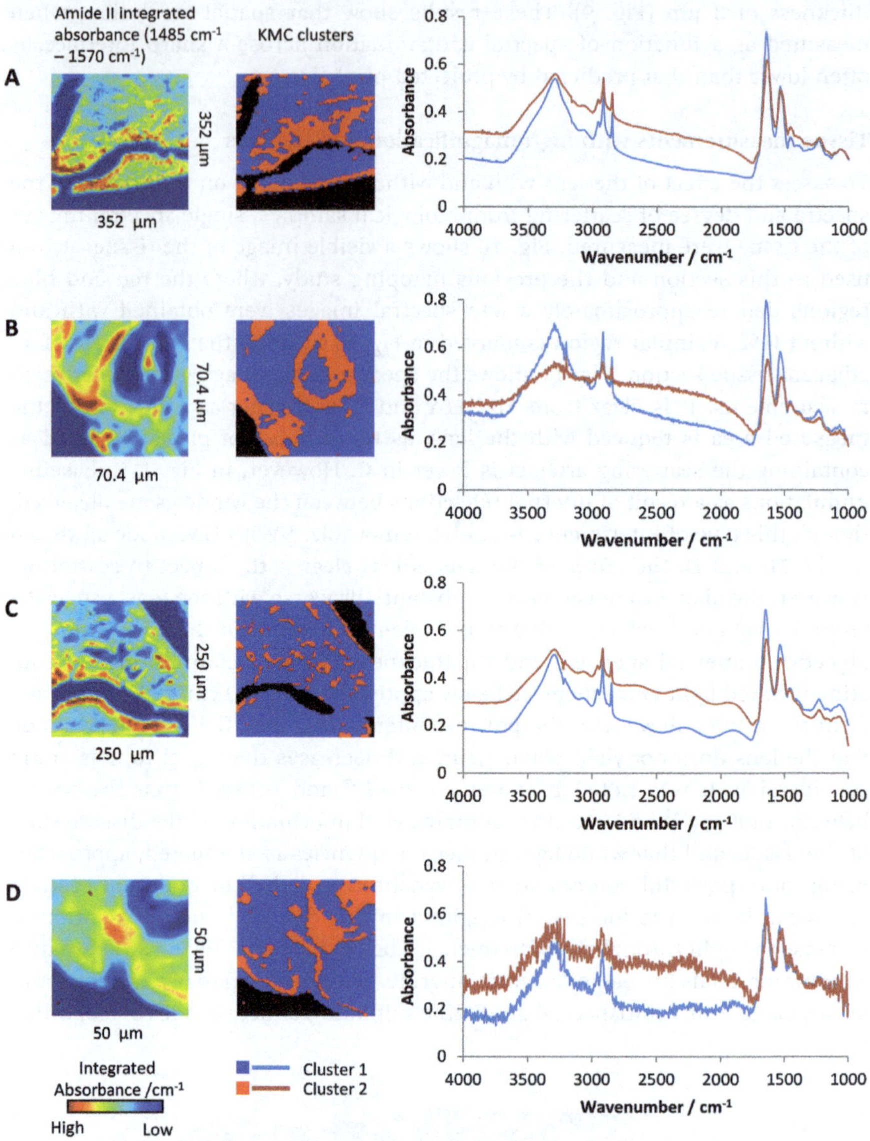

Fig. 11 Integrated absorbance images of tissue and clusters, and average spectra from each cluster imaged (A) without HM, without lens, (B) with HM, without lens, (C) without HM, with lens and (D) with HM, with lens.

would not only increase spatial resolution but still allow for alignment using visible light and increase transmission below 1200 cm^{-1}. However, throughput was severely diminished, as predicted by the Fresnel equations which give a total transmittance of approximately 54.5% for 4 surfaces, *i.e.* window and lens surfaces. Interference from internal reflections also severely hindered usability such that the quality of the spectra was significantly worse compared with using calcium fluoride. Transmittance through calcium fluoride for the same configuration is approximately 90%. Optimising the thickness of the lenses and windows would also reduce attenuation of light at the mentioned frequencies.

Micro-ATR spectroscopic imaging of tissue

Micro-ATR mode uses a crystal (commonly germanium) in contact with the sample through which infrared light undergoes total internal reflection. Scattering effects are eliminated and the spatial resolution is enhanced due to the high refractive index of the ATR crystal. Here, a section of tissue with edges and holes was chosen as in transmission, and it was expected that the dispersion artefact would be present. Using the same analytical approach, it can be seen that the dispersion artefact is completely removed and the second cluster identified is simply poorer quality spectra from bad detector pixels rather than the spectra containing scattering (Fig. 12). Within the fingerprint region, the depth of penetration of the evanescent wave is approximately 0.5 µm,[40] and is independent of sample thickness, allowing a variety of tissue samples to be studied without the need for microtoming. For biopsies, the ATR mode does not require extensive preparation such as microtoming and deparaffinisation, which could affect the chemical properties of the sample.

Imaging of live cells

Imaging live HEK 293A cells within microfluidic devices benefits greatly from the use of the lens as shown in Fig. 13 and our previous work.[43] Spatially smoothing the spectra using a 3 pixel Gaussian filter and producing integrated absorbance images from the amide-II band shows an increase in sharpness using the lens, making comparison with the visible images relatively straightforward. Cell morphology is easily assessed from spectroscopic images which is an indication of the state of health of the cells. For live biological systems and other microfluidics, the use of two windows is often unavoidable and the use of lenses to correct refraction is a practical way to significantly improve the spectral quality.

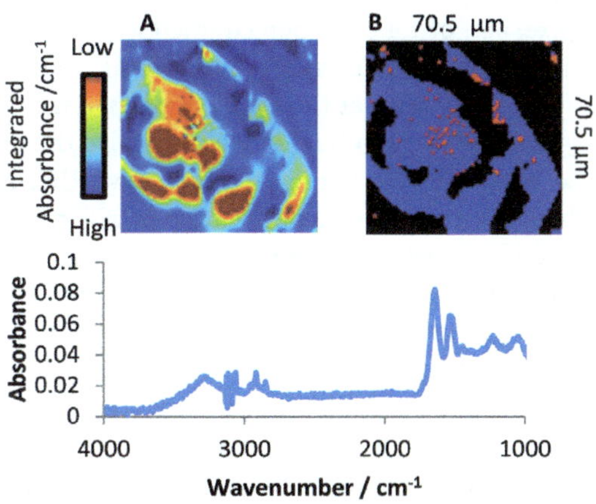

Fig. 12 Micro-ATR image of oesophageal tissue (A) and KMC clusters (B). The graph shows the average spectrum from cluster 1 (blue regions), indicating the complete removal of scattering.

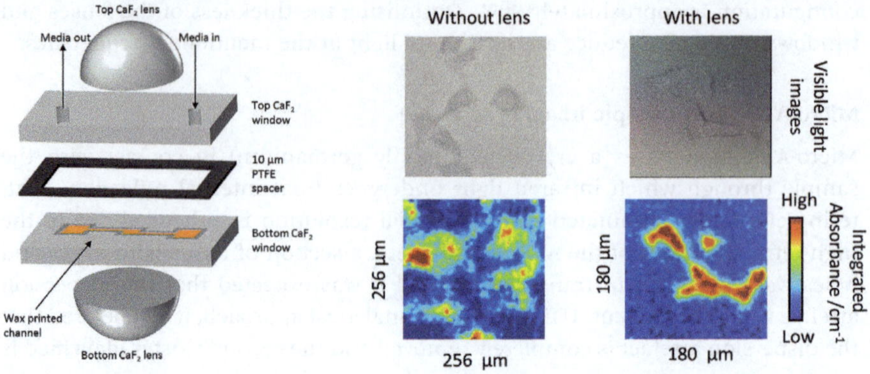

Fig. 13 Microfluidic device with live cells imaged with and without the lens through 4 mm thick calcium fluoride windows. The spectroscopic images are generated using the amide II band between 1485 cm^{-1} and 1570 cm^{-1}, with the benefit of the lens apparent compared to without.

Conclusions

Combining mapping and spectroscopic imaging of tissue mounted on a calcium fluoride window, with an added calcium fluoride lens has been demonstrated. Using this method, it is possible to image large sections of tissue, improving spatial resolution, correcting for chromatic aberration and reducing scattering, as shown using *k*-means clustering classification. As each tile is scanned separately and assembled post-scan, the method is reliable, allowing for maps of any size to be acquired before stitching and computational analysis. The reduction of scattering would assist in further computational processing of spectra to remove the remaining artefacts. HCA was able to correctly classify high-grade dysplasia from normal tissue for the map obtained without the lens but was unable to do so for the map obtained with the lens due to the decreased throughput of infrared light at wavenumbers below 1200 cm^{-1}.

The Agilent Cary 620 microscope contains high-magnification (HM) optics, giving a projected pixel size of 1.1 μm when coupled with a 15× objective. In non-HM mode, the projected pixel size is larger (5.5 μm) due to the use of expanding optics to give a larger field of view at the expense of spatial resolution. We show using the USAF target the change in field of view and also the performance of the added lens coupled with the HM optics. It was shown that in cases with and without the lens, the last elements of group 7 are partially resolved. However, the contrast is slightly reduced using the lens due to internal reflections from the increased number of surfaces through which the infrared light passes. With a projected pixel size of 785 nm, the noise is also increased, thus the full potential of using the lens with HM optics could only be realised using more powerful sources such as a synchrotron or quantum cascade laser.

The USAF target represents a highly idealised system not similar to real biomedical samples. Therefore, a laminate sample containing a sharp interface between two different polymers was imaged in transmission using both a top and a bottom lens. This method using two lenses corrected chromatic aberration for the condenser and the objective, maximising the potential spatial resolution that could be obtained. Using the integrated absorbance of a spectral band present in one polymer but not the other, an absorbance profile was obtained across the interface of the obtained image. The change in integrated absorbance from one polymer to the other indicates the useful spatial resolution of the system, as the objective of spectroscopic imaging is to spatially resolve spectrally dissimilar regions *i.e.* where the spectral features of one are absent in the other. The width of the interface was shown to be 8.5 μm, similar to that observed in previous work and nevertheless, lower than the theoretical maximum resolution. The over-sampling capability of the HM optics thus provides no discernible increase in spatial resolution when used with the lens. Importantly, micro-ATR spectroscopic imaging of the same interface show higher resolution is obtained (3 μm) in spite of the larger projected pixel size, demonstrating the importance of measuring real samples to assess spatial resolution rather than assuming the projected pixel size is the spatial resolution.

Spectral images of the tissue section with and without the lens, and with and without HM optics were analysed using *k*-means clustering. Without HM optics, the reduction in scattering due to the addition of the lens was evident from the reduction in the number of pixels classified as containing the dispersion artefact. However, with HM optics, scattering is not reduced significantly using the added lens, and the signal to noise ratio is significantly worse, making analysis for diagnostic purposes more challenging and, should mapping be performed, increasing the time to map an area significantly. The highest performance of the lens can therefore be obtained using an objective without HM optics, reducing scattering and achieving sufficient signal to noise. To address the loss of throughput at the lower frequencies, which are important for clinical diagnostics, thinner windows and lenses could be used.[14] For comparison with transmission images, spectra were obtained from the same sample using a micro-ATR germanium crystal. Spectra from *k*-means clustering showed the absence of scattering and demonstrated the feasibility of using ATR mode to obtain high-quality spectra and that it can be used for similar studies without the need for microtoming.

This approach could also be applied to other devices such as microfluidic systems, which often operate at quasi-steady state. With these systems, sandwiching of samples between spectroscopic windows is required for the transmission measurements, hence the lens is particularly useful to remove chromatic aberration. We present a microfluidic device containing live HEK cells, imaged with and without the lens. With the lens, the correlation with the visible image is straightforward and the cell boundary is clear, aiding in the assessment of their morphology. The use of lenses and the mapping approach allows the entire device to be studied at high resolution.

Note added after first publication

This article replaces the version published on 08 April 2016, which contained an error in Fig. 4.

References

1 *Infrared and Raman Spectroscopic Imaging*, ed. R. Salzer and H. W. Siesler, Wiley-VCH Verlag GmbH & Co. KGaA, Weinheim, 2nd edn, 2014.

2 *Raman, Infrared, and near-infrared chemical imaging*, ed. S. Šašić and Y. Ozaki, John Wiley & Sons, Inc., Hoboken, New Jersey, 2010.
3 S. G. Kazarian and K. L. A. Chan, *Appl. Spectrosc.*, 2010, **64**, 135A–152A.
4 G. Steiner and E. Koch, *Anal. Bioanal. Chem.*, 2009, **394**, 671–678.
5 T. P. Wrobel, A. Vichi, M. Baranska and S. G. Kazarian, *Appl. Spectrosc.*, 2015, **69**, 1170–1174.
6 A. Gupper and S. G. Kazarian, *Macromolecules*, 2005, **38**, 2327–2332.
7 C. Vogel, E. Wessel and H. W. Siesler, *Biomacromolecules*, 2008, **9**, 523–527.
8 J. L. Koenig, *Microspectroscopic Imaging of Polymers*, American Chemical Society, 1998.
9 A. A. Gabrienko, C. H. Lai and S. G. Kazarian, *Energy Fuels*, 2014, **28**, 964–971.
10 F. H. Tay and S. G. Kazarian, *Energy Fuels*, 2009, **23**, 4059–4067.
11 K. L. A. Chan and S. G. Kazarian, *Anal. Chem.*, 2012, **84**, 4052–4056.
12 K. L. A. Chan, X. Niu, A. J. deMello and S. G. Kazarian, *Anal. Chem.*, 2011, **83**, 3606–3609.
13 E. Carter, K. Tam, R. Armstrong and P. Lay, *Biophys. Rev.*, 2009, **1**, 95–103.
14 K. L. A. Chan and S. G. Kazarian, *Chem. Soc. Rev.*, 2016, **45**, 1850–1864.
15 L. M. Miller, M. W. Bourassa and R. J. Smith, *Biochim. Biophys. Acta, Biomembr.*, 2013, **1828**, 2339–2346.
16 S. G. Kazarian and K. L. A. Chan, *Analyst*, 2013, **138**, 1940–1951.
17 P. Gelfand, R. J. Smith, E. Stavitski, D. R. Borchelt and L. M. Miller, *Anal. Chem.*, 2015, **87**, 6025–6031.
18 S. G. Kazarian and K. L. A. Chan, *Biochim. Biophys. Acta, Biomembr.*, 2006, **1758**, 858–867.
19 L. S. Leslie, A. Kadjacsy-Balla and R. Bhargava, in *Medical Imaging 2015: Digital Pathology*, ed. M. N. Gurcan and A. Madabhushi, Spie-Int Soc Optical Engineering, Bellingham, 2015, vol. 9420.
20 B. Bird, M. Miljkovic, S. Remiszewski, A. Akalin, M. Kon and M. Diem, *Lab. Invest.*, 2012, **92**, 1358–1373.
21 C. Hughes and M. J. Baker, *Analyst*, 2016, **141**, 467–475.
22 D. C. Fernandez, R. Bhargava, S. M. Hewitt and I. W. Levin, *Nat. Biotechnol.*, 2005, **23**, 469–474.
23 L. M. Miller and P. Dumas, *Biochim. Biophys. Acta, Biomembr.*, 2006, **1758**, 846–857.
24 J. A. Dougan, K. L. A. Chan and S. G. Kazarian, in *Infrared and Raman Spectroscopic Imaging*, Wiley-VCH Verlag GmbH & Co. KGaA, 2014, pp. 397–444, DOI: 10.1002/9783527678136.ch9.
25 K. L. A. Chan and S. G. Kazarian, *Anal. Chem.*, 2013, **85**, 1029–1036.
26 C. R. Findlay, R. Wiens, M. Rak, J. Sedlmair, C. J. Hirschmugl, J. Morrison, C. J. Mundy, M. Kansiz and K. M. Gough, *Analyst*, 2015, **140**, 2493–2503.
27 S. Inoué, in *Handbook Of Biological Confocal Microscopy*, ed. J. B. Pawley, Springer, US, 2006, ch. 1, pp. 1–19, DOI: 10.1007/978-0-387-45524-2_1.
28 D. Mayerich, T. van Dijk, M. J. Walsh, M. V. Schulmerich, P. S. Carney and R. Bhargava, *Analyst*, 2014, **139**, 4031–4036.
29 A. Kohler, J. Sulé-Suso, G. D. Sockalingum, M. Tobin, F. Bahrami, Y. Yang, J. Pijanka, P. Dumas, M. Cotte, D. G. van Pittius, G. Parkes and H. Martens, *Appl. Spectrosc.*, 2008, **62**, 259–266.
30 P. Bassan, A. Kohler, H. Martens, J. Lee, H. J. Byrne, P. Dumas, E. Gazi, M. Brown, N. Clarke and P. Gardner, *Analyst*, 2010, **135**, 268–277.

31 R. K. Reddy, M. J. Walsh, M. V. Schulmerich, P. S. Carney and R. Bhargava, *Appl. Spectrosc.*, 2013, **67**, 93–105.

32 V. Zohdi, D. R. Whelan, B. R. Wood, J. T. Pearson, K. R. Bambery and M. J. Black, *PLoS One*, 2015, **10**, e0116491.

33 B. Mohlenhoff, M. Romeo, M. Diem and B. R. Wood, *Biophys. J.*, 2005, **88**, 3635–3640.

34 P. Bassan, A. Sachdeva, A. Kohler, C. Hughes, A. Henderson, J. Boyle, J. H. Shanks, M. Brown, N. W. Clarke and P. Gardner, *Analyst*, 2012, **137**, 1370–1377.

35 P. Bassan, A. Sachdeva, J. H. Shanks, M. D. Brown, N. W. Clarke and P. Gardner, *Analyst*, 2013, **138**, 7066–7069.

36 A. Sroka-Bartnicka, J. Kimber, L. Borkowski, M. Pawlowska, I. Polkowska, G. Kalisz, A. Belcarz, K. Jozwiak, G. Ginalska and S. Kazarian, *Anal. Bioanal. Chem.*, 2015, **407**, 7775–7785.

37 L. Foreman, J. A. Kimber, K. V. Oliver, J. M. Brown, S. M. Janes, T. Fearn, S. G. Kazarian and P. Rich, Optical Diagnostics and Sensing Xv: Toward Point-of-Care Diagnostics, *Proc. SPIE*, 2015, **9332**, 93320S.

38 T. D. Wang, G. Triadafilopoulos, J. M. Crawford, L. R. Dixon, T. Bhandari, P. Sahbaie, S. Friedland, R. Soetikno and C. H. Contag, *Proc. Natl. Acad. Sci. U. S. A.*, 2007, **104**, 15864–15869.

39 L. Quaroni and A. G. Casson, *Analyst*, 2009, **134**, 1240–1246.

40 K. L. A. Chan and S. G. Kazarian, *Appl. Spectrosc.*, 2003, **57**, 381–389.

41 A. M. A. Pistorius and W. J. DeGrip, *Vib. Spectrosc.*, 2004, **36**, 89–95.

42 P. Bassan, M. J. Weida, J. Rowlette and P. Gardner, *Analyst*, 2014, **139**, 3856–3859.

43 K. L. A. Chan and S. G. Kazarian, *Analyst*, 2013, **138**, 4040–4047.

Faraday Discussions

PAPER

Multi-centre Raman spectral mapping of oesophageal cancer tissues: a study to assess system transferability

M. Isabelle,*[a] J. Dorney,[b] A. Lewis,[c] G. R. Lloyd,[a] O. Old,[a] N. Shepherd,[a] M. Rodriguez-Justo,[c] H. Barr,[a] K. Lau,[d] I. Bell,[d] S. Ohrel,[d] G. Thomas,[c] N. Stone[b] and C. Kendall[a]

Received 18th November 2015, Accepted 17th December 2015
DOI: 10.1039/c5fd00183h

The potential for Raman spectroscopy to provide early and improved diagnosis on a wide range of tissue and biopsy samples *in situ* is well documented. The standard histopathology diagnostic methods of reviewing H&E and/or immunohistochemical (IHC) stained tissue sections provides valuable clinical information, but requires both logistics (review, analysis and interpretation by an expert) and costly processing and reagents. Vibrational spectroscopy offers a complimentary diagnostic tool providing specific and multiplexed information relating to molecular structure and composition, but is not yet used to a significant extent in a clinical setting. One of the challenges for clinical implementation is that each Raman spectrometer system will have different characteristics and therefore spectra are not readily compatible between systems. This is essential for clinical implementation where classification models are used to compare measured biochemical or tissue spectra against a library training dataset. In this study, we demonstrate the development and validation of a classification model to discriminate between adenocarcinoma (AC) and non-cancerous intraepithelial metaplasia (IM) oesophageal tissue samples, measured on three different Raman instruments across three different locations. Spectra were corrected using system transfer spectral correction algorithms including wavenumber shift (offset) correction, instrument response correction and baseline removal. The results from this study indicate that the combined correction methods do minimize the instrument and sample quality variations within and between the instrument sites. However, more tissue samples of varying pathology states and greater tissue area coverage (per sample) are needed to properly assess the ability of Raman spectroscopy and system transferability algorithms over multiple instrument sites.

[a]*Biophotonics Research Unit and Pathology Department, Gloucestershire Hospitals NHS Foundation Trust, Gloucester, UK. E-mail: martini_30@hotmail.com*
[b]*Biomedical Spectroscopy, School of Physics, University of Exeter, UK*
[c]*Department of Cell and Developmental Biology, University College London, London, UK*
[d]*Spectroscopy Products Division, Renishaw plc, Wotton-Under-Edge, Gloucestershire, UK*

Introduction

Oesophageal cancer is the fifth biggest cancer killer, with rising incidence rates.[1] Once patients are found to have symptoms such as trouble swallowing the disease is advanced and usually incurable (15% of patients survive 5 years[1]). Individuals with a precursor condition called Barrett's oesophagus (caused by reflux disease), have an increased risk of developing cancer.[2,3] These patients can be diagnosed at an early stage using endoscopic surveillance, which includes biopsy. Most gastrointestinal cancers occur on the mucosal surface, however dysplastic changes in the oesophagus are invisible to the naked eye during endoscopy.[4] Therefore patients with Barrett's oesophagus are subjected to routine endoscopy with collection of biopsy samples for analysis by histopathology. However, the gold standard histopathology is not able to accurately and reproducibly identify those patients with early disease or those who may be harbouring more advanced disease in unsampled areas of the oesophagus,[5] therefore improved and objective diagnostic tools are needed.[6]

Raman spectroscopy (RS) makes use of biochemical or molecular analysis of target tissues rather than an individual's assessment of cellular and tissue appearance,[7] therefore RS provides an objective, non-destructive, label-free, non-invasive, rapid and economical method for diagnosis. RS has the potential to be more automated and adaptable to routine screening and provides a spectroscopic signature of all molecular constituents of the tissue sample, which facilitates the analysis and use of subtle molecular changes to classify different pathology and tissue states. Due to these pathology and tissue-specific spectroscopic signatures, RS has been used as a tool in histopathology,[8–14] and cytology.[15–18] It is based on inelastic light scattering and occurs when the biological sample is illuminated by monochromatic laser light. The interactions between the incident photons and molecules in the sample result in scattering of the light to produce a characteristic biomolecular fingerprint of the sample.[19]

Kendall *et al.* demonstrated the ability of RS to discriminate between 8 pathological groups in oesophageal cancer: normal squamous, three subtypes of Barrett's oesophagus (intestinal metaplasia, cardiac and fundic type mucosa), high-grade dysplasia, adenocarcinoma, squamous dysplasia and squamous cell carcinoma with sensitivities between 73% and 100% and specificities of 90–100%.[8] Hutchings *et al.* and Shetty *et al.* have demonstrated the advanced use of chemometrics (principal component analysis score maps) to generate biochemical maps. These maps were used to identify cellular constituents associated with malignancy,[20] and to discriminate between different pathology and tissue types.[21]

Development of a miniature confocal fibre optic probe for Raman spectroscopic measurements (Day *et al.* 2009 (ref. 22)) enabled the use of fibre optic Raman probe spectral measurements *ex vivo* in which Almond *et al.* (Almond *et al.* 2012 (ref. 23)) performed Raman spectroscopic measurements on tissue samples from 28 patients using a custom-built fibre-optic Raman probe, in conjunction with multivariate classification models, to differentiate between benign and neoplastic oesophageal cancer and pre-cancer. The Raman probe system was able to differentiate between normal squamous, Barrett's oesophagus and neoplasia with sensitivities of (83% to 86%) and specificities of (89% to 99%). In another study by the same group (Almond *et al.* 2014 (ref. 24)), using tissue from 62 patients, the authors demonstrated that the Raman probe system, in conjunction with principal component-fed linear discriminant analysis (PC-LDA), was able to achieve a sensitivity of 86% and a specificity of 88% for detecting high-grade dysplasia and adenocarcinoma. An additional study using a Raman fibre optic probe was conducted by Bergholt and coworkers who used, in clinic during endoscopic examination, a multimodal image-guided Raman endoscopy technique for real-time *in vivo* diagnosis of cancer in the oesophagus.[25] 75 oesophageal tissue sites from 27 patients were measured, and using a LDA diagnostic model they were able to achieve an accuracy of 96%, sensitivity of 97% and specificity of 95% in the *in vivo* diagnosis of oesophageal cancer.

There is clearly a clinical need for these types of validated multivariate disease-specific classification models, developed to classify and discriminate pathology classes. However, due to instrument artefacts and environmental differences between locations (mentioned below), a successful classification model constructed using spectra from one instrument system may not have the ability to accurately predict spectra measured on another second instrument.

Physical and chemical constitution (*e.g.* viscosity, particle size, and surface texture) of the tissue samples and standards can impact this transfer process. The instrument response function can affect signal intensity values (these differences/changes could be due to part replacement, aging of parts and sources, and gradual change in quantum efficiency of the detector over time) and wavenumber registration shifts. Ideally these factors can be prevented by pre-calibrating prior to sample measurement using calibration standards. Lastly, instrument environment can affect the validity of the classification model on the different instruments. Absolute temperature changes, both short and long term, may possibly induce instrument changes due to thermal expansion, leading to misalignment of optical components and the shifting of spectral peaks along the wavelength axis.[26] There are a number of standardization methods that can be used to attempt to correct the intensity and wavenumber shifts including single wavelength standardization (SWS), direct standardization (DS) and piecewise direct standardization (PDS).[26] These methods involve determining the instrumental differences from a subset of samples measured on the primary site instrument, and regressing against the same subset measured on the secondary site instrument. This determines a spectral response that can be used to correct the model differences for future measurements.

The ability to implement models created with data collected on a number of machines at various times that can classify data measured on different instruments at different sites is critical for large-scale implementation of Raman spectroscopy in the clinic. In this work we investigate the use of calibration methods (wavenumber offset correction, instrument response correction and baseline fluorescence correction) to develop transferable classification models capable of discriminating between intraepithelial metaplasia (IM) and adenocarcinoma (AC) pathologies in oesophagus tissue using Raman spectroscopic mapping. Stone *et al.*[27] first attempted system transferability between two similar Raman systems (at the same site) by determining the energy transfer function of each system from a calibrated tungsten filament lamp and a secondary NPL (National Physical Laboratory) standard, and then correcting the measured oesophageal tissue spectra by multiplying this by the transfer function. The authors noted that by calibrating for energy sensitivity in each of the systems, there were very few differences in the mean tissue spectra between the two systems. This current study, however, is the first paper that has attempted to build a classification model, from mapped Raman tissue spectra, and validate in three different geographical instrument sites with the intention of creating a transferable diagnostic tool suitable for clinical application. Additionally it is the first paper to document the use of Raman map spectra (compared to commonly point-measured spectra) collected from oesophageal tissue samples to develop and validate a pathology classification model.

Materials and methods

Tissue collection and preparation

All tissue used in this work was obtained from patients undergoing scheduled endoscopy (for Barrett's surveillance or dysplasia) or from patients undergoing surgery for oesophageal cancer. These procedures were performed under local (endoscopic resection) or general (oesophageal resection) anaesthetic in accordance with an approved ethical proposal [Gloucestershire Local Research Ethics Committee]. At all times the General Medical Council (GMC) Guidelines on good clinical practice were followed.

20 oesophageal samples were collected in total, 10 IM and 10 AC. Informed written consent was provided prior to collection of the tissue. Each specimen was placed in a labelled cryovial (Thermo Scientific, Waltham, MA, USA) and the specimens were immediately snap-frozen in liquid nitrogen, before storage at −80 °C.

For every specimen, tissue sections were cut and stained with H&E (haematoxylin and eosin) and reviewed by a consultant histopathologist. Acetate paper on which the specimen was mounted in the operating theatre acted not only to orientate the sample but also allowed the specimens to adhere to the microtome chuck without the need for cutting agents such as Optimal Cutting Temperature compound (OCT).

A freezing microtome was used to obtain a 5–6 micron section that was mounted on a plain glass slide for H&E staining, while three 8-micron-thick tissue sections were mounted on three (one for each instrument site) circular 20 mm × 1 mm Raman-grade calcium fluoride (CaF_2) discs (Crystran, Poole, UK) and stored at −80 °C in 20 mm round coin capsules (Leuchtturm, Geesthacht, Germany). Before being used in all of the Raman experiments, samples were allowed to passively thaw at 23 °C for up to 15 minutes prior to measurement.

Raman spectral tissue measurement

All three sites (Biophotonics Research Unit, Gloucestershire Hospitals NHS Foundation Trust, Gloucester, UK; University of Exeter, Exeter, UK; and University College London, London, UK) used a Renishaw RA800 series benchtop Raman system configured for pathology use; 785 nm laser excitation with a ×50 objective, and motorised XYZ stage. The system is equipped with transmitted and reflected white light imaging for sample location. StreamLine™ Raman imaging, a fast mapping method, was used for data collection. The spectral resolution was 2.5 cm^{-1} with a range of 150–2100 cm^{-1}. The Raman spatial resolution is ~1 μm. The system comes with fully-automated alignment and calibration routines to ensure optimal data reproducibility and transferability. An acquisition time of 60 s per line was used to collect a 50 × 50 pixel Raman image of two selected regions of interest from the tissue section (determined by reviewing the H&E slide).

The spectrometers were regularly calibrated and checked using an additional set of external standards including silicon, green glass and polystyrene, to automatically calibrate the wavenumber axis and to ensure optimal signal.

Data processing and analysis

Streamline™ maps were loaded into Matlab R2014b (The Mathworks Inc., Natick, Massachusetts, USA) for data pre-processing. The spectral range was cropped to the fingerprint region (400–1800 cm^{-1}).

Three spectral correction transfer methods were applied and tested to determine their influences on model classification performance: instrument response correction, wavenumber offset correction and EMSC fluorescence baseline correction.

Instrument response correction

Instrument response correction was achieved by comparing the ratio of the daily measured and calibrated standard spectra from each of the instrument systems measured from a piece of standardized green glass (three pieces from the same source were used, one on each of the three systems) to provide an instrument response profile that was then used to correct spectra obtained from the tissue specimens.

Wavenumber offset correction

Spectra were wavenumber offset corrected using the phenylalanine peak of the average spectrum of each map. Peak fitting a pseudo-Voigt profile was used to correct the phenylalanine position to 1003 cm^{-1}.

Extended multiplicative scatter correction (EMSC) fluorescence baseline correction

Extended Multiplicative Scatter Correction (EMSC), a least squares modelling procedure, was used in a manner already described in detail in previous publications,[28,29] including an investigation for electronic removal of paraffin peaks from infrared spectra.[30] EMSC was used to account for scaling and artefacts using a polynomial baseline (3rd order), non-tissue measured instrument spectrum for each of the instrument systems (measuring without CaF_2 substrate or tissue sample present) and a representative mean tissue spectrum for each instrument spectral dataset. In brief, EMSC is adapted in order to neutralize any background (*e.g.* fluorescence) using a least squares modelling procedure. This involves an interference matrix introduced into the EMSC algorithm composed of the non-tissue instrument spectrum and the mean tissue spectrum for each instrument site. This function removes baseline fluorescence arising from any tissue fluorophores and instrument optics and normalises the dataset to reduce the effect of total signal variation (*e.g.* due to variable focussing and thickness of the sample).

Cosmic ray and fluorescence saturation removal

It is crucial to either remove or replace sharp spikes due to cosmic rays and saturated spectra due to fluorescence in order to prevent them from unduly influencing the data analysis. These artefacts were filtered from the spectral images by applying a 5 × 5 moving window two-dimensional median filter to each wavenumber. For the 5 × 5 window used, the centre pixel was replaced with the median value of all pixels in the window for the chosen wavenumber.

Principal component analysis (PCA)

Prior to creating and validating a classification model, PCA was performed on the complete tissue dataset prior to and after implementing the correction algorithms. This determined whether the correction methods were able to minimize instrument system differences created as a result of different instrument environment and instrument response function, as well as potential sample quality differences.

PCA reduces the spectral data to a smaller number of components that describe the major variations in the dataset, in this case instrument, sample quality and pathology variations.

Partial least squares followed by discriminant analysis (PLS-DA)

A total of 219 367 individual Raman spectra were collected, across all three systems (intraepithelial metaplasia = 114 456; adenocarcinoma = 104 911).

The data matrix was mean-centred prior to the application of Partial Least Squares Discriminant Analysis (PLS-DA). PLSDA reduces the spectral data to a smaller number of components that describe the major variations in the dataset, while simultaneously correlating the captured variance with the consultant pathology classification obtained for each tissue sample.[31]

The number of components used for discriminant analysis was determined by increasing the number included until the ability of the algorithm to accurately predict the pathology group of spectra in the second leave-some-out test dataset no longer increased.[31]

Model validation using leave-one-map-out (LOMOCV) cross-validation

To fully assess the suitability of Raman spectroscopy for distinguishing between the two pathology groups (intraepithelial metaplasia and adenocarcinoma) and to test the system transferability performance across the three instrument sites, the dataset was split into training and validation sets by leaving one map out of the tissue spectra.

The training dataset was used to generate a statistical model and the validation dataset was projected onto the model to assess its performance.

Results and discussion

In order to assess the ability of the wavenumber offset correction (WOC) algorithm on the tissue spectral data and whether it will adequately correct for any potential wavenumber shift at any of the instrument sites, the polymer spectra recorded every day, prior to tissue measurements, was analysed prior and post WOC. Fig. 1 shows the mean spectra of the polymer data calculated from a 5 s

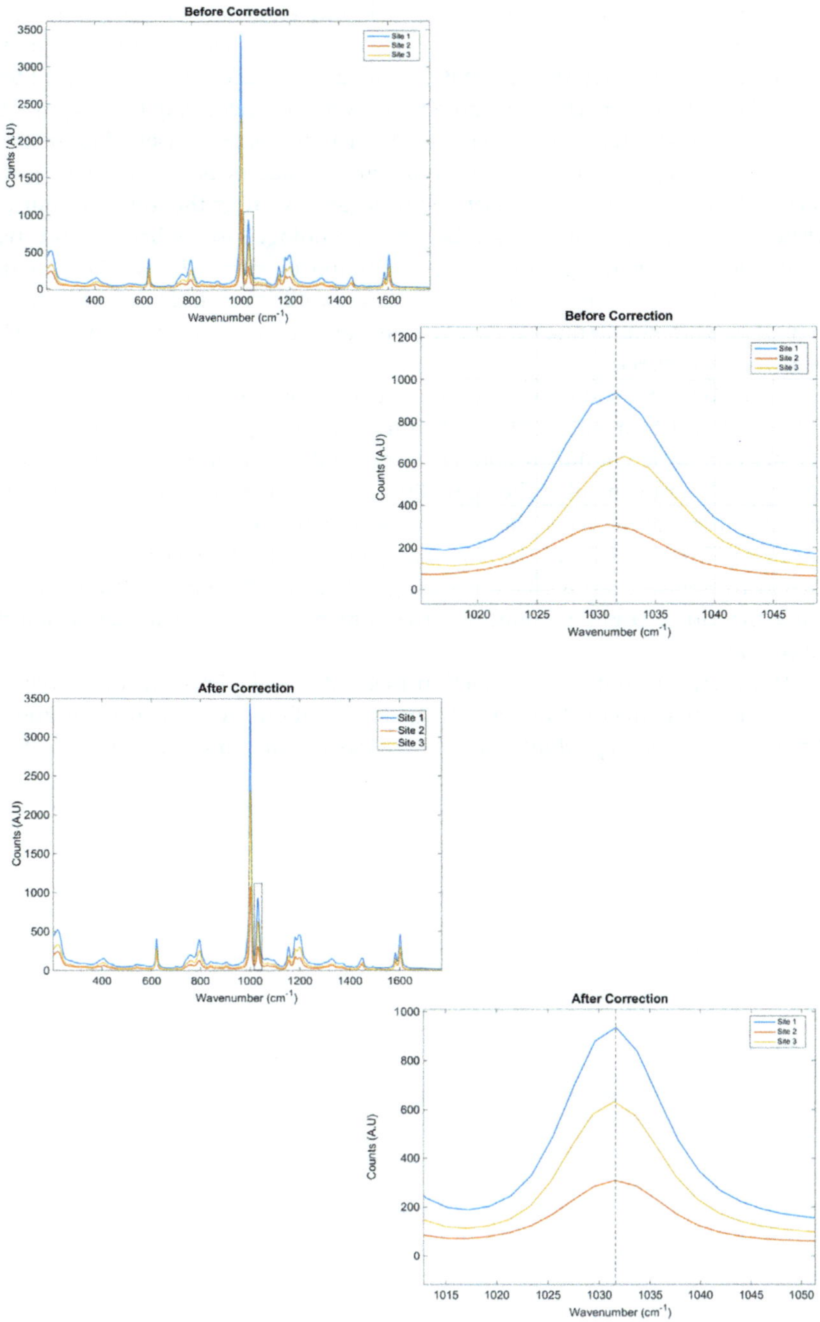

(a)

(b)

Fig. 1 Plot of mean spectra of polymer standard (a) before and (b) after corrections for offset wavenumber correction. Right panels represent zoom-enhanced areas of the spectra for the 1033 cm^{-1} peak.

time-acquired 50 × 50 pixel map. In Fig. 1a, it is clear that there is a slight shift in the ~1033 cm^{-1} peak (and therefore rest of the spectrum) between each of the sites. Three polymer peaks were initially selected across the fingerprint range and a second order polynomial correction, including an offset, was applied to correct for wavenumber shifts between sites. However, on inspection of the polynomial coefficients, it was determined that the values were effectively zero for all terms apart from the offset. This indicated that it was only necessary to apply an offset correction to the tissue data set. For the actual tissue dataset, this offset was corrected using the phenylalanine 1003 cm^{-1} tissue peak. Not correcting these slight shifts in the tissue dataset could negatively affect the ability of our algorithm to accurately differentiate between pathology groups because the tissue peaks accounting for pathology state differences would be masked by these spectral shifts. By applying the pseudo-Voigt profile peak-fitting algorithm to the data, Fig. 1b demonstrates that it is possible to correct for these spectral shifts and align the spectral data.

Using WOC in addition to response correction and EMSC on the complete tissue data set from each of the instrument sites, Fig. 2 demonstrates the effectiveness of these combined correction methods to reduce the variation in the standard deviation at all of the sites. Site 1 still seems to show increased variation in comparison to the other sites. Although similar tissue areas were selected to be measured from the H&E by the site 1 user, it is possible that each user at the other two instrument sites could be measuring a slightly different area from site 1, and this, therefore, could account for the difference in the calculated standard deviation.

However, the combined correction methods (as depicted in the flowchart in Fig. 3) help to correct for any sample thickness differences as well as variations in baseline, caused by fluorophores in tissue and any instrument optical

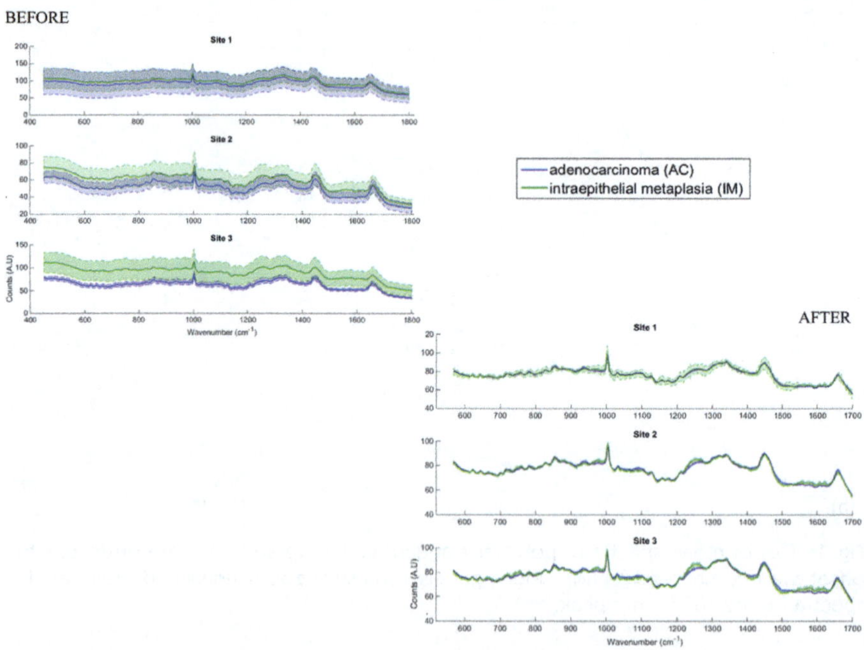

Fig. 2 Plot of mean spectra of pathologies (by instrument site) with ±0.5STD before and after offset wavenumber, instrument response correction, and fluorescence baseline correction and normalisation applied.

fluorescence. In addition, the instrument response function is corrected by comparing the ratio of the daily green glass standard measurement and the calibrated standard spectra to decipher the instrument response profile, which is then applied to the same day collected map tissue spectra.

In order to illustrate the impact of the correction methods to minimise the instrument site variation within the entire dataset, PCA was applied to the mean-centred tissue dataset and the first two PCA components are plotted in Fig. 4. PC1 and PC2 account for the greatest variance in the tissue data, independent of any pathology supervision or prior knowledge. It is clear in Fig. 4a, that there is potentially an instrument and sample quality variation between and within the instrument sites. The individual sites are separating and site 3 is seems to have 3 clusters of data, which could be due to either pathology or sample quality differences.

After applying the correction algorithms in Fig. 4b, which will reduce the variation due to sample thickness and quality, fluorescence baselines as well as spectral offset and instrument transfer function, it is clear that between and within the individual instrument sites, these differences are reduced and the data

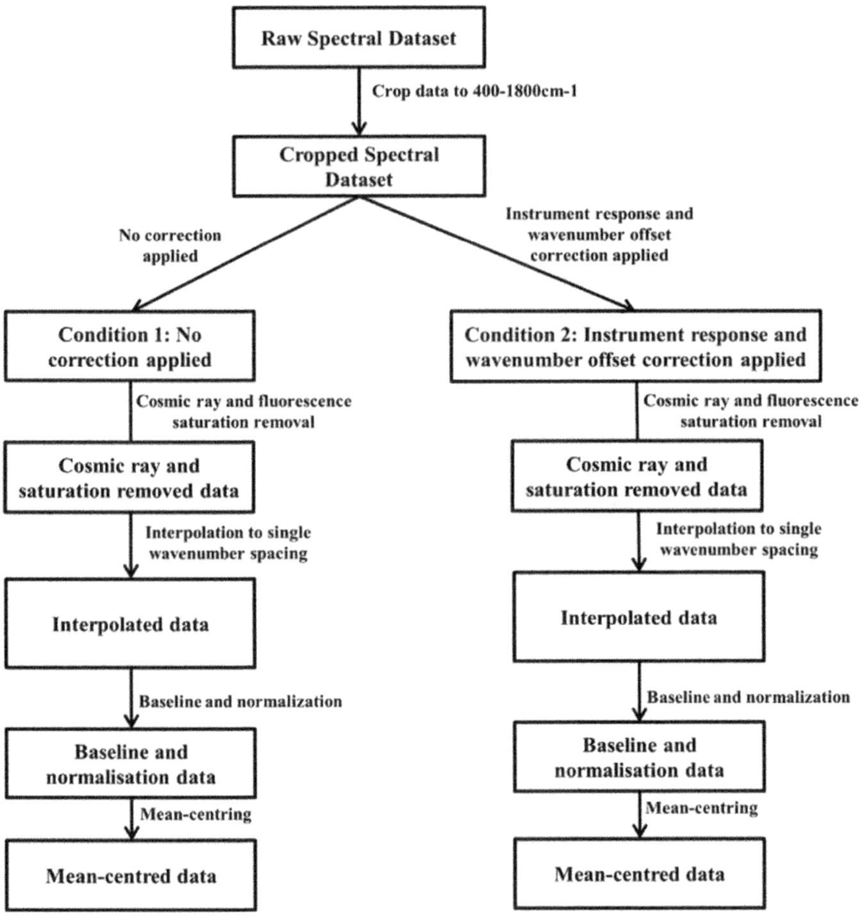

Fig. 3 Steps applied for correction and processing of tissue data.

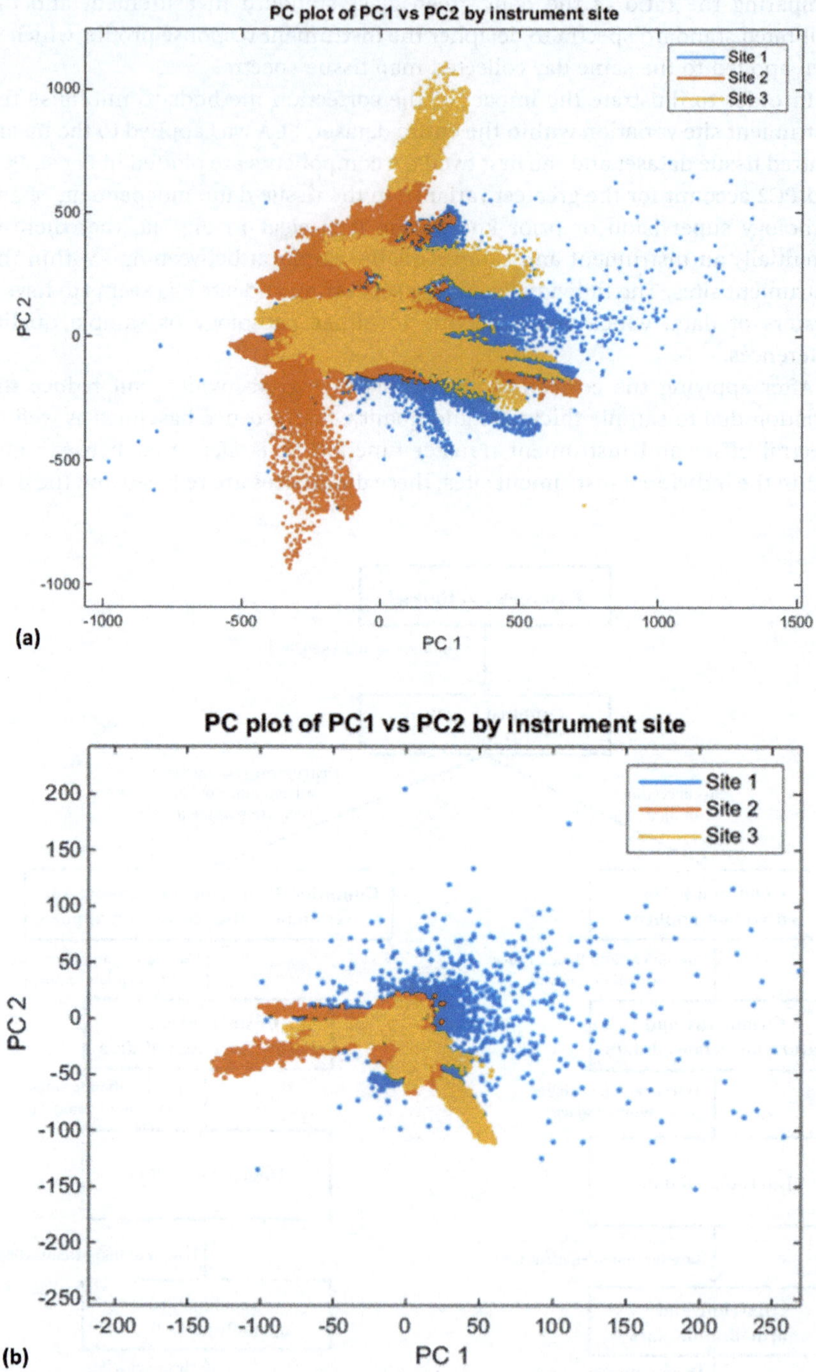

Fig. 4 Plot of PCA scores of PC1 *vs.* PC2 of mean-centred tri-centre data to demonstrate the instrument site dependence of data and need for correction algorithms. (a) Before (b) after correction.

seems to be more tightly clustered. However there still appears to be some instrument site variation even after the correction has been applied. Due to the small sample size (10 adenocarcinoma and 10 intraepithelial metaplasia tissue samples) as well as the small region of area actually mapped on each tissue section (two 50 × 50 mini-maps at ~1.1 μm spatial resolution), we do not expect there to be a great variance within the pathology states in the existing limited tissue dataset. Therefore, in this current tissue dataset, it is assumed that any instrument and sample quality variation would become more prominent. We believe that these remaining site variations could be due to two different factors: (1) instrument and (2) sampling area.

Correction for instrument differences (*i.e.* wavenumber offset correction, instrument response and background correction) have been applied, therefore the remaining differences seen in Fig. 4b between the sites could be either due to deficiencies in the correction methods or remaining instrument differences that we have not addressed. For example, the signal and noise levels were higher in site 1 (due to higher laser power at objective), which would have increased both the Raman peak signals from tissue as well the instrument noise, enough to separate some of the site 1 data from the site 2 & 3 datasets, as demonstrated in the PC plot. Additionally, point spread function (PSF) could also play a part in instrument differences as it is already a challenge to maintain a constant PSF imaging performance over the full range of one detector, never mind three different detectors on the three different systems.

The second source of difference is possibly due to sampling differences between the sites. After further analysis of the PC plot, it was discovered that a majority of the data points with the most variation were in the IM pathology. After checking with the areas selected from each of the tissue sections from all the sites, it was concluded that there was a small difference in how the IM tissue sampling areas were selected between site 1 and sites 2 & 3. The Barrett's epithelium in IM, compared with adenocarcinoma, contains a number of tissue features: goblet cells and intervening non-goblet columnar cells in crypts with irregularity, budding and distortion. IM areas selected by sites 2 & 3 concentrated mostly on dense areas of goblet cells showing dysplasia, while site-1-selected areas included more tissue feature coverage: goblet cells, non-goblet columnar cells in the crypts as well as some of the luminal surface. We believe it is the inclusion of a broader region of tissue features at site 1 measurements compared to sites 2 & 3 that is responsible for the increase in data variation. For future measurements, there will be more agreement of the sampling area and features being measured. These considerations will hopefully help reduce some of the variation between the measurements sites.

Table 1 documents the training performance of the site 1 tissue dataset after applying PLS-DA and the impact of applying the correction algorithms. The tissue data from site 1 was mean-centred and then PLS was applied to reduce the spectral data to a smaller number of components, the major variation in the dataset relevant to the pathology states. Although the correction method improved the resulting specificity (99.5% *vs.* 95.9%) in discriminating adenocarcinoma from intraepithelial metaplasia, the sensitivity reduced from 99.5% to 94.7% following correction, which indicates little impact on the training performance for a single site. However this is not greatly surprising as the dataset is

Table 1 Table for training performance of PLS-DA on site 1 before and after correction applied to differentiate intraepithelial metaplasia (IM) from adenocarcinoma (AC)

	Site 1 as training set	
	Sensitivity (%)	Specificity (%)
Before any correction methods applied	99.5	95.9
After all correction methods applied	94.7	99.5

from one instrument site and therefore the instrument transfer function corrections we applied will have minimal impact compared to multi-site analysis.

In order to assess some of the spectral differences between the two pathology groups of interest (adenocarcinoma and intraepithelial metaplasia), the corrected tissue spectra from all three instrument sites were combined and the mean spectrum for each of the two pathology groups was calculated (Fig. 5). By analysing the spectral differences in these two spectra, adenocarcinoma has higher intensities in nucleic acids (cytosine and uracil at 783 cm^{-1} and 840 cm^{-1} O–P–O stretch) but reduced protein content (amino acid peaks at 754, 1002, 1121, 1603–1622 cm^{-1}, and amide III at 1246 cm^{-1}). Shetty *et al.*[20] documented similar changes and attributed this to cells undergoing dysplastic or malignant change in carcinogenesis in oesophageal tissue. This observation is also in agreement with Stone *et al.*,[10] who observed an increase in nucleic acid content in increasing carcinogenesis in laryngeal cancer tissue samples using similar peak positions for reference as did Haka *et al.*[32] in breast cancer tissue using Raman spectroscopy.

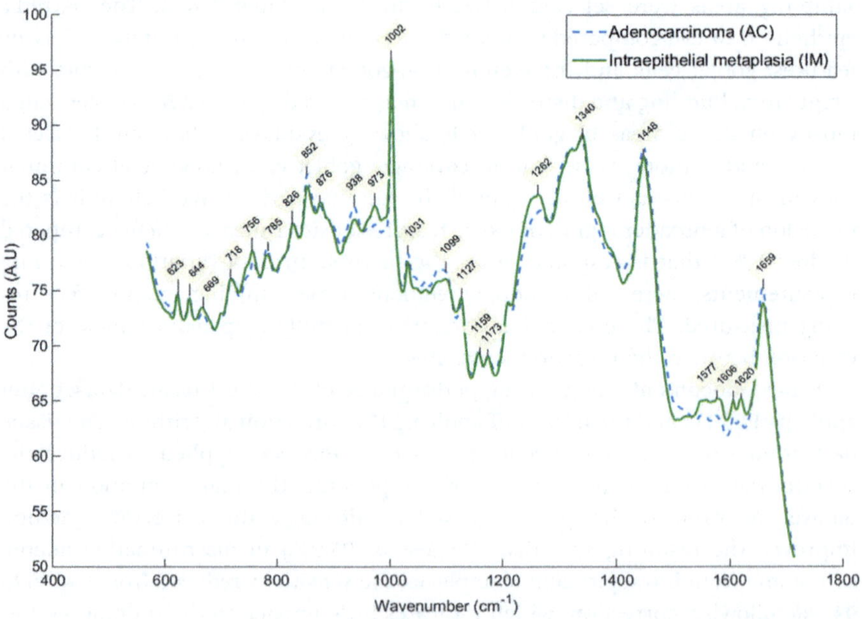

Fig. 5 Mean spectra for intraepithelial metaplasia (IM) *vs.* adenocarcinoma (AC) for all sites.

Finally, to assess the classification performance of measured tissue spectra from all three instrument sites with and without correction applied, leave-one-sample-out cross-validation (LOOCV) and using data from one instrument site as the training site (site 1) analysis was performed. Table 2 describes the classification performance achieved. Again, although the correction methods improved the sensitivity of the classification model to differentiate between adenocarcinoma and intraepithelial metaplasia (from 86% to 96% when site 2 was the test set and 73% to 79% when site 3 was the test set) the specificity was reduced (from 84% to 77% for site 2 as the test set and 85% to 81% when site 3 was the test set). This suggests that although the correction method improves the classification sensitivity of the model (*i.e.* the ability of the model to correctly identify those samples that are adenocarcinoma correctly) there is a reduction in the specificity (ability of the model to correctly identify those samples that are not classified as adenocarcinomas).

Overall, the results together indicate that although the correction models do minimize the instrument and sample quality variations within and between the instrument sites, they do not currently improve the overall performance of the classification model when taking into account both the specificity and sensitivity performance. In order to further improve the classification performance of these models, potentially further standardisation procedures, calibrations, transfer corrections and modelling options need to be implemented in order to achieve similar results to the within-site model. Additionally this classification model was constructed from only 20 tissue samples over two pathology groups with limited sampling area from each tissue sample. A larger sample collection with more tissue samples of varying pathology states and greater tissue area coverage (per sample) is needed to properly assess the ability of Raman spectroscopy and system transferability algorithms over multiple instrument sites in oesophageal cancer diagnostics.

A few studies in the last 20 years have investigated model transferability in the field of spectroscopy.[33–36] Myles *et al.*[37] published a study on a multivariate classification model constructed from near-infrared spectra measured in order to discriminate between green *Arabica* and *Robusta* coffee beans. This model was transferred to a second instrument and investigated for classification performance. In order to allow this classification model to perform optimally on this second system, multiple correction algorithms were investigated including: a modified version of slope/bias correction (SBC), orthogonal signal correction

Table 2 Table of classification performance for LOOCV with and without all correction methods applied to differentiate intraepithelial metaplasia (IM) from adenocarcinoma (AC)

	Site 1 as training set (site 2 as test set)		Site 1 as training set (site 3 as test set)	
	Sensitivity (%)	Specificity (%)	Sensitivity (%)	Specificity (%)
Before any correction methods applied	86.2	84.4	72.8	85.3
After all correction methods applied	95.7	76.9	78.8	80.8

(OSC), and model updating (MU). These corrections resulted in misclassification errors in the second instrument site between 5–10%. Anibal *et al.*[38] assessed the performance of multivariate calibration transfer methods (such as Piecewise Direct Standardization (PDS)) used to discriminate between unadulterated samples and samples adulterated with four Sudan dyes. These were measured using UV-visible spectroscopy by different operators and varying the time in which the samples were analysed to represent the different experimental parameters/conditions that need correcting. By using PDS, the authors were able to determine transfer functions and apply them to improve the classification performance for the other measurement conditions to yield results comparable to the original measurement condition. Rodriguez and coworkers[39] used standardisation models on Raman spectra of a set of pharmaceutical chemical samples. These chemicals were measured on five commercially available 785 nm Raman spectrometers (different laboratory-based and portable platforms). As part of the standardisation process, the authors applied shift correction, intensity correction, and resolution matching and interpolation. The results indicated that the differences in Raman spectra from these chemical standards acquired on different spectrometers can be corrected, with satisfactory results, using their recommended standardisation protocol. They concluded that the most dramatic effects came from correcting for instrument–instrument variations in Raman shift axis and resolution matching.

All these previous studies were performed on commercial chemicals and food products. Human tissue samples measurements are subject to greater variations in sample quality, thickness, density, composition and preparation, which require a greater level of standardisation and system transfer spectral correction algorithms.

A larger study involving both oesophageal and colon cancer tissue samples using the existing correction algorithms documented in this current study as well as a number of multivariate calibration transfer methods (such as Piecewise Direct Standardization (PDS) model updating (MU)) is currently being investigated within our collaboration.

Of interest to future clinical implementation of Raman spectroscopy is the need to generate a classification model from data collected on a number of different machines and to then use the system transferability correction algorithms to predict pathology samples collected from a completely different set of sites and Raman spectroscopy systems. This current study used instruments of similar parameter designs, however, the true test of the implementation of this technology in the clinic and worldwide is being able to implement this model on independent machine(s) of different design specifications. This would however require significant collaboration between an increased number of institutes, much larger sample sizes and more advanced multivariate analysis.

Conclusions

Raman spectroscopy combined with advanced multivariate analysis is able to decipher a wealth of biochemical information from tissue samples of different pathology states. This information can be used to identify biomarkers associated with pre-malignant and carcinogenic change. This is important for non-invasive detection of disease as well as its classification and grading to help in patient diagnosis and prognosis.

Streamline™ Raman mapping and imaging in conjunction with rapid spectral analysis has significant potential to be able to collect more tissue spectra data from tissue samples, increasing the amount of information available to spectroscopists to develop validated classification models.

In order to implement this technology in the clinic and to in general introduce Raman spectroscopy as a clinical tool for more global use in medical diagnostics, classification models constructed from spectra from one or more instruments must to be transferable between instruments and centres. Studies such as this help to investigate the potential of these models with their system transfer correction algorithms to transfer between different Raman spectroscopy systems at geographically different locations. There are a number of challenges ahead in order to develop the classification and optimal system transfer models, however collaborative studies such as this introduce the means and foundation for these further investigations.

Acknowledgements

This work was funded and supported by Innovate UK – 'SMART: Stratified Medicine through Advanced Raman Technologies'. We thank Renishaw Plc for system support and development. The histopathology technicians at Gloucestershire Hospitals NHS Trust in Gloucester, especially Joanne Motte and Sarah Brown, have worked extremely hard to complete the additional sample processing and preparation.

References

1 Cancer Research UK, *Cancer Res.*, UK, 2012.
2 H. Pohl and H. G. Welch, *J. Natl. Cancer Inst.*, 2005, **97**, 142–146.
3 J. Lagergren, R. Bergstrom, A. Lindgren and O. Nyren, *N. Engl. J. Med.*, 1999, **340**, 825–831.
4 S. M. Wildi and M. B. Wallace, *Barrett's Esophagus and Esophageal Adenocarcinoma*, John Wiley & Sons, 2008.
5 B. J. Reid, R. C. Haggitt, C. E. Rubin, G. Roth, C. M. Surawicz, G. van Belle, K. Lewin, W. M. Weinstein, D. A. Antonioli and H. Goldman, *Hum. Pathol.*, 1988, **19**, 166–178.
6 C. L. Booth and K. S. Thompson, *J. Gastrointest. Oncol.*, 2012, **3**, 232–242.
7 L. M. Almond, J. Hutchings, N. Shepherd, H. Barr, N. Stone and C. Kendall, *J. Biophotonics*, 2011, **4**, 685–695.
8 C. Kendall, N. Stone, N. Shepherd, K. Geboes, B. Warren, R. Bennett and H. Barr, *J. Pathol.*, 2003, **200**, 602–609.
9 N. Stone, C. Kendall, N. Shepherd, P. Crow and H. Barr, *J. Raman Spectrosc.*, 2002, **33**, 564–573.
10 N. Stone, P. Stavroulaki, C. Kendall, M. Birchall and H. Barr, *Laryngoscope*, 2000, **110**, 1756–1763.
11 P. Crow, N. Stone, C. A. Kendall, J. S. Uff, J. A. M. Farmer, H. Barr and M. P. J. Wright, *Br. J. Cancer*, 2003, **89**, 106–108.

12 J. Smith, C. Kendall, A. Sammon, J. Christie-Brown and N. Stone, *Technology in Cancer Research & Treatment*, 2003, **2**, 327–332.
13 R. Kast, G. Auner, S. Yurgelevic, B. Broadbent, A. Raghunathan, L. M. Poisson, T. Mikkelsen, M. L. Rosenblum and S. N. Kalkanis, *J. Neuro-Oncol.*, 2015, **125**, 287–295.
14 F. M. Lyng, D. Traynor, I. R. M. Ramos, F. Bonnier and H. J. Byrne, *Anal. Bioanal. Chem.*, 2015, **407**, 8279–8289.
15 Z. Farhane, F. Bonnier, A. Casey, A. Maguire, L. O'Neill and H. J. Byrne, *Analyst*, 2015, **140**, 5908–5919.
16 S. J. Harder, Q. Matthews, M. Isabelle, A. G. Brolo, J. J. Lum and A. Jirasek, *Appl. Spectrosc.*, 2015, **69**, 193–204.
17 T. Tolstik, C. Marquardt, C. Matthäus, N. Bergner, C. Bielecki, C. Krafft, A. Stallmach and J. Popp, *Analyst*, 2014, **139**, 6036–6043.
18 Q. Matthews, A. Brolo, J. Lum, X. Duan and A. Jirasek, *Phys. Med. Biol.*, 2011, **56**, 19–38.
19 C. Kendall, M. Isabelle, F. Bazant-Hegemark, J. Hutchings, L. Orr, J. Babrah, R. Baker and N. Stone, *Analyst*, 2009, **134**, 1029–1045.
20 G. Shetty, C. Kendall, N. Shepherd, N. Stone and H. Barr, *Br. J. Cancer*, 2006, **94**, 1460–1464.
21 J. Hutchings, C. Kendall, B. Smith, N. Shepherd, H. Barr and N. Stone, *J. Biophotonics*, 2009, **2**, 91–103.
22 J. C. C. Day, R. Bennett, B. Smith, C. Kendall, J. Hutchings, G. M. Meaden, C. Born, S. Yu and N. Stone, *Phys. Med. Biol.*, 2009, **54**, 7077–7087.
23 L. M. Almond, J. Hutchings, C. Kendall, J. C. C. Day, O. A. C. Stevens, G. R. Lloyd, N. A. Shepherd, H. Barr and N. Stone, *J. Biomed. Opt.*, 2012, **17**, 081421.
24 L. M. Almond, J. Hutchings, G. Lloyd, H. Barr, N. Shepherd, J. Day, O. Stevens, S. Sanders, M. Wadley, N. Stone and C. Kendall, *Gastrointest. Endosc.*, 2014, **79**, 37–45.
25 M. S. Bergholt, W. Zheng, K. Lin, K. Y. Ho, M. Teh, K. G. Yeoh, J. B. So and Z. Huang, *Technol. Cancer Res. Treat.*, 2011, **10**, 103–112.
26 D. Ozdemir, M. Mosley and R. Williams, *Appl. Spectrosc.*, 1998, **52**, 1203–1209.
27 N. Stone, C. Kendall and H. Barr, in *Handbook of Vibrational Spectroscopy*, John Wiley & Sons, Ltd, 2006.
28 N. K. Afseth and A. Kohler, *Chemom. Intell. Lab. Syst.*, 2012, **117**, 92–99.
29 H. Martens and E. Stark, *J. Pharm. Biomed. Anal.*, 1991, **9**, 625–635.
30 J. Nallala, G. R. Lloyd and N. Stone, *Analyst*, 2015, **140**, 2369–2375.
31 G. R. Lloyd, L. E. Orr, J. Christie-Brown, K. McCarthy, S. Rose, M. Thomas and N. Stone, *Analyst*, 2013, **138**, 3900–3908.
32 A. S. Haka, K. E. Shafer-Peltier, M. Fitzmaurice, J. Crowe, R. R. Dasari and M. S. Feld, *Proc. Natl. Acad. Sci. U. S. A.*, 2005, **102**, 12371–12376.
33 C. K. Mann and T. J. Vickers, *Appl. Spectrosc.*, 1999, **53**, 856–861.
34 H. Nguyen Quang, M. Jouan and N. Quy Dao, *Anal. Chim. Acta*, 1999, **379**, 159–167.
35 H. Swierenga, A. P. de Weijer, R. J. van Wijk and L. M. C. Buydens, *Chemom. Intell. Lab. Syst.*, 1999, **49**, 1–17.
36 H. Swierenga, A. P. de Weijer and L. M. C. Buydens, *J. Chemom.*, 1999, **13**, 237–249.

37 A. J. Myles, T. A. Zimmerman and S. D. Brown, *Appl. Spectrosc.*, 2006, **60**, 1198–1203.

38 C. V. Di Anibal, I. Ruisánchez, M. Fernández, R. Forteza, V. Cerdà and M. Pilar Callao, *Food Chem.*, 2012, **134**, 2326–2331.

39 J. D. Rodriguez, B. J. Westenberger, L. F. Buhse and J. F. Kauffman, *Analyst*, 2011, **136**, 4232–4240.

Faraday Discussions

PAPER

Label-free classification of colon cancer grading using infrared spectral histopathology

C. Kuepper,[a] F. Großerueschkamp,[a] A. Kallenbach-Thieltges,[a] A. Mosig,[a] A. Tannapfel[b] and K. Gerwert*[a]

Received 10th November 2015, Accepted 11th December 2015
DOI: 10.1039/c5fd00157a

In recent years spectral histopathology (SHP) has been established as a label-free method to identify cancer within tissue. Herein, this approach is extended. It is not only used to identify tumour tissue with a sensitivity of 94% and a specificity of 100%, but in addition the tumour grading is determined. Grading is a measure of how much the tumour cells differ from the healthy cells. The grading ranges from G1 (well-differentiated), to G2 (moderately differentiated), G3 (poorly differentiated) and in rare cases to G4 (anaplastic). The grading is prognostic and is needed for the therapeutic decision of the clinician. The presented results show good agreement between the annotation by SHP and by pathologists. A correlation matrix is presented, and the results show that SHP provides prognostic values in colon cancer, which are obtained in a label-free and automated manner. It might become an important automated diagnostic tool at the bedside in precision medicine.

Introduction

Colorectal cancer

Colorectal cancer is a major cause of morbidity and mortality throughout the world, and the third most common cancer worldwide. It affects men and women almost equally, with just over 1 million new cases every year.[1] In the majority of cases the colorectal carcinoma originates in polyps, referred to as benign adenomas. About 80% of colorectal carcinomas are sporadic with no hereditary deposition. The two most common hereditary risk factors are "familial adenomatous polyposis" (FAP)[2,3] and "hereditary non polyposis colon cancer" (HNPCC or Lynch Syndrome).[3] Patients with FAP suffer from hundreds or thousands of polyps in the colorectal mucosa, with a manifested near 100% risk of malignancy. Lynch syndrome shows only a small number of polyps. Non-hereditary risk factors

[a]*Chair of Biophysics, Faculty of Biology and Biotechnology Ruhr University Bochum, Germany. E-mail: gerwert@bph.rub.de*
[b]*Institute of Pathology, Ruhr University Bochum, Germany*

are aging, high fat nourishment in combination with a lack of physical exercise, and smoking.

The first level of detecting and characterizing colon cancer is visual inspection during colonoscopy. A diagnosis is performed on a biopsy by pathologists *via* histopathological examination using haematoxylin and eosin (H&E) stained tissue thin sections.[4] The current guidelines of the UICC for classification of colorectal carcinoma follow the TNM (tumour, lymph nodes, metastasis) system.[5] The staging is a measure of how the cancer has spread through the organism. The TNM system characterizes the local infiltration of the primary tumour, the lymph node status, and potential distant metastasis in other organs. In contrast to staging, the WHO GRADING system addresses the differentiation of tumour cells.[6] The grade score reaches from G1 (well-differentiated), to G2 (moderately differentiated), G3 (poorly differentiated) and in rare cases to G4 (undifferentiated). Well and moderately differentiated tumours are summarised as "low grade", and poorly and undifferentiated tumours as "high grade" carcinomas. The grading of cancer tissue samples is important for the prognosis of cancer patients.[7–9] Fig. 1 shows exemplary images of H&E stained colorectal cancer tissue at different differentiation states, and for comparison a sample of normal colon mucosa. Here we illustrate that spectral histopathology not only distinguishes between tumour and healthy tissue, but in addition provides the grading of the tumour.

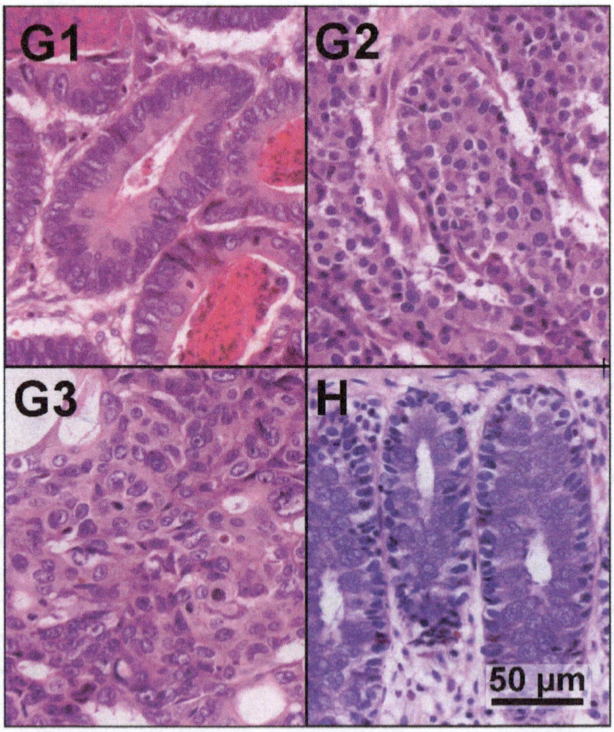

Fig. 1 Different dedifferentiation states of colon carcinoma. G1, G2 and G3 show the grading states during tumour genesis and H represents a H&E stained sample of healthy colonic tissue with normal cells.

Spectral histopathology *via* infrared imaging

In the last decade, many studies have shown that spectral histopathology is capable of classifying tissue[10,11] and especially diseased tissue.[12–20] The IR spectra measured per pixel represent mostly an integral signal of the proteome and genome. Each spectrum is assigned a specific colour. This results in an index colour image, by which the tissue is classified and tumour is identified. In particular, colorectal carcinoma is identified in this way by IR imaging.[21,22] Even though these previous studies have shown that SHP can differentiate between healthy and cancer tissue label-free, it is of limited value for clinical diagnosis, because this question is easily and very quickly answered by histopathology already. Therefore here the approach is significantly extended, and in addition the grading of the tumour is determined. The grading is much more difficult to classify and depends critically on the expertise of the respective pathologist and the time taken for the diagnosis. Biomax tissue micro arrays (TMA) were measured with an Agilent Cary 620 FTIR microscope and subsequently H&E stained. The resulting index colour images were analysed and compared with the morphological characteristics provided by H&E staining. These results exhibited good correlation between the annotation by SHP and the annotation by pathologists, as shown in a correlation matrix. This shows that the method is a useful tool for label-free automated and precise colon cancer tissue grading.

Bioinformatics workflow

We established a workflow that integrates FTIR microscopy, bioinformatics and histopathology (Fig. 2). Primarily, the tissue thin sections were measured with FTIR imaging using a focal plane array detector (FPA) with 128 × 128 MCT elements. The measured spectral map is clustered by an unsupervised algorithm in the training stage (hierarchical clustering, HCA, *k*-means). The resulting index colour image represents the spectral distribution over the examined tissue section. In parallel the tissue is still accessible for H&E and/or immunohistochemical staining due to the marker-free character of the SHP. In collaboration with pathologists, the index colour image based on the spectral map is correlated with the classically stained image of the sample. A database of spectral "fingerprints" is generated for different tissue and disease types from this expert annotation.

The spectral database enables us to train supervised classification algorithms like the artificial neural networks (ANN), support vector machines (SVM), or random forests (RF).[23–25] As previously shown in our approach, we are using Random Forest (RF) classifiers, which have proven to be accurate, easy to use and robust. The workflow was established in our lab previously for colon and lung cancer.[22,31] Unknown spectral maps of tissue thin sections can be automatically annotated with the trained RF (validation stage). The accuracy of the trained supervised classifier is determined on an independent pool of patients to ensure that no over fitting is occurring.

Experimental

Sample preparation

Tissue micro arrays (TMA, Table 1) displaying samples of colorectal carcinomas with different gradings were purchased from US Biomax Inc. (Rockville, MD,

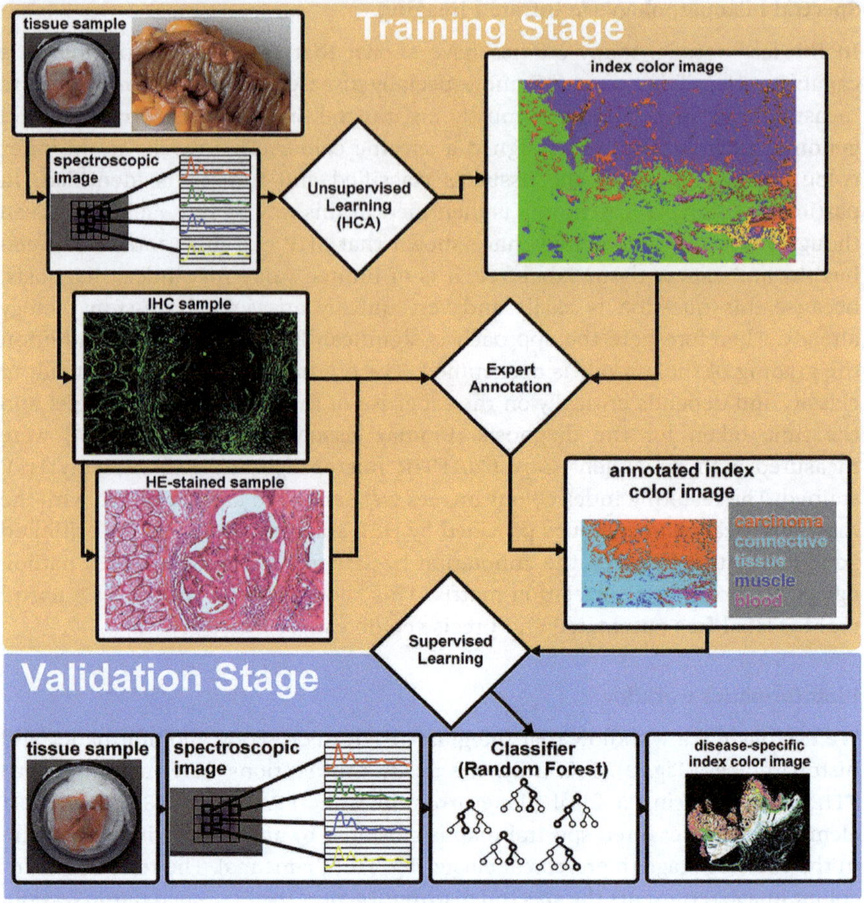

Fig. 2 Workflow of the training and validation stage. In the training stage the spectral maps were correlated to classical histopathological annotation by an expert. The resulting spectral database is used for the training of a supervised classification algorithm that is validated on independent samples in the validation stage.

USA). The samples were 5 μm thick and were placed on LowE slides [Kevley Technologies, Chesterland, OH, USA]. Before the spectral measurements, they were deparaffinised using standard protocols.[26] Afterwards the samples were stored and measured under dry air.

Table 1 Summary of the measured samples by their grading, the used TMA slides and patient number used overall and for training

Biomax ID	Chosen *patients* (*cases*)	G1	G2	G3
CO1002b	32 (*32*)	6 (6)	18 (18)	8 (8)
BCO51111	70 (136)	6 (9)	55 (110)	9 (17)
CO722	23 (23)	4 (4)	17 (17)	2 (2)
No. independent patients – validation		*14*	*87*	*17*
No. patients – training		*2*	*3*	*2*

Data acquisition

Infrared hyperspectral data acquisition was performed in transflection mode using an Agilent system (Santa Clara, California, USA), consisting of a Cary 620 infrared microscope in combination with a Cary 670 FTIR spectrometer. Spectral data were collected by a mounted liquid nitrogen cooled focal plane array (FPA) MCT detector with 128 × 128 elements, providing a field of view (FOV) of approximately 715 μm × 715 μm. The Fourier transformation was performed with the Agilent Resolution Pro Software with Mertz phase correction, a Blackman-Harris-4-term apodization and a zero filling of 2. The spectra were saved between 3700–950 cm^{-1} with a spectral resolution of 4 cm^{-1}. For the transflection (reflection–absorption) measurements, the tissue sections had been prepared on LowE slides. An inherent problem of the occurrence of a standing wave electric field in the transflection mode was described for infrared microscopy, which leads to shifts of and variances in the ratio of absorption bands, especially between the amid I and amid II bands.[32,33] However, simulations have shown that the resulting intensity artefact is minimized when objectives with high numeric apertures are used.[34] Therefore, here a high numeric aperture of 0.62 was used. In addition we tested the second derivative, which minimizes the effects of the standing wave artefact. This resulted in the same supervised classification of the colon cancer grading. Therefore, we used the vector normalized spectra.

The resulting raw spectral maps were pre-processed using the previously described workflow.[22] Strong artefacts possibly arising from cracks or folds in the tissue were eliminated by quality control based on the signal-to-noise ratio and the intensity of the amid I band. The remaining spectra were subjected to a Mie and resonance-Mie correction based on EMSC[27–29] in the wavenumber range from 2300 to 950 cm^{-1}. The correction was performed with only one iteration step, but a higher number of iteration steps (up to 20) were tested due to low scattering effects, as this does not alter the final classification. During the last step the spectra were smoothed by a 9 point Savitzky Golay filter,[30] providing second derivative spectra for the unsupervised multivariate methods like hierarchical and *k*-means clustering. For the RF classification, the spectra were only corrected for resonance Mie scattering as previously described. For both methods the analysis was performed on the fingerprint region from 1800–950 cm^{-1}.

Data selection and training of a supervised classifier

As we have shown before, the RF classifier is capable of distinguishing tissue types and identifying cancerous regions in colorectal tissue sections.[22] Here we established a hierarchical application of two consecutive RFs (Fig. 3). Spectra identified as cancerous were isolated and furthermore analysed by a second RF trained for recognizing the grade of differentiation of colorectal carcinomas. Training data were acquired from two spots of G1 and three spots of G2 and G3. Compared to the first RF using 100 data points in the fingerprint region, the second RF is supplied with 385 data points on an equidistant wavenumber scale from the wavenumber interval of 1800–950 cm^{-1}. All computations were performed in MATLAB (TheMathWorks Inc., Natick, MA, USA).

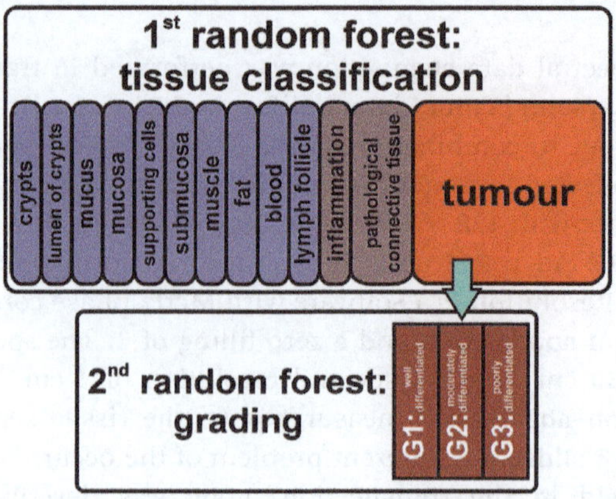

Fig. 3 The first RF detects different tissue types and pathological regions. The spectra of tumorous regions were transferred to the second RF, which determines the grading of the cancer cells.

For this study 191 sample spots collected from 125 patients, covering different carcinoma grades (1, 2 and 3), were analysed. 19 samples from 16 patients were identified as G1, 145 samples from 90 patients were identified as G2, and G3 was analysed from 27 samples from 19 patients (Table 1). The TMAs provided by US Biomax were standardised and annotated by two clinical pathologists. The microarrays were H&E stained after IR data collection. This allowed us to compare the morphological characteristics of the tissue spots with the index colour images provided by IR-SHP, leading to good correlation not only of tissue types but even regarding the grading of colon carcinomas. This study was performed in two phases. During the training stage, samples of each grade were selected randomly, spectra were analysed and training spectra representative of each grade were identified by visual inspection supported by the expertise of a Biomax independent clinical pathologist. The distribution of the measured patients over the three TMAs is shown in Table 1. The number of patients in the training set is balanced (2 for G1, 3 for G2 and 2 for G3). From these we established a training dataset of 987 representative spectra – 355 spectra for G1, 285 spectra for G2, and 347 spectra for G3. The training data set is well balanced among the three dedifferentiation grades.

In the validation stage the trained grading RF classifier was validated on 14 patients with a well-differentiated (G1), 87 patients with a moderately differentiated (G2), and 17 patients with a poorly differentiated (G3) tumour. This distribution represents the clinical occurrence of colorectal adenocarcinomas with around 10% well (G1), 70% moderately (G2) and 20% poorly differentiated carcinomas.[35] In the validation the data set does not need to be balanced between the three grades. The prediction of the grading RF was performed on the tumour spectra identified by the previously published first RF.[22] The grading RF was trained with 5000 trees and 16 features randomly chosen from the spectral data points per decision in the trees. We present the H&E stained images of the core samples, combined with corresponding spectral images that are the basis of the spectral grading of the tumour.

Results and discussion

Classification of colon tissue and colon carcinoma by infrared spectral histopathology

In the past we reported on the automated label-free classification of colon cancer tissue sections.[22] Such classification by the first RF is shown in Fig. 4 in comparison with the H&E stained sample image. We reached an accuracy of 96% combined with a high sensitivity of 94% and a specificity of 100%. This analysis was performed on 46 randomly chosen independent samples. The spectra that were classified as tumorous (see Fig. 4D) were further analysed in the second RF, which determined the grading of the cancer cells.

SHP yields reliable classification of well-differentiated colonic carcinomas

For each grade one microarray tissue sample is shown as an example in Fig. 5. In total 191 tissue spots were measured and analysed. The validation samples are presented, while the training data originates from independent samples. Fig. 5 shows the H&E stained sample of a colorectal well-differentiated cancer overlaid with tumour class spectra of the RF-based IR image. The cancerous regions *via* IR-SHP are clearly identified. The grading is given by the colour codes: G1 – well-differentiated – in red, G2 – moderately differentiated – in green, and G3 – poorly differentiated – in blue pixels.

We have chosen a threshold value of 5% of all tumour spectra for the highest grade to be taken into account for the final classification, in order to prevent false classification. The occasionally visible pixels annotating higher grades than G1 are isolated and mostly located at the edge of tumour tissue. They represent between 3.5% (G3) and 4.2% (G2) of all spectra that were annotated as "tumour" by the first level RF. The majority of 92.3% of tumour spectra were classified as G1, matching the annotation given by two clinical pathologists at US Biomax. Thus, the lowest grade with cells showing the least dedifferentiation of cancerous cells is reliably recognized by the algorithm. The pixels of the SHP index colour image match with the tumour cells quite well even at higher magnification of the H&E stained image. Small parts of the samples' cancerous regions are shown at a higher magnification in Fig. 5A–C. Tumour cells with barely visible changes in the morphology of the nuclei were assigned by the classifier as G1 cancer. The RF-based spectral image derived from a marker-free method leads to a precise classification of G1 colonic cancer tissue. This is promising, because tumour grading provides information on treatment and prognosis of colon cancer patients. Patients suffering from an early detected well-differentiated cancer have a better prognosis and may be treated with a less aggressive medication, providing better and also individual treatment with a positive outcome and the consequence of a better health-related quality of life. For 19 of 191 samples, these were characterized as well-differentiated (G1). For 18 of these 19, the annotation by our method was correct, leading to a sensitivity of 94%. The lack of false positive predictions results in a specificity of 100% regarding G1. Note that the indicators of sensitivity and specificity for the evaluation of classifiers can only be specified for binary classifiers. Thus we present these indicators (Fig. 8A) for each individual grade against the remaining other grades, rather than providing comprehensive indicators for the overall

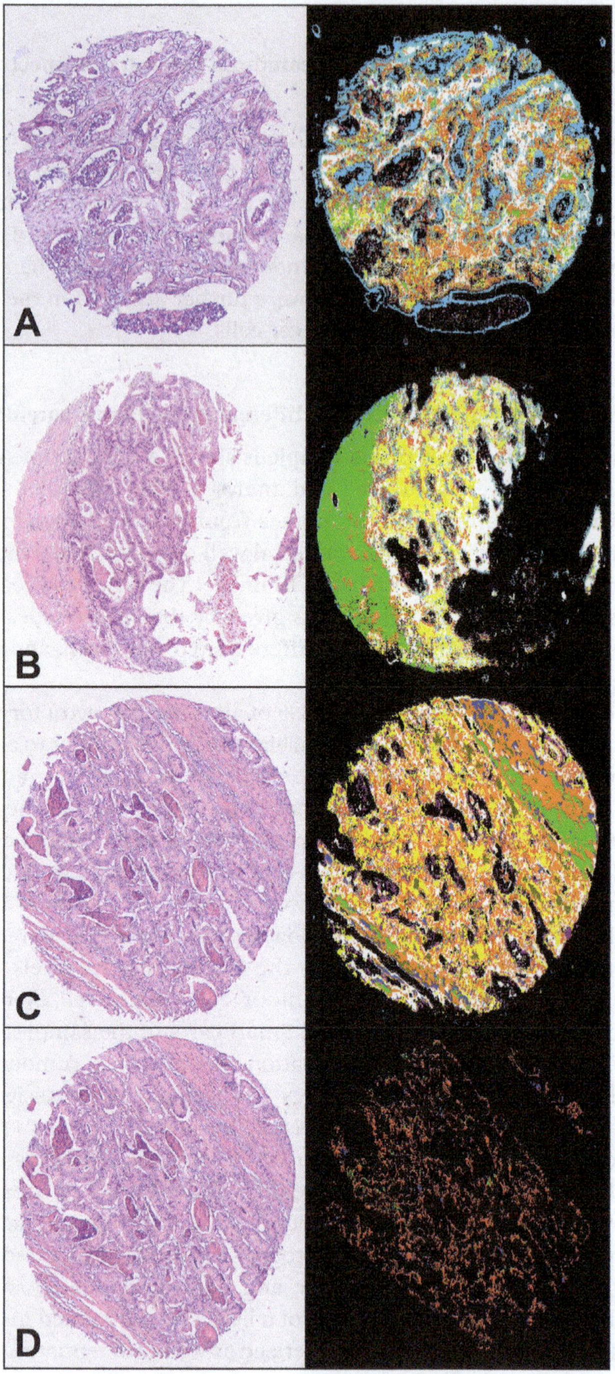

Fig. 4 From A to C exemplary index colour images of the first RF are presented. The high correlation between the H&E and the SHP can be seen. The colour code is as follows: green and yellow hues denote connective tissue, white indicates musculature, cyan is connective tissue with supporting cells, pink is the lumen of the crypts, olive indicates blood, light blue is mucus (this occurs in whole tissue and is highly influenced by scattering effects, as seen in A, but it does not affect the tumour detection), blue is pathological connective tissue, orange is inflammatory tissue, and red is the tumour region. (D) The same spot as C but only the pathological regions are presented.

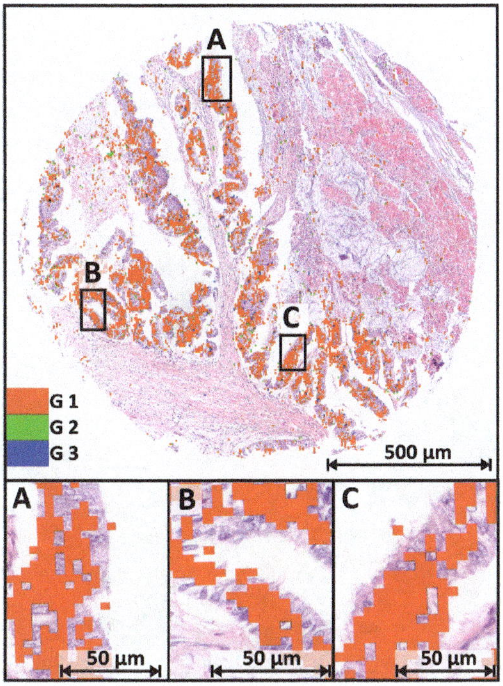

Fig. 5 H&E stained image overlaid with spectral false colour image representing the annotation of tumours. The three grades are shown with different colours. G1 is marked red, green was used for G2 and blue pixels represent G3.

prediction of all three grades. As a comprehensive indicator for all three classes, we present a confusion matrix in Fig. 8B.

A significant fraction of cancerous regions in G2 samples is classified as G1

Fig. 7 shows one exemplary tissue core sample annotated as G2 carcinoma by the pathologists' annotation. The IR image is again enlarged to the same resolution as the H&E stained image. The overlay shows a compliance of the tumour identified with SHP and cancerous regions visibly highlighted through the H&E staining. The majority of spectra in this case have been classified as G1 or G2, represented by red and green pixels. Again a small amount of isolated G3 (blue) pixels (approximately 1%) are scattered throughout the index colour image. They have not been taken into account for the analysis. In particular, 60% of the tumour class spectra were annotated G1, while 39% were classified as moderately differentiated (G2). Thus the sample is annotated as a G2 tumour, even though the large amount of over 60% were annotated G1. This is reasonable due to the fact that a tumour is never a homogenous mass of cells in the exact same state of dedifferentiation. In routine histopathological work up, the grading was performed according to the less differentiated part of the tumour. There are always cells or whole regions present, still barely differing from their tissue of origin. In the enlarged areas shown in Fig. 6, tumour cells with enlarged nuclei are visible. The changes in shape and morphology during the progress of dedifferentiation

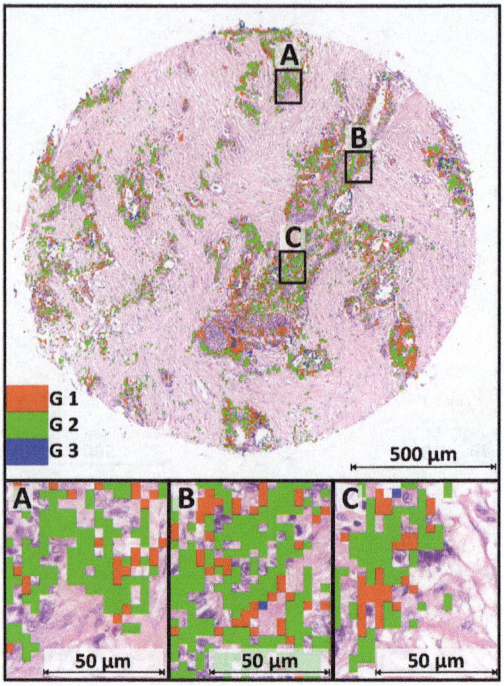

Fig. 6 H&E stained sample of grade 2 tumour overlaid with IR spectral image. The three colours red, green and blue cover again the three grades in ascending order.

are more pronounced as compared to the G1 sample. In total, 145 core samples, annotated as moderately differentiated (G2) carcinoma by clinical pathologists, were in the dataset. 119 out of these 145 samples were correctly predicted by SHP. The remaining 26 samples were annotated as G3, but none was spectrally graded lower than the grading by a pathologist.

G3 colorectal carcinoma shows the whole range of the dedifferentiation progress

Fig. 7 shows a Biomax TMA sample annotated by Biomax pathologists as a G3 carcinoma, exhibiting diffuse regions of poorly differentiated cells. The enlarged areas show selected regions with a variety of cancer cells. The nuclei are big in relation to the cell bodies and their shape differs distinctly from the healthy cells in their tissue of origin. The morphology of tumour cells is prominently illustrated in the H&E stained sample, due to the massively upregulated transcription activity necessary for a high proliferation rate. About one third (33.9%) of all spectra annotated as tumour in this sample were classified as poorly differentiated (G3) carcinoma. The rest is annotated as moderately differentiated (G2 – 44.4%) and 21.6% of tumour spectra were classified as well-differentiated (G1). With this outcome of the prediction the sample is annotated as a G3 carcinoma, matching the pathologists' diagnosis. In the dataset there were 27 core samples that were given a diagnosis of G3 adenocarcinoma. The classifier predicted 25 out of 27 correctly. A further 24 samples were annotated as G3 but didn't actually

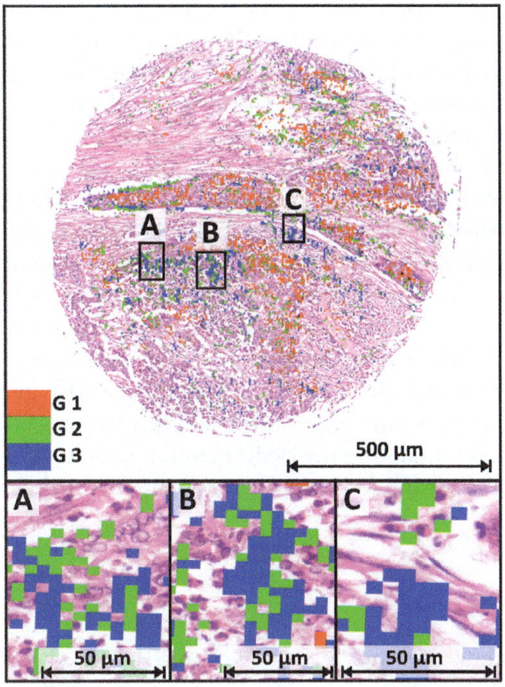

Fig. 7 H&E stained sample of a G3 colorectal carcinoma. The colour code is used as before. All three grades are determined in this sample, with a high amount of G3 (approx. 34%), leading to the annotation matching the diagnosis given by clinical pathologists.

match the pathologists' vote (false positive). These 24 samples are exactly predicted as false negatives regarding G2. This might be due to the fact that the intermediate G2 and G3 are broadly similar in the biochemical status of the tissue. Tumour cells from G1 carcinomas, which are still well-differentiated, are better distinguished from the higher grades than moderately (G2) and poorly (G3) differentiated carcinomas are from each other. None of the microarrays includes a sample of a G4 carcinoma, therefore no analysis was possible of tissue and cells showing anaplasia.

SHP analysis leads to a reproducible annotation of colorectal carcinoma grading

Overall SHP predicted the given diagnosis of cancer grading in 85% of 191 cases of colorectal cancer tissue samples. For each grade the sensitivity, specificity and accuracy was determined using standards for evaluating a binary classifier. The prediction was assigned to true positives (TP), true negatives (TN), false positives (FP) and false negatives (FN). These four basic values, corresponding to the actual diagnosis and classification outcome, built the basis of the evaluation of the classifier. In Fig. 8B the sensitivity, specificity and accuracy for each grade is demonstrated in a diagram. The sensitivity reaches from 83% to 94%, giving the peak ratio of the automated annotation. The specificity is determined with values from 87% up to 100% for well-differentiated (G1) carcinomas. Both values combined lead to the accuracy of the classifier, which reaches from 86% to 99%, as shown in Fig. 8A. Again this shows the improved capabilities of our SHP approach in detecting G1 tumours. Summing up these evaluations in a confusion matrix, the emphasis on the identification of well-differentiated (G1) is even more visible. Fig. 8B shows the confusion matrix of our classifier, with columns corresponding to the actual values (diagnosis by pathologists), and the rows correspond to the classification value (predicted grade via SHP). The matrix has a colour scheme for better understanding, beginning with blue for low values and ending in red for the high values. The desirable intrinsic diagonal pattern can be seen in the three red fields, referring to the fact that the classifier achieves high agreement rates. It illustrates also that well-differentiated (G1) has the best identification rates. In summary, the presented work demonstrates a workflow for fast, accurate and reproducible annotation of colorectal carcinoma. Earlier we presented that SHP is capable of distinguishing different tissue types and disease patterns like cancerous tissue regions in colorectal tissue sections, and now we advance our work for analysing the dedifferentiation state of tumour cells. This paves the way to precise and individual care for patients suffering from colorectal carcinoma.

Conclusions

This follow-up study presents a new level in SHP for classification of colon carcinoma. We demonstrate that FTIR imaging may not just classify tissue morphology and identify tumour, as has been demonstrated in numerous previous studies, but is also able to distinguish cell differentiation and thereby tumour grading. The grading, in addition to the detection of the tumour, paves the way to a better and more precise characterization of colon carcinoma. The approach of using a hierarchy of two (or potentially even more) spectral classifiers seems promising and should also lead to good results in annotation and characterization of other diseases.[31] Furthermore, our study utilizes higher order

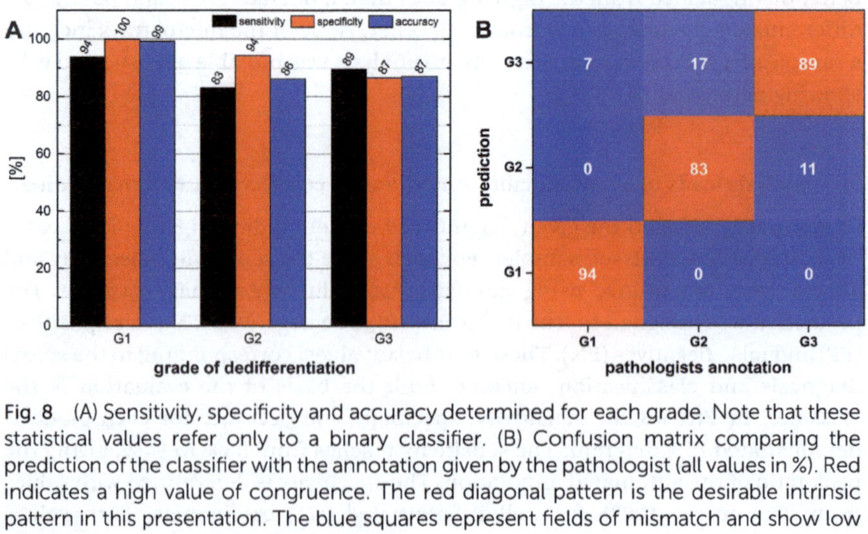

Fig. 8 (A) Sensitivity, specificity and accuracy determined for each grade. Note that these statistical values refer only to a binary classifier. (B) Confusion matrix comparing the prediction of the classifier with the annotation given by the pathologist (all values in %). Red indicates a high value of congruence. The red diagonal pattern is the desirable intrinsic pattern in this presentation. The blue squares represent fields of mismatch and show low values.

features – namely relative proportions of areas associated with different grades – for characterizing the samples. In our case, a label-free, robust, reliable, operator-independent and reproducible method for the identification and characterization of colon cancer and its grading is presented.

Acknowledgements

This research was supported by the Protein Research Unit Ruhr within Europe (PURE) – project 233-1.08.03.03-031.68079 – Ministry of Innovation, Science and Research of North-Rhine Westphalia, Germany. Furthermore we want to thank Prof. Dr M. Heise for English corrections.

Notes and references

1 American Cancer Society. Colorecatal Cancer, http://www.cancer.org/cancer/colonandrectumcancer/detailedguide/colorectal-cancer-key-statistics, Accessed 09/2015.

2 A. E. de Jong, M. van Puijenbroek, Y. Hendriks, C. Tops, J. Wijnen, M. G. Ausems, H. Meijers-Heijboer, A. Wagner, T. van Os, A. H. Brocker-Vriends, H. F. Vasen and H. Morreau, *Clin. Cancer Res.*, 2004, **10**, 972–980.

3 A. de la Chapelle, *N. Engl. J. Med.*, 2003, **349**, 209–210.

4 G. Avwioro, *Int. J. Pharmacol. Clin. Sci.*, 2011, **1**, 24–34.

5 Ch. Wittekind and H.-J. Meyer, *TNM Klassifikation maligner Tumoren. 7*, Auflage, Wiley-VCH, Weinheim, 2010.

6 National Cancer Institute of the US National Institutes of Health, http://www.cancer.gov/cancertopics/factsheet/Detection/tumor-grade, Accessed 09/2015.

7 C. Hassan, A. Zullo, M. Risio, F. P. Rossini and S. Morini, *Dis. Colon Rectum*, 2005, **48**, 1588–1596.

8 H. Ueno, H. Mochizuki, Y. Hashiguchi, H. Shimazaki, S. Aida, K. Hase, S. Matsukuma, T. Kanai, H. Kurihara, K. Ozawa, K. Yoshimura and S. Bekku, *Gastroenterology*, 2004, **127**, 385–394.

9 W. Schmiegel, A. Reinacher-Schick, D. Arnold, U. Graeven, V. Heinemann, R. Porschen, J. Riemann, C. Rödel, R. Sauer, M. Wieser, W. Schmitt, H. J. Schmoll, T. Seufferlein, I. Kopp and C. Pox, *Z. Gastroenterol.*, 2008, **46**, 799–840.

10 M. Diem, M. Miljkovic, B. Bird, T. Chernenko, J. Schubert, E. Marcsisin, A. Mazur, E. Kingston, E. Zuser, K. Papamarkakis and N. Laver, *Spectroscopy*, 2012, **27**, 463–496.

11 L. Chiriboga, P. Xie, H. Yee, V. Vigorita, D. Zarou, D. Zakim and M. Diem, *Biospectroscopy*, 1998, **4**, 47–53.

12 M. Diem, A. Mazur, K. Lenau, J. Schubert, B. Bird, M. Miljković, C. Krafft and J. Popp, *J. Biophotonics*, 2013, **6**, 855–886.

13 B. R. Wood, L. Chiriboga, H. Yee, M. A. Quinn, D. McNaughton and M. Diem, *Gynecol. Oncol.*, 2004, **93**, 59–68.

14 R. Bhargava, D. C. Fernandez, M. D. Schaeberle and I. W. Levin, *PittCon, Paper 211*, 2002.

15 W. Steller, J. Einenkel, L.-C. Horn, U.-D. Braumann, H. Binder, R. Salzer and C. Krafft, *Anal. Bioanal. Chem.*, 2006, **384**, 145–154.

16 C. Krafft and R. Salzer, *Handbook vibrational spectroscopy*, 2008.
17 M. Diem, P. R. Griffiths and J. M. Chalmers, *Vibrational Spectroscopy for Medical Diagnosis*, John Wiley & Sons, Chichester, UK, 2008.
18 N. Amharref, A. Beljebbar, S. Dukic, L. Venteo, L. Schneider, M. Pluot, R. Vistelle and M. Manfaita, *Biochim. Biophys. Acta, Biomembr.*, 2006, **1758**(7), 892–899.
19 M. J. Romeo and M. Diem, *Vib. Spectrosc.*, 2005, **38**, 115–119.
20 B. Bird, M. Miljkoviæ, S. Remiszewski, A. Akalin, M. Kon and M. Diem, *Lab. Invest.*, 2012, **92**, 1358–1373.
21 P. Lasch, M. Diem, W. Hänsch and D. Naumann, *J. Chemom.*, 2006, **20**, 209–220.
22 A. Kallenbach-Thieltges, F. Großerüschkamp, A. Mosig, M. Diem, A. Tannapfel and K. Gerwert, *J. Biophotonics*, 2013, **6**, 88–100.
23 R. Diaz-Uriarte and S. A. de Andres, *BMC Bioinformatics*, 2006, **7**(1), 3.
24 A. Statnikov, L. Wang and C. F. Aliferis, *BMC Bioinformatics*, 2008, **9**(1), 319.
25 L. Breiman, *Mach. Learn.*, 2001, **45**(1), 5–32.
26 DCS-Innovative Diagnostik, http://www.dcs-diagnostics.de/data/IHC_Entparaffinierung_web.pdf, Accessed 06/2015.
27 P. Bassan, A. Kohler, H. Martens, J. Lee, H. J. Byrne, P. Dumas, E. Gazi, M. Brown, N. Clarke and P. Gardner, *Analyst*, 2010, **135**, 268–277.
28 P. Bassan, H. J. Byrne and F. Bonnier, *Analyst*, 2009, **134**, 1586–1593.
29 P. Bassan, A. Sachdeva, A. Kohler, J. Lee, P. Dumas and P. Gardner, *Analyst*, 2012, **137**(6), 1370–1377.
30 A. Savitzky and M. J. E. Golay, *Anal. Chem.*, 1964, **36**(8), 1627–1639.
31 F. Großerüschkamp, A. Kallenbach-Thieltges, T. Behrens, T. Brüning, M. Altmayer, G. Stamatis, D. Theegarten and K. Gerwert, *Analyst*, 2015, **140**, 2114–2120.
32 P. Bassan, J. Lee, A. Sachdeva, J. Pissardini, K. M. Dorling, J. S. Fletcher, A. Henderson and P. Gardner, *Analyst*, 2013, **138**, 144.
33 M. Miljkovic, B. Bird and M. Diem, *Analyst*, 2012, **137**, 3954–3960.
34 T. P. Wrobel, B. Wajnchold, H. J. Byrne and M. Baranska, *Vib. Spectrosc.*, 2013, **69**, 84–92.
35 M. Fleming, S. Ravula, S. F. Tatishchev and H. L. Wang, *J. Gastrointest. Oncol.*, 2012, **3**(3), 153–173.

Faraday Discussions

PAPER

Advancements in quantum cascade laser-based infrared microscopy of aqueous media

K. Haase,† N. Kröger-Lui,†* A. Pucci, A. Schönhals and W. Petrich

Received 17th November 2015, Accepted 14th December 2015
DOI: 10.1039/c5fd00177c

The large mid-infrared absorption coefficient of water frequently hampers the rapid, label-free infrared microscopy of biological objects in their natural aqueous environment. However, the high spectral power density of quantum cascade lasers is shifting this limitation such that mid-infrared absorbance images can be acquired *in situ* within signal-to-noise ratios of up to 100. Even at sample thicknesses well above 50 μm, signal-to-noise ratios above 10 are readily achieved. The quantum cascade laser-based microspectroscopy of aqueous media is exemplified by imaging an aqueous yeast solution and quantifying glucose consumption, ethanol generation as well as the production of carbon dioxide gas during fermentation.

1 Introduction

Spectroscopy in the mid-infrared spectral region can be used for label-free and quantitative measurements of biomolecules, such as carbohydrates, DNA or proteins. However, when trying to investigate biological objects in their natural habitat the high mid-infrared absorption coefficient of water constitutes a challenging obstacle.

Mid-infrared microspectroscopic studies of individual live cells in aqueous environments were reported more than a decade ago.[1] These measurements were performed using a FT-IR spectrometer with a globar source and a water layer thickness of 10 μm. Further studies of live cells in aqueous media as well as microfluidic events based on FT-IR-spectroscopy with a globar were conducted at a path length of 10–25 μm.[2–5] Shortly after the first studies of cells in aqueous media using globar-based FT-IR microspectroscopy, Moss *et al.* investigated cell cultures and single cells under aqueous media using synchrotron radiation-based FT-IR microspectroscopy, with a measured path length of 11–12 μm. For the first time, cell division and motility were observed during the FT-IR experiment.[6]

Heidelberg University, Kirchhoff Institute for Physics, Im Neuenheimer Feld 227, 69120 Heidelberg, Germany. E-mail: niels.kroegerlui@kip.uni-heidelberg.de

† These authors contributed equally to this work.

Microfabricated sample holders have been developed to investigate single cells in aqueous solutions using synchrotron-based FT-IR microspectroscopy at path lengths of up to 12 μm. For larger cells, measurements were performed using a conventional flow cell with spacers for nominal path lengths of 7.6 μm and 12.7 μm.[7] To monitor arsenic-induced changes to intracellular biomolecules in live leukemia cells, synchrotron-based FT-IR microspectroscopy was performed at a path length of less than 10 μm.[8] In more recent studies, synchrotron-based FT-IR microspectroscopy was used to observe the progression of fibroblast cells through the cell cycle.[9] Further information is provided in, for example, ref. 10 and 11.

The recent progress in quantum cascade laser (QCL)[12] technology allows for the rapid investigation of samples at water layer thicknesses well above 10 μm. QCLs have been successfully used, for example, in non-imaging mid-infrared spectroscopy-based sensor systems to determine the concentration of glucose in aqueous solutions *in vitro*.[13] Furthermore, measurements of glucose concentration in human blood serum were performed using a flow cell with a liquid layer thickness of 140 μm together with a highly sensitive, liquid-nitrogen-cooled detector, whereby a root mean square error of prediction of 6.9 mg dL^{-1} was achieved.[14] More recently, protein concentration and secondary structure in water have been determined using a QCL-based setup with a flow cell with a 38 μm liquid layer thickness.[15] Finally, fibre-based QCL sensors enabled the continuous, minimally invasive quantification of glucose in the interstitial fluid of subcutaneous tissue of rats *in vivo*.[16]

QCLs have also been used for infrared microscopy in order to increase measurement speed whilst maintaining a reasonable signal-to-noise ratio.[17–20] The rapid investigation of thin tissue sections using QCL-based infrared microscopes has been demonstrated for selected spectral regions.[21,22] The investigation of infrared absorption at pre-selected, discrete frequencies can speed up the hyperspectral imaging of thin tissue sections even further.[22–24] Microscopic imaging at a single mid-infrared wavelength can be performed even in real time using QCL-based microscopes, as was recently demonstrated using living microorganisms in a measurement chamber with a water layer thickness of 25 μm and at a frame rate of 50 Hz.[25]

Ultimately, the feasibility and performance of QCL-based microscopy is governed by the ratio between the obtainable signal and the noise. Hence, in this manuscript the aspect of signal-to-noise is defined, analysed and interpreted. QCL-based microspectroscopy of a biologically active aqueous sample is illustrated on the example of a fermentation process.

2 Materials and methods

All infrared measurements described here are conducted with a custom built, QCL-based infrared microscope (Fig. 1). For illumination of the sample a QCL (type CW-MHF, Daylight Solutions Inc., San Diego, USA) can be tuned over the spectral region of 1027–1087 cm^{-1} within less than 12 seconds. The calibration of laser output power as well as wavelength can be performed in parallel with infrared microscopy. The microscope uses an uncooled microbolometer array detector (VarioCam hr head, InfraTec GmbH, Dresden, Germany) with 640 × 480 pixel elements and a pixel pitch of 25 μm for infrared measurements in

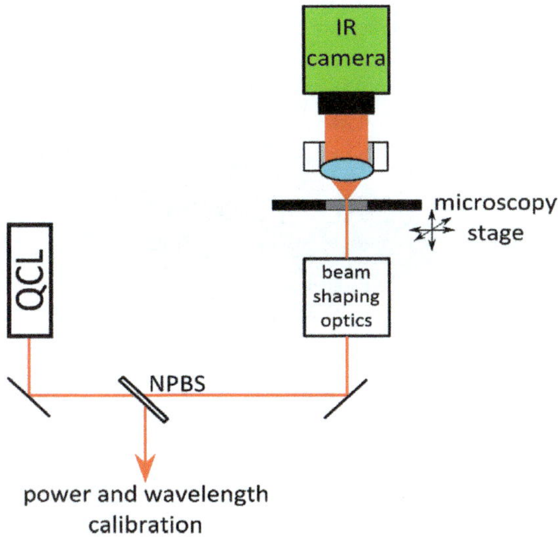

Fig. 1 Schematic of the QCL-based infrared microscope. The laser beam originates from a tunable QCL (1030–1090 cm^{-1}) and is directed towards a non-polarizing beam splitter (NPBS). The power and wavelength of the QCL can be monitored parallel to microscopy. The magnification of the microscope was chosen to be 1 : 1 (with a 10.6 × 11.7 mm^2 field of view and a 23.0 ± 3.5 μm pixel pitch) by selection of a macro lens in front of the microbolometer array detector. Adjustable beam shaping optics can be used to, for example, limit the QCL illumination to the size of the field of view of the detector.[20]

transmission mode. For large objects, lens optics for a nominal magnification of 1 : 1 can be chosen. The optical resolution in 1 : 1 magnification was determined to be 23.0 ± 3.5 μm for a field of view of 10.6 × 11.7 mm^2. A more detailed description of the setup can be found in ref. 20.

For infrared transmission measurements of samples in aqueous solutions, a custom built cuvette (Fig. 2) was mounted on the microscopy stage of the infrared microscope. The cuvette consists of two barium fluoride windows (Korth Kristalle GmbH, Altenholz, Germany) and an exchangeable mylar or PTFE spacer (Specac Ltd, Kent, England) determining the water layer thickness of the measurement chamber. The cuvette is equipped with a heating element (Omega Engineering Limited, United Kingdom) and a temperature sensor (PT100, Heraeus, Germany) enabling temperature control by a digital PID controller (ETC8420, ENDA, Germany).

In order to discuss signal-to-noise levels, the transmission of the water-filled cuvette is referenced to the transmission signal of the empty cuvette, thus enabling the calculation of absorbance. By fitting a model for the absorbance of water to the spectra in each pixel, one can estimate the noise in each pixel by the root mean square deviation between the data points and the fit. We followed the quantification of noise provided in ref. 26. More precisely, the 'single-pixel noise' can be estimated by dividing the mean transmission by the root mean square error of the fit. It is important to note that pixel-to-pixel variations among the spectra of neighbouring pixels in an image will not contribute to the single-pixel noise. Hence, for a more detailed analysis of noise in imaging applications, analysis of the deviation of pixel count rates within an image is important.[24]

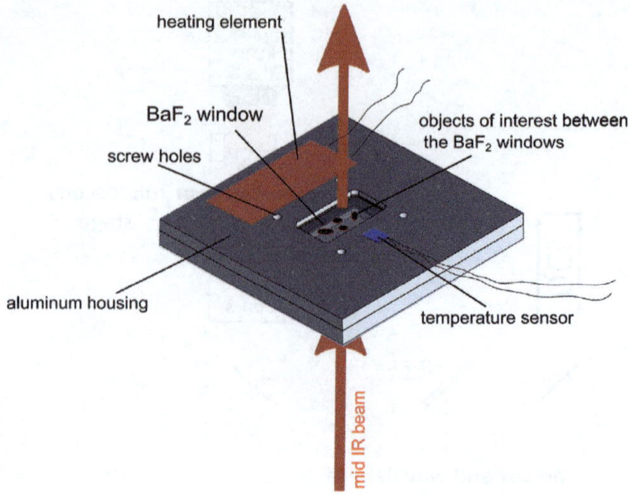

Fig. 2 Schematic of the cuvette design. Barium fluoride windows are separated by a Mylar or PTFE spacer. The volume of the cuvette is sealed using an O-ring, which is placed between the barium fluoride windows and the aluminium housing. The cuvette temperature is controlled using a temperature sensor and a heater plus the respective electronic circuitry.

The *in situ* measurements of biologically active fluids were performed in two different configurations:

(1) In the first configuration 2 μL of solution was dispensed onto the lower BaF_2 window of the chamber. A 1 mm thick PTFE spacer was used such that the droplet did not come into contact with the upper BaF_2 window. The primary sealing of the cuvette by the spacer is supposed to prevent the solution from evaporating out of the chamber. The chamber was investigated for about 9 h taking data every 30 min while the temperature was constantly held at 35.0 ± 0.5 °C. Each acquisition cycle was performed averaging over 20 laser sweep scans, leading to a measurement time < 10 min. The image data was referenced to a measurement without any object in the beam path. The inhomogeneous thickness of the solution droplet due to its free shape was taken into account by spatially distinguishing pixels containing solution and pixels that are background *via* an upper ($T = 0.5$) and lower ($T = 0.08$) threshold in the transmission T. The spectra were then averaged over the region assigned to solution and investigated using a pure component analysis.

(2) In the second configuration a Mylar spacer with a thickness of 12 μm was used and the chamber was completely filled with solution. In this setting, the CO_2 created during the fermentation process generates gas bubbles within the liquid. In a later stage, smaller bubbles agglomerate to form larger areas of CO_2 such that the volume of CO_2 can be derived from the infrared transmission image by means of thresholding followed by a convex hull segmentation. In this way, a simultaneous monitoring of glucose, ethanol as well as CO_2 can be performed.

For either of the configurations we monitored a fermentation process as an example in that the measurement chamber was filled with a solution of the following constituents: 0.5 g dL^{-1} yeast (Saccharomyces cerevisiae, Alcotec,

United Kingdom), 1 g dL^{-1} LB medium (LB Broth, Sigma Aldrich, Germany), 2 g dL^{-1} glucose (D(+)glucose, Merck KGaA, Germany).

3 Results and discussion

In order to elucidate the feasibility of and conditions for imaging living matter, two series of measurements were performed in the course of the experiments described here, one targeting the signal-to-noise aspects of the QCL-based microspectroscopy and a second one to illustrate the application capabilities.

3.1 Signal and noise

In a first series of measurements the cuvette was filled with water of varying thickness. The average transmission of water in the spectral range of the QCL ranges from 54.1% ± 0.8% to 2.8% ± 0.3% for water thicknesses between 6 μm and 50 μm.‡ The analysis of the thickness dependence yields an extinction coefficient of 685 cm^{-1}. Since scattering of the MIR radiation in the thin water layer is negligible, the extinction coefficient is identical to the absorption coefficient and agrees with the data reported in the literature (see *e.g.* ref. 27) within the error margins.

The spectral dependence of the raw signal of *e.g.* the IR camera's center pixel is shown for the case of the empty cuvette in Fig. 3. The comparison between this signal and the raw signal obtained through a cuvette filled with a water layer of 6 μm thickness reveals that the spectral shape of the raw signal is governed by the wavenumber dependence of the laser output power ("gain profile").

When assuming that the spectral dependence of transmission through water can be approximated by a moderate, but linear spectral dependence of *e.g.* 0.0002 cm^{-1} (dashed line in Fig. 3b), the difference between the actual transmission at

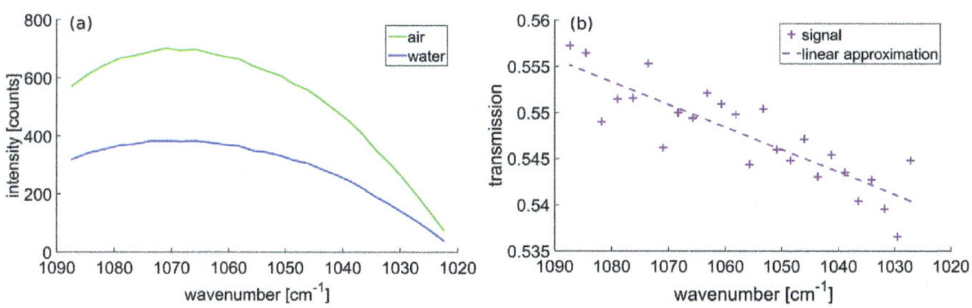

Fig. 3 (a) Single-pixel raw signals originating from the transmitted laser radiation during a laser sweep for the case of an air-filled (green) and a water-filled (blue) cuvette with layer thickness 6 μm. (b) The ratio of these two measurements reveals the transmission of approximately 54% in agreement with the expectations from the literature. The dashed line indicates the very moderate wavenumber dependence of the transmission through water, which may be approximated by a simple linear function.

‡ Note that the transmission of the filled cuvette is hereby compared to the empty cuvette with just one BaF_2 window and corrected for the thickness-independent, systematic deviation of 7.4% caused by the difference in Fresnel reflections for the BaF_2–water and BaF_2–air interfaces.

any wavenumber and this linear fit may be used as an estimate for the noise as detected by the pixel.

We here define "noise" as the deviation between a measured signal y_i and a reference y_{ref} according to

$$N = \sqrt{\frac{1}{n-1}\sum_{i=1}^{n}\left(y_i - y_{ref}\right)^2}$$

In the case of single-pixel noise N_s, we refer to the transmission signal at any given wavenumber as y_i and the corresponding value of the linear approximation to the wavenumber dependence, *i.e.* the dashed line in Fig. 3b, as y_{ref}. The sum then averages the squares of these deviations over n wavenumber intervals within a given wavenumber range under consideration. For the data in Fig. 3, for example, the corresponding single-pixel noise N_s of the transmission amounts to 0.37% for the full spectral tuning range of the QCL (1027–1087 cm^{-1}) and to 0.32% for the more intense and central part of the spectrum, *i.e.* between 1041 cm^{-1} and 1079 cm^{-1}.

In contrast to the single-pixel analysis of the noise, the pixel-to-pixel variation has to be considered in an imaging scenario, too: even if the single-pixel noise would be close to zero, an image may still look "noisy" under constant illumination. Hence, the single-pixel noise provides a lower limit for the overall noise of an image and the pixel-to-pixel variation might add to the imaging noise. We therefore evaluate the noise among the pixels as a further measure for noise and hereby refer to it as pixel-to-pixel noise N_{pp}.§ The pixel-to-pixel noise is calculated according to the above formula, whereby a field of m pixels is considered as follows: given the negligible wavelength dependence of the transmission through water over the given tuning ranges, we take the spectral average of the transmission at any pixel as y_i and the common average of y_i over all m pixels as y_{ref}. For the data shown in Fig. 3, for example, the pixel-to-pixel noise N_{pp} of the transmission amounts to 0.45% for the full and 0.40% for the central part of the spectral tuning range.

Finally, the combination of the inherent noise in each pixel N_s and the deviations among the pixels N_{pp} provides an estimate for what we dubbed the total noise N_t, which is calculated by comparing the signal of each pixel at any given wavenumber (y_i) to the overall mean of these signals averaged among the m pixels under consideration (y_{ref}). The total noise N_t of the transmission equals 0.59% and 0.51% for the spectral ranges of 1027–1087 cm^{-1} and 1041–1079 cm^{-1}, respectively, for the data shown in Fig. 3.

The spectrally averaged transmission as well as the single pixel noise, the pixel-to-pixel noise and the total noise are shown in Fig. 4 for the various thicknesses of the water layer.

The transmission noise decreases with decreasing signal (*i.e.* increasing thickness of the absorbing water layer) down to a level of ~0.2% for the single pixel noise. The heterogeneity among the pixels leads to a noise floor of ~0.3%,

§ If it is unknown whether this single-pixel noise will be the same for all pixels, one may wish to average this single-pixel noise over a selection of or all pixels of a transmission image. We here provide the analysis for a central area of 9 × 9 pixels unless noted otherwise.

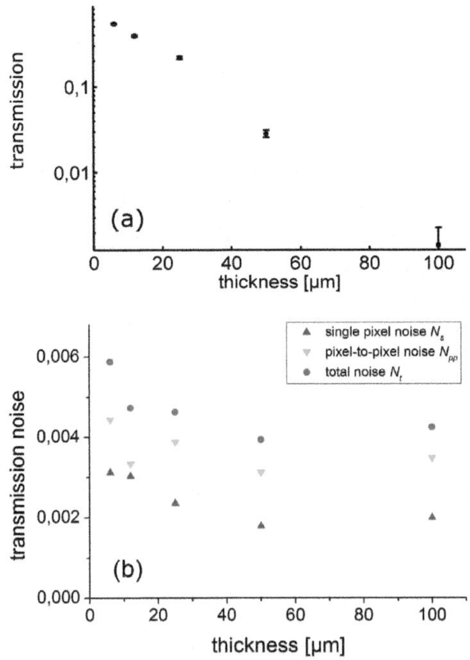

Fig. 4 (a) Transmission and (b) its various noise contributions as a function of layer thickness. The data shown is taken from a 9 × 9 pixel region in the centre of the measurement window and is averaged over the spectral range from 1041–1079 cm^{-1}.

which overall can be summarized as a total noise background for transmission around 0.4%. Further analysis shows that the noise approximately follows the expected square root dependence on the transmission (plus the constant noise floor) and provides sufficient signal as compared to the noise, even for thicknesses well above 50 µm (Fig. 5). This, in turn, means that in our setup transmission measurements can be performed in aqueous solutions for

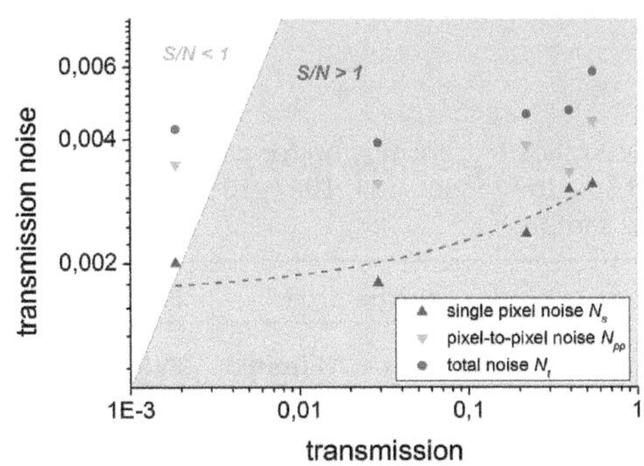

Fig. 5 The transmission noise as a function of the transmission based on the data shown in Fig. 4. The shaded area marks the region for which the signal-to-noise ratio is larger than 1. A square root relationship above the background noise is sketched for the example of single-pixel noise (dashed line).

thicknesses well above 50 μm, whereby even at 50 μm the signal is still approximately 10 times larger than the noise. The complete list of average transmission signals as well as the signal-to-noise ratios for the various noise sources are provided as a function of thickness of the water layer in Table 1.

While the analysis of the transmission T enables a better understanding of the origin of the noise and the fundamental limitations, the investigation of the noise in terms of the absorbance A is of a more practical importance. As an example for the absorbance noise, the spectral dependence of the single-pixel noise and the total noise are displayed in Fig. 6 for the case of a 6 μm thick water layer. In either case, the absorbance noise reaches its respective minima of 0.002 and 0.004 in the wavenumber range at which the gain profile is at its maximum (see also Fig. 3). This finding suggests that the noise is intensity-dependent, as already speculated from the above discussion of the transmission data.

In the light of the fundamental differences between absorbance noise and transmission noise, it is important to point out that the loss of transmission is of most interest in practice rather than the transmission itself. In other words, and as is well known in this field, the highest transmission (short path lengths) is therefore not desirable since it offers no absorbance signal of the analyte under investigation, while at the same time investigations at very low transmission (large path lengths) are impaired by noise. Hence, the optimum balance between signal gain and noise increase, *i.e.* the optimum signal-to-noise ratio (S/N), is sought. The absorbance S/N (A/dA) and the transmission S/N (T/dT) are related by:

$$\frac{A}{\mathrm{d}A} = \frac{-\log(T)}{-\mathrm{d}T/T} = \log(T)\frac{T}{\mathrm{d}T}$$

While both the transmission S/N and the absorbance S/N decrease with increasing sample thickness for large thicknesses, it is of note that transmission S/N decreases while the absorbance S/N increases with increasing sample thickness for small thicknesses. The corresponding values for absorbance and the various signal-to-noise ratios are provided in Table 2 for the respective sample thicknesses.

Table 1 Signal-to-noise ratios of transmission for different layer thicknesses, methods of noise calculation, and spectral ranges (full: 1027–1087 cm^{-1}; limited: 1041–1049 cm^{-1}) based on a 9 × 9 pixel sample

Thickness [μm]	Average transmission	Single pixel		Pixel-to-pixel		Total	
		Full	Limited	Full	Limited	Full	Limited
6	0.54	145.4	168.1	121.2	137.2	91.3	106.6
12	0.39	109.4	125.1	115.8	125.8	81.8	94.4
25	0.22	73.5	89.9	55.7	65.3	46.6	56.0
50	0.029	11.5	15.6	9.0	10.5	7.2	8.9
100	0.0019	0.5	0.9	0.4	0.6	0.3	0.5

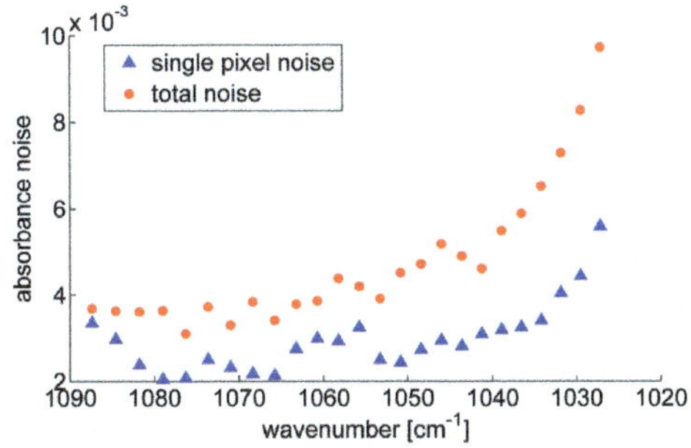

Fig. 6 Spectral dependence of the single-pixel noise and the total noise of the absorbance. The noise reaches its minimum in the spectral range of the maximum of the laser gain profile, suggesting that the noise is intensity-dependent.

An example for the experimental data is given in Fig. 7. The optimum balance between gain in absorbance signal and increase in noise with increasing sample thickness appears to be around 20 μm for our setup, which agrees well with our prior estimates.[28] The single-pixel as well as the pixel-to-pixel noise also peak around that thickness.

Since the signal-to-noise ratio is dependent on laser intensity and, thus, on the overall counts c detected by each individual pixel, the S/N may be further improved by averaging over successive frames for applications, in which the time resolution is less critical than the S/N. When averaging over *e.g.* 5 successive frames, the absorbance noise is found to decrease by a factor of around 2. In this context, it is interesting to investigate the noise (*e.g.* the total absorbance noise) as a function of the overall count rate, which changes over the tuning range as well as when averaging over successive measurement frames. Since the spectral dependence of the water absorption in this range is negligible, the main effect upon tuning will be the laser intensity variation over the gain profile. Indeed, the (*e.g.* total absorbance) noise depends on the laser intensity as well as on the number of frames used for averaging as expected; $\sim(c + c_0)^{-0.5}$, where c_0 is the dark count rate mentioned above (Fig. 8).

Table 2 Signal-to-noise ratios of absorbance for different layer thicknesses, methods of noise calculation, and spectral ranges (full: 1027–1087 cm^{-1}; limited: 1041–1049 cm^{-1}) based on a 9 × 9 pixel sample

Thickness [μm]	Average absorbance	Single pixel		Pixel-to-pixel		Total	
		Full	Limited	Full	Limited	Full	Limited
6	0.27	88.4	101.9	74.1	83.7	55.9	65.1
12	0.41	101.0	115.2	108.6	117.6	76.7	88.3
25	0.66	108.9	132.8	84.6	98.8	70.6	84.7
50	1.54	33.2	49.8	29.3	36.4	23.0	30.8
100	2.71	10.4	10.7	9.8	9.6	6.7	6.9

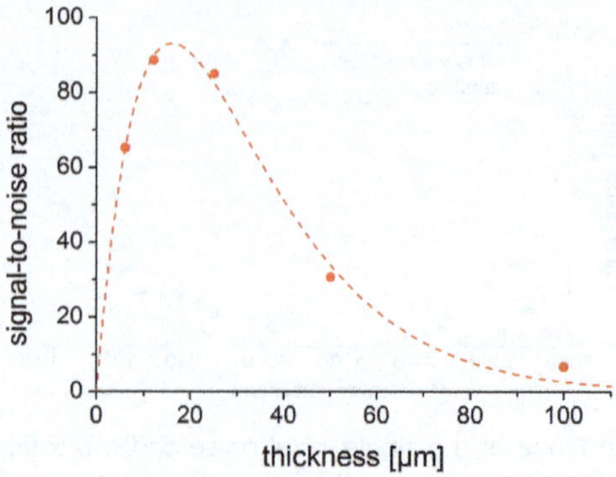

Fig. 7 Signal-to-noise ratio in terms of absorbance signal and total noise as a function of thickness of the water layer. The dashed line indicates the theoretical trend.

3.2 Application example

In order to exemplify the capability of QCL-based mid-infrared micro-spectroscopy, a fermentation process was studied with respect to glucose consumption, ethanol generation and CO_2 production according to

$$C_6H_{12}O_6 \rightarrow 2C_2H_5OH + 2CO_2.$$

A Fourier-transform Infrared (FT-IR) spectrum of the fermentation solution (Fig. 9) was acquired using a standard FT-IR spectrometer (Tensor 27, Bruker Optik GmbH, Germany), and it indicates that our QCL's tuning range may already suffice for the quantitative investigation of the fermentation process. This tuning range in particular includes the C–O–H vibration band of glucose as well as the C–O stretching vibration of ethanol.

The QCL-based experiment was performed in the two ways mentioned above. For the case in which a droplet of solution was allowed to exchange vapour and gas with the surrounding, it is important to know that the absorbance of the IR-

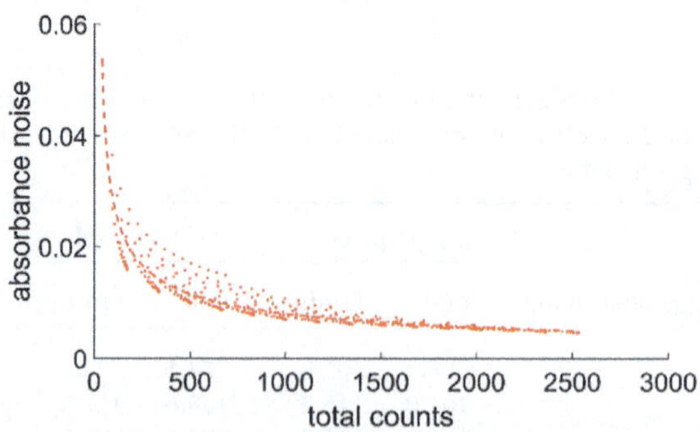

Fig. 8 The absorption signal-to-noise ratio shows a clear dependence on the mean intensity. The counts are varied along the gain profile of the laser as well as by averaging over successive frames. The dashed line indicates a $\sim (c + c_0)^{-0.5}$ dependence.

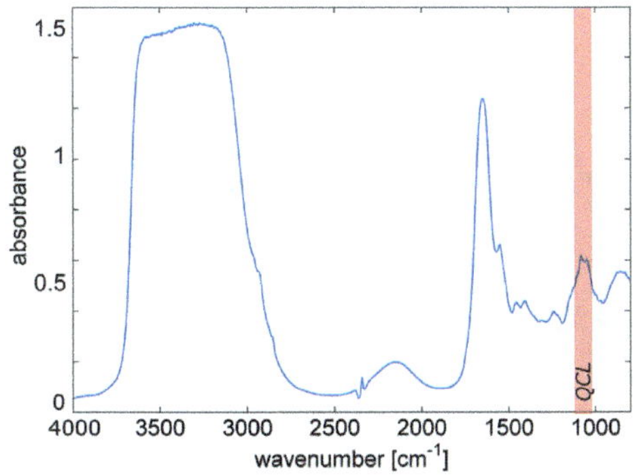

Fig. 9 FT-IR spectrum of the fermentation solution.

active absorption band of gaseous CO_2 within the tuning range of the QCL is well below 10^{-3} due to the short layer thickness and the low density of the gas. Hence, it can be assumed that contributions of CO_2 to the overall signal are negligible in terms of infrared absorption, and it can be shown that this finding holds true also in terms of pressure increase and water displacement. At the same time, the initial evaporation of water from the liquid droplet will came to a halt very rapidly since the small volume inside the cuvette will be saturated with water vapour already by 4 nL of liquid water (35 °C).

The pure component spectra of glucose, ethanol, yeast and LB medium were determined in an independent series of measurements (Fig. 10) with the

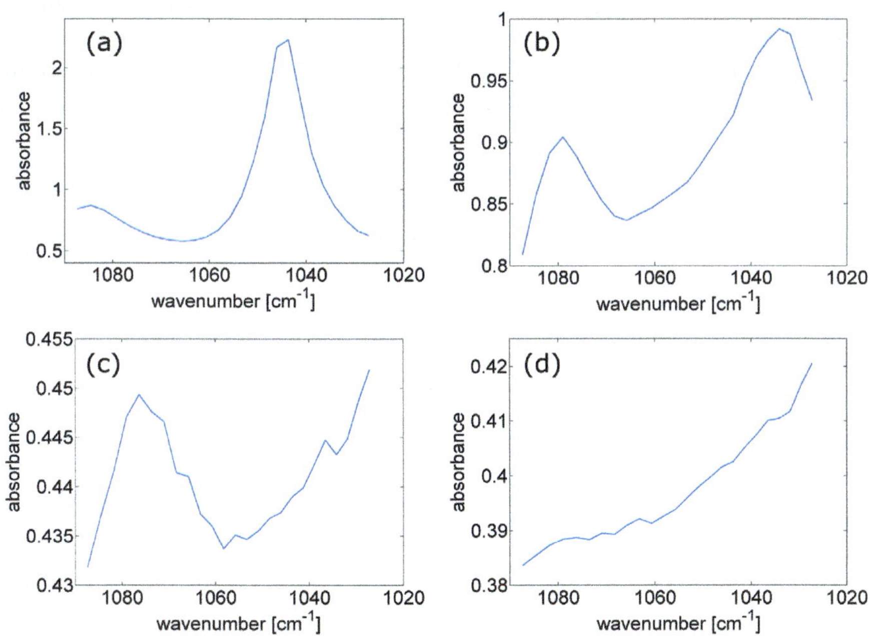

Fig. 10 Pure component spectra of (a) ethanol, (b) glucose, (c) yeast and (d) LB medium. The error bars amount to 0.05 for (a), 0.06 for (b), 0.03 for (c) and 0.03 for (d).

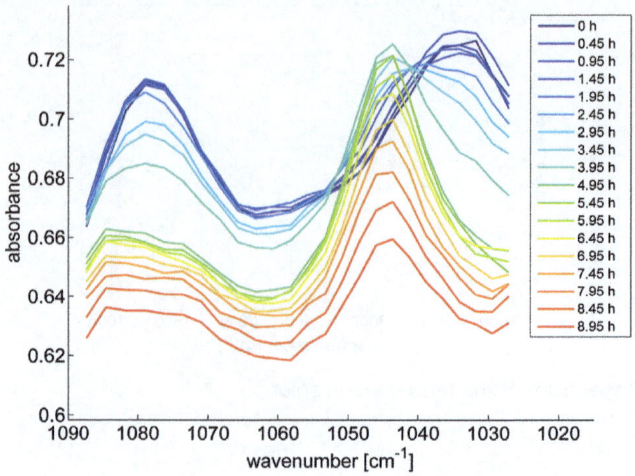

Fig. 11 Time evolution of the infrared spectrum of the fermentation droplet.

microscope setup. These spectra of the relevant constituents can be used to analyse the time evolution of the fermentation solution (Fig. 11) in which the changes from a glucose-like spectrum towards an ethanol-like spectrum are evident. The result of a pure component analysis is shown in Fig. 12. Since the height of the droplet is not known, no absolute values for the concentrations can be provided other than average mass and volume concentration per area for glucose and ethanol, respectively. The increase in ethanol concentration follows the decrease in glucose concentrations for the first 5 hours of the experiment, as expected. All glucose appears to be consumed after 5 hours, which is in agreement with the estimation of the glucose consumption rate of the fermentation at a temperature of 35 °C. It is of note, however, that the cuvette seems to become leaky during these experiments such that the ethanol vapour could escape. This,

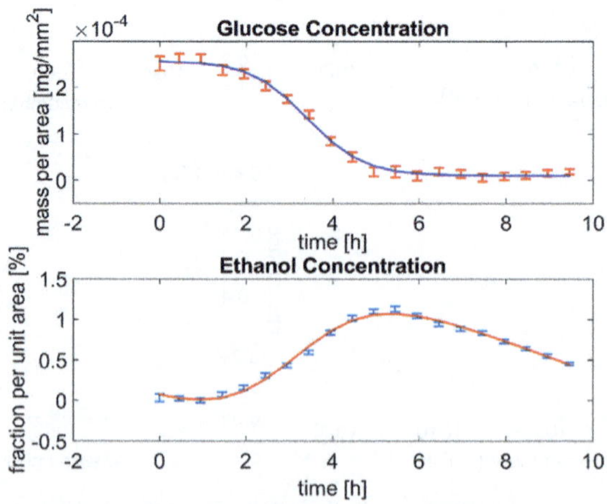

Fig. 12 Results of the pure component analysis of the data shown in Fig. 11.

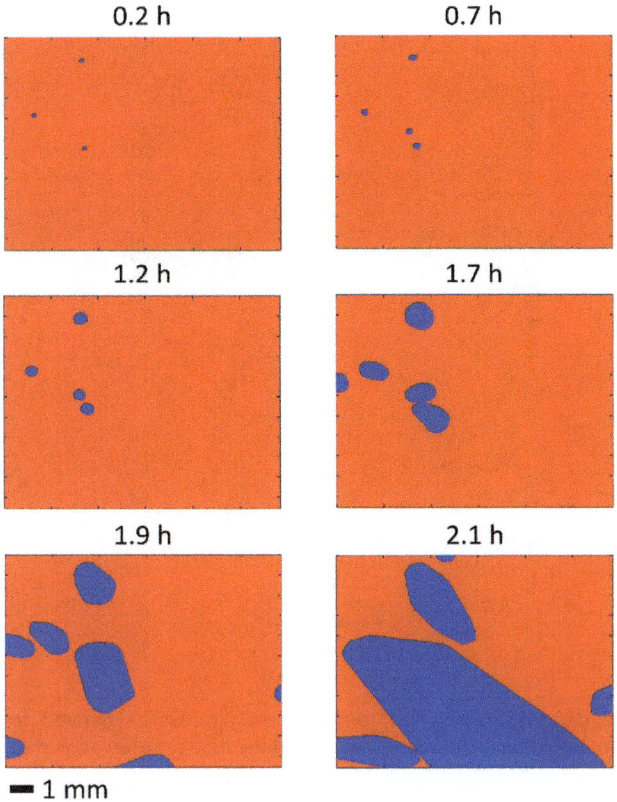

Fig. 13 Evolution of the CO_2-filled bubbles (blue) over time as derived by the segmentation process.

in turn, leads to an observable loss of liquid ethanol by continued evaporation with a rate of 0.2% h^{-1}.

When targeting a more quantitative analysis, the sample thickness must be known. Hence, we performed the second series of experiments in which we used a thin and fully filled cuvette. We thus could achieve a well-defined thickness of the liquid and were also able to quantitatively estimate the volume of CO_2 produced: a thresholding followed by a convex hull segmentation analysis of the absorbance images easily reveals the CO_2 bubble locations and their sizes (Fig. 13). The data shows that a measurable amount of ethanol is starting to be generated 1.5 h after the start, which nicely follows the expected behaviour¶ according to the above reaction equation. After 2.5 hours, the area covered by the bubbles becomes too large and prevents the further analysis of the gas volume. No ethanol leakage was evident in this experiment.

A joint analysis of the glucose, ethanol and CO_2 concentrations follows the expected behaviour (Fig. 14) and leads to a maximum glucose consumption rate of 120 mg per hour per 1 g yeast.

¶ In this experiment, the temperature had not fully stabilized during the initial 20 min.

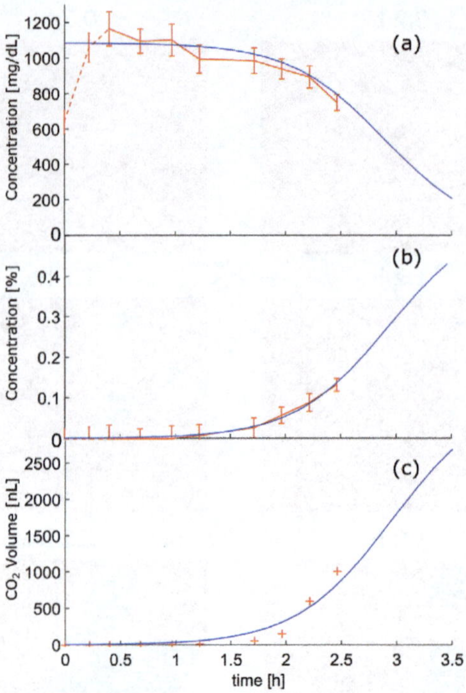

Fig. 14 Time dependence of the fermentation process as derived from the pure component analysis ((a) glucose, (b) ethanol) and the area segmentation method ((c) CO_2).

4 Conclusions

Hyperspectral imaging in the mid-infrared region is often hampered by signal-to-noise considerations, which finally lead to issues in the speed of data acquisition, the size of the object, and/or instrument cost. Since biological processes mostly take place in liquid water and since water has a large extinction coefficient in this spectral region, hyperspectral imaging of biological processes or specimens *in situ* is difficult to achieve.

The high spectral power density of tunable quantum cascade lasers has thus opened up novel access to mid-infrared hyperspectral imaging *in situ* and *in vivo*.

It is thus worthwhile to consider the obtainable signals as well as the noise limits in QCL-based mid-infrared microspectroscopy. For our setup, the optimum S/N is achieved at path lengths around 20 μm, while S/N ratios above 10 are still maintained for water layer thicknesses well above 50 μm. The S/N improves with laser power as well as integration time and follows the expected theoretical dependence such that the whole field will benefit from *e.g.* future improvements in laser power.

The QCL-based microspectroscopy is exemplified by observing the fermentation process. Quantitative results for the concurrent observation of glucose consumption, alcohol production and CO_2 generation could readily be obtained and are in agreement with expectations.

Acknowledgements

The project was funded by the Baden-Württemberg Stiftung. A. Schönhals acknowledges support by the Konrad-Adenauer-Stiftung and K. Haase and A. Schönhals acknowledge support by the Heidelberg Graduate School of Fundamental Physics.

Notes and references

1 M. Miljkovic, M. Romeo, C. Matthäus and M. Diem, Infrared Microspectroscopy of Individual Human Cervical Cancer (HeLa) Cells Suspended in Growth Medium, *Biopolymers*, 2004, **74**, 172–175.
2 E. J. (Swain) Marcsisin, C. M. Uttero, M. Miljković and M. Diem, Infrared microspectroscopy of live cells in aqueous media, *Analyst*, 2010, **135**, 3227–3232.
3 K. L. Andrew Chan, X. Niu, A. J. deMello and S. G. Kazarian, Generation of Chemical Movies: FT-IR Spectroscopic Imaging of Segmented Flows, *Anal. Chem.*, 2011, **83**, 3606–3609.
4 K. L. Andrew Chan and S. G. Kazarian, FT-IR Spectroscopic Imaging of Reactions in Multiphase Flow in Microfluidic Channels, *Anal. Chem.*, 2012, **84**, 4052–4056.
5 K. L. Andrew Chan and S. G. Kazarian, Aberration-free FTIR spectroscopic imaging of live cells in microfluidic devices, *Analyst*, 2013, **138**, 4040–4047.
6 D. A. Moss, M. Keese and R. Pepperkok, IR microspectroscopy of live cells, *Vib. Spectrosc.*, 2005, **38**, 185–191.
7 M. J. Tobin, L. Puskar, R. L. Barber, E. C. Harvey, P. Heraud, B. R. Wood, K. R. Bambery, C. T. Dillon and K. L. Munro, FTIR spectroscopy of single live cells in aqueous media by synchrotron IR microscopy using microfabricated sample holders, *Vib. Spectrosc.*, 2010, **53**, 34–38.
8 K. L. Munro, K. R. Bambery, E. A. Carter, L. Puskar, M. J. Tobin, B. R. Wood and C. T. Dillon, Synchrotron radiation infrared microspectroscopy of arsenic-induced changes to intracellular biomolecules in live leukemia cells, *Vib. Spectrosc.*, 2010, **53**, 39–44.
9 D. R. Whelan, K. R. Bambery, L. Puskar, D. McNaughton and B. R. Wood, Synchrotron Fourier transform infrared (FTIR) analysis of single living cells progressing through the cell cycle, *Analyst*, 2013, **138**, 3891–3899.
10 M. J. Baker, J. Trevisan, P. Bassan, R. Bhargava, H. J. Butler, K. M. Dorling, P. R. Fielden, S. W. Fogarty, N. J. Fullwood, K. A. Heys, C. Hughes, P. Lasch, P. L. Martin-Hirsch, B. Obinaju, G. D. Sockalingum, J. Sulé-Suso, R. J. Strong, M. J. Walsh, B. R. Wood, P. Gardner and F. L. Martin, Using Fourier transform IR spectroscopy to analyze biological materials, *Nat. Protoc.*, 2014, **9**, 1771–1791.
11 L. Quaroni and T. Zlateva, Infrared spectromicroscopy of biochemistry in functional single cells, *Analyst*, 2011, **136**, 3219–3232.
12 J. Faist, F. Capasso, D. L. Sivco, C. Sirtori, A. L. Hutchinson and A. Y. Cho, Quantum Cascade Laser, *Science*, 1994, **264**, 553–556.
13 C. Vrančić, A. Fomichova, N. Gretz, C. Herrmann, S. Hoecker, A. Pucci and W. Petrich, Continuous glucose monitoring by means of mid-infrared transmission laser spectroscopy *in vitro*, *Analyst*, 2011, **136**, 1192–1198.

14 M. Brandstetter, L. Volgger, A. Genner, C. Jungbauer and B. Lendl, Direct determination of glucose, lactate and triglycerides in blood serum by a tunable quantum cascade laser-based mid-IR sensor, *Appl. Phys. B*, 2013, **110**, 233–239.

15 M. R. Alcaráz, A. Schaighofer, C. Kristament, G. Ramer, M. Brandstetter, H. Goicoechea and B. Lendl, External-Cavity Quantum Cascade Laser Spectroscopy for Mid-IR Transmission Measurements of Proteins in Aqueous Solution, *Anal. Chem.*, 2015, **87**, 6980–6987.

16 C. Vrančić, N. Kröger, N. Gretz, S. Neudecker, A. Pucci and W. Petrich, A Quantitative Look Inside the Body: Minimally Invasive Infrared Analysis *in vivo*, *Anal. Chem.*, 2014, **86**, 10511–10514.

17 M. C. Phillips and N. Hô, Infrared hyperspectral imaging using a broadly tunable external cavity quantum cascade laser and microbolometer focal plane array, *Opt. Express*, 2008, **16**, 1836–1845.

18 M. C. Phillips and B. E. Bernacki, Hyperspectral microscopy of explosives particles using an external cavity quantum cascade laser, *Opt. Eng.*, 2012, **52**, 061302.

19 M. R. Kole, R. K. Reddy, M. V. Schulmerich, M. K. Gelber and R. Bhargava, Discrete frequency infrared microspectroscopy and imaging with a tunable quantum cascade laser, *Anal. Chem.*, 2012, **84**, 10366–10372.

20 N. Kröger, A. Egl, M. Engel, N. Gretz, K. Haase, I. Herpich, S. Neudecker, A. Pucci, A. Schönhals and W. Petrich, Rapid hyperspectral imaging in the mid-infrared, *Proc. SPIE-Int. Soc. Opt. Eng.*, 2014, **8939**, 89390Z.

21 N. Kröger, A. Egl, M. Engel, N. Gretz, K. Haase, I. Herpich, B. Kränzlin, S. Neudecker, A. Pucci, A. Schönhals, J. Vogt and W. Petrich, Quantum cascade laser-based hyper-spectral imaging of biological tissue, *J. Biomed. Opt.*, 2014, **19**, 111607.

22 N. Kröger-Lui, N. Gretz, K. Haase, B. Kränzlin, S. Neudecker, A. Pucci, A. Regenscheit, A. Schönhals and W. Petrich, Rapid identification of goblet cells in unstained colon thin sections by means of quantum cascade laser-based infrared microspectroscopy, *Analyst*, 2015, **140**, 2086–2092.

23 P. Bassan, M. J. Weida, J. Rowlette and P. Gardner, Large scale infrared imaging of tissue micro arrays (TMAs) using a tunable Quantum Cascade Laser (QCL) based micro-scope, *Analyst*, 2014, **139**, 3856.

24 K. Yeh, S. Kenkel, J.-N. Liu and R. Bhargava, Fast Infrared Chemical Imaging with a Quantum Cascade Laser, *Anal. Chem.*, 2015, **87**, 485–493.

25 K. Haase, N. Kröger-Lui, A. Pucci, A. Schönhals and W. Petrich, Real-time mid-infrared imaging of living microorganisms, *J. Biophotonics*, 2016, **9**, 61–66.

26 P. Lasch and W. Petrich, Data Acquisition and Analysis in Biomedical Vibrational Spectroscopy, in *Biomedical Appli-cations of Synchrotron Infrared Microspectroscopy: A Practical Approach, RSC Analytical Spectroscopy Series*, ed. D. Moss, Royal Society of Chemistry, Cambridge, UK, 2010, vol. 11, pp. 192–225.

27 G. M. Hale and M. R. Querry, Optical Constants of Water in the 200 nm to 200 μm Wavelength Region, *Appl. Opt.*, 1973, **12**, 555–563.

28 C. Herrmann, C. Vrančić, A. Fomichova, N. Gretz, S. Hoecker, A. Pucci and W. Petrich, *In vitro* characteristics of a mid-infrared, continuous glucose sensor, *Proc. SPIE-Int. Soc. Opt. Eng.*, 2010, **8219**, 75600E.

Faraday Discussions

PAPER

High-throughput quantum cascade laser (QCL) spectral histopathology: a practical approach towards clinical translation

Michael J. Pilling,[a] Alex Henderson,[a] Benjamin Bird,[b] Mick D. Brown,[c] Noel W. Clarke[c] and Peter Gardner*[d]

Received 17th November 2015, Accepted 4th January 2016
DOI: 10.1039/c5fd00176e

Infrared microscopy has become one of the key techniques in the biomedical research field for interrogating tissue. In partnership with multivariate analysis and machine learning techniques, it has become widely accepted as a method that can distinguish between normal and cancerous tissue with both high sensitivity and high specificity. While spectral histopathology (SHP) is highly promising for improved clinical diagnosis, several practical barriers currently exist, which need to be addressed before successful implementation in the clinic. Sample throughput and speed of acquisition are key barriers and have been driven by the high volume of samples awaiting histopathological examination. FTIR chemical imaging utilising FPA technology is currently state-of-the-art for infrared chemical imaging, and recent advances in its technology have dramatically reduced acquisition times. Despite this, infrared microscopy measurements on a tissue microarray (TMA), often encompassing several million spectra, takes several hours to acquire. The problem lies with the vast quantities of data that FTIR collects; each pixel in a chemical image is derived from a full infrared spectrum, itself composed of thousands of individual data points. Furthermore, data management is quickly becoming a barrier to clinical translation and poses the question of how to store these incessantly growing data sets. Recently, doubts have been raised as to whether the full spectral range is actually required for accurate disease diagnosis using SHP. These studies suggest that once spectral biomarkers have been predetermined it may be possible to diagnose disease based on a limited number of discrete spectral features. In this current study, we explore the possibility of utilising discrete frequency chemical imaging for acquiring high-throughput, high-resolution chemical images. Utilising a quantum cascade laser imaging microscope with discrete frequency collection at key

[a]*Manchester Institute of Biotechnology, University of Manchester, 131 Princess Street, Manchester, M1 7DN, UK*

[b]*Daylight Solutions, 15378 Avenue of Science, Suite 200, San Diego, CA 92128-3407, USA*

[c]*Genito Urinary Cancer Research Group, Institute of Cancer Sciences, Paterson Building, The University of Manchester, Wilmslow Road, Manchester, M20 4BX, UK*

[d]*Manchester Institute of Biotechnology, University of Manchester, 131 Princess Street, Manchester, M1 7DN, UK. E-mail: Peter.Gardner@Manchester.ac.uk; Fax: +44 (0) 161 306 5201; Tel: +44 (0) 161 306 4463*

diagnostic wavelengths, we demonstrate that we can diagnose prostate cancer with high sensitivity and specificity. Finally we extend the study to a large patient dataset utilising tissue microarrays, and show that high sensitivity and specificity can be achieved using high-throughput, rapid data collection, thereby paving the way for practical implementation in the clinic.

1 Introduction

Histopathology is currently the gold standard for identifying the manifestation of disease in tissue. Principally relying on changes in morphology and architecture highlighted through selective staining,[1,2] a highly trained pathologist can diagnose disease, suggest possible treatments and even provide information on likely prognosis. Microscopic examination of stained tissue biopsy sections presents the pathologist with a high degree of information, and histopathology is currently unsurpassed in its diagnostic accuracy. However, manual examination of individual tissue biopsies is extremely time-consuming, with each section being individually interrogated for the presence of abnormalities. Limited throughput inevitably results in significant delays between the time a biopsy is obtained and a diagnosis being made, with clear implications for patient care and treatment. Furthermore disease diagnosis based on tissue morphology and architecture is inherently subjective, often resulting in intra- and inter-observer error.[3] This situation has been exacerbated by national cancer screening programs, with the number of tissue biopsies being harvested increasing annually. Desire for increased throughput, improved accuracy and a reduction in repeat biopsies are clear drivers for the implementation of complementary methods for disease diagnosis.

Over the last decade, spectral histopathology (SHP) has demonstrated great promise for the diagnosis of the diseased state. Fourier transform infrared chemical imaging has gained attention in the biomedical field as a rapidly emerging technology for disease diagnosis.[4,5] Biological material can be interrogated without the need for exogenous labels, little or no sample preparation, and in a non-destructive manner. The technique exploits the high chemical sensitivity of infrared spectroscopy, in combination with microscopy, to provide spatially resolved measurements that are rich in biochemical content. Whereas conventional histology relies on the subjective interpretation of tissue architecture and cellular morphology, this approach relies on reproducible physical measurements of sample chemistry, and the potential to reduce misdiagnosis.

In partnership with machine learning methods, FTIR chemical imaging has demonstrated the ability to distinguish between normal and cancerous tissue with high sensitivity and specificity,[6–9] and also to determine cancer grade[10] and staging.[11] However, clinical translation has been inhibited until recently by technological advancements failing to deliver what is required to make it competitive with current histological methods. Developments in focal plane array (FPA) detector technology[12] have drastically reduced acquisition times, but until recently[13] could not compete with the high-resolution images obtainable in bright field imaging. Early signs of invasive cancer are often manifested in the basement membrane, the basal layer in prostate[14] and myoepithelium in breast cancer.[15,16] Conclusive early diagnosis requires detection of subtle changes on the subcellular level across microscopic membranes only obtainable through high-quality high-resolution chemical images.

Recent technological advancements have resulted in commercially available infrared microscopes utilising 0.62 NA 15× magnification optics with a 128 × 128 FPA, enabling imaging of a 140 μm × 140 μm area to be imaged as a single measurement with 1.1 μm pixel size and a diffraction-limited spatial resolution of about 6 μm at 1667 cm^{-1}. However, the inherent trade-off between high resolution and acquisition times inevitably makes high-resolution imaging impractical due to excessive measurement times. Obtaining high-resolution (1.1 μm) chemical images from a single 1 mm tissue microarray (TMA) core will typically take between 5–6 hours to acquire, followed by a further 40–50 minutes to process the interferograms and stitch the tiles together.[17] The time taken to record single cores generally makes FTIR chemical imaging, even using 128 × 128 FPA, unsuitable for high-throughput imaging of tissue biopsies and full TMAs. The problem lies with the vast quantities of data an FTIR chemical imaging system acquires when using an FPA detector. A single infrared tile consists of 16 384 pixels (for a 128 × 128 FPA) and each pixel itself consists of an entire infrared spectrum. Imaging of a full TMA core with 1 mm diameter at 1.1 μm pixel resolution is typically performed using 64 infrared tiles, resulting in a large spectral data cube requiring over 13 GB to store. Since FTIR relies on the Fellgett advantage and collects all wavelengths simultaneously, restricting the spectral range does not reduce the acquisition time. Speed of acquisition and data management issues are rapidly becoming a significant barrier to clinical translation.

Recently, doubts have been raised as to whether entire infrared spectra are necessary for disease diagnosis using SHP. Studies suggest that once spectral biomarkers have been identified, it may be possible to use a selection of key wavenumbers for diagnosing disease.[18–20] In this paper we report on a novel study using discrete frequency imaging utilising a Spero Quantum Cascade Laser (QCL)-based full-field imaging infrared microscope for disease diagnosis. We investigate the practicalities of utilising high-resolution, high-throughput chemical imaging using discrete frequencies, and consider implications for improved disease diagnosis.

2 Materials and methods

2.1 Sample preparation

Formalin-fixed, paraffin-embedded prostate tissue samples were obtained following informed consent and ethical approval (Trent Multi-centre Research Ethics Committee 01/4/061). A 12 μm-thick section was taken from each paraffin block and fixed to a BaF_2 slide (75 mm × 25 mm × 1 mm) for infrared transmission measurements. BaF_2 was chosen since it has a better low wavenumber cut-off than CaF_2 (950 cm^{-1} compared with 1000 cm^{-1}) and does not suffer from the electric field standing wave effect,[21–24] which can be a problem for low-e infrared reflecting slides. Serial sections from each block were fixed to glass and underwent Haematoxylin and Eosin (H&E) staining for bright field imaging. The samples mounted on BaF_2 were left in wax and did not undergo deparaffinization. This reduces the risk of further chemical alterations from clearing solvents, and reduces Mie scattering *via* refractive index matching.[25,26]

2.2 Infrared chemical imaging

Infrared chemical images were acquired with a Spero infrared microscope (Daylight Solutions Inc., San Diego, CA, USA) utilising quantum cascade laser technology.[27] Employing four separate high-brightness QCL modules in a single multiplexed source enables continuous access to the fingerprint region between 900 and 1800 cm^{-1}. The system is equipped with a high-pixel-density (480 × 480) uncooled microbolometer FPA. A 0.7 NA, 12.5× compound refractive objective was used in transmission mode, providing a large field of view of 650 μm × 650 μm with a corresponding pixel size of 1.35 μm, yielding a diffraction-limited spatial resolution of about 5 μm at 1667 cm^{-1}.

The tissue used in the study arise from 29 separate cancer patients consisting of 50 unique 1 mm diameter cores spread over two separate TMAs. Each core is assigned as either cancerous (containing malignant tissue), or normal-associated (from a cancer patient but containing no malignant tissue). Wherever possible a normal core and a cancerous core were measured for each patient. However, this was not always possible due to some cores being missing from the array. The sample set consisted of an equal number of 25 normal-associated and 25 cancerous cores. Background images were collected prior to each TMA core, taken from a clean area of the sample that was free of tissue or paraffin. Chemical images of each TMA core were collected using the mosaic method, with each core measured individually as a 2 × 2 mosaic. A single core using 27 discrete wavenumbers consisting of 921 600 pixels took approximately 5 minutes and 30 seconds to collect. Each sample tile is ratioed to its background in real time and, upon completion of the collection, automatically exported as a datacube in MATLAB format ready for stitching post collection.

2.3 Data pre-processing

Data pre-processing was performed using MATLAB 2013a (The MathWorks Inc., Natick, MA, USA) and the ProSpect Toolbox (London Spectroscopy Ltd., London, UK). Infrared tiles were stitched together using software written in house and saved as a 960 × 960 × 27 hyperspectral datacube, and also as a chemical image based on the intensity of the amide I band. Stitching together 4 tiles to form a hyperspectral data cube using a dual core Intel i7-2600 with 16 GB RAM took on average just 6 seconds per core and required only 80 MB of storage space. Spectra were quality tested to remove areas of the images where no tissue was present, or where there was a high degree of scattering. Quality testing was based on the intensity of the amide I band, with those spectra having amide I absorbance between 0.1–2.0 being retained. Each spectrum was baseline corrected using a linear rubber band correction at 1000 cm^{-1} and 1734 cm^{-1}. Finally the spectra were normalised to the intensity of the amide I band to account for different thicknesses of the tissue sample.

3 Results and discussion

3.1 Wavenumber selection for discrete frequency imaging

Successful exploitation of discrete frequency chemical imaging for high-throughput disease diagnosis requires the intelligent selection of salient frequencies that provide the greatest discriminatory power between diseased and healthy states. Failure to choose the correct wavenumbers could result in crucial spectral biomarkers being missed and could directly impact diagnostic accuracy. In addition, not all wavenumbers are suitable biomarkers and often provide little or no useful biochemical information. Acquiring too many wavenumbers increases measurement times and therefore reduces throughput. Numerous examples exist in the literature of well-established biomarkers[28,29] determined using FTIR chemical imaging. However, to date no studies have been performed on the transferability of biomarkers obtained using FTIR to discrete frequency IR spectroscopy. We have addressed this by acquiring full band spectra and subsequently identifying key biomarkers at sparsely located frequencies.

Chemical images were acquired in the spectral range 1000–1800 cm^{-1} from tissue cores from two patients who had been diagnosed with prostate cancer. The first patient core was histologically classified as Normal-Associated Tissue (NAT) and contained normal tissue components only. The second patient core was classified as cancerous and contained morphological features consistent with a Gleason grade of 4. Since the cores available for the study had Gleason grades ranging between 3 and 5, choosing a core with a Gleason grade of 4 encompasses the middle of the cancer severity range. In principle, utilising a larger patient set for acquiring continuous infrared spectra would enable improved identification of the key wavenumbers. However, for the scope of this proof-of-concept study and due to the limited time available, we elected to choose a normal core and a cancerous core in the middle of the cancer severity range. Chemical images for each core based on the intensity of the amide I band are shown in Fig. 1.

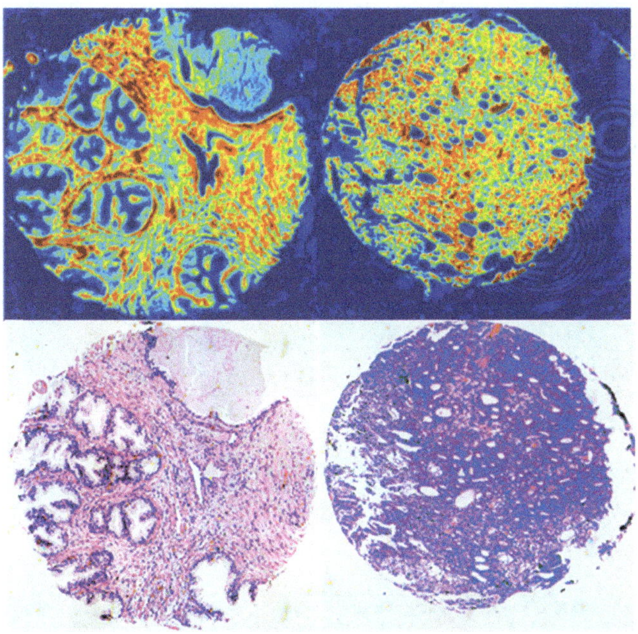

Fig. 1 QCL chemical images of the amide I band intensity and H&E-stained serial section (bottom) for normal-associated tissue (left) and cancerous tissue (right) used to identify the key wavenumbers for discrete frequency classification.

Employing similar methods as Fernandez,[20] a database was constructed consisting of 5000 spectra each for cancerous epithelium and normal-associated epithelium. The spectra were quality tested, truncated between 1350 and 1500 cm^{-1} (to remove spectral regions describing bands of paraffin), and normalised to the amide I band. Mean spectra of the normal-associated and cancerous epithelium tissue are displayed in Fig. 2. Upon first inspection the spectra appear relatively similar, although some subtle differences can be discerned between 1000 and 1300 cm^{-1}.

Half of the spectra from each class were selected at random and fed into a random forest[30] algorithm (software available from http://code.google.com/p/randomforest-matlab/). Random forests have the advantage that, unlike other supervised classifiers, they do not require feature selection prior to use. A random forest will return a measure for variable importance and identify the most important wavenumbers for classification. Alternative methods for wavenumber selection are available, such as partial least squares discriminant analysis (PLS-DA) and variable importance for projection (VIP), as described by Lloyd.[31] The classifier was trained using 500 trees, with the number of wavenumbers selected at random to try and split each node (mtry) set to 2. The remainder of spectra in the database that had not been used for training were used to test the model.

Receiver operator curves present an effective way to visualise the performance of the classifier. Each tree votes to classify a spectrum to a specific class, and the number of votes provides a probability estimate to each spectrum belonging to a particular class. Varying the probability acceptance thresholds adjusts the trade-off between sensitivity and specificity and produces a receiver operator curve (ROC). The ROCs obtained using the random forest classifier are displayed in Fig. 3.

The optimal situation would be for curves to be situated at the top left hand corner of the plot, which indicates both high sensitivity and high specificity.

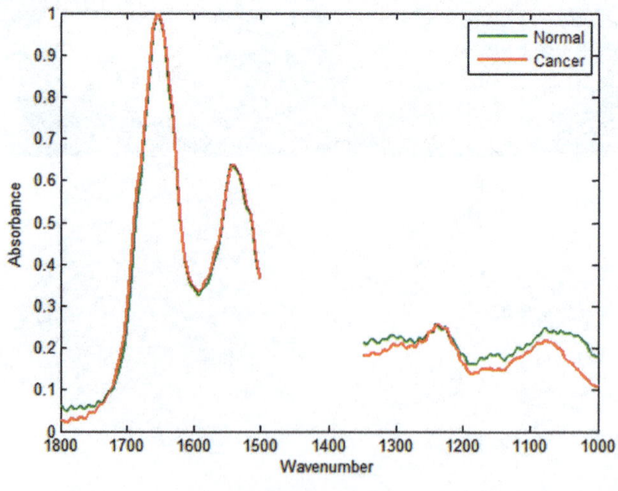

Fig. 2 Mean spectra for normal-associated epithelium and cancerous epithelium from the database constructed from two prostate tissue cores, following truncation to remove the spectral regions describing wax, and normalisation to the 1652 cm^{-1} band.

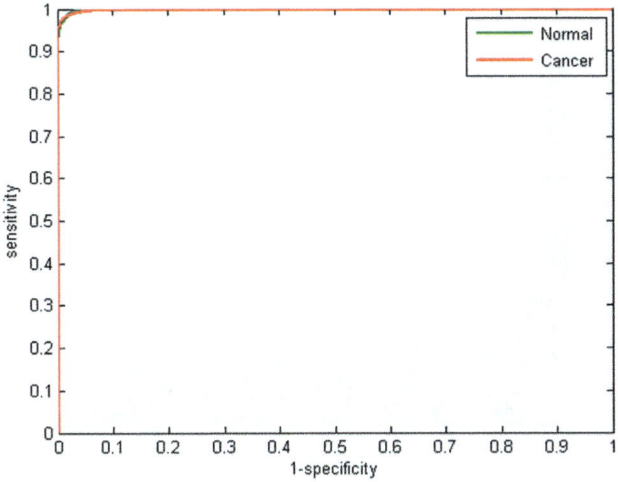

Fig. 3 Receiver operator curves for normal and cancerous epithelium spectra using 2500 from each class for training and testing. AUC = 0.9991.

Conversely, a poor classifier would be shown as a plot close to a diagonal line between the origin and top right corner. Area under the curve (AUC) is a widely accepted measure of classifier performance. The AUC for the plot shown in Fig. 3 is 0.991, demonstrating high performance of the classifier. Setting a probability of acceptance threshold of 0.5 enables a confusion matrix to be calculated and the ability to determine the proportion of each class that is correctly classified. Table 1 shows that normal-associated epithelium spectra are correctly classified with an accuracy of 97.25%, and cancerous epithelium with an accuracy 97.19%.

Wavenumbers were then ranked in order of variable importance using a GINI importance plot to determine which were most important in distinguishing between normal and cancerous epithelium. Fig. 4a and b show typical GINI plots used to select the 25 most important features. The top 25 wavenumbers from a single GINI plot were selected for data collection. Subsequent repetition of the analysis shows that the first 14 wavenumbers are consistently in the top 16, but the remaining 11 wavenumbers selected can be ranked as far down as 58. This is not too surprising given that the difference in importance starts to drop off significantly after 20.

The twenty five discriminating wavenumbers that were originally used in order of variable importance are shown in Table 2. The selected wavenumbers broadly overlap absorption bands centred at 1032 cm^{-1} ν (C–O) glycogen, 1080 cm^{-1} ν_s (PO_2^-), 1236 cm^{-1} ν_{as} (PO_2^-), 1540 cm^{-1} (amide II), and 1656 cm^{-1} (amide I).

Table 1 Confusion matrix showing classification accuracy for normal-associated and cancerous epithelium using the random forest classifier with 500 trees

	Normal	Cancer
Normal	**97.25**	2.25
Cancer	2.81	**97.19**

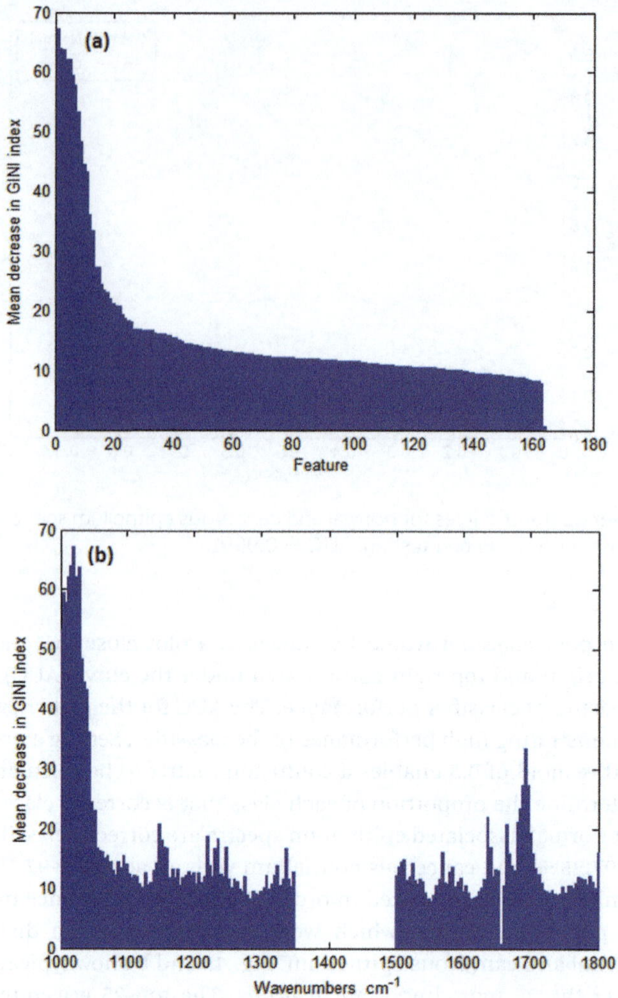

Fig. 4 GINI importance plot (a) as a function of wavenumber and (b) ranked in order of variable importance.

3.2 Discrete frequency imaging and classification

Discrete frequency chemical images were acquired from each of the 50 prostate tissue biopsy cores. Two wavenumbers at 1652 cm^{-1} and 1734 cm^{-1} in addition to

Table 2 25 key wavenumbers (cm^{-1}) ranked in order of variable importance, as identified by the random forest classifier. Figures in parentheses indicate the variable importance ranking, with the lowest number being the highest ranking

Wavenumber (cm^{-1})							
(1) 1024	(2) 1020	(3) 1028	(4) 1016	(5) 1012	(6) 1072	(7) 1000	(8) 1004
(9) 1032	(10) 1636	(11) 1068	(12) 1088	(13) 1684	(14) 1092	(15) 1008	(16) 1640
(17) 1688	(18) 1692	(19) 1064	(20) 1648	(21) 1236	(22) 1696	(23) 1096	(24) 1524
(25) 1044							

the twenty five key wavenumbers (Table 2) were also acquired. The difference in absorbance at these two wavenumbers enabled the height of the amide I band to be determined, and this was used to quality check the spectra with spectra having an amide I intensity of between 0.1–2 being retained. Fig. 5 shows chemical images from a single prostate tissue core based on the intensity of the 1652 cm^{-1}, 1524 cm^{-1}, and 1236 cm^{-1} bands and the H&E-stained serial section. The chemical images shown have been quality tested and spectra with amide I peak absorbance intensity between 0.1 and 2.0 retained. The image illustrates that rapid chemical imaging using discrete frequencies enables different types of tissue to be highlighted depending on the chosen frequency. Chemical images obtained at 1652 cm^{-1} and 1524 cm^{-1} enable differentiation between epithelium and stroma, while the 1236 cm^{-1} chemical image highlights regions of stroma.

Chemical images from each of the 50 cores were compared to the corresponding H&E-stained serial sections to identify regions of cancerous and normal-associated epithelium. The patients were then randomly divided into two separate libraries to form a training cohort (15 patients) and a testing cohort (14 patients). The patients in each cohort were fairly evenly distributed across the two separate TMA slides. The training cohort had a split of 8 patients on one slide and 7 on the other. While the testing cohort was split with 8 patients on one slide and 6 on the second slide. Using the methods previously described by Fernandez,[20] two spectral databases were constructed from these cohorts, consisting of a training data set and an independent test set. Dividing the patients (and the data) prior to building the classifier ensures that the test set is completely

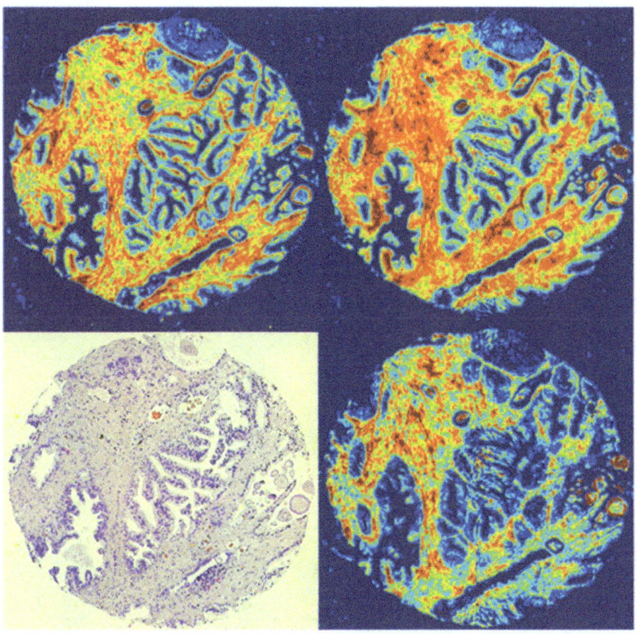

Fig. 5 Discrete frequency chemical images of a prostate tissue single core obtained using (clockwise from top left) 1652 cm^{-1}, 1524 cm^{-1}, and 1236 cm^{-1} band intensity and H&E-stained serial section.

independent, since no spectra used in training the model will be used for testing. Equal numbers of spectra from each class (normal-associated and cancerous epithelium) were extracted from the training database. Spectra were quality tested, baseline corrected and normalised to the amide I band. The mean cancerous epithelium and normal epithelium spectra based on 207 505 measurements each are shown in Fig. 6. Despite the limited number of data points in each spectrum, subtle spectral differences between the two classes are discernible, particularly between 1000 and 1240 cm^{-1}.

Half the spectra contained in the training database were randomly selected to train the model, with the remainder forming a validation test set. Metrics fed into the classifier were based on the absorbance values for each of the 25 discrete frequencies, and also all possible ratio combinations for the discrete frequency dataset, which yielded 325 features in total. The random forest classifier was then trained on the 207 505 partitioned spectra using 200 trees, which enabled the classifier to be constructed in approximately 90 minutes. The remaining spectra in the training data base were used to validate the model. The receiver operator curves obtained are displayed in Fig. 7. AUC values for the classifier are close to 1 (0.9895), indicating that the classifier can easily differentiate between normal and cancerous epithelium spectra. Despite utilising only 25 wavenumbers, the correctness of classification is high, with sensitivity and specificity of 93.39% and 94.72% respectively, as shown by the confusion matrix in Table 3.

The large number of features used to train the random forest classifier, and the substantial size of the data set, are the main factors responsible for lengthy training times. In an attempt to speed up training, the classifier was also trained using only the absorbance values at each of the 25 discrete frequencies. Training using 207 505 spectra per class using 200 trees enabled the random forest classifier to be constructed in just 8 minutes. The ROCs obtained using 25 features are shown in Fig. 8. The reduction in the features used in training has an impact on the performance of the classifier, the AUC decreasing from 0.9895 to 0.9625.

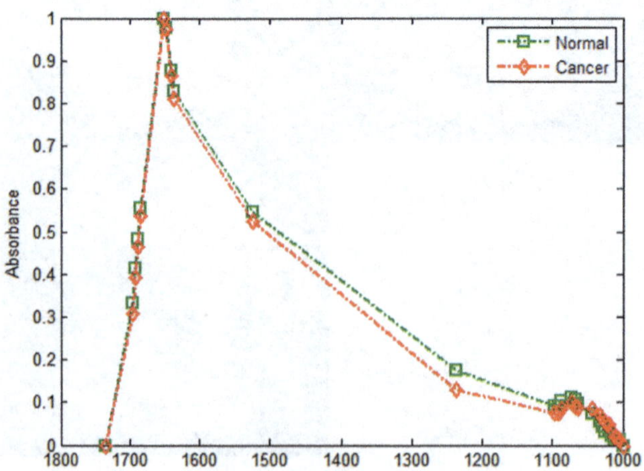

Fig. 6 Discrete frequency mean spectra utilising 27 wavenumbers for cancerous and normal-associated epithelium. Dashed lines are present as a guide to the eye.

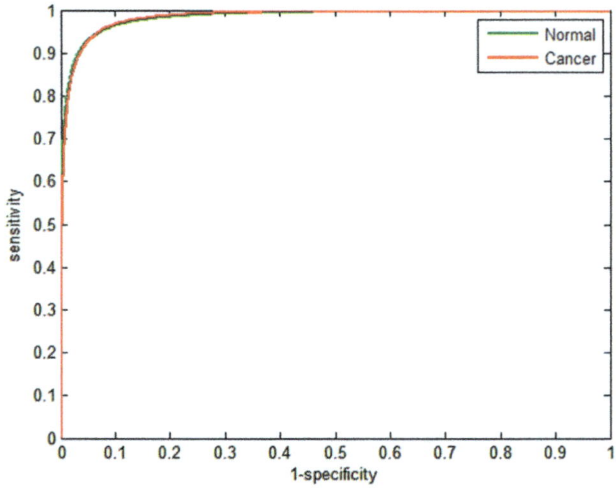

Fig. 7 Receiver operator curves (ROC) with 25 wavenumbers (325 features), using validation data for normal-associated and cancerous epithelium. Area under the curve values (AUC) are normal = 0.9851, cancer = 0.9851.

Furthermore the sensitivity and specificity decreases to 89.14% and 90.32% respectively, suggesting that despite the increased processing times, using 325 features constructed from the 25 discrete frequencies is more effective.

The training data was then subjected to repeated random sub-sampling validation using ten repeats. In each case half the spectra in the database were randomly selected and used for training, while the remainder served as validation spectra. Table 4 shows the mean and standard deviation for the calculated sensitivity and specificity of the ten classifiers trained. The mean sensitivity and specificity from the repeated sub-sampling is high and provides a very small standard deviation, indicating that the classifier accuracy is not dependent on the spectra used to train and test the model.

3.3 Discrete frequency classification with restricted numbers of wavenumbers

While it is evident that 25 discrete wavenumbers allows good classification accuracy on the validation data set, the effect of the number of discrete wavenumbers measured on classification accuracy is a key question. Clinical translation of discrete frequency infrared imaging requires high-throughput, high-resolution imaging utilising as few discrete wavenumbers as possible. Naturally, there will be a trade-off between the number of wavenumbers acquired and the classification accuracy. We have addressed this by reducing the number of

Table 3 Confusion matrix showing correctness of classification using 25 wavenumbers for normal and cancerous epithelium

	Normal	Cancer
Normal	**93.39%**	6.61%
Cancer	5.28%	**94.72%**

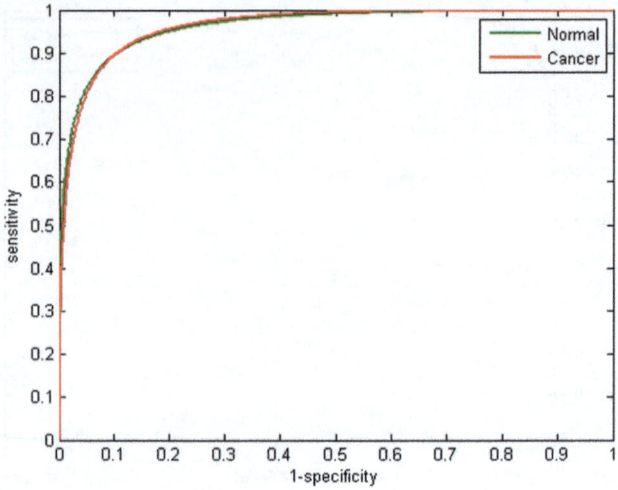

Fig. 8 Receiver operator curves (ROC) with 25 wavenumbers (25 features), using validation data for normal-associated and cancerous epithelium. Area under the curve values (AUC) are normal = 0.9625, cancer = 0.9625.

wavenumbers used to train the model. We elected to reduce the number of wavenumbers used in classification rather than re-measuring all the cores with the respective number of wavenumbers due to time considerations and the desire for better comparability. In each case the subset of wavenumbers used was those with the highest variable importance values (Table 2), to ensure optimal classifier performance. Six separate experiments were performed on the training database using varying numbers of discrete wavenumbers. Table 5 details the discrete wavenumbers used for training and validating the random forest classification model.

The performance of each classifier is shown in the ROCs in Fig. 9(a–f). Decreasing the number of wavenumbers used in classification reduces the performance of the classifier, as observed in the AUC values of 0.9780 and 0.9739 for 20 and 18 wavenumbers respectively. Reducing the number of discrete frequencies is expected to reduce classifier performance, since less information is being used during training. Surprisingly, training the random forest with just 18 wavenumbers still enables excellent discrimination between normal-associated and cancerous epithelium tissue. Reducing the numbers of wavenumbers further to 16 discrete frequencies only has a marginal effect on classifier performance (AUC = 0.9772). However, when using 12 or 10 discrete frequencies, the classifier performance begins to deteriorate with AUC values of 0.9557 and 0.9421 respectively.

Table 4 Mean and standard deviation of sensitivity and specificity, obtained using repeated random sub-sampling validation of ten trained classifiers

	Sensitivity	Specificity
Mean	94.60%	93.39%
Standard deviation	0.0012	0.0010

Table 5 Discrete wavenumbers used for training the classifier with 20, 18, 16, 14, 12, and 10 different wavenumbers

Number of discrete wavenumbers	Discrete wavenumbers (cm^{-1}) used for random forest classification								
20	1024	1020	1028	1016	1012	1072	1000	1004	1032
	1636	1068	1088	1684	1092	1008	1640	1688	1692
	1064	1648							
18	1024	1020	1028	1016	1012	1072	1000	1004	1032
	1636	1068	1088	1684	1092	1008	1640	1688	1692
16	1024	1020	1028	1016	1012	1072	1000	1004	1032
	1636	1068	1088	1684	1092	1008	1640		
14	1024	1020	1028	1016	1012	1072	1000	1004	1032
	1636	1068	1088	1684	1092				
12	1024	1020	1028	1016	1012	1072	1000	1004	1032
	1636	1068	1088						
10	1024	1020	1028	1016	1012	1072	1000	1004	1032
	1636								

AUC values provide a good comparison of classification accuracy, however a more meaningful measure is the proportion of correctly classified spectra. Table 6 shows the proportion of correctly classified cancerous (sensitivity) and normal-associated epithelium (specificity) as a function of the discrete frequencies used in classification. The values for sensitivity and specificity are the mean values based on repeated random sub-sampling using ten repeats.

The sensitivity and specificity are broadly in line with the AUC values, and using all 25 wavenumbers enables high classification accuracy. Reducing the number of discrete frequencies to 16 still results in good classification accuracy with sensitivity and specificity of 91.88% and 91.03% respectively. Performance of the classifier becomes poorer when using 12 or fewer discrete frequencies. However using only 10 wavenumbers still enables surprisingly good classification accuracy, with sensitivity and specificity of 87.15% and 86.80%. Inspection of Table 5 reveals that, when using 10 discrete frequencies, the majority of the wavenumbers are in the range 1000–1072 cm^{-1}, indicating that important spectral biomarkers are located here.

The number of discrete frequencies chosen when acquiring chemical images is a key parameter. However, the time penalty associated with collecting increasing numbers of discrete frequencies is also an important consideration. Furthermore, as the number of discrete frequencies increases, so does the time required to train the random forest classifier. The performance of the random forest classifier as a function of AUC, sensitivity, specificity, acquisition time per core, and training time are shown in Table 7.

The resulting sensitivity and selectivity are excellent when using the full 25 discrete frequencies, and a single core can be measured in 5.5 minutes, which is a reasonable timescale. However, constructing the classifier takes the longest time at *ca.* 90 minutes. Utilising only 10 discrete frequencies enables fast data acquisition (3.27 minutes), and the random forest classifier can be constructed in just 17 minutes. However, the improved throughput and analysis time is offset by the reduced sensitivity and specificity of 87.15% and 86.80% respectively. To put this into perspective it is crucial to understand what timescales would be

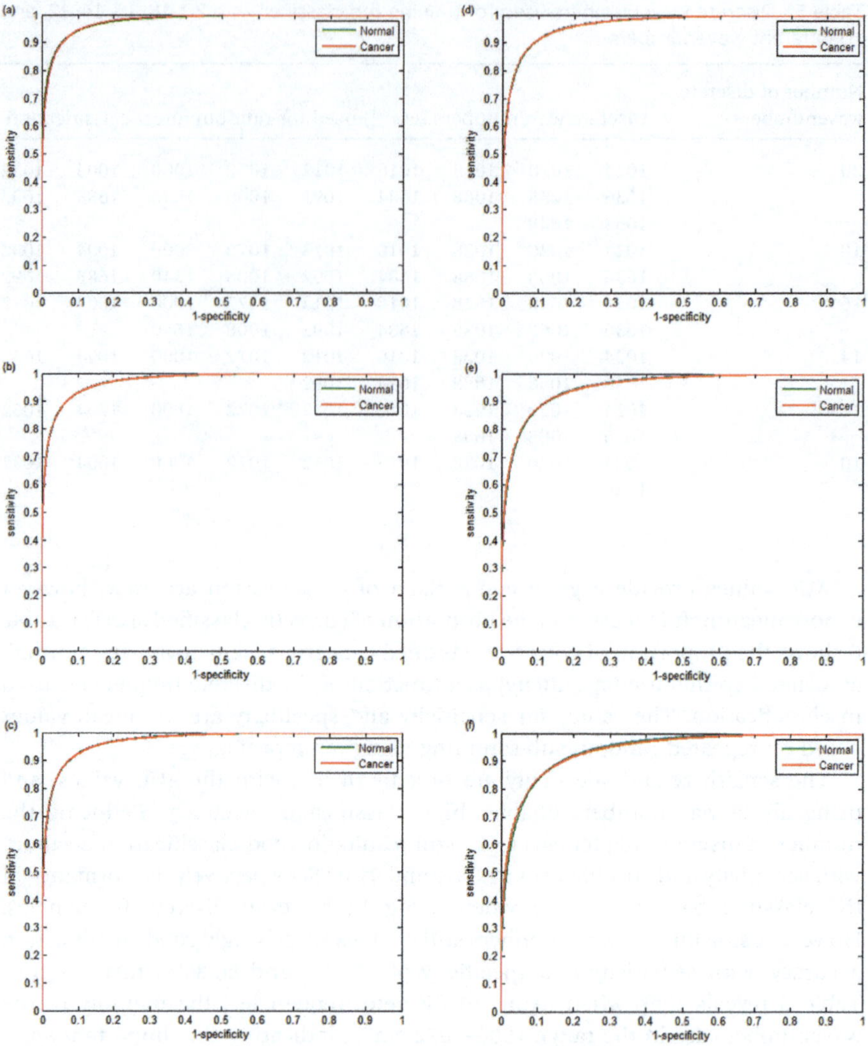

Fig. 9 Receiver operator curves using validation data for normal-associated and cancerous epithelium with (a) 20, (b) 18, (c) 16, (d) 14, (e) 12 and (f) 10, discrete frequencies. AUC values are 0.9780, 0.9739, 0.9720, 0.9669, 0.9557 and 0.9421 respectively.

clinically acceptable. Once the classifier has been trained and robustly validated there would not be a requirement to retrain the classifier on a regular basis. Therefore, provided that the classifier can be trained within reasonable time-scales, then the key parameter is the collection time per core. Utilising between 14

Table 6 Table showing sensitivity and specificity for the validation data using random subset sampling using ten repeats

Number of discrete frequencies	25	20	18	16	14	12	10
Sensitivity (%)	94.60	93.02	92.27	91.88	91.13	89.11	87.15
Specificity (%)	93.39	91.71	91.16	91.03	90.05	88.53	86.80

Table 7 Table showing AUC, sensitivity, specificity, collection time per core, and classifier training time as a function of the number of discrete frequencies used with the random forest classifier

No. of frequencies	AUC	Sensitivity	Specificity	Collection time per core (min)	Training time (min)
25	0.9851	94.60	93.39	5.5	90
20	0.9780	93.02	91.71	4.47	60
18	0.9739	92.27	91.16	4.33	48
16	0.9772	91.88	91.03	4.13	37
14	0.9669	91.13	90.05	4	32
12	0.9557	89.11	88.53	3.6	24
10	0.9421	87.15	86.80	3.27	17

and 16 discrete frequencies enables each core to be measured in approximately 4 minutes while maintaining sensitivity and specificity >90%. Although there is a slight reduction in sensitivity and specificity compared to utilising the full 25 discrete frequencies, there is a considerable time saving of approximately 90 seconds per core. We would suggest that acquiring high-resolution images of a single TMA acquired in just four minutes, while maintaining high sensitivity and specificity, would be clinically acceptable.

As QCL-based, full-field imaging technology continues to advance over the coming years, this tradeoff will become less apparent to the clinician. The underlying technology employed in this work is scalable and has the potential to reach data collection times 1–2 orders of magnitude shorter, limited by the thermal time constant of the bolometer (typically 0.33/fps) and the time required to step the stage a single FOV when building mosaic images. Even today, if a slightly lower pixel resolution of 4.25 μm is deemed acceptable for the application, a 9.5× increase in throughput could be achieved simply by using the 0.3 NA 4× objective with a 2 mm × 2 mm FOV. In this configuration, tissue cores with diameters up to 2 mm could be imaged in a quarter of the times reported in this work. Current and expected future trends in data acquisition times as a function of the number of discrete wavenumbers employed in the diagnostic for two different area–pixel resolution configurations are shown in Fig. 10.

Estimating future throughput trends assumed two camera frames (at 30 fps) are used per discrete wavenumber to ensure adequate settling and 125 msec stage mosaic step times. Based on these results, it becomes immediately apparent that whole-slide diagnostic imaging could eventually be completed in a matter of minutes using the protocols developed in this work.

3.4 Discrete frequency classification: independent test set

Testing classifier performance using the same patients for training and testing is likely to produce favourable results, since inter-patient variability does not become a factor. Implementation of SHP in the clinic requires that good classification of disease state can be achieved as new patients are introduced. Confidence in SHP using discrete frequency imaging can only be achieved if it performs well on patients in an independent test set. Each random forest classifier was used to classify epithelium spectra from the 14 patients in the independent test

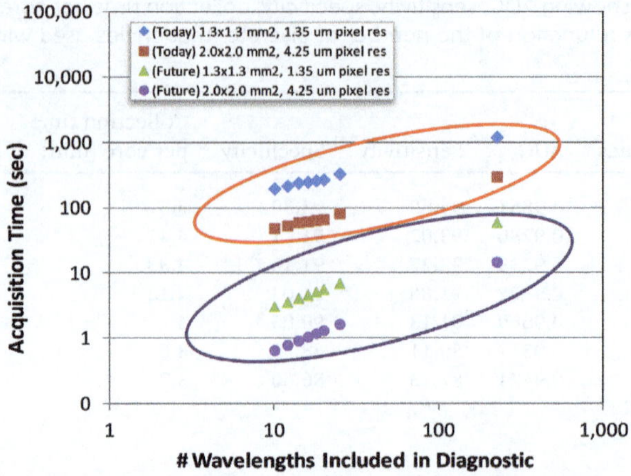

Fig. 10 Data acquisition times *vs.* the number of discrete wavenumbers included in the diagnostic for the current and future QCL full-field imaging technology used in this work. Two different imaging configurations were used in this analysis: (1) 1.3 mm × 1.3 mm (2 × 2 mosaic) at 1.35 μm pixel size (this work), and 2.0 mm × 2.0 mm (single FOV) at 4.25 μm pixel size.

set. The ROCs obtained in each case are shown in Fig. 11(a–f) and 12. Fig. 11(a) shows the ROC obtained when using 25 discrete frequencies for training and classification on the independent test set. The AUC values that were obtained for the validation set were observed to be all close to 1, indicating good discrimination between classes when training and testing occurs on the same patients. However, testing the classifier on the independent test set reduces the AUC values from 0.9851 for the validation data to 0.8395 for the independent test set. Reduced classification accuracy is expected to occur for the independent test, since the data used to test the model are from new patients and therefore completely independent. Reducing the number of discrete frequencies decreases AUC values for the independent test set, in a similar manner observed for the training data set. The AUC value of 0.8396 obtained using 20 discrete frequencies instead of 25 (0.8395) is very similar, indicating that classification performance has not deteriorated significantly. Although there is a slight reduction in AUC (0.8163) when using 16 wavenumbers, each ROC plot appears broadly similar. Classification performance only appears to deteriorate significantly when utilising 14 or fewer discrete frequencies. Using only 10 discrete frequencies (Fig. 12), the AUC value decreases to 0.7808, which is in stark contrast to the validation set, which had an AUC value of 0.9421.

The effect of reducing the number of discrete frequencies on sensitivity and specificity is shown in Table 8. Utilising the full 25 discrete frequencies enables reasonable classification accuracy rates of 72.14% and 80.23% for sensitivity and specificity respectively. Reducing the number of discrete frequencies to 16 only has a limited impact on classification, with sensitivity and specificity values of 70.46% and 78.10%. In contrast to the validation set, the sensitivity does not appear to deteriorate significantly when reducing the number of discrete frequencies. Specificity, however, does appear to be strongly correlated to the number of discrete frequencies, and performance drops off sharply when less

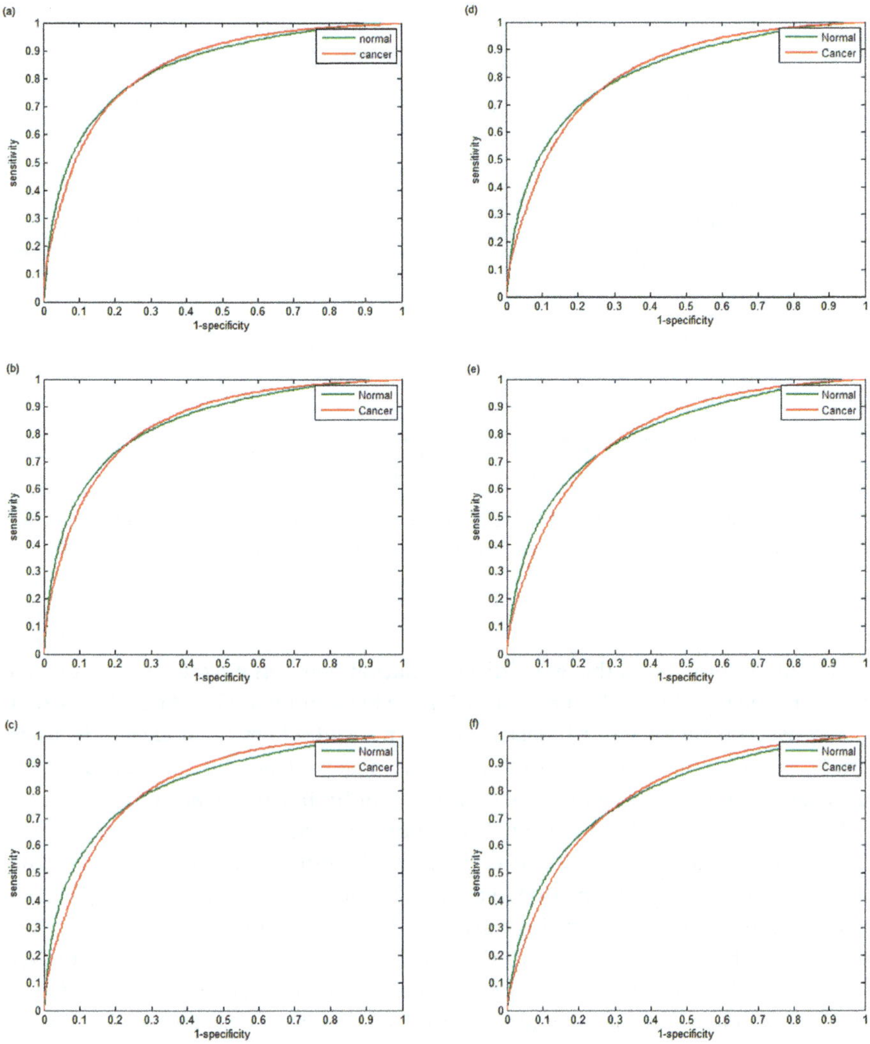

Fig. 11 Receiver operator curves using the independent test set for normal-associated and cancerous epithelium with (a) 25, (b) 20, (c) 18, (d) 16, (e) 14, and (f) 12 discrete frequencies. AUC values are 0.8395, 0.8396, 0.8261, 0.8163, 0.8044 and 0.7876 respectively.

than 14 wavenumbers are used. When using only 10 wavenumbers the classification of the independent test set is poorer, with a mean sensitivity and specificity of 68.73% and 73.51% respectively.

The poorer performance of the classifiers on the independent test set is surprising considering the excellent classifier performance using the training data. Since all patients in this study have been diagnosed with prostate cancer, there is likely to be considerable biochemical variability between patients. To perform well on new patients, the model needs to be trained on a dataset that encompasses this variation. Given the limited patient numbers available in this study for training and testing, it is likely that the model did not have sufficient

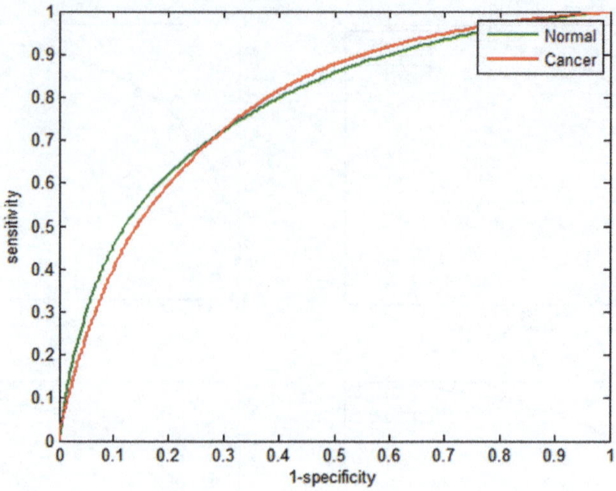

Fig. 12 Receiver operator curves (ROC) with 10 wavenumbers using independent test set for normal-associated and cancerous epithelium. Area under the curve (AUC) values are 0.7808.

variability built in to enable good discrimination between normal and cancerous tissue for new patients. Similar findings have been published by Pounder[32] when using spectral histology of breast tissue using FTIR chemical imaging. In their study, good classification performance was observed on the training data for classifying epithelium, lymphocytes and myofibroblast-rich stroma with AUC values of 0.94. Upon classifying an independent test set there was a deterioration in classifier performance, with AUC values in the range of 0.8–0.88. The authors described this effect as being due to the limited number of cores (50) and patients used in the study. These findings are broadly in line with the classifier performance which we have detailed in this paper. We have also considered whether instrumental or sample preparation parameters could be a contributory factor to the poorer classification accuracy of the independent test set. Variability in sample and substrate thickness, and whether the samples are left in wax or dewaxed are all parameters that could potentially affect classifier performance. However, a much larger study investigating the effect of each parameter will be required to determine the optimum parameters for classification performance. Another possibility for the poorer classification on the independent test set is the selection of the salient spectral frequencies. In this proof-of-concept study only two patients were used for selecting the spectral frequencies used to train and test the model. Given the biochemical variability within a patient population, it is unlikely that two patients are a sufficiently large dataset for identifying the key

Table 8 Table showing sensitivity and specificity for the independent test set using random subset sampling for ten trained classifiers

Discrete frequencies	25	20	18	16	14	12	10
Sensitivity (%)	72.14	71.29	71.13	70.46	69.48	68.25	68.73
Specificity (%)	80.23	80.83	78.86	78.10	76.69	75.07	73.51

biomarkers. In the future it is recommended that a larger patient population is used for frequency collection, and this is planned to be conducted. Although our preliminary results are promising, larger studies using a more diverse patient database would be required to fully evaluate the full potential of discrete frequency imaging for disease diagnosis.

4 Conclusions

Discrete frequency infrared chemical imaging has the potential to provide high-resolution, high-throughput chemical images on a timescale that could revolutionise spectral histopathology. In this study, we have demonstrated that high-quality chemical images of tissue biopsy cores composed of almost a million pixels can be obtained in a matter of minutes. Comparable chemical images obtained on a state-of-the-art FTIR system using an FPA detector would have taken several hours. We have clearly demonstrated on a validation set that excellent classifier performance can be achieved by careful selection of discrete frequencies. We have further shown that significant time advantages can be achieved by using just 16 discrete frequencies while maintaining good classification accuracy. Testing the classifier on an independent test set produced mixed results, with poorer accuracy than on the validation set. However, reasonable classification accuracy could still be achieved when using 16 or more discrete frequencies. Classifier performance may have been compromised by only using two patients for selecting the optimal wavenumbers. Utilising a larger patient population for determining the key biomarkers will be important in any future studies. Limitations on the number of patient tissue core biopsy samples available are the most likely cause of the reduced accuracy when testing on new patients. Prospects for this new and exciting technology are bright. However, further work needs to be performed on significantly larger patient numbers to fully understand its potential for successful implementation in the clinic.

Acknowledgements

We would like to thank Daylight Solutions for making the Spero infrared microscope available to us for the duration of the study. We would also like to acknowledge Jeremy Rowlette, Edeline Fotheringham, Miles Weida, Bill Mohar and Matthew Barre for their help on the project and also for their roles during informative discussions. PG and MP would like to acknowledge the EPSRC for funding (EP/K02311X/1, EP/L012952/1).

Notes and references

1 L. G. Luna, *Manual of Histologic Staining Methods of the Armed Forces Institute of Pathology*, McGraw Hill, New York, 1968.
2 W. P. Michael and H. Ross, *Histology a text and atlas*, Lippincott Williams & Wilkins, 6th edn, 2010.
3 J. B. Lattouf and F. Saad, *BJU Int.*, 2002, **90**, 694–698, discussion 698–699.
4 H. Fabian, P. Lasch, M. Boese and W. Haensch, *Biopolymers*, 2002, **67**, 354–357.
5 B. R. Wood, L. Chiriboga, H. Yee, M. A. Quinn, D. McNaughton and M. Diem, *Gynecol. Oncol.*, 2004, **93**, 59–68.

6 B. Bird, M. Miljkovi, S. Remiszewski, A. Akalin, M. Kon and M. Diem, *Lab. Invest.*, 2012, **92**, 1358–1373.

7 N. Bergner, B. F. M. Romeike, R. Reichart, R. Kalff, C. Krafft and J. Popp, *Analyst*, 2013, **138**, 3983–3990.

8 M. J. Baker, E. Gazi, M. D. Brown, J. H. Shanks, N. W. Clarke and P. Gardner, *J. Biophotonics*, 2009, **2**, 104–113.

9 A. Akalin, B. Bird, X. Mu, M. A. Kon, A. Ergin, S. H. Remiszewski, C. M. Thompson, D. J. Raz and M. Diem, *Lab. Invest.*, 2015, **95**, 406–421.

10 E. Gazi, M. Baker, J. Dwyer, N. P. Lockyer, P. Gardner, J. H. Shanks, R. S. Reeve, C. A. Hart, N. W. Clarke and M. D. Brown, *Eur. Urol.*, 2006, **50**, 750–761.

11 N. Wald and E. Goormaghtigh, *Analyst*, 2015, **140**, 2144–2155.

12 K. M. Dorling and M. J. Baker, *Trends Biotechnol.*, 2013, **31**, 437–438.

13 M. J. Walsh, D. Mayerich, A. Kajdacsy-Balla and R. Bhargava, 2012.

14 A. Liu, L. Wei, W. A. Gardner, C.-X. Deng and Y.-G. Man, *Int. J. Biol. Sci.*, 2009, **5**, 276–285.

15 P. R. Pandey, J. Saidou and K. Watabe, *Front. Biosci.*, 2010, **15**, 226–236.

16 M. J. Walsh, S. E. Holton, A. Kajdacsy-Balla and R. Bhargava, *Vib. Spectrosc.*, 2012, **60**, 23–28.

17 L. S. Leslie, A. Kadjacsy-Balla and R. Bhargava, *Medical Imaging: Digital Pathology*, 2015, 9420.

18 P. Bassan, J. Mellor, J. Shapiro, K. J. Williams, M. P. Lisanti and P. Gardner, *Anal. Chem.*, 2014, **86**, 1648–1653.

19 R. Bhargava, *Anal. Bioanal. Chem.*, 2007, **389**, 1155–1169.

20 D. C. Fernandez, R. Bhargava, S. M. Hewitt and I. W. Levin, *Nat. Biotechnol.*, 2005, **23**, 469–474.

21 H. Brooke, B. V. Bronk, J. N. McCutcheon, S. L. Morgan and M. L. Myrick, *Appl. Spectrosc.*, 2009, **63**, 1293–1302.

22 J. Filik, M. D. Frogley, J. K. Pijanka, K. Wehbe and G. Cinque, *Analyst*, 2012, **137**, 853–861.

23 P. Bassan, J. Lee, A. Sachdeva, J. Pissardini, K. M. Dorling, J. S. Fletcher, A. Henderson and P. Gardner, *Analyst*, 2013, **138**, 144–157.

24 M. J. Pilling, P. Bassan and P. Gardner, *Analyst*, 2015, **140**, 2383–2392.

25 P. Bassan, A. Sachdeva, J. H. Shanks, M. D. Brown, N. W. Clarke and P. Gardner, 2014, **9041**, 90410D.

26 M. J. Baker, J. Trevisan, P. Bassan, R. Bhargava, H. J. Butler, K. M. Dorling, P. R. Fielden, S. W. Fogarty, N. J. Fullwood, K. A. Heys, C. Hughes, P. Lasch, P. L. Martin-Hirsch, B. Obinaju, G. D. Sockalingum, J. Sulé-Suso, R. J. Strong, M. J. Walsh, B. R. Wood, P. Gardner and F. L. Martin, *Nat. Protoc.*, 2014, **9**, 1771–1791.

27 P. Bassan, M. J. Weida, J. Rowlette and P. Gardner, *Analyst*, 2014, **139**, 3856–3859.

28 Y. Yang, J. Sulé-Suso, G. D. Sockalingum, G. Kegelaer, M. Manfait and A. J. El Haj, *Biopolymers*, 2005, **78**, 311–317.

29 E. Gazi, J. Dwyer, P. Gardner, A. Ghanbari-Siahkali, A. P. Wade, J. Miyan, N. P. Lockyer, J. C. Vickerman, N. W. Clarke, J. H. Shanks, L. J. Scott, C. A. Hart and M. Brown, *J. Pathol.*, 2003, **201**, 99–108.

30 L. Breiman, *Journal of Machine Learning*, 2001, **45**, 5–32.

31 G. R. Lloyd and N. Stone, *Appl. Spectrosc.*, 2015, **69**, 1066–1073.

32 F. N. Pounder and R. Bhargava, 2009, vol 7182, p. 718206.

Faraday Discussions

DISCUSSIONS

Spectral Pathology: general discussion

Christopher Sammon, Zachary D. Schultz, Sergei Kazarian, Hugh Barr, Royston Goodacre, Duncan Graham, Matthew J. Baker, Peter Gardner, Bayden Wood, Colin J. Campbell, Richard Dluhy, Samir El-Mashtoly, Christopher Phillips, Jonathan Frost, Max Diem, Achim Kohler, Parvez Haris, Alexander Apolonskiy, Hemmel Amrania, Peter Lasch, Zhe Zhang, Wolfgang Petrich, Ganesh Dhruvananda Sockalingum, Nick Stone, Klaus Gerwert, Ioan Notingher, Rohit Bhargava, Niels Kröger-Lui, Martin Isabelle and Michael Pilling

DOI: 10.1039/C6FD90011A

Parvez Haris opened the discussion of the introductory lecture by Max Diem: Varying degrees of accuracy have been obtained for discrimination between cancerous and non-cancerous tissues using vibrational spectroscopic methods. What are the explanations for these variations in accuracy between cancerous and non-cancerous tissues and how do they correlate with accuracy from other techniques including histopathology? It was reported that changes in a specific protein (PDL-1) in tissue sections were detected by FTIR spectroscopic imaging, to compare cancerous with non-cancerous tissues. This is very interesting but needs to be proven with additional data as peaks in the spectra of complex tissue samples could arise from not only chemical differences but physical differences associated with changes in tissue stiffness, dynamics *etc.* Therefore, combining histopathology, proteomics, metabolomics and ultrasound elastography with vibrational spectroscopic imaging could pave the way for a better understanding of the changes associated with cancer and could also improve our understanding of the vibrational spectra of cells and tissues. Basic studies on understanding the factors influencing spectral changes in tissues need to be better understood for the technique to be widely accepted as a tool for cancer diagnosis.

Rohit Bhargava replied: Variations in accuracy arise from at least the following factors:

1. Measurement noise, unless it is reduced to a level where it does not affect classification (see ref. 1).

2. Population variance, person to person variance, tumor heterogeneity and clinic-to-clinic variations that likely arise from sample processing. For a complete discussion of all factors see ref. 2.

3. Shortcomings in the design of experiments, especially in assessing confidence intervals as related to sample sizes (see the article by Pounder, Reddy and

Bhargava in this issue (DOI: 10.1039/c5fd00199d) for confidence intervals and for sample size discussions, please see ref. 3 by Beleteis *et al.*).

4. Combined optical effects arising from the optics, sample (especially thickness), substrate and interfaces. There is no general description or solution to this problem though there are ongoing attempts to resolve these effects *via* theory, models and chemometric approaches.

The accuracy of histopathology is not usually assessed as there is no secondary standard. However, there are examples of several studies that discuss the error rates of pathologists, concordance, grading accuracy *etc.* These are organ and disease dependent.

There can be instances of a protein being measured directly (as for PDL-1) but there could be another reason. For a specific protein, it is not necessary for the concentration to be directly measured. More than likely, in tissue, the protein and some features of the IR spectrum correlate to the same underlying physiology, disease or cellular state that is characterized by the protein expression. Hence, it is possible to correlate protein expression with a specific cellular identity or transformation.

Improving our understanding of the origins of spectra and their changes is indeed of great importance. The combined multimodal approach suggested may prove to be useful.

1 R. Bhargava, Towards a practical Fourier transform infrared chemical imaging protocol for cancer histopathology, *Anal. Bioanal. Chem.*, 2007, **389**(4), 1155–1169.
2 J. T. Kwak, R. Reddy, S. Sinha and R. Bhargava, Analysis of variance in spectroscopic imaging data from human tissues, *Anal. Chem.*, 2012, **84**(2), 1063–1069.
3 C. Beleites, U. Neugebauer, T. Bocklitz, C. Krafft and J. Popp, Sample size planning for classification models, *Anal. Chim. Acta*, 2013, **760**, 25–33.

Max Diem responded: I disagree with the question in that the accuracy in classification of normal and cancerous tissue has uniformly been above 95% for most of the recent studies from different laboratories. However, classification between normal and cancerous tissue should not be the goal of spectral histopathology (SHP), since this is a task any pathologist can do. The question rather should be: what are the accuracies for the sub-classification, grading and staging of cancers? It is in this area that SHP adds value.

In terms of the cancer marker results reported in my talk, I emphasized that we may not see the cancer marker per se, but a surrogate. For example, the co-localization of regions of high TTF-1 concentration, as detected by immunohistochemistry, combined with the SHP image may imply that we see regions in the infrared that have high transcription activity (TTF-1 is a transcription factor), and that IR spectroscopy detects the high rate of cell division that accompanies transcription activity.

Sergei Kazarian added: Quantifying the accuracy of histopathology *vs.* spectroscopic classification of tissue is an ongoing challenge and is not something we specifically address in our paper here. However, we recently published a paper[1] in which the progression of all disease stages to cancer within a single biopsy were identified spectroscopically and verified independently by histopathology. As such, FTIR spectroscopic classification of cancerous/non-cancerous tissue, and even assessment of the degree of dysplasia, is possible and work is being done by us and others to improve the sensitivity and specificity of this approach, as well as to correlate spectral band changes with additional data from alternative methods.

1 L. Foreman J. A. Kimber, K. V. Oliver, J. M. Brown, S. M. Janes, T. Fearn, S. G. Kazarian, P. Rich, Assessing Dysplasia of a Bronchial Biopsy with FTIR Spectroscopic imaging, *Proc. of SPIE*, 2015, **9332**, 93320S

Klaus Gerwert commented: In the talk individual band assignments in the spectra of cells and tissue were discussed. I disagree in general to those individual band assignments. I have a lot of experience with band assignments in recombinant proteins using either site-directed mutagenesis or isotopic labeling. Thereby we assigned clear cut bands to individual protein groups.

However, in cells or tissue the concentration of an individual protein is much smaller and the bands of thousands of proteins overlap. In addition lipids, metabolites, RNA and DNA bands overlap. The IR spectra can only be used as a fingerprint. They have to be analysed by bioinformatic tools but not by hand. These tools are able to distinguish objectively between the different patterns in the spectra. However, by combining the label-free imaging with molecular resolving techniques like proteome analysis and Next Generation Sequencing individual proteins or mutations can be identified. We have already classified different subclasses of mesothelioma cancer by IR Imaging. We cut the different classes out by laser microdissection. The cut out tissue was than analyzed by proteome analysis. 143 differently expressed proteins were thereby identified, which causes the differences in the spectra of the different tissue classes. By this approach detailed molecular information is obtained. I do think this is the path we should follow in the future to provide detailed molecular information to the clinician which is relevant for his therapeutic decision.

Matthew J. Baker asked: For spectral cytopathology and other uses of spectroscopic analysis (*e.g.* tissue) can you comment on the usefulness of using low *e* substrates particularly in light of recently discovered artefacts and what is the optimum substrate for the future?

Max Diem answered: The effects of a standing electromagnetic wave on reflective surfaces is now well understood, thanks to the work by Peter Gardner.[1] We also have recently commented on these effects in an article[2] and pointed out that by using 2nd derivative spectra in a restricted spectral range, the intensity distortions can be minimized. Furthermore, the effect is small when a monolayer of thin, exfoliated epithelial cells are investigated, but is more significant in thick layers of cultured cells.

1 P. Bassan, J. Lee, A. Sachdeva, J. Pissardini, K. M. Dorling, J. S. Fletcher, A. Henderson and P. Gardner, The inherent problem of transflection-mode infrared spectroscopic microscopy and the ramifications for biomedical single point and imaging applications, *Analyst*, 2013, **138**, 144–157.
2 M. Miljković, B. Bird, K. Lenau, A. I. Mazur and M. Diem, *Analyst*, 2013, **138**, 3975–3982.

Nick Stone questioned: How do you define a spectral diagnostic decision on a particular sample? Do you provide a pathologist with a digitally stained image and allow them to make a decision on individual abnormally labelled cells? Or do you use a voting/polling scheme, such as that used by Rohit and others (DOI: 10.1039/c5fd00199d) to provide the pathologist with an abnormality score for the sample as a whole?

Max Diem replied: The SHP results are rendered as a report that contains an image of the sample tissue coded by tissue type, *e.g.* connective tissue with inflammatory response, cancer main type, and sub-type/grade, all as color-coded images. in addition, H&E as well as IHC images may be shown. A pie chart of percentages of the abundance of each tissue type/disease type is provided, along with the patient information available.

Duncan Graham opened a general discussion of Max Diem's, Rohit Bhargava's and Sergei Kazarian's papers: This work from Max, Rohit and Sergei on using IR for clinical pathology looks great and if I were a patient I'd want to have this used to help with my diagnosis. How soon can we benefit from this in the clinic and what are the barriers to this happening? How can these barriers be overcome?

Rohit Bhargava responded: There are several technological and scientific barriers that need to be overcome:

1. Demonstration of value. Laboratory studies that conclusively demonstrate the solution of a problem of interest to clinicians, healthcare professionals or patients are not yet widely available. There are some examples in the grading of lung and colon tissue, for example, as mentioned in this meeting by Diem *et al.* (DOI: 10.1039/c6fd00023a) and Gerwert *et al.* (DOI: 10.1039/c5fd00157a). There is also recent work on prognosis that directly competes with existing standards and shows the power of IR imaging.[1] Significantly more work is needed to establish an extensive set of "solved" problems.

2. Instrumentation. Imaging today takes far too long to measure samples of clinical interest. ~ 1 cm x 1 mm is a typical size that needs to be analyzed in a short time period, for example, in 10 min. At 1 micron pixel^{-1}, in accordance with the theory for high definition IR imaging, this implies a pixel rate of ~ 1 megapixel min^{-1}. The data acquisition rate by itself is insufficient, as the signal to noise ratio of the data are also critical for classification. Hence, signal averaging may be needed to attain that performance level, leading to increased pressure on acquisition time. It appears that new laser based discrete frequency instrumentation can progress towards this performance level.[2] I anticipate that further progress in the field will permit fast enough data acquisition to make the technology practical.

3. Relation to biomedical knowledge. It is very important for the field to relate to biomedical knowledge but also provide new insights or studies that are not possible. A direct attempt to provide H&E and immune-images using IR imaging is a step in this direction.[3] Several examples from the meeting, including studies on metabolism and drug-ligand interactions can provide new scientific insights whereas cancer grading and detection can provide capabilities that relate to knowledge in a significantly improved manner.

4. Understanding the origins of signals and strategies to extract more information. While a theory of image formation is now available for IR microscopy,[4,5] the relation of sample morphology to recorded data still remains to be completely resolved. The forward problem, *i.e.* prediction of spectra if the structure and

refractive index are known, can now be solved. The inverse problem, *i.e.* determining the local spectral and structural properties of tissue from recorded data, remains to be solved. It has been solved for simple objects like spheres and cylinders.[6, 7] While one approach to practical translation is to use model-based corrections as demonstrated by the work of Kohler *et al.* (DOI: 10.1039/c5fd00171d), other approaches can and should be developed. It is likely that different samples/problems may require different methods or need to be optimized for different conditions (*e.g.*, speed of correction). Hence, it is very important for the community to develop methods to address this need. In addition to these scientific accomplishments, a commercial product needs to be developed. This includes the development of a design that is efficient in obtaining high signal to noise ratios while making the trade-offs needed to keep costs low, components manufactured in a cost effective manner, software that is user friendly and results in which confidence is assured. Finally, and perhaps most crucially, a tremendous effort needs to be made in educating the biomedical community who will be the users of the technology as well as other physical and computational scientists who provide a robust community to further develop the technology.

1 J. T. Kwak, A. Kajdacsy-Balla, V. Macias, M. Walsh, S. Sinha and R. Bhargava, *Sci. Rep.*, 2015, **5**, 8758.
2 K. Yeh, S. Kenkel, J.-N. Liu and R. Bhargava, *Anal. Chem.*, 2015, **87**, 485–493.
3 D. Mayerich, M. J. Walsh, A. Kadjacsy-Balla, P. S. Ray, S. M. Hewitt and R. Bhargava, *Technology*, 2015, **3**(1), 27–31.
4 B. J. Davis, P. S. Carney and R. Bhargava, *Anal. Chem.*, 2010, **82**, 3487–3499.
5 B. J. Davis, P. S. Carney and R. Bahrgava, *Anal. Chem.*, 2010, **82**, 2474–2486.
6 T. van Dijk, D. Mayerich, P. S. Carney and R. Bhargava, *Appl. Spectrosc.*, 2013, **67**, 546–552.
7 B. J. Davis, P. S. Carney and R. Bhargava, *Anal. Chem.*, 2011, **83**, 525–532.

Max Diem answered: The technological problems largely have been overcome. The main challenge now is to have a system and method in place that fits into the workflow, and is accepted in the diagnostic community. Cireca Theranostics, LLC, is heavily involved in these steps, and unveiling of a platform to introduce SHP to the clinical community is planned for later this year.

Sergei Kazarian responded: The gap between implementation in the research laboratory and practical application in the clinic is formidable. Many more samples would be required to be analysed by the new and current practice methods in order provide the necessary statistical validation. Clear add-on value and/or cost-effectiveness *versus* standard methods would also be required. Such studies would take several years and will require substantial financial support and commercial involvement.

Jonathan Frost asked Max Diem: One of the significant potential benefits of vibrational spectroscopy is the ability to go beyond replicating the routine histopathological diagnosis by offering prognostic or other information not currently available to the histopathologist. In the lung cancer study presented there are a significant number of patients with a relatively short disease course and quantifiable response to treatment. Has the ability of the spectroscopic analysis to determine disease course or response to treatment been evaluated within this study?

Max Diem answered: This is being investigated in the presently ongoing study with a national cancer center in the US, for which patient outcome and other metadata are available. The original study reported here used commercial lung cancer samples, and the aim was the accurate classification of the main lung cancer subtypes.

Royston Goodacre said: The problem seems to be that the wrong samples are being used. For example in Max's study he reminds us that the 5 year survival of Small Cell Lung Carcinoma (SCLC) is still far too low (reported as 7%; and has been for decades). But the patients being analysed already have cancer and are at the end stage of the disease process. Can we learn from improvements in breast cancer survival where screening has been highly successful in reducing mortality? What samples can be used for an early screening of lung cancer?

Max Diem replied: Lung cancer often is, indeed, asymptomatic until the cancer is more than a few cm in size. At this stage, it is often fatal. For lung cancer screening, chest X-rays can be used, but this is not widely performed due to the low specificity of lung X-rays, and radiation dose concerns. For spectral screening, cytology can be performed on sputum or lung lavages, but cytology has, in general, low accuracy. Spectral cytopathology could be used here, but no studies on lung spectral cytopathology have been carried out. The main difficulty, in all forms of cytology, is the "gold standard", since cytology produces few abnormal cells, even in the case of disease. Thus, for most cytological work, biopsies are the gold standard, and that, of course, defeats the purpose of screening.

Wolfgang Petrich addressed Max Diem and Rohit Bhargava: To me there appear to be diverging opinions concerning the question whether hyperspectral pathology will or will not replace pathologists. While Max Diem stressed the point that we won't replace the pathologist, Rohit Bhargava said – with reference to the objectivity and reproducibility of the hyperspectral "judgement" – we will replace them. Frankly, I believe that the truth lies in between. And, beyond what has been discussed so far, my question "*can (and should) we replace the pathologist?*" relates to the fact that, as of today, pathologists set the gold standard. On the one hand, the immediate "true diagnosis" related to a specific image or tissue sample is therefore always provided by a pathologist. Hence, the best an alternative method could deliver is: to be as good as the gold standard. A straight forward example along these lines would be the infrared-based replacement of H&E staining in a first step. As I just said: we could possibly show that we can nicely color our images like if it were an H&E stain, but it will be very difficult to show that this particular surrogate staining would be any "better" than an H&E stain. If, on the other hand, you want to replace the pathologist then you need to change the established gold standard. You will only be able to do that if you can show that the medical outcome based on a hyperspectral assessment (solely) is better than the outcome based on the histopathologist's diagnosis. In cancer, for example, you will need to show that IR- or Raman-approaches are more sensitive and specific with respect to the detection of cancer than the histopathologist. In this case, you start redefining the gold standard with reference to the medical outcome which for cancer necessarily means that you have to do longitudinal studies. These usually take a long time and are quite costly. Hence, if we are targeting cancer with the aim to replace the histopathologist, my concern is that our focus on cancer at the same time substantially raises the hurdles for a conceivable replacement claim. So, generally speaking, is cancer diagnostics simply the wrong example?

Max Diem answered: I will stick to my guns and state that attempting to replace pathology at this point is counterproductive and will make enemies. After all, are there any professionals out there who want to be replaced by a machine?

The other point raised in this question is a very important one: of course I can take my hyperspectral image and make it look like an H&E stained tissue. But this is a naïve approach to what a pathologist does. A pathologist looks at details, such as chromatin distribution in the nucleus, the existence of intercellular bridges (to classify squamous cell carcinoma), the presence of fibroblasts and connective tissue in adenocarcinomas, and the abundance of immune cells. This detail is absent in an image that just produces red and blue hues to mimic H&E stained tissue. We as spectroscopists must abandon our "we are better then them" attitude and really learn what a pathologist needs, and learn from the enormous knowledge a pathologist has. So, what can we offer? We can classify poorly differentiated or undifferentiated cancers where classical pathology has issues. We can offer a subjective sub-classification and grading of tumours that is now very subjective. We may supplement immunohistochemistry (IHC) as SHP often co-localizes with areas of IHC response. We can predict outcomes from longitudinal studies. Those are aspects where we can interact with pathologists, and advance the field of diagnostics.

Rohit Bhargava responded: This is a very interesting point. Part of my comment was specifically targeted to promote discussion. It does not seem that an absolute answer of "replace" or "cannot replace" will be true. While it is difficult to predict the future, technology has typically replaced human functions where productivity is limited, repetition is burdensome or the need for capabilities beyond average human ability has arisen. In most instances, further, it is not that the human is replaced but that there is significant time freed up to focus on other aspects. One could imagine a situation in pathology where routine tasks are conducted using spectroscopic imaging whereas difficult cases are examined by a pathologist. At the other end, it is difficult for even the most experienced pathologist to observe subtle changes in morphology that can have profound effects on the disease course. If spectroscopy can provide an answer, then it can be the standard test and "replace" the pathologist. The replacement, however, is of dubious value since the contrast mechanism in pathology is not sufficient for this task to begin with. I do agree with the comment that we need to show results that outperform the current state of the art; otherwise, like any other technology, we will not be utilized. Cancer is no doubt a difficult problem but one that is immediately recognizable. Without convincing hypotheses of why other diseases would be more amenable to our technologies, in my opinion, cancer is a great challenge and one of high impact that is rightly the focus of many groups.

Hemmel Amrania opened discussion of the paper by Rohit Bhargava: Having spent last few years working very closely with breast pathologists, it is clear in my mind that the detection of tumours using H&E staining is fairly trivial and is well established. The more pertinent unmet need exists in the sub-classification of tumours into high or low risk cases and patient prognosis from biopsy examinations. Please comment on the possible translation of your work to the clinic and how it could be implemented in existing clinical workflows.

Rohit Bhargava responded: Indeed, the detection of tumors is easy. However, it does require time and effort, usually very tedious work to wade through the majority of biopsies that do not have tumors or where real-time decisions are needed (surgery). Hence, automated technologies that can accomplish the same are needed. Further, it is a very high technological bar to achieve almost perfect discrimination of tumors in tissue. Hence, it is a worthy goal for the community. Even though it may not be a great clinical need, it is a confidence-boosting demonstration and can potentially be useful as an adjunct to prevent the (small number) of errors pathologists do make in recognizing tumors. The work demonstrated in this volume can potentially lead to a very fast and easy method to recognize disease. We anticipate that it can be used for both adjunct examinations of routine specimens as well as intra-operatively. Obviously, much more work will be needed to define the protocol and develop an instrument *etc.*

Ioan Notingher continued the general discussion of the papers by Max Diem and Rohit Bhargava: For using spectroscopy in the clinic, the performance of spectroscopy is always compared to histopathology. While histopathology is very good, in many cases analysis cannot be carried out on the entire tissue specimen. For example in the case of cancer resections, typically only 1-5% of the margins are looked at by histopathology. Such sampling errors can lead to many false negative diagnoses, and to sub-optimal treatment. IR/Raman can provide advantages in these cases, if it can look at the entire surface of the tissue specimens.

Christopher Phillips added: We feel that it is not constructive to talk of replacing clinicians with these spectroscopy-based technologies. In the field of cancer, for example, H&E biopsies are the "gold standard" and they will always continue to be relied upon by the clinicians. It is best to work with the pathologists themselves, to develop ways of using our spectroscopic techniques to augment the existing protocols, in a way that the pathologists can understand and thus will be able to adopt.

Alexander Apolonskiy opened discussion of Sergei Kazarian's paper: Spatial resolution and dispersion - what is the relation between them? Did you mean the dispersion of the biomaterial or the dispersion of the optics? If you are discussing the dispersion of the optics, you can play with different materials. How do you compensate for the dispersion of the bio-sample?

Sergei Kazarian responded: Dispersion in the form of refraction (appearing as chromatic aberration) causes infrared light at different frequencies to be focused on different focal planes. This produces images that are in focus only at one frequency of light. In our paper, the added lenses remove chromatic aberration making the images focused at all wavelengths. Then, the spatial resolution is limited by the diffraction of light, depending only on the wavelength of light and the numerical aperture (notwithstanding optical aberrations). The lenses also remove the effects of scattering from the sample as shown in our spectra, so scattering from the bio-sample (particularly from the areas with significant irregularities present) is corrected. We also used zinc sulphide lenses and windows as a different material, with the hope that magnification would be further increased. However, the reflection losses when using them were too high for any meaningful result due to the high refractive index of that material and the number of interfaces through which the light travelled. So far, we have successfully demonstrated that an added CaF_2 lens, which creates a pseudo hemisphere, removes the effect of chromatic aberration on the obtained images, reduces scattering and increases magnification and spatial resolution. This was applied to cells and tissues, including the combination of imaging with mapping to image large areas using this approach.

Christopher Sammon asked: Considering the nature of the medical imaging challenge, how critical is the driver towards improving spectral resolution, given the generally adopted approach of binning pixels from IR images to both improve spectral quality (*i.e.* S/N) and reduce the volume of these vast data sets?

Sergei Kazarian replied: In our paper, we applied 4x4 binning (average of 16 pixels) to perform the HCA, which was necessary to increase the signal to noise ratio. The use of new laser based sources (QCLs) will increase the signal to noise ratio and render binning unnecessary, as will the use of more powerful computers for analysis.

However, for localised detection of cancer,[1] high resolution is necessary, especially as spectral signatures would be averaged out if binning were used. Additionally, live cell studies benefit from high spatial resolution, for example with the ability to view drug distribution within cells. Micro ATR-FTIR spectroscopic imaging with a Ge crystal, demonstrates truly increased spatial resolution due to the increase of the numerical aperture by a factor of 4, which has been assessed by the stringent edge profile test of the polymer interface. The benefits of the high spatial resolution achieved are evident in measurements of live cells for sub-cellular resolution as has been demonstrated in our paper here (DOI: 10.1039/c5fd00158g) and previous publications.[2] The important issue is how the achieved spatial resolution is assessed and I was advocating over the years that it should be assessed *via* the measurement of the absorbance profile for the spectral bands across a sharp interface between two dissimilar polymers as this provides a realistic test for actual chemical spatial resolution in spectroscopic imaging.

1 S. G. Kazarian and K. L. A. Chan, *Appl. Spectrosc.*, 2010, **64**, 135A–152A.
2 S. G. Kazarian and K. L. A. Chan, *Analyst*, 2013, **138**, 1940–1951.

Bayden Wood opened discussion of Martin Isabelle's paper: You state that adenocarcinoma has higher intensities in nucleic acids (cytosine and uracil at 783 cm-1 and 840 cm-1 (O–P–O stretch) compared to intraepithelial metaplasia. Yet when I look at Figure 5 in the paper there appears to be no apparent difference between the band at 785 cm-1, which I assume you mean 783 cm-1. Moreover there is no band at 840 cm-1 so I assume you mean 852 cm-1 or 826 cm-1. In either case there is no difference in any of these bands or if there is it is less than 1 Raman count, which when compared to Figure 2 is far less than the standard deviation. So how can you possibly state unequivocally that there is a difference in the amount of DNA between these two cancer types?

Martin Isabelle responded: A lot of discussion from this conference has centred on the caution required when assigning peaks and bands to biomolecules, due to the overlapping nature of the complex biomolecular functional groups in biological tissue and cells. We acknowledge this despite it being consistently done by many groups in vibrational spectroscopy and in publications in many journals for many years. We need to develop the capability to make reliable band assignments by using additional experimental methods, such as isotopic labelling and mutagenesis, to confirm many of the assignments. Additionally, it cannot be ruled out that *in vivo* at least part of the spectral contribution to the region where DNA bands absorb is due to other molecules.

Although the difference spectrum, Fig. 1 and 2 attached, (which we used to assess peak differences but did not include as a figure in the manuscript) indicates a small peak intensity difference in 783cm^{-1} (the value mentioned in the main text of the manuscript) and between 840-860cm^{-1} between the adenocarcinoma and intraepithelial metaplasia pathology groups, I would therefore agree that we cannot unequivocally state there is a difference in the amount of DNA between these two cancer types without performing these additional experimental methods. These experimental methods are not available to us at this time, but may be something we look to include in future studies.

Matthew J. Baker said: Following processing your specificity decreases. Are you throwing away information during the processing procedures and which value is more important for your analysis, sensitivity or specificity?

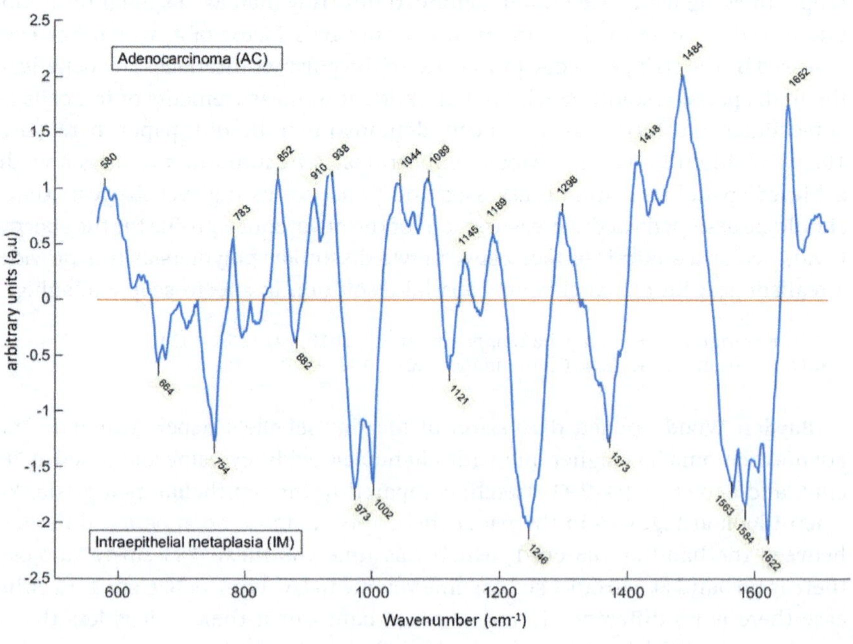

Fig. 1

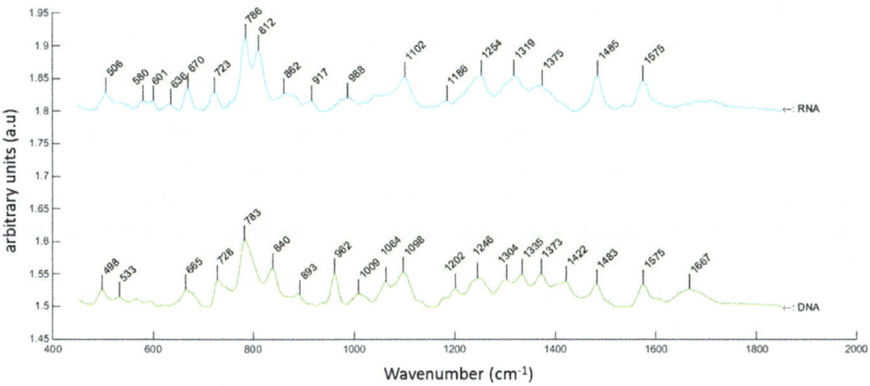

Fig. 2

Martin Isabelle replied: Caution is always required when analysing spectral data to try not to over-process the raw spectral data and lose important spectral information. This concern has been repeated particularly regarding the "subtraction" or "removal" of the fluorescence baseline (contributed by the tissue fluorophore from samples, fluorescence from the instrument optics and the substrate) which could actually be an important parameter for pathology classification/discrimination due to differences in the fluorophore composition in the samples between pathology states.

In this work, we are trying to only eliminate or minimize the intra- and inter-system differences (wavenumber shifts, system response differences, instrument-specific fluorescence) so they do not skew the pathology classification when using data from all three systems. Ultimately the lessons we have learned from this preliminary stage of the project will enable us to anticipate or compensate prior to future measurements, and understand which factors (instrument, environment, user *etc.*) we need calibrate or control better so that we do not have to perform some of the post-processing correction measures. However the correction methods explained and demonstrated here will help researchers attempt to correct their multi-system spectral datasets where they have encountered unexpected intra- and inter-system variations that their pre-calibration procedures have not taken into consideration. Obviously if we can share these observations, correction methods as well as pre-calibration procedures and considerations as a community, it will enable future larger clinical and research studies to measure richer and cleaner spectral datasets without the need to post-process and risk the chance of removing important spectral biomarker information.

Whether higher sensitivity/lower specificity or lower sensitivity/higher specificity is more important depends very much on the clinical problem specific to the pathology or tissue type under consideration. Higher sensitivity corresponds to a high negative predictive value, the ideal property of a "rule-out" test, while high specificity corresponds to high positive predictive value and is the ideal property of a "rule-in" test. In the case of differentiating oesophageal intestinal metaplasia (IM) *vs.* adenocarcinoma (AC), we want to ensure that we do not misdiagnose any AC as IM and therefore miss removing any cancerous lesions, so we would prefer the "rule-out" test (higher sensitivity), but this situation could change as we include more pathology groups and subtypes in the project where some of the differentiation becomes more difficult (*e.g.* low *vs.* high grade dysplasia).

Colin J. Campbell commented: I particularly like the aspect of your paper that addressed multi-centre comparisons. As we enter an era where large-scale IR/Raman datasets are going to be made open-access online, is there a role for the community or journal editors to recommend or dictate standards for the community on how we collect data and report the metadata associated with these datasets?

Do you think that such standards would make it easier for us to compare datasets in a meaningful way?

Martin Isabelle responded: Absolutely and we plan to share our experience, applied correction methods and recommendations for calibration standards and procedures with these communities (*e.g.* CLIRSPEC, eCOST Raman4clinics) and technical and scientific protocol journals (Some of this has already been done *e.g.* ref. 1). We strongly believe that the ability to implement models created with data collected on a number of machines at different times that can classify data measured on different instruments at different sites is critical for large scale implementation of Raman spectroscopy in the clinic. Without this, it may not be possible to use or apply large sample datasets and classification models clearly demonstrated to work on one single instrument outside of that research or hospital site, severely reducing the significance of the Raman spectroscopy pathology discrimination for that particular disease type on a global scale. We have to be able to clearly demonstrate that these disease-specific classification models developed on one instrument can be used to classify data measured on other different instruments without the need for having to create a brand new large dataset every time specifically for that machine, that takes many months to years. We need repeatability, validation and reproducibility on a multi-platform and large scale before realistic clinical integration.

1 H. J. Butler, L. Ashton, B. Bird, G. Cinque, K. Curtis, J. Dorney, K. Esmonde-White, N. J. Fullwood, B. Gardner, P. L. Martin-Hirsch, M. J. Walsh, M. R. McAinsh, N. Stone and F. L. Martin, *Nat. Protoc.*, 2016, **11**(4), 664–687.

Max Diem asked: Have you done instrument calibration on a homogeneous sample? If you run a homogeneous sample you don't have a focusing issue. In IR we have created a slide that can be used as a standard with a sharp edge to check resolution independent of position. Should this be the first step in Raman instrument calibration?

Martin Isabelle answered: Agreed, we have just started a number of experiments to measure and determine image quality and resolution between the three systems using a homogeneous polymer sample. White light image quality and Raman sampling resolution were tested on all three instruments using a set of samples (*e.g.* including silicon on a chrome resolution target) that aided the differentiation of signal profiles across the edges. We plan to incorporate a standard routine in-system calibration to ensure resolution is calibrated.

Nick Stone continued the discussion of the paper by Rohit Bhargava: Traditionally instrument transferability between instruments has been seen as more simple for IR instruments when compared to Raman. However, we don't see many studies demonstrating training and testing models between instruments. Can you comment on the key elements we must consider in IR transferability and how you correct for them?

Rohit Bhargava answered: This is a major consideration and should be a focus of attention for our community. The reason for instrument transferability has been the use of the interferometer, which provides exceptional ordinate accuracy. The only issues which then remain are the apodization (which is easily controlled) and the transmission of the instrument itself. These are further mitigated if absorption spectra are compared. However, that is true only for non-imaging, homogeneous samples. For imaging heterogeneous samples, the same optical configuration should have the same reproducibility as with conventional FT-IR spectrometers. The challenge arises when the optics of the system are changed. For two completely different optical systems, the key elements in IR imaging (let us consider FT-IR with a globar for simplicity) are:

1. Image formation optics
2. Detector configuration and fore optics
3. Sample

The sample becomes crucial as light scattered from the sample is collected differently by lenses of a different numerical aperture (NA) and with different dimensions, for example. In principle, the optics of a system can be characterized and a sample of known structure can be measured. There is now a framework to simulate a microscope to match this data (forward problem) that is available *via* our work, for example. Using the parameters of the second instrument, we can predict the measured spectrum. However, there is unlikely to be a general solution for a simple function that can "normalize" the differences between two instruments. For tissue classification problems, however, whether the changes in the spectra arise from a change in optics is unclear. A similar effect likely exists in spontaneous Raman microscopy but is largely masked as the Rayleigh and Raman scattered photons are distinct. The effect of Rayleigh scattering may show up in confocal Raman but will be difficult to separate from laser noise for thin objects.

Ganesh Dhruvananda Sockalingum continued the discussion of the paper by Martin Isabelle: The efficiency of classifiers has been tested on data from the same samples recorded on two IR machines from the same manufacturer. A loss of sensitivity and specificity has been observed when a database built on one machine is used to identify spectra recorded on the second machine. The issue of data transfer has not been solved and there is still a lot of work to be done in that sense. For Raman, the issue seems even more complicated due to different kinds of microscopes. When micro-Raman data or images are recorded on two different systems, how can the spatial information be restored and compared? This would indeed help cytologists and pathologists to look at the same objects that have been recorded with Raman.

Martin Isabelle replied: Agreed and we believe that there is a lot more work required for system data transferability, especially as you correctly point out that there are different kinds of microscopes and laser wavelengths used in Raman microspectroscopy. We have already mentioned the influence of inter-system optical specific fluorescence baseline variances (due in part to the microscope objectives as well as the other optical components) in the dataset and correction methods used to attempt to correct for this. As well as the instrument factors, we have many other environmental, physical and user factors to consider (Fig. 3).

We have not done a lot of spatial image comparison or correction so far as we have only been attempting to collect spectral information from roughly the same areas of the tissue in each system to develop a classification model rather than any in depth spectral image comparison. However it is a very good point and, although not included in the manuscript, we have performed white light image quality and Raman sampling resolution experiments on all three instruments using a set of samples (*e.g.* USAF target, biological samples). The spatial correlation between the subsequently collected white light images and Raman map was optimised on each instrument and is typically within 1 to 2 μm . This is definitely a lot more important for cell spectroscopy when dealing with measurements from whole and sub-cellular locations. However I believe assessing the impact of image resolution matching on tissue measurements is an important part of our current project and we will continue to explore this to understand its impact on our inter-system data and classification performance as well as to investigate the comparison of the white light images against the Raman spectral images produced.

Richard Dluhy commented: NIST-referenced broad band tungsten lamps with regulated power supplies that are used as primary intensity standards for CCD calibration exist. Would the use of these kinds of primary intensity standards help in the transferability issues you mentioned?

Martin Isabelle responded: It is our understanding that most optical spectroscopy research groups use some form of NIST or NPL or other well-established type of standard for signal intensity calibration in addition to the standards for

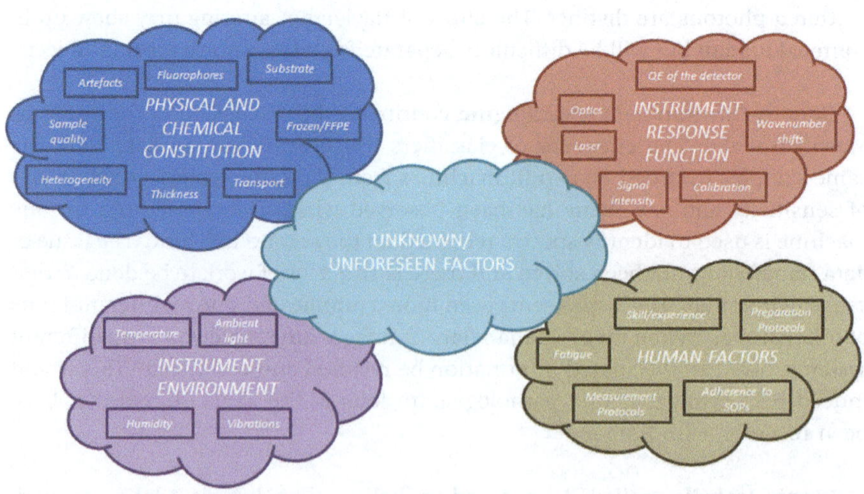

Fig. 3

wavenumber calibration. We have used secondary W-filament lamp standards calibrated to primary NPL standards for calibration. However, with microscope systems, the beam path of the light can impact on the signal measured. Diffuse scattering surfaces are required beneath the microscope objectives. A simpler approach is to use fluorescent green glass with a calibrated spectrum, which sits at the focus of the microscope objective and light follows the same route as that when measuring samples of interest. In this study we used our own tertiary glass standard and a NIST calibrated standard (NIST Standard Reference Material 2241), which we find has worse spatial shape variance.

However this small study shows that these standards did not improve model transferability between systems, which implies that other unaccounted for factors such as those mentioned in the paper and the presentation (physical, environmental, user, *etc.*) are important for model transferability. However, the standard measured and the corrections applied certainly help to minimise the majority of the signal intensity and shape differences between instruments during pre-measurement calibration.

Achim Kohler said: I would first like to give you a comment and then ask a question. 1. You showed that pre-processing decreased the overall performance of the classification model. This happens only if relevant information is removed by the preprocessing procedure. A way to investigate this is to check if the Extented Multiplicative Signal Correction (EMSC) parameters contain valuable information (see ref. 1) 2. I have a question about how the EMSC was performed: Was the same model used for the training set and the test set?

1 A. Kohler, C. Kirschner, A. Oust and H. Martens, Extended multiplicative signal correction as a tool for separation and characterisation of physical and chemical information in Fourier transform infrared microscopy images of cryo-sections of beef loin, *Appl. Spectrosc.*, 2005, **59**(6), 707–716.

Martin Isabelle replied: Yes, as in response to Matthew J. Baker's comment, if we do not take care when using pre-processing procedures such as EMSC, we can run the risk of removing important tissue and pathology information. Ultimately the lessons we have learnt from this first part of the project will enable us to anticipate or compensate ahead of future measurements, what factors (instrument, environment and user) we need to calibrate or control better so that we do not have to perform some of the post-processing correction measures that we documented in this manuscript.

For our purposes to remove the fluorescence baseline and account for scaling and artefacts, a representative mean tissue spectrum for each instrument spectral dataset was calculated as the reference spectrum, a non-tissue measured instrument spectrum for each of the instrument systems (measuring without the CaF2 substrate or tissue sample present) was used as the interference spectrum. This modified function removes baseline fluorescence arising from any variance in the tissue fluorophores and instrument optics and normalises the dataset to reduce the effect of total signal variation (*e.g.* due to variable focusing and the thickness of the sample).

In our data analysis, the EMSC is applied to both the training and test sets, however the reference spectrum is only calculated for the training set data and so the EMSC applied to the training and test sets uses the tissue mean spectrum reference spectrum from the training set, this way the test data is truly independent). However as stated in response to a point raised by Matthew J. Baker about the worry of removing important tissue fluorophore information when removing baseline fluorescence and the risk of removing important tissue biomarker information when using EMSC, we are continuing to investigate and assess what parameters we use as the reference and interference spectra to minimize the risk of over processing and removal of important tissue fluorophore information, but to still allow us to minimize inter-system optical instrument-specific fluorescence baseline contribution variation from our datasets. We have attempted to standardise the choice, manufacture and testing precision of the optics used in the systems. Usually we obtained the same optics coatings (inc. objectives) when ordered from the same suppliers and ordered at the same. However this is very difficult to control and we have to address the realistic situation where we will not be able to do this on a wider scale in other research groups which means that realistically we will need to accommodate for these variations by some degree of spectral correction during data processing.

Max Diem remarked: Have you done instrument comparison by correlation of images?

Martin Isabelle answered: Although this is not documented or investigated in this manuscript, image and resolution comparisons were performed on all three instruments prior to the start of the tissue measurements. White light image quality and Raman sampling resolution were tested on all three instruments using a set of samples (*e.g.* USAF target, biological samples). The spatial correlation between the subsequently collected white light images and Raman maps was optimised on each instrument and is typically within 1 to 2 μm.

Samir El-Mashtoly continued discussion of the introductory lecture by Max Diem: Most researchers focus on substrate calibration but the majority of us ignore some information about the patient, *e.g.* oncogenic mutations which are important for therapy.

Did you consider mutation in genes or receptors such as the EGFR in your lung study?

Max Diem replied: Correlation against patient metadata, such as age, gender, smoking and overexpression of certain cancer markers is presently performed at Cireca. However, the mutation of the EGFR has not yet been investigated.

Royston Goodacre continued discussion of the paper by Martin Isabelle: This comment is directed at Martin. One of the problems with referring to single sensitivities and specificities is that one does not know the true variance. Max and Rohit included confidence intervals for these (and other values; *e.g.* AUROC), and this is a much better way of approximating the true answer (*i.e.*, to try and approximate the central limit theory).

Martin Isabelle responded: Yes I agree, although sensitivities and specificities are often given without confidence intervals as a measure of model classification performance in vibrational spectroscopy. However, I agree in principle that the use of confidence intervals and AUROC are important to include where possible.

It is important to mention that this is a preliminary study. We have emphasised several times in the manuscript that only a small number of samples were available at this time as we are only half-way through the project and tissue collection is ongoing. The use of confidence intervals (and approximation of the central limit theory or true answer) would be more informative when there are larger sample sizes per pathology group (*i.e.* more representative variance of the subtypes in the groups *etc.*). We prefer to use "leave sample out" cross-validation rather than "leave spectra out" cross validation as we feel it is more representative of a true validation set, where a set of samples are completely excluded until prediction. In a "leave samples out" scenario, confidence intervals would require us to set aside a number of samples simultaneously, and while this may be the ideal situation it is of questionable benefit for small datasets such as the one presented in our paper. It is important, of course, that conclusions drawn from the analysis of small datasets are carefully moderated in view of the limitations. Once this larger sample collection is complete, we plan to estimate confidence intervals and AUROC as suggested since it will enables us to suggest the best decision threshold for clinical purposes; higher sensitivity corresponds to a high negative predictive value and the ideal property of a "rule-out" test while high specificity corresponds to high positive predictive value and is the ideal property of a "rule-in" test.

Achim Kohler continued discussion of the paper by Sergei Kazarian: As I understand, the chromatic aberration is caused due to the refractive index change that occurs when the infrared radiation enters the window from air. The result is that a sample position cannot be in focus for every wavelength acquired. For sample positions that are out of focus, the scattered light is being directed into the wrong pixel on the detector. This introduces a blurring. You show that you can correct for this effect in an elegant way, by using a lens which corrects for the chromatic aberration. That is a great achievement. Still, I have problems to understand how this can strongly reduce dips in your spectrum which arise from fluctuations of the real refractive index due to absorption resonances. This was especially visible close to the Amide I band. Do you have an explanation for this? Could the reduction of the dip simply be an effect of the reduction of the blurring. If the blurring is reduced and the resolution is increased, edge positions inside the mask may not be contaminated from highly scattering positions that are outside the region that is selected by the mask.

Sergei Kazarian replied: You are correct that the increased spatial resolution will reduce the number of pixels containing the scattering artefact. However, we believe that its intensity is reduced due to the increased numerical aperture of the system with the lens, collecting more of the scattered light, in a similar way to the paper by Dazzi *et al.*[1] However, while the mask may have had some effect, our other publications in which a mask was not used showed a similar reduction in scattering.[2,3] Therefore the mask and increase in spatial resolution cannot account for the observed reduction in scattering alone.

1 A. Dazzi, A. Deniset-Besseau and P. Lasch, *Analyst*, 2013, **138**, 4191–4201.
2 K. L. A. Chan and S. G. Kazarian, *Anal. Chem.*, 2013, **85**, 1029–1036.

3 A. Sroka-Bartnicka, J. A. Kimber, L. Borkowski, M. Pawlowska, I. Polkowska, G. Kalisz, A. Belcarz, K. Jozwiak, G. Ginalska and S. G. Kazarian, *Anal. Bioanal. Chem.*, 2015, **407**(25), 7775–7785.

Parvez Haris opened a general discussion of the paper by Klaus Gerwert: Due to the "black box" approach being widely employed in the diagnosis of cancer, using Infrared and Raman spectroscopy, valuable information regarding spectral changes that could be related to molecular alterations are being ignored. This is a great pity, since each spectrum contains valuable structural and environmental information that can be explored to obtain a better understanding of the changes associated with the development of cancer. Previous studies have shown that bands in the FTIR spectra of proteins can change significantly due to differences in hydration and environmental factors (see ref. 1–4). Do differences in the stiffness/density of the tissues contribute to spectral differences that could partially explain the discrimination between cancerous and non-cancerous tissues using FTIR and Raman spectroscopy? It is known that the stiffness of cancerous tissues differs from that of normal tissues.[5] Therefore, correlating tissue stiffness with spectral changes may yield important information that could help advance the field of vibrational spectroscopy in biomedical research. Differences in tissue stiffness have been utilised for cancer diagnosis using ultrasound elastography.[5] Is it not possible that changes in tissue/stiffness will alter the vibration of molecules and these alterations will be manifested in spectral changes? Related to this is the hydration of the tissues. Cancer and non-cancerous tissues are known to differ in their hydration status and this may also be used to identify factors that may explain spectral differences that help in the discrimination between cancerous and non-cancerous tissues. Therefore, obtaining a good understanding of the sample hydration status, stiffness and also molecular composition data will be valuable for identifying the most important factors that can help discriminate between normal and cancer tissues. It would be helpful to address these issues so that cancer discrimination using FTIR/Raman spectroscopy goes beyond the "black box" approach.

1 M. Jackson, P. I. Haris, and D. Chapman, Fourier transform infrared spectroscopic studies of calcium-binding proteins, *Biochemistry*, 1991, **30**(40), 9681–9686.
2 P. I. Haris, M. Coke, and D. Chapman, Fourier transform infrared spectroscopic investigation of rhodopsin structure and its comparison with bacteriorhodopsin, *Biochimica et Biophysica Acta (BBA) - Protein Structure and Molecular Enzymology*, 1989, **995**(2), 160–167.
3 E. L. Karjalainen, and A. Barth, Vibrational coupling between helices influences the amide I infrared absorption of proteins: Application to bacteriorhodopsin and rhodopsin, *J. Phys. Chem. B*, 2012, **116**(15), 4448–4456.
4 M. van de Weert, P. I. Haris, W. E. Hennink, and D. J. Crommelin, Fourier transform infrared spectrometric analysis of protein conformation: effect of sampling method and stress factors, *Anal. Biochem.*, 2001, **297**(2), 160–169.
5 G. Salomon, J. Köllerman, I. Thederan, F. K. Chun, L. Budäus, T. Schlomm, H. Isbarn, H. Heinzer, H. Huland, and M. Graefen, Evaluation of prostate cancer detection with ultrasound real-time elastography: a comparison with step section pathological analysis after radical prostatectomy, *Eur. Urol.*, 2008, **54**(6), 1354–1362.

Klaus Gerwert responded: Differences in the density of the tissues contribute to spectral differences. If you compare the raw spectral data of cancer tissue with non-cancer tissue you detect differences in the protein concentration which could be related to higher density but also the higher proliferation rate. The effect of stiffness will have no specific effect on the spectra and will hardly be observable.

The influence of hydration is minimized because the samples are measured under dry air conditions which leads to low hydration states of the tissue. This is necessary to minimize the influence of water in the IR spectra. However, density, stiffness and hydration are unspecific. These effects might contribute to spectral differences between cancer and non-cancer tissue however, these are not specific. But today in clinics, detailed molecular information is needed for targeted therapy. This can be obtained by combination with molecular resolving techniques as outlined before.

Max Diem commented: I don't know how the macroscopic stiffness variations between normal and cancerous tissue affects the observed vibrational spectra of the processed and thinly sliced tissue biopsies used here. In addition, the samples are often gross mixtures of tissue types (connective , glandular, vascular *etc.*), and I don't believe that all these tissue types are equally affected in terms of their mechanical properties.

But I agree that a biochemical interpretation of observed changes in tissue spectra is useful; in my talk, I emphasized the spectral changes due to necrosis and fibrosis. I disagree with the comments by Prof. Gerwert, who stated that he thinks such changes cannot be assigned. It is, of course, impossible to detect single molecules in SHP, but classes of peptide structures, such as aggregated β-sheets, or triple helical (collagen) structures can be detected, even with the naked eye.

Parvez Haris added: It is inconceivable that differences in stiffness (and hydration) variations between normal and cancerous tissue will not result in any changes in the vibrational spectra of samples. Stiffness and hydration variations between cancerous and non-cancerous tissues may be highly specific and can be combined with other markers for a better understanding of the disease and its diagnosis. The impact of these differences is likely to be carried forward when the tissues are cut into thin sections. These differences may manifest in spectral changes. I am confident that in due course it would be possible to correlate stiffness/hydration maps of tissue sections with corresponding vibrational spectroscopic maps. By combining mechanical and chemical parameters it may be possible to improve the sub-classification, grading and staging of cancers. This multi-pronged strategy needs to be explored in the future.

Max Diem asked: Did you compare cancer in glandular epithelial tissue and normal glandular tissue, or just cancer and normal tissue (which could be normal connective tissue)?

Klaus Gerwert answered: We compared different tissue types already in a previous paper.[1] The presented classifier in this paper differentiates between cancer, mucosa and crypts as glandular epithelial tissue, and other normal tissue types such as muscle or connective tissue. The work presented here analyses the tumor grading of adenocarcinomas. The different cancer grades are compared with each other, but not against healthy normal tissue.

1 A. Kallenbach-Thieltges, F. Großerüschkamp, A. Mosig, M. Diem, A. Tannapfel and K. Gerwert, *J. Biophotonics*, 2013, **6**(1), 88–100.

Peter Lasch remarked: Klaus' study nicely illustrates that IR-SHP in combination with multivariate pattern recognition approaches ("fingerprint approach") can be used not only for cancer detection, but also for cancer grading. I entirely agree with Klaus' statement that only the analysis of IR fingerprints will be successful. Nevertheless, I would like to add that we should not forget that IR spectroscopy is at the same time a sensitive analytical technique which provides structural and compositional information of the samples studied. Despite the very high degree of overlap it is sometimes possible to assign IR marker bands in biomedical IR spectra without using isotope labelling techniques. For example, marker bands from collagen and DNA are highly characteristic and sometimes allow unequivocal assignment in the IR spectra of biomedical samples. I believe that the two examples below (Fig. 4 and Fig. 5) nicely illustrate that band assignment is indeed possible without the use of isotopic labels.

Klaus Gerwert responded: These are nice, but rare examples in which a highly concentrated protein such as collagen in skin can be observed. But in general single proteins cannot be identified in tissue. Today spectral histopathology has to

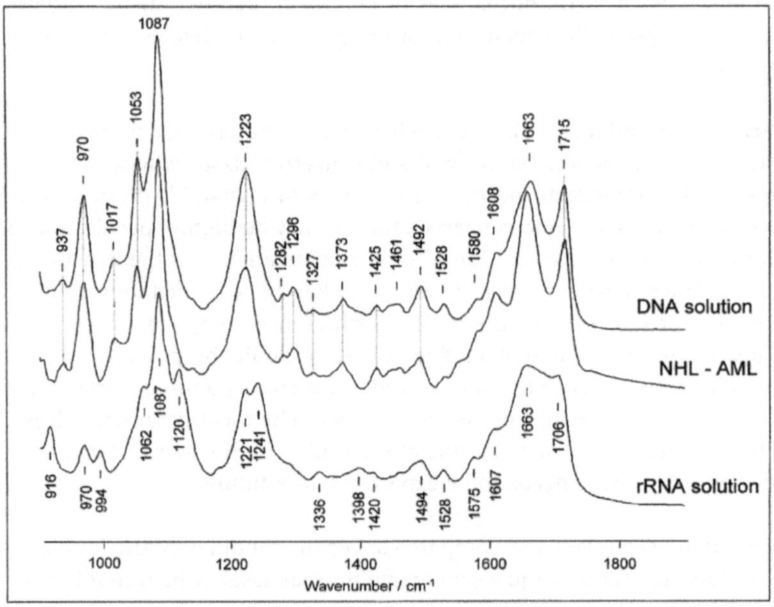

Fig. 4 Buffer-corrected IR absorbance spectra (min-max normalized) from DNA and rRNA solutions and a double difference spectrum "NHL minus AML". NHL: difference spectrum between a spectrum from non-Hodgkin lymphoma cells measured in aqueous solution and a buffer spectrum (optical path length 10 μm). AML: difference spectrum between a spectrum from acute myeloic leukemia cells measured in aqueous solution and a buffer spectrum (optical path length 10 μm). NHL-AML: weighted double difference spectrum of the NHL and the AML spectra. Note the striking similarity between this and the IR spectrum of the DNA solution. From a biological point of view, the DNA-like double difference spectrum makes perfect sense: NHL tumor cells usually have larger nuclei than AML cells and the DNA-to-protein ratio is also higher in NHL cells. Source: Heinz Fabian (not published, mid-1990s). Measurements were carried out during Heinz's visit at the Institute for Biodiagnostics (IBD) of the National Research Council (Canada) in Winnipeg.

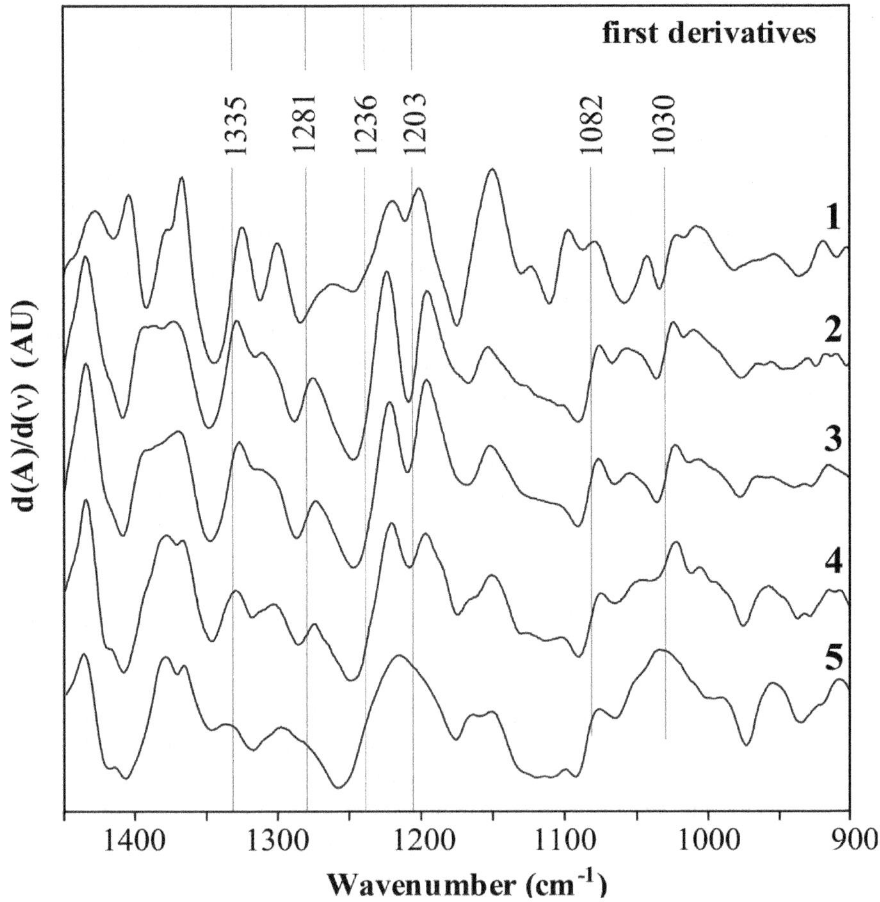

Fig. 5 Comparison of FTIR spectra (first derivatives, vector-normalized) obtained from the two structural proteins elastin (1) and collagen type I (2) and FTIR microspectra from the human skin: Stratum reticulare (3), Stratum papillare (4) and amelanotic melanoma (5). Collagen type I exhibits characteristic absorption features in the amide III region at 1203, 1236, 1281, 1335 cm^{-1} and around 1030 and 1082 cm^{-1}. Note the striking similarity of the IR spectrum from isolated collagen I (2) and from Stratum reticulare (3). Stratum papillare contains more cells so that the spectral features of collagen are less evident in spectrum (3), cf. the characteristic M-shape feature in spectra (2–4) around 1205 cm^{-1} which indicates the presence of collagen in the first derivative spectra. The spectra suggest also that collagen cannot be detected in the tumor region. (Taken from Lasch, 1999 (page 65) http://www.ftir-workshop.org/publications/diss_lasch.pdf).

compete in diagnostics with molecular resolving techniques like proteome analysis and Next Generation Sequencing (NGS). NGS for example can determine the Ras status; wild type (WT) or oncogenic mutations. Such information is directly used in targeted therapy against the EGF receptor upstream of Ras. Only for WT Ras, an expensive antibody therapy against the EGF-receptor is applied. Such detailed information can never be directly extracted out of the tissue spectra and we do have to be careful with over interpretation of the tissue spectra. Only in combination with these molecular resolving techniques we can provide detailed information in precision medicine otherwise we will not be taken seriously by the clinicians.

Nick Stone said: Pathologists have 50–60% agreement with each other for many early cancers. How do you get them to make a decision that agrees for each sample? Can you get pathologists to give precise calibration for each pixel?

Klaus Gerwert replied: To overcome the problem with different annotations by different pathologists in our studies we always add additional immunochemistry staining and go back then with these results if there is a conflict between the pathologists and usually end up with an agreement after intense discussion. Pathologists cannot give precise calibration for each pixel. For a pathologist the annotation of a larger tissue region is more important than the annotation of single pixel in the order of a cell size. Furthermore the error of the calibration is in the order of the pixel size and so you cannot be sure that the annotated pixel overlaps with the spectral image pixel.

Alexander Apolonskiy opened discussion of the paper by Niels Kröger-Lui: Regarding the noise analysis in your measurement, to properly analyse it, you need to analyse the contributions from the laser, the detector and the measuring system. What are the contributions in terms of noise from each of these?

Niels Kröger-Lui replied: We fully agree in that a further analysis will require us to identify the individual noise contributions. However, as a first step it was important for us to quantify the overall noise of the system – and we were happy to see that we are not so far away from standard FT-IR even at the much shorter times used for QCL-based MIR microspectroscopy. As far as the individual contributions are concerned, we have shown that, for our microscope, noise is clearly Poissonian (see Fig. 8 in the paper). Hence, an increase in laser power and/or optical throughput of the measuring system will increase the signal to noise ratio.

Christopher Phillips opened a general discussion of the paper by Michael J. Pilling: Suppose one were to first do a training run to choose the best wavelengths, then repeat the training run, but with a larger data averaging time, and therefore a better signal to noise ratio (SNR), does the software necessarily choose the same set of wavelengths both times?

Michael Pilling answered: The software used to acquire the infrared data does not have the functionality to co-add scans directly, and we cannot control how many frames the micro-bolometer collects at each discrete frequency. We had considered acquiring a chemical image of a core, then repeating this, and averaging the discrete frequency images obtained in each case. However we have found that the noise we observe in the QCL spectra is correlated rather than random, therefore co-adding scans does not serve to improve the signal to noise ratio. Although we could not directly compare the wavelength selection for spectra of differing SNR, we did perform repeats of the wavenumber selection using random subsets of spectra. We found that the first 14 wavenumbers of those we selected as key features, consistently appeared in the top 16 ranked wavenumbers.

Zachary D. Schultz said: The bands chosen by your algorithm and presented in the paper (Table 5) appear to cluster around certain frequencies. Can you comment on how more evenly distributed bands, for example fixed intervals across the spectral window, might affect the results observed?

Michael Pilling responded: The QCL infrared imaging system used in the study utilises four independent laser modules, enabling complete access to the fingerprint region between 900–1800 cm^{-1}. Each QCL covers a different spectral range and there is some degree of overlap between the lasers. Tuning a given QCL module is relatively fast and if all the discrete frequencies required are contained within the spectral range of a single QCL, then they can all be accessed rapidly. Switching between different QCL modules is more time consuming and in our experience can take several seconds. Therefore if the desired discrete frequencies span all four modules then this would increase acquisition time. However it is also important to remember that the bands were not chosen arbitrarily, and that the wavenumbers were chosen using the random forest algorithm based on which we enabled the greatest discrimination from normal and cancerous tissue. Similar clustering has been reported by Lloyd[1] by using a variable importance projection for the detection of dysplasia in human colon polyp sections.

1 G. R. Lloyd and N. Stone, Method for Identification of Spectral Targets in Discrete Frequency Infrared Spectroscopy for Clinical Diagnostics, *Appl. Spectrosc.*, 2015, **69**, 1066–1073.

Peter Gardner continued the discussion of Klaus Gerwert's paper: In your presentation you stressed the importance of determining cancer grading, rather than just cancer *vs.* non-cancer. I completely agree with this but in doing this we are trying to correlate the spectral changes (*i.e.* changes in biochemistry) with changes in tissue architecture. However the changes in tissue architecture are part of a continuum and the boundaries between grades have been drawn somewhat arbitrarily. It is quite possible that the biochemical changes do not occur at the same point in the progression of the disease as the architectural changes. In the future therefore, rather than trying to correlate one with the other you could envisage a new grading system with boundaries based on the chemical changes observed. Is this something that you have considered?

Klaus Gerwert replied: Yes, this is exactly what we have in mind. In the future a disease should be classified by the underlying molecular events and not by the morphological changes alone, as has been done for about 100 years in pathology. Especially, vibrational spectroscopy could contribute to such a new classification of diseases.

Christopher Sammon continued the discussion of the paper by Michael Pilling,: In your paper you describe improving the time resolution of collecting your images, but you describe improvements from 5 : 30 min to 3 : 30 min, these small changes seem somewhat arbitrary. What is the ultimate time resolution target for this approach?

Followed up with; as mentioned in earlier presentations the goal of distinguishing cancerous *versus* non-cancerous tissue is considered fairly trivial, how suitable is the QCL methodology described to the determination of cancer grade? How far could the envelope of data point reduction be pushed whilst still facilitating grading, what should our expectations be for this kind of approach for rapid cancer identification? Is QCL the solution? (Given sufficient time and resources to fully explore its potential).

Michael Pilling answered: In the paper we have tried to consider what time-scales would be acceptable in the clinic. Whereas a time saving of 2 min per core may seem trivial, extrapolating this to acquiring a full TMA (which could consist of up to 200 patients) results in significantly higher throughput. The optimal situation would be to acquire the infrared data as quickly as possible, however it is important to consider the trade off between sensitivity, specificity and throughput. A further possibility is the use of a low magnification objective which allows a full core to be imaged as a single tile in under a minute. We would envisage that an appropriate use of QCL technology in its current state of maturity would be for cancer screening. The majority of tissue biopsies undergoing pathological assessment are normal, and each one needs to be manually examined by a pathologist to rule out cancer. QCL imaging could be used to identify which tissue biopsies are normal and these would then not need further pathological review. Only in cases where there is any risk of the tissue being abnormal, or highly suspect would there be the need to have pathological assessment.

The determination of cancer grade is significantly more challenging than differentiating between normal and cancerous tissue. Differences in spectra for various cancer grades are subtle, which necessitates good signal to noise ratios and will presumably require either a continuous spectrum or an increased number of discrete frequencies. Even if the full spectral range is required than QCL imaging operating under high spatial resolution has the potential to offer considerable time savings compared to FTIR. Typical FTIR acquisition times at 1.1 μm pixel size, for a single 1 mm diameter core are between 4–6 h. Using QCL imaging the same core can be acquired at 1.3 μm pixel size in approximately 30 min. Another possibility is to use the lower magnification optics providing a pixel size of 4.16 μm which would enable a fourfold reduction in acquisition times. Since the majority of work to date has focused on speed rather than diagnostic accuracy further work would need to be performed in assessing the potential of this new technology.

Sergei Kazarian asked: Would it be possible to achieve a quantitative study of different severity ranges of cancerous tissues (disease stage) using this approach?

Michael Pilling replied: Numerous studies using FTIR microscopy have demonstrated that it is indeed possible to identify disease stage from an infrared spectrum. Spectral differences between cancer stages are more subtle than those between normal tissue and cancerous tissue. Therefore in order to successfully differentiate between different stages of cancer it is likely that a larger number of discrete frequencies would be required. In principle this would be possible but a more detailed study investigating the feasibility of disease staging and the impact on throughput and classification accuracy is required.

Achim Kohler commented: For wavelength selection only samples from two patients have been used. Thus, it is difficult to validate the selected wavelengths. One possibility to validate the selection of wavelengths is a cross-model validation approach. See for example ref. 1.

1 I. Karaman, E. M. Qannari, H. Martens, M. S. Hedemann, K. E. B. Knudsen, A. Kohler, Comparison of Sparse and Jack-knife partial least squares regression methods for variable selection, *Chemom. Intell. Lab. Syst.*, 2013, **122**, 65–77.

Michael Pilling responded: For this proof of concept study wavelength selection was achieved using just two patients, and we agree that this is not ideal. If the study was to be repeated using a larger number of patients we agree that this would be a better approach to robustly validate the wavenumber selection as suggested.

Klaus Gerwert asked: How accurate is the system in the 3–5 min time frame?

Michael Pilling replied: We have investigated the reproducibility of the system by recording a background using the full spectral range 1000–1800 cm^{-1} with no sample on the stage. We have then acquired spectra immediately after this, again with no sample present. Using this approach the time between the background and spectrum for each wavenumber is approximately 6 min. Over this time period the absorbance peak to peak noise is approximately 0.005, and over a time period of 36 min this increases to approximately 0.015. We have also found that the variability in the wavenumber positional stability is imperceptible over a 3–5 min time frame. However there is sometimes some noticeable variability especially at the high wavenumber end, between day to day measurements.

Hugh Barr continued the discussion of the paper by Klaus Gerwert: In clinical medicine we find that we need to analyse historical tissue to identify the expression of various molecules that may be responsive to new chemotherapy agents, that have become available. In addition some biopsy samples from very vulnerable areas may be small. Histopathology is destructive so the use of the available tissue is very precious. Using spectroscopy are you able to target specific molecules to identify whether the cancer would be responsive to drugs such as Herceptin?

Klaus Gerwert answered: Yes, by combining it with molecular resolving techniques like proteome-analysis and next generation sequencing we can quantify the response to specific drugs in tissues by FTIR imaging. However, using cancer cell lines we can monitor the drug response directly using Raman imaging. We have published this Raman approach already for kinase inhibitors of EGF receptors and also antibodies.

Royston Goodacre addressed everyone: From Klaus we've heard about proteomics and NGS (next generation sequencing), but it's worth reminding people that cancer is a disease of *metabolism* and so it makes sense to measure these small molecules (*e.g.* using mass spectrometry). An important one in non-oxidative phosphorylation is lactic acid (part of the Warburg effect). Has anyone observed pH changes in cancer grading or tumor *vs.* normal? Are you simply measuring local pH changes with vibrational spectroscopy (one could do that a lot cheaper)?

Rohit Bhargava replied: Indeed. This is a very important point. Unfortunately, most of the tissue studies done with IR imaging are on fixed and embedded tissue that has been washed numerous times with different solvents. Hence, metabolites may not be available. Most certainly, fresh, snap frozen or live cells will provide the information. While it does not seem that we are measuring pH at this point, its measurement will likely be important when the appropriate samples are analyzed.

Nick Stone responded: When studying the composition of breast calcifications associated with breast disease with FTIR,[1] we noted that the carbonate ion substitution in the calcium hydroxyapaptite reduced with advancing malignancy. We believe this was due to the lower pH in malignant tissues increasing the carbonate ion substitution. So by using the surrogate marker of carbonate ion signals in the calcifications we believe we were in effect probing the local pH of the surrounding tissue fluid.

1 R. Baker, K. D. Rogers, N. Shepherd, N. Stone, New relationships between breast micro-calcifications and cancer, *Br. J. Cancer*, 2010, **103**, 1034–1039.

Klaus Gerwert answered: With vibrational spectroscopy we measure much more than local pH changes. By combining it with molecular resolving techniques we obtain a deep insight in the molecular changes caused by the cancer. These molecular differences are important for the therapeutic decision made by the clinician. For example an oncogenic Ras Mutation determines if an antibody therapy will be successful or not. For precise personalised therapy we need much more than the determination of a pH change, but molecular details for successful targeted therapy.

Bayden Wood continued the discussion of Michael Pilling's paper: It would take approximately 3 min to record a Focal Plane Array (FPA) image of a tissue section of 1 cm × 1 cm using 16 cm^{-1} spectral resolution with 4 co-added interferograms and using a 4× magnification IR objective. With that image you would be able to pre-process the spectra to correct for Mie scattering (or resonance Mie, R-Mie, scattering), vector normalise the spectra and take the second-derivative. The time frame over the same area using a QCL with say 16 discrete points would be even slower at ~10 mins but you don't get any opportunity to pre-process and correct the spectra for Mie or R-Mie scattering using the QCL approach and you miss out on the high wavenumber end. So I don't really see a time advantage in using QCL and a number of disadvantages in terms of pre-processing and limited spectral range. So what are the advantages of QCL imaging over FPA imaging?

Michael Pilling responded: We agree that using discrete frequency imaging limits the types of pre-processing available since most multivariate methods aren't suitable due to the limited number of data points in the spectra. For this reason we chose to work with paraffinised tissue rather than dewaxed, to reduce Mie scattering since this could not be corrected for. The acquired data only had minimal pre-processing which was a quality check based on the amide I intensity, baseline correction at each end of the spectrum followed by normalisation to the amide I band. Utilising a ×4 objective on a FTIR microscopy system does indeed enable large areas of tissue to be imaged, but at the price of significantly poorer spatial resolution. Some pathologists will argue that the chemical images obtained using IR microscopy, have poor spatial resolution that are acquired on a system which is considerably more expensive than they currently have. High spatial resolution is important for the early diagnosis of disease, which often occurs at the basement membrane (in prostate and breast cancer). While operating at 16 cm^{-1} the spectral resolution is perfectly fine for basic histology, more investigation is required on whether this is suitable for disease diagnosis which would require the differentiation of relatively subtle spectral features. We feel that QCL has the potential to augment FTIR microscopy since it enables chemical images with high spatial resolution to be rapidly acquired.

Christopher Phillips addressed Michael Pilling and Rohit Bhargava: We find that informed choices can be better than unsupervised software protocols when choosing the appropriate wavelengths for sparse datasets like these. This way we can *e.g.* use the known disease pathology, *e.g.* to isolate known biomarkers like the phosphate-to-amide concentration ratio. Also, one can implement bespoke background subtraction and ratiometric methods that make the measurement more robust and reproducible.

Rohit Bhargava replied: Yes, indeed. This is very true. There are two approaches to spectral analysis. The first is "whole spectrum" methods that essentially select large contiguous spectral regions and were the method of choice in most approaches. In 2005, the approach of selecting specific, individual quantities from the spectrum (termed metrics by the authors to differentiate from just spectral features) was introduced.[1] This, in turn, sparked the development of filters for imaging in a so called discrete frequency fashion.[2] The current use of QCL and other laser-based instruments is a major advance and completely complements the wavelength selection approach.

1 R. Bhargava, D. C. Fernandez, S. M. Hewitt and I. W. Levin, High throughput assessment of cells and tissues: Bayesian classification of spectral metrics from infrared vibrational spectroscopic imaging data, *Biochim. Biophys. Acta, Biomembr.*, 2006, **1758**(7), 830–845.
2 A. K. Kodali, M. Schulmerich, J. Ip, G. Yen, B. T. Cunningham and R. Bhargava, Narrow-band midinfrared reflectance filters using guided mode resonance, *Anal. Chem.*, 2010, **82**(13), 5697–5706.

Michael Pilling responded: We found that the selected wavenumbers broadly agreed with known disease pathology, with wavenumbers associated with glycogen, phosphate, amide I and amide II being identified. The wavenumbers used were not chosen arbitrarily, they are the key features identified by random forests. Therefore we are fairly confident that the wavenumbers selected are appropriate for differentiation between normal and cancerous tissue.

Nick Stone addressed Wolfgang Petrich, Niels Kröger-Lui and Michael Pilling: In relation to QCL coherence effects, in the spectra you see ripples in the data. Are these artefacts the same for every sample, do they vary, should we be concerned about these and how do you overcome them?

Wolfgang Petrich responded: As you know one of the key differences between a broad-band thermal source and a narrow-band laser source (such as a quantum cascade laser) is given by the coherence properties of the emitted radiation. What you may call "ripples" or "artefacts" are in fact the speckle pattern arising from various opportunities for position-dependent phase shifts of the electromagnetic waves. In a histopathological application of QCL-based mid-IR microscopy it is crucial to understand that in particular the tissue sample itself will give rise to local phase shifts, which results in a sample and position-dependent speckle pattern, thus deteriorating the desired, "pure" absorbance information of the image. I believe that any reasonable QCL-based microscopy will have to consider this effect in detail prior to building the microscope and, yes, there are ways to mitigate the impact of the high degree of coherence. We have implemented one option in our microscope in Heidelberg and, personally, I think understanding this was crucial to our image quality. From my point of view, we were able to achieve excellent image quality even for a single frame, *i.e.* a raw mid-IR image taken within 20 ms and displayed without any averaging or smoothing or such.

Niels Kröger-Lui replied: When using QCLs for infrared microscopy, coherence effects can deteriorate the chemical information. However, there are ways to overcome this. So-called fringes, caused by interference at the optical elements in the beam path can be mitigated by using anti-reflective coatings. It is important to keep in mind that AR coatings for a broad spectral range (*e.g.* the full fingerprint region), will not yield perfect results. Also, since these fringes are characteristic for the measurement system, they can be addressed with computational methods. Using AR-coated optics and/or computational methods for fringe reduction are strong methods to remove coherence effects that are characteristic for the measurement system, which do not vary with the sample. The remaining problems to solve are coherence effects which are caused by interference at the sample itself (so-called speckles). These effects vary from sample to sample and are much harder to correct. For samples of known dimensions, coherence effects contain a lot of useful information. However, when investigating a sample with unknown shape (*e.g.* a thin tissue slice), extracting pure chemical information from a mix of chemical and physical information can be nearly impossible. For spectral histopathology this is a huge problem, since tissue classification is usually based on chemical information (which is obtainable using FTIR-techniques). Coherence effects, which are typical for QCL-based infrared microscopes will cause distortions of the spectra and may severely affect classification results.

One option to mitigate this problem is to perform point mapping of the sample. Since our custom built QCL-based infrared microscope is targeting the fast chemical imaging of thin tissue slices, we had to find a solution to remove fringes and speckles. Hence, a point mapping setup was not our first choice because of the speed advantage of an imaging setup. One possible solution for removing coherence effects in an imaging setup was implemented in our custom-built setup. We believe that we managed to mitigate coherence effects to such a degree that the dominating source of error is Poissonian/random noise (as indicated by the results in our paper). This means that spectra acquired with our custom-built QCL-based microscope in Heidelberg compare very well with FTIR spectra and can be used for spectral histopathology. We also filed a patent for our solution, so technical details will soon be public knowledge.

Michael Pilling answered: We have observed these artefacts in spectra obtained when measuring different samples, and found that the intensity of the ripples is quite variable. In the absence of a sample the artefacts are still present and are observed to increase in intensity as a function of time from when the background was collected. The influence of the coherence effects depends on how intense they are relative to the intensity of the absorption peaks within the spectra. Provided that the absorption peaks are sufficiently intense, then the effect of the artefacts on classification performance is likely to be limited. Removal of coherence effects experimentally is a potential solution, which has been demonstrated by Kole using a rotating diffuser plate consisting of a scratched salt disk.[1]

1 M. R. Kole, R. K. Reddy, M. V. Schulmerich, M. K. Gelber and R. Bhargava, Discrete frequency infrared microspectroscopy and imaging with a tunable quantum cascade laser, *Anal. Chem.*, 2012, **84**(23), 10366–10372.

Zhe Zhang continued this general discussion of Niels Kröger-Lui's and Michael Pilling's papers: Q1: As mentioned before, the so called noise is due to the correlated laser. Thus, the so called noise is not random noise. Is there a method to remove it, especially for discrete wavenumbers?

Q2: In Niels' paper, the best S/N ratio is achievable using a sample thickness of 20 μm, and in both papers which utilize the QCL microscope, the samples you used were above 12 μm. However, it is more common to cut biopsy tissue or TMAs at about 4 μm thick. Will the S/N ratio be sufficient to be able to measure this kind of sample? What is the influence of the noise on the classification of these thinner samples?

Q3: The optimum 20 μm thickness was for a homogeneous sample. Is this optimum different for a heterogeneous sample for example tissue or cell?

Niels Kröger-Lui replied: Q1: We were aware of this before we started to design our custom built QCL-based infrared microscope and implemented one possible solution for the problem. From the results presented in this paper, we conclude that the measurement error of our system is indeed dominated by Poissonian noise. These findings translate directly to imaging at discrete wavenumbers. In fact, by restricting the analysis to one or a few discrete wavenumbers, generally more integration time will be available to further minimize noise.

Q2: In previous publications, we investigated tissue samples of 8 μm thickness. For example, we measured in the carbohydrate spectral fingerprint region (1030–1090 cm^{-1}), where the average absorbance of our sample (colonic mucosa) was around 0.09–0.18 a.u. for colonic epithelium. We were able to subdivide colonic mucosa into its main tissue components using a random forest classifier. Furthermore, when using only two wavelengths, we were able to highlight the distribution of mucins, with an integration time of less than 5 s. We did not investigate whether we would have been able to achieve similar results with thinner (or thicker) slices. I am very confident that our system would be capable of imaging 4 μm thick tissue samples at the amide I and II bands, as the absorbance values are much higher than those in the carbohydrate spectral fingerprint region. Also, since the measurement error of our system is dominated by Poissonian noise, we could increase integration time in order to minimize noise to an acceptable level to image 4 µm thick slices, even for spectroscopy in spectral regions with low average absorbance.

From a conceptual standpoint, noise will, on the one hand affect classification as soon as the noise level becomes comparable to the spectral differences between the classes. On the other hand, thinner samples might reveal more information on morphology, as you will not only measure the absorbance of the projection of several cell layers but the absorbance of single cell layers (see next answer).

Q3: For a heterogeneous sample like a thin tissue slice you should be aware that when investigating slices of 20 µm thickness, one will typically obtain absorbance information from the transmission through multiple cell layers. Even though signal to noise ratio might be better compared to thin slices, one loses information about morphology. I would speculate that, for classification, thin slices would yield better results despite the lower signal to noise. If needed, it may also be permissible for thinner slices to enlarge the integration time (if, like in the case of our microscope in Heidelberg, the measurement error is dominated by Poissonian noise) in order to obtain the best results.

Michael Pilling responded: Since the noise is correlated and not random it is not possible to reduce the noise by using longer acquisition times or an increased number of scans. Discrete wavenumber collection prevents the utilisation of most multivariate techniques for removing noise, therefore the noise would need to be removed experimentally. Kole *et al.* have demonstrated that coherent effects can be reduced by using a rotating diffuser plate consisting of a scratched calcium fluoride plate.[1]

Determining the signal to noise ratio when using discrete wavenumbers is problematic because there are insufficient data points to get a meaningful measure of the SNR. In a different study we have acquired data using the full spectral range for 5 µm tissue and have found that the correlated noise is almost imperceptible, which suggests that the SNR is sufficient for classification. Using thinner samples reduces the intensity of the absorption bands and therefore for a constant level of noise has the effect of reducing the signal to noise ratio. As samples become thinner then the SNR decreases which is likely to reduce the classification accuracy. The optimum thickness depends on having a sufficient path length through the sample, while ensuring that sufficient light reaches the infrared detector. Mie scattering at the edges of cells or tissue will further reduce the amount of light reaching the detector so will change the optimum thickness compared to a homogeneous sample.

1 M. R. Kole, R. K. Reddy, M. V. Schulmerich, M. K. Gelber and R. Bhargava, Discrete frequency infrared microspectroscopy and imaging with a tunable quantum cascade laser, *Anal. Chem.*, 2012, **84**(23), 10366–10372.

Alexander Apolonskiy continued the discussion of the paper by Niels Kröger-Lui: There are 2 remedies to avoid fringing, both by means of suppressing the interaction of light transmitted or reflected from the borders of a sample. One way is to use wedged substrates, another is to use an anti-reflective coating on all surfaces.

Wolfgang Petrich responded: I fully agree with this and, yes, we do use anti-reflection coatings *etc.* However, in the case of quantum cascade laser-based microscopy, the high degree of coherence of the infrared radiation causes interference patterns not only from the optics *etc.* but also, in a very fundamental way, from the sample itself. In the visible range, this phenomenon is referred to as speckle patterns, meaning nothing but stable local interference minima and maxima. Hence, the desired "absorbance" image is obscured by bright dots at seemingly random positions. This phenomenon is indeed vastly different from what people are used to when using broad band sources. Hence, intrinsically, our switching from a globar source to a quantum cascade laser implied that we have to mitigate the effect of fringing or speckle patterns for that part. We found a workable solution for this and, hence, our microscope was designed from scratch around the fundamental concept of mitigating the impact of the high degree of coherence (spatial as well as temporal, by the way) of the laser.

Niels Kröger-Lui answered: The use of wedged substrates will definitely remove fringes from the substrate. Also, anti-reflective coatings should always be used on all surfaces. In the special case of histopathology, it is important to note that interference does also originate from the sample itself and it is precisely this type of interference which contributes to fringing or, more precisely, speckle patterns.

Sergei Kazarian commented: This is an interesting demonstration of the application of QCL, although for the case of a fermentation reaction, it appears to be like using a sledgehammer to crack a nut, because the need for high signal to noise is not particularly substantiated here, and the approach of multiple reflection ATR using a conventional globar source would have worked equally well. You mention that the cuvette became leaky in your experiments and that ethanol vapour may escape too. It seems that CO_2 released during the fermentation causes an increase in the pressure within the cuvette that results in this leak. So, if you were able to collect CO_2 instead and measure the amount of released CO_2 by volume (in a capillary tube attached to the cuvette) then you would be able to correlate the volume obtained by imaging the CO_2 bubbles with the volume measured by such a simple approach. This would also allow you to measure CO_2 amount even when the bubbles become too large for imaging experiments.

Christopher Phillips asked: Concerning interference artefacts in the laser images, they can be removed by using a broader band illumination source. We find that bespoke narrow-band IR filters, used with a simple globar, can give similar (~mW) powers for spectral imaging, without interference problems. (See DOI: 10.1039/c5fd00150a and references therein).

Niels Kröger-Lui responded: I agree that interference artefacts can effectively be mitigated by going back to a broad band, large area illumination source. However, in the field of biomedical vibrational spectroscopy you should be able to spectrally resolve at least 4 cm^{-1} and you should concentrate the emitted infrared power into this bandwidth. It is precisely these benefits of narrow bandwidth and high spectral power density which triggered our efforts to use quantum cascade lasers. These properties of a QCL enable, for example, very fast discrete frequency imaging (with integration times in the millisecond range when using microbolometer array detectors), or performance of spectral histopathology using narrow-band spectral markers. Since, in addition, broad QCL tuning ranges are now becoming available, a change in wavelength does not require any hardware changes within a wavenumber span of, say, 200 cm^{-1}.Taking this example, tuning over this range with the above mentioned spectral resolution would require 50 different filters. Hence, I believe it is a question of the specific application. For a dedicated application with a limited number of wavenumbers (bands) to be used, a globar-plus-filter setting may suffice. The more flexibility you need, the more likely you are going to end up with either a FT-IR or a QCL-based microscope. Having said this, I would, however, like to reiterate that our custom built QCL-microscope was designed with guidelines to eliminate interference artefacts while preserving the key benefits of narrow band QCL sources of high spectral power density.

Ganesh Dhruvananda Sockalingum questioned: Are there any limitations in doing polarisation measurements using discrete frequencies? Can this inform about collagen structure for example?

Niels Kröger-Lui replied: Since QCL radiation is linearly polarized, it should be possible to observe optical rotation with a single polarizer used as an analyzer.

Klaus Gerwert said: What do you see when you take spectra of *C. elegans*? Could you identify marker bands in the spectra that indicate changes in the membrane potential?

Niels Kröger-Lui answered: In case of the *C. elegans* we performed discrete frequency imaging with a 50 Hz frame rate, using a microbolometer array detector. For a single frequency, chemical information is obviously limited. Also, the displacement of water by the moving *C. elegans* needs to be considered to obtain chemical information. Taking full spectra of *C. elegans* will, at this point, be more time consuming. When using a microbolometer array detector with a 50 Hz frame rate, we speculate that imaging at a small number of discrete frequencies (*e.g.* by rapidly switching between two QCLs) might be a good solution to correct for the effects of water displacement while maintaining a high acquisition speed. Also, if cooling requirements and cost aspects allow for the use of a MCT focal plane array, this detector might help to further improve acquisition speed and acquire full spectra which could be baseline corrected, in order to account for the effects of water displacement.

Faraday Discussions

PAPER

Raman spectroscopy for cytopathology of exfoliated cervical cells

I. R. Ramos,[ab] A. D. Meade,[b] O. Ibrahim,[ab] H. J. Byrne,[c] M. McMenamin,[d] M. McKenna,[d] A. Malkin[e] and F. M. Lyng*[ab]

Received 27th November 2015, Accepted 21st December 2015
DOI: 10.1039/c5fd00197h

Cervical cancer is the fourth most common cancer affecting women worldwide but mortality can be decreased by early detection of pre-malignant lesions. The Pap smear test is the most commonly used method in cervical cancer screening programmes. Although specificity is high for this test, it is widely acknowledged that sensitivity can be poor mainly due to the subjective nature of the test. There is a need for new objective tests for the early detection of pre-malignant cervical lesions. Over the past two decades, Raman spectroscopy has emerged as a promising new technology for cancer screening and diagnosis. The aim of this study was to evaluate the potential of Raman spectroscopy for cervical cancer screening using both Cervical Intraepithelial Neoplasia (CIN) and Squamous Intraepithelial Lesion (SIL) classification terminology. ThinPrep® Pap samples were recruited from a cervical screening population. Raman spectra were recorded from single cell nuclei and subjected to multivariate statistical analysis. Normal and abnormal ThinPrep® samples were discriminated based on the biochemical fingerprint of the cells using Principal Component Analysis (PCA). Principal Component Analysis – Linear Discriminant Analysis (PCA-LDA) was employed to build classification models based on either CIN or SIL terminology. This study has shown that Raman spectroscopy can be successfully applied to the study of routine cervical cytology samples from a cervical screening programme and that the use of CIN terminology resulted in improved sensitivity for high grade cases.

Introduction

Cervical cancer is the fourth most common cancer in women worldwide, accounting for an estimated 528 000 new cases and 266 000 deaths in 2012.[1]

[a]*DIT Centre for Radiation and Environmental Science, FOCAS Research Institute, Dublin Institute of Technology, Kevin St, Dublin 8, Ireland. E-mail: fiona.lyng@dit.ie; Fax: +353 1 4027904; Tel: +353 1 4027972*

[b]*School of Physics, Dublin Institute of Technology, Kevin St, Dublin 8, Ireland*

[c]*FOCAS Research Institute, Dublin Institute of Technology, Kevin St, Dublin 8, Ireland*

[d]*Cytopathology Department, Altnagelvin Hospital, Western Health and Social Care Trust, Derry, Northern Ireland, UK*

[e]*School of Biological Sciences, Dublin Institute of Technology, Kevin St, Dublin 8, Ireland*

However, the mortality associated with cervical cancer can be significantly reduced if this disease is detected at the early stages of development or at the pre-cancer stage, termed cervical intraepithelial neoplasia (CIN). Cervical cancer mainly affects younger women, about 60% of cases occurring in women under 50 years of age. Persistent infection with human papillomavirus (HPV) (such as high risk HPV types 16 and 18) is accepted as the major cause for the development of cervical pre-cancer and cancer.[2] Other risk factors include smoking, immuno-suppression, long term use of oral contraceptives and socioeconomic status.[3]

Cervical cancers are usually preceded by a long phase of pre-invasive disease. This phase is characterised microscopically as a sequence of events progressing from cellular atypia to various grades of dysplasia or CIN before progression to invasive carcinoma. Introduced in 1968, CIN is the most common terminology for cervical histology. It is a three-tiered system, divided into grades 1, 2 and 3, whereby CIN1 corresponds to mild dysplasia, CIN2 to moderate dysplasia, and CIN3 corresponds to both severe dysplasia and carcinoma *in situ*.[4] In cytology, the British Society for Clinical Cytology (BSCC) reporting system for cervical cancer refers to different grades of dyskaryosis which relate to the three-tiered CIN terminology with CIN1 corresponding to mild dyskaryosis, CIN2 corresponding to moderate dyskaryosis and CIN3 to severe dyskaryosis. Since then, advances in HPV research and liquid based cytology led to the introduction of the Bethesda System in the United States of America.[5] According to this nomenclature, squamous intraepithelial lesion (SIL) encompasses a range of non-invasive cervical epithelial abnormalities, comprising low grade (LSIL) and high grade lesions (HSIL). Low grade lesions correspond to cellular changes associated with the HPV cytopathic effect (koilocytotic atypia) and mild dysplasia (CIN1), whereas high grade lesions correspond to moderate dysplasia, severe dysplasia, and carcinoma *in situ* (CIN2 and 3). The two-tiered terminology of SIL is the standard reporting system for cervical cytology used in many screening programmes across developed countries. Since 2012, the College of American Pathologists and the American Society for Colposcopy and Cervical Pathology has recommended a uniform terminology to describe the histology of HPV-associated squamous disease across all anogenital sites. The two tiered terminology of LSIL and HSIL is recommended, as it reflects the biology of transient HPV infections and persistent pre-cancerous lesions.[6] More recently, the BSCC terminology was also revised to

Table 1 Summary of the different cervical cytology/histology reporting systems

Cervical cytology/histology reporting systems			
CIN[a]	BSCC[b] (1996)	SIL[c] (Bethesda)/CAP[d]/ASCCP[e]	BSCC (2013)
1	Mild dyskaryosis	LSIL[f]	Low grade dyskaryosis
2	Moderate dyskaryosis	HSIL[g]	High grade dyskaryosis (moderate)
3	Severe dyskaryosis		High grade dyskaryosis (severe)

[a] Cervical intraepithelial neoplasia. [b] British Society for Clinical Cytology. [c] Squamous intraepithelial lesion. [d] College of American Pathologists. [e] American Society for Colposcopy and Cervical Pathology. [f] Low grade squamous intraepithelial lesion. [g] High grade squamous intraepithelial lesion.

comprise the terms low grade and high grade, with the option of high grade dyskaryosis being further reported as moderate or severe.[7] Table 1 summarises the different classification systems.

The gradual progression of cervical cancer can allow the detection of dysplastic changes before invasive cancer develops through cervical cancer screening programmes. These screening programmes are common in developed countries, greatly reducing the mortality rates due to cervical cancer, but are not yet implemented in developing countries due to a lack of infrastructure and funding. The Papanicolaou (Pap) test is the most common screening method for cervical cancer and its precursor lesions.[8] The advantages of the Pap test are that it is non-invasive, inexpensive and widely accepted. However, although it can have high specificity of up to 95–98%, sensitivity rates can vary from 74 to 96%, due to sampling, technical and/or interobserver errors which are associated with the subjective nature of cytological screening.[9] There is, therefore, a need for new objective screening tests for cervical cancer.

Over the past 15 years, excellent sensitivity and specificity values have been reported using vibrational spectroscopy, Infrared (IR) and Raman, for the diagnosis of a wide range of cancers, including breast, prostate, oesophageal, colon, lung, oral and cervical cancer.[10–12] IR spectroscopy is based on the absorption of infrared radiation by the sample and the fact that molecules absorb specific frequencies of the incident light which are characteristic of their structure. Raman spectroscopy is an optical method based on inelastic light scattering. The sample is illuminated by monochromatic laser light and the interactions between the incident photons and molecules in the sample result in scattering of the light. The coupling of the light generates vibrations within the material which are characteristic of the chemical structure and the energy of the scattered light is reduced by an amount equal to the energy of the vibrational energy. Thus, the positions, relative intensities and shapes of the bands in a Raman spectrum carry detailed information about the molecular composition of the sample.

A number of papers by Wong and co-workers in the early 1990's showed significant differences in the cervical cytology of cells between normal, pre-cancer and cancer samples using IR spectroscopy.[13–16] However, these initial studies recorded spectra from cell pellets rather than from individual cells and a number of confounding factors such as neutrophils, endocervical columnar cells, metaplastic cells, cervical mucus and debris were subsequently identified.[17] Similar confounding factors, such as inflammation, metaplasia, hormonal changes, metabolic activity, blood and mucus, were identified by other groups.[18–22] Nevertheless, an early study by Cohenford and Rigas[23] reported an important finding that morphologically normal exfoliated cells from women with dysplasia or cancer exhibited extensive IR spectral changes. This finding was confirmed by Schubert *et al.*,[24] who showed spectral changes in cytologically normal cells in dysplastic samples, most likely due to HPV infection. Significant overlap was observed between negative, LSIL and HSIL cases using ATR—FTIR spectroscopy with maximal overlap between negative and LSIL cases.[25] But a more recent study showed that cervical pre-cancer is more accurately identified when histology rather than cytology is used as the gold standard to classify the samples.[26]

There are relatively fewer studies on cervical cytology using Raman spectroscopy, most likely because of the issues with confounding factors. Vargis *et al.*[27] showed that Raman spectroscopy could classify HPV-positive and HPV-negative cytology samples with an accuracy of 98.5%. Rubina *et al.*[28] reported a classification accuracy of ~80% using Raman spectroscopy to distinguish between normal and cervical cancer cytology samples. Cytology samples were treated with a red blood cell lysis buffer prior to Raman acquisition as the spectra of cervical cancer samples were dominated by blood features. Both of these studies used cell pellets rather than recording the Raman spectra from individual cells and this probably resulted in the relatively low classification accuracy in the study by Rubina *et al.* due to sample heterogeneity. A recent study by Bonnier *et al.*[29] presented new methods for recording Raman spectra from ThinPrep® cervical cytology samples. Pre-treatment of the slides with hydrogen peroxide to clear blood residue contamination before Raman recording was shown to minimise the variability within the data sets resulting in the collection of highly reproducible data with clear discrimination between negative cytology and CIN cytology. All data was recorded on glass ThinPrep® slides which are currently used for clinical cervical cytology rather than spectroscopic substrates such as calcium fluoride substrates which are commonly used in biospectroscopy research studies. Although these substrates reduce the presence of confounding contributions of the substrate, they are significantly more expensive which may prohibit clinical applications.[30]

The aim of this study was to evaluate the potential of Raman spectroscopy for cervical cancer screening using routine cervical cytology samples from a cervical screening programme. Cytology samples were classified according to both CIN and SIL terminology and Raman classification models compared.

Experimental

Sample collection and slide preparation

166 unstained smear samples in ThinPrep® slides were obtained from the Western Health & Social Care Trust Altnagelvin Hospital, Northern Ireland, with the approval of the Research Ethics Committee Northern Ireland.

Smears were collected, processed *via* the ThinPrep® method, Papanicolaou (Pap) stained and screened in the hospital by specialised personnel according to the guidelines in practice. A total of 88 negative, 35 CIN1 (or 35 LSIL) and 21 CIN2 and 22 CIN3 (or 43 HSIL) cases were randomly selected and included in this study. Each case represents a sample from an individual patient.

One duplicate slide of each selected case was prepared using the ThinPrep® method, fixed in 100% ethanol and air-dried. The samples were sent to the Dublin Institute of Technology for Raman analysis. Before recording, each slide was pretreated with hydrogen peroxidase (H_2O_2), as per an in-house protocol[29] to remove any contaminating blood and debris.

Raman microspectroscopy

Raman measurements were performed using a HORIBA Jobin Yvon XploRA™ system (Villeneuve d'Ascq, France), which incorporates an Olympus microscope BX41 equipped with a 100× objective (MPlanN, Olympus, N.A. 0.9) and a 532 nm diode laser source. To avoid any photo damage to the sample, the power of the laser was set at 50%. The confocal hole was set at 100 μm and a 1200 lines per mm grating was used, which gave a spectral dispersion of ~3 cm^{-1} per pixel. The backscattered light was collected using an air-cooled CCD detector

(Andor, 1024 × 256 pixels) and the spectrometer was controlled by Labspec V5.0 software. The system was calibrated to the 520.7 cm^{-1} spectral line of silicon.

Raman signals from each cell nucleus were integrated twice for 30 seconds in the spectral range of 400–1800 cm^{-1}. Spectra from a minimum of 10 cell nuclei were recorded per sample, depending on the quality of each slide. The data is presented as the average of all 10 cellular spectra recorded from each individual patient.

Data analysis

Data analysis was performed in Matlab [Mathworks, CA, USA] according to protocols developed in house.[31] The pre-processing of the raw Raman spectra included the application of a Savitsky–Golay filter (5th order, 13 points) to smooth the spectra and the subtraction of the glass background according to an in house non-negative least squares (NNLS) model.[32] The data set has also been corrected for baseline and vector normalized to facilitate comparison before principal component analysis (PCA) was employed to highlight the variability existing in the spectral data set recorded. Principal component analysis – linear discriminant analysis (PCA-LDA) was also employed to generate a classification model based on the features highlighted by PCA analysis. The optimal number of principal components (PCs) to generate the PCA-LDA model was established and 10 fold cross-validation used to test it. Furthermore, leave one out cross validation (LOOCV) was then used to evaluate the performance of the PCA-LDA classification model and the sensitivity and specificity rates were calculated for both SIL and CIN classifications.

Results

The mean Raman spectra of cervical cytology samples in the fingerprint region of 400–1800 cm^{-1} are shown in Fig. 1a. The samples are presented according to the SIL classification system; negative, LSIL and HSIL. The main peaks are indicated and assignments are listed in Table 2.[33] The mean spectra of negative, LSIL and HSIL samples show similar features, the main differences being observed around the 1318/1339 cm^{-1} region, in which the ratio of the intensities of these two peaks increases from negative to LSIL and HSIL samples as shown in Fig. 1b.

To further highlight any differences between the spectral profiles of the samples, PCA was employed. Fig. 2a shows the PCA scatterplot for all the SIL

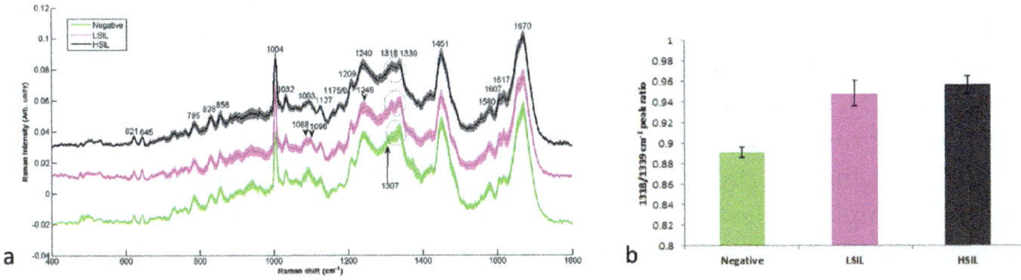

Fig. 1 (a) Mean Raman spectra from negative, LSIL and HSIL cervical cytology samples. Shading denotes the standard deviation, (b) 1318/1339 cm^{-1} peak ratios for negative, LSIL and HSIL cervical cytology samples.

Table 2 Tentative peak assignments[28] for Raman spectra shown in Fig. 1

Raman shift (cm^{-1})	Assignment
621	C–C twisting mode of phenylalanine (proteins)
645	C–C twisting mode of phenylalanine (proteins)
722	Adenine
786	DNA: O–P–O; pyrimidine ring breathing mode
828	Phosphodiester; O–P–O stretching DNA/RNA
856	Amino acid side chain vibrations of proline & hydroxyproline, as well as a (C–C) vibration of the collagen backbone
1032	CH_2CH_3 bending modes of collagen & phospholipids; phenylalanine of collagen; proline (collagen assignment)
1093	Symmetric PO_2^- stretching vibration of the DNA backbone–phosphate backbone vibration as a marker mode for the DNA concentration C–N of proteins
1127	ν(C–N)
1175/76	Cytosine, guanine
1209	Tryptophan & phenylalanine
1240	Amide III; differences in collagen content; asymmetric phosphate stretching modes
1246	Amide III (of collagen)
1307	CH_3/CH_2 twisting, wagging &/or bending mode of collagens & lipids
1318	Guanine (ring breathing modes of the DNA/RNA bases)–C–H deformation (protein); amide III (α-helix)
1339	CH_2/CH_3 wagging & twisting mode in collagen, nucleic acid & tryptophan
1451	CH_2CH_3 deformation
1580	C–C stretching
1607	C=C phenylalanine, tyrosine
1617	C=C phenylalanine, tyrosine
1670	Amide I; C=C stretching vibrations; amide I (anti-parallel β-sheet); n(C=C) trans, lipids, fatty acids

classified samples. From the PCA scatterplot, it can be seen that negative (green), LSIL (magenta) and HSIL (black) samples are separated according to the first (PC1) and second (PC2) principal components which account respectively for 76.81% and 7.548% of the variance explained in the dataset. The loadings for PC1 and PC2 are shown in Fig. 2b and c. The negative samples seem to separate from the LSIL samples according to PC1, with negative samples having more DNA (~814 cm^{-1}), protein and lipids (1307, 1446, 1453 cm^{-1}), amide III (1242 cm^{-1}) and I (1690 cm^{-1}) also featuring prominently. Furthermore, the negative samples separate to a large extent from the HSIL samples according to PC2. Negative samples show stronger amide III (1243, 1375 cm^{-1}) and protein/lipid (1339 cm^{-1}) features, whereas HSIL samples display stronger amide I (1606 cm^{-1}) and amide II (1544 cm^{-1}). In addition, the PC2 loading also highlights differences in nucleic acids, the features at 1458 and 1485/7 cm^{-1} being more prominent in the spectra of negative samples. DNA features at 481 and 786 cm^{-1} are more prominent in negative samples, whereas the feature at 893 cm^{-1} is more prominent in HSIL samples. Similarly, phosphate and phosphodiester bonds at 812 cm^{-1} are more prominent in HSIL samples, whereas those at 1087–9 cm^{-1} are more prominent in negative samples. The separation between LSIL and HSIL samples results from a combination of PC1 and PC2. Taking the PC1 and PC2 assignments for the

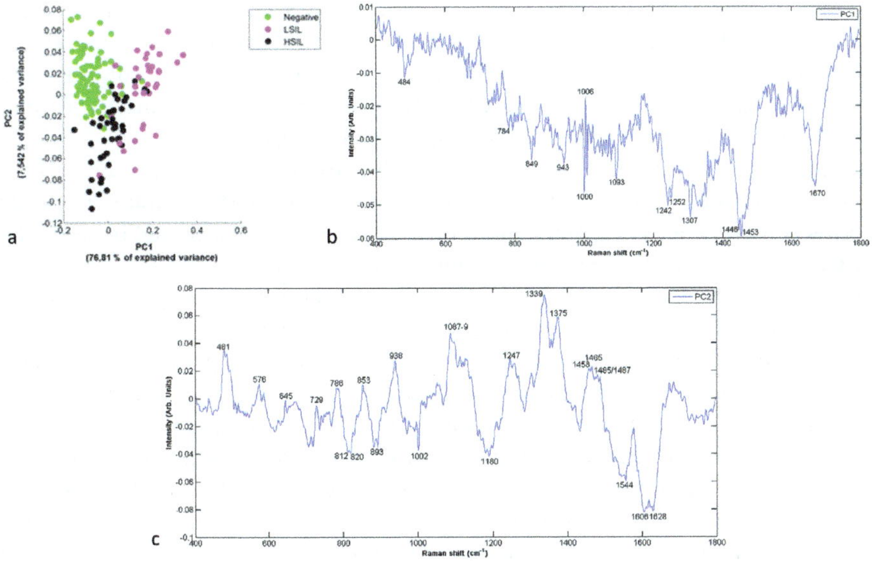

Fig. 2 (a) PCA scatterplot, (b) PC1 loading and (c) PC2 loading of negative, LSIL and HSIL cervical cytology samples.

negative samples as a reference, the LSIL samples have a similar PC2 profile to the negative samples whereas the HSIL samples have a similar PC1 profile to the negative samples.

The PCA results suggest that significant differences can be found in the Raman spectral profile of cell nuclei to distinguish between negative, LSIL and HSIL samples. PCA-LDA was therefore used to generate a classification model based on the features highlighted by PCA analysis. The results showed that a model with 12 PCs was best for LDA. Leave one out cross validation (LOOCV) was then used to evaluate the performance of the PCA-LDA classification model and the sensitivity and specificity rates are shown in Table 3.

Apart from the HSIL sensitivity, the performance of the PCA-LDA model is quite encouraging, with all the sensitivity and specificity values being above 90%. Sensitivity is also called the true positive fraction and accounts for the number of reported positives that are correctly identified as such. As the HSIL group consists of samples previously classed by the CIN classification system as CIN2 and CIN3, it was decided to re-evaluate the data according to the CIN classification.

Table 3 Sensitivity and specificity from Principal Component Analysis – Linear Discriminant Analysis (PCA-LDA) classification of negative, LSIL and HSIL cervical cytology samples

	SIL classification		
	Negative	LSIL	HSIL
Sensitivity	100.00%	94.29%	86.05%
Specificity	97.22%	95.42%	100.00%

Fig. 3a shows the mean Raman spectra of the CIN2 and CIN3 samples. Only small differences can be observed; a peak at 1246 cm^{-1} assigned to amide III and one at 1417 cm^{-1} assigned to C=C stretching in a quinoid ring are only observed in the CIN2 spectra whereas the CH_3/CH_2 twisting or bending mode of lipid/collagen at 1309 cm^{-1} is only evident in the CIN3 spectra.

PCA was performed on the dataset according to the CIN classification and the PCA scatterplot is shown in Fig. 3b. CIN2 (red) is not clearly separated from the other sample groups but, along PC1, it can be seen to be distributed between the CIN3 (black) and CIN1 (magenta) samples. To further elucidate the differences between all the CIN samples, pairwise PCA was performed on spectra from the CIN1, CIN2 and CIN3 samples and the results are shown in Fig. 4. The CIN1 and CIN2 samples show some overlap but there is some separation along PC1 (Fig. 4a). PC1 is positively dominated by amide I (1663 cm^{-1}) and other protein features (~1309, 1449 cm^{-1}), as shown in Fig. 4b, indicating that these are more prominent in CIN2 samples. Similarly, CIN1 and CIN3 show some overlap but there is some separation along PC1 (Fig. 4c). In this case, PC1 is also positively dominated by amide I (1669 cm^{-1}) and other protein features (~1308, 1453 cm^{-1}), as shown in Fig. 4d, indicating that these are more prominent in CIN3 samples compared to CIN1 samples. CIN2 and CIN3 samples show reasonably good separation with some overlap along PC3 (Fig. 4e). PC3 is positively dominated by nucleic acid features at 722, 786, 810 and 850 cm^{-1} and amide III at 1242 cm^{-1}, which are more intense in the spectra of the CIN3 samples, whereas amide I features at 1651 cm^{-1} and the C–H vibration of proteins and lipids at 1449 cm^{-1} are more prominent in the Raman spectra of the CIN2 samples.

For the PCA-LDA of the dataset according to the CIN classification, a model with 14 PCs was best for LDA and, similar to the SIL classification analysis, LOOCV was then used to evaluate the performance of the PCA-LDA classification model and the sensitivity and specificity rates are shown in Table 4. All the sensitivity and specificity values of the PCA-LDA model were greater than 90%.

Discussion

In this study Raman spectra were acquired from the nuclei of single cells of ThinPrep® cervical cytology specimens on glass slides. Recording spectra from the cell nuclei may explain the similarity between the average spectrum of the

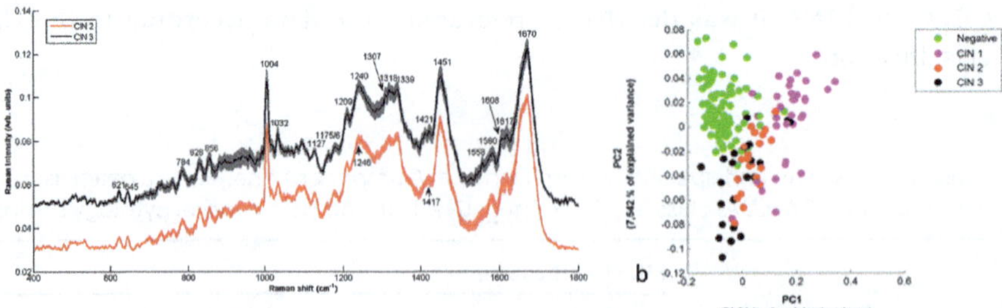

Fig. 3 (a) Mean Raman spectra from CIN2 and CIN3 cervical cytology samples. Shading denotes the standard deviation, (b) PCA scatterplot for negative, CIN1, CIN2 and CIN3 cervical cytology samples.

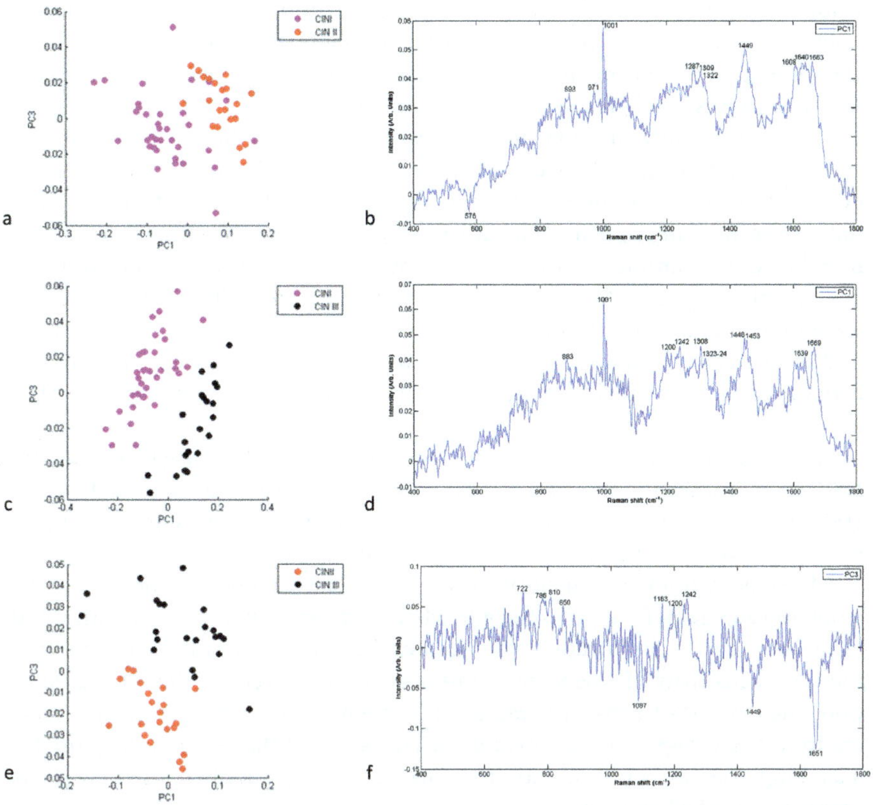

Fig. 4 (a) PCA scatterplot and (b) PC1 loading for CIN1 and CIN2 cervical cytology samples, (c) PCA scatterplot and (d) PC1 loading for CIN1 and CIN3 cervical cytology samples, (e) PCA scatterplot and (f) PC3 loading for CIN2 and CIN3 cervical cytology samples.

different sampling groups, as biochemically the nucleus of a normal cell and the nucleus of an abnormal cell are similar. Several studies have reported increases in nucleic acids in abnormal tissue samples compared to normal tissue samples.[34–37] Abnormal tissue is characterised by increased cell proliferation rates and therefore more cells are expected to be detected/scanned (per area) on abnormal samples, resulting in substantial differences compared to normal tissue samples. In the present study, despite targeting the nuclei of the cells, it is mainly protein

Table 4 Sensitivity and specificity from Principal Component Analysis – Linear Discriminant Analysis (PCA-LDA) classification of negative, CIN1, CIN2 and CIN3 cervical cytology samples

	CIN classification			
	Negative	CIN1	CIN2	CIN3
Sensitivity	100.00%	94.29%	100.00%	90.91%
Specificity	97.37%	98.47%	97.24%	100.00%

features which seem to discriminate the groups, although some differences in nucleic acid features were also observed. It should be noted that the Raman signals will have some degree of contribution from the cytoplasm of the cells as the laser passes through the cell to reach the nucleus.

A previous study of cervical cytology specimens, although conducted on cellular pellets rather than cell monolayers, reported amide I (1660 cm^{-1}), ∂CH_2 (1450 cm^{-1}) and phenylalanine (1002 cm^{-1}) as the main features dominating the Raman spectra of negative samples.[28] After a treatment to remove blood from the samples an increase in protein content (at 1006, 1450 and 1660 cm^{-1}) and changes in their secondary structure due to positive amide III bands were found. This adds to the case that instead of concentrating on increases and/or decreases in DNA and nucleic acids, Raman spectroscopy profiling for cervical cancer diagnosis can benefit from a better understanding of protein assignments.

A change in ratio at 1318/1339 cm^{-1} was also observed for negative, LSIL and HSIL cases, suggesting a decrease in the lipid/protein to guanine ratio in LSIL and HSIL samples, which may result from either a reduction of lipids/proteins and/or an increase in the nucleic acid (guanine) content of these samples.

The sensitivity and specificity values for the PCA-LDA models generated in this study were extremely high, especially when compared with Pap screening.[9] All values in the CIN classification were greater than 90% and with the exception of the HSIL sensitivity of ~86%, the same was also observed for the SIL classification. The lower HSIL sensitivity value of 86% seems to suggest that the CIN2 cases are from a heterogeneous group and perhaps some of the samples diagnosed as such might be biochemically closer to CIN1 than CIN3 and therefore may not be correctly classified as HSIL.

In fact, a recent study by Doorbar *et al.*[38] which investigated the correlation of CIN classification and HPV infection status suggested that some reported CIN2 cases, when analysed by an immunohistochemistry panel of P16INK4a, MCM and HPV-encoded E4, in fact group with CIN1 rather than CIN3 cases. The study showed that the combination of the identification of surrogates of high-risk HPV E6/E7 activity (P16INK4a and MCM), together with the detection of the abundant HPV-encoded E4 protein, was able to identify both transient and transforming lesions. This approach not only allowed distinguishing of true papillomavirus infections from similar pathologies but also the division of the heterogeneous CIN2 category into those that are CIN1-like with transient HPV infection expressing E4, and those that do not express E4 and therefore are more closely related to CIN3 cases with transforming HPV infection.

Conclusions

This study has shown that Raman spectroscopy can be successfully applied to the study of ThinPrep® cervical cytology samples from a cervical screening programme. Samples were prepared according to standard protocols and good quality Raman spectra were recorded from unstained single cervical cells on glass slides. Excellent sensitivity and specificity values were obtained particularly when CIN rather than SIL terminology was used to classify the samples. This suggests that the HSIL category is quite heterogeneous with CIN2 cases being biochemically different to CIN3 cases.

Acknowledgements

Financial support from the Dublin Institute of Technology Fiosraigh Research Excellence Scheme is gratefully acknowledged.

References

1 L. A. Torre, F. Bray, R. L. Siegel, J. Ferlay, J. Lortet-Tieulent and A. Jemal, *Ca-Cancer J. Clin.*, 2015, **65**, 87–108.

2 J. M. Walboomers, M. V. Jacobs, M. M. Manos, F. X. Bosch, J. A. Kummer, K. V. Shah, P. J. Snijders, J. Peto, C. J. Meijer and N. Munoz, *J. Pathol.*, 1999, **189**, 12–19.

3 E. L. Franco, N. F. Schlecht and D. Saslow, *Cancer J.*, 2003, **9**, 348–359.

4 H. M. Shingleton, R. M. Richart, J. Wiener and D. Spiro, *Cancer Res.*, 1968, **28**, 695–706.

5 D. Solomon, D. Davey, R. Kurman, A. Moriarty, D. O'Connor, M. Prey, S. Raab, M. Sherman, D. Wilbur, T. Wright Jr and N. Young, *JAMA, J. Am. Med. Assoc.*, 2002, **287**, 2114–2119.

6 T. M. Darragh, T. J. Colgan, J. T. Cox, D. S. Heller, M. R. Henry, R. D. Luff, T. McCalmont, R. Nayar, J. M. Palefsky, M. H. Stoler, E. J. Wilkinson, R. J. Zaino and D. C. Wilbur, *Journal of Lower Genital Tract Disease*, 2012, **16**, 205–242.

7 N. C. S. Programmes, Achievable standards, Benchmarks for reporting, and Criteria for evaluating cervical cytopathology, *NHS CSP*, Publication No 1, Sheffield, 3rd edn, 2013.

8 L. G. Koss and M. R. Melamed, *Koss' Diagnostic Cytology and Its Histopathologic Bases*, Lippincott Williams & Wilkins, Philadelphia, fifth edn, 2006.

9 H. C. Kitchener, R. Blanks, H. Cubie, M. Desai, G. Dunn, R. Legood, A. Gray, Z. Sadique and S. Moss, *Health Tech. Assess.*, 2011, **15**(iii–iv, ix–xi), 1–170.

10 M. Diem, A. Mazur, K. Lenau, J. Schubert, B. Bird, M. Miljkovic, C. Krafft and J. Popp, *J. Biophotonics*, 2013, **6**, 855–886.

11 K. Kong, C. Kendall, N. Stone and I. Notingher, *Adv. Drug Delivery Rev.*, 2015, **89**, 121–134.

12 F. M. Lyng, D. Traynor, I. R. Ramos, F. Bonnier and H. J. Byrne, *Anal. Bioanal. Chem.*, 2015, **407**, 8279–8289.

13 M. Fung Kee Fung, M. Senterman, P. Eid, W. Faught, N. Z. Mikhael and P. T. Wong, *Gynecol. Oncol.*, 1997, **66**, 10–15.

14 S. Neviliappan, L. Fang Kan, T. Tiang Lee Walter, S. Arulkumaran and P. T. T. Wong, *Gynecol. Oncol.*, 2002, **85**, 170–174.

15 P. T. Wong, R. K. Wong, T. A. Caputo, T. A. Godwin and B. Rigas, *Proc. Natl. Acad. Sci. U. S. A.*, 1991, **88**, 10988–10992.

16 H. M. Yazdi, M. A. Bertrand and P. T. Wong, *Acta Cytol.*, 1996, **40**, 664–668.

17 P. T. T. Wong, M. K. Senterman, P. Jackli, R. K. Wong, S. Salib, C. E. Campbell, R. Feigel, W. Faught and M. Fung Kee Fung, *Biopolymers*, 2002, **67**, 376–386.

18 L. Chiriboga, P. Xie, V. Vigorita, D. Zarou, D. Zakim and M. Diem, *Biospectroscopy*, 1998, **4**, 55–59.

19 M. A. Cohenford, T. A. Godwin, F. Cahn, P. Bhandare, T. A. Caputo and B. Rigas, *Gynecol. Oncol.*, 1997, **66**, 59–65.

20 M. Diem, L. Chiriboga, P. Lasch and A. Pacifico, *Biopolymers*, 2002, **67**, 349–353.

21 M. J. Romeo, M. A. Quinn, F. R. Burden and D. McNaughton, *Biopolymers*, 2002, **67**, 362–366.

22 B. R. Wood, M. A. Quinn, B. Tait, M. Ashdown, T. Hislop, M. Romeo and D. McNaughton, *Biospectroscopy*, 1998, **4**, 75–91.

23 M. A. Cohenford and B. Rigas, *Proc. Natl. Acad. Sci. U. S. A.*, 1998, **95**, 15327–15332.

24 J. M. Schubert, B. Bird, K. Papamarkakis, M. Miljkovic, K. Bedrossian, N. Laver and M. Diem, *Lab. Invest.*, 2010, **90**, 1068–1077.

25 N. C. Purandare, I. I. Patel, J. Trevisan, N. Bolger, R. Kelehan, G. von Bunau, P. L. Martin-Hirsch, W. J. Prendiville and F. L. Martin, *Analyst*, 2013, **138**, 3909–3916.

26 K. Gajjar, A. A. Ahmadzai, G. Valasoulis, J. Trevisan, C. Founta, M. Nasioutziki, A. Loufopoulos, M. Kyrgiou, S. M. Stasinou, P. Karakitsos, E. Paraskevaidis, B. Da Gama-Rose, P. L. Martin-Hirsch and F. L. Martin, *PLoS One*, 2014, **9**, e82416.

27 E. Vargis, Y.-W. Tang, D. Khabele and A. Mahadevan-Jansen, *Clin. Transl. Oncol.*, 2012, **5**, 172–179.

28 S. Rubina, M. Amita, K. D. Kedar, R. Bharat and C. M. Krishna, *Vib. Spectrosc.*, 2013, **68**, 115–121.

29 F. Bonnier, D. Traynor, P. Kearney, C. Clarke, P. Knief, C. Martin, J. J. O'Leary, H. J. Byrne and F. Lyng, *Anal. Methods*, 2014, **6**, 7831–7841.

30 H. J. Byrne, M. Baranska, G. J. Puppels, N. Stone, B. Wood, K. M. Gough, P. Lasch, P. Heraud, J. Sulé-Suso and G. D. Sockalingum, *Analyst*, 2015, **140**, 2066–2073.

31 P. Knief, Dublin Institute of Technology, 2010.

32 O. Ibrahim, A. Maguire, A. D. Meade, S. Flint, H. J. Byrne and F. M. Lyng, manuscript in preparation.

33 Z. Movasaghi, S. Rehman and I. U. Rehman, *Appl. Spectrosc. Rev.*, 2007, **42**, 493–541.

34 C. M. Krishna, G. D. Sockalingum, B. M. Vadhiraja, K. Maheedhar, A. C. K. Rao, L. Rao, L. Venteo, M. Plutot, D. J. Fernandes, M. S. Vidyasagar, B. V. B. Kartha and M. Manfait, *Biopolymers*, 2006, **85**, 214–221.

35 F. M. Lyng, E. O. Faolain, J. Conroy, A. D. Meade, P. Knief, B. Duffy, M. B. Hunter, J. M. Byrne, P. Kelehan and H. J. Byrne, *Exp. Mol. Pathol.*, 2007, **82**, 121–129.

36 N. Rashid, H. Nawaz, K. W. Poon, F. Bonnier, S. Bakhiet, C. Martin, J. J. O'Leary, H. J. Byrne and F. M. Lyng, *Exp. Mol. Pathol.*, 2014, **97**, 554–564.

37 K. M. Tan, C. S. Herrington and C. T. Brown, *J. Biophotonics*, 2011, **4**, 40–48.

38 H. Griffin, Y. Soneji, R. van Baars, R. Arora, D. Jenkins, M. van de Sandt, Z. Wu, W. Quint, R. Jach, K. Okon, H. Huras, A. Singer and J. Doorbar, *Mod. Pathol.*, 2015, **28**, 977–993.

Faraday Discussions

PAPER

Towards quantitative molecular mapping of cells by Raman microscopy: using AFM for decoupling molecular concentration and cell topography†

Radu Boitor,[a] Faris Sinjab,[a] Stephanie Strohbuecker,[b] Virginie Sottile[b] and Ioan Notingher*[a]

Received 16th November 2015, Accepted 26th November 2015

DOI: 10.1039/c5fd00172b

Raman micro-spectroscopy (RMS) is a non-invasive technique for imaging live cells *in vitro*. However, obtaining quantitative molecular information from Raman spectra is difficult because the intensity of a Raman band is proportional to the number of molecules in the sampled volume, which depends on the local molecular concentration and the thickness of the cell. In order to understand these effects, we combined RMS with atomic force microscopy (AFM), a technique that can measure accurately the thickness profile of the cells. Solution-based calibration models for RNA and albumin were developed to create quantitative maps of RNA and proteins in individual fixed cells. The maps were built by applying the solution-based calibration models, based on partial least squares fitting (PLS), on raster-scan Raman maps, after accounting for the local cell height obtained from the AFM. We found that concentrations of RNA in the cytoplasm of mouse neuroprogenitor stem cells (NSCs) were as high as 25 ± 6 mg ml^{-1}, while proteins were distributed more uniformly and reached concentrations as high as $\sim 50 \pm 12$ mg ml^{-1}. The combined AFM–Raman datasets from fixed cells were also used to investigate potential improvements for normalization of Raman spectral maps. For all Raman maps of fixed cells ($n = 10$), we found a linear relationship between the scores corresponding to the first component (PC1) and the cell height profile obtained by AFM. We used PC1 scores to reconstruct the relative height profiles of independent cells ($n = 10$), and obtained correlation coefficients with AFM maps higher than 0.99. Using this normalization method, qualitative maps of RNA and protein were used to obtain concentrations for live NSCs. While this study demonstrates the potential of using AFM and RMS for measuring concentration maps for individual NSCs *in vitro*, further studies are required to establish the robustness of the normalization method based on principal component analysis when comparing Raman spectra of cells with large morphological differences.

[a]*School of Physics and Astronomy, University of Nottingham, University Park, Nottingham NG7 2RD, UK. E-mail: ioan.notingher@nottingham.ac.uk*

[b]*Wolfson STEM Centre, School of Medicine, University of Nottingham, University Park, Nottingham NG7 2RD, UK*

† Electronic supplementary information (ESI) available. See DOI: 10.1039/c5fd00172b

Introduction

Cells are the fundamental units of any living organism. Their functions rely on highly orchestrated, spatially and time-dependent molecular interactions that underpin all biological processes. However, because of their small size and fragile nature, non-invasive detection and quantification of molecular interactions in living cells has always been a challenge. Various techniques for quantitative molecular analysis of cells do exist (*e.g.* mass spectrometry,[1] polymerase chain reaction (PCR)[2]), yet they require destructive procedures and usually apply to whole cell populations instead of single cells. Therefore, these techniques provide only single time-points and cannot provide insight into discrete and dynamic molecular events in living cells. Similar limitations apply to microscopy techniques based on fluorescence labeling of cells. As most molecules of interest are found inside the cells, cell fixation and membrane permeabilisation are required in order to allow the fluorescent molecules, which often are large dye molecules, to enter the cell cytoplasm and attach to the molecules of interest. To circumvent this antibody-based detection approach, transgenic strategies to express markers such as green fluorescent protein (GFP) have been developed. However, these techniques cannot be readily applied to primary cells, and genetic manipulation of cells requires laborious protocols to be developed for each cell type. Another limitation common to all fluorescence imaging methods is the difficulty in quantifying the results due to intensity variations of the fluorescence emission caused by photobleaching as well as variations in staining protocols.

Raman micro-spectroscopy (RMS) is a well-established analytical technique based on interaction of light with molecules present in the sample. Research during the last two decades showed that RMS can provide detailed molecular information on complex biological samples such as tissues and cells.[3] The main advances in this field have been recently reviewed.[4–6] An important feature of RMS is that it can measure time- and spatially resolved molecular processes in live cells maintained under physiological conditions *in vitro*, and without requiring labeling.[7–10]

Recently, attempts have been reported for quantitative molecular analysis of individual cells, as well as mapping the concentration of molecular components in live cells. The use of Raman spectral calibration models developed using samples with known concentrations (microparticles or solutions) has been reported for quantifying glycogen content in stem cell populations[11] and RNA in neuroprogenitor stem cells.[12] However, one main challenge when attempting such quantitative analysis arises from the fact that the intensity of a Raman band is proportional to the number of molecules in the sampled volume, which depends on the local concentration and the local thickness of the cell, as well as the laser focusing conditions. Furthermore, most biological processes are also accompanied by morphological changes of the cells. Therefore, it is often difficult to extract and quantify the molecular concentration in live cells because the 3D morphology of the cells is not known (Fig. 1).

When attempting to discriminate between two groups of cells based on their Raman spectra, it is common to normalize the Raman spectra, either using the intensity of certain bands that are expected to remain constant (*e.g.* 1450 cm^{-1} [13] or 1004 cm^{-1} [14]), the total area under the baseline-corrected spectrum,[15,16] or

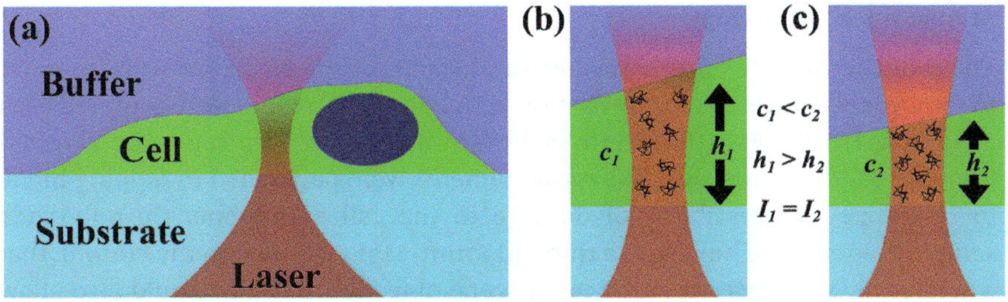

Fig. 1 Coupling between concentration and cell thickness in Raman micro-spectroscopy measurements. A schematic of the Raman sampling geometry for cells is shown in (a). The situations shown in (b) and (c) will give the same Raman scattered intensity, despite different molecular concentrations.

using the standard normal variance method.[17–19] Such normalization procedures have also been combined with solution-based calibration models to obtain molecular maps (*e.g.* RNA) in individual cells.[12] However, no studies on the validity of these normalization methods have been reported to date, thus the ability of these techniques to truly account for the morphology of the cells remains unclear.

In this paper we used confocal Raman micro-spectroscopy in conjunction with atomic force microscopy (AFM) in order to separate the effects of cell morphology and molecular concentration in the measured Raman spectra of cells. AFM was used to measure the topography maps of each individual cell, and this information was used to correct the intensity of the Raman spectra. Although in principle the topography of the cell could be obtained by recording a 3D map using a confocal Raman microscope, such measurements are time-consuming, and are prone to errors due to uncertainties related to the sampling volume of the microscope (variations in refractive index within the cell and the interfaces at the cell surface). Here, we use solution-based calibration models for proteins and nucleic acids, and combined AFM–Raman mapping to obtain quantitative molecular concentration maps for fixed cells (neuroprogenitor stem cells, NSCs). We also observed that principal component analysis (PCA) may be used for normalization of Raman spectral maps to take into account variations in cell thickness. This normalization method may be particularly useful for normalizing the Raman spectra of live cells, since AFM measurements are typically impractical because of long acquisition times for images and cell motility.

Materials and methods

Cell culture

Cell culture reagents were from Life Technologies (UK) unless otherwise stated. Mouse neural stem cells were cultured as previously described.[12] Briefly, cells were maintained in NSC medium made of DMEM/F12 and Neurobasal medium (1 : 1), N2, B27, antibiotics and supplemented with bFGF and EGF (Sigma-Aldrich, UK) at 20 ng ml^{-1} each. Cells were seeded onto low Matrigel-coated chambers (Becton Dickinson, UK) at 10^5 cells per ml using Accutase followed by a PBS wash. Cells were fixed with 4% paraformaldehyde (PFA, VWR) where indicated.

Atomic force microscopy

The cells that were fixed were washed in phosphate buffered saline (PBS) in order to remove cell debris caused by fixation with PFA. AFM images were acquired with a Nanowizard II AFM (JPK, Germany) that was mounted on top of an inverted Olympus IX-71 microscope. The measurements were performed in tapping mode to minimize the possibility of damage to the cell membrane. Custom-made sample holders were built with a thin (0.17 mm) MgF_2 window on the bottom, that would easily allow access to the microscope objective, but that would also allow space for the AFM scan head from above. Liquid-mode AFM tips (Multi75GB-G, BudgetSensors) were used for all measurements.

Raman micro-spectroscopy

The sample holders containing the cells were moved to a home-built Raman instrument, which was based on another inverted Olympus IX-71 microscope with an Olympus water immersion 60× objective (NA 1.2, a droplet of water was placed between the objective and coverslip) and with an automated stepper-motor stage (Prior Scientific) used to raster scan the sample, using custom-built software. A 785 nm Ti:Sapphire laser beam was focused to a laser spot, which provided 230 mW of power at the sample. The Raman back-scattered light was directed to a spectrometer (Andor Shamrock 303i with iDus CCD) *via* a fiber connection to obtain the Raman spectra. The spectrometer was calibrated using a polystyrene sample using peaks in the fingerprint region. Each fixed cell was scanned for 20 to 60 minutes, depending on its size. For Raman mapping, the integration time for each spectrum was 3 seconds and the step size in X and Y directions was 1 μm. For the axial scans, Raman spectra were recorded at regular 1.1 μm z intervals.

Live cell Raman measurements were performed on the same setup but with different experimental parameters. In order to facilitate live cell measurements, the setup was also equipped with an environmental enclosure (Solent Scientific), which kept the cells under physiological conditions during the measurements (37 °C, 5% CO_2). The same 785 nm laser line was used as for fixed cells, but the power was decreased to 170 mW and the acquisition time was reduced to 1 s. These parameters have been shown to have a minimal impact on cell physiology, while at the same time providing usable Raman spectra.[12]

Raman calibration models

Quantification of the two investigated cellular components was performed with the use of partial least squares regression (PLS). In order to build a PLS calibration model from Raman spectra, 14 known concentration solutions of RNA (yeast tRNA, lyophilized powder, Sigma UK) and protein (bovine serum albumin, lyophilized powder, Sigma UK) were created. All solutions were created by dissolving the two analytes in high purity deionized water (resistivity of 18.2 MΩ cm). In order to ensure the best correlation between the calibration model and the spectra of fixed cells, the same experimental parameters were used as for the acquisition of spectra from fixed cells: laser power at the sample was 230 mW and the acquisition time for each spectrum was 3 seconds. The Raman spectra of the calibration samples were measured using 6 μm-thick spacers (Omni-cell spacers from Specac).

Data processing

All data processing was carried out using custom-built MATLAB functions. Cosmic ray spikes were automatically removed from each spectrum before processing. Singular-value decomposition (SVD) was used to remove noise from the spectra whilst preserving 80–90% of the spectral information. Principal component analysis (PCA) was also used for the quick identification of various cell components and for the removal of the background (composed of contributions from the substrate, objective and the medium surrounding the cell). For the z profile data, PC1 values at different relative z positions had a linear baseline subtracted and were offset to 0 minimum value before fitting a Gaussian curve. The top-hat function representing the AFM height, used for deconvolution with the PC1 z profile had a width specified to an accuracy of 0.1 μm. The background was linearly subtracted from each spectrum used in the PLS calibration and prediction, in order to limit the spectral features to the ones that are specific to each investigated cellular component. For these spectra, the background was identified automatically with the use of k-means clustering analysis.

For AFM image processing, the background was levelled by fitting a linear plane between 3 points outside the cells and subtracting it from the dataset. In order to correlate the AFM images with the Raman maps the higher resolution AFM images were transformed and cropped in an automated routine to align the AFM image with the frame of the Raman image. The resolution of the resulting AFM image was then reduced to match that of the Raman maps, allowing the data from each to be approximately matched point-for-point.

Results and discussion

Fig. 2 presents typical examples of AFM and Raman spectral images for a fixed NSC. Fig. 2(b) shows that the selected Raman spectra at various positions inside the cell represent the molecular composition of the NSC and indicate a high chemical heterogeneity. Raman bands associated to nucleic acids include backbone vibrations at 788 cm^{-1} (O–P–O) and 1095 cm^{-1} (PO_2–), while vibrations specific to nucleotides are detected at 782 cm^{-1} (thymine, cytosine and uracil) and

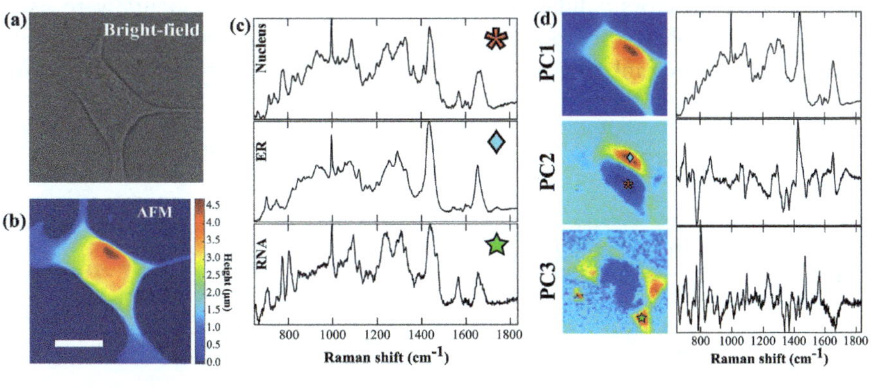

Fig. 2 Bright-field (a) and AFM (b) images of the same NSC (scale bar: 20 mm). Typical spectra from regions indicated by markers in PC score images (c) (SVD from PCs 2 : 10). (d) PCA images and loadings.

1578 cm^{-1} (guanine and adenine).[20,21] In agreement with previous reports, spectra with an intense band assigned to RNA (813 cm^{-1}, typically assigned to O–P–O vibrations in single-stranded nucleic acids) were observed at positions inside the cytoplasm of the NSC cells.[12] Raman bands corresponding to proteins are found in the ranges 1660–1670 cm^{-1} (amide I), 1450 cm^{-1} (C–H bending), 1200–1300 cm^{-1} (amide III), 1005 cm^{-1} (phenylalanine), 854 cm^{-1} (tyrosine) and 760 cm^{-1} (tryptophan).[22] All lipids are characterized by intense Raman bands at 1449 cm^{-1} (C–H bending vibrations), 1301 cm^{-1} (CH_2 twisting), 1000–1100 cm^{-1} spectral range (C–C stretching), while Raman bands characteristic to unsaturated lipids can be found at 1658 cm^{-1} (C=C stretching).[23] Membrane phospholipids also exhibit Raman bands in the 700–900 cm^{-1} spectral range assigned to different residues at the phosphate ester headgroup.[23] High-resolution Raman spectral images of cells can be created by plotting the scores obtained after principal component analysis (PCA) of all Raman spectra (Fig. 2(d)). These images can identify tentatively cellular structures, such as the nucleus (PC2 – negative scores), region rich in lipids and proteins surrounding the nucleus – potentially corresponding to the endoplasmatic reticulum (PC2 – positive scores), as well as the RNA rich regions at the edges of the cells (PC3). Indeed, bands corresponding to the molecular species found in these regions can be identified in the PCA loadings (Fig. 2(d)).

Nevertheless, PCA analysis can provide only qualitative information regarding the molecular composition of cells. To obtain quantitative information from the Raman spectra, calibration models based on water solutions of RNA and albumin (model for proteins) were developed (Raman spectra are presented in ESI Fig. S1†). To reduce the uncertainties related to the laser depth of field, the calibration solutions were placed in a sample chamber of 6 μm thickness in between two MgF_2 coverslips (Fig. 3(a)). As the thickness of the cells was 1–6 μm, the spectra recorded using this sample chamber are more representative for the Raman spectra of cells.

The PLS calibration models were built using fourteen RNA and albumin solutions with known concentrations spaced evenly over the expected concentrations in cells.[24] For the RNA solutions, the calibration model provided a good linear fit with an R^2 value of 0.961. Given the relatively low number of samples used for calibration, the prediction performance of the model was estimated by leave-one-out cross validation. When the first four components were used for the PLS model, the root mean squared error for cross-validation was RMSECV = 2.18 mg ml^{-1} (relative error 13%). Similarly, thirteen Raman spectra

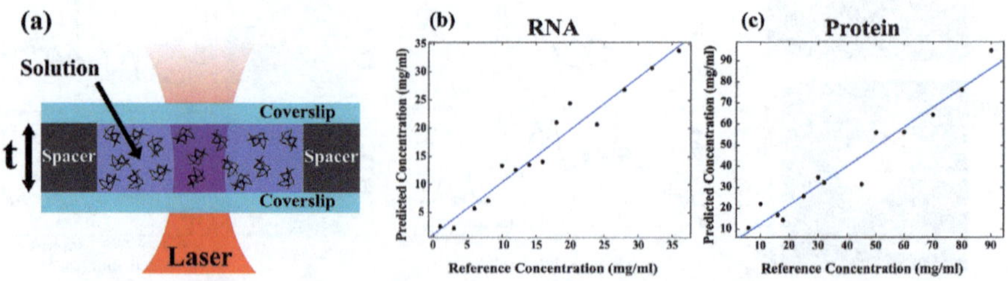

Fig. 3 PLS calibration of albumin and RNA. (a) Schematic of the experiment utilizing $t = 6$ μm spacers. Calibration curves for RNA (b) and albumin (c) in phosphate buffered saline solutions. The symbols represent the samples used for independent validation.

of albumin solutions were used for the protein model, producing an R^2 value for the calibration curve of 0.942 and a RMSECV of 6.29 mg ml^{-1} (relative error 15%).

Next, we applied the solution-based calibrations for RNA and albumin on 2D Raman maps of fixed cells in order to predict the concentration distributions of RNA and proteins. For each pixel in the Raman map (~1 µm × 1 µm), the value of cellular height obtained from the AFM map was used to calculate the local concentrations of RNA and proteins. Fig. 4 presents typical examples of quantitative maps for RNA and proteins for fixed NSCs. Individual Raman spectra from selected positions in the cells are presented in the ESI.†

At each pixel in the concentration maps in Fig. 4, the values for the RNA and protein concentrations were calculated by applying the calibration models described in Fig. 3 and then scaling using the cell thickness from the AFM map. For cell height values smaller than 1 µm in the AFM images, the corresponding Raman spectral intensity approached the limit of detection for the instrument. This resulted in the large values of concentration for the background, thus a threshold was applied based on the AFM height, allowing concentration maps only where cell spectra were detectable by the Raman instrument. When considering the accuracy of the calibration models (~15%) and errors in aligning the AFM and Raman maps, we estimate the errors in the concentration maps to be

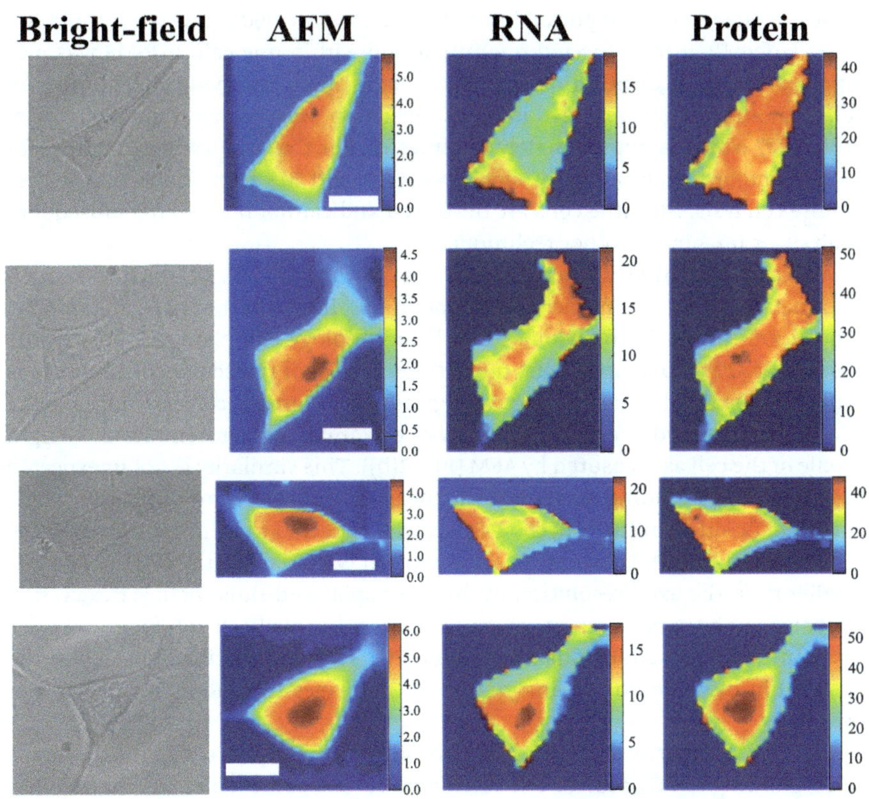

Fig. 4 Quantitative Raman mapping. Bright field images, transformed and matched AFM images (scale bar 10 µm, color scale in µm) with corresponding concentration maps for RNA and proteins (color scale in mg ml^{-1}) for three NSCs.

~25%. These errors may be reduced by increasing the number of samples in the calibration models and by using an integrated AFM–Raman instrument to reduce misalignment between AFM and Raman maps. Colocalized AFM–Raman measurements could eliminate the need for the alignment procedure in post-processing, providing more accurate results.

Fig. 4 shows that regions with high concentrations of RNA can be identified in the cytoplasm of NSCs, with maximum concentrations of typically 25 ± 6 mg ml^{-1} consistently between cells. We found that the concentration of protein can reach values as high as 50 ± 12 mg ml^{-1}, and was more widespread throughout the cell. The AFM and Raman spectral maps can also be used to quantify the total quantity of RNA and proteins in a single cell. As such, for typical NSC, the total quantity of RNA was found to be 12 pg (with values ranging from 11.1 pg to 13.1 pg), and 35 pg for proteins (values again in the range of 32.8–40.2 pg). The sensitivity level of Raman microscopy compares favorably with other single-cell analytical techniques, such as mass spectrometry and PCR. Real-time PCR (RT-PCR) is routinely used for quantification of microRNAs from cell cytoplasm, and achieves detection levels of ~15 pg of RNA.[25] Mass spectrometry (MS) is a very sensitive analysis technique that allows analysis of even single proteins.[26] However, a key challenge when analyzing cellular components is the preparation and isolation of these components. Detection of RNA was achieved by inductively coupled plasma mass spectrometry (ICP-MS), by quantifying the amount of $^{31}P^{+}$ in the investigated sample.[27] However, the reported limit of detection was ~60 pg, which is roughly 5 times larger than the expected quantity found within one cell. ICP-MS was also reported for quantification of proteins in cells, by quantifying the number of sulfur atoms.[28] However, the limit of detection was higher than 300 pg of proteins,[29] which is approximately 10 times higher than the values obtained for RMS. A major disadvantage of single-cell PCR and MS techniques is that they require cell lysis; thus, the cells are destroyed and cannot be used for subsequent analysis or imaging by other techniques.

Although the AFM allows accurate measurements of cell thickness, such information is often impractical to obtain for live cells because of the long acquisition times, during which the cell morphology can change. Thus, in the absence of AFM topography data, the absolute values of molecular concentrations cannot be obtained. However, Fig. 2 suggests that the image corresponding to the PC1 scores obtained from the PCA analysis of Raman maps resembles the height profile of the cell as measured by AFM (Fig. 2(b)). This similarity is not unexpected because the PC1 loading represents the mean Raman spectrum of the cell. Thus, at a given position in the cell, the PC1 score is expected to be proportional to the cell mass and height. However, this assumption is valid when the cell thickness is smaller than the axial resolution of the laser spot, and thus the laser excitation intensity can be considered constant across the sampled cell volume.

To investigate the relationship between PC1 scores and cell height (measured by AFM) in more depth, we analyzed the PC1 images from Raman maps of five fixed cells.

Fig. 5(a) compares the AFM height profiles of five fixed cells and the maps corresponding to PC1 scores. The calculated correlation coefficients for the AFM and PC1 maps for the individual cells was above 0.99, indicating that the PC1 scores from a Raman raster scan can provide an accurate relative height profile for the cells. Fig. 5(a) also shows that the map of PC1 scores is a more accurate representation of the cell topography that the previously used intensity of the 1450 cm^{-1} band.[12] The use of the 1450 cm^{-1} map can be convenient to use for normalization because it corresponds mainly to CH_2 vibrations found in most biomolecules, but also it is an intense band for which the intensity can be calculated accurately. Nevertheless, Fig. 5 indicates that the images corresponding to the 1450 cm^{-1} over-represent the lipid-rich regions within the cells while under-representing the protein-rich regions. Normalization to the 1001 cm^{-1} band (Fig. 5) shows a better correlation to the AFM height map than the 1450 cm^{-1} normalization, however there is also significant noise from the background (errors in calculation of local baseline), which is avoided when using PC1. A regression analysis of the PC1 scores and AFM height at 9–11 randomly selected points within each of the five cells (Fig. 5(b)) shows that, at each sampling point, the height of the cell is proportional to the PC1 score. However, Fig. 5(b) shows that the slopes of the regression lines vary from cell to cell, therefore the maps of PC1 scores can provide only a relative profile of the cell topography.

We also investigated whether it is possible to obtain an absolute height profile of a cell by measuring the height of the cell at a single point using a Raman *z* scan, and use this height value to scale the height profile obtained from PC1. In order to achieve this, we first calculated the axial point spread function (PSF_z) of the Raman instrument by deconvolution of the Raman *z* scans on fixed cells and the cell thickness values measured by AFM (modeled as a top-hat function). Fig. 6(a) presents the 2D Raman maps corresponding to PC1, PC2 and PC3 at various values of *z*. These results show that when the *z* position along the axial direction is changed, the Raman spectral images obtained by PCA analysis change in intensity but the general profiles remain the same. This indicates that the PSF_z is significantly larger than the thickness of the cell.

Fig. 6(d) presents typical *z* profiles of the PC1 scores, obtained at two positions in a fixed cell. These *z* profiles represent the convolution of the PSF_z of the instrument and the thickness of the cell at the corresponding positions. As the heights of the cells at these positions were measured by AFM, the PSF_z of the instrument was calculated by deconvolution, obtaining a Gaussian function with FWHM of 10.7 ± 0.2 μm (data from 3 cells were used, with 13 Raman *z* scans for

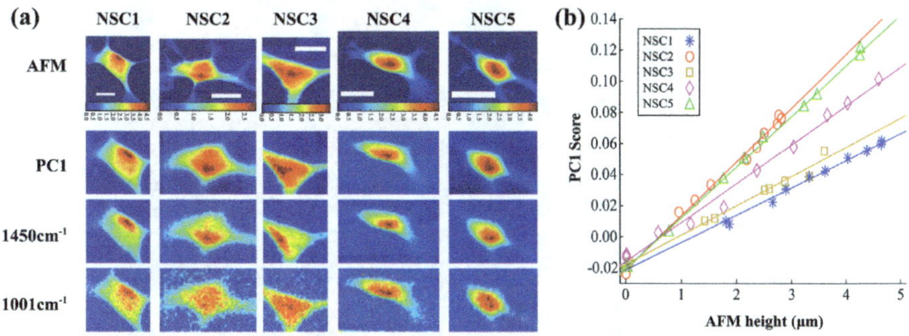

Fig. 5 Cell normalization comparison. (a) Shows AFM maps (scale bar 20 μm) for five different NSCs, with corresponding maps from Raman data, based on PC1 scores, and normalization by peak area for 1450 cm^{-1} and 1001 cm^{-1} peaks. (b) Correlation of AFM height with co-localized PC1 scores.

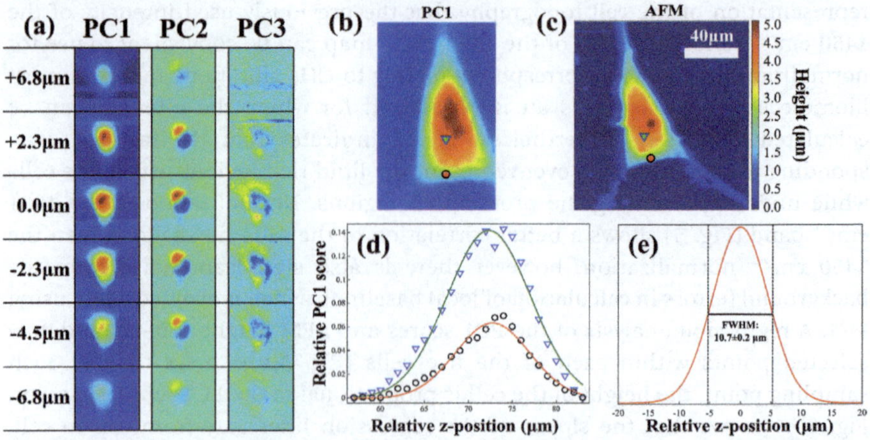

Fig. 6 (a) Raman *z*-profiling images. (b) PC1 map and (c) AFM for an NSC for which 25 line scans at different *z* positions were obtained. (d) Shows the variation in the value of the PC1 score as the sampling *z* position is varied at two locations on the cell of different thickness (indicated in (c) and (d)). Several PC1 *z* profiles were fitted to a Gaussian function, matched to a top-hat function of width given by the corresponding AFM height, and deconvolved to obtain an estimate for the PSF_z. (e) Shows the resulting average PSF_z measured from 13 deconvolution measurements.

each cell). This measured FWHM was larger than the theoretical value of 1.45 μm; discrepancies likely to be due to the use of MgF_2 rather than a glass/quartz coverslip, as well as potential differences in refractive index variations within the

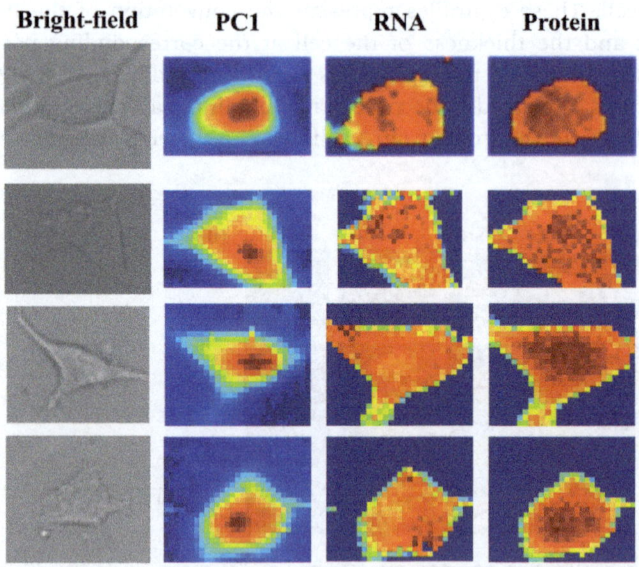

Fig. 7 Relative concentrations of RNA and proteins in live NSCs. The concentration maps were obtained by applying the calibration models described in Fig. 3 on Raman maps normalized using the scores of PC1.

cells. After the calculation of the PSF_z, Raman z scans obtained from five new cells were deconvolved in order to obtain the local height of the cells. However, comparison between the height values obtained by deconvolution and the values measured by AFM indicated typical errors of $\sim$50%. The main factor affecting these errors is related to the fact that the FWHM of the PSF_z is much larger than the typical heights of the cells ($\sim$1–5 μm). Using a confocal Raman microscope with a narrower PSF_z may improve the measurement of the cell height. However, measuring 2D Raman maps with a narrower PSF_z would create PC1 maps that are no longer representative of cell thickness as the laser intensity in the axial direction is no longer constant throughout the cell thickness. Thus, a full 3D Raman scan would be required in order to measure the cell topography and obtain absolute molecular concentration maps of the cell.

Because the acquisition of 3D Raman maps is impractical, in particular for live cells, we measured 2D Raman maps of live NSCs and obtained relative concentration maps for RNA and proteins by applying the calibration models in Fig. 3 and normalizing the Raman spectra using the PC1 scores. Fig. 7 shows that, after normalization, the concentration of proteins in the cells is homogenous, which is in agreement with the results for fixed cells. This result is to be expected, especially considering that the acquisition time of the Raman maps was 5–15 minutes, averaging any potential short-lived heterogeneities. On the other hand, RNA distribution indicates a higher concentration of RNA in the cytoplasm than in the nucleus and the endoplasmic reticulum, in agreement with the results for fixed cells, and expected biologically.

Conclusion

RMS and AFM are non-invasive label-free techniques that can be used on cells to provide complementary information. RMS mapping allows qualitative measurements of the spatial distribution of particular biomolecules in living cells. However, when comparing cells with different morphologies, it is difficult to know whether spectral differences arise from molecular differences (*i.e.* differences in concentration) or are due to cell topography. Here, we combined RMS with AFM to understand these effects. AFM was used to map the topography of fixed cells, and thus provide a representation of cell volume. Using solution-based calibration models for RNA and albumin, the RMS and AFM data were used to create quantitative maps of RNA and proteins in individual fixed cells. Because the acquisition times of AFM and Raman measurements are too long to be compatible with live cell measurements, we also used the combined AFM–Raman dataset from fixed cells to investigate potential improvements for normalization of Raman spectral maps. We found that if the PSF of the Raman instrument is significantly wider than the thickness of the cells, the maps of Raman PC1 scores can be correlated with the AFM map, and thus the cell volume. Using this technique, qualitative maps of RNA and protein concentrations were obtained for live cells. While this study demonstrates the potential of using AFM and RMS for measuring concentration maps for individual NSCs *in vitro*, further studies are required to establish the robustness of the normalization method based on principal component analysis when comparing Raman spectra of cells with large morphological differences.

Acknowledgements

The development of the Raman micro-spectrometer was supported by the Biotechnology and Biological Sciences Research Council UK research grant BB/G010285/1. The NSC culture was available thanks to support from the Alzheimer's Society (S. S. and V. S.).

References

1 L. Mueller, H. Traub, N. Jakubowski, D. Drescher, V. I. Baranov and J. Kneipp, Trends in single-cell analysis by use of ICP-MS, *Anal. Bioanal. Chem.*, 2014, **406**(27), 6963–6977.

2 A. Ståhlberg and M. Kubista, The workflow of single-cell expression profiling using quantitative real-time PCR, *Expert Rev. Mol. Diagn.*, 2014, **14**(3), 323–331.

3 G. J. Puppels, F. F. M. de Mul, C. Otto, J. Greve, M. Robert-Nicoud, D. J. Arndt-Jovin and T. M. Jovin, Studying single living cells and chromosomes by confocal Raman microspectroscopy, *Nature*, 1990, **347**, 301–303.

4 A. F. Palonpon, M. Sodeoka and K. Fujita, Molecular imaging of live cells by Raman microscopy, *Curr. Opin. Chem. Biol.*, 2013, **17**(4), 708–715.

5 T. Huser and J. Chan, Raman spectroscopy for physiological investigations of tissues and cells, *Adv. Drug Delivery Rev.*, 2015, **89**, 57–70.

6 B. Kann, H. L. Offerhaus, M. Windbergs and C. Otto, Raman microscopy for cellular investigations—From single cell imaging to drug carrier uptake visualization, *Adv. Drug Delivery Rev.*, 2015, **89**, 71–90.

7 M. Okada, N. Isaac Smith, A. Flotildes Palonpon, H. Endo, S. Kawata, M. Sodeoka and K. Fujita, Label-free Raman observation of cytochrome c dynamics during apoptosis, *Proc. Natl. Acad. Sci. U. S. A.*, 2012, **109**(1), 28–32.

8 H. N. Noothalapati Venkata and S. Shigeto, Stable isotope-labeled Raman imaging reveals dynamic proteome localization to lipid droplets in single fission yeast cells, *Chem. Biol.*, 2012, **19**(11), 1373–1380.

9 F. C. Pascut, S. Kalra, V. George, N. Welch, C. Denning and I. Notingher, Non-invasive label-free monitoring the cardiac differentiation of human embryonic stem cells *in vitro* by Raman spectroscopy, *Biochim. Biophys. Acta, Gen. Subj.*, 2013, **1830**(6), 3517–3524.

10 A. Naemat, H. M. Elsheikha, A. Al-Sandaqchi, K. Kong, A. Ghita and I. Notingher, Analysis of interaction between the apicomplexan protozoan Toxoplasma gondii and host cells using label-free Raman spectroscopy, *Analyst*, 2015, **140**(3), 756–764.

11 S. O. Konorov, H. Georg Schulze, C. G. Atkins, J. M. Piret, S. A. Aparicio, R. FB Turner and M. W. Blades, Absolute quantification of intracellular glycogen content in human embryonic stem cells with Raman microspectroscopy, *Anal. Chem.*, 2011, **83**(16), 6254–6258.

12 A. Ghita, F. C. Pascut, M. Mather, V. Sottile and I. Notingher, Cytoplasmic RNA in undifferentiated neural stem cells: a potential label-free Raman spectral marker for assessing the undifferentiated status, *Anal. Chem.*, 2012, **84**(7), 3155–3162.

13 I. Notingher, S. Verrier, S. Haque, J. M. Polak and L. L. Hench, Spectroscopic study of human lung epithelial cells (A549) in culture: living cells *versus* dead cells, *Biopolymers*, 2003, **72**(4), 230–240.

14 Y. Chen, L. Su, G. N. Zhang, L. H. Wang, A. G. Shen, X. D. Zhou, X. H. Wang and J. M. Hu, *In vivo* and *in situ* monitoring of the nitric oxide stimulus response of single cancer cells by Raman spectroscopy, *Laser Phys. Lett.*, 2013, **10**(4), 045608.

15 Q. Matthews, A. Jirasek, J. Lum, X. Duan and A. G. Brolo, Variability in Raman spectra of single human tumor cells cultured *in vitro*: correlation with cell cycle and culture confluency, *Appl. Spectrosc.*, 2010, **64**(8), 871–887.

16 M. Chen, N. McReynolds, E. C. Campbell, M. Mazilu, J. Barbosa, K. Dholakia and S. J. Powis, The Use of Wavelength Modulated Raman Spectroscopy in Label-Free Identification of T Lymphocyte Subsets, Natural Killer Cells and Dendritic Cells, *PLoS One*, 2015, **10**, e0125158.

17 F. C. Pascut, H. T. Goh, N. Welch, L. D. Buttery, C. Denning and I. Notingher, Noninvasive detection and imaging of molecular markers in live cardiomyocytes derived from human embryonic stem cells, *Biophys. J.*, 2011, **100**(1), 251–259.

18 T. Tolstik, C. Marquardt, C. Matthäus, N. Bergner, C. Bielecki, C. Krafft, A. Stallmach and J. Popp, Discrimination and classification of liver cancer cells and proliferation states by Raman spectroscopic imaging, *Analyst*, 2014, **139**(22), 6036–6044.

19 S. F. El-Mashtoly, H. K. Yosef, D. Petersen, L. Mavarani, A. Maghnouj, S. Hahn, C. Kötting and K. Gerwert, Label-Free Raman Spectroscopic Imaging Monitors the Integral Physiologically Relevant Drug Responses in Cancer Cells, *Anal. Chem.*, 2015, **87**(14), 7297–7304.

20 S. C. Erfurth, E. J. Kiser and W. L. Peticolas, Determination of the backbone structure of nucleic acids and nucleic acid oligomers by laser Raman scattering, *Proc. Natl. Acad. Sci. U. S. A.*, 1972, **69**(4), 938–941.

21 J. M. Benevides and G. J. Thomas, Characterization of DNA structures by Raman spectroscopy: high-salt and low-salt forms of double helical poly(dG-dC) in H_2O and D_2O solutions and application to B, Z and A-DNA*, *Nucleic Acids Res.*, 1983, **11**(16), 5747–5761.

22 T. G. Spiro and B. P. Gaber, Laser Raman scattering as a probe of protein structure, *Annu. Rev. Biochem.*, 1977, **46**(1), 553–570.

23 C. Krafft, L. Neudert, T. Simat and R. Salzer, Near infrared Raman spectra of human brain lipids, *Spectrochim. Acta, Part A*, 2005, **61**(7), 1529–1535.

24 R. Milo, What is the total number of protein molecules per cell volume? A call to rethink some published values, *BioEssays*, 2013, **35**(12), 1050–1055.

25 C. Chen, D. A. Ridzon, A. J. Broomer, Z. Zhou, D. H. Lee, J. T. Nguyen, M. Barbisin, N. L. Xu, V. R. Mahuvakar, M. R. Andersen, K. Q. Lao, K. J. Livak and K. J. Guegler, Real-time quantification of microRNAs by stem–loop RT-PCR, *Nucleic Acids Res.*, 2005, **33**(20), e179.

26 G. A. Valaskovic, N. L. Kelleher and F. W. McLafferty, Attomole protein characterization by capillary electrophoresis-mass spectrometry, *Science*, 1996, **273**, 1199–1202.

27 S.-I. Fujii, K. Inagaki, S.-I. Miyashita, A. S. Groombridge, K. Nagasawa, K. Chiba and A. Takatsu, Separation and quantification of RNA molecules using size-exclusion chromatography hyphenated with inductively coupled plasma-mass spectrometry, *Electrophoresis*, 2014, **35**(9), 1315–1318.

28 M. Wind, A. Wegener, A. Eisenmenger, R. Kellner and W. D. Lehmann, Sulfur as the Key Element for Quantitative Protein Analysis by Capillary Liquid

Chromatography Coupled to Element Mass Spectrometry, *Angew. Chem., Int. Ed.*, 2003, **42**(29), 3425–3427.

29 N. Zinn, R. Krüger, P. Leonhard and J. Bettmer, μLC coupled to ICP-SFMS with post-column isotope dilution analysis of sulfur for absolute protein quantification, *Anal. Bioanal. Chem.*, 2008, **391**(2), 537–543.

Faraday Discussions

PAPER

Vibrational spectroscopy in sensing radiobiological effects: analyses of targeted and non-targeted effects in human keratinocytes†

Aidan D. Meade,*[ab] Orla Howe,[bc] Valérie Unterreiner,[d] Ganesh D. Sockalingum,[ef] Hugh J. Byrne[g] and Fiona M. Lyng[ab]

Received 3rd December 2015, Accepted 14th December 2015
DOI: 10.1039/c5fd00208g

Modern models of radiobiological effects include mechanisms of damage initiation, sensing and repair, for those cells that directly absorb ionizing radiation as well as those that experience molecular signals from directly irradiated cells. In the former case, the effects are termed targeted effects while, in the latter, non-targeted effects. It has emerged that phenomena occur at low doses below 1 Gy in directly irradiated cells that are associated with cell-cycle-dependent mechanisms of DNA damage sensing and repair. Likewise in non-targeted bystander-irradiated cells the effect saturates at 0.5 Gy. Both effects at these doses challenge the limits of detection of vibrational spectroscopy. In this paper, a study of the sensing of both targeted and non-targeted effects in HaCaT human keratinocytes irradiated with gamma ray photons is conducted with vibrational spectroscopy. In the case of directly irradiated cells, it is shown that the HaCaT cell line does exhibit both hyperradiosensitivity and increased radioresistance at low doses, a transition between the two effects occurring at a dose of 200 mGy, and that cell survival and other physiological effects as a function of dose follow the induced repair model. Both Raman and FTIR signatures are shown to follow a similar model, suggesting that the spectra include signatures of DNA damage sensing and repair. In bystander-irradiated cells, pro- and anti-apoptotic signalling and mechanisms of ROS damage were inhibited in the mitogen-activated protein kinase (MAPK)

[a]*School of Physics, Dublin Institute of Technology, Kevin Street, Dublin 8, Ireland. E-mail: aidan.meade@dit.ie*
[b]*DIT Centre for Radiation and Environmental Science, Focas Research Institute, Dublin Institute of Technology, Camden Row, Dublin 8, Ireland*
[c]*School of Biological Sciences, Dublin Institute of Technology, Kevin Street, Dublin 8, Ireland*
[d]*Plateforme en Imagerie Cellulaire et Tissulaire (PICT), Université de Reims Champagne-Ardenne, 51 rue Cognacq-Jay, 51095 Reims Cedex, France*
[e]*Université de Reims Champagne-Ardenne, MéDIAN-Biophotonique et Technologies pour la Santé, UFR de Pharmacie, 51 rue Cognacq-Jay, 51095 Reims Cedex, France*
[f]*CNRS UMR 7369, Matrice extracellulaire et Dynamique Cellulaire, MEDyC, 51095 Reims Cedex, France*
[g]*Focas Research Institute, Dublin Institute of Technology, Camden Row, Dublin 8, Ireland*

† Electronic supplementary information (ESI) available. See DOI: 10.1039/c5fd00208g

transduction pathway. It is shown that Raman spectral profiles of bystander-irradiated cells are correlated with markers of bystander signalling and molecular transduction. This work demonstrates for the first time that both targeted and non-targeted effects of ionizing radiation damage are detected by vibrational spectroscopy *in vitro*.

Introduction

The central dogma of radiation biology is that ionizing radiation must interact with the DNA of the cell in order to produce *targeted* radiobiological effects. This view has been challenged in recent decades by the discovery of a host of *non-targeted effects*, including radiation-induced bystander effects, genomic instability and abscopal effects, in which similar radiobiological effects are observed in unirradiated cells in the vicinity of cells that have been irradiated.[1–3] Experimental evidence of this has been widely observed using either X-ray microbeams[4] or high linear energy transfer (LET) sources such as alpha particles for irradiation of single cells,[5] and in 'medium-transfer' experiments in which medium from irradiated donor cells is used to expose unirradiated recipient cells.[3,6] The precise mechanisms governing how this effect occurs remain unknown, but it is thought to occur *via* secretion of factors from the irradiated cell that interact with the unirradiated cell.[7,8] It has also been demonstrated that the effect has an approximately linear dose response up to a dose of 500 mGy, and saturates from 500 mGy to 5 Gy.[9]

In tandem with the discovery of non-targeted effects, a new view of targeted effects has emerged for directly irradiated cells. It has recently emerged that, below 1 Gy, a region of high cell radiosensitivity exists, followed by a region of radioresistance. The former effect is termed low dose radiation hypersensitivity (HRS) and the latter increased radioresistance (IRR). In the traditional linear-quadratic (LQ) model of cell survival with dose (eqn (1)), these effects are not taken into account. This model has been modified as the induced repair (IR) model of cell survival (eqn (2)).[10]

$$S = \exp(-\alpha D - \beta D^2) \tag{1}$$

$$S = \exp\left\{-\left[\alpha_H\left(1 + \left[\frac{\alpha_L}{\alpha_H} - 1\right] \times \exp\left(-\frac{D}{D_C}\right)\right)\right] \times D - \beta D^2\right\} \tag{2}$$

Within eqn (1), the α and β terms are the linear and quadratic coefficients describing the variation in cell survival at low and high doses respectively (where dose is denoted as D). In eqn (2), the α_L and α_H terms are incorporated to describe the biphasic dose–response at doses up to 1 Gy, and D_C is the dose point at which the transition from low dose hyperradiosensitivity (HRS at $D < D_C$) to increased radioresistance (IRR for dose, D, between D_C and ~1 Gy) occurs.[11]

The sensing of DNA damage is thought to have a threshold in dose, D_C, at which the molecular machinery sensing the damage is activated, and this subsequently results in a transition between HRS and IRR.[11–13] It has recently emerged that two molecular checkpoints exist in the G_2 phase of the cell cycle,[14,15] an early one that is associated with the ataxia telangiectasia mutated (ATM) gene and whose activation appears to only occur at doses >200 mGy, and a late one that is dose-dependent. It appears that the activation of the early checkpoint is associated with the HRS to IRR transition, such that doses below 200 mGy are insufficient to activate the DNA damage sensing machinery.[11,12] The activation of the G_2/M checkpoints is also dependent on the point in the cell cycle at which the cell is irradiated, as well as the radiation dose.[12,16] It has also emerged recently that HRS is dependent on the cell's ability to undergo p53-dependent apoptosis,[17,18] such that the HRS phenomenon may be a mechanism by which cells that sustain low levels of radiation damage at or below $D_C \sim$ 200 mGy are allowed to die rather than proceed to mitosis with unrepaired DNA damage.[17,18]

The present study, which is in two parts, aims to demonstrate that spectroscopic signatures of both targeted effects and non-targeted effects are observable in a human keratinocyte cell line (HaCaT). In the first part, an analysis of FTIR and Raman spectroscopic data as a function of radiation dose and time after irradiation is conducted *in vitro*, in parallel with measurements of various characteristics of the physiological activity of the cell. Recent data has shown the sensitivity of Raman spectroscopy to low dose irradiation of lymphocytes.[19] The dose range analysed here includes both the low dose region, below 1 Gy, and above this dose point, in contrast to previous work that has focused on analysing the spectroscopic effect of exposure to high dose ionizing photonic radiation.[20–22] It is demonstrated for the first time that the HaCaT cell line exhibits characteristics of both hyperradiosensitivity and increased radioresistance, both in terms of variation in its cell survival and physiological activity with radiation dose. It is also demonstrated, for the first time, that a wide range of spectroscopic signatures vary in accordance with the induced repair model with radiation dose.

In the second part, Raman spectroscopy is used to identify molecular changes occurring in recipient bystander HaCaT cells exposed to irradiated cell-conditioned medium (ICCM) harvested from irradiated donor HaCaT cells. Modifications of the bystander effect are then analysed spectroscopically using inhibitors of the mitogen-activated protein kinase (MAPK) pathway and inhibitors of reactive oxygen species (ROS). The MAPK signalling transduction pathways are related to the initiation of multiple and distinct cellular processes in response to activation by extracellular stresses or molecular factors,[23–26] including the regulation of tumour cell growth.[27,28] Within the MAPK superfamily, sub-pathways, including the ERK, JNK and p38 pathways, exist that control individual cellular responses, and that have a degree of dynamic balance.[24–26] The pro-apoptotic c-Jun NH_2-terminal kinase (JNK) 1/2/3 pathway is activated by radiation-induced ceramide generation and pro-apoptotic death receptors (such as tumour necrosis factor-α (TNF-α) and the Fas and TRAIL ligands) on the plasma membrane of cells.[23,24,29] Activation or inhibition of the ERK pathway appears to inhibit or enhance JNK activation, respectively, leading to a dynamic balance between ERK and JNK activation after radiation exposure.[23–26] This complex interplay between pro- and anti-survival enzyme signalling is shown to be observable spectroscopically for the first time, according to signatures that match reference biological measurements of pro- and anti-apoptotic signalling.

Materials and methods

Cell culture and sample preparation

Human keratinocytes (HaCaT) were cultured in Dulbecco's MEM : F12 (1 : 1) whole medium (Sigma, Dorset, UK) supplemented with 10% foetal calf serum (Gibco, Irvine, UK), 1% penicillin–streptomycin solution 1000 IU (Gibco, Irvine, UK), 2 mM L-glutamine (Gibco, Irvine, UK) and 1 μg ml^{-1} hydrocortisone (Sigma, Dorset, UK). Cells were cultured and maintained in an incubator at 37 °C with 95% relative humidity and 5% CO_2. Cells were routinely subcultured at 80% confluency using a 1 : 1 solution of 0.25% trypsin and 1 mM versene at 37 °C.

FTIR was conducted on samples placed on MirrIR (Kevley Technologies) slides that were cut into 20 mm × 25 mm pieces and sterilised in 70% industrial methylated spirits before being placed in 6-well plates. These slides were then coated in a sterile gelatin layer that was prepared from a 2% (b/w) in deionized water solution that has been previously shown to aid the attachment of cells to these substrates.[30] Cell suspensions in DMEM–F12 were added at densities of 1.5 × 10^4 cells per substrate for study at 96 hours after direct irradiation. All cell samples were allowed to attach to the substrate for 24 hours before irradiation. Samples for direct irradiation and subsequent Raman microspectroscopy were also prepared for analysis at 96 hours after irradiation by depositing suspensions of 2.5 × 10^4 HaCaT cells onto fused quartz disks, also coated in a sterile aqueous gelatin solution and cultured for 24 hours in DMEM–F12 with all supplements.[30]

Irradiation was performed using a cobalt-60 teletherapy source with doses of 5 mGy, 20 mGy, 50 mGy, 200 mGy, 500 mGy, 750 mGy, 1 Gy, 2.5 Gy and 5 Gy in addition to a sham-irradiated control, as detailed elsewhere.[31] All samples for FTIR and Raman of directly irradiated cells were prepared in triplicate and were fixed in 4% formalin after irradiation, and stored in dH_2O at 4 °C until analysis.

Preparation of irradiated cell-conditioned medium for bystander irradiation

HaCaT donor cells were irradiated with γ-radiation from a cobalt 60 teletherapy unit. Doses of 0.05 Gy and 0.5 Gy were employed for the measurement of parallel protein expression profiles. This allowed the establishment of the dose dependence of the target protein expression profile. Donor cells were irradiated at room temperature with a concentration of 7.5 × 10^5 cells per flask. Control flasks were sham-irradiated (0 Gy).

A dose of 0.5 Gy was used to prepare ICCM for the Raman spectroscopy experiment. The cells for subsequent Raman spectroscopy were irradiated in T25 flasks containing 5 ml DMEM–F12 medium that did not contain the phenol red indicator (so as not to optically interfere with or attenuate the illuminating source during subsequent Raman measurements).

All MAPK inhibitors for the Raman experiment were added to the flasks in their respective concentrations 15 to 45 minutes prior to the irradiation (inhibitors were added to flasks for irradiation and their respective controls at the same time). The MEK inhibitors U0126 (inhibitor of ERK1) and PD98059 (inhibitor of ERK1 + ERK2) (A. G. Scientific Inc., San Diego CA) were used at a final concentration of 20 μM. The JNK inhibitor (SP600125) and the p38 inhibitor (SB203580) (A. G. Scientific Inc.) were both used at final concentrations of 10 μM. Catalase (Sigma-Aldrich) was prepared at a concentration of 20 μg ml^{-1} in whole medium and used immediately. Each of the inhibitors prevent phosphorylation of the respective signalling protein within the MAPK pathway, and thus prevent their activation.[32]

ICCM was decanted from donor flasks one hour after irradiation and was filtered through a 0.22 μm filter to ensure that no irradiated cells were contained within the ICCM. The medium was then divided into 5 ml aliquots, stored at −80 °C, and subsequently thawed when required for experiments.

Sample preparation for exposure to irradiated cell culture medium and Raman spectroscopy

A total of 1×10^6 recipient HaCaT cells for exposure to ICCM and subsequent Raman spectroscopy were plated onto 25 × 25 mm quartz coverslips that had been precoated to improve biocompatibility with 2% gelatin solution in dH_2O, and the cells were allowed to adhere to the substrates for 24 hours prior to the beginning of the exposure.[30] The HaCaT cells on coated quartz substrates were placed in 35 mm Petri dishes for exposure to ICCM, and a volume of 3 ml ICCM was then added to cover the cells. The cells were exposed to ICCM for either 30 minutes or 24 hours, at which point they were used for Raman spectroscopy.

FTIR and Raman spectroscopy

FTIR was performed on directly irradiated cells using a Perkin-Elmer GX-II spectrometer, over the 4000 to 720 cm^{-1} wavenumber range with an aperture size of 100 μm × 100 μm, at a spectroscopic resolution of 4 cm^{-1} and with 64 scans per spectrum. All spectra were recorded in transreflection mode; approximately 300 point spectra were recorded at each dose and time point. Spectra were vector normalised before analysis.

Raman spectra were acquired using two Horiba-Jobin Yvon HR-800 Raman spectrometers, both equipped with a 785 nm source laser. Spectra were acquired using a confocal hole diameter of 100 μm on both instruments. For the measurement of spectra from directly irradiated cells, a 300 lines per mm grating was used to provide dispersion of the Raman signal, while a 950 lines per mm grating was used to provide dispersion for the bystander-irradiated cells. Both instruments were calibrated using the 520.7 cm^{-1} line of silicon. A spectrum of a neon lamp source was also taken as a reference for wavelength calibration of the spectrometer CCD detectors. A water immersion objective with ×100 magnification was used for all spectroscopic measurements, whereby directly irradiated cells were immersed in deionized water and bystander-irradiated cells were immersed in ICCM during spectroscopic measurements. Spectra of the quartz substrate were acquired in triplicate prior to, and at the end of, each measurement.

Raman spectra of approximately 40 directly irradiated HaCaT cells at each dose point were acquired in a line scan across the cells with a step interval of 3 μm, such that spectra included contributions from the nucleus, cytoplasm and membrane.

Raman spectra were recorded from ICCM-exposed cells in their live state after exposure to ICCM for the necessary exposure time. A total of approximately 30 spectra per sample in the 30 minutes exposure category and approximately 100 spectra per sample in the 24 hours ICCM exposure category were acquired from the nuclei of randomly selected cells. All measurements were performed in triplicate. Three spectra were recorded of the quartz background and the ICCM for each exposure condition.

Spectroscopic processing was performed using Matlab v8.5 (The MathWorks Inc., USA) with the PLS-Toolbox v8 (Eigenvector Research Inc.). All Raman spectra were preprocessed according to procedures detailed elsewhere.[33] Spectra were calibrated using an alignment procedure involving the fitting of a polynomial to the relationship between peak positions in the spectrum of a neon lamp. The signature of the quartz background (in the case of directly irradiated cells) or the quartz and medium background (in the case of bystander-irradiated cells) was subtracted from spectra before removing any residual baseline using an in-house rubberband algorithm.[33] Spectra were then vector normalised. Outlying spectra were removed by a repeated Grubb's test of Mahalanobis distances between the scores of the first three principal components of each spectroscopic dataset.

To aid spectral interpretation, spectra of pure samples of various biochemical components of the cell were acquired. For this purpose these samples were purchased directly from Sigma-Aldrich and used without further modification.

Linear discriminant analysis

Linear discriminant analysis (LDA) develops a linear classification hyperplane between multivariate objects on the basis of a set of features that describe the objects, such that the classification error is minimized. LDA has previously been used with vibrational spectra of various cell and tissue types to assign spectra to different biological classes.[19,34,35] In the present work, principal components analysis (PCA) was used as a dimensionality reduction methodology, and the principal component scores were then used as inputs to the LDA classification algorithm.

Cellular physiological assays

In parallel with the spectroscopic measurements on directly irradiated cells, quantitative measurements of various aspects of the cellular physiology were performed in triplicate experiments. Cellular proliferation and viability were quantified using parallel measurements of Alamar Blue (AB) and Neutral Red (NR) fluorescence, which quantify cellular metabolism and viability, respectively, and whose measurement is detailed elsewhere.[30] Briefly, cells were seeded at a concentration 1.9×10^3 cells per well in 96 well plates for the measurements of proliferation and viability at 96 hours post-irradiation.

Further measurements were conducted of the absorbance of MTT (3-(4,5-dimethylthiazol-2-yl)-2,5-diphenyltetrazolium bromide) at 96 hours after irradiation using the same cell concentrations as detailed above. MTT is a yellow-coloured molecule that is converted to purple formazan by mitochondrial reductases in living cells, and thus measurements of its absorbance are indicative of the mitochondrial activity in the cell. A stock MTT solution at a concentration of 5 mg ml^{-1} was prepared by dissolving MTT in sterile PBS, with a working solution prepared as a 1 : 10 dilution of the stock solution in DMEM–F12 medium without supplements.[36] After incubation of the cells with the working solution for 3 hours at 37 °C, the formazan was released from the cells using DMSO, and its absorbance was measured at 560 nm.

A clonogenic assay was also performed according to a published protocol.[37] Cells were seeded at concentrations of 200 cells per well in 6-well tissue culture plates and γ-irradiated after a 24 hour period of attachment. Ten days after irradiation, the cells were stained using carbol fuchsin, and the number of colonies formed were counted manually.

Protein extraction

HaCaT bystander/recipient cells were centrifuged at 1200 rpm for 10 minutes and washed once with phosphate-buffered saline (PBS). 100–200 μl of protein lysis buffer (Tris-buffer 1 with NP-40 detergent supplemented with dithiothreitol (DTT), phenylmethylsulfonylfluride (PMSF), sodium fluoride (NaF) and sodium vanidate (NaVa)) were added to the cell pellet and the cells were incubated on ice for 45 minutes to allow proteins to be released into the cellular environment. The lysed cells were spun at 12 000 rpm and the supernatant containing the proteins was transferred into new tubes. Protein estimation was carried out on the supernatant using a multiwell plate (Falcon) with a Bio-Rad D_C protein assay kit (Bio-Rad) according to the manufacturer's instructions (Falcon, St. Louis, MO, USA). The protein concentration was obtained by reading the samples at 595 nm wavelength on a micro-plate reader (Tecan Genios). The protein samples were prepared at a concentration of 50 μg μl^{-1} and protein loading dye (1 mM Tris·Cl (pH 6.8), 200 mM dithiothreitol, 4% SDS (electrophoresis grade), 0.2% bromophenol blue and 20% glycerol) was added to a volume suitable for the wells in the PAGE gel (25 μl max). The samples were then placed in a heat block at 100 °C for 10 minutes to denature and linearize the proteins.

Sodium dodecyl sulphate–polyacrylamide gel electrophoresis (SDS-PAGE) and western blotting

Depending on the kDa size of the proteins to be detected, 10% or 12% polyacrylamide gels were used to separate the proteins in the above samples. The separating gel was made up with 30% acrylamide (Bio-Rad laboratories), 1.5 M Tris solution pH 8.8 (BDH, Poole, UK), 10% Sodium Dodecyl Sulphate (SDS) (Sigma-Aldrich), 10% APS, *N*′-tetramethylethylenediamine (TEMED) (Sigma-Aldrich) and water, to a final desired volume for the electrophoresis apparatus. A 4% stacking gel was then made up with the same constituents as above, except the 1.5 M Tris solution (BDH) was replaced with a 1 M Tris solution at pH 6.8. The samples were loaded carefully onto the gel, leaving the end lanes free for loading commercial protein markers. The main protein marker used was Benchmark™ pre-stained ladder (Invitrogen) because of the required range of proteins from 10–200 kDa. The samples were run through the gel with 1× PAGE running buffer (made from 5 × 30 g Tris per l, 144 g glycine per l and 100 ml 10% SDS per l) at a constant electrical current of 20 mA per gel.

The separated immobilised proteins on the PAGE gel were then transferred/blotted onto a hybond C nitrocellulose membrane (Amersham-Pharmacia Biotech). The gel was placed on top of the membrane and two pieces of blotting paper were sandwiched each side. The proteins were transferred from the gel onto the membrane through an electrical field of 100 volts for 1 hour in a tank transfer apparatus (Bio-Rad laboratories) with 1× high molecular weight (HMW) transfer buffer, which was 100 ml of 10× HMW buffer (of 60 g l^{-1} Tris, 30 g l^{-1} glycine and

36 ml of 10% SDS) added to 200 ml of methanol and made up to 1 litre and pre-cooled, fresh for every western blot transfer.

Immunoblotting and chemiluminescent detection of JNK-1, ERK-3, caspase 8 and caspase 3 proteins

The western blot membranes were incubated with 5% non-fat milk powder in Tris-buffered saline with 0.1% Tween 20 solution ((TBS-T) 2.42 g l^{-1} Tris, 8.0 g l^{-1} $NaCl_2$, 0.1% Tween and adjusted to pH 7.6 with HCl) for 1 hour at RT, to block non-specific protein binding to the membrane. Each of the primary antibodies (IgG) (R&D systems) was diluted according to manufacturers' instructions in a 3% milk protein solution in TBS-T. For human c-jun N-terminal Kinase-1 (JNK-1) monoclonal mouse (MAB1387 clone) antibody and human extracellular signal-regulated kinase 4 (ERK4) monoclonal rat (MAB3914 clone), a concentration of 0.2 μg ml^{-1} was recommended by the manufacturers and was therefore applied. Secondary antibodies were conjugated to horseradish peroxidase for chemiluminescent detection. Anti-mouse IgG-HRP antibody (HAF007) and anti-rat IgG-HRP antibody (HAF005) were used with JNK-1 and ERK-4 primary antibodies respectively.

For human caspase-8, affinity purified polyclonal rabbit (AF1650 clone) antibody and human caspase-3, affinity purified polyclonal goat (AF-605-NA) antibody, a concentration of 0.5 μg ml^{-1} was recommended by the manufacturers and was therefore applied. Similarly, anti-rabbit IgG-HRP antibody (HAF008) and anti-goat IgG-HRP (HAF017) antibody were used with caspase-8 and caspase-3 primary antibodies, respectively. The western blots were exposed to the primary antibodies diluted in 3% protein solution and incubated overnight and were then washed three times with TBS-T buffer for 10 minutes each to remove unbound antibody. The secondary antibodies were all diluted 1 : 1000 in 5% milk protein solution and incubated on the membranes for 1 hour at room temperature. The membrane was washed three times with TBS–T buffer as above.

Protein bands were visualised by using an ECL chemiluminescence detection kit (Amersham), which detects the HRP-tagged secondary antibodies conjugated to the primary antibodies for JNK-1, ERK-4, caspase-8 and caspase-3, and the blot was imaged using an X-ray film (Bio-rad), which was developed by an X-ray developer machine (Agfa).

Results and discussion

A. Spectroscopic analyses of directly irradiated cells

A. 1. Analysis of variation in physiological endpoints with dose. As the variation in spectroscopic content is likely to be associated with variation in physiological activity in the cell, the results of the physiological assays at 96 hours after irradiation were interrogated with respect to their agreement with the Linear Quadratic (LQ – eqn (1)) and Induced Repair (IR – eqn (2)) models of survival *versus* radiation dose (see the introduction for full equations of each).

The cell surviving fraction (survival) at each dose was calculated at each dose point using the clonogenic assay data. It was found that the survival of the HaCaT keratinocytes as a function of radiation dose (measured at 10 days after irradiation) was in better agreement with the IR model of cell survival in comparison with the LQ model of survival. The variation in cell survival as a function of dose is shown in Fig. 1 and is non-monotonic with a local minimum in the region of 0.1 Gy (100 mGy). This observation implies that HaCaT keratinocytes exhibit low dose hyperradiosensitivity (HRS) and increased radioresistance (IRR). An estimate of the location of this local minimum, signifying the threshold for transition from HRS to IRR (D_C) in HaCaT cells, was ascertained from the fitting parameters of eqn (2) to be 89 mGy (with 95% confidence intervals (CI) from 26 mGy to 151 mGy). This figure is in agreement with the 100–200 mGy level noted previously.[11] HRS and IRR have not previously been observed in HaCaT cells using the clonogenic assay,[38,39] although HRS has been observed in epidermal and epithelial keratinocytes *in vivo*.[40,41] The other physiological assay results (AB, NR, MTT) at 96 hours after irradiation also agree well with the IR model (Fig. 2), and the value of D_C determined from this data was 220 mGy (95% CI: 60 mGy, 380 mGy), 152 mGy (95% CI: 61 mGy, 240 mGy) and 277 mGy (95% CI: 0 mGy, 1100 mGy) (from the AB, NR and MTT data respectively). Each of these physiological parameters is slightly different to that measured by the clonogenic assay (reproductive capacity), and they are measured at an earlier time point (96 hours after irradiation as compared to 10 days after irradiation in the case of the clonogenic assay), which may account for the slight differences in the estimates of D_C obtained from the IR fit to each. Each assay is in reasonable agreement with the IR model, and, importantly, the presence of a local minimum at approximately 200 mGy in each IR model fit confirms the presence of a HRS/IRR response in HaCaT cells.

A. 2. Analysis of variation in spectroscopic features of DNA with dose at 96 hours after irradiation. Sample FTIR and Raman spectra of the sham-irradiated HaCaT cells are depicted in Fig. 3. These are in agreement with spectra of this cell line measured previously.[31,42,43]

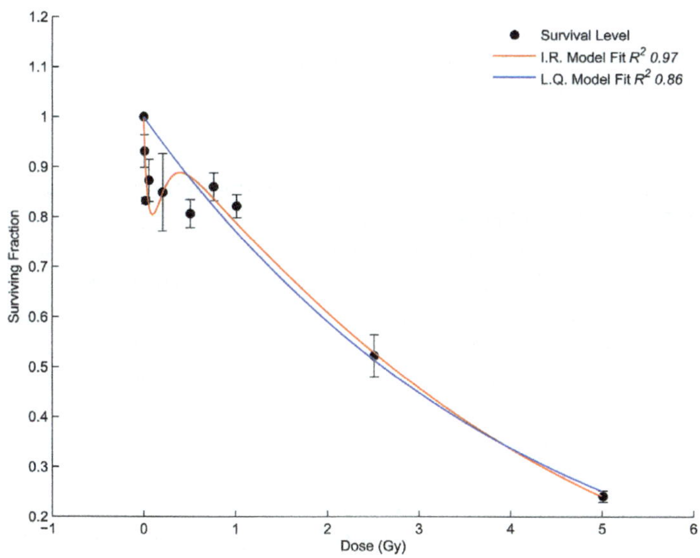

Fig. 1 Variation in survival of HaCaT keratinocytes with dose, with data fitted using the Linear Quadratic (LQ) and Induced Repair (IR) models of cellular survival. Error bars depict the standard error on the mean at each dose point, where survival measurements have been performed three times in triplicate.

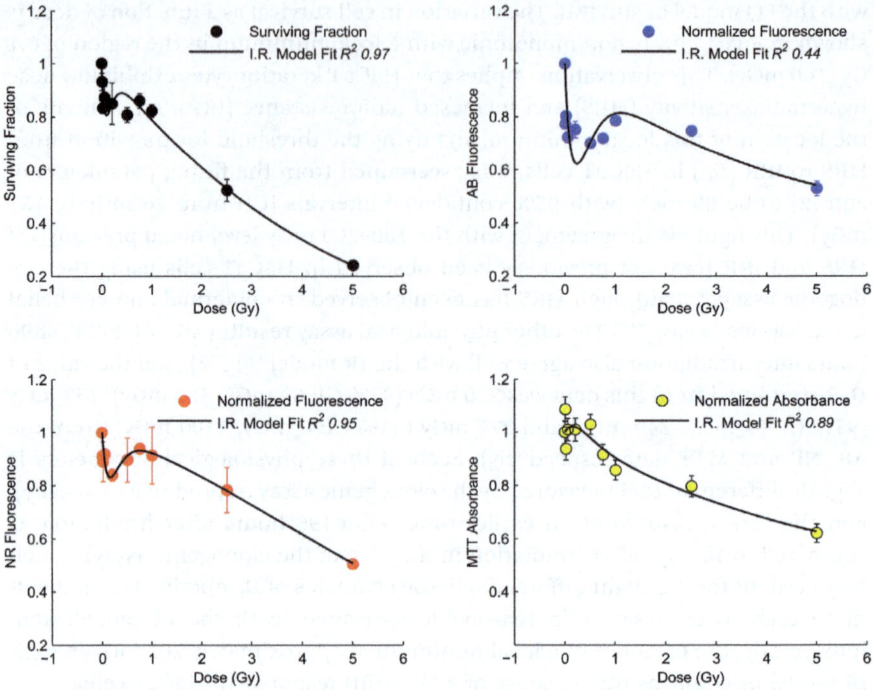

Fig. 2 Variation in (a) clonogenic response, (b) Alamar Blue (AB) fluorescence, (c), Neutral Red (NR) fluorescence and (d) MTT fluorescence with dose, with data fitted using the Induced Repair (IR) model of cellular survival. Error bars depict the standard error on the mean at each dose point, where all measurements have been performed three times in triplicate and normalized to their sham-irradiated controls.

The variation in the spectroscopic absorbance (for FTIR data) or intensity (for Raman data) of various bands with dose was examined at 96 hours after irradiation, with a view to determining whether the variation in spectral features with dose followed the IR model. As the IR model is designed for use in fitting survival data that decrease with increasing dose relative to the sham-irradiated control, as shown in Fig. 1, the following procedure was used to express band areas as a function of those in the control spectrum. Mean spectra were calculated at each dose point and the difference spectrum between the spectrum at any dose point and control spectrum was calculated. The band area of interest in the difference spectrum was then calculated and expressed as a fraction of the band area in the control. While this means that band areas will appear to decrease compared to those seen in the control, it will be noted in the following analysis whether band areas were in fact seen to increase with dose relative to those in the control.

Fig. 4a depicts the variation of the phosphodiester bond ($\nu - PO_4^-$ of deoxyribose) in DNA measured *via* FTIR and the stretching vibration of the DNA backbone (895 to 898 cm^{-1}) measured *via* Raman spectroscopy. It is clear that the variation in the vibrational amplitude of these features in both FTIR and Raman spectra agrees well with the IR model, and overall, the amplitudes are found to decrease relative to the control as a function of dose. Interestingly, the values of D_C predicted by the fits to the vibrations of the DNA backbone and the

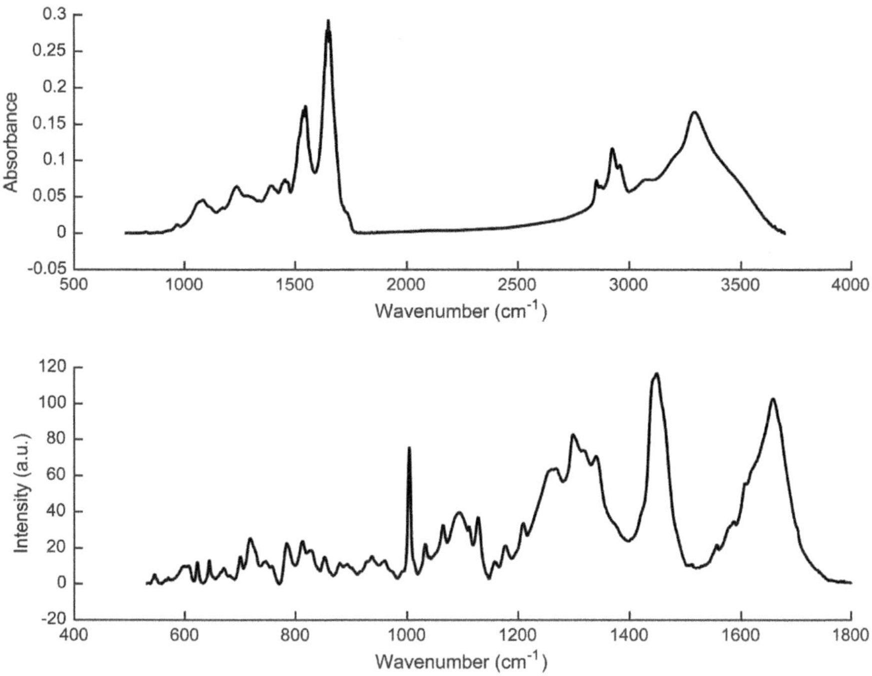

Fig. 3 Sample FTIR (top panel) and Raman spectra (bottom panel) of HaCaT keratinocytes.

phosphodiester stretching in the Raman spectra, and that predicted by the asymmetric stretching vibration in the FTIR spectra, are all in excellent agreement with each other and the value of D_C of 213 mGy (95% CI 115 mGy to 312 mGy) is within the bounds of confidence predicted by the physiological assays discussed earlier. It is therefore likely that the variation in amplitude of the DNA vibrational intensity is due to structural modifications in the DNA molecule as a result of radiation-induced DNA lesions and associated damage repair processes.

The transcription of RNA from cellular DNA and the associated synthesis of protein depends on the integrity of the DNA, and therefore on the level of DNA damage and the extent of DNA repair. It would therefore be expected that the trend in the vibrational intensities of both species, as a function of dose, should follow that seen already in respect of DNA vibrational modes. The change in absorbance of the RNA backbone stretching vibration at 996 cm^{-1} in the FTIR spectra (which overlaps with the ring bending mode of uracil) and the change in intensity of the RNA phosphodiester symmetric stretching vibration at 811 cm^{-1} in the Raman spectra are shown in Fig. 4b. Both follow the IR model in their variation with dose, and decrease with respect to the control. In Fig. 4c, the change in the absorbance of the amide I α-helix mode in both the FTIR (a) and Raman (b) spectra is also presented. Again, in both instances, the variation with dose follows the IR model, but in this case increases in the absorbance of the spectral features relative to the control sample are observed, as a function of dose.

There is no evidence of substantial changes to the structure of protein, as there are no significant differences in the levels of α-helix, β-sheet, or disordered structural vibrations (data not shown) in either the FTIR or Raman spectra, as

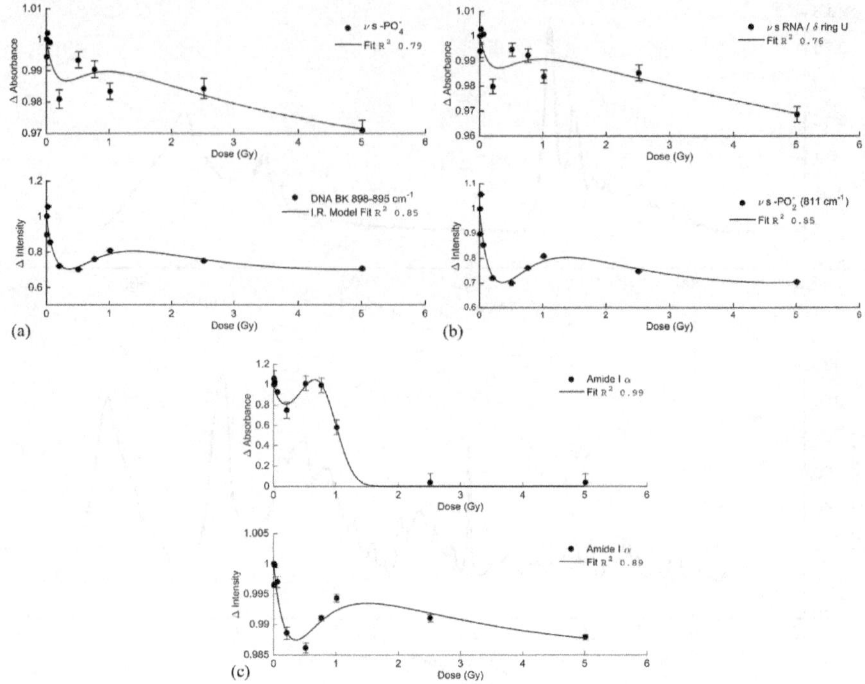

Fig. 4 (a) Change in the absorbance of the phosphodiester stretching vibration ($\nu_s - PO_4^-$) in DNA measured *via* FTIR (top) and the intensity of the DNA backbone measured *via* Raman spectroscopy (bottom) with dose at 96 hours after irradiation. Error bars depict the extent of the standard error on the mean at each dose point. The error bars on the Raman data are very slight (<1%) and so are not visible. (b) Change in the absorbance of the overlapping stretching vibration in RNA and ring bending mode in uracil measured *via* FTIR (top) and the intensity of the RNA phosphate stretching vibration measured *via* Raman spectroscopy (bottom) with dose at 96 hours after irradiation. Error bars depict the extent of the standard error on the mean at each dose point. The error bars on the Raman data are very slight (<1%) and so are not visible. (c) Change in the absorbance of the amide Iα vibration in FTIR spectra (top) and the intensity of the same vibration in Raman spectra (bottom) of HaCaT cells with dose at 96 hours after irradiation. Error bars depict the extent of the standard error on the mean at each dose point. The error bars on the Raman data are very slight (<1%) and so are not visible.

a function of dose. Overall, these features suggest decreases to the total concentration of protein within the cell, relative to the sham-irradiated control, rather than widespread structural changes. It has been previously observed by Zelig[44] and Liu[45] and their colleagues that an increase in the amide I β-sheet conformation of protein in FTIR spectra was associated with cells undergoing apoptotic or necrotic death. There is no spectral evidence of this effect either in the FTIR or Raman data shown here.

The intensities of the vibrational modes of lipids, within the fingerprint region of both the FTIR and Raman spectra, were not observed to follow the trend of the IR model (data not shown). The processes leading to lipid damage as a result of ionizing radiation exposure are not associated with damage to nuclear DNA, but rather lipid peroxidation processes mediated in the cytoplasm and mitochondria.[46] In this context, it would not be expected that lipid structural vibrations would have a correlation with the dose dependence of the IR model.

Typically, the cell senses DNA damage within an hour of irradiation and repair begins through the formation of foci around the site of damage, where large molecular complexes bind and clamp the damage site to effect repair. As spectra were measured at a time that is beyond the point (typically within 1 hour of irradiation) at which DNA damage sensing and repair occurs through standard processes, the variation in the amplitude of these features is unlikely to be associated with DNA repair. The repair of these lesions is critical to cell survival and they are thought to be repaired through parallel 'fast' and 'slow' repair processes, the latter of which can lead to DNA lesions persisting up to 128 hours after irradiation.[47] In addition, delayed DNA repair can occur, by which repair does not proceed as normal. The latter two processes account for up to 20% of DNA repair. Given that the mechanisms of HRS/IRR are dominated by DNA repair processes, the variation of the amplitude of these vibrations is likely to be governed by the combined effect of both persistent radiation-induced lesions to the backbone of the DNA molecule and restrictions to its vibrational intensity through DNA repair processes. It is known that DNA double-strand break repair processes involve binding of various large molecular complexes to the site of damage, which may restrict the vibration of the molecule, and these may account for the reduction in the intensity of the DNA bands in the Raman and FTIR spectra.

Previously it had been shown that, in regression of spectroscopic measurements against radiation dose, a non-linear multivariate regression model outperformed a multivariate linear regression.[31,43] The present study has demonstrated that this non-linearity has its origin in the spectroscopic detection of HRS and IRR and ionizing radiation-mediated DNA damage sensing mechanisms that result in cell survival or death.

B. Spectroscopic analyses of bystander-irradiated cells

B.1. MAPK and caspase protein expression from western blotting. The expression of proteins involved in the MAPK pathway and in a caspase-dependent apoptosis pathway were measured after exposure to bystander medium for 1, 6, 12 and 24 hours. Fig. 5 depicts the expression of each of the target proteins (JNK-1, ERK-4, caspase 8 and 3), as measured using western blotting. The antibody to JNK-1 mostly detects JNK1 but also the JNK 2 and 3 isoforms. The antibody to ERK-4 detects this isoform of the protein, and ERK-4 is an isoform that complexes with ERK3 to activate MAPK-activated protein kinase 5 (MEK5) in the pro-survival pathway.

Caspases exist as inactive proenzymes that undergo proteolytic processing at conserved aspartic residues to produce two subunits, large and small, that dimerize to form the active enzyme.[48,49] Caspase 8 is an initiator of the extrinsic apoptotic pathway and is known to interact with JNK for apoptosis and also inflammatory pathways and switch on caspase 3. JNK activates caspase 8 and then caspase 3 in the pro-apoptotic cascade, while also having the potential to produce Bid (a pro-apoptotic protein like Bax from the bcl2 family), which then initiates the intrinsic apoptotic cascade downstream.

JNK protein expression has a transient expression level at 0 Gy that decreases with dose at early time points (1–6 h) but increases at later time points (12–24 h).

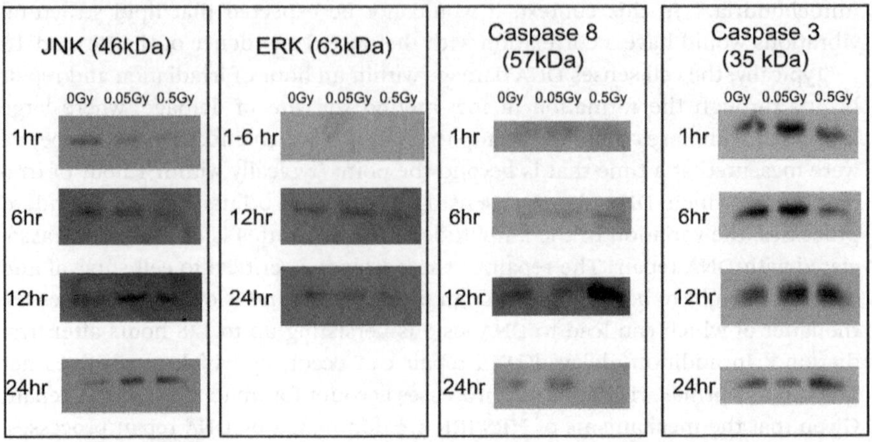

Fig. 5 Western blots of target proteins showing expression levels with dose and duration of exposure to ICCM.

The ERK 1/2 protein could not be seen at either 1 h or 6 h time points, but at 12 and 24 h, ERK 2 seems to be expressed at 0 Gy and reduces with dose in the 0.05 Gy and 0.5 Gy samples. Caspase 8 expression is seen to increase with dose for 1, 6 and 12 hours post addition of bystander medium, but at 24 hours, caspase 8 expression decreases with dose. At 12 hours post exposure to bystander medium, caspase 3 increases with dose, and this is consistent up to 24 hours. Overall, these results are consistent with those observed previously[32,50] of MAPK signalling in human keratinocytes.

B.2. Analysis of difference spectra at 30 min and 24 hours. Fig. 6a and b depict the difference Raman spectrum between the control and ICCM-exposed sample (0.5 Gy) at, respectively, 30 minutes and 24 hours. A two-sided Student's t-test was conducted to identify spectroscopic regions in the difference spectra that vary with statistical significance ($p < 0.005$) in the bystander sample with respect to the control. These are highlighted in each figure (note that the t-test treats each data point independently and therefore identifies individual points at 1 cm^{-1} intervals rather than bands). There are only very slight changes to the spectroscopic content of the ICCM-exposed cells after a 30 minutes exposure period. After 24 hours of exposure to 0.5 Gy ICCM, substantial spectroscopic modifications have occurred throughout the spectrum. Both findings are in agreement with the observations of Lyng *et al.*,[32] who demonstrated that, while no bystander effects were seen after 30 minutes exposure to ICCM, significant effects were observed after a 24 hour exposure period.

The mean spectrum of each sample after 24 hours exposure to ICCM is supplied as ESI Fig. S1.† In this figure, the mean spectrum for each sample is supplied, together with that of the medium in which it is measured. The medium spectrum itself possesses a broad band, centered at approximately 1637 cm^{-1}, which is mainly attributed to a weak water vibration, normally centered at ~1640 cm^{-1}, and small contributions from protein amide vibrations in the medium. A series of vibrations between 1498 cm^{-1} and 1096 cm^{-1} are due to CH deformation vibrations with proteins and carbohydrates (glucose, galactose), and

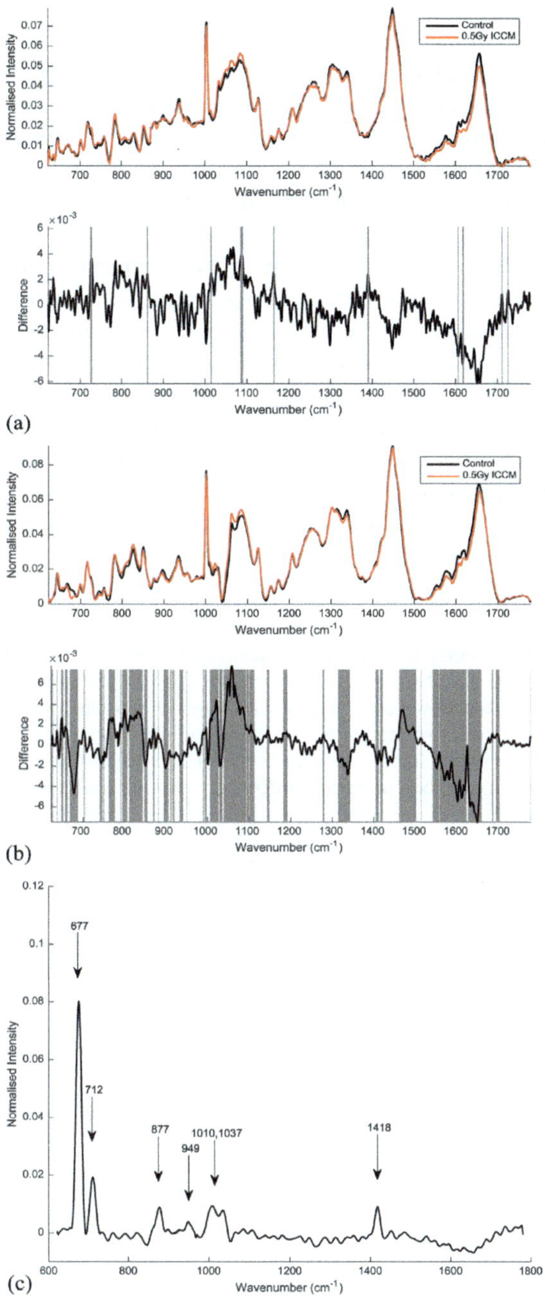

Fig. 6 (a) Mean spectra of control and 0.5 Gy ICCM-exposed sample at 30 minutes (top panel), plus difference spectrum between exposed and control sample (bottom panel). Vertical shaded regions show regions of significant difference between control and bystander-irradiated sample ($p > 0.005$). (b) Mean spectra of control and 0.5 Gy ICCM-exposed sample at 24 hours (top panel), plus difference spectrum between exposed and control sample (bottom panel). The vertical shaded regions show regions of significant difference between control and bystander-irradiated sample ($p > 0.005$). (c) Difference spectrum between bystander ICCM (0.5 Gy) and control medium (0 Gy). This is the first observation of the possible Raman spectrum of the secreted bystander factor or factors in ICCM.

characteristic ring vibration modes of amino acids and nucleic acid bases (such as phenylalanine, tryptophan, tyrosine, thymine and adenine), contained within the medium and its supplements. A strong vibration at 1044 cm^{-1} (attributed to tyrosine) and vibrations centered at 714 cm^{-1} (tyrosine), 771 cm^{-1} (glucose) and 831 cm^{-1} (phenylalanine) are also prominent. The corrected mean spectra of the sample in Fig. S1† do not contain visible spectroscopic features originating from the medium, providing evidence of the effectiveness of the correction methodology.

In Fig. 6c, a difference spectrum between the sham-irradiated control and 0.5 Gy ICCM is presented. This difference spectrum, in principle, contains the signature of the bystander effect, or molecules mediating the effect. Despite much analysis over the past number of decades, the identity of the bystander factor has never been determined. It is thought that the species may be a low-weight, heat-labile protein. Although no identification of the species is attempted here, the spectrum indicates that the molecular structure of this species potentially includes a C–S–C (stretching vibration at 677 cm^{-1} (ref. 51)) and P–C moiety (stretching vibration at 712 cm^{-1} (ref. 52)) together with C–C skeletal structure (stretch at 877 to 949 cm^{-1}),[30] C–O (1010 cm^{-1}),[30] C–O–P (1037 cm^{-1})[30] and C–H (deformation at 1418 cm^{-1}).[30]

Table 1 Confusion matrices for PCA-LDA classification with spectra taken from cells after a 30 min exposure to ICCM. In each LDA, control spectra are paired with their corresponding ICCM-exposed counterparts. The low levels of classification accuracy suggest that the spectra from cells exposed to ICCM for 30 minutes cannot be distinguished from their controls. In each sub-table rows indicate proportions in the actual class, with columns indicating proportions predicted into each class by the LDA

	Control	0.5 Gy ICCM
Control	58	42
0.5 Gy ICCM	46	54
	Control + JNK Inh.	0.5 Gy ICCM + JNK Inh.
Control + JNK Inh.	49	51
0.5 Gy ICCM + JNK Inh.	36	64
	Control + catalase	0.5 Gy ICCM + catalase
Control + catalase	63	37
0.5 Gy ICCM + catalase	32	68
	Control + ERK 1/2 Inh.	0.5 Gy ICCM + ERK 1/2 Inh.
Control + ERK 1/2 Inh.	60	40
0.5 Gy ICCM + ERK 1/2 Inh.	41	59

In the following section, a comparative analysis between the spectral features in the bystander-medium-exposed samples and their MAPK-inhibited counterparts is provided using PCA-LDA classification rates and associated loadings.

B.3. Linear discriminant analysis. HaCaT recipient cells were exposed to bystander medium for periods of either 30 minutes or 24 hours, at which point Raman spectra were acquired from them in their live forms.

Classifiers were then prepared with PCA-LDA using spectra from cells exposed to bystander medium with inhibitors for 30 minutes, whereby each bystander-irradiated sample was classified with respect to its corresponding control. In the present analysis, a 70% classification rate is considered a minimum indicating successful classification.[53] Classification rates were established using a leave-one-out-cross-validation (LOOCV) including the first four principal components. The PCA-LDA classifier was unable to distinguish between spectra of the control and ICCM-exposed samples at 30 minutes, returning an average classification accuracy of 58% (calculated using leave-one-out cross-validation and shown in Table 1).

However, after exposure to bystander medium for a period of 24 hours, the classification rates rise substantially. In Table 2, the results of LOOCV of PCA-LDA classifiers are shown, whereby each inhibited sample is classified with respect to

Table 2 Confusion matrices for PCA-LDA classification with spectra taken from cells after 24 hours exposure to ICCM. In each LDA, control spectra are paired with their corresponding 0.5 Gy ICCM-exposed sample to examine the effect of inhibition of the bystander effect on the ability of PCA-LDA to identify the bystander effect. In each sub-table, rows indicate proportions in the actual class, and columns indicate proportions predicted into each class by the LDA

	0 Gy control	0.5 Gy ICCM
0 Gy control	78	22
0.5 Gy ICCM	10	90
	0.5 Gy ICCM	**0.5 Gy ICCM + JNK Inh.**
0.5 Gy ICCM	87	13
0.5 Gy ICCM + JNK Inh.	18	82
	0.5 Gy ICCM	**0.5 Gy ICCM + catalase**
0.5 Gy ICCM	100	0
0.5 Gy ICCM + catalase	0	100
	0.5 Gy ICCM	**0.5 Gy ICCM + ERK 1/2 Inh.**
0.5 Gy ICCM	100	0
0.5 Gy ICCM + ERK 1/2 Inh.	0	100

the 0.5 Gy bystander-irradiated sample (ICCM) at this time point. The classification rates do vary with the inhibitor used with the ICCM, but are all relatively high, whereby the explained variance by the PCA totalled approximately 60% in the first 6 PCs. These observations demonstrate that the Raman signal includes molecular evidence of the effect of the ICCM exposure on the cell, and can discriminate as to whether the exposure is attenuated by inhibition of MAPK transduction and reactive oxygen species (ROS) attack. In addition, the small level of explained variance also demonstrates the complexity of the spectroscopic changes mediated by the bystander effect.

The PC scores and loadings plots for each individual classification are shown in Fig. 7a to d. In each PC scores plot, the location of the PCA LDA classification hyperplane is used to determine the combination of loadings that differentiate the samples, and therefore explain the spectroscopic origin of the classification.[54] In Fig. 7a it is shown that the differentiation between the control and ICCM-exposed sample is through a combination of PC1 and PC2. PC1 is dominated by the spectrum of a protein strongly resembling histone, whereas PC2 is dominated

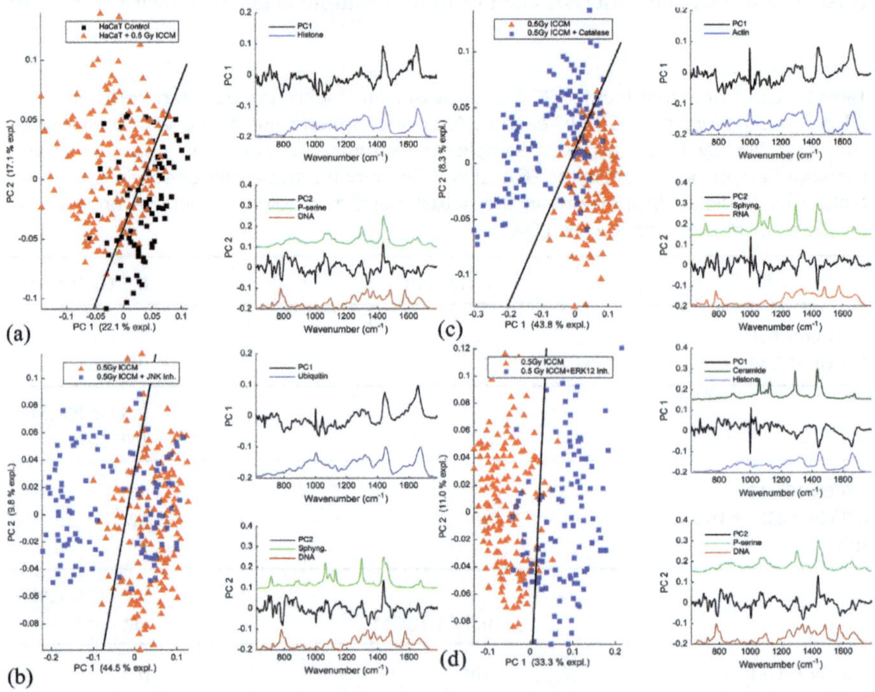

Fig. 7 The black line in each scores plot depicts the location of the classification hyperplane and the percentage labels on the score plot axes indicate the variance explained by that principal component. (a) Principal component scores plot (left) and loadings (right) in the differentiation of control from ICCM-exposed HaCaT cells at 24 hours. (b) Principal component scores plot (left) and loadings (right) in the differentiation of ICCM-exposed HaCaT cells from JNK-inhibited cells at 24 hours. (c) Principal component scores plot (left) and loadings (right) in the differentiation of ICCM-exposed HaCaT cells from catalase-inhibited cells at 24 hours. (d) Principal component scores plot (left) and loadings (right) in the differentiation of ICCM-exposed HaCaT cells from ERK12-inhibited HaCaT cells at 24 hours.

by modes originating from both phosphatidylserine (P-serine) and DNA. The proportions of variance described by each component suggest that the loadings of PC1 are the dominant spectroscopic effect of exposure to ICCM, whereas the loading of PC2 describes the within class variance of the spectra in both datasets, while also contributing to the PCA-LDA discrimination. In Fig. 5, it is shown that bystander signalling in HaCaT cells exposed to 0.5 Gy ICCM includes upregulation of JNK, ERK and caspase 3 and 8 over a 24 hour period. Although the expression is shown to be temporally variant, in particular caspase 3 signalling is strongly sustained to 24 hours. This suggests that the observed loadings differentiating control and ICCM-exposed cells may be associated with apoptotic effects including externalisation of phosphatidylserine, DNA cleaving and protein coagulation.

Loadings similar to those in PC2 have previously been observed after exposure of prostate,[21] lung and breast[20] to X-rays (6 MeV), with doses between 15 Gy and 50 Gy. In these studies, this component was linked to variability between cell lines due to cell cycle position. In the present instance, PC2 may therefore be partially associated with cell cycle variability arising from bystander-induced G2 phase arrest in DNA repair. No evidence of signatures of elevated glycogen as a marker of radiobiological effect were found here, however, in contrast to those previously observed in directly irradiated lung, breast and prostate tumour cells at doses above 2 Gy.[55] However, as the present work has been conducted with a cell line from non-cancerous tissue cultured *in vitro* with full supplements and non-hypoxic conditions, glycogen synthesis may not be required to ensure cell survival.[56]

In Fig. 7b, the effect of inhibition of the pro-apoptotic JNK enzyme is seen, in this case through the emergence of signatures of signalling proteins (ubiquitin) in PC1, and overlapping components of the cellular membrane (sphyngomyelin) and DNA in PC2. This may be evidence of increased cellular proliferation and DNA synthesis in the JNK-inhibited sample relative to the ICCM-exposed sample. The classification rate in Table 2 would suggest that, although the inhibition of JNK signalling contributes to cell survival after ICCM exposure, this process is incomplete, a proportion of cells receiving ICCM-mediated stress and pro-apoptotic signalling by an alternative pathway. Again the proportions of variance described by each component suggest that the differentiation between classes is due to a substantial increase in pro-survival signalling in the JNK-inhibited sample represented by the signature in PC1, with a weak effect due to pro-apoptotic signalling in PC2.

In Fig. 7c, the effect of inhibition of reactive oxygen species and caspase-dependent apoptosis is investigated. Here, the emergence of signatures of structural proteins within the cytoplasm (actin) and membrane (sphyngomyelin) are seen together with evidence of transcription (RNA). Again, these are suggestive of anti-apoptotic signalling through inhibition of stress transduction signals that are present in the ICCM. As catalase is an effective inhibitor of ROS in keratinocytes, it will abrogate the ROS stress signalling, which leads to JNK phosphorylation and pro-apoptotic signalling; the high level of classification relative to the ICCM-exposed sample indicates that this inhibition is seen in all of the cells analysed here.

In Fig. 7d, the effect of inhibition of the phosphorylation of pro-survival enzymes ERK1 and ERK2 is investigated. In this case, PC1 provides the bulk of the discrimination between the ICCM-exposed sample and the ERK-inhibited sample, with PC2 providing a slight contribution. The dominant loading that provides the discrimination is PC1, which contains a mixture of signatures from histone and ceramide. In PC2, evidence again of the potential externalisation of phosphatidylserine and DNA condensation occurring during apoptosis is apparent, though this is a weak source of common variance in the ICCM-exposed sample and ERK-inhibited sample, and therefore demonstrates that this effect occurs in both cell lines. The loadings in PC1 are consistent with an increase in apoptotic signalling effects, which may exceed the levels in the 0.5 Gy ICCM-exposed sample, an observation supported by the high level of classification accuracy.

These results suggest strongly that Raman spectroscopy can detect, and differentiate, bystander-induced stresses and their effects on pro- and anti-apoptotic signalling in the cell. Together with classification rates, the Raman signatures provide a picture of the interplay between ROS stress and signalling from pro- and anti-apoptotic enzymes in contributing to the fate of the cell.

Conclusions

This study demonstrates, for the first time, the sensitivity of vibrational spectroscopy to both direct targeted radiation-induced effects across the continuum of doses from low to high, and to non-targeted effects induced by bystander signalling. In practical exposure situations, in which both effects contribute to cell stress, their combined impact renders spectral interpretation challenging. However, the results here demonstrate that separating the spectroscopic evidence of each effect is possible, and that this evidence aligns well with parallel biological metrics of radiobiological effects. The ability to detect both directly mediated damage and stress-induced effects is an important observation. The results are encouraging for the future of vibrational spectroscopy in radiobiological research.

Acknowledgements

This research was supported by the National Biophotonics and Imaging Platform (NBIP) Ireland funded under the Higher Education Authority PRTLI (Programme for Research in Third Level Institutions) Cycle 4, co-funded by the Irish Government and the European Union Structural Fund and by the FP6 Integrated Project, *Non-targeted effects of ionising radiation* (*NOTE*) FI6R 036465. The authors also acknowledge funding through the Irish Research Council for Science Engineering and Technology 'Ulysses' France–Ireland mobility programme.

References

1 W. F. Morgan and M. B. Sowa, *Cancer Lett.*, 2015, **356**, 17–21.
2 C. Mothersill and C. B. Seymour, *Nat. Rev. Cancer*, 2004, **4**, 158–164.
3 K. M. Prise, G. Schettino, M. Folkard and K. D. Held, *Lancet Oncol.*, 2005, **6**, 520–528.
4 G. Schettino, M. Folkard, B. D. Michael and K. M. Prise, *Radiat. Res.*, 2005, **163**, 332–336.

5 H. Zhou, G. Randers-Pehrson, C. A. Waldren, D. Vannais, E. J. Hall and T. K. Hei, *Proc. Natl. Acad. Sci. U. S. A.*, 2000, **97**, 2099–2104.
6 C. Mothersill and C. B. Seymour, *Nat. Rev. Cancer*, 2004, **4**, 158–164.
7 W. F. Morgan, *Radiat. Res.*, 2003, **159**, 567–580.
8 W. F. Morgan, *Radiat. Res.*, 2003, **159**, 581–596.
9 L. J. Wu, G. Randers-Pehrson, A. Xu, C. A. Waldren, C. R. Geard, Z. Yu and T. K. Hei, *Proc. Natl. Acad. Sci. U. S. A.*, 1999, **96**, 4959–4964.
10 M. C. Joiner, B. Marples, P. Lambin, S. C. Short and I. Turesson, *Int. J. Radiat. Oncol., Biol., Phys.*, 2001, **49**, 379–389.
11 B. Marples, B. G. Wouters, S. J. Collis, A. J. Chalmers and M. C. Joiner, *Radiat. Res.*, 2004, **161**, 247–255.
12 B. Marples and S. J. Collis, *Int. J. Radiat. Oncol., Biol., Phys.*, 2008, **70**, 1310–1318.
13 L. Martin, B. Marples, M. Coffey, M. Lawler, D. Hollywood and L. Marignol, *Radiat. Res.*, 2009, **172**, 405–413.
14 C. J. Bakkenist and M. B. Kastan, *Nature*, 2003, **421**, 499–506.
15 B. Xu, S. T. Kim, D. S. Lim and M. B. Kastan, *Mol. Cell. Biol.*, 2002, **22**, 1049–1059.
16 M. Fernet, F. Megnin-Chanet, J. Hall and V. Favaudon, *DNA Repair*, 2010, **9**, 48–57.
17 L. Enns, K. T. Bogen, J. Wizniak, A. D. Murtha and M. Weinfeld, *Mol. Cancer Res.*, 2004, **2**, 557–566.
18 S. A. Krueger, M. C. Joiner, M. Weinfeld, E. Piasentin and B. Marples, *Radiat. Res.*, 2007, **167**, 260–267.
19 A. Maguire, I. Vegacarrascal, L. White, B. McClean, O. Howe, F. M. Lyng and A. D. Meade, *Radiat. Res.*, 2015, **183**, 407–416.
20 Q. Matthews, A. Jirasek, J. J. Lum and A. G. Brolo, *Phys. Med. Biol.*, 2011, **56**, 6839–6855.
21 Q. Matthews, A. Brolo, J. Lum, X. Duan and A. Jirasek, *Phys. Med. Biol.*, 2011, **56**, 19–38.
22 N. Gault, O. Rigaud, J. L. Poncy and J. L. Lefaix, *Int. J. Radiat. Biol.*, 2005, **81**, 767–779.
23 K. Valerie, A. Yacoub, M. P. Hagan, D. T. Curiel, P. B. Fisher, S. Grant and P. Dent, *Mol. Cancer Ther.*, 2007, **6**, 789–801.
24 P. Dent, A. Yacoub, P. B. Fisher, M. P. Hagan and S. Grant, *Oncogene*, 2003, **22**, 5885–5896.
25 M. Raman, W. Chen and M. H. Cobb, *Oncogene*, 2007, **26**, 3100–3112.
26 L. Chang and M. Karin, *Nature*, 2001, **410**, 37–40.
27 E. F. Wagner and A. R. Nebreda, *Nat. Rev. Cancer*, 2009, **9**, 537–549.
28 K. Matsuda, T. Idezawa, X. J. You, N. H. Kothari, H. Fan and M. Korc, *Cancer Res.*, 2002, **62**, 5611–5617.
29 P. Dent, A. Yacoub, J. Contessa, R. Caron, G. Amorino, K. Valerie, M. P. Hagan, S. Grant and R. Schmidt-Ullrich, *Radiat. Res.*, 2003, **159**, 283–300.
30 A. D. Meade, F. M. Lyng, P. Knief and H. J. Byrne, *Anal. Bioanal. Chem.*, 2007, **387**, 1717–1728.
31 A. D. Meade, C. Clarke, H. J. Byrne and F. M. Lyng, *Radiat. Res.*, 2010, **173**, 225–237.
32 F. M. Lyng, P. Maguire, B. McClean, C. Seymour and C. Mothersill, *Radiat. Res.*, 2006, **165**, 400–409.

33 A. Maguire, I. Vega-Carrascal, J. Bryant, L. White, O. Howe, F. M. Lyng and A. D. Meade, *Analyst*, 2015, **140**, 2473–2481.

34 K. Das, N. Stone, C. Kendall, C. Fowler and J. Christie-Brown, *Lasers Med. Sci.*, 2006, **21**, 192–197.

35 A. Nijssen, S. Koljenovic, T. C. Bakker Schut, P. J. Caspers and G. J. Puppels, *J. Biophotonics*, 2009, **2**, 29–36.

36 M. Davoren, E. Herzog, A. Casey, B. Cottineau, G. Chambers, H. J. Byrne and F. M. Lyng, *Toxicol. in Vitro*, 2007, **21**, 438–448.

37 N. A. Franken, H. M. Rodermond, J. Stap, J. Haveman and C. van Bree, *Nat. Protoc.*, 2006, **1**, 2315–2319.

38 L. A. Ryan, C. B. Seymour, M. C. Joiner and C. E. Mothersill, *Int. J. Radiat. Biol.*, 2009, **85**, 87–95.

39 C. Mothersill, C. B. Seymour and M. C. Joiner, *Radiat. Res.*, 2002, **157**, 526–532.

40 I. Turesson, J. Nyman, F. Qvarnstrom, M. Simonsson, M. Book, I. Hermansson, S. Sigurdardottir and K. A. Johansson, *Radiother. Oncol.*, 2010, **94**, 90–101.

41 M. Simonsson, F. Qvarnstrom, J. Nyman, K. A. Johansson, H. Garmo and I. Turesson, *Radiother. Oncol.*, 2008, **88**, 388–397.

42 A. D. Meade, H. J. Byrne and F. M. Lyng, *Mutat. Res., Rev. Mutat. Res.*, 2010, **704**, 108–114.

43 A. D. Meade, C. Clarke, H. J. Byrne and F. M. Lyng, in *2014 IEEE International Conference on Bioinformatics and Biomedicine (BIBM)*, IEEE, 2014, pp. 254–260.

44 U. Zelig, J. Kapelushnik, R. Moreh, S. Mordechai and I. Nathan, *Biophys. J.*, 2009, **97**, 2107–2114.

45 K. Z. Liu, L. Jia, S. M. Kelsey, A. C. Newland and H. H. Mantsch, *Apoptosis*, 2001, **6**, 269–278.

46 E. I. Azzam, J.-P. Jay-Gerin and D. Pain, *Cancer Lett.*, 2012, **327**, 48–60.

47 A. Noda, Y. Hirai, K. Hamasaki, H. Mitani, N. Nakamura and Y. Kodama, *J. Cell Sci.*, 2012, **125**, 5280–5287.

48 E. M. Creagh, *Trends Immunol.*, 2014, **35**, 631–640.

49 M. Leist and M. Jäättelä, *Nat. Rev. Mol. Cell Biol.*, 2001, **2**, 589–598.

50 H. Furlong, C. Mothersill, F. M. Lyng and O. Howe, *Mutat. Res.*, 2013**741–742**, 35–43.

51 P. J. Caspers, A. C. Williams, E. A. Carter, H. G. Edwards, B. W. Barry, H. A. Bruining and G. J. Puppels, *Pharm. Res.*, 2002, **19**, 1577–1580.

52 S. D. Christesen, *Appl. Spectrosc.*, 1988, **42**, 318–321.

53 A. Nijssen, T. C. Bakker Schut, F. Heule, P. J. Caspers, D. P. Hayes, M. H. Neumann and G. J. Puppels, *Journal of Investigative Dermatology*, 2002, **119**, 64–69.

54 F. L. Martin, J. G. Kelly, V. Llabjani, P. L. Martin-Hirsch, I. I. Patel, J. Trevisan, N. J. Fullwood and M. J. Walsh, *Nat. Protoc.*, 2010, **5**, 1748–1760.

55 Q. Matthews, M. Isabelle, S. J. Harder, J. Smazynski, W. Beckham, A. G. Brolo, A. Jirasek and J. J. Lum, *PLoS One*, 2015, **10**, e0135356.

56 M. C. Brahimi-Horn, G. Bellot and J. Pouysségur, *Curr. Opin. Genet. Dev.*, 2011, **21**, 67–72.

Faraday Discussions

PAPER

Mie scatter corrections in single cell infrared microspectroscopy

Tatiana Konevskikh,*[a] Rozalia Lukacs,[a] Reinhold Blümel,[b] Arkadi Ponossov[a] and Achim Kohler[a]

Received 16th November 2015, Accepted 16th December 2015
DOI: 10.1039/c5fd00171d

Strong Mie scattering signatures hamper the chemical interpretation and multivariate analysis of the infrared microscopy spectra of single cells and tissues. During recent years, several numerical Mie scatter correction algorithms for the infrared spectroscopy of single cells have been published. In the paper at hand, we critically reviewed existing algorithms for the correction of Mie scattering and suggest improvements. We developed an iterative algorithm based on Extended Multiplicative Scatter Correction (EMSC), for the retrieval of pure absorbance spectra from highly distorted infrared spectra of single cells. The new algorithm uses the van de Hulst approximation formula for the extinction efficiency employing a complex refractive index. The iterative algorithm involves the establishment of an EMSC meta-model. While existing iterative algorithms for the correction of resonant Mie scattering employ three independent parameters for establishing a meta-model, we could decrease the number of parameters from three to two independent parameters, which reduced the calculation time for the Mie scattering curves for the iterative EMSC meta-model by a factor of 10. Moreover, by employing the Hilbert transform for evaluating the Kramers–Kronig relations based on a FFT algorithm in Matlab, we further improved the speed of the algorithm by a factor of 100. For testing the algorithm we simulate distorted apparent absorbance spectra by utilizing the exact theory for the scattering of infrared light at absorbing spheres, taking into account the high numerical aperture of infrared microscopes employed for the analysis of single cells and tissues. In addition, the algorithm was applied to measured absorbance spectra of single lung cancer cells.

Introduction

Since the invention of infrared microscopes in the 90s, infrared microscopy has become an attractive tool for the analysis of single cells and connective tissues in biological and medical sciences.[1] *Via* infrared microscopy, tissues and cells can be analysed chemically without destroying the chemical structure of the material.

[a]*Department of Mathematical Sciences and Technology (IMT), Norwegian University of Life Sciences, 1430 Ås, Norway. E-mail: tatiana.konevskikh@nmbu.no; Fax: +47 64 96 54 01; Tel: +47 67 23 15 84*
[b]*Department of Physics, Wesleyan University, Middletown, Connecticut 06459-0155, USA*

Unfortunately, strong scatter effects that hamper the chemical interpretation of spectra occur, which have been interpreted as Mie-type scattering.[2] According to Mie theory,[3,4] single cells are highly efficient scatterers in the infrared, since the wavelength of the infrared radiation is the same order as the size of the cells.

The correction of Mie type scattering in the infrared spectra of single cells is difficult, since scattering and absorption are highly entangled. The reason for this is that due to the strong scattering, a large part of the scattered light does not reach the detector: only scattered light that is collected by the Schwarzschild optics reaches the detector, while a substantial part of the light is scattered into solid angles, which are not covered by the Schwarzschild optics. The scattered light that does not reach the detector leads to apparent absorption signatures in the measured absorbance spectrum, and the researcher is at a loss to decide which part of the measured absorbance spectrum is due to scattering and which part is due to chemical absorption. Thus, in the FTIR microspectroscopy of single cells, the measured absorbance spectrum is also termed the apparent absorbance spectrum, since it contains both absorption and scattering signatures. Due to the difficulties in the interpretation of absorption peaks, the scattering and absorption of infrared radiation at cells have been studied during recent years and algorithms have been established to separate scattering and absorption in the apparent absorbance spectra from the infrared spectroscopy of single cells.[5–11] All established algorithms, are based on Mie-theory, an exact theory for the scattering of infrared light at absorbing spheres developed by Gustav Mie in 1905.[3] The theory describes the scattering of electromagnetic radiation at spheres rigorously. Based on the exact Mie theory and approximation formulas thereof, a number of approaches have been developed, which aim at disentangling the scattering and absorption signatures in the infrared spectra of single cells. In 2008, Kohler *et al.* developed an algorithm based on extended multiplicative signal correction (EMSC), which corrects Mie scattering in the infrared spectra of cells.[9] The algorithm employs an approximation formula for the extinction efficiency and takes into account a constant refractive index. Since the exact size, morphology and refractive index of a cell are unknown in all practical situations, the algorithm employs a meta-model that comprises a range of cell sizes and refractive indices. The algorithm involves several approximations. (1) It considers only the extinction efficiency in the forward direction when calculating the absorbance, while infrared microscopes use Schwarzschild optics and collect light over a large numerical aperture. (2) The extinction in the forward direction is calculated by an approximation formula developed by v. de Hulst.[4] (3) The algorithm considers the refractive index as constant and real in the employed wavelength range, while it is known that for absorbing scatterers, the refractive index has a non-zero imaginary part and a fluctuating real part, *i.e.*, is dispersive (4). The algorithm approximates the apparent absorbance by the scattering efficiency. Despite the approximations employed, the algorithm corrects the broad Mie scatter oscillations in the apparent absorbance of single cell spectra, while so-called dispersive artefacts due to the resonant Mie effect remain uncorrected. The term 'dispersive artefact' circumscribes the fact that absorption resonances lead to fluctuations in the real part of the refractive index, which affect the extinction efficiency and thereby the measured absorbance spectrum. In 2010, Bassan *et al.*[6] developed an iterative algorithm based on EMSC, which allows correction of the dispersive artefact. The algorithm developed by Bassan *et al.*[6] involves several approximations. (1) As in

Kohler's algorithm,[9] the apparent absorbance is approximated by the extinction efficiency in the forward direction developed by v. de Hulst.[4] (2) While Bassan's algorithm[6] takes into account a fluctuating real refractive index, it does not use the imaginary part of the refractive index. Recently, van de Dijk *et al.*[10] have introduced an iterative algorithm taking into account the high numerical aperture of infrared microscopes. The algorithm employs exact Mie formulas for the absorbance and treats the refractive index as complex. While it was shown that the algorithm corrects sharp dispersive Mie signatures in the spectra of PMMA spheres,[10] it has also been demonstrated that the algorithm is not applicable to biological samples, which have a complex composition and a shape that often deviates substantially from a sphere.[11]

Although the algorithm of Bassan *et al.*[6] does not take into account a complex refractive index, it has been widely employed during recent years, since it provides a stable extraction of the pure absorbance spectrum. It is currently considered as the best existing method for correcting the Mie scatter distortions in the infrared microspectroscopy of cells and tissues.

The aim of the present paper is to further develop the algorithm by Bassan *et al.*[6] by taking into account a complex refractive index according to Mie theory. The algorithm of Bassan *et al.*[6] requires the calculation of the Kramers–Kronig transform iteratively. Since this turns out to be a very time consuming process, we further aimed at improving the speed of the algorithm. In order to test the newly developed algorithm, we created a test set of apparent absorbance spectra. The test set consists of spectra that were distorted according to exact Mie theory. For the creation of the test set, the optical set-up including the Schwarzschild optics with focusing and collecting optics has been taken into account. The algorithm presented in this paper is based on the algorithm of Kohler *et al.*[9] and the algorithm of Bassan *et al.*[6] based on a meta-model taking into account a broad parameter range for parameters such as the refractive index, size of the cell and effective sample thickness. While the resonant Mie scatter correction algorithm developed by Bassan *et al.*[6] uses three independent parameters for building the meta-model, we will show in the present paper that the meta-model can be set up with two independent parameters, which further improves the speed of the algorithm and the usage of memory in the modelling step.

Simulation of pure absorbance spectra

To validate the Mie scatter correction algorithm developed in this paper, a simulated data set of pure absorbance spectra was created. The spectra were simulated such that the obtained absorbance spectra resembled a matrigel spectrum of Bassan *et al.*,[6] *i.e.* a spectrum of an artificial base membrane consisting mainly of proteins. The matrigel spectrum is considered to be a nearly scatter-free pure absorbance spectrum. A random number generator was used to change the heights ($\pm$20%), and shift band positions ($\pm$1 cm^{-1}) of peaks by superimposing Lorentz lines for 50 spectra. This data set was divided into two data sets of 25 spectra. For data set one, we systematically changed the amplitudes at the peak positions 1116, 1127, 1172, 1233, 1294, 1388, 1544 and 1648 cm^{-1} by making them either higher or lower for each respective band. For data set two, we changed the amplitudes at the positions 1075, 1155, 1192, 1243, 1315, 1404, 1456, 1551 and 1656 cm^{-1} by again making them either higher or lower for each

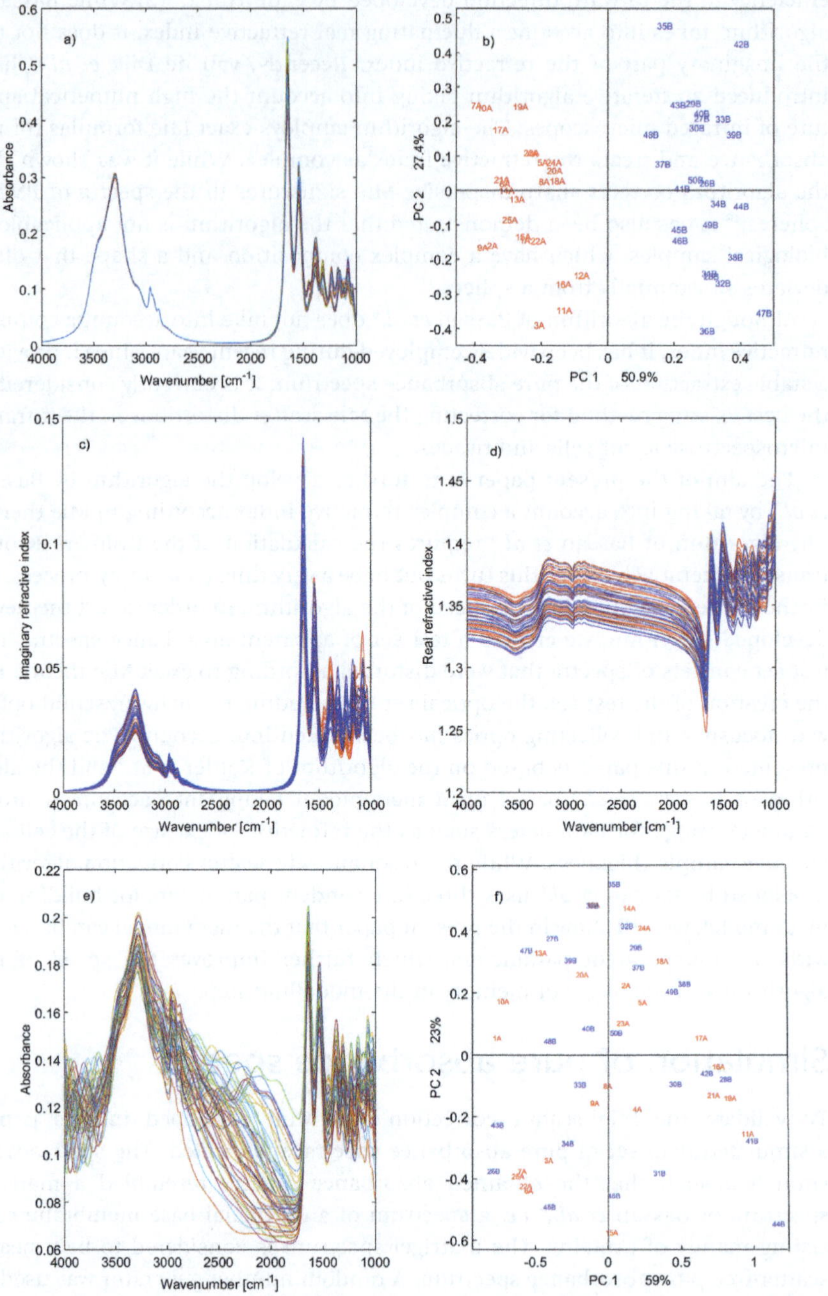

Fig. 1 (a) A set of simulated pure absorbance spectra. (b) PCA of simulated pure absorbance spectra. (c) Imaginary part of the refractive index obtained from the simulated absorbance spectra (d) real part of the refractive index obtained from the simulated absorbance spectra. (e) Apparent absorbance spectra obtained by employing exact Mie theory taking into account the optical setup of an infrared microscope. (f) PCA of simulated apparent absorbance spectra.

respective band. When analysing these data sets by PCA, two clusters of samples were obtained according to the design of the simulation. The simulated spectra are shown in Fig. 1a. The corresponding score plot for the first two PCA components is shown in Fig. 1b. The first two components account in total for 78.3% of the variance in these data. From the simulated pure absorbance spectra, the imaginary parts of refractive indices $n'(\tilde{\nu})$ were calculated according to

$$n'(\tilde{\nu}) = \frac{A(\tilde{\nu})\ln(10)}{4\pi d_{\text{eff}}\tilde{\nu}}, \tag{1}$$

where d_{eff} is the effective thickness of the cell[11] and $\tilde{\nu}$ is the wavenumber. The real part of the fluctuating part of the refractive index was calculated using the Kramers–Kronig transform

$$n(\tilde{\nu}) = n_0 + \frac{2}{\pi}P\int_0^{\infty}\frac{sn'(s)}{s^2 - \tilde{\nu}^2}\text{d}s, \tag{2}$$

where n_0 is the constant part of the real refractive index, P denotes the Cauchy principal value integral and n' is the imaginary part of the refractive index.[12]

Simulation of apparent absorbance spectra according to exact Mie theory

In order to simulate apparent absorbance spectra, exact Mie theory was used. In the simulations, the optical setup of an infrared microscope with a numerical aperture NA was taken into account. The apparent absorbance was calculated according to

$$A_{\text{app}} = -\log_{10}\left(1 - \frac{\pi a^2}{G}Q_{\text{ext}} + \frac{1}{4\pi G\tilde{\nu}^2}\int_0^{\theta_{\text{NA}}}[i_1(\theta) + i_2(\theta)]\sin(\theta)\text{d}\theta\right), \tag{3}$$

where a is the radius of the spherical scatterer, G is the size of the aperture and $\tilde{\nu}$ is the wavenumber. The integration is performed over the numerical aperture θ_{NA} with integrand functions $i_{1,2}(\theta)$, which are calculated from the scattering amplitudes $S_{1,2}(\theta)$. Details are given in the ESI material S.2 of a paper by Lukacs *et al.*[11] For each simulated pure absorbance spectrum (see description in the previous section), the imaginary and real parts of the refractive index were calculated according to eqn (1) and (2) and displayed in Fig. 1c and d, respectively. For each simulated apparent absorbance spectrum A_{app}, random values were chosen for n_0 and a from the intervals $1.1 < n_0 < 1.4$ and $2\ \mu\text{m} < a < 5.5\ \mu\text{m}$. Following this procedure, a set of 50 simulated apparent absorbance spectra was obtained. The apparent absorbance spectra are shown in Fig. 1e. The score plot of the first two components of the corresponding PCA analysis is shown in Fig. 1f. We can see that due to the scatter distortions, the two groups that were observed in the score plot of the first two components of the PCA of the pure absorbance spectra (Fig. 1b) are now mixed. It is important to note that visualization of the score plots of higher components also reveals the grouping in the apparent absorbance spectra. This is expected, since the scatter distortions are simply leading to a distortion of the main variation pattern, but the information related to the grouping according to the chemical difference is still obtained in the simulated apparent absorbance spectra. It is important to mention that the simulated apparent absorbance spectra as shown in Fig. 1e include the so-called dispersive artefact. When the apparent absorbance spectra in Fig. 1e are corrected according to the EMSC algorithm developed by Kohler *et al.*,[9] the dispersive artefact can be clearly seen (results not shown). The dispersive artefact is due to the fluctuations of the real refractive index caused by the absorption resonances and is not corrected by the algorithm developed by Kohler *et al.*,[9] since the model assumes a real and constant refractive index.

Further, we would like to mention that the apparent absorbance spectra in Fig. 1e contain ripples which are higher frequency oscillations. Ripples can be seen clearly in the region from 2800 cm^{-1} to 1800 cm^{-1}. These ripples also appear in other spectral regions, but are less visible since they overlap with chemical absorbance bands. While we have observed ripples in spectra of pollen,[11] ripples are usually not dominant in spectra of human and animal cells and tissues.

Extended multiplicative signal correction and meta-modelling

For the extraction of the pure absorbance spectra from simulated apparent absorbance spectra, an iterative algorithm based on EMSC was developed, which is a further extension of the algorithms presented by Kohler *et al.*[9] and Bassan *et al.*[7] Multiplicative signal correction (MSC) and EMSC have been introduced for pre-processing of near-infrared spectra.[13,14] It has been shown in the past that both methods are versatile tools for correcting infrared spectra of biological materials.[6,9,15–17] When EMSC is used for estimating and correcting Mie scattering, a measured absorbance spectrum $A_{\text{app}}(\tilde{\nu})$ is approximated by a reference spectrum $Z_{\text{ref}}(\tilde{\nu})$ times a multiplicative effect b, plus deviations from this reference spectrum expressed by a constant baseline c plus a sum of components $p_i(\tilde{\nu})$ times respective parameters g_i

$$A_{\text{app}}(\tilde{\nu}) = bZ_{\text{ref}}(\tilde{\nu}) + c + \sum_{i=1}^{A_{\text{opt}}} g_i p_i(\tilde{\nu}) + \varepsilon(\tilde{\nu}). \tag{4}$$

The un-modelled part is captured by the residual term $\varepsilon(\tilde{\nu})$. Further baseline effects may be included in the model by adding polynomials to the EMSC model.[15,16] Since EMSC models an apparent absorbance spectrum around a reference spectrum, the estimation of the model parameters has turned out to be a very stable process. This is because the infrared spectra of biological materials have very similar spectral signatures deriving from protein, fat and carbohydrate absorptions, leading to a visually very similar overall shape of the spectrum. When the parameters are estimated according to eqn (4), the apparent absorbance spectrum is corrected according to

$$A_{\text{corr}}(\tilde{\nu}) = \frac{\left(A_{\text{app}}(\tilde{\nu}) - c - \sum_{i=1}^{A_{\text{opt}}} g_i p_i(\tilde{\nu})\right)}{b}. \tag{5}$$

In an EMSC Mie model, the component spectra $p_i(\tilde{\nu})$ are obtained from a meta-model based on Mie theory. In the meta-model used by Kohler *et al.*[9] and Bassan *et al.*,[7] the scatter extinction was approximated by the formula derived by van de

Hulst,[4] which was originally developed for a constant and real refractive index and is written as

$$Q_{\text{ext}}(\tilde{\nu}) \approx 2 - \frac{4}{\rho}\sin\rho + \left(\frac{4}{\rho^2}\right)(1 - \cos\rho), \tag{6}$$

where ρ is given by

$$\rho = 4\pi a\tilde{\nu}(n - 1) \tag{7}$$

and a and n are the radius of the spherical particle and the real refractive index, respectively. In order to allow for the correction of the dispersive artefact, Bassan *et al.*[7] introduced in eqn (6) n, a non-constant real refractive index, which was calculated from an estimate of the pure absorbance according to the Kramers–Kronig transform (eqn (2)), while the imaginary part $n'(\tilde{\nu})$ of the refractive index may be calculated from an estimate of the pure absorbance spectrum $A(\tilde{\nu})$ according to eqn (1). This relation was simplified by Bassan *et al.*[7] assuming that the imaginary part $n'(\tilde{\nu})$ of the refractive index is proportional to the absorbance $A(\tilde{\nu})$, *i.e.*

$$n'(\tilde{\nu}) \approx sA(\tilde{\nu}) \tag{8}$$

where s is a proportionality factor. The proportionality is only a rough approximation since the wavenumber varies over the range under consideration. When the imaginary part of the refractive index, $n'(\tilde{\nu})$, is known, the real part, $n(\tilde{\nu})$, can be calculated according to the Kramers–Kronig relation in eqn (2).

The iterative algorithm developed by Bassan *et al.* works as follows

The algorithm of Bassan corrects a scatter distorted apparent absorbance spectrum $A_{\text{app}}(\tilde{\nu})$ according the following iterative procedure, which is based on the EMSC model in eqn (4).

Initialization. Eqn (4) requires the initialization of a reference spectrum $Z_{\text{ref}}(\tilde{\nu})$, which ideally should be closed to the pure absorbance spectrum. To initialize the algorithm, a matrigel spectrum[6] was used as the pure absorbance spectrum $A(\tilde{\nu})$.

Iterative algorithm. (1) The reference spectrum $Z_{\text{ref}}(\tilde{\nu})$ in eqn (4) is replaced by an estimate of the pure absorbance spectrum.

(2) The estimate of the pure absorbance spectrum is further used to calculate the imaginary part of the refractive index $n'(\tilde{\nu})$ according to eqn (1). From the imaginary part of the refractive index $n'(\tilde{\nu})$, the real part of the refractive index, $n(\tilde{\nu})$, is calculated by the Kramers–Kronig transform according to eqn (2). The real part of the refractive index is then used to calculate the extinction efficiency $Q_{\text{ext}}(\tilde{\nu})$ according to eqn (6) for a wide range of parameters a, n_0 and the proportionality factor of eqn (8), denoted by s. For each parameter typically its range is covered by 10 different values. The apparent absorbance $A_{\text{app}}(\tilde{\nu})$ is then calculated for the range of parameters a, n_0 and s by assuming that $A_{\text{app}}(\tilde{\nu}) \approx Q_{\text{ext}}(\tilde{\nu})$, resulting in a set of $10 \times 10 \times 10 = 1000$ apparent absorbance spectra.

(3) The set of apparent absorbance spectra $A_{\text{app}}(\tilde{\nu})$ is approximated by a meta-model using PCA, resulting in a set of principal components $p_i(\tilde{\nu})$ that are used as components in eqn (4). The EMSC parameters in eqn (4) are estimated by ordinary

least square fits. After estimation of the EMSC parameters, the spectrum is corrected according to eqn (5), resulting in a corrected spectrum $A_{corr}(\tilde{\nu})$.

(4) The estimate of the pure absorbance spectrum is replaced by the corrected spectrum $A_{corr}(\tilde{\nu})$ and the algorithm is reiterated starting with step (1).

The iterative algorithm by Bassan *et al.*[6] deserves some comments. It is important to mention that the assumed proportionality $A_{app}(\tilde{\nu}) \sim Q_{ext}(\tilde{\nu})$ that is employed in step (1), is a rough approximation of eqn (3).[11] It involves neglecting the third term in eqn (3) resulting in

$$A_{app} = -\log_{10}\left(1 - \frac{\pi a^2}{G} Q_{ext}\right) \tag{9}$$

where a is the radius of the sphere and G is the detector area, and further the expansion of the logarithm of eqn (9) up to linear order, resulting in

$$A_{app} \approx \frac{\pi a^2}{G} Q_{ext}, \tag{10}$$

explaining the proportionality between $A_{app}(\tilde{\nu})$ and $Q_{ext}(\tilde{\nu})$.

Improved algorithm suggested in this paper

The iterative algorithm suggested in this paper involves several improvements. First, the iterative algorithm in this paper is based on a meta-model, which involves a complex refractive index according to Mie theory. We used an approximation formula for the extinction efficiency, that has been derived by van de Hulst[4]

$$Q_{ext}(\tilde{\nu}) \approx 2 - 4e^{-\rho \tan\beta} \frac{\cos\beta}{\rho} \sin(\rho - \beta) - 4e^{-\rho \tan\beta} \left(\frac{\cos\beta}{\rho}\right)^2 \cos(\rho - 2\beta) + 4\left(\frac{\cos\beta}{\rho}\right)^2 \cos(2\beta) \tag{11}$$

with

$$\rho = 4\pi a\tilde{\nu}(n - 1) \text{ and } \tan\beta = n'/(n - 1) \tag{12}$$

where n and n' are the real and imaginary parts of the refractive index, respectively. The differences between eqn (6) and (11) are illustrated in ref. 11. The differences are significant, both with respect to the position of the band and the absolute values of the estimated extinction and absorbance.

The second improvement relates to the complexity of the meta-model. While in the previous work of Bassan *et al.*[6] a parameter model was used including ranges of three parameters (3 dimensions), it can be shown that a 2-dimensional parameter model is sufficient for parameter estimation. Details about the reduction from the 3-parameter model to the 2-parameter model are given in the Appendix A. The application of a 2-dimensional parameter model decreases the computation time and the required amount of computer memory. This can be illustrated by an example. When the three parameter ranges represented by the size of the cell, the background refractive index and the scaling parameter s of eqn (8) are covered and 10 values are used to cover each parameter range, 1000 Mie extinction curves need to be simulated and used for the establishment of the meta-model. Employing a two-parameter model, 100 spectra are sufficient to cover the same range. Thus, memory usage is reduced by a factor 10 and the computation time for the establishment of the meta-model is reduced as well.

A third improvement of the algorithm was achieved by speeding up the Kramers–Kronig transform of eqn (2). Kramers–Kronig relations are mainly used in optical spectroscopy to determine the complex refractive index $\hat{n} = n + in'$ of the medium from the measured absorption, transmission or reflection spectrum. The refractive index is an important quantity when considering the scattering and absorption of light at biological materials. The real part n of this index describes the refractive properties of the material; the imaginary part n' of it, determines the absorptive properties of the material. The most employed method for obtaining n from n' and *vice versa* is the Kramers–Kronig transform, which expresses the real part n in terms of the imaginary part n' according to

$$n(\tilde{\nu}) = \frac{2}{\pi} P \int_0^{+\infty} \frac{sn'(s)}{s^2 - \tilde{\nu}^2} \mathrm{d}s, \tag{13a}$$

and the imaginary part n' in terms of the real part n by

$$n'(\tilde{\nu}) = -\frac{2\tilde{\nu}}{\pi} P \int_0^{+\infty} \frac{n(s)}{s^2 - \tilde{\nu}^2} \mathrm{d}s. \tag{13b}$$

In Appendix B, we show that these relations are equivalent to the Hilbert transform, since the real part of the refractive index is an even function of the wavenumber and the imaginary part of the refractive index is an odd function of wavenumber, *i.e.* $n(\tilde{\nu}) = n(-\tilde{\nu})$ and $n'(\tilde{\nu}) = -n'(-\tilde{\nu})$. This can be seen from the Lorentz model. Details are given in the ESI material S.1 of ref. 11. The Hilbert transform is written as

$$n(\tilde{\nu}) = \frac{1}{\pi} P \int_{-\infty}^{+\infty} \frac{n'(s)}{s - \tilde{\nu}} \mathrm{d}s = -\frac{1}{\pi \tilde{\nu}} * n'(\tilde{\nu}) \tag{14a}$$

and

$$n'(\tilde{\nu}) = -\frac{1}{\pi} P \int_0^{+\infty} \frac{n(s)}{s - \tilde{\nu}} \mathrm{d}s = \frac{1}{\pi \tilde{\nu}} * n(\tilde{\nu}) \tag{14b}$$

respectively, where * denotes convolution. The Hilbert transform can be calculated *via* Fast Fourier Transform (FFT). The FFT method is based on the fact that both real and imaginary parts of the complex refractive index defining the Hilbert transform are proportional to the convolution product between $n(\tilde{\nu})$ or $n'(\tilde{\nu})$ and the convolution kernel function $\frac{1}{\pi \tilde{\nu}}$. In eqn (14) there is a singularity, when s is equal to $\tilde{\nu}$. The singularity problem is theoretically bypassed by introducing the Cauchy principal value.[18]

The new iterative algorithm works as follows

The algorithm corrects a scatter distorted apparent absorbance spectrum $A_{\mathrm{app}}(\tilde{\nu})$ according to the following iterative procedure which is based on the EMSC model in eqn (4).

Initialization. The reference spectrum $Z_{\mathrm{ref}}(\tilde{\nu})$ in eqn (4) is initialized by a matrigel spectrum[6] or another appropriate non-distorted spectrum.

Iterative algorithm. (1) The reference spectrum $Z_{\text{ref}}(\tilde{\nu})$ in eqn (4) may be either updated[6] or the same reference spectrum may be used in each iteration.

(2) The estimate of the pure absorbance spectrum is further used to calculate the imaginary part of the refractive index $n'(\tilde{\nu})$ according to eqn (1). The real part of the refractive index, $n(\tilde{\nu})$, is calculated from the imaginary part of the refractive index, $n'(\tilde{\nu})$, by the Hilbert transform (function Hilbert in Matlab) according to eqn (14a). The calculations of the Hilbert function are based on the FFT algorithm. The real part of the refractive index is then used to calculate the extinction efficiency $Q_{\text{ext}}(\tilde{\nu})$ according to eqn (11) involving a complex refractive index. In order to cover the relevant parameter range, the scaled parameters γ and α_0 according to eqn (A20) and (A21) are used (see Appendix A) with the ranges $\alpha_0 \in [0.2\ \mu\text{m}, 2.2\ \mu\text{m}] \times 4\pi$ and $\gamma \in \left[5 \times 10^4 \frac{1}{m},\ 6 \times 10^5 \frac{1}{m}\right]$. The corresponding ranges of n_0, a and f are given in (A13). A set of 50 apparent absorbance spectra $A_{\text{app}}(\tilde{\nu})$ is then calculated according to eqn (9) uniformly covering the parameter ranges for α_0 and γ, and stored as rows in a matrix M.

(3) The set of apparent absorbance spectra $A_{\text{app}}(\tilde{\nu})$, *i.e.* the matrix M is approximated by a meta-model using PCA resulting in a set of principal components $p_i(\tilde{\nu})$ that are used as components in eqn (4). Prior to approximation by PCA, the matrix M is orthogonalized with respect to the reference spectrum $Z_{\text{ref}}(\tilde{\nu})$. This is to avoid competition between the parameters b and g_i (see eqn (4)) in the subsequent parameter estimation. The EMSC parameters in eqn (4) are estimated by ordinary least squares fits. After estimation of the EMSC parameters, the

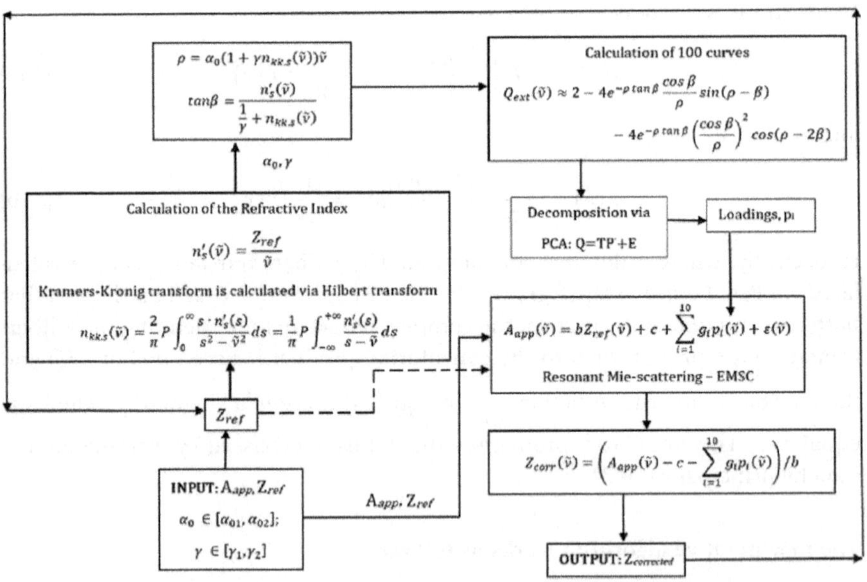

Fig. 2 Schematic illustration of the iterative algorithm for the retrieval of the pure absorbance spectrum from the apparent absorbance spectrum. After each iteration, a new estimate of the pure absorbance spectrum is obtained. The next iteration is initialized by using the estimate of the pure absorbance spectrum for updating the real and imaginary parts of the refractive index for the iterative algorithm.

spectrum is corrected according to eqn (5) resulting in a corrected spectrum $A_{corr}(\tilde{\nu})$.

(4) The estimate of the pure absorbance spectrum is replaced by the corrected spectrum $A_{corr}(\tilde{\nu})$ and the algorithm is reiterated starting with step (1).

This iterative algorithm is described schematically in Fig. 2. We suggest to call the algorithm fast resonant Mie scatter correction.

EMSC correction of simulated and measured apparent absorbance spectra

The correction of an apparent absorbance spectrum according to the algorithm suggested in this paper is demonstrated in Fig. 3. In Fig. 3a an example of a simulated apparent absorbance spectrum is shown. The spectrum was simulated employing eqn (3). As parameters we used $n_0 = 1.35$ and $a = 3.81$ µm. Corrected spectra that were obtained with the first three iterations of the correction algorithm proposed in this paper are shown in Fig. 3b together with the reference spectrum and the pure absorbance spectrum. As a reference spectrum we used the average spectrum of all the simulated pure absorbance spectra. Alternatively, the matrigel spectrum could be used. It can be seen that the algorithm converges quickly and that the corrected spectra are close to the pure absorbance spectra used for the simulation. Ripples that can be observed in the region between 4000 cm^{-1} and 3500 cm^{-1} were not corrected successfully. This is expected, since the van de Hulst formula used for the correction (see eqn (11)) does not describe the Mie ripples. In Fig. 4a, a simulated apparent absorbance spectrum containing strong ripples is shown. As parameters, we used $n_0 = 1.36$ and $a = 5.12$ µm. Ripples with these intensities are usually not observed in the

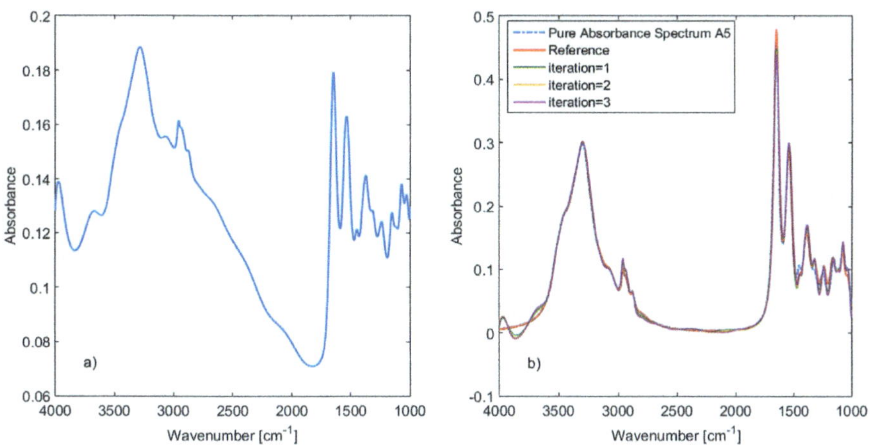

Fig. 3 (a) An example of a simulated apparent absorbance spectrum is shown. (b) Spectra corrected by an iterative EMSC meta-model employing the van de Hulst approximate formula with complex refractive index are shown. Corrected spectra for iterations 1, 2 and 3 of the correction algorithm (green, brown and violet, respectively) are close to the underlying pure absorbance spectrum (blue dashed line) and different from the reference spectrum (red line) employed. For simulation of the apparent absorbance spectrum in (a) we used the parameters $n_0 = 1.35$ and $a = 3.81$ µm.

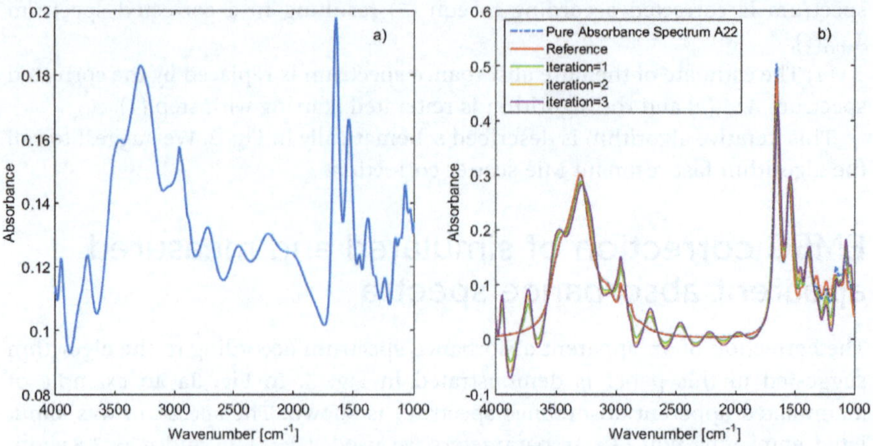

Fig. 4 (a) Apparent absorbance spectrum containing ripples. For the simulation of the apparent absorbance spectrum the parameters $n_0 = 1.36$ and $a = 5.12$ μm were used. (b) The corrected spectra for iterations 1, 2 and 3 of the correction algorithm (green, brown and violet, respectively). They still contain the ripple structure. As expected, the EMSC meta-model employing the van de Hulst approximate formula with complex refractive index does not correct the ripples.

infrared microspectroscopy of single cells. An explanation for this will be given elsewhere. In the corrected spectra in Fig. 4b it can be seen clearly that the van de Hulst algorithm is not capable of correcting the ripples.

The complete set of corrected spectra for the simulated apparent absorbance spectra from Fig. 1e is shown in Fig. 5. The spectra that correspond to the red

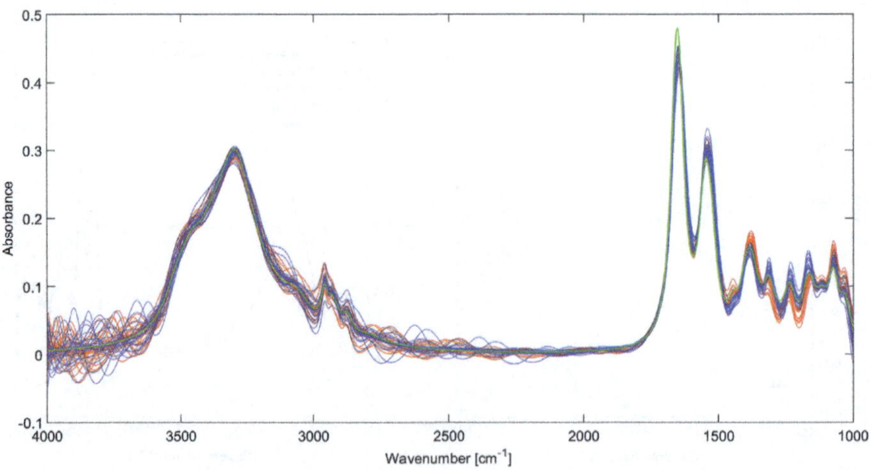

Fig. 5 The complete set of simulated apparent absorbance spectra shown in Fig. 1e is corrected by the algorithm presented in this paper. The spectra that correspond to red scores of the pure absorbance spectra in Fig. 1b are drawn in red; the spectra that correspond to blue scores of the pure absorbance spectra in Fig. 1b are marked blue. The reference spectrum is shown in green. Some corrected absorbance spectra contain ripples. As expected, the EMSC meta-model employing the van de Hulst formula with complex refractive index does not remove ripples.

scores of the pure absorbance spectra in Fig. 1b are plotted in red in Fig. 5, while the spectra that correspond to the blue scores of the pure absorbance spectra in Fig. 1b are plotted in blue in Fig. 5. The reference spectrum is shown in green. It can be seen that the correction of the spectra worked well. The two groups of spectra can be separated visually after correction. As mentioned before, ripples are not well corrected, which does not pose a major problem, since strong ripples are usually not observed in the infrared microspectroscopy of single cells. The computation time for the correction of 50 spectra (one iteration in the iterative algorithm) using a 4th generation Intel®Core™i7-4702HQ quad core processor, 16GB memory and Matlab 2015 is 77 seconds. This is a substantial improvement to the algorithm developed by Bassan *et al.*[6] which uses 765 seconds on the same computer (employing numerical integration and a parameter model with 3 parameters). No speed optimization of the algorithm by vectorization or parallel programming has been done so far. It is important to mention that the Kramers–Kronig part of the algorithm is improved by a factor of approximately 200 when using the algorithm based on the Hilbert transform compared to numerical integration. The improvement of the Kramers–Kronig part shows advantages especially when more than 1 iteration is applied in the iterative algorithm.

In order to demonstrate that the chemical information that was distorted in the apparent absorbance spectra can be restored, we investigated the peak ratios before and after correction. In Fig. 6, the peak ratios for the simulated pure absorbance spectra of the band at 1546 cm^{-1} and the band at 1387 cm^{-1} are shown in blue. The peak ratio of these bands is simulated such that the 50 spectra can be clearly separated into two groups of 25 spectra each. Spectra 1–25 have a lower peak ratio than spectra 26–50. The corresponding peak ratios for the apparent absorbance spectra are shown in magenta. It is obvious that the scatter distorted apparent absorbance spectra do not allow the separation of these two groups with the help of the peak ratio of the band at 1546 cm^{-1} and the band at 1387 cm^{-1}. The apparent absorbance spectra were then corrected by the iterative

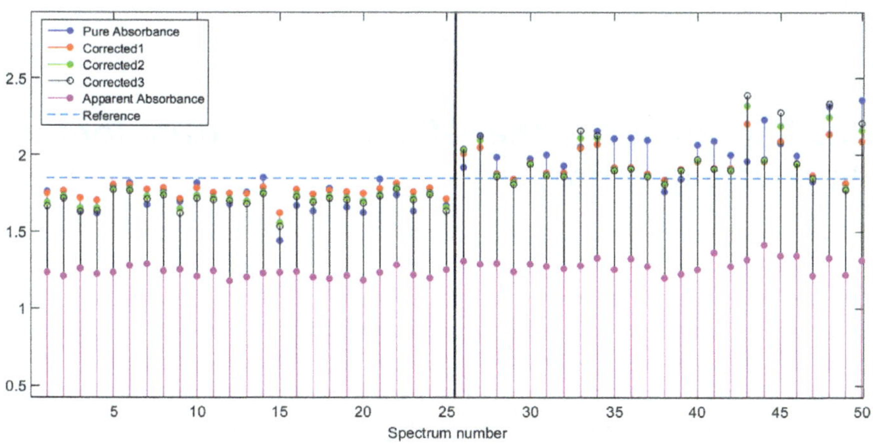

Fig. 6 The peak ratio is shown for the bands at 1546 cm^{-1} and 1387 cm^{-1} for the simulated pure absorbance spectra (blue), for iterations 1, 2 and 3 of the correction algorithm (red, green and black, respectively) and for the apparent absorbance (magenta). The ratio for the reference spectra is plotted as a blue dashed line for comparison.

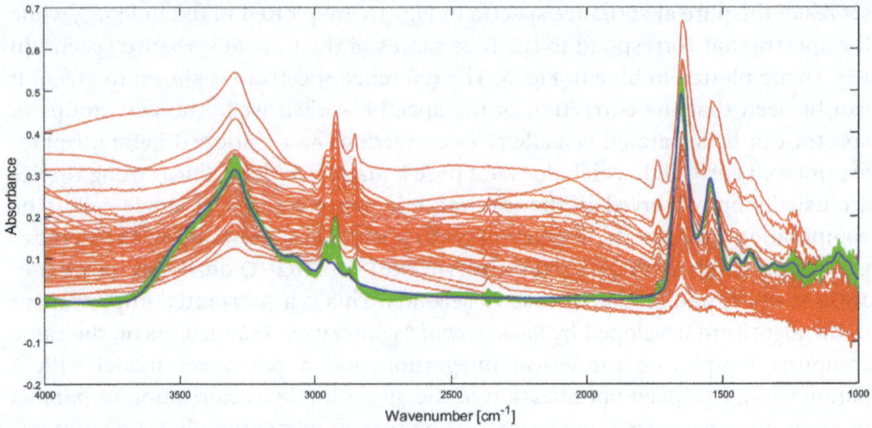

Fig. 7 The measured infrared microspectroscopy spectra of lung cancer cells are shown in red together with the corrected spectra in green. For the correction one iteration step was used. The matrigel spectrum was used as the reference spectrum (in blue).

algorithm. Corrected spectra are shown in red (one iteration), green (two iterations) and black (three iterations). The ratio obtained from the average spectrum of all the simulated pure absorbance spectra that was used as the reference spectrum for the EMSC model is shown as a blue dashed line for comparison. It can be seen that the iterative algorithm retrieves a good estimation of the ratio of the simulated pure absorbance spectrum.

We further tested the algorithm on a set of measured infrared microspectroscopy spectra of lung cancer cells. Details about the data set can be found in the work of Kohler *et al.*[9] The measured spectra are shown in Fig. 7 (red) together with the corrected spectra (green). For the correction one iteration step was used. The matrigel spectrum was used as the reference spectrum.[6] In Fig. 7, it can be seen that correction works visually very well. The resonant Mie effect that was not corrected by the algorithm presented by Kohler *et al.*[9] is now successfully corrected.

The importance of the reference spectrum for the iterative algorithm

An important comment relates to the updating of the EMSC reference spectrum in the iterative algorithm. The dashed line in Fig. 2 indicates, that there are two options: either the reference spectrum is updated by the new estimate of the pure absorbance spectrum obtained after each iteration or the initial reference spectrum is kept and only the imaginary part of the refractive index and, after Kramers–Kronig transformation, the real part of the refractive index is updated. We have observed that an update of the reference spectrum may lead to instabilities in the iterative algorithm and the result obtained may completely depend on the number of iterations used, *i.e.* the iterative algorithm may drift completely apart from reasonable solutions. Thus, we do not suggest to update the reference spectrum for EMSC in the iterative algorithm.

The use of a good reference spectrum is crucial for the success of the correction algorithm. In Fig. 7 it is obvious that the corrected spectra become in their overall shape very close to the reference spectrum. This can be clearly seen when comparing the reference spectrum (for example in Fig. 4b) with the corrected spectra in Fig. 7. The same observation can be made in the paper of Bassan *et al.*,[6] where the same reference spectrum was used as in the current paper. In ref. 6, after correction all spectra are in their overall shape similar to the matrigel spectrum used as a reference in EMSC. This is due to the high flexibility of the meta-model used for the correction. Notwithstanding, although the corrected spectra tend overal in their shape towards the reference spectrum, we could clearly show that chemical information can be restored by the suggested algorithm.

Conclusions

In this work we presented an improved iterative EMSC algorithm for correcting Mie scattering in the infrared microspectroscopy of single cells and tissues. The iterative EMSC algorithm employs a meta-model based on an approximate formula by van de Hulst, taking into account a complex refractive index for correcting Mie scattering. The new iterative algorithm was tested using a simulated set of apparent absorbance spectra and a set of measured apparent absorbance spectra. The simulated apparent absorbance spectra were obtained by first simulating pure absorbance spectra and then generating scatter distorted apparent absorbance spectra by full Mie theory. For the simulations, the full optical set up of an infrared microscope including the collecting and focusing Schwarzschild optics were taken into account. The set of simulated spectra used in this paper is thus more difficult to correct than the spectra simulated in ref. 6, where the spectra were simulated using exactly the same approximation formula that was afterwards used for the correction algorithm. In addition, the spectra simulated in ref. 6 were obtained by the addition of pure absorbance spectra to simulated scatter spectra. This facilitated the correction of the spectra in ref. 6, since scatter distortions were only added on top of the pure absorbance spectra. In our simulations the scatter distorted spectra where obtained according to eqn (3) and do not contain additive contributions of the pure absorbance spectra. It is important to mention that according to Mie theory, apparent absorbance spectra are not obtained by adding pure absorbance spectra to scatter contributions. Thus, the retrieval of chemical information from the simulated apparent absorbance spectra taking into account the full optical setup as employed in the current paper, demonstrates the capability of the algorithm presented in this paper well.

The measured spectra used for correction in the current paper were obtained from single lung cancer cells.[9] The correction of the measured spectra was successful, while it is obvious that the corrected spectra tend in their overall shape towards the matrigel spectrum employed. The high flexibility of the meta-model results in corrected spectra that are in their overall shape very similar to the reference spectrum used in the EMSC model, while we showed by employing a simulated data set that the chemical characteristics of the pure absorbance spectra could be restored.

The iterative algorithm developed by Bassan *et al.*[6] involves a numerical integration in order to perform a Kramers–Kronig transform. In the algorithm presented in this paper, we replaced the integral of the Kramers–Kronig transform by a fast Fourier transform (FFT) algorithm. This reduced the computation time of the Kramers–Kronig transform approximately by a factor of 100. Moreover we have shown that two independent parameters α_0 and γ (each parameter contains 10 equidistant values in its respective range) are sufficient for compressing 100 Mie scattering curves into a small number of principal components loading spectra to estimate the scattering contributions in the EMSC meta-model.

While we have shown that the new algorithm retrieves the pure absorbance spectra from highly distorted apparent absorbance spectra, ripples that are present in the simulated apparent absorbance spectra could not be corrected, since the van de Hulst approximation does not account for ripples. The ripples that are visible in the simulated apparent absorbance spectra can be explained by diffractive surface waves. While we have observed that the appearance of ripples is not present or suppressed in measured spectra, we believe that the correction of ripples is in general not required for the correction of the infrared microspectroscopic spectra of single cells and tissues. Ripples may be absent in measured spectra because the apertures used in infrared microspectroscopy are comparable to the size of the cells, while the exact Mie theory which assumes an incoming plane wave assumes apertures that are much bigger than the probed cell. Thus, diffractive surface waves that are causing ripples in exact Mie theory may be suppressed in practical measurement situations.

Appendix A

Complexity of meta model

For the purpose of building a meta-model we take into account a suitable parameter space for the approximation formula employed for the extinction efficiency. The approximation formula given by eqn (11) contains the parameters ρ and β. We define

$$\alpha = 4\pi a(n - 1), \tag{A1}$$

where n is the real refractive index and a is the size of the sphere. It follows that $\rho = \tilde{\nu}\alpha$.

When the pure absorbance spectrum $A(\tilde{\nu})$ is known, the imaginary part of the refractive index can be calculated according to

$$n'(\tilde{\nu}) = \frac{A(\tilde{\nu})\ln(10)}{4\pi d_{\text{eff}}\tilde{\nu}}, \tag{A2}$$

where d_{eff} is the effective thickness of the cell.[11] The real part of the refractive index $n(\tilde{\nu})$ can then be calculated by the Kramers–Kronig relation according to

$$n(\tilde{\nu}) = n_0 + \frac{2}{\pi}P\int_0^{\infty}\frac{sn'(s)}{s^2 - \tilde{\nu}^2}\text{d}s, \tag{A3}$$

where n_0 is the constant part of the real refractive index and the integral term

$$n_{\text{kk}}(\tilde{\nu}) = \frac{2}{\pi}P\int_0^{\infty}\frac{sn'(s)}{s^2 - \tilde{\nu}^2}\text{d}s \tag{A4}$$

is the fluctuating part of the real refractive index.

It follows that

$$\alpha = 4\pi a(n_0 + n_{kk} - 1). \tag{A5}$$

If n_{kk} is calculated from the pure absorbance spectrum $A(\tilde{\nu})$ according to eqn (A2) and (A4), the effective thickness d_{eff} of the cell is in general not known. We therefore define a scaled imaginary part of the refractive index, $n'_s(\tilde{\nu})$, according to

$$n'_s(\tilde{\nu}) = \frac{A(\tilde{\nu})}{\tilde{\nu}}, \tag{A6}$$

where

$$n'(\tilde{\nu}) = fn'_s(\tilde{\nu}), \tag{A7}$$

with the scaling constant

$$f = \frac{\ln(10)}{4\pi d_{eff}}. \tag{A8}$$

The fluctuating part is consequently calculated according to

$$n_{kk}(\tilde{\nu}) = fn_{kk,s}(\tilde{\nu}) \tag{A9}$$

where

$$n_{kk,s}(\tilde{\nu}) = \frac{2}{\pi} P \int_0^\infty \frac{sn'_s}{s^2 - \tilde{\nu}^2} ds. \tag{A10}$$

Thus, we obtain

$$\alpha = 4\pi a(n_0 + fn_{kk,s} - 1) \tag{A11}$$

and

$$\tan \beta = \frac{fn'_s}{n_0 + fn_{kk,s} - 1}. \tag{A12}$$

Generally the exact numbers for a, n_0 and f are not known. Only ranges for these values can be given. These parameters have to be estimated in the modeling process. The following ranges were found to be optimal

$$n_0 \in [1.1, 1.4], \tag{A13a}$$

$$f \in \left[10^4 \frac{1}{m}, \ 3 \times 10^4 \frac{1}{m}\right], \tag{A13b}$$

$$a \in [2\ \mu m, 5.5\ \mu m]. \tag{A13c}$$

Since the parameters a and n_0 are not independent (see (A5)), these parameter ranges are not absolute. We include for example considerably higher values of a than indicated by the range in (A13c), when the refractive index in an apparent absorbance spectrum is below the maximum value of (A13a), since the two parameters enter eqn (A5) as a product. The parameters in eqn (A13) are not independent and one parameter can be omitted by rescaling as we will show in the following. We can write eqn (A11) as

$$\alpha = 4\pi a(n_0 - 1)\left(1 + \frac{f n_{\mathrm{kk,s}}}{(n_0 - 1)}\right), \tag{A14}$$

and eqn (A12) as

$$\tan\beta = \frac{n'_s}{\dfrac{n_0 - 1}{f} + n_{\mathrm{kk,s}}}. \tag{A15}$$

By defining

$$\alpha_0 = 4\pi a(n_0 - 1), \tag{A16}$$

and

$$\gamma = \frac{f}{(n_0 - 1)}, \tag{A17}$$

we obtain

$$\alpha = \alpha_0(1 + \gamma n_{\mathrm{kk,s}}), \tag{A18}$$

and

$$\tan\beta = \frac{n'_s}{1/\gamma + n_{\mathrm{kk,s}}} \tag{A19}$$

with the respective ranges that correspond to the values in eqn (A13):

$$\gamma \in \left[5\times 10^4 \frac{1}{m},\ \ 6\times 10^5 \frac{1}{m}\right] \tag{A20}$$

$$\alpha_0 \in [0.2\ \mu\mathrm{m},\ 2.2\ \mu\mathrm{m}] \times 4\pi \tag{A21}$$

It follows that the obtained model contains two independent parameters γ and α_0. The parameter γ is the new scaling value for the non-constant refractive index. A correct estimate of this parameter is important for the biochemical interpretation of the FTIR spectra. The parameter α_0 corresponds to the α-value of the non-resonant case used in the work of Kohler *et al.*[9] A good estimation of the α_0-value is important for the estimation of the physical parameters as the size of the cells a and the constant part of the real refractive index n_0.

Appendix B

Hilbert transform and Kramers–Kronig relations for a calculation of the complex refractive index

In order to define the complex refractive index $\hat{n}(\tilde{\nu}) = n(\tilde{\nu}) + in'(\tilde{\nu})$ of materials from the measured absorption, transmission or reflection spectra in optical spectroscopy, Kramers–Kronig relations are used. The complex refractive index is an important physical quantity when considering the scattering and absorption of infrared light at biological materials. The real part $n(\tilde{\nu})$ of this index describes the refractive properties of the material; the imaginary part $n'(\tilde{\nu})$ of it determines the absorptive properties of the material. A method for obtaining $n(\tilde{\nu})$ and $n'(\tilde{\nu})$ is the Kramers–Kronig transform, which relates the real part $n(\tilde{\nu})$ to the imaginary part $n'(\tilde{\nu})$ of the complex refractive index:

$$n(\tilde{\nu}) = \frac{2}{\pi} P \int_0^{+\infty} \frac{s n'(s)}{s^2 - \tilde{\nu}^2} \mathrm{d}s, \tag{B1a}$$

$$n'(\tilde{\nu}) = -\frac{2\tilde{\nu}}{\pi} P \int_0^{+\infty} \frac{n(s)}{s^2 - \tilde{\nu}^2} \mathrm{d}s, \tag{B1b}$$

where P denotes Cauchy's principal value.

These relations are equivalent to the Hilbert transform, provided that the real part is an even function of wavenumber and the imaginary part is an odd function of wavenumber, *i.e.* $n(\tilde{\nu}) = n(-\tilde{\nu})$ and $n'(\tilde{\nu}) = -n'(-\tilde{\nu})$:

$$n(\tilde{\nu}) = \frac{1}{\pi} P \int_{-\infty}^{+\infty} \frac{n'(s)}{s - \tilde{\nu}} \mathrm{d}s = -\frac{1}{\pi \tilde{\nu}} * n'(\tilde{\nu}), \tag{B2a}$$

$$n'(\tilde{\nu}) = -\frac{1}{\pi} P \int_{-\infty}^{+\infty} \frac{n(s)}{s - \tilde{\nu}} \mathrm{d}s = \frac{1}{\pi \tilde{\nu}} * n(\tilde{\nu}). \tag{B2b}$$

Let us consider the connections between eqn (B1) and (B2). Because $n'(\tilde{\nu})$ is an odd function of wavenumber, this property permits the conversion of the Hilbert transform pair (eqn (B2a)) into the Kramers–Kronig relations. The Hilbert transform pair (eqn (B2a)) is expanded as

$$n(\tilde{\nu}) = \frac{1}{\pi} P \int_{-\infty}^{+\infty} \frac{n'(s)}{s - \tilde{\nu}} \mathrm{d}s = \frac{1}{\pi} P \int_{-\infty}^{0} \frac{n'(s)}{s - \tilde{\nu}} \mathrm{d}s + \frac{1}{\pi} P \int_{0}^{+\infty} \frac{n'(s)}{s - \tilde{\nu}} \mathrm{d}s$$

with $n'(\tilde{\nu}) = -n'(-\tilde{\nu})$ we get

$$n(\tilde{\nu}) = \frac{1}{\pi} P \int_{0}^{+\infty} \frac{n'(s)}{s + \tilde{\nu}} \mathrm{d}s + \frac{1}{\pi} P \int_{0}^{+\infty} \frac{n'(s)}{s - \tilde{\nu}} \mathrm{d}s = \frac{2}{\pi} P \int_{0}^{+\infty} \frac{s n'(s)}{s^2 - \tilde{\nu}^2} \mathrm{d}s.$$

Because $n(\tilde{\nu})$ is an even function of wavenumber, this property permits the conversion of the Hilbert transform pair (eqn (B2b)) into the Kramers–Kronig relations. The Hilbert transform pair (eqn (B2b)) is expanded as

$$n'(\tilde{\nu}) = -\frac{1}{\pi} P \int_{-\infty}^{+\infty} \frac{n(s)}{s - \tilde{\nu}} \mathrm{d}s = -\frac{1}{\pi} P \int_{-\infty}^{0} \frac{n(s)}{s - \tilde{\nu}} \mathrm{d}s - \frac{1}{\pi} P \int_{0}^{+\infty} \frac{n(s)}{s - \tilde{\nu}} \mathrm{d}s$$

with $n(\tilde{\nu}) = n(-\tilde{\nu})$

$$n'(\tilde{\nu}) = \frac{1}{\pi} P \int_{0}^{+\infty} \frac{n(s)}{s + \tilde{\nu}} \mathrm{d}s - \frac{1}{\pi} P \int_{0}^{+\infty} \frac{n(s)}{s - \tilde{\nu}} \mathrm{d}s = -\frac{2\tilde{\nu}}{\pi} P \int_{0}^{+\infty} \frac{n(s)}{s^2 - \tilde{\nu}^2} \mathrm{d}s.$$

The Hilbert transform can be calculated *via* the Fast Fourier Transform (FFT). The FFT method is based on the fact that both the real and imaginary parts of the complex refractive index defining the Hilbert transform are proportional to the convolution product between $n(\tilde{\nu})$ or $n'(\tilde{\nu})$ and the convolution kernel function $\frac{1}{\pi\tilde{\nu}}$.

In eqn (B2) there is a singularity, when s is equal to $\tilde{\nu}$. The singularity problem is theoretically bypassed by introducing the Cauchy principal value.

Derivations of Hilbert transform and Kramers–Kronig relations

If $\hat{n}(\tilde{\nu})$ is an analytic function, then $\frac{\hat{n}(s)}{s-\tilde{\nu}}$ is an analytic function too except at the pole $s = \tilde{\nu}$. Cauchy's theorem states that

$$\oint f(s)\mathrm{d}s = 0,$$

provided that the closed contour encloses no poles of the analytic function $f(s)$. Let us apply Cauchy's theorem to the function $\frac{\hat{n}(s)}{s-\tilde{\nu}}$, where $\tilde{\nu}$ is a point on the real axis, and the contour, shown in Fig. 8, is the union of four curves with parametric representations

$$\mathrm{C}_1\text{: } s = \Omega, \text{ where } -R \leq \Omega \leq \tilde{\nu} - \rho;$$

$$\mathrm{C}_2\text{: } s = \tilde{\nu} + \rho\mathrm{e}^{\mathrm{i}\varphi}, \text{ where } 0 \leq \varphi \leq \pi;$$

$$\mathrm{C}_3\text{: } s = \Omega, \text{ where } \tilde{\nu} + \rho \leq \Omega \leq R;$$

$$\mathrm{C}_4\text{: } s = R\mathrm{e}^{\mathrm{i}\varphi}, \text{ where } 0 \leq \varphi \leq \pi.$$

From Cauchy's theorem we have

$$\int_{-R}^{\tilde{\nu}-\rho} \frac{\hat{n}(s)}{s-\tilde{\nu}}\mathrm{d}s + \int_{\pi}^{0} \frac{\hat{n}(\tilde{\nu}+\rho\mathrm{e}^{\mathrm{i}\varphi})}{\tilde{\nu}+\rho\mathrm{e}^{\mathrm{i}\varphi}-\tilde{\nu}}\mathrm{i}\rho\mathrm{e}^{\mathrm{i}\varphi}\mathrm{d}\varphi + \int_{\tilde{\nu}+\rho}^{R} \frac{\hat{n}(s)}{s-\tilde{\nu}}\mathrm{d}s + \int_{0}^{\pi} \frac{\hat{n}(R\mathrm{e}^{\mathrm{i}\varphi})}{R\mathrm{e}^{\mathrm{i}\varphi}-\tilde{\nu}}\mathrm{i}R\mathrm{e}^{\mathrm{i}\varphi}\mathrm{d}\varphi = 0.$$

Therefore

$$\int_{-R}^{\tilde{\nu}-\rho} \frac{\hat{n}(s)}{s-\tilde{\nu}}\mathrm{d}s + \mathrm{i}\int_{\pi}^{0} \hat{n}(\tilde{\nu}+\rho\mathrm{e}^{\mathrm{i}\varphi})\mathrm{d}\varphi + \int_{\tilde{\nu}+\rho}^{R} \frac{\hat{n}(s)}{s-\tilde{\nu}}\mathrm{d}s + \mathrm{i}\int_{0}^{\pi} \frac{\hat{n}(R\mathrm{e}^{\mathrm{i}\varphi})}{R\mathrm{e}^{\mathrm{i}\varphi}-\tilde{\nu}}R\mathrm{e}^{\mathrm{i}\varphi}\mathrm{d}\varphi = 0.$$

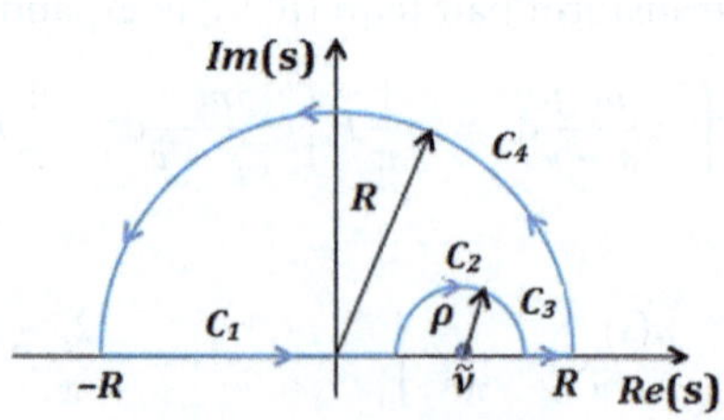

Fig. 8 Contour in the complex plane used to derive the Kramers–Kronig relations. The integration path, which is skirting the singularity point $s = \tilde{\nu}$, is indicated by arrows. The radius R of the outer semicircle tends to infinity. The radius ρ of the small semicircle about $\tilde{\nu}$ tends to zero.

Putting $\rho \to 0$ and $R \to \infty$, we have

$$\lim_{\substack{\rho\to 0,\\ R\to\infty}}\left(\int_{-R}^{\tilde{\nu}-\rho}\frac{\hat{n}(s)}{s-\tilde{\nu}}\mathrm{d}s+\int_{\tilde{\nu}+\rho}^{R}\frac{\hat{n}(s)}{s-\tilde{\nu}}\mathrm{d}s+\mathrm{i}\int_{\pi}^{0}\hat{n}(\tilde{\nu}+\rho\mathrm{e}^{\mathrm{i}\varphi})\mathrm{d}\varphi+\mathrm{i}\int_{0}^{\pi}\frac{\hat{n}(R\mathrm{e}^{\mathrm{i}\varphi})}{R\mathrm{e}^{\mathrm{i}\varphi}-\tilde{\nu}}R\mathrm{e}^{\mathrm{i}\varphi}\mathrm{d}\varphi\right)=0,$$

$$\lim_{\rho\to 0}\left(\int_{-\infty}^{\tilde{\nu}-\rho}\frac{\hat{n}(s)}{s-\tilde{\nu}}\mathrm{d}s+\int_{\tilde{\nu}+\rho}^{+\infty}\frac{\hat{n}(s)}{s-\tilde{\nu}}\mathrm{d}s\right)-\mathrm{i}\pi\hat{n}(\tilde{\nu})=0.$$

We obtain the last line since the fourth term vanishes as $R \to \infty$ if $\lim_{|s|\to\infty}\hat{n}(s)=0$.

According to the definition of the Cauchy principal value of an integral we get

$$\lim_{\rho\to +0}\left(\int_{-\infty}^{\tilde{\nu}-\rho}\frac{\hat{n}(s)}{s-\tilde{\nu}}\mathrm{d}s+\int_{\tilde{\nu}+\rho}^{+\infty}\frac{\hat{n}(s)}{s-\tilde{\nu}}\mathrm{d}s\right)=P\int_{-\infty}^{+\infty}\frac{\hat{n}(s)}{s-\tilde{\nu}}\mathrm{d}s, \tag{1}$$

and therefore, from (1), we have

$$\frac{1}{\pi i}P\int_{-\infty}^{+\infty}\frac{\hat{n}(s)}{s-\tilde{\nu}}\mathrm{d}s=\hat{n}(\tilde{\nu}).$$

Taking the real and imaginary parts of $\hat{n}(\tilde{\nu}) = n(\tilde{\nu}) + \mathrm{i}n'(\tilde{\nu})$, we get

$$H[n(\tilde{\nu})]=n'(\tilde{\nu})=-\frac{1}{\pi}P\int_{-\infty}^{+\infty}\frac{n(s)}{s-\tilde{\nu}}\mathrm{d}s,$$

$$H^{-1}[n'(\tilde{\nu})]=n(\tilde{\nu})=\frac{1}{\pi}P\int_{-\infty}^{+\infty}\frac{n'(s)}{s-\tilde{\nu}}\mathrm{d}s.$$

The last two expressions are called the *Hilbert transform pair*.

Now we focus on the numerical solution of the Hilbert transform pair.

By means of the Fourier transform definition an analytic signal $x(t)$ can be represented in the following way

$$x(t)=F^{-1}[X(f)]=\int_{-\infty}^{+\infty}X(f)\exp(\mathrm{i}2\pi ft)\mathrm{d}f,$$

where

$$X(f)=F[x(t)]=\int_{-\infty}^{+\infty}x(t)\exp(-\mathrm{i}2\pi ft)\mathrm{d}t.$$

According to the definition of the Hilbert transform for a function $x(t)$ we have

$$\tilde{x}(t)=H(x(t))=\frac{1}{\pi}P\int_{-\infty}^{+\infty}\frac{x(s)}{s-t}\mathrm{d}s=\frac{1}{\pi t}*x(t).$$

Since

$$F\left(\frac{1}{\pi t}\right) = -i[\text{sign}(f)] = \begin{cases} -i, & f > 0; \\ 0, & f = 0; \\ i, & f < 0. \end{cases}$$

Then

$$F[\tilde{x}(t)] = \tilde{X}(f) = F\left[\frac{1}{\pi t} * x(t)\right] = F[x(\text{t})] \times F\left(\frac{1}{\pi t}\right) = -i[\text{sign}(f)] \times X(f).$$

Multiplying both parts of the last expression by $i[\text{sign}(f)]$, we get

$$i[\text{sign}(f)]\tilde{X}(f) = X(f).$$

This equation is used for the numerical solution of the Hilbert transform pair. The algorithm of calculating the Hilbert transform consists of three steps:

(1) To calculate the spectrum $X(f)$:

$$X(f) = F[x(t)];$$

(2) To apply the expression $i[\text{sign}(f)] \times \tilde{X}(f) = X(f)$ for calculating $\tilde{X}(f)$:

$$\tilde{X}(f) = -i[\text{sign}(f)]X(f)$$

(3) To calculate $\tilde{x}(t)$:

$$\tilde{x}(t) = F^{-1}[\tilde{X}(f)].$$

These three steps are used in Matlab for the calculation of eqn (B2) *via* the function *Hilbert*.

More details about the numerical solution of the Kramers–Kronig transforms based on the FFT algorithm are presented by P. Bruzzoni *et al.*[19]

Abbreviations

EMSC	Extend multiplicative signal correction
MSC	Multiplicative signal correction
FFT	Fast Fourier transform
PCA	Principal component analysis

Acknowledgements

The authors wish to thank Peter Gardener and Paul Bassan for providing the matrigel and Josep Sule Suso for providing the set of measured lung cancer cell spectra.

References

1 D. Wetzel and J. Reffner, *Cereal Foods World*, 1993, **38**, 9–20.
2 B. Mohlenhoff, M. Romeo, M. Diem and B. R. Woody, *Biophys. J.*, 2005, **88**, 3635–3640.
3 G. Mie, *Ann. Phys.*, 1908, **25**, 377–452.
4 H. C. van de Hulst, *Light scattering by small particles*, Courier Corporation, 1957.
5 P. Bassan, H. J. Byrne, F. Bonnier, J. Lee, P. Dumas and P. Gardner, *Analyst*, 2009, **134**, 1586–1593.
6 P. Bassan, A. Kohler, H. Martens, J. Lee, H. J. Byrne, P. Dumas, E. Gazi, M. Brown, N. Clarke and P. Gardner, *Analyst*, 2010, **135**, 268–277.
7 P. Bassan, A. Kohler, H. Martens, J. Lee, E. Jackson, N. Lockyer, P. Dumas, M. Brown, N. Clarke and P. Gardner, *J. Biophotonics*, 2010, **3**, 609–620.
8 P. Bassan, A. Sachdeva, A. Kohler, C. Hughes, A. Henderson, J. Boyle, J. H. Shanks, M. Brown, N. W. Clarke and P. Gardner, *Analyst*, 2012, **137**, 1370–1377.
9 A. Kohler, J. Sulé-Suso, G. D. Sockalingum, M. Tobin, F. Bahrami, Y. Yang, J. Pijanka, P. Dumas, M. Cotte, D. G. van Pittius, G. Parkes and H. Martens, *Appl. Spectrosc.*, 2008, **62**, 259–266.
10 T. van Dijk, D. Mayerich, P. S. Carney and R. Bhargava, *Appl. Spectrosc.*, 2013, **67**, 546–552.
11 R. Lukacs, R. Blümel, B. Zimmerman, M. Bağcıoğlu and A. Kohler, *Analyst*, 2015, **140**, 3273–3284.
12 L. D. Landau, J. Bell, M. Kearsley, L. Pitaevskii, E. Lifshitz and J. Sykes, *Electrodynamics of continuous media*, Elsevier, 1984.
13 J. L. Ilari, H. Martens and T. Isaksson, *Appl. Spectrosc.*, 1988, **42**, 722–728.
14 H. Martens and E. Stark, *J. Pharm. Biomed. Anal.*, 1991, **9**, 625–635.
15 A. Kohler, C. Kirschner, A. Oust and H. Martens, *Appl. Spectrosc.*, 2005, **59**, 707–716.
16 N. K. Afseth and A. Kohler, *Chemom. Intell. Lab. Syst.*, 2012, **117**, 92–99.
17 B. Zimmermann and A. Kohler, *Appl. Spectrosc.*, 2013, **67**, 892–902.
18 R. P. Kanwal, *Linear integral equations*, Springer Science & Business Media, 2013.
19 P. Bruzzoni, R. Carranza, J. C. Lacoste and E. Crespo, *Electrochim. Acta*, 2002, **48**, 341–347.

Faraday Discussions

PAPER

Infrared imaging of small molecules in living cells: from *in vitro* metabolic analysis to cytopathology

Luca Quaroni,*[a] Theodora Zlateva,[b] Katia Wehbe[b] and Gianfelice Cinque[b]

Received 9th November 2015, Accepted 14th December 2015
DOI: 10.1039/c5fd00156k

A major topic in InfraRed (IR) spectroscopic studies of living cells is the complexity of the vibrational spectra, involving hundreds of overlapping absorption bands from all the cellular components present at detectable concentrations. We focus on the relative contribution of both small-molecule metabolites and macromolecules, while defining the spectroscopic properties of cells and tissue in the middle IR (midIR) region. As a consequence, we show the limitations of current interpretative schemes that rely on a small number of macromolecules for IR band assignment. The discussion is framed specifically around the glycolytic metabolism of cancer cells because of the potential pharmacological applications. Several metabolites involved in glycolysis by A549 lung cancer cells can be identified by this approach, which we refer to as Correlated Cellular Spectro-Microscopy (CSM). It is noteworthy that the rate of formation or consumption of specific molecules could be quantitatively assessed by this approach. We now extend this analysis to the two-dimensional case by performing IR imaging on single cells and cell clusters, detecting variations of metabolite concentration in time and space across the sample. The molecular detail obtained from this analysis allows its use in evaluating the pharmacological effect of inhibitors of glycolytic enzymes with potential consequences for *in vitro* drug testing. Finally we highlight the implications of the spectral contribution from cellular metabolites on applications in IR spectral cytopathology (SCP).

Introduction

Over the last twenty years IR spectromicroscopy has seen increasing applications in the study of cells and tissue because of its high sensitivity to molecular properties and its capability to perform a spatially resolved analysis of the sample. An IR absorption measurement can simultaneously provide information about the identity, concentration and biochemical properties of sample components. In

[a]*Institute of Nuclear Physics, Polish Academy of Sciences, Kraków, Poland. E-mail: quaroni@bluewin.ch*
[b]*Diamond Light Source, Didcot, Oxfordshire, OX11 0DE, UK*

spectromicroscopy the measurement can be performed on microscopic samples, including cells and tissue portions, with a spatial resolution which, in some optical configurations (*e.g.* coupled to a Synchrotron Radiation (SR) source), is limited only by light diffraction.[1] IR mapping or imaging experiments provide the spatial distribution of molecules, without resorting to staining and, in principle, with minimal or no previous knowledge of sample composition. All these features have led to increased interest in using IR spectromicroscopy for the characterization of biological systems. The compositional sensitivity of IR absorption measurements has also generated enthusiasm about the diagnostic potential of the technique, leading to a new discipline that has been termed spectral cytopathology (SCP). In SCP IR absorption spectra from biological samples are used to discriminate between healthy and pathological states by relying on the changes in molecular composition that accompany the onset of disease.[2–6]

An important development in biological applications of IR spectromicroscopy is its use to characterize biochemical reactions in living cells, using a combination of bench-top and synchrotron radiation sources. Early investigations in this field aimed at performing compositional studies of cells in an environment that ensures their viability.[7–13] In recent years this effort has further shaped itself into two broad directions. One direction of activity aims at developing the competence to perform classical and quantitative biochemical experiments *in vivo*, such as kinetics, structure–function correlations and mechanistic studies.[14–17] A different, although related, direction of activity aims at taking advantage of the broad sensitivity of IR absorption spectroscopy to study the activity of cells as a whole, such as in the quantitative description of metabolic networks.[18,19] Both directions share a common challenge in the need to identify and assign patterns of absorption bands amid the complexity of cellular IR absorption spectra.

To date the canonical approach to the interpretation of IR absorption spectra from biological samples follows a minimalistic scheme that reduces the spectrum to a combination of bands from a few classes of macromolecule families (proteins, polysaccharides, nucleic acids) plus acyl-containing lipids.[20] Developed during the early days of cellular IR studies, this scheme works as a general guide for the assignment of the most intense bands observed in tissue spectra.[21,22] Although very useful, it was pointed out early that such interpretation was a preliminary but not exhaustive approach, and it was suggested that more specific and detailed analysis should be performed in the context of each individual problem. Despite these cautionary notes, such minimalistic interpretation has become the standard approach for biological spectral analysis so far in IR literature. Reports that identified considerable contributions from other molecules in cellular IR spectra[23,24] did not substantially change this paradigm. Such neglect can be ascribed to the relatively low spectral contribution of most individual molecular components other than the canonical ones (proteins, polysaccharides, nucleic acids, acyl lipids), to the difficulty in making reliable band assignments in the complex cellular environment, and to the low signal-to-noise ratio in spectromicroscopy measurements.

In an effort to clarify the spectral contribution of small molecules and to address the problem of band assignment, we have recently introduced a method called correlated cellular spectromicroscopy (CSM). The method relies on two-dimensional correlation spectroscopy (2D COS) to identify clusters of bands belonging to the same molecule in a complex evolving system. The analysis identifies several bands from specific molecules, allowing identification by spectral fingerprinting. CSM is suitable for real-time metabolomic analysis, including the measurement of rates of metabolite formation and consumption. Its application to the study of glycolysis in hypoxic A549 cells can detect several molecules associated to glycolysis.[18]

In this work we further develop the issues that led us to introduce CSM and discuss in general terms the role of small molecule metabolites in cellular IR absorption spectra. We want to show how a CSM analysis can be further integrated with 2D IR imaging to provide a description of spatially resolved metabolic activity at the single cell level, finally drawing general conclusions on the implications of these results in SCP.

Experimental section

Cell culture

A549 cells (Human Caucasian Lung Carcinoma; ECACC Catalogue Number 86012804) were cultured directly on CaF_2 windows in Dulbecco's Modified Eagle Medium (DMEM) high glucose 4.5 mg ml^{-1}, no pyruvate, containing sodium bicarbonate (3.7 g l^{-1}) and supplemented with 10% foetal bovine serum and 1% penicillin–streptomycin antibiotic. Cultures were kept at 37 °C with 5% CO_2 in an incubator saturated with water vapour. The initial culture medium was replaced with a fresh medium after 33 hours and the cells in the exponential growth phase were allowed to grow for an additional 1 hour before using for metabolic analysis. For FTIR spectromicroscopy measurements the cell-coated window (final cell coverage 40–60%) was transferred to a custom-built solution sample holder kept at 37 °C.

Sample holder for FTIR imaging on cells

The sample holder is a variation of a sandwich design with two optical windows separated by a spacer and has been described previously.[18] A 15 μm gold-coated metal spacer is used to avoid damaging the cells. The temperature of the sample holder is controlled by the thermostatic fluid from a circulating bath. The cells are grown on an optical window as described above. Prior to the experiment the cell-coated window is placed into the holder with the growth medium. The spacer is superimposed and the second window is placed on top of the stack, followed by gentle tightening. No perfusion of the sample is performed to ensure the build-up of hypoxic conditions.

Beamline setup

IR experiments were performed at MIRIAM beamline B22 of the Diamond Light Source, using a Hyperion 3000 microscope coupled to a Vertex 80v interferometer (Bruker UK, Coventry). A summary of beamline parameters and performances for diffraction-limited FTIR microanalysis can be found in ref. 25.

Synchrotron based IR images and microspectra were acquired on the upstream IR microscope by means of a liquid N_2-cooled FPA photovoltaic MCT detector 64 × 64 pixels. By using a 36× objective and a 20× condenser optics in transmission mode, similarly to the work described in ref. 26, the corresponding field of view is circa 70 × 70 μm^2. A broadband KBr beamsplitter and a Ge filter provide the mid-

IR range up to 4000 cm^{-1}, while the detector cut-off was 950 cm^{-1}, as per current FPA technology.

FTIR measurements

Single-channel spectra were obtained by performing a Fourier Transform of the interferogram after apodization with a Blackman–Harris 3-Term function, using a zero-filling factor of 2 and a power phase correction mode.

A background single-channel measurement was performed on a portion of DMEM without cells. The IR beam was then moved to the location of a cell cluster for the sample single-channel measurements described below. 512 scans were averaged for sample and background measurements. Sequential measurements were recorded at 10 min intervals. Pixel response was averaged according to a 2 × 2 binning scheme.

Spectra are reported as differences from the first measurement in each pixel location and converted to their second derivative by using 13 points Savitzky–Golay smoothing. IR images are built by mapping the distribution of the peak of the second derivative for bands of interest.

All data processing and image construction are performed using OPUS 7.1 from Bruker Optics (Ettlingen, Germany).

2D Correlation spectroscopy

2D Correlation Spectroscopy (2D COS) was introduced as a method to quantify and parametrize the relationship between spectral properties that vary as a function of time or another physical quantity.[27] The algorithm involves calculation of the correlation function of the evolving spectral profiles, followed by application to the latter of a 2D Fourier transform algorithm. The result is a complex function of two wavenumber variables. The real component is called the synchronous plot, and the imaginary component is called the asynchronous plot. The synchronous plot represents correlations between pairs of wavenumbers that are changing fully in phase or with opposite phase. Two peaks that change in phase have positive "correlation peaks" in the synchronous plot, on both sides of the diagonal. Peaks that change with opposite phase have negative correlation peaks. The asynchronous plot represents correlations between pairs of wavenumbers that have an out-of-phase relationship.

Bands from the same molecule must change perfectly in phase, with positive correlation peaks in the synchronous plot and no correlation peaks in the asynchronous plot. The correlation properties of spectral changes arising from the same molecule allow us to assign the full set of bands belonging to that molecule. These properties are the basis of correlated cellular spectromicroscopy (CSM) for the analysis of reactions in living cells.[18]

Results and discussion

The application of CSM to metabolic turnover in hypoxic A549 cells has recently allowed us to identify several metabolites involved in glycolysis, including the end product of fermentation, lactate.[18] We are now building on this work by expanding these results into the second dimension and reconstructing the distribution in space of metabolite formation around individual cells.

By using FTIR imaging with a focal plane detector and a synchrotron light source we are able to track the distribution of metabolite formation in a cluster of hypoxic A549 cells adherent on a CaF_2 optical window. To maximize the spectral quality (figure of merit signal-to-noise ratio or s/n), we performed IR imaging experiments using MIRIAM beamline B22 at Diamond Light Source. The advantage of using a bright synchrotron source and coupling it to a two-dimensional FPA detector is more effective illumination for full-field IR imaging at high magnification.[28] More specifically for the MIRIAM beamline optics, the 36× objective demagnifies the 40 μm FPA pixel to an effective size of ~1.1 μm at the sample, and the theoretical signal gain expected per pixel is 20 times with respect to a broadband conventional IR source like a globar. Recent experimental work at this beamline[26] showed an overall IR signal gain up to 6 times per FPA pixel on living cells in a 10 micron water layer. Of note, the specific beamline design and multiple optics systems allow effective illumination of the FPA detector within the microscope specification for spectroscopy (illumination variation within 50% of the peak intensity on the area). An additional difficulty that can be addressed by synchrotron-based IR imaging is to overcome the strong background due to an aqueous sample environment. This is very challenging for IR spectroscopy because of the strong absorption from water that causes a significant throughput reduction. In the optical configuration used for these experiments, synchrotron illumination gives a clear advantage in terms of increased photon flux density relative to the globar, which allows enough signal at the single pixel level for proper absorption calculation, including water correction and second derivative analysis. In so doing, we follow an approach that has already proven useful in identifying the metabolic turnover of isotopically labelled water molecules by functional fibroblasts.[26]

Cells are enclosed in the chamber of the sample holder and brought to the IR microscope for measurement while maintaining a temperature $T = 37$ °C. No flushing of the cell with medium is performed after sealing, thus allowing the establishment of hypoxic conditions. A background measurement is performed on a region of the chamber where only medium is present. The field of view is then moved to a group of cells where spectra are measured sequentially.

Fig. 1 shows the visible light transmission image of a cluster of A549 cells and the corresponding distribution of several absorption bands throughout the cluster. The bands were recorded as difference of absorption from the beginning of the measurement. The maps in Fig. 1 correspond to the total variation over a 50 min interval. The maps were constructed by plotting the intensity of the 2nd derivative of absorption at the given wavenumber. Assignments to bands of specific molecules are based on the CSM analysis of this cellular system under identical conditions.[18] The vertical scale of each map has been expanded differently to enhance visibility of the distribution of interest.

The periphery of the full field of view is not usable because of the low light intensity that illuminates the detector. This is due to the inhomogeneous illuminations of the sample when using the synchrotron light source. The light source was aligned to increase flux in the central portion of the detector, while excluding the periphery. Images have been cropped relative to the original size to exclude most of the unusable regions from the field of view. Despite this, some portions of the cropped image still display high noise levels and high apparent baselines and are of little use for a sensitive measurement. These regions

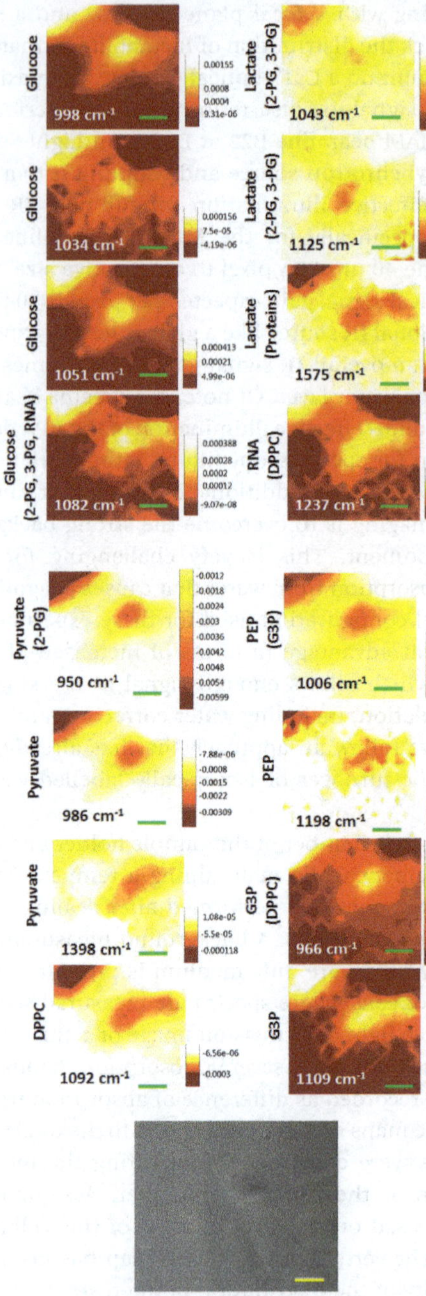

Fig. 1 IR images from a cluster of A549 cells at different wavelengths. Images show the distribution of the second derivative peak of specific absorption bands. Spectral images are reported as differences from the first image, recorded 50 min after the beginning of the measurement. The maps correspond to selected absorption bands associated to specific molecules involved in glycolysis. Each map includes the assignment to a specific molecule as determined in ref. 18. The presence of overlapping bands from other molecules is represented by listing in brackets the molecules giving rise to the overlap. Inset: visible image of the cell cluster collected soon after these IR images.

correspond to the corners of the image and are generally sent out of scale when adjusting the intensity interval to be displayed.

It is immediately apparent from Fig. 1 that changes in metabolite absorption are observed in space and in time. Most of the changes are localized in the proximity of one large cell in the upper centre of the map and give rise to clear gradients around this cell. Another location where a clear concentration gradient is observed is the upper left corner of the map, because of the presence of other cells located at the very edge of the field of view. More cells are seen in the lower half of the image, but without any detectable concentration gradients. It is notable that the latter are seen shifting their location during the course of this measurement, thus confirming their viability.

The high level of smoothing associated to the Savitzky–Golay algorithm, 13 points, is required for extracting the weaker spectral changes from the noisy baseline and to allow mapping. However, this improvement comes at the cost of spectral resolution. Several overlapping bands cannot be resolved and some of the maps in Fig. 1 must be assigned to spectral contributions from more than one molecule. Specific cases of overlap are highlighted both in the figure and in the text. Presently, the need to decrease spectral resolution to compensate for a low s/n ratio is the main drawback in the application of IR imaging to metabolic analysis *in vivo*. In comparison, spectral analysis by CSM in a sample larger than 50×50 μm can be performed by using 7 or 9 smoothing points in the Savitzky–Golay algorithm.[18]

Some discrimination is possible between events occurring in the exterior and in the interior of the cell, although only on a qualitative basis. The presence of headspace between the top of the cell and the upper optical window means that some contribution from components of the medium is present in every measurement location and cannot be separated without accurate knowledge of the topography of the cell.

Gradients in space are observed for bands identified by CSM. Glucose and lactate provide the strongest contribution to the spectral variations and the associated maps. This is in agreement with their role as substrate and final product of glycolysis and fermentation. In the case of lactate, maxima in the concentration gradients for all bands are observed both in the central part of the cell and in its periphery, extending into the extracellular space. This is expected since lactate is known to be released by the cell into its surroundings. In contrast, glucose bands show a negative gradient, with a minimum of concentration only in the central region of the cell. This is also expected since glucose is consumed within the cytoplasm. Glycolysis is fully catalysed by enzymes that are distributed in the cytoplasm itself, as opposed to oxidative phosphorylation, which is localized in the mitochondria. Therefore the decrease in glucose concentration is expected to track the volume of cytoplasm that is probed by the beam. The quality of the cellular image in the inset of Fig. 1 is degraded by chromatic aberrations introduced by the optical windows, preventing us from making an accurate comparison. Nonetheless, this is at least qualitatively confirmed by comparison of maps from glucose bands at 998 cm^{-1}, 1034 cm^{-1} and 1051 cm^{-1} (which do not include major overlap from other bands) and with the visible light image. Glucose gradients track the topography of the cell and appear negligible outside of the cell. The implication is that cellular uptake, as opposed to diffusion or enzymatic degradation, is a rate-limiting step in glucose consumption.

One interesting observation is the presence of clear gradients associated to bands that were only weakly detected in extended clusters of A549 cells under identical conditions.[18] Such metabolites include pyruvate, phosphoenolpyruvate and the phosphoglycerates. All of them are present at absolute concentrations within the detection limit of IR spectroscopy, and have been observed in macroscopic measurements of metabolic turnover, for example in yeast lysates.[29] The lack of spectroscopic evidence for these species in the difference spectra from A549 cell clusters was attributed to steady state conditions that maintain the concentration of intermediate metabolites relatively stable throughout the cluster. Under these conditions only reactants and final products are observed in difference spectra. Intermediate species are observed only as far as they are displaced from homeostasis by a transient perturbation.[18] In the current imaging measurements the local concentration of a molecule varies in response to both local turnover and diffusion, making it easier to avoid steady state conditions. The resulting concentration gradients extend in space and also change over time, facilitating detection of intermediate species compared to an averaged bulk measurement.

Spectromicroscopy measurements on A549 cell clusters[18] failed to detect appreciable changes in the concentrations of ATP and inorganic phosphate (PO_4), despite the fact that the absolute cellular concentrations of these molecules are well within the detection limits of the technique. Imaging measurement in this work also failed to detect any unequivocal contributions from these two species. No gradients can be observed for the strongest absorption band specific to aqueous ATP, at 1075 cm^{-1}, nor for the strongest absorptions from PO_4 at pH values close to neutral, at 1077 cm^{-1} and 1159 cm^{-1}. The reasons for this absence are not clear. It is known from bulk measurements that cellular variations in ATP concentration are very rapid following perturbation of a system,[30] and a stationary state is quickly reached. Nonetheless, the absence of observable concentration gradients in space is surprising. Possible explanations will be tested in future experiments. One possibility is that the spectra of these species in the cellular environment differ from those of the species in bulk aqueous solution. Complexation to H-bonding donors or to metal ions can shift vibrational transitions by a few, up to tens, of wavenumber units. In this case we would need to identify the correct spectral bands for the cellular forms before being able to locate them in cellular spectra and maps. An alternative or additional possibility is that the intracellular localization of these species prevents the formation of gradients driven by passive or active transport across the cellular membrane. The combination of this effect with the rapid achievement of steady-state conditions would prevent the observation of concentration changes in the present experiment. Nonetheless, the latter explanation would be surprising given the current understanding of PO_4 and ATP transport.

Current results confirm that the detectability of small molecules in cellular spectra is viable. In an attempt to generalize this conclusion and encourage the design of future experiments, we have searched available information about small molecules that are potentially identifiable in an FTIR experiment. Table 1 summarizes the results of a search performed in the Human Metabolome Database[31] (HMDB), maintained at the University of Alberta, Canada (http://www.hmdb.ca). The table lists the cytoplasmic concentration values reported by the HMDB for several human metabolites. Each entry is the outcome of

Table 1 Reported cytoplasmic concentrations of human metabolites from the Human Metabolome Database (HMDB). The red line shows a recently reported limit for the detectability of a small molecule in an FTIR measurement[32]

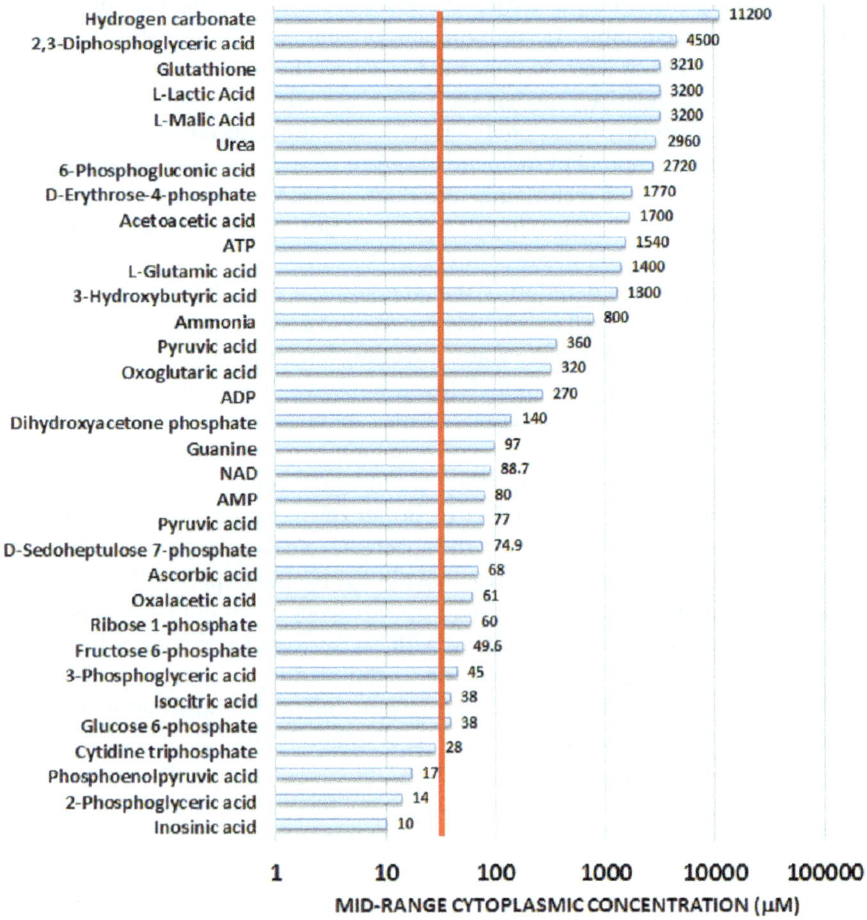

measurements from an individual publication. The individual HMDB entry provides a link to the original publication and corresponding information about the method, source cell type, data analysis and associated uncertainties. The table lists only the most concentrated metabolites, expected to be detectable in an FTIR bulk measurement. As detection limit we tentatively take the value of 20 μm, recently reported by Chan *et al.*[32] Detection limits with a microscopy configuration are expected to be at least one order of magnitude worse, depending on aperture size, optical configuration and light source. Due to the variability and limited selection of cell type, sample preparation, and analytical method, Table 1 is not meant to represent the composition of any real cellular state. It is rather a general guide to what specific molecules are potentially accessible to investigation with FTIR spectroscopy or microscopy measurements. The list of entries is expected to increase in the future, as accurate cytoplasmic concentrations of more metabolites are reported. One outstanding absence is the concentration value for phosphate species, which is currently being debated, but which is expected to be at upper end of the table. The detection limit is also expected to shift to lower values as the sensitivity of FTIR and other IR spectroscopic methods is improved.

Table 1 shows that only a small fraction of known human metabolites, of the order of 1%, is detectable. Despite the low percentage value, in absolute terms this corresponds to at least 30 molecular species. The spectral complexity associated even to this relatively modest set is daunting, when considering that $3N-6$ ($N=$ number of atoms) vibrational normal modes are associated to each molecular species, even if not all of them are observed in an absorption experiment. In this respect, the use of techniques like CSM, coupled to isotopic substitution experiments, will prove critical in mining the wealth of biochemical information associated to cellular and tissue samples.

One important consideration is that the spectral contribution of individual metabolites varies from cell to cell. It is expected to depend on all the factors that influence the cellular metabolic state, including the phase of the cell cycle, the cellular environment and the availability of nutrients. A related consideration is the verification that several of these molecules are not a stable complement of the cell, but are constantly exchanged with the medium and with adjoining cells over the scale of minutes and seconds. This is unlike the complement of macromolecules, which on average are subject to slower turnover. The immediate consequence of these observations is that the spectroscopic properties of a cell (we are talking about IR absorption spectroscopy, but this could apply to Raman spectroscopy as well, and to other vibrational spectroscopy techniques) are dependent on how the cell has been interacting with its surroundings shortly before its spectroscopic probing. The composition of the medium, the local temperature and the level of cellular confinement are of course all obvious and major factors in determining the metabolic complement of the cell that is probed by spectromicroscopy experiments. However, many other factors also affect the composition of the system, such as the shape of the cell relative to the confocal aperture used for the measurement (which determines the ratio of cellular volume to the medium volume that is present in the spectra), the flow rate of the medium in a microincubator, the diffusion rate of the various nutrients through the sample, the retention time of the cell in a sampling loop or sampling capillary while being transferred during a biopsy, the delay in the transfer of a cell from one holder to another, the duration of a centrifugation or filtration step during cellular harvesting, the method and skill of the executor of a biopsy in separating a cell from its surrounding tissue, *etc.* All these factors can affect the resulting complement of small molecules of a cellular sample, while they have little effect on the contribution of major macromolecular components, which are more stably associated to the cell and show less variability in relative concentration. Many of these factors can be reproducibly controlled when operating *in vitro*, *e.g.* when comparing the spectra of cell lines cultivated under identical conditions and using the same sample holder. This variability is also much less appreciated when working on tissue sections. Most sections are obtained from biopsies or tissue harvest that are immediately followed by cryopreservation, fixation and/or sectioning. This ensures that the complement of metabolites of the tissue is either fully retained with appropriate spatial distribution (as in cryosectioning) or is removed in a reproducible way (as in fixation followed by sectioning). The possibility to control the loss of metabolites is one likely reason for the early success of SCP applications based on the analysis of tissue sections. However, major variability is expected when comparing cellular samples of different origin and measured in different laboratories. Loss or variation of the metabolite complement of single cells is difficult to control, and this uncertainty will be the main barrier in developing reliable diagnostic applications for SCP in a clinical setting, where small procedural variations are expected between laboratories, between different practitioners and between different operations of the same practitioner.

Conclusions

By performing time-resolved and spatially resolved measurements of cellular systems we have observed and identified a number of small molecule metabolites in A549 cells maintained under hypoxic conditions. This work demonstrates the viability to use CSM and IR imaging with synchrotron radiation to study metabolic turnover with single cell resolution.

To our knowledge, IR imaging is the first technique that allows discriminating cellular metabolic production in a non-destructive way, retaining cellular function, with single cell resolution. These results open the way to the use of IR imaging in a CSM analysis scheme to perform specific studies of the biochemistry of individual cells in multicellular systems. Techniques for the co-culture of different cell types and the use of organotypic cultures can allow extension to systems of increased complexity. The role of fluctuations of local concentrations in metabolic networks and the interaction between metabolic activities in adjoining cells can all be measured and hypothesis can be tested.

One potential offshoot is the use of the technique to assess the activity of compounds of pharmacological interest in multicellular assemblies that act as models for more complex living organs. Local compound turnover, partition between different cell types and intracellular exchange of metabolic products can all be tested, qualitatively and quantitatively, in an IR imaging experiment and analysed by CSM.

An additional direction for future development is the transfer of the CSM approach to other spectroscopic techniques that provide correspondingly rich and complex spectra of cellular samples. Raman spectromicroscopy and imaging is the obvious first candidate, being also a vibrational spectroscopic technique.[33] However, any set of spectroscopic data can be potentially analysed using a 2D COS approach, including the ones based on UV/vis and X-ray photon absorption and scattering.[27] One appealing development will be the integration of different techniques into a single approach that follows the CSM scheme. The resulting multispectral method would be more than the sum of its parts. Correlations between widely separated spectral regions may help to associate electronic and vibrational transitions to the same molecule. A correlation approach that involved both IR and Raman spectra would be especially powerful. It would take advantage of the complementary selection rules for the two techniques to provide more solid band assignments and more accurate identification of molecules.

The main limitations are the ones that generally affect all experiments of IR spectromicroscopy and imaging. One disadvantage is the relatively short penetration depth of mid-IR radiation in tissue, of the order of 10–50 μm, depending on the specific spectral interval, which restricts most experiments to single or double cellular layers. The second disadvantage is the limited sensitivity, which, despite the extensive list of detectable molecules presented in Table 1, still excludes the vast majority of metabolites. Concerning the latter aspect, an increase in the sensitivity of 2D COS-based methods, such as CSM, is obtained because of the reduced noise contributions in synchronous plots,[34] although the improvement has not been quantified yet.

The present work confirms that the detectability of small molecules in IR cellular spectra is far from being an exception; it is rather the rule. Although in absolute terms their contribution is often smaller than that of the canonical components of cellular spectra, it is far from negligible. Future work on cellular and tissue IR spectra aiming at molecular characterization of the sample will need to keep the contribution from small molecules in consideration, and conclusions from past work that neglected this contribution should be reassessed.

Acknowledgements

We are grateful to Prof. Frank Martin, Lancaster University, and Dr Endre Laczko, Functional Genomics Centre Zurich, for helpful discussion, and to Blagoj Sarafimov, Paul Scherrer Institute, for construction of the sample holder. Some of the authors (LQ and TZ) acknowledge Diamond Light Source for beam time allocation at the MIRIAM beamline, B22, (proposal SM12221) and support, which contributed to the results presented here.

References

1 P. Dumas and L. Miller, *Vib. Spectrosc.*, 2003, **32**, 3–21.

2 M. Diem, A. Mazur, K. Lenau, J. Schubert, B. Bird, M. Miljkovic, C. Krafft and J. Popp, *J. Biophotonics*, 2013, **6**, 855–886.

3 P. Lasch, J. Kneipp and D. Naumann, *Spectroscopy of Biological Molecules: New Directions*, 1999, pp. 509–510.

4 D. C. Fernandez, R. Bhargava, S. M. Hewitt and I. W. Levin, *Nat. Biotechnol.*, 2005, **23**, 469–474.

5 G. Bellisola and C. Sorio, *Am. J. Cancer Res.*, 2012, **2**, 1–21.

6 N. C. Purandare, I. I Patel, J. Trevisan, N. Bolger, R. Kelehna, G. von Bünau, P. L. Martin-Hirsch, W. J. Prendiville and F. L. Martin, *Analyst*, 2013, **138**, 3909–3916.

7 L. Quaroni and T. Zlateva, *Analyst*, 2011, **136**, 3219–3232.

8 J. R. Mourant, R. R. Gibson, T. M. Johnson, S. Carpenter, K. W. Short, Y. R. Yamada and J. P. Freyer, *Phys. Med. Biol.*, 2003, **48**, 243–257.

9 H.-Y. N. Holman, K. A. Bjornstad, M. P. McNamara, M. C. Martin, W. R. McKinney and E. A. Blakely, *J. Biomed. Opt.*, 2002, **7**, 417–424.

10 M. Romeo, C. Matthaus, M. Miljkovic, S. Boydston-White and M. Diem, *Biomedical Vibrational Spectroscopy and Biohazard Detection Technologies*, 2004, vol. 5321, pp. 124–129.

11 M. K. Kuimova, K. L. Chan and S. G. Kazarian, *Appl. Spectrosc.*, 2009, **63**, 164–171.

12 P. Heraud, J. Beardall, D. McNaughton and B. Wood, *Phycologia*, 2005, **44**, 104.

13 M. J. Nasse, S. Ratti, M. Giordano and C. J. Hirschmugl, *Appl. Spectrosc.*, 2009, **63**, 1181–1186.

14 L. Quaroni, T. Zlateva, D. Bedolla, S. Massaro and V. Torre, *ChemPhysChem*, 2008, **9**, 1380–1382.
15 L. Quaroni, T. Zlateva and E. Normand, *Anal. Chem.*, 2011, **83**, 7371–7380.
16 D. Ami, A. Natalello, M. Lotti and S. M. Doglia, *Microb. Cell Fact.*, 2013, **12**.
17 L. Chen, H.-Y. N. Holman, Z. Hao, H. A. Bechtel, M. C. Martin, C. Wu and S. Chu, *Anal. Chem.*, 2012, **84**, 4118–4125.
18 L. Quaroni and T. Zlateva, *Anal. Chem.*, 2014, **86**, 6887–6895.
19 D. I. Ellis, G. G. Harrigan and R. Goodacre, *Metab. Profiling*, 2003, 111–124.
20 M. J. Baker, J. Trevisan, P. Bassan, R. Bhargava, H. J. Butler, K. M. Dorling, P. R. Fielden, S. W. Fogarty, N. J. Fullwood, K. A. Heys, C. Hughes, P. Lasch, P. L. Martin-Hirsch, B. Obinaju, G. D. Sockalingum, J. Sule-Suso, R. J. Strong, M. J. Walsh, B. R. Wood, P. Gardner and F. L. Martin, *Nat. Protoc.*, 2014, **9**, 1771–1791.
21 M. Diem, S. Boydston-White and L. Chiriboga, *Appl. Spectrosc.*, 1999, **53**, 148A–161A.
22 L. M. Miller and P. Dumas, *Biochim. Biophys. Acta, Biomembr.*, 2006, **1758**, 846–857.
23 M. Z. Kastyak, M. Szczerbowska-Boruchowska, D. Adamek, B. Tomik, M. Lankosz and K. M. Gough, *Neuroscience*, 2010, **166**, 1119–1128.
24 K. L. Goff, L. Quaroni, T. Pedersen and K. E. Wilson, in *Wirms 2009: 5th International Workshop on Infrared Microscopy and Spectroscopy with Accelerator Based Sources*, ed. A. Predoi-Cross and B. E. Billinghurst, 2010, vol. 1214, pp. 54–56.
25 G. Cinque, M. Frogley, K. Wehbe, J. Filik and J. Pijanka, *Synchrotron Radiat. News*, 2011, **24**, 24–33.
26 L. Quaroni, T. Zlateva, B. Sarafimov, H. W. Kreuzer, K. Wehbe, E. L. Hegg and G. Cinque, *Biophys. Chem.*, 2014, **189**, 40–48.
27 I. Noda, *Appl. Spectrosc.*, 1993, **47**, 1329–1336.
28 M. J. Nasse, M. J. Walsh, E. C. Mattson, R. Reininger, A. Kajdacsy-Balla, V. Macias, R. Bhargava and C. J. Hirschmugl, *Nat. Methods*, 2011, **8**, 413–416.
29 T. Mair, L. Zimanyi, P. Khoroshyy, A. Muller and S. C. Muller, *BioSystems*, 2006, **83**, 188–194.
30 F. L. Martin and A. E. M. McLean, *Toxicology*, 1995, **104**, 91–97.
31 D. S. Wishart, C. Knox, A. C. Guo, R. Eisner, N. Young, B. Gautam, D. D. Hau, N. Psychogios, E. Dong, S. Bouatra, R. Mandal, I. Sinelnikov, J. Xia, L. Jia, J. A. Cruz, E. Lim, C. A. Sobsey, S. Shrivastava, P. Huang, P. Liu, L. Fang, J. Peng, R. Fradette, D. Cheng, D. Tzur, M. Clements, A. Lewis, A. De Souza, A. Zuniga, M. Dawe, Y. Xiong, D. Clive, R. Greiner, A. Nazyrova, R. Shaykhutdinov, L. Li, H. J. Vogel and I. Forsythe, *Nucleic Acids Res.*, 2009, **37**, D603–D610.
32 K. L. A. Chan and P. L. V. Fale, *Anal. Chem.*, 2014, **86**, 11673–11679.
33 K. Kong, C. Kendall, N. Stone and I. Notingher, *Adv. Drug Delivery Rev.*, 2015, **89**, 121–134.
34 L. Quaroni and E. Normand, in *Wirms 2009: 5th International Workshop on Infrared Microscopy and Spectroscopy with Accelerator Based Sources*, 2010, vol. 1214, pp. 66–68.

Faraday Discussions

PAPER

Chemotherapeutic response to cisplatin-like drugs in human breast cancer cells probed by vibrational microspectroscopy†

A. L. M. Batista de Carvalho,*[a] M. Pilling,[b] P. Gardner,[b] J. Doherty,[bc] G. Cinque,[c] K. Wehbe,[c] C. Kelley,[c] L. A. E. Batista de Carvalho[a] and M. P. M. Marques*[ad]

Received 30th October 2015, Accepted 11th January 2016

DOI: 10.1039/c5fd00148j

Studies of drug–cell interactions in cancer model systems are essential in the preclinical stage of rational drug design, which relies on a thorough understanding of the mechanisms underlying cytotoxic activity and biological effects, at a molecular level. This study aimed at applying complementary vibrational spectroscopy methods to evaluate the cellular impact of two Pt(II) and Pd(II) dinuclear chelates with spermine (Pt_2Spm and Pd_2Spm), using cisplatin (*cis*-$Pt(NH_3)_2Cl_2$) as a reference compound. Their effects on cellular metabolism were monitored in a human triple-negative metastatic breast cancer cell line (MDA-MB-231) by Raman and synchrotron-radiation infrared microspectroscopies, for different drug concentrations (2–8 μM) at 48 h exposure. Multivariate data analysis was applied (unsupervised PCA), unveiling drug- and concentration-dependent effects: apart from discrimination between control and drug-treated cells, a clear separation was obtained for the different agents studied – mononuclear *vs.* polynuclear, and Pt(II) *vs.* Pd(II). Spectral biomarkers of drug action were identified, as well as the cellular response to the chemotherapeutic insult. The main effect of the tested compounds was found to be on DNA, lipids and proteins, the Pd(II) agent having a more significant impact on proteins while its Pt(II) homologue affected the cellular lipid content at lower concentrations, which suggests the occurrence of distinct and unconventional pathways of cytotoxicity for these dinuclear polyamine complexes. Raman and FTIR microspectroscopies were confirmed as powerful non-invasive techniques to obtain unique spectral signatures of the biochemical impact and physiological reaction of cells to anticancer agents.

[a]"Química-Física Molecular", Univ. Coimbra, 3004-535 Coimbra, Portugal. E-mail: pmc@ci.uc.pt; analmbc@gmail.com

[b]Manchester Institute of Biotechnology, Univ. Manchester, Manchester, M1 7DN, UK

[c]Diamond Light Source, Harwell Science and Innovation Campus, Didcot, Oxfordshire OX11 0DE, UK

[d]Dep. Life Sciences, Univ. Coimbra, 3000-456 Coimbra, Portugal

† Electronic supplementary information (ESI) available: Raman, FTIR and INS spectra (600–1800 cm^{-1}) of the complexes under study. See DOI: 10.1039/c5fd00148j

A Introduction

Chemotherapy is one of the main approaches in the management of cancer (alongside radiotherapy and surgery) and aims at killing neoplastic cells with minimal damage to healthy tissue. Breast cancer is the second most common type of cancer worldwide and the most frequent among women,[1] being the leading cause of female mortality. Particularly for the more aggressive triple negative invasive breast carcinoma (lacking the oestrogen, progesterone and human epidermal growth factor receptors (ER, PR and HER2)), the prognosis is still poor due to its heterogeneity and high metastatic capacity.[2,3] Development of new anticancer agents and therapeutic strategies against this type of tumour is therefore of paramount importance, aiming at a higher efficacy regarding cell growth inhibition and decreased invasiveness, coupled to lower acquired resistance and deleterious side effects.[4]

Since Rosenberg's serendipitous discovery of cisplatin (*cis*-$Pt(NH_3)_2Cl_2$)[5–7] and in spite of the high number of studies in this field,[7–10] only three platinum-based compounds are currently approved worldwide as anticancer drugs for clinical use: cisplatin, carboplatin and oxaliplatin, that act through covalent binding to DNA forming predominantly intrastrand 1,2- and 1,3-crosslinks, which induce cell cycle arrest and apoptosis. However (apart from being administered intravenously) they are associated with severe side effects (*e.g.* nephrotoxicity, hepatotoxicity or myelosuppression) as well as acquired resistance, which strongly limit their application, particularly in single agent chemotherapy. In addition, they display a very low efficiency against metastatic cancers, which are those responsible for a higher mortality worldwide. For the last two decades, particular attention has been paid to polynuclear Pt(II) and Pd(II) chelates (comprising 2 or 3 metal centres) with flexible polyamine ligands, aiming at an increased activity and higher selectivity.[11] These display a higher cytotoxicity towards several types of cancer (*e.g.* oestrogen-independent (ER^-) breast carcinoma) as compared to the classic Pt(II) agents, due to more severe and less repairable DNA damage upon formation of long-range interstrand drug–DNA adducts, not available to conventional drugs.[12,13] However, their mechanism of action is not yet fully understood at the molecular level. Furthermore, their Pd(II) analogues have recently arisen as innovative and promising alternatives to Pt(II) compounds,[9–11,14,15] particularly against metastatic breast cancer,[16] namely when using strongly coordinating polydentate ligands such as polyamines, which minimise displacement by sulfhydryl groups from glutathione and other cellular thiols (responsible for drug inactivation and deleterious side effects).[17] A careful design of palladium complexes may thus allow targeting strategies leading to different profiles of drug activity and to a lack of cross-resistance as compared with platinum-based agents.[12]

In the last few years, the authors have been involved in the study of polynuclear platinum and palladium-based complexes with polyamine ligands as promising anticancer agents towards several human cancer cells (*e.g.* metastatic breast carcinoma, castration-resistant prostate cancer and melanoma). A thorough knowledge of these compounds and of their interactions with their main pharmacological target (DNA), as well as with other biomolecules involved in the drug pharmacokinetics (*e.g.* glutathione) and pharmacodynamics, has been attained through vibrational spectroscopy (Raman, FTIR and Inelastic Neutron Scattering (INS))[18–24] coupled to synchrotron-based Extended X-ray Absorption Fine Structure (EXAFS),[23] as well as by NMR metabolomics[22,25] and electrochemical techniques.[26,27] In order to design such effective chemotherapeutic agents, it is essential not only to determine their effect on cellular growth, metabolism and function, but also to understand the mechanisms (at a molecular level) associated with this biological impact. This can be accomplished through *in vitro* experiments in cells, followed by *ex vivo* and *in vivo* studies, coupled to high resolution spectroscopic techniques. In the last decades, vibrational spectroscopy has arisen as a suitable and extremely accurate analytical tool for the characterisation of biological systems, as it provides detailed data on the investigated sample. Its high sensitivity, non-invasive nature and real-time molecular imaging capability, without the need for dyes or external probes, allows the attainment of extremely accurate biochemical information and images of cells and tissues, in both *in vitro* and *in vivo*/*in situ* conditions. The interplay between the vibrational profile and spectra–structure correlation yields quantitative and readily comprehensible data even for inhomogeneous biological samples. In particular, Raman confocal microspectroscopy of cell cultures constitutes a unique probing technique, yielding precise information on their functionality and metabolic profile under distinct conditions, namely as a response to chemotherapeutic agents.[28,29] Additionally, coupling the complementary technique Fourier transform infrared (FTIR) microspectroscopy to the brilliance of synchrotron radiation (SR) sources leads to a massive increase in the quality of the results (relative to benchtop globar sources), yielding broadband spectra with subcellular resolution and unsurpassed signal-to-noise ratio. This constitutes an unmatched approach for routine chemical profiling of cells, allowing accurate examination of cellular environments and monitoring of drug bioavailability, biodistribution and target site accumulation after administration, as well as drug delivery across biological barriers and cellular response to treatment.[30–32] Since cells are known to have specific vibrational spectroscopic signatures for different physiological conditions (regarding viability, cell cycle, apoptosis or stress), a correlation between a particular vibrational profile and a definite effect or mode of action should predictably be attained.[33]

The present study aimed at probing the cellular impact of two Pt(II) and Pd(II) dinuclear chelates with the biogenic polyamine spermine (Spm = $H_2N(CH_2)_3NH(CH_2)_4NH(CH_2)_3NH_2$), synthesised and fully characterised by the authors[18–24,34–36] – Pt_2SpmCl_4 and Pd_2SpmCl_4 (hereafter denominated Pt_2Spm and Pd_2Spm). Raman microspectroscopy and synchrotron-radiation infrared microspectroscopy (SR-IRMS) were applied to probe the effects of their interaction with the cellular components, on a human triple-negative breast cancer cell line, as compared to the well established anticancer drug cisplatin. This was achieved relying on the microRaman equipment from the Vibrational Spectroscopy Laboratory for Bioprobing of the Research Group QFM/University of Coimbra (Portugal) (VIBIMA/QFM-UC),[37] as well as on the state-of-the-art fingerprint microprobe analysis, detection and imaging capabilities of B22 beamline from Diamond Light Source (DLS, UK) (MIRIAM – Multimode InfraRed Imaging and Microspectroscopy).[38,39]

The data obtained within this work, coupled to conformational analysis of the metal complexes and cytological assays (for evaluation of growth-inhibition, cytotoxic and anti-metastatic activities) previously carried out at QFM-UC,[14–16,22,25–27,40–45] will pave the way for future studies on other platinum and palladium-based polyamine agents prepared by the authors. This approach should provide valuable clues for the rational design of novel cisplatin-like drugs, displaying a higher efficiency, lower toxicity, decreased acquired resistance and possible oral administration.

B Methods

B.1 Chemicals

Antibiotics (penicillin–streptomycin 100× solution), Dulbecco's Modified Eagle's Medium-high glucose (DMEM-HG, 4500 mg L^{-1} glucose), *cis*-dichlorodiammine platinum(II) (cisplatin, $Pt(NH_3)_2Cl_2$, >98%), ethylenediaminetetraacetic acid (EDTA, disodium salt, dihydrate), formalin (10% neutral-buffered formalin, *ca.* 4% formaldehyde), phosphate buffered saline (PBS), potassium tetrachloropalladate(II) (K_2PdCl_4, >99.9%), potassium tetrachloroplatinate(II) (K_2PtCl_4, >99.9%), spermine (*N*,*N'*-bis(3-aminopropyl)-1,4-diaminobutane), trypan blue (0.4%), trypsin–EDTA (1×), and inorganic salts and acids (of analytical grade) were purchased from Sigma-Aldrich Chemical S.A. (Sintra, Portugal). Fetal bovine serum (FBS) was obtained from Gibco-Life Technologies (Porto, Portugal).

B.2 Synthesis and characterisation of the complexes

Synthesis of the Pt_2Spm and Pd_2Spm complexes were carried out according to published procedures,[46] optimised by the authors.[24,41] Briefly, 2 mmol of K_2PtCl_4 or K_2PdCl_4, respectively, was dissolved in a minimal amount of water, and an aqueous solution containing 1 mmol of spermine was added dropwise under continuous stirring. The reaction was allowed to occur for 24 h, after which the resulting yellow powder was filtered off and washed with acetone. Recrystallization of these products from water afforded yellow-orange needle-shaped crystals. The newly synthesised compounds were fully characterised (and tested as to their purity) by elemental analysis and vibrational spectroscopy[18,24] (FTIR, Raman and INS, coupled to *ab initio* theoretical calculations, Fig. S1/ESI†).

Pt_2Spm (yield 60%) – elemental analysis, found (for $Pt_2(C_{10}N_4H_{26})Cl_4$): C: 16.2%; H: 3.5%; N: 7.6%, Cl: 19.1%; calculated – C: 16.4%; H: 3.6%; N: 7.5%, Cl: 19.3%. Pd_2Spm (yield 68%) – elemental analysis, found (for $Pd_2(C_{10}N_4H_{26})Cl_4$): C: 21.2%; H: 4.7%; N: 9.6%, Cl: 25.9%; calculated – C: 21.5%; H: 4.7%; N: 9.9%, Cl: 25.6%.

The FTIR spectra of the solid complexes were obtained (at room temperature, in KBr disks *ca.* 0.5% (w/w)) using a Bruker Optics Vertex 70 FTIR spectrometer, coupled to a KBr beamsplitter and a liquid nitrogen cooled Mercury Cadmium Telluride (MCT) detector. Each spectrum was the sum of 50 scans, at a 2 cm^{-1} resolution. The error in wavenumbers was estimated to be less than 1 cm^{-1}. The Raman spectra of Pt_2Spm and Pd_2Spm were acquired (at room temperature) with a Bruker RFS-100 Fourier transform Raman spectrometer, equipped with an InGaAs detector, with near-infrared excitation (in order to overcome fluorescence) provided by the 1064 nm line of a Nd:YAG laser (Coherent, model Compass-1064/500N) yielding *ca.* 150 mW at the sample position. A 180° geometry was employed, and each spectrum was the average of three repeated measurements of 50 scans each, the resolution being set to 2 cm^{-1}. The neutron measurements were carried out at the ISIS Pulsed Neutron Source of the Rutherford Appleton Laboratory (UK). INS data were recorded (at 10 K) on the TOSCA spectrometer,[47,48] an indirect geometry time-of-flight, high resolution (($\Delta E/E$) *ca.* 1.25%), broad range spectrometer.

B.3 Preparation of solutions

Stock solutions were prepared in PBS, at concentrations of 1 mM for cisplatin, and 500 μM for Pt_2Spm and Pd_2Spm, sterile-filtered and stored at −20 °C. Further dilutions to the final desired concentrations were made in PBS, immediately before addition to the cells.

B.4 Cell culture

The epithelial human breast cancer cell line MDA-MB-231 (human Caucasian triple-negative, claudin-low, breast carcinoma)[49] was purchased from the European Collection of Cell Cultures (ECACC, UK). The cells were cultured as monolayers, at 37 °C, in a humidified atmosphere of 5% CO_2. The cultures were maintained in a DMEM-HG medium supplemented with 10% (v/v) heat-inactivated FBS, 1% (v/v) penicillin (100 U mL^{-1})/streptomycin (100 mg mL^{-1}) and sodium bicarbonate-20 mM (pH 7.4). Cells were subcultured at 80% confluence, using trypsin–EDTA (1×). Under these conditions, the duplication time was 26 h. The cells were always in the logarithmic phase of growth when the tested agents were added.

B.5 Sample preparation for spectroscopic analysis

Upon harvesting by trypsinization, cells were centrifuged and the pellet was resuspended in culture medium and seeded, at a concentration of 3×10^4 cells per cm^2, on optical substrates suitable for either Raman[50] or FTIR data collection:[51] respectively MgF_2 (2 × 20 mm) and CaF_2 (IR-grade, 1 × 13 mm) disks, previously cleaned with ethanol-70%. Upon incubation for 24 h (allowing the cells to attach), cells were treated with different concentrations of each tested compound (2, 4 and 8 μM) and left to culture for a further 48 h. The growth medium was then removed, cells were washed twice with NaCl-0.9%, fixed in 4% formalin (diluted in NaCl-0.9% from the commercial 10% neutral-buffered formalin solution) for 10 min (ref. 51) and washed several times with pure water (to remove any residual salt). The disks were allowed to air-dry prior to spectroscopic analysis. Spectral collection was performed for the dry samples at B22/DLS, and within NaCl-0.9% (w/v) solution at VIBIMA/QFM-UC.

All samples were prepared in triplicate, in a single experiment.

B.6 Vibrational spectroscopy

B.6.1 Raman microspectroscopy. Raman spectra of MDA-MB-231 fixed cells on MgF_2 disks, both untreated and drug-treated, were recorded (at room temperature) at the VIBIMA laboratory of the University of Coimbra (Portugal)[37] and at the peripheral laboratories of the B22 beamline from Diamond Light Source (UK).[38,39] In all cases, samples of untreated cells were used as a control.

At VIBIMA/QFM-UC, the spectra were obtained in the 600–1800 cm^{-1} range, using a Horiba Jobin-Yvon T64000 spectrometer in direct configuration mode (focal distance 0.640 m, aperture *f*/7.5), equipped with a holographic grating of 1800 grooves per mm. The entrance slit was set to 200 μm. Rayleigh elastic scattering was rejected by a Notch filter, which reduces its intensity by a factor of 10^6. The detection system was a liquid nitrogen cooled non-intensified 1024 × 256 pixels (1″) CCD. The 514.5 nm line of an Ar^+ laser (Coherent, model Innova 300-05) was used as the excitation radiation, yielding *ca.* 10 mW at the sample position. All the spectra were recorded using an Olympus 60× water immersion objective (Olympus LUMPLFLN 60XW, NA 1.0, wd 2 mm). A 200 μm confocal pinhole rejected signals from out-of-focus regions of the sample. Apart from the control (untreated cells), samples containing 2, 4 and 8 μM of the tested drugs (48 h exposure) were analysed. 30 to 40 spectra were collected per sample (in different cells), with 4 accumulations and 60 s of exposure, at <1 cm^{-1} spectral resolution.

At B22/DLS, Raman data for samples of fixed cells exposed to drugs at 2, 4 and 8 μM (48 h exposure) and the controls were acquired in the 600–1800 cm^{-1} and 2600–3600 cm^{-1} regions, using a Bruker Senterra dispersive Raman micro-spectrometer with a CCD detector (1024 × 265 pixels), using a 100× objective (Olympus BX, NA 0.9, Japan), a 50 μm diameter confocal aperture and a 532 nm laser line providing *ca.* 25 mW at the sample. 150 to 200 spectra were collected per sample (from cytoplasm and the nucleus for each cell, Fig. 1), with 5 s of exposure and 6 co-added scans per chosen point, sampling approximately 75 to 100 cells per sample, at 9–18 cm^{-1} spectral resolution.

B.6.2 SR-IRMS. Synchrotron-radiation infrared microspectroscopy data of MDA-MB-231 fixed cells on CaF_2 IR optical substrates, both untreated and drug-treated, were acquired (at room temperature) at the infrared beamline B22 (MIRIAM) from Diamond Light Source (UK).[38,39]

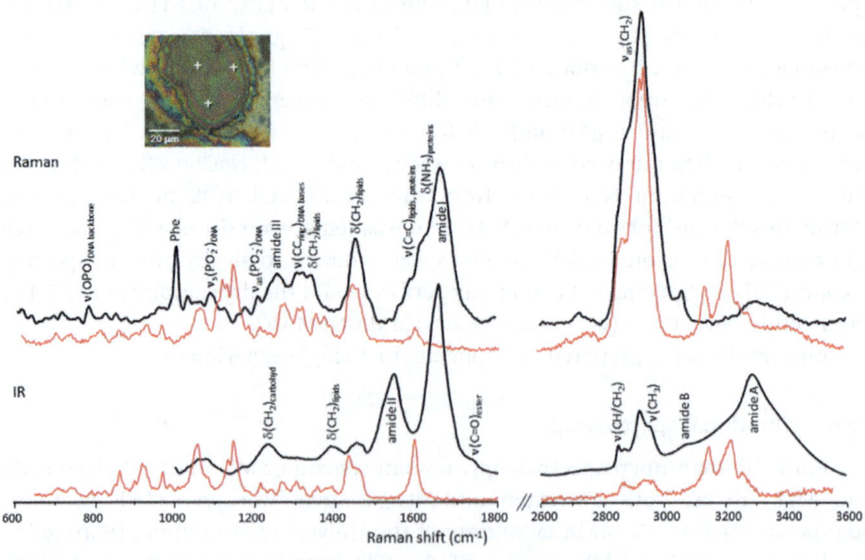

Fig. 1 Mean Raman (saline-dipped fixed cells, Horiba Jobin-Yvon T64000/514.5 nm) and infrared spectra (fixed cells, MIRIAM) for untreated MDA-MB-231 cells, and microscopic image of a cell (×100 magnification) showing some of the points at which data were captured. The spectra for Pd_2Spm are shown in red.

Acquisition in the mid-IR region 650–3900 cm^{-1} (Ge filter cut off) was carried out in transmission mode using a Hyperion 3000 microscope coupled to a Bruker Vertex 80 FTIR spectrometer, equipped with liquid nitrogen cooled Mercury Cadmium Telluride (MCT) high sensitivity detection (50 × 50 μm^2 area) and a synchrotron radiation IR source yielding 300 mA at the sample position (microbeam fwhm *ca.* 15 × 15 μm^2 and detected area between 5 × 5 to 15 × 15 μm^2 selected *via* slits). An area of 10 × 10 μm^2 at the sample (NA 0.5) matched with a 36× cassegrain condenser for transmission measurement.

Samples containing drug at 4 and 8 μM concentrations (for a 48 h exposure) were analysed, as well as the controls (drug-free). 150 to 200 spectra were collected per sample (from at least 2 points per cell sampling 75 to 100 cells per sample, Fig. 1), with 256 co-added interferograms per chosen point (as well as background every 10 spectra), *i.e. ca.* 40 s per spectrum in order to obtain the signal-to-noise ratio for further data analysis. Three replicates were acquired for each sample.

B.7 Data preprocessing and analysis

Raman spectra collection was carried out using the LabSpec 5.0 (Horiba, France) and OPUS 7.2 (Bruker Optik, Germany) packages, respectively in VIBIMA/QFM-UC and B22/DLS. Data pre-processing was performed with The Unscrambler® X software version 10.2 (CAMO, Norway). Spurious peaks caused by cosmic rays were removed and the spectra were selected for data treatment by discarding those with a very poor S/N ratio. The spectra were baseline corrected (second order polynomial in the 600–1800 cm^{-1} region and linear baseline correction for the 2650–3450 cm^{-1} interval), smoothed by the Savitsky–Golay filter (order 1, 9 points) which enabled the reduction of noise without band distortion, and vector normalised (minimising the effects of variable thickness within and among the cell samples).

FTIR data acquisition was performed using OPUS 7.2, which compensated for atmospheric and water vapour interferences, and corrected for background intensity changes within the chosen wavenumber interval. Infrared absorbance spectra were obtained by ratioing to a background measured from a clean area of the sample where no cells were present. Since for spectral acquisition in transmission mode the sample is illuminated through the substrate, reflection losses related to the relative refractive indices at the substrate–sample interface must be considered in the data treatment, as well as optical contributions from sample–air and air–substrate boundaries. This was achieved by RMieS correction (with the Extended Multiplicative Signal Correction (EMSC) algorithm,[52,53] 20 iterations) of the single cell spectra, vector normalised and mean centered. The spectra were corrected independently before computing the average spectrum per cell, as each spectrum may have experienced a different spectral distortion due to the heterogeneous nature of the cells distributed along the two-dimensional arrangement of pixels over which the sample was illuminated. The data were quality checked to remove any spectra which still exhibited high levels of scattering. Quality checking was based on the amide I band intensity, spectra with absorbance between 0.1 and 1 being retained. Initial pre-processing was based on a PC-based noise reduction algorithm: noise reduction was achieved by retaining a selected number of principal components and then recombing the dataset. Careful selection of the retained components enables random noise to be removed while still preserving the chemical information in the spectra. Utilising PC-based noise reduction of the data with 40 PCs enabled good improvements to be observed in the signal-to-noise ratio.

For both FTIR and Raman, the spectra were plotted using Origin 8.0 (OriginLab, USA). Multivariate analysis of the results – unsupervised principal components analysis (PCA) – was carried out with The Unscramber X 10.2 and using Matlab 2012a (The MathWorks Inc., Natick, MA) and the ProSpect Toolbox (London Spectroscopy Ltd., London, UK). The order of the PCs denotes their importance to the dataset, PC1 corresponding to the highest amount of variation. Analysis was carried out separately for each spectral region probed. For the Raman data, PCA was performed using the cross validation method, while for the infrared results the nonlinear iterative partial least squares algorithm and the leverage correction validation method were applied.

C Results and discussion

Following previous biological studies by the authors on the effect of the tested Pt(II) and Pd(II) spermine complexes on the growth and viability of several human cancer cells,[14–16,22,26,27,40,41,43,45] including the hormone-unresponsive breast carcinoma currently investigated, a cellular response at the biochemical level is presently sought, through analysis of the vibrational pattern of drug-treated and untreated cells. This multidisciplinary approach aims to link biological response to changes in cellular chemistry and metabolism, thus identifying the spectral markers of drug activity and paving the way for an understanding of the drug mode of action at the molecular level, hopefully helping to predict drug efficacy under specific conditions. According to former evaluation of the anti-tumour effect of Pt_2Spm and Pd_2Spm towards MDA-MB-231 cells, at different time-points and doses,[16] a 2 to 8 μM concentration range was chosen for the current study, for a drug-exposure period of 48 h, as these conditions correspond to an optimal drug effect, the concentration leading to a 50% cell growth inhibition (IC_{50}) at 48 h being 5.4 μM for Pt_2Spm and 4.1 μM for Pd_2Spm. Hence, since IC_{50} will induce cell growth inhibition in half the population of the probed cells, higher concentrations (*ca.* twice the IC_{50} value) were also tested in order to better identify the drug-induced biochemical changes.

Spectral biochemical signatures of the cell, at each condition tested (type of drug and concentration), were generated by averaging the spectra acquired from different cellular locations. The data (both FTIR and Raman) were captured as single point spectra (Fig. 1), across a cell population comprising randomly distributed cells displaying a high spectral heterogeneity (also due to cells being in different points of the cell cycle). Therefore, a large number of data points is required, for the mean resulting spectrum to be representative of the sample. This procedure, as opposed to single cell spectral acquisition, minimizes the effect of the cell cycle profile (predominantly G_1 under normal conditions). Cisplatin, a DNA major groove binder with a well characterised mechanism of action[54] was used as a reliable control, its effect on the MDA-MB-231 cells being compared and correlated to that of the cisplatin-like dinuclear Pt(II) and Pd(II) chelates. Similarly to cisplatin, these are expected to be cell-cycle non specific agents, allowing their cellular impact to be accurately retrieved from analysis of the mean spectra currently recorded, which average out the possibly different chemical fingerprints from cells at different cell cycle stages within the probed population. The MDA-MB-231 cell line showed to be sensitive to both tested spermine complexes, which induced clear spectral changes (complementary Raman and infrared) in the wide range presently analysed (600–3600 cm^{-1}), enabling the identification of specific features (biomarkers) associated with the biochemical effect of each investigated agent and the subsequent physiological response from the cell – the spectral signature of drug action.

The cells were analysed upon chemical fixation with a 4% formaldehyde solution (formalin). This type of biocompatible preservation method is intended to retain the biochemical conditions as similar as possible to the native physiological state, for prolonged time periods, whilst minimising auto-fluorescence and avoiding contamination from the fixative. The effects of chemical fixation are known to be cell line- and fixative-dependent, but are still not fully characterised in the literature. Formalin is a widespread chemical fixative that causes cross-linking between the aldehyde and the primary and secondary amine groups of cellular proteins (*via* methylene bridges), creating a gel and thus retaining the cellular constituents in a close to *in vivo* relationship. While no considerable effects of this fixation procedure have been observed in the IR spectra of single cells,[55] it is expected to induce some reduction in vibrational band intensity, resulting from the disruption of lipid assembly and from changes in protein conformation (mainly reflected in the amide I and II bands). Nevertheless, formalin fixation has been found to be the best method for preserving cellular integrity with a weak impact on the overall molecular content and yielding the spectral profile closest to that observed in the live state, while hardly affecting discrimination between cells under distinct conditions (namely control and drug-exposed).[31,56–58] No formalin contamination was verified in the measured spectral data, with no signals being detected for its typical $t(CH_2)$, $\nu(CO)$ and $\delta(CH_2)$ modes at *ca.* 907, 1040 and 1490 cm^{-1}.

Despite the intense vibrational band characteristics of the tested Pt(II) and Pd(II) agents within the currently probed spectral window (Fig. 1 and S1/ESI†),[18,24] they were indiscernible intracellularly (either by FTIR or Raman), which can be explained by the low concentrations used (not exceeding 8 μM), coupled to intracellular chloride hydrolysis and the strong interaction with their main pharmacological target (DNA) and other biomolecules.[18,23,24] Also, the vibrational modes involving the metal center lie in the low frequency region of the spectrum (below 550 cm^{-1}) which does not fall within the interval currently measured (>600 cm^{-1}). It should be noted, however, that the 1000–1800 cm^{-1} interval may comprise features from the polyamine ligand within the dinuclear complexes,[24] but these are overruled by the strong signals due to the cellular components (Table 1, Fig. 1).

Notwithstanding the complexity of the biochemical processes taking place within the cell, the use of chemometric techniques (PCA), providing further insight into the source of the observed spectral variability, enabled discrimination between spectral patterns obtained for distinct levels of drug exposure. From the loading plots corresponding to each principal component it was possible to derive biochemical information based on the differentiation between the tested conditions, and to evidence inter-group variance. Additionally, loading plots could be matched with the corresponding experimental spectra, and the positive and negative peaks in the loadings were assigned to the different groups observed

Table 1 Raman and infrared bands for live human breast cancer cells (MDA-MB-231). Features from specific drug-prompted DNA and protein conformational rearrangements are presented in red. The signals exclusively detected by infrared are shaded in grey

Band (cm^{-1})	Assignment[a]			
	Nucleic acids	Proteins	Lipids	Carbohydrates
622	Z-DNA/dG (ν(CC)$_{ring}$)	Phe (t(CC))		
645		ν(CS), Tyr (t(CC))		
668	A-DNA/dG (ν(CC)$_{ring}$)			
669	B-DNA/G,T (ν(CC)$_{ring}$)			
678	B-DNA/A,G,T,C (ν(CC)$_{ring}$)	Trp (ν(CC)$_{ring}$)		
697-900			phosphate esters (ν_s(OPO))	
698	B-DNA/dG (ν(CC)$_{ring}$)	Met (ν(CS))		
717			phosphocholine (ν(C–$(CH_3)_3N^+$)	
728	B-DNA/A (ν(CC)$_{ring}$)	Trp (ν_s(CC)$_{ring}$)		
746	B-DNA/T (ν(CC)$_{ring}$)	porphyrin (ν(CC)$_{ring}$)		
749	Z-DNA (ν(OPO)$_{backbone}$)			
757	B-DNA/dT (ν(CC)$_{ring}$)	Trp (ν_s(CC)$_{ring}$)		
782-795	C,T,U (ν(CC)$_{ring}$)			
785	B-DNA (ν(OPO)$_{backbone}$)			
792	Z-DNA (ν(OPO)$_{backbone}$)			
805	A-DNA (ν(OPO)$_{backbone}$)			
812	RNA (ν(OPO)$_{backbone}$)	Pro, Tyr (ν(CC))		
830	B-DNA (ν(OPO)$_{backbone}$)	Pro, Tyr (ν(CC))		
835-840	Z-DNA (ν(OPO)$_{backbone}$)			Glu (δ(CCO))
855		Pro, Tyr, Val (ν(CC), δ(CCH))		polysaccharides (ϒ(COC))
868	Z-DNA (ν(OPO)$_{backbone}$)	Pro (ν(CC))		mono, disaccharides (ν(COC))
879	RNA (ribose, ν(CC)$_{ring}$)	Tyr (ν(CC)$_{ring}$), Hyp (ν(CC))	phosphocholine (ν_s(CCN^+))	ν(CC), ν(C-O)$_{ring}$
883		ρ(CH_2)		
901	deoxyribose (ν(CC)$_{ring}$)	ν(CC)	fatty acids (ν(CC), ν(CO))	ν(COC)
915	RNA (ribose, ν(CC)$_{ring}$)			
920-940	B-DNA/deoxyribose (ν(CC)$_{ring}$)	Pro (ν(CC))		Glu (ν(CC))
925-930	Z-DNA (ν(OPO)$_{backbone}$)			
935-940		ν(CC) (α-helix)	glycolipids (ν(COC))	ν(COC)$_{glycosidic}$
940	RNA/ribose (ν(CC)$_{ring}$)			polysaccharides (skeletal modes)
950-1250		phosphorylated proteins (ν(OPO))		
951		ν_s(CH_3) (α-helix)		
965-973	DNA (ν(CC)/ν(CO)$_{backbone}$)			polysaccharides (δ(C=O))
974	RNA (ribose, ν(CC)$_{ring}$)			
980		ν(CC) (β-sheet)	δ(C=CH)	
1003		Phe (ν_s(CC)$_{ring}$)		
1032		Phe (δ(CH)), ν(O–CH_3)	ν(CC), phospholipids (δ(CH))	ν(CC), ν(CO), ν(C–OH)
1050	RNA (ν(OPO))		phosphate esters (ν_{as}(OPO))	ν(C–OH)
1063	B-DNA/deoxyribose (ν(CO))	ν(CC), ν(CN)	ν(CC), ν(CO)	ν(CC), ν(CO), δ(OCH)
1082		ν(CC), ν(CN)	phospholipids (ν_s(PO_2^-))	glycogen (ν(CC), ν(CO))
1092	B-DNA (ν_s(PO_2^-))			
1095	Z-DNA (ν_s(PO_2^-))		ν(CC), ν(CN)	ν(CC), ν(CO)
1099	A-DNA (ν_s(PO_2^-))			
1127	RNA/ribose (ν(CO))	ν(CN)	ν(CC)$_{acyl}$ (*trans* conformation)	ν(CO), ν(CC)
1157		ν(CC), ν(CN), δ(CH_2)	δ(CH_2), ν(=C–C=)$_{conjugated}$	δ(CH_2)
1158		ν(CC), ν(CO), ν(CN)	phosphocholine (ν_s($N(CH_3)_3^+$))	ν(CO)
1175	C,G,T (ν(CC)$_{ring}$)	Tyr, Phe (δ(CH))		
1207		Hyp, Phe, Tyr (ν(CC))		
1220-1300		amide III (δ(NH), ν(CN))		
1229-1235		amide III/β-sheet		
1237	B-DNA (ν_{as}(PO_2^-))			
1240	RNA (ν(PO_2^-))			
1243-1259		amide III/random coil		
1242-1266	RNA/dT (ν(CC)$_{ring}$)			
1247-1252	C (δ(NH_2))			
1252-1264	A,T (ν(CC)$_{ring}$)	δ(CH_2), δ(C=C–H)	δ(CH_2), δ(C=C–H)$_{phospholipids}$	δ(CH_2), ω(CH_2), t(CH_2)
1254	dC (ν(CC)$_{ring}$)			
1270	RNA/C,U (ν(CC)$_{ring}$)		phospholipds (δ(C=C))	

1272		amide III/α-helix	δ(CH_2), ω(CH_2), t(CH_2)	δ(CH_2), ω(CH_2), t(CH_2)
1300	RNA/A,C ($\nu(CC)_{ring}$)			
1318	G ($\nu(CC)_{ring}$)	δ(CH_2)	δ(CH_2)	δ(CH_2)
1323	G ($\nu(CC)_{ring}$)			
1339	G ($\nu(CC)_{ring}$)			
1340	A,G ($\nu(CC)_{ring}$)	δ(CH_2), Trp ($\nu(CC)_{ring}$)		δ(CH_2)
1375	A,G,T ($\nu(CC)_{ring}$)	glycoproteins (δ(CH_3))	lipids/acyl chains (δ(CH_3))	saccharides (δ(CH_2))
1396		δ(CH_2), ρ(CH_2)	membrane lipids (δ(CH_2))	δ(CH_2), ρ(CH_2)
1405		δ(NH_2)		
1420	A-DNA (δ(CH_2))			
1420-1480		δ(CH_2), δ(CH_3)	δ(CH_2), δ(CH_3), aromatic lipids	δ(CH_2)
1421	A,G ($\nu(CC)_{ring}$)			
1423	B-DNA/A,G (δ(CH_2))			
1425	Z-DNA (δ(CH_2))			
1440-1450			δ(CH_2)	
1483		δ(NH_3)		
1487	A,G ($\nu(CC)_{ring}$)	δ(NH_3^+)		
1512		Tyr ($\nu(CC)_{ring}$)		
1513	C ($\nu(CC)_{ring}$)			
1518		$\nu(C{=}C)_{porphyrin}$		
1550-1560		amide II ((δ(CN-H)/ν(CN))		
1555	G ($\nu(CC)_{ring}$)	Trp ($\nu(CC)_{ring}$), $\nu(C{=}C)_{porphyrin}$		
1579	A,G ($\nu(CC)_{ring}$)			
1582		ν(C=C), ν(C=N)	ν(C=C), ν(C=N)	
1604	T ($\nu(CC)_{ring}$)	Phe (δ(C=CH))		
1604-1618	A ($\nu(CC)_{ring}$), C (δ(NH_2))	Phe, Tyr, Trp (ν(C=C)), δ(NH_2)		
1660	DNA (δ(NH))	Amide I (ν(C=O))/α-helix	ν(C=C)	
1612-1690		amide I/β-sheet, antiparallel		
1626-1640		amide I/β-sheet, parallel		
1640-1650		amide I/random coil		
1690		$\nu(C{=}O)_{amino\ acid\ side\ chain}$		
1712	A-DNA (ν(C=O))			
1717	B-DNA (ν(C=O))			
1730-1742			phospholipids ($\nu(C{=}O)_{ester}$)	
2850-2875		$\nu_s(CH_2)$	$\nu_s(CH)$, $\nu_s(CH_2)$	$\nu_s(CH)$, $\nu_s(CH_2)$
2880		$\nu_s(CH_3)$	$\nu_s(CH_3)$	$\nu_s(CH_3)$
2900-2935		$\nu_{as}(CH_2)$	$\nu_{as}(CH_2)$	$\nu_{as}(CH_2)$
2960		$\nu_{as}(CH_3)$	$\nu_{as}(CH_3)$	$\nu_{as}(CH_3)$
3060		amide B		
3250-3290		amide A (ν(NH))		
3300-3350			ν(OH)	ν(OH)

[a] A – adenine; C – cytosine; dG – deoxyguanine; dT – deoxythymine; G – guanine; Glu – glucose; Hyp – hydroxyproline; Met – methionine; Phe – phenylalanine; Pro – proline; T – thymine; Trp – tryptophan; Tyr – tyrosine; U – uracil; Val – valine. δ – in-plane deformation; ϒ – out-of-plane deformation; ν – stretching; ρ – rocking; t – twisting; ω – wagging. s – symmetric; as – anti-symmetric.

in the score plots. As discrimination was attained according to PCs dominated by spectral feature characteristics of the cellular components (and not the drugs), it may be concluded that the detected spectral variance is not due to the presence of the external agents but instead to their distinct effects on the cell's biochemical profile coupled to specific physiological reactions – the impact of the drugs is therefore being monitored.

C.1 Raman microspectroscopy

The Raman spectra of the MDA-MB-231 fixed cells, prior to drug administration (Fig. 1), reflected a vibrational pattern consistent with a healthy state as opposed to dead cells,[28,59] mainly evidenced by the DNA $\nu(O{-}P{-}O)_{backbone}$ and $\nu(PO_2^-)$ modes (respectively at 785, and 1092/1237 cm^{-1}), the ν(O–P–O) features assigned to phosphorylated proteins (from 950 to 1250 cm^{-1}) and the protein amide I band (ν(C=O) at 1660 cm^{-1}). The DNA ν(O–P–O)$_{backbone}$ is the most sensitive peak to recognise cell death, since this is accompanied by a breakdown of the phosphodiester bonds in the double helix, leading to a significant reduction (up to 80%) of the band intensity. The main Raman signals observed for untreated MDA-MB-231 cells are encompassed in Table 1. Assignment of the signals relied on reported studies on several cell lines and biomolecules.[28,60,61] Two different visible lasers were used for the acquisition of the Raman spectra, with 514 and 532 nm wavelengths (at QFM-UC and DLS, respectively). This very small energy difference in the incident radiation did not lead to significant changes in the intensity of the observed bands (the efficiency of the Raman scattering process depending on the fourth power of the excitation frequency). Moreover, data collection either in dry samples or saline-dipped cells did not elicit any significant variations in the spectral profile (data not shown).

The spectra of drug-exposed cells after 48 h of incubation with cisplatin and the Pt- and Pd–spermine agents displayed marked differences relative to the non-exposed (control) cells, reflected in vibrational features associated with specific biomolecules. Additionally, a slight blue shift (*ca.* 2–3 cm^{-1}) was detected for some features assigned to proteins (ν(CN), amide I) and lipids ($\delta(CH_2/CH_3)$), in agreement with previously reported studies.[62,63] Upon unsupervised PCA analysis, a more detailed picture of the drug-impact on the cellular biochemical profile was

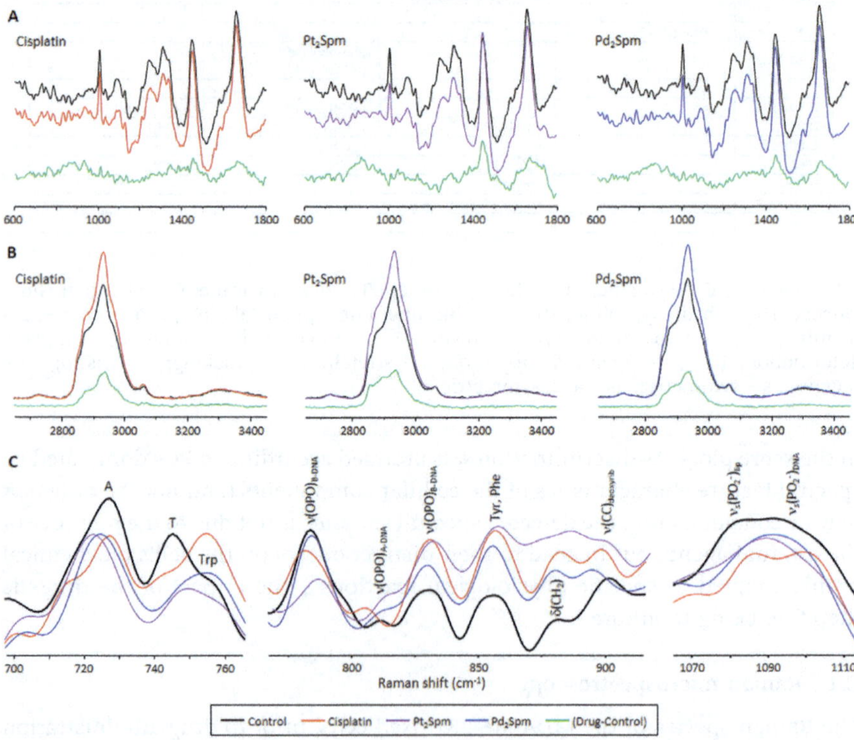

Fig. 2 Mean Raman spectra (fixed cells, Bruker Senterra/532 nm, 600–1800 cm^{-1} (A) and 2650–3450 cm^{-1} (B)), and difference spectra to the control, for drug-treated (4 μM) MDA-MB-231 cells. (C) – Particular spectral regions affected by drug exposure.

attained. A drug-dependent response was verified, as depicted in Fig. 2, which represents the mean Raman spectra measured for cells upon incubation with cisplatin, Pt_2Spm and Pd_2Spm, at 4 μM, and the difference spectra to the control, the corresponding PCA loading and score plots being shown in Fig. 3.

The most affected DNA bands upon drug exposure were found to be those from deoxyguanine (dG, 698 cm^{-1}), thymine (746, 782, 1375, 1604 cm^{-1}), adenine (728, 1252, 1340–1487, 1579, 1604–1618 cm^{-1}), cytosine (782–795 cm^{-1}), guanine

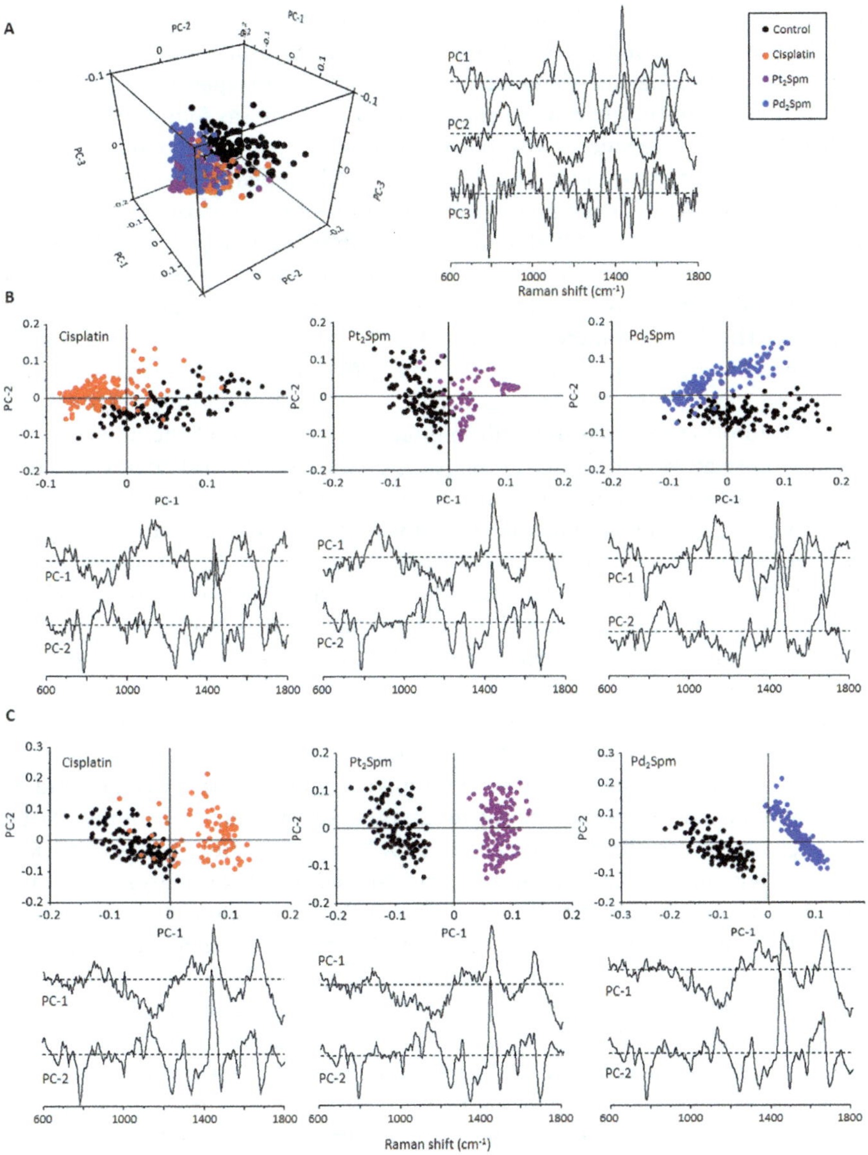

Fig. 3 PCA score and loading plots of Raman data (fixed cells, Bruker Senterra/532 nm, 600–1800 cm^{-1}) for drug-treated MDA-MB-231 cells *versus* the control: all tested agents, 4 μM (A). Cisplatin, Pt_2Spm and Pd_2Spm – 4 μM (B) and 8 μM (C) (for clarity the loadings are offset, the dashed horizontal lines indicating zero loading).

(669–678, 1175, 1318–1375, 1421–1487, 1555–1579 cm^{-1}) and deoxyribose (901–940, 1063 cm^{-1}, more perturbed by the polynuclear agents than by cisplatin) (Table 1). These variations evidence the occurrence of drug–DNA crosslinks, predominantly intrastrand 1,2 GG, 1,2 AG, 1,3 GTG,[64] leading to local unwinding of the native helix. While the adenine ring breathing vibration at 728 cm^{-1} decreased in intensity and was slightly red-shifted upon treatment with Pt_2Spm and Pd_2Spm (to 723 and 725 cm^{-1}, respectively), the thymine band at 746 cm^{-1} was deviated to higher frequencies in the presence of cisplatin or Pt_2Spm (755 cm^{-1}) and Pd_2Spm (749 cm^{-1}) (Fig. 2(C)). The guanine signal at 1487 cm^{-1} (detected as a shoulder of the strong $\delta(CH_2)_{lipids}$ band at 1448 cm^{-1}) is particularly sensitive to the binding of electrophilic agents (such as cisplatin-like compounds) at the N_7 acceptor, having decreased in intensity (nearly disappearing) upon drug-interaction. As for the guanine band at *ca.* 1579 cm^{-1}, it was found to be slightly reduced in the presence of the polynuclear Spm agents, as opposed to the cisplatin-induced blue shift (to 1585 cm^{-1}). Additionally, the strong signal centered at 1660 cm^{-1}, encompassing contributions from protein $\nu(C{=}O)$ (amide I), lipid $\nu(C{=}C)$ and DNA $\delta(N\text{–}H)$ modes, (the latter being highly sensitive to the breakage of Watson–Crick H-bonds), raised in intensity in the presence of all three metal complexes studied (as evidenced by the difference spectra in Fig. 2(A)). This intensity increase observed for DNA Raman bands in drug-exposed cells may be explained by a hyperchromic effect upon base unstacking and unpairing. Changes in the bands associated with the DNA phosphate groups are specially informative regarding a drug-prompted distortion of the double helix, namely those at 785 cm^{-1} ($\nu(O\text{–}P\text{–}O)_{backbone}$ characteristics of B-DNA, combined to cytosine ring breathing), 1092 cm^{-1} ($\nu_s(PO_2^-)$) and 1237 cm^{-1} ($\nu_{as}(PO_2^-)$). The $\nu(O\text{–}P\text{–}O)_{backbone}$ band suffered an intensity decrease, more evident in the presence of Pt_2Spm (main discrimination from the control in PC1) as compared to Pd_2Spm (mostly differentiated in PC2) (Fig. 2(C) and 3). The symmetric stretching mode of the PO_2^- groups external to the helix backbone (at 1092 cm^{-1}, partially overlapped with the $\nu_s(PO_2^-)$ band from phospholipids at 1082 cm^{-1}) showed slight deviations for Pt_2Spm (1090 cm^{-1}) and Pd_2Spm (1091 cm^{-1}), as compared to cisplatin (1099 cm^{-1}) (Fig. 2(C)). These spectral changes are in agreement with a B- to A-DNA (or ultimately Z-DNA) transition due to covalent drug-binding,[23] yielding strong crosslinks responsible for disruption of the base stacking and/or base pairing. Apart from the intensity decrease upon drug exposure of the $\nu(O\text{–}P\text{–}O)_{backbone}$ bands ascribed to the native double helix structure (at 785 and 830 cm^{-1}), a red shift of the 698 cm^{-1} signal assigned to B-DNA deoxyguanine was detected, for all tested agents, to about 668 cm^{-1} (compatible with a B- to A-DNA transition) and 622 cm^{-1} (reflecting a B- to Z-form rearrangement). Moreover, apart from the B-DNA marker at 830 cm^{-1}, a typical band for A-DNA[65] was detected at 805, 809 and 810 cm^{-1} respectively for the cisplatin, Pt_2Spm and Pd_2Spm-containing cells (Fig. 2(C)). These are reliable spectral biomarkers of the metal-based drugs' interaction with DNA and consequent distortion of the DNA backbone, the $\nu(PO_2^-)$ modes being particularly responsive to variations in the electrostatic environment of the phosphate group within the double helix.

Changes in the vibrational modes ascribed to cellular proteins (Table 1) were also observed, reflecting a drug effect either by direct interaction (*e.g.* binding to cytoplasmic proteins) or indirectly (namely *via* proteolysis during the apoptotic process). In addition, the recognised cytotoxic mechanism of metal-based agents involves the binding of high mobility group (HMG) proteins to DNA upon platination- (or palladation)-induced conformational rearrangement of the double helix (precluding repair mechanisms). The most relevant markers associated with protein variations were those at 728/757 cm^{-1} (tryptophan ring breathing), 855 cm^{-1} (Tyr), 935–940 cm^{-1} ($\nu(CC)_{protein}$, α-helix), 1003 cm^{-1} (phenylalanine symmetric ring stretching), 1207 cm^{-1} (ν(CC) from phenylalanine, tryptophan and hydroxyproline), 1127–1158 cm^{-1} (ν(CN)), 1220–1300 cm^{-1} (amide III, ν(CN)/(δ(CN–H))), 1448 cm^{-1} (δ(CH) overlapped with similar modes from lipids) and 1660 cm^{-1} (predominantly amide I). While the impact on the protein skeletal stretching modes (*e.g.* ν(CN) at 1127 and 1157 cm^{-1}) was found to be noteworthy for the cells incubated with cisplatin (PC1), it was much less marked for Pt_2Spm and Pd_2Spm (PC1 and PC2), for which variance was more significant in the amide III band (Fig. 3(B)). The Trp ring breathing feature, detected at 757 cm^{-1} in the untreated cells, suffered a slight red shift (to 755 cm^{-1}) upon exposure to cisplatin and the Spm complexes (for which this signal was partially overlapped with the one from the thymine ring, Fig. 2(C)). As for the $\rho(CH_2)$ signal from the peptide chains, it was shifted from 883 to 885 cm^{-1} only in the presence of Pt_2Spm. Additionally, direct binding of the drugs to membrane and cytoplasmic proteins is anticipated to cause variations in their secondary and tertiary structures, reflected in the characteristic features of β-sheet, random coil and α-helix conformations (at 1229–1235 cm^{-1}, 1243–1253 cm^{-1} and 1270–1300 cm^{-1}, respectively, Table 1). These were perceived in the amide III band, namely through changes in the signal characteristics of the α-helix signal at 1272 cm^{-1}, primarily in the presence of Pt_2Spm (1260 cm^{-1}) and Pd_2Spm (1254 cm^{-1}), evidencing a partial conversion to a random coil structure. Moreover, exposure of side chains containing aromatic residues as a consequence of protein unfolding is suggested to be responsible for the very slight intensity increase of some of the bands ascribed to their ring breathing modes,[63] which was currently observed only for Phe (Fig. 2(A)). The drug impact on the protein amide I band, in turn, showed to be noticeable for cisplatin, as clearly evidenced by the respective PC1 loading (Fig. 3(B)). In turn, Pt_2Spm and Pd_2Spm were found to have a higher effect on the $\delta(CH_2)$ and $\delta(NH_2)$ modes from the peptide chains (at *ca.* 1440 and 1618 cm^{-1}, Table 1), as shown in the PC2 loading plot. Analysis of the high frequency region of the spectra (Fig. 2(B) and 4) allowed the conclusion that the amide A band at 3290 cm^{-1} (exclusively due to protein ν(NH) modes), as well as the amide B signal at 3060 cm^{-1} (arising from a Fermi resonance between amide I's first overtone and ν(NH) and hence strongly dependent on intramolecular H-bonds), did not undergo significant changes upon drug exposure, as opposed to the CH stretching modes of the cellular biomolecules (2850–2960 cm^{-1}).

MDA-MB-231 cells are especially rich in lipids (*e.g.* glycerophospholipids, known to be increased in cancer cells[66]). The direct drug impact on these cell constituents was reflected as an increase of the intense Raman feature centered at 1448 cm^{-1} ($\delta(CH_2)$), particularly for Pt_2Spm, as well as of the bands at 2850–2960 cm^{-1}, clearly evidenced in the difference spectra obtained for all the tested drugs (Fig. 2(A) and (B)). This may also be inferred by analysis of the loading plots corresponding to Pt_2Spm and Pd_2Spm (respectively in PC1 and PC2, Fig. 3(B)). However, since CH_2 deformations from cellular lipids occur in the same wavenumber interval as similar modes from proteins and carbohydrates (Table 1), as well as from the spermine polyamine ligand (Fig. 1), PCA differentiation should

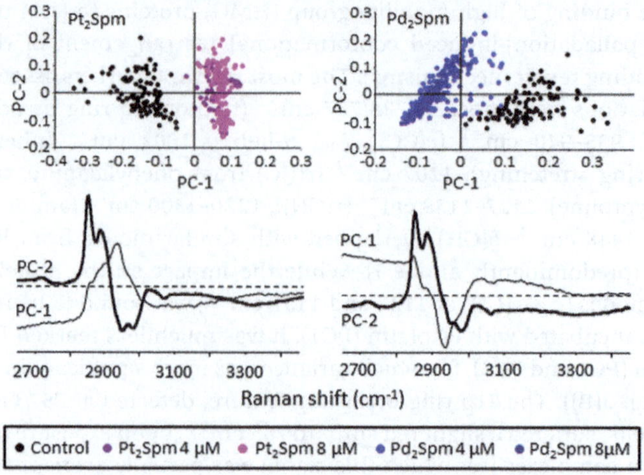

Fig. 4 PCA score and loading plots of Raman data (fixed cells, Bruker Senterra/532 nm, 2650–3450 cm^{-1}) for Pt_2Spm- and Pd_2Spm-treated (4 and 8 μM) MDA-MB-231 cells *versus* the control (the dashed horizontal lines indicates zero loading).

be interpreted with care in this region. The PC2 loading plot for Pd_2Spm, for instance, also showed a very marked discrimination according to protein and carbohydrate CH_2 deformations (*ca.* 1430 cm^{-1}). The highest frequency spectral region is therefore more suited for this analysis, the drug-induced changes observed in the typical $\nu(CH_2)$ (2850–2935 cm^{-1}) and $\nu(CH_3)$ (2960 cm^{-1}) vibrational modes from lipids being reflected in the clear separation from the untreated cells along PC1, for all agents investigated: the loadings corresponding to the (CH) stretching modes from methyl groups being positive for cisplatin and negative for both spermine complexes (Fig. 4). This evidences different biochemical impacts induced by these mono- and dinuclear agents on lipid terminal moieties (CH_3) *versus* $-(CH_2)_n-$/$-(HC{=}CH{-}CH_2)_n-$ chains, which is in accordance with the presence of the alkyl moieties within the ligand in the Pt_2Spm and Pd_2Spm complexes that are prone to interact efficiently with the hydrophobic chain of lipid molecules (as opposed to $Pt(NH_3)_2Cl_2$). Regarding membrane lipids, it was possible to identify changes in the vibrational mode ascribed to their phosphate-ester headgroups, at 1732 cm^{-1}, that was shifted to 1739 cm^{-1} for Pt_2Spm but hardly detectable (at *ca.* 1742 cm^{-1}) for cisplatin- and Pd_2Spm-treated cells. The $\nu(COC)$ modes from glycosidic bonds in glycolipids, centered at 938 cm^{-1}, were found to be deviated to 934 cm^{-1} for cisplatin and Pd_2Spm, and to 931 cm^{-1} for Pt_2Spm. Moreover, the signal at 1158 cm^{-1}, characteristic of the $N(CH_3)_3^+$ choline group (*e.g.* from phosphatidylcholine and sphingomyelin, two main cell membrane constituents), experienced a shift to 1163 cm^{-1} for the cisplatin-exposed cells and an intensity decrease for the samples exposed to the dinuclear Spm chelates. This may be indicative of the formation of membrane vesicles, as a way for the cell to remove toxins (*e.g.* the administered compounds) by exocytosis. Overall, the drug effect on cellular lipids results from an interplay of several (sometimes disparate) factors, such as an increase in neutral lipids (associated with cisplatin's nephrotoxicity), an enhanced lipid peroxidation, and a reduction in the total lipid content due to membrane degradation and decreased cell viability.

Fig. 3(A) displays the unsupervised PCA analysis considering all agents under study, allowing the identification of discrimination according to PC2 that represents variance in protein, lipids and DNA (specifically $\delta(NH_2)$, and ring modes from bases and deoxyribose), with a slight overlap between drug-exposed and control cells (mostly for cisplatin). Apart from this distinction between control and treated cells, it was possible to separate each of the three drugs currently tested based on their effect on the cell chemical profile, as compared to the control (Fig. 2 and 3), reflecting the previously described drug-dependent response. Differentiation was achieved in PC1 for cisplatin and Pt_2Spm, while for Pd_2Spm it was attained in PC2 for the lowest doses and in PC1 for 8 μM (Fig. 3(B) and (C)). For cisplatin-treated samples the main impact was observed on DNA and proteins, with an effect on lipids being detected only for 8 μM. The dinuclear Pt(II) complex, in turn, was responsible for a strong effect on lipids, already at the lowest concentration tested (2 μM), also evidenced through changes in the corresponding high energy $\nu(CH_2)$ modes. In turn, the effect on proteins (noticeable in PC1 and PC2) was only verified for concentrations equal to 4 μM or higher. At 8 μM there was a small effect on RNA (bands centered at 879 cm^{-1}). Pd_2Spm produced a noteworthy impact on proteins, for all concentrations tested. As to the Pd_2Spm-induced variance on lipids and DNA, it was significant for 4 μM and did not increase at 8 μM (Fig. 3(B) and (C)).

A dose-dependent effect was uncovered by separate analysis of the data measured for each compound, a significantly improved discrimination having been achieved for higher concentrations, as expected (Fig. 3(B) and (C)). Comparing the results for the Pt_2Spm- and Pd_2Spm-treated cells *versus* the

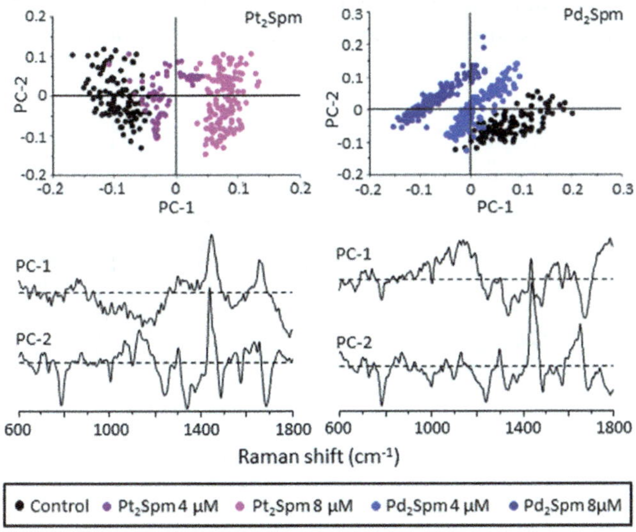

Fig. 5 PCA score and loading plots of Raman data (fixed cells, Bruker Senterra/532 nm, 600–1800 cm^{-1}) for Pt_2Spm- and Pd_2Spm-treated (4 and 8 μM) MDA-MB-231 cells *versus* the control (for clarity the loadings are offset, the dashed horizontal lines indicating zero loading).

control, for 4 and 8 μM (Fig. 5), allowed the conclusion that separation from the control as well as between both Pt_2Spm concentrations occurred along PC1, the discriminant peaks corresponding predominantly to DNA and lipids. For Pd_2Spm, this distinction was not as noticeable (increasing with drug concentration), and occurred in PC1 reflecting DNA (at 785 and *ca.* 1320 cm^{-1}) and protein components (at *ca.* 1150 and 1660 cm^{-1}), with a very small variance for the lipids (at 1448 cm^{-1}). Interestingly, while in the fingerprint region discrimination was achieved between 4 and 8 μM concentrations for both Pt- and Pd-complexes (Fig. 5), in the highest frequency range, mainly reflecting the impact on lipids, no differentiation was observed between distinct drug doses for Pt_2Spm and a very small separation was detected for Pd_2Spm (Fig. 4). Hence, the effect on lipids appears to become dose-independent above 4 μM, which implies a higher sensitivity of the cellular lipid content to the administered drugs as compared to the other biomolecules, a maximum effect being attained for lower concentrations of the chemotherapeutic agent, particularly for Pt_2Spm.

Discrimination between the two dinuclear spermine complexes was also attained, through PC2 (Fig. 6) that predominantly represents the cellular DNA (at 678, 728, 746, 785, 974, *ca.* 1180, 1315, 1420, 1490 and 1620–1660 cm^{-1}) and RNA (at 812 and 920–940 cm^{-1}). This reflects the distinct impact of each chelate on nucleic acids, suggesting the occurrence of different pathways for cytotoxicity according to the nature of the metal centre.

In summary, the Raman spectra presently obtained for the MDA-MB-231 cells may be resolved into distinct regions of statistically significant spectral biomarkers which were found to undergo variations upon drug treatment, corresponding to different biochemical impacts. The overall effect of the tested agents on the vibrational features from the main cellular components is consistent with changes in DNA, as well as in cytoplasmic proteins and lipids, and the onset of apoptosis involving chromatin condensation, DNA fragmentation, membrane blebbing, protein proteolysis (*e.g.* in the cytoskeleton) and lipid accumulation. The distinct outcomes on the cell chemical content currently

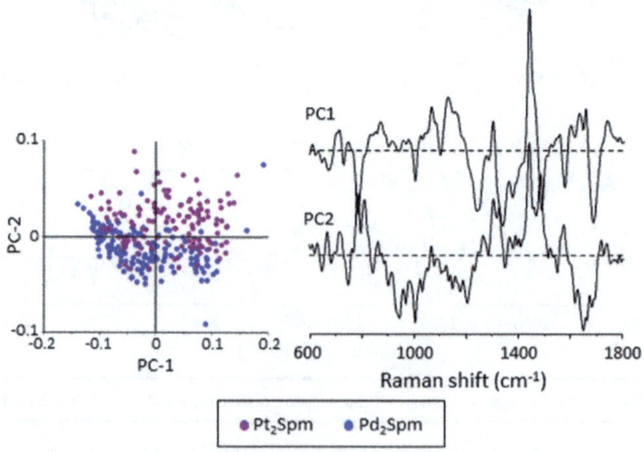

Fig. 6 PCA score and loading plots of Raman data (fixed cells, Bruker Senterra/532 nm, 600–1800 cm^{-1}) for drug-treated MDA-MB-231 cells: Pt_2Spm *versus* Pd_2Spm (8 μM) (for clarity the loadings are offset, the dashed horizontal lines indicating zero loading).

unveiled for Pt_2Spm and Pd_2Spm may be indicative of the occurrence of alternative pathways of cytotoxicity for the palladium complex, as opposed to a preferentially apoptotic mechanism of cell death induced by its platinum homologue (as is widely accepted for the Pt-drugs in clinical use – cisplatin, carboplatin and oxaliplatin). The results presently obtained for cisplatin are in accordance with reported Raman studies on the cisplatin effect on human lung adenocarcinoma cells,[62,63] despite the longer drug exposure time considered by these authors – 96 h as compared to the 48 h incubation period used in the current work. Additionally, the chemical variations associated with the apoptotic process presently described agree with time-course Raman imaging experiments during apoptosis performed in the same breast cancer cell line, maintained under physiological conditions.[50]

C.2 SR-infrared microspectroscopy

The main infrared bands presently observed for the MDA-MB-231 cells are depicted in Fig. 1, clearly showing the signals assigned to the protein characteristic modes amide A and amide II (δ(CN–H)/ν(CN) at *ca.* 1550 cm^{-1}), as well as to the ester groups from phospholipids (*ca.* 1730 cm^{-1}, Table 1). Infrared signatures from untreated and drug-treated cells were successfully retrieved from the data, evidencing a good PCA discrimination on the basis of the distinct biochemical profile observed for each condition. Since the main conclusions are in accordance with the Raman results previously described in detail, the following discussion will only emphasize the specific effects unveiled by IRMS analysis.

In view of the optical substrate transmission range (0.35–10 μm wavelength) it was not possible to reliably explore the data below *ca.* 1100 cm^{-1}, *i.e.* the nucleic acid region was strongly affected by the cut off of the CaF_2 substrate at 1000 cm^{-1}.[51] Spectra were therefore truncated between 1100 and 1800 cm^{-1} for baseline correction prior to PCA analysis. When applying the PCA, the spectral interval from 1100 to 1250 cm^{-1} was not found to yield significant information. PCA applied to the spectral range 1250–1800 cm^{-1} showed the largest variability in the data, allowing us to distinguish the impact of each drug type effectively. Actually, the score and loading plots extracted for PC2 and PC3 (Fig. 7) evidenced significant differences between cisplatin and the platinum and palladium–spermine complexes. In each case, good discrimination was observed along PC3, enabling separation between the untreated control and the drug-treated cells: while there was some overlap in the scatter plots, two centroids could be clearly distinguished. No other principal components were able to successfully discriminate between the control and drug-exposed cells. The loading plot along PC3 for the cisplatin-containing samples displayed an intense peak at 1659 cm^{-1} (amide I) and a relatively weak band at 1556 cm^{-1} (amide II), reflecting a clear effect on proteins. Furthermore, two distinct signals ascribed to ν(C=O) modes appeared, respectively from amino acid side chains and antiparallel β-sheet conformations (at 1690 cm^{-1}), and from phospholipid ester groups (at 1741 cm^{-1}, Table 1). In addition to these peaks observed for the cisplatin-exposed samples, further bands were detected upon incubation with the spermine complexes, at 1460 and 1628 cm^{-1}, due to anti-symmetric CH_2 scissoring from lipids and to the amide I protein mode (β-sheet parallel), respectively. The band at 1741 cm^{-1} for the cisplatin-treated cells was found to be slightly shifted to 1728 cm^{-1} in the presence of Pt_2Spm and Pd_2Spm, possibly as a consequence of drug interaction with the cell

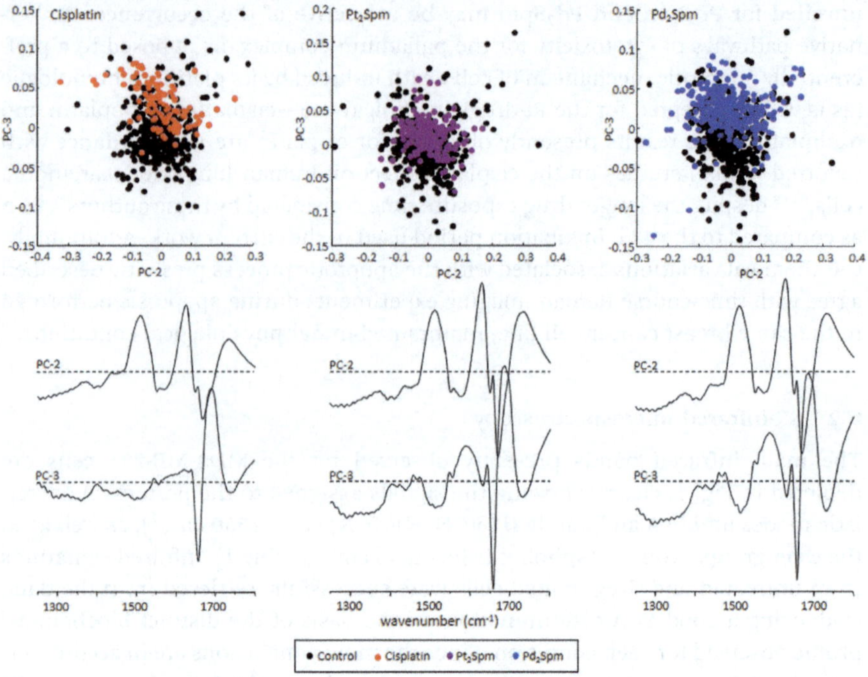

Fig. 7 PCA score and loading plots of FTIR data (fixed cells, MIRIAM, 1250–1800 cm^{-1}) for cisplatin-, Pt$_2$Spm- and Pd$_2$Spm-treated (4 μM) MDA-MB-231 cells *versus* the control (for clarity the loadings are offset, the dashed horizontal lines indicating zero loading).

membrane (phospholipid moieties). These results indicate the significant impact of the Pt(II) and Pd(II)-polyamine agents on the cellular proteins and lipids, as verified by Raman. Strong similarities were apparent in the loading plots for PC3 for both Pt$_2$Spm and Pd$_2$Spm, which suggests an identical impact of these drugs on the MDA-MB-231 cells, differing from cisplatin mostly in the bands ascribed to the CH$_2$ stretching modes from lipids (*ca.* 1450 cm^{-1}). Also, the differences exposed by comparison with the cisplatin loading plot indicate that the activity of Pt$_2$Spm and Pd$_2$Spm follows a different pathway, which is to be expected since the biochemical interaction of these dinuclear highly flexible complexes to DNA and other cellular components is likely to occur *via* a somewhat distinct mechanism to that of mononuclear cisplatin.

The effect of drug exposure on the cellular lipid constituents was also assessed by applying PCA to the 2650–3450 cm^{-1} spectral region (comprising the amide A and lipid CH stretching vibrational modes). Fig. 8 depicts the score and loading plots for the cells exposed to each of the three tested drugs *versus* the control, showing some separation along PC2. For each compound, these plots display predominantly positive scores while the untreated controls are chiefly negative, reflecting different modes of action. Similar profiles were obtained in the PC2 loadings, the main differentiating signal from the control being the one centered at 2960 cm^{-1} ascribed to methyl group stretching (mainly from the lipids). The increase in this band's intensity upon drug exposure is in agreement with the PCA plots obtained from the Raman spectra (Fig. 4). The amide A (3290 cm^{-1}) and amide B (3060 cm^{-1}) protein modes were also found to undergo drug-induced

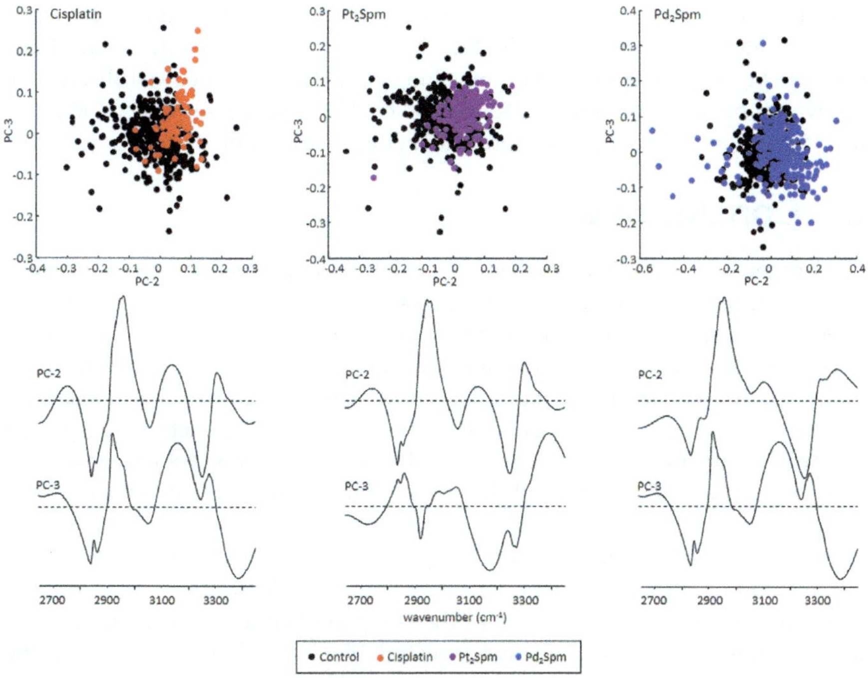

Fig. 8 PCA score and loading plots of FTIR data (fixed cells, MIRIAM, 2650–3450 cm^{-1}) for cisplatin-, Pt_2Spm- and Pd_2Spm-treated (4 μM) MDA-MB-231 cells *versus* the control (for clarity the loadings are offset, the dashed horizontal lines indicating zero loading).

changes (namely an intensity decrease) for both Spm complexes, supporting an impact on proteins. The effect on these particular features could only be clearly evidenced by infrared, since these are weak signals in Raman (Fig. 2).

The effect of changing the metal centre within the spermine chelates was also probed, in the high frequency spectral range. The biological effect on the treated

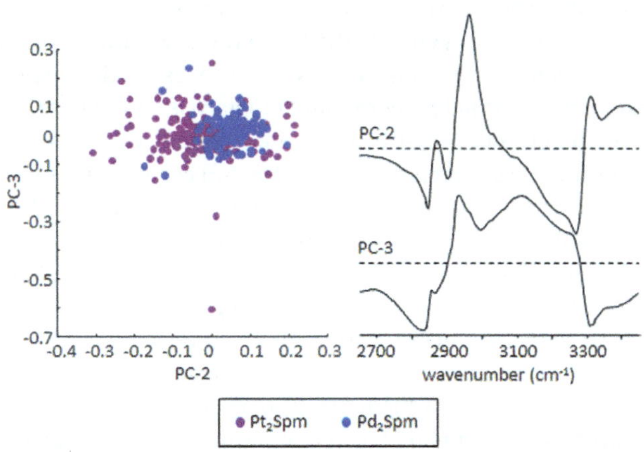

Fig. 9 PCA score and loading plots of FTIR data (fixed cells, MIRIAM, 2650–3450 cm^{-1}) for drug-treated MDA-MB-231 cells: Pt_2Spm *versus* Pd_2Spm (4 μM) (for clarity the loadings are offset, the dashed horizontal lines indicating zero loading).

cells (for a 4 μM drug concentration) was brought out on examination of the score and loading plots represented in Fig. 9, a good discrimination having been achieved between Pt_2Spm and Pd_2Spm along PC2. Major variance was observed in the typical bands from lipids (2850 to 2960 cm^{-1}), as well as in a broad features due to the amide A mode, once more evidencing an impact on proteins and lipids.

D Conclusions

The present study aimed at applying optical vibrational spectroscopy methods – Raman and synchrotron-radiation infrared microspectroscopy (SR-IRMS) – for the investigation and measurement of the metabolic impact of and cellular response to two potential cisplatin-like anticancer agents varying only in the nature of the metal centre – Pt_2Spm and Pd_2Spm. These were administered separately to human metastatic breast cancer cells (MDA-MB-231), and distinctive spectral signatures were obtained as a function of drug concentration (2–8 μM), for a 48 h exposure period. The widely used chemotherapeutic drug cisplatin (with a known mode of action) was chosen as a reference compound. Although the tested chemotherapeutic agents were not tracked intracellularly, this approach yielded an accurate description of variations in cellular biochemistry, with molecular specificity, in the presence of each of the compounds under study, allowing the identification of vibrational bands (spectral biomarkers) assigned to specific drug-elicited effects and to the cell response to this perturbation.

The MDA-MB-231 cell line showed to be sensitive to the tested compounds, being prompted to apoptosis by incubation with the metal complexes (for 48 h). Drug exposure triggered both a direct biochemical effect and a physiological response from the cell, which were reflected in the cellular vibrational signature but not always clearly distinguished. Actually, while changes in the vibrational features from DNA could be unmistakably ascribed to direct drug binding, variations in the bands assigned to lipids and proteins were the combined outcome of direct drug interaction and the cellular reaction to the chemotherapeutic insult (*e.g.* formation of membrane vesicles for exocytosis or biochemical events following apoptotic stimuli such as protein degradation).

The detected changes in the infrared and Raman signatures of particular intracellular constituents triggered by the Pt– and Pd–spermine complexes under study were compared to the control cells, as well as to the cisplatin-exposed samples. Unsupervised principal components analysis of the data enabled us to clearly distinguish between drug-treated and untreated cells. Additionally, it was possible to discriminate between groups of data and correlate them with different chemotherapeutic conditions, regarding concentration and type of drug – either Pt(II)- or Pd(II)-based, mononuclear (cisplatin) or dinuclear (Pt_2Spm and Pd_2Spm).

Both infrared and Raman data reflected drug- and dose-dependent responses, the main effect on intracellular metabolism being on lipids, DNA and proteins. Bands associated with the B-DNA native form suffered a decrease in intensity compatible with drug-induced conformational rearrangements, upon formation of stable adducts within the double helix. In particular, the vibrational modes of the DNA deoxyribose–phosphate backbone, which are very sensitive to base stacking and base-pairing, were suitable markers of drug activity for the metal-based agents currently investigated. When compared to cisplatin and Pt_2Spm, Pd_2Spm elicited a significant effect on the DNA backbone conformation (detected by Raman) and on proteins (in agreement with previously reported effects on the cytoskeleton[16]). The cellular lipid components, in turn, appeared to be more sensitive to the Pt–spermine complex, not only relative to its palladium homologue but also to cisplatin. This effect on lipids is noteworthy, in view of the close link between lipid accumulation and breast cancer aggressiveness and metastatic spread.[67] These findings are consistent with the mode of action generally accepted for cisplatin and cisplatin-like DNA groove-binding agents, but also unveiled alternative pathways of cytotoxicity for the spermine dinuclear complexes, that also appear to differ according to the nature of the metal centres (Pt(II) or Pd(II)).

The results obtained along this work were related with a thorough conformational analysis of the newly synthesised metal-based agents (by Raman, IR and INS coupled to theoretical approaches)[18–24,34–36] and with biological assays for the evaluation of cytotoxicity performed alongside on the same systems – cell growth and viability inhibition, as well as anti-angiogenic and anti-metastatic abilities[14–16,27,40–43,45] – allowing us to establish a reliable correlation between structural/conformational features, cytotoxic activity and impact on cellular metabolism. This data, coupled to previous INS and EXAFS studies on drug–DNA adducts (carried out at ISIS and DLS[18,21,23,24]), will pave the way for future work on other promising cisplatin-like anticancer agents (both Pt(II) and Pd(II)) which have been studied by the team in the last few years, as well as on distinct human cancer cells. In addition, experiments are envisaged to monitor the drugs biodistribution within the cell, by differentiating the responses in different subcellular regions as well as through the use of synchrotron radiation FTIR microspectroscopy for obtaining high spatial resolution chemical images on single cells. Furthermore, quasi-elastic neutron spectroscopy studies are under way for probing water within living human cancer cells, with a view to ascertain differences upon drug incubation (possibly unveiling secondary targets of drug action).

This type of study is expected to contribute to the elucidation of potentially novel pathways of cytotoxicity for tailored Pt(II) and Pd(II) polynuclear compounds, leading to diverse and improved pharmacological properties coupled to minimal acquired resistance, and hopefully allowing the identification and prediction of the cellular response to chemotherapy.

Acknowledgements

The Portuguese Foundation for Science and Technology – UID/MULTI/00070/2013, PhD fellowship SFRH/BD/72851/2010 and Project PTDC/QEQ-MED/1890/2014. CICECO, the University of Aveiro (PT), for access to the FT-Raman spectrometer. The STFC Rutherford Appleton Laboratory (UK), for access to neutron beam facilities (RB 520092, 1320035), work supported by the European Commission – 7th Framework Programme/Key Action *Strengthening the European Research Area, Research Infrastructures* (CP-CSA_INFRA-2008-1.1.1 No. 226507-NMI3). Diamond Light Source (UK), for time on B22/MIRIAM (SM10065), under the European programme Calypso FP7.

References

1 J. Ferlay, I. Soerjomataram, M. Ervik, R. Dikshit, S. Eser, C. Mathers, M. Rebelo, D. M. Parkin, D. Forman and F. Bray, *GLOBOCAN 2012 v1.0, Cancer Incidence*

and Mortality Worldwide: IARC CancerBase No. 11, International Agency for Research on Cancer, Lyon, France, 2013, http://globocan.iarc.fr, accessed 14 October 2014.
2 B. Weigelt, J. L. Peterse and L. J. van't Veer, *Nat. Rev. Cancer*, 2005, **5**, 591.
3 H. Doppler and P. Storz, *Cell Metab.*, 2015, **22**, 536.
4 M. P. Decatris, S. Sundar and K. J. O'Byrne, *Cancer Treat. Rev.*, 2004, **30**, 53.
5 B. Rosenberg, L. Vancamp and T. Krigas, *Nature*, 1965, **205**, 698.
6 B. Rosenberg, L. VanCamp, J. E. Trosko and V. H. Mansour, *Nature*, 1969, **222**, 385.
7 N. J. Wheate, S. Walker, G. E. Craig and R. Oun, *Dalton Trans.*, 2010, **39**, 8113.
8 L. Kelland, *Nat. Rev. Cancer*, 2007, **7**, 573.
9 N. P. Farrell, *Curr. Top. Med. Chem.*, 2011, **11**, 2623.
10 J. Reedijk, *Pure Appl. Chem.*, 2011, **83**, 1709.
11 S. Komeda, *Metallomics*, 2011, **3**, 650.
12 M. P. M. Marques, *ISRN Spectrosc.*, 2013, **2013**, 29.
13 C. S. Matos, A. L. M. Batista de Carvalho, R. P. Lopes and M. P. M. Marques, *Curr. Med. Chem.*, 2012, **19**, 4678.
14 T. M. Silva, S. Andersson, S. K. Sukumaran, M. P. Marques, L. Persson and S. Oredsson, *PLoS One*, 2013, **8**, e55651.
15 T. M. Silva, S. M. Fiuza, M. P. M. Marques, L. Persson and S. Oredsson, *Amino Acids*, 2014, **46**, 339.
16 S. M. Fiuza, J. Holy, L. A. E. Batista de Carvalho and M. P. M. Marques, *Chem. Biol. Drug Des.*, 2011, **77**, 477.
17 A. M. Florea and D. Busselberg, *Cancers*, 2011, **3**, 1351.
18 L. A. E. Batista de Carvalho, M. P. M. Marques, C. Martin, S. F. Parker and J. Tomkinson, *ChemPhysChem*, 2011, **12**, 1334.
19 T. M. Silva, S. Oredsson, L. Persson, P. Woster and M. P. M. Marques, *J. Inorg. Biochem.*, 2012, **108**, 1.
20 A. L. M. Batista de Carvalho, S. M. Fiuza, J. Tomkinson, L. A. E. Batista de Carvalho and M. P. M. Marques, *Spectrosc. Int. J.*, 2012, **27**, 403.
21 M. P. M. Marques, R. Valero, S. F. Parker, J. Tomkinson and L. A. E. Batista de Carvalho, *J. Phys. Chem. B*, 2013, **117**, 6421.
22 I. Lamego, I. F. Duarte, M. P. M. Marques and A. M. Gil, *J. Proteome Res.*, 2014, **13**, 6033.
23 M. P. M. Marques, D. Gianolio, G. Cibin, J. Tomkinson, S. F. Parker, R. Valero, R. P. Lopes and L. A. E. Batista de Carvalho, *Phys. Chem. Chem. Phys.*, 2015, **17**, 5155.
24 S. M. Fiuza, A. M. Amado, S. F. Parker, M. P. M. Marques and L. A. E. Batista de Carvalho, *New J. Chem.*, 2015, **39**, 6274.
25 I. F. Duarte, I. Lamego, J. Marques, M. P. M. Marques, B. J. Blaise and A. M. Gil, *J. Proteome Res.*, 2010, **9**, 5877.
26 O. Corduneanu, A. M. Chiorcea-Paquim, S. M. Fiuza, M. P. M. Marques and A. M. Oliveira-Brett, *Bioelectrochemistry*, 2010, **78**, 97.
27 O. Corduneanu, A. M. Chiorcea-Paquim, V. Diculescu, S. M. Fiuza, M. P. M. Marques and A. M. Oliveira-Brett, *Anal. Chem.*, 2010, **82**, 1245.
28 I. Notingher, *Sensors*, 2007, **7**, 1343.
29 Z. Farhane, F. Bonnier, A. Casey and H. J. Byrne, *Analyst*, 2015, **140**, 5212.
30 A. Marcelli, A. Cricenti, W. M. Kwiatek and C. Petibois, *Biotechnol. Adv.*, 2012, **30**, 1390.

31 M. J. Baker, J. Trevisan, P. Bassan, R. Bhargava, H. J. Butler, K. M. Dorling, P. R. Fielden, S. W. Fogarty, N. J. Fullwood, K. A. Heys, C. Hughes, P. Lasch, P. L. Martin-Hirsch, B. Obinaju, G. D. Sockalingum, J. Sule-Suso, R. J. Strong, M. J. Walsh, B. R. Wood, P. Gardner and F. L. Martin, *Nat. Protoc.*, 2014, **9**, 1771.

32 M. Jimenez-Hernandez, M. D. Brown, C. Hughes, N. W. Clarke and P. Gardner, *Analyst*, 2015, **140**, 4453.

33 A. Mignolet, A. Derenne, M. Smolina, B. R. Wood and E. Goormaghtigh, *Biochim. Biophys. Acta, Proteins Proteomics*, 2016, **1864**, 85.

34 S. M. Fiuza, A. M. Amado, M. P. M. Marques and L. A. E. Batista de Carvalho, *J. Phys. Chem. A*, 2008, **112**, 3253.

35 A. M. Amado, S. M. Fiuza, M. P. M. Marques and L. A. E. Batista de Carvalho, *J. Chem. Phys.*, 2007, **127**, 185104.

36 S. M. Fiuza, A. M. Amado, H. F. Dos Santos, M. P. M. Marques and L. A. E. Batista de Carvalho, *Phys. Chem. Chem. Phys.*, 2010, **12**, 14309.

37 VIBIMA, http://www.uc.pt/en/iii/infrastructures/, accessed 18 September 2015.

38 Diamond Light Source, http://www.diamond.ac.uk/, accessed 18 September 2015.

39 G. Cinque, M. Frogley, K. Wehbe, J. Filik and J. Pijanka, *Synchrotron Radiat. News*, 2011, **24**, 24.

40 M. P. M. Marques, T. Girao, M. C. P. De Lima, A. Gameiro, E. Pereira and P. Garcia, *BBA, Biochim. Biophys. Acta, Mol. Cell Res.*, 2002, **1589**, 63.

41 L. J. Teixeira, M. Seabra, E. Reis, M. T. G. da Cruz, M. C. P. de Lima, E. Pereira, M. A. Miranda and M. P. M. Marques, *J. Med. Chem.*, 2004, **47**, 2917.

42 S. M. Fiuza, A. M. Amado, P. J. Oliveira, V. A. Sardao, L. A. E. Batista de Carvalho and M. P. M. Marques, *Lett. Drug Des. Discovery*, 2006, **3**, 149.

43 A. S. Soares, S. M. Fiuza, M. J. Goncalves, L. A. E. Batista de Carvalho, M. Paula, M. P. M. Marques and A. M. Urbano, *Lett. Drug Des. Discovery*, 2007, **4**, 460.

44 A. Tassoni, N. Bagni, M. Ferri, M. Franceschetti, A. Khomutov, M. P. Marques, S. M. Fiuza, A. R. Simonian and D. Serafini-Fracassini, *Plant Physiol. Biochem.*, 2010, **48**, 496.

45 R. Tummala, P. Diegelman, S. M. Fiuza, L. A. E. Batista de Carvalho, M. P. M. Marques, D. L. Kramer, K. Clark, S. Vujcic, C. W. Porter and L. Pendyala, *Oncol. Rep.*, 2010, **24**, 15.

46 G. Codina, A. Caubet, C. Lopez, V. Moreno and E. Molins, *Helv. Chim. Acta*, 1999, **82**, 1025.

47 ISIS-STFC, http://www.isis.stfc.ac.uk/instruments/tosca/, accessed 18 September 2015.

48 S. F. Parker, C. J. Carlile, T. Pike, J. Tomkinson, R. J. Newport, C. Andreani, F. P. Ricci, F. Sacchetti and M. Zoppi, *Phys. B*, 1997, **241–243**, 154.

49 ECACC, https://www.phe-culturecollections.org.uk/products/celllines/generalcell/detail.jsp?refId=92020424&collection=ecacc_gc, accessed 18 September 2015.

50 A. Zoladek, F. C. Pascut, P. Patel and I. Notingher, *J. Raman Spectrosc.*, 2011, **42**, 251.

51 K. Wehbe, J. Filik, M. D. Frogley and G. Cinque, *Anal. Bioanal. Chem.*, 2013, **405**, 1311.

52 P. Bassan, A. Kohler, H. Martens, J. Lee, H. J. Byrne, P. Dumas, E. Gazi, M. Brown, N. Clarke and P. Gardner, *Analyst*, 2010, **135**, 268.

53 P. Bassan, A. Sachdeva, A. Kohler, C. Hughes, A. Henderson, J. Boyle, J. H. Shanks, M. Brown, N. W. Clarke and P. Gardner, *Analyst*, 2012, **137**, 1370.

54 S. Dasari and P. B. Tchounwou, *Eur. J. Pharmacol.*, 2014, **740**, 364.

55 E. Gazi, J. Dwyer, N. P. Lockyer, J. Miyan, P. Gardner, C. Hart, M. Brown and N. W. Clarke, *Biopolymers*, 2005, **77**, 18.

56 M. M. Mariani, P. Lampen, J. Popp, B. R. Wood and V. Deckert, *Analyst*, 2009, **134**, 1154.

57 A. D. Meade, C. Clarke, F. Draux, G. D. Sockalingum, M. Manfait, F. M. Lyng and H. J. Byrne, *Anal. Bioanal. Chem.*, 2010, **396**, 1781.

58 F. Draux, C. Gobinet, J. Sule-Suso, A. Trussardi, M. Manfait, P. Jeannesson and G. D. Sockalingum, *Anal. Bioanal. Chem.*, 2010, **397**, 2727.

59 I. Notingher, S. Verrier, S. Haque, J. M. Polak and L. L. Hench, *Biopolymers*, 2003, **72**, 230.

60 J. De Gelder, K. De Gussem, P. Vandenabeele and L. Moens, *J. Raman Spectrosc.*, 2007, **38**, 1133.

61 S. Verrier, A. Zoladek and I. Notingher, *Methods Mol. Biol.*, 2011, **740**, 179.

62 H. Nawaz, F. Bonnier, P. Knief, O. Howe, F. M. Lyng, A. D. Meade and H. J. Byrne, *Analyst*, 2010, **135**, 3070.

63 H. Nawaz, F. Bonnier, A. D. Meade, F. M. Lyng and H. J. Byrne, *Analyst*, 2011, **136**, 2450.

64 O. Vrana, V. Masek, V. Drazan and V. Brabec, *J. Struct. Biol.*, 2007, **159**, 1.

65 J. M. Benevides and G. J. Thomas, *Nucleic Acids Res.*, 1983, **11**, 5747.

66 A. Z. Fernandis and M. R. Wenk, *J. Chromatogr. B: Anal. Technol. Biomed. Life Sci.*, 2009, **877**, 2830.

67 C. Nieva, M. Marro, N. Santana-Codina, S. Rao, D. Petrov and A. Sierra, *PLoS One*, 2012, **7**, e46456.

Faraday Discussions

DISCUSSIONS

Single cell analysis/data handling: general discussion

Matthew J. Baker, Royston Goodacre, Christopher Sammon, Maria Paula Marques, Peter Gardner, William Tipping, Josep Sulé-Suso, Bayden Wood, Hugh J. Byrne, Michael Hermes, Pavel Matousek, Colin J. Campbell, Samir El-Mashtoly, Jonathan Frost, Christopher Phillips, Max Diem, Achim Kohler, Katherine Lau, Sergei Kazarian, Wolfgang Petrich, Gavin Lloyd, Ines Delfino, Gianfelice Cinque, Martin Isabelle, Nick Stone, Catherine Kendall, Lauren Jamieson, David Perez-Guaita, Louise Clark, Klaus Gerwert, Ioan Notingher, Luca Quaroni, Rohit Bhargava, Aidan Meade and Fiona Lyng

DOI: 10.1039/C6FD90012G

Jonathan Frost opened the discussion of the paper by Fiona Lyng: A clinically significant question within this work is whether Raman spectroscopy and multivariate analysis are able to identify those women with histologically proven abnormalities (i.e. CIN 1–3). The standard screening test (routine cytopathology) does not perform flawlessly and it is valuable to have a direct comparison of the classification accuracy of Raman spectroscopy and routine cytopathology in identifying histological abnormalities. In addition, analysis of the role of Raman spectroscopy in augmenting standard cytopathology to increase sensitivity and specificity would be interesting. Have these analyses been undertaken?

Fiona Lyng answered: For the samples reported in the present paper, we do not have histological confirmation of CIN 1–3. However, we are currently undertaking a study of 1000 patients where we will have histology follow up so we will be able to compare Raman spectroscopy and routine cytology in identifying histologically confirmed CIN1–3. Also an analysis of the role of Raman spectroscopy in augmenting standard cytopathology to increase sensitivity and specificity is planned.

Katherine Lau remarked: The cytopathologists I've spoken to said that they have to find the cancer cells among hundreds of normal cells, similar to finding a needle in a haystack. Do you need to screen every cell using your Raman-based method? And how does it compare with histology in terms of analysis time? Would a pre-screen step be implemented to shorten the overall analysis time?

Fiona Lyng answered: No, for this study we did not screen every cell. The slide was first examined under a microscope and abnormal cells were identified based on nucleus to cytoplasm ratio. Raman spectra were recorded from every abnormal cell identified for the CIN 1–3 cases. Each slide typically took 20 min to record. This is slower than cytology where typically slides would be screened in 5-10 minutes. However, for abnormal cases, the slide would be screened again by two further people before being reported so the overall time might be comparable.

Max Diem asked: Have you considered the hormonal level (*i.e.*, time since the last menstrual period) in the discrimination between different disease states?

Fiona Lyng responded: Yes we have just completed a study looking at hormonal changes due to the menstrual cycle, the use of hormone based contraceptives and the onset of menopause.[1] We have shown that, similar to previous work by Romeo *et al.*,[2] hormonal changes do not interfere with the ability of Raman spectroscopy to discriminate between normal and abnormal cervical smear samples. A previous *in vivo* study by Kanter *et al.*[3] showed that hormonal status does influence the Raman spectra from the cervix and that incorporating hormonal status into their classification algorithm increased the accuracy from 88% to 94%. It should be noted, however that the main spectral differences found in the study by Kanter *et al.* were at 1250 cm^{-1}, 1300 cm^{-1} and 1320 cm^{-1}, most likely due to changes in the proteins collagen and elastin, from connective tissue underlying the cervical epithelium.

1 D. Traynor, P. Kearney, I. Ramos, C. M. Martin, J. J. O'Leary and F. M. Lyng, A study of variability due to hormonal effects in cervical smear samples using Raman spectroscopy, *J. Biophotonics*, 2016, *submitted*.
2 M. J. Romeo, B. R. Wood and D. McNaughton, Observing the cyclical changes in cervical epithelium using infrared microspectroscopy, *Vib. Spectrosc.*, 2002, **28**(1), 167–175
3 E. M. Kanter, S. Majumder, G. J. Kanter, E. M. Woeste and A. Mahadevan-Jasen, Effect of hormonal variation on Raman spectra for cervical disease detection, *Am. J. Obstet. Gynecol.*, 2009, **200**, 512.e1–512.e5.

Colin J. Campbell said: I appreciate that you often have little role to play in defining the way that clinical samples will be collected, and I note that yours were fixed and treated with peroxide. Given that you're carrying out chemical imaging, does this type of treatment degrade any of the chemical information in your sample and would it be better to carry out the analysis on (for example) freshly frozen samples.

Fiona Lyng responded: We have previously carried out a study specifically looking at the effects of fixation and other processing steps[1] which showed biochemical changes due to sample processing. However, as you say, for these clinical samples, it is not possible to change the way the samples are collected as they are collected according to cytology SOPs into PreservCyte, a methanol based fixative.

The slides are then treated with hydrogen peroxide to remove any contamination due to blood residues. We have found this to be essential to minimise variability in the Raman spectra collected from these Thinprep cervical cytology samples.[2]

1 E. Ó Faoláin, M. Hunter, J. Byrne, P. Kelehan, M. McNamara, H. J. Byrne and F. M. Lyng, A study examining the effects of tissue processing on human tissue sections using vibrational spectroscopy, *Vib. Spectrosc.*, 2005, **38**(1–2), 121–127.
2 F. Bonnier, D. Traynor, P. Kearney, C. Clarke, P. Knief, C. Martin, J. J. O'Leary, H. J. Byrne and F. Lyng, Processing ThinPrep cervical cytological samples for Raman spectroscopic analysis, *Anal. Methods*, 2014, **6**, 7831.

Klaus Gerwert asked: Do you think your spectral cytopathological approach could substitute the conventional cervix test? How long does it take to make a spectral test and how long does the conventional test last in comparison? Could the spectral test be automated?

Fiona Lyng replied: I think that the approach could certainly be used in addition to the conventional Pap test. Slides could be examined firstly using Raman spectroscopy, followed by Pap staining and cytological evaluation perhaps just for the abnormal cases. As the majority of cases are normal, this would save time and resources.

As for the previous question, each slide typically took 20 min to record. Typically slides can be screened in 5–10 minutes by cytologists but, for abnormal cases, the slide is screened again by two further people before being reported.

Yes the spectral test could certainly be automated by using cell recognition to find the cells to record.

Bayden Wood said: You mentioned the problems of achieving a representative sample from the transformation zone of the cervix for cervical cancer detection using Raman spectroscopy. For Pap smears cells are often collected using both an Ayre spatula and Cytobrush and a mixture of ectocervical cells and mucus secreting columnar cells indicate the sample has actually been collected from the transformation zone of the cervix. Do you get mixtures of ectocervical cells and columnar cells in your samples? If not then is there anyway to ensure that the cells have been collected form the transformation zone of the cervix where cervical cancer often first manifests?

Fiona Lyng replied: The samples for this study were residual liquid based cytology samples from a cervical screening programme and they had already been assessed and reported by the cytology laboratory. As such they would have contained a representative cell sample from either the endocervix and/or the transformation zone as well as ectocervical cells. As the samples had already been screened and reported, we did not specifically look for endocervical cells in our unstained slides for Raman spectroscopy. Also as the focus is on squamous abnormality, we target ectocervical cells as opposed to endocervical cells.

Josep Sulé-Suso asked: If you take the data to cytologists and show them you can discriminate between cancer and non-cancer, what feedback do you get when asking them to use this technique in clinical practice?

Fiona Lyng answered: Our clinical collaborators are very enthusiastic. Together with our collaborators in the Coombe Women and Infants University Hospital in Dublin and funded by the Health Research Board, we are currently undertaking a study of 1000 patients.

Michael Hermes questioned: Which information was used to subset the data set for cross validation? Is the approach capable to account for patient to patient variation?

Fiona Lyng replied: Leave one patient out cross validation was used to account for patient to patient variation.

Christopher Sammon opened the discussion of the paper by Ioan Notingher: There are conflicting reports about the suitability of Raman for conducting multiple measurements on live cells, for example Goodacre's group in Manchester observed cell death after ~7 measurements over a period of time (Lorna Ashton, personal communication). So my question relates to the practical aspects of such measurements. How critical is the laser excitation wavelength, the laser power density at the sample, cell phenotype, sampling set up *etc.*? What guidance can you give to people working in this field? Are there best practices that can be adopted?

Ioan Notingher answered: See reply to Roy Goodacre's question about the energy on live cells during Raman measurement.

Royston Goodacre asked: How much energy is on the live cell during the Raman measurements? Have validation tests been made to check that the phenotype is not modified? For example have you looked for the expression of heat shock transcription factor 1 (HSF1). This can be measured by single cell PCR. For more on HSF1 for monitoring stress see ref. 1.

1 A. Mitchell, L. Ashton, X. B. Yang, R. Goodacre, A. Smith and J. Kirkham, Raman spectroscopy using an aseptic method can characterise osteo-induction in dental pulp stromal cells, *Analyst*, 2015, **140**, 7347–7354.

Ioan Notingher replied: On live cells we typically use ~100mW. For this particular cell type, viability staining was done to ensure the cells were viable after the Raman measurements (reported in ref. 1). Although the viability of the cells was not compromised, the undifferentiated cells had high motility during the measurements (see ref.1 for details). The phenotype of the cells was checked after the measurements by staining for nestin. Doing single cell PCR to look at HSF1 is a very good idea to check for other adverse effects of the laser.

It is worth mentioning also that the effect of the laser was investigated on other cell types: there was no effect on ability of human embryonic stem cells to differentiate towards the cardiomyocyte phenotype,[2] it did not affect the beating frequency of cardiomyocyte a,[3] and it did not affect the ability of protozoan parazytes to invade host mammalian cells.[4]

1 A. Ghita, F. C. Pascutt, M. Mather, V. Sottile and I. Notingher, *Anal. Chem.*, 2012, **84**, 3155.
2 F. C. Pascut, S. Kalra, V. George, N. Wlech, C. Denning and I. Notingher, *Biochim. Biophys. Acta, Gen Subj.*, 2013, **1830**, 3517.
3 F. C. Pascut, H. T. Goh, N. Welch, L. D. Buttery, C. Denning and I. Notingher, *Biophys. J.*, 2011, **100**, 251.
4 A. Naemat, H. M. Elsheikha, R. A. Boitor and I. Notingher, *Sci. Rep.*, 2016, **6**, 20811

Nick Stone continued the discussion of Fiona Lyng's paper: In the gut we can explore changes in mucins with spectroscopy and the relationship with disease. Is this possible in the cervix and therefore spectroscopy of a mucin smear would reduce the need to find the small number of abnormal cells available?

Fiona Lyng answered: Yes this is definitely an interesting area. However, this hasn't been possible in our study to date as the clinical samples are collected as part of the cervical screening programme in GP clinics so it is not possible to influence how a sample is collected.

Catherine Kendall commented: Fiona you mentioned HPV. Is the HPV status of the patients in your study known? Have you correlated HPV status data with the spectral results? Would this be useful clinically?

Fiona Lyng responded: Yes the HPV status is known for the abnormal (CIN) cases in this study. The Cobas HPV test is performed routinely at our collaborating hospital and this provides specific genotyping information for HPV types 16 and 18, and concurrently detects 12 other high-risk HPV types. We are currently working on correlating this HPV data with the spectral data. Yes this would be of great interest clinically as HPV testing is currently being added to cervical screening programmes.

Samir El-Mashtoly asked: HPV is a major risk factor for cervical cancer. Have you looked at samples infected with HPV but not diagnosed with cervical cancer? This might help in early diagnosis of cervical cancer.

Fiona Lyng responded: Our group and others have previously looked at HPV infection using cell line models.[1–3] For the cytology samples reported in the present paper, we are currently correlating HPV status with Raman spectral profiles.

1 P. R. T. Jess, D. D. W. Simth, M. Mazilu, K. Dholakia, A. C. Riches and C. S. Herrington, Early detection of cervical neoplasia by Raman spectroscopy, *Int. J. Cancer*, 2007, **121**, 2723–2728.
2 K. M. Ostroswka, A. Malkin, A. Meade, J. O'Leary, C. Martin, C. Spillane, H. J. Byrne and F. M. Lyng, Investigation of the influence of high-risk human papillomavirus on the biochemical composition of cervical cancer cells using vibrational spectroscopy, *Analyst*, 2010, **135**, 3087–3093.
3 E. Vargis, Y.-W.Tang, D. Khabele and A. Mahadevan-Jansen, Near-infrared Raman Microspectroscopy Detects High-risk Human Papillomaviruses, *Transl. Oncol.*, 2012, **5**, 172–179

Matthew J. Baker continued the discussion of the paper by Ioan Notingher: Have you taken any other physical measurements (e.g. Young's modulus) from the AFM to correlate with spectral measurements and how would this be implemented?

Ioan Notingher responded: It is possible but we did not measure this. Such experiments are worth doing when studying live cells. In our application, we believe that AFM is not viable to be applied on live cells, simply because the morphology can change by the time the Raman experiments are completed. We did AFM only on the fixed cells, and for these, stiffness values are not truly relevant.

Sergei Kazarian asked: Have you measured the depth of field in your Raman study? This is important both for extracting specific chemical information from within the cell and for comparing with other data. Quantitative studies would require the depth resolution achieved to be assessed for the system used in this investigation. For example, Neil Everall has highlighted the importance of such issues and in particular, of assessment of depth resolution for an system optimised for a specific study.[1]

1 N. J. Everall, Confocal Raman microscopy: common errors and artefacts, *Analyst*, 2010, **135**(10), 2512–2522.

Ioan Notingher responded: This is a very good point. Yes, the depth of field measurements are included in the paper (Fig. 6): the value for the FWHM was 10 μm.

Pavel Matousek asked: In order to determine the thickness of the cells in your experiments did you consider using conventional imaging microscopy? That is imaging sharply the top and bottom of the cell and from the required movement of the microscope stage in the z-direction between the two positions deriving the thickness of the cell.

Ioan Notingher replied: Conventional microscopy is not sufficiently accurate. However, there are several optical techniques that may provide information regarding cell topography.

Katherine Lau said: There are many examples in the literature showing AFM has been performed on live cells. What was the reason you did not want to perform AFM on live cells?

How did you assign PC2 to the endoplasmic reticulum?

Ioan Notingher responded: AFM: in principle, indeed, the AFM-Raman can be applied on live cells. However, in our study, we wanted to validate the methods, thus we used fixed cells. The time for AFM and Raman mapping of a cell is too long (~30 minutes), therefore cell topography can change, thus inducing additional errors. The long term is indeed to achieve quantitative measurements on live cells.

Max Diem asked: What RNA is most likely be detected by your measurements? In cytoplasm, it may be r-RNA, but in the nucleus, most likely t-RNA. Can you use that for a surrogate measurement of protein synthesis? Do you know where cells are in the cell cycle?

Ioan Notingher answered: I agree, it is likely that we measure r-RNA. We tried to confirm this by combining Raman imaging with fluorescence imaging for r-RNA, but the results were not very reliable. Yes, quantitative measurements of RNA concentration may be related to protein synthesis – it would be interesting to try it out!

Samir El-Mashtoly said: From our experience in the Raman imaging of cells, the cell thickness is dependent on the cell cycle.

Have you checked this point? Also how many cells have you measured in terms of statistics to estimate the cell thickness?

Ioan Notingher replied: This is a single cell study - AFM and Raman were used on each cell.

Achim Kohler remarked: I have two questions/comments. 1. My first comment concerns the way you estimate cell thickness. The effective optical path length, which correlates with the thickness of the cell, appears as the first principal component when spectra are not preprocessed. In this case the explained variance of the first principal component is very high, since the effective optical path length of biological samples creates a huge variation. What is the explained variance of the PCA components? When spectra are preprocessed by MSC or EMSC, the EMSC parameter that relates to the effective optical path length can be correlated with the cell thickness. See images of EMSC parameters in ref. 1. This is a more straight-forward approach than using the PCA component. The PCA component may contain other spectral information that co-varies with the effective optical path length.

2. In your paper, regression models are established on the basis of pure solutions of RNA and proteins with different concentrations and then used to estimate the RNA and proteins in the cells. The regression models based on the pure solutions cannot "not learn" about the subtle differences needed to estimate the RNA and proteins in cells. What do your regression coefficients look like? In order to establish regression coefficients that can estimate cell components, regression coefficients need to know everything about the interfering cell components. Thus, regression models based on pure components cannot perform well when being applied to cells. In order to establish regression models for predicting cell components, models need to be established by using cells with known component concentrations. This is in general difficult to achieve, since it is difficult to perform reference analyses. In addition, the spectral difference may be tiny and strongly depending on the interfering components. This makes it very difficult to establish models.

1 A. Kohler, C. Kirschner, A. Oust, H. Martens, Extended multiplicative signal correction as a tool for separation and characterisation of physical and chemical information in Fourier transform infrared microscopy images of cryo-sections of beef loin, *Appl. Spectrosc.*, 2005, **59**(6), 707–716.

Ioan Notingher responded: Point 1: The first component in PCA can be used for the estimation of cell thickness, as shown in our paper, (confirmed by calculating the correlation with AFM images). Agreed, there may be also other ways of obtaining similar information, such as MSC or EMSC. Point 2: I fully agree. Currently, the errors are ~25%, we are working on improving the calibration methods to reduce these errors. However, it is a difficult task.

Pavel Matousek commented: The point about the difficulty of applying PLS predictions to other cells if the training set was derived from a single cell appears to be valid. To mitigate the issue of inter-cell protein content variability could you not use 10 or 100 cells in your calibration step instead of one to expose PLS to the existing inter-cell variability?

Ioan Notingher replied: Indeed, the more cells are used in the training, the more accurate PLS is. In this study we just wanted to investigate the feasibility of using the combined AFM-Raman for obtaining quantitative data related to the molecular composition of cells.

Hugh J. Byrne said: In terms of possible sample photodegradation during Raman measurement, it is important to differentiate between photochemical and photothermal effects. The former are usually due to photooxidation and therefore need absorption by a chromophore at the source wavelength and that absorption should be at higher energy than the critical activation energy of the creation of singlet oxygen. This equates to ~650 nm. Photooxidation is a linear process and G. Pupples has demonstrated the differences between live cells and fixed cells.[1] Photothermal effects do not require a chromophore (although they help build up heat if they are not fluorescent) and are intensity dependent processes (of the order cubic). They can be alleviated by providing good thermal contact through the substrate or for example by measuring in water immersion.[2]

1 G. J. Puppels, J. H. Olminkhof, G. M. Segers-Nolten, C. Otto, F. F. de Mul and J. Greve, *Exp. Cell Res.*, 1991, **195**, 361.
2 F. Bonnier, S. M. Ali, P. Knief, H. Lambkin, K. Flynn, V. McDonagh, C. Healy, T. C. Lee, F. M. Lyng and H. J. Byrne, *Vib. Spectrosc.*, 2012, **61**, 124–132.

Ioan Notingher responded: Yes, I agree. Most of the damage caused by visible lasers (wavelength less than 650 nm) is due to photochemical effects, that is why we use 785 nm lasers in all our studies. There has also been lot of work in the optical tweezers community, see for example ref. 1 and 2. These show that cell damage is minimised around 785 nm compared to lower and higher wavelengths. We also confirmed this by measuring the beating frequency of live cardiomyocytes before and after laser irradiation, showing that the laser did not induce any differences.[3]

1 K. Neuman, E. H. Chass, G. F. Liou, K. Bergman and S. M. Block, *Biophys. J.*, 1999, 77, 2856.
2 H. Liang, K. T. Vu, P. Krishnan, T. C. Trang, D. Shin, S. Kimel and M. W. Berns, *Biophys. J.*, 1996, **70**, 1529.
3 F. C. Pascut, H. T. Goh, N. Welch, L. D. Buttery, C. Denning and I. Notingher, *Biophys. J.*, 2011, **100**(1), 251–259

Royston Goodacre commented: On the point of using PLSR for quantification (this may or not be appropriate from external mixtures) there are other chemoemtric methods. There is a nice paper in *Nature Protocols* from Roma Tauler's group on MCR-ALS for image analysis where they extract spectral features directly from IR images.[1] This might be worth considering.

I also have a comment on the choice of cells. Some cells are "shy" and for example we have found that neuroblastoma cells move away from the laser during Raman, making measurements difficult (785 nm, ~30 mW, on a inVia, Renishaw microscope).

1 J. Felten, H. Hall, J. Jaumot, R. Tauler, A. de Juan, and A. Gorzsás, Vibrational spectroscopic image analysis of biological material using multivariate curve resolution-alternating least squares (MCR-ALS), *Nat. Protoc.*, 2015, **10**, 217–240.

Sergei Kazarian opened a general discussion of the paper by Aidan Meade: The effect of inhibitors on the presence of reactive oxygen species in irradiated cells was evaluated. However, the effect of inhibitors should also be assessed independently, *i.e.* spectroscopically assessing non-irradiated cells in presence of inhibitors and then comparing them with irradiated samples. Was this attempted?

Aidan Meade responded: The effect of the inhibitors on control cells was studied and the data is supplied as supplementary figures to this paper.

Colin J. Campbell asked: Can you explain the mechanism by which the signal is passed from the irradiated cells to the bystander cells and are there spectroscopic differences between the directly irradiated cells and the bystander cells that correlate with different mechanisms of death?

Aidan Meade answered: At present the molecular identity of the bystander signal remains unknown, though it has been confirmed that it does comprise a heat-labile low molecular weight protein, reactive oxygen species, and perhaps some involvement by exosomes. The data presented in this paper (Fig. 6c) show that the molecular signal is observable spectroscopically in the difference spectrum between the unconditioned and irradiated-cell conditioned medium, though interpretation of this spectrum was not performed.

Martin Isabelle remarked: Following on from the previous question that was asked and Aidan's response, I can further comment that as co-author on the publication from the Canadian group (UVIC/BC Cancer Agency) that was mentioned,[1,2] we also observed the complexity of the biomolecular interactions taking place in radiobiology. As the audience member correctly pointed out, we did observe a change in glycogen content as a result of changes in radiation dose. However I agree that this is only one of the biomolecular changes that are occurring in cells and tissues as a result of radiation dose and the molecular interplay in radiobiology is a lot more complicated than one single biomolecule. When investigating vibrational spectroscopy in the scope of radiobiology, we need to bear this mind and look at possible changes in lipids, nucleic acids, proteins and other biomolecules as well. I actually have two unrelated questions for Aidan. The first question is: Did you attempt any form of DNA fragmentation, double-strand damage strand break or radiation-induced foci molecular biology analysis that was correlated with the spectroscopic observations? For example, TUNEL assay and/or Comet (Single-Cell Gel Electrophoresis) assay? If so, did these results correlate with the nucleic acid changes observed in the spectra?

The second question: With the complexity of radiobiology and the lack of knowledge about the detailed interplay of molecular mechanisms involved in radiobiology (DNA damage, apoptosis, DNA repair, mitochondrial damage, ROS, bystander effect, abscopal effect *etc.*), do you think vibrational spectroscopy has a future role in clinical diagnostics or theradiagnostics in radiation treatment planning and personalised treatment? What about prognostics? Could vibrational spectroscopy be used as a tool to assess treatment response to radiation treatment (or even chemotherapy or hormone therapy)? Or even be able to predict tumour response to radiotherapy?

1 Q. Matthews, M. Isabelle, S. J. Harder, J. Smazynski, W. Beckham, A. G. Brolo, A. Jirasek and J. J. Lum, Radiation-Induced Glycogen Accumulation Detected by Single Cell Raman Spectroscopy Is Associated with Radioresistance that Can Be Reversed by Metformin, *PloS One*, 2015, **10**(8), e0135356.
2 S. J. Harder, M. Isabelle, L. DeVorkin, J. Smazynski, W. Beckham, A. G. Brolo, J. J. Lum and A. Jirasek, Raman spectroscopy identifies radiation response in human non-small cell lung cancer xenografts, *Sci. Rep.*, 2016, **6**, 21006.

Aidan Meade replied: In answer to the first question we did not specifically perform DNA damage analysis in directly or bystander irradiated cells, though this data does exist in our laboratory in the form of gamma-H2AX foci analyses. We have previously seen a correlation in spectral features within Raman spectra and increases in numbers of gamma-H2AX foci within irradiated lymphocytes[1] although it has been difficult to identify spectral markers of this damage as alterations of the whole cellular spectrum are seen.

In answer to the second question, it may be possible to investigate this with carefully chosen *in vitro* and *in vivo* models. The complexity of the cellular response to treatment, and its variability between individuals, will present difficulties in associating a given type of response with specific spectral markers, and in making determinations on any aspect of the patient response phenotype. These difficulties are already seen in the various clinically available genetic tests for segregating cancer patients by response and will therefore be challenges in any spectroscopic study; the issue of whether vibrational spectroscopy has a role in this area remains an open question.

1 A. Maguire, I. Vegacarrascal, L. White, B. McClean, O. Howe, F. M. Lyng and A. D. Meade, *Radiat. Res.*, 2015, **183**(4), 407–416.

Maria Paula Marques said: You used HaCaT cells in your study. Have you considered using cancer cells (e.g. a skin cancer cell line)? Which distinct results, if any, would you expect had you probed cancer cells?

Aidan Meade responded: There was no consideration given to using a cancer cell line here as the overall intention in this and linked studies[1,2] was to examine the potential for detection of the effects of external radiation on the skin with vibrational spectroscopy. Depending on the whether the cell line selected commits to p53 dependent apoptosis in preference to DNA repair after irradiation it might be expected that an increase in spectral signatures relatable to apoptosis would be seen. Biological characterisation of the cell line response to radiation would therefore be key to understanding the spectral response within this experiment.

1 A. D. Meade, C. CLarke, H. J. Byrne and F. M. Lyng, *Radiat. Res.*, 2010, **173**(2), 225–237.
2 A. D. Meade, H. J. Byrne and F. M. Lyng, *Mutat. Res., Rev. Mutat. Res.*, 2010, **704**(1–3), 108–114

Sergei Kazarian commented: The accuracy of two models with different numbers of variables are compared. Clearly, the model with more variables would give a better fit, so is it a fair comparison?

Aidan Meade replied: The use of the induced-repair (IR) model here is intended to highlight the variation in spectroscopic signatures at low doses. The linear quadratic (LQ) model does not describe this behaviour. Given the clear deviation of the data from the LQ model it is necessary to use the IR model to provide a more complete modelling of the spectral and biological reference data over the dose range.

Nick Stone asked: When plotting the changing DNA peaks with radiation dose, have you started with a hypothesis of what you're looking for or mined the data and identified correlations to plot, ignoring peaks which don't fit/make sense?

Aidan Meade answered: The variation in all of the spectroscopic bands with dose were interrogated, as mentioned in the paper. Some agreement was seen between the induced repair model and the variation in the intensities of the DNA, RNA and certain protein bands, which were referenced to variations in the parallel biological metrics with dose. However, many of the protein and lipid bands displayed very different variations, some increasing, and some decreasing with dose. These may be signatures of protein lysis effects and of free radical damage to the cell membrane, though further work is required here to fully explain this behaviour and reference appropriate biological metrics that were not used in this study.

Bayden Wood said: Can you assess the recovery rate of the cells after irradiation?

Have you put irradiated cells back into the media, allowed them to recover and determined if the spectral changes revert back to the spectra of the control samples?

Aidan Meade responded: This is a very interesting question, and although the aim of the study was not specifically to examine this response, the data in the paper does allow the question to be considered. The surviving fraction after irradiation (see Fig. 1 from this paper) was measured 10 days after irradiation and showed that the extent of the repair response and associated cell survival is dose dependent. In this study the Raman and FTIR spectra were measured in directly irradiated cells which were incubated at 37 °C for 96 hours (4 days) after irradiation and it is shown that the cellular spectra do show dose dependent alterations in cells which have normal morphology and which have survived the irradiation to this time point (Fig. 4 in the paper). The evidence suggests, therefore, that the spectra of irradiated cells do not revert to the control spectra in surviving cells after irradiation.

Ines Delfino asked: 1) What was the laser power at the sample used for performing Raman measurements?

2) Regarding the study of the bystander effect, different durations of exposure to the irradiated cell-conditioned medium were selected for observing target proteins (after 1, 6, 12 and 24 hours of exposure) and detecting Raman spectra (after 30 min and 24 h of exposure). Can you comment this selection?

Aidan Meade replied: 1) The laser power at the sample was 70 mW for the spectrometer used at the University of Reims, and 100 mW for the spectrometer used at DIT. 2) Previous studies performed by Lyng *et al.*[1] demonstrated that the time course of bystander signalling in keratinocytes (HPV-G line) showed evidence of signalling but little cellular effect at 30 minutes after exposure, with strong evidence of exposure at 24 hours after exposure to the irradiated cell-cultured medium. The present study was designed to match Lyng's study in regards to the times at which Raman spectra were acquired. However, it was necessary to establish whether HaCaT cells expressed the MAPK proteins in bystander signalling, how this varied with dose, and to examine the temporal variation in expression in order to plausibly argue that the spectral variations seen at 24 hours were due to MAPK signalling and caspase-dependent apoptosis.

1 F. M. Lyng, P. Maguire, B. McClean, C. Seymour and C. Mothersill, *Radiat. Res.*, 2006, **165**(4), 400–409.

Lauren Jamieson asked: Cells in the body exists in a three dimensional environment. When a cancer tumour develops the central regions become hypoxic as they get further from the well organised vasculature of the body. These hypoxic regions are known to develop radioresistance.[1] Have you looked at how hypoxia effects the response you get in the Raman to radiation? This is very nice work and it would be interesting to see if you pick up the differences with increasing levels of hypoxia for both targeted and non-targeted effects.

1 L. E. Jamieson, D. J. Harrison, and C. J. Campbell, Chemical analysis of multicellular tumour spheroids, *Analyst*, 2015, **140**, 3910–3920.

Aidan Meade answered: This is a very interesting idea and I am not aware of the study having been done. I would hope that spectral alterations in normoxic versus hypoxic irradiated cells would be observable though this has yet to be proven using unstimulated Raman spectroscopy.

Louise Clark asked: At the moment you are using gamma radiation, would you be looking towards using X-radiation in the future?

Aidan Meade responded: We have conducted experiments with both X-ray radiation and gamma radiation and will be publishing some data from the former experiments in the near future. However, given that the relative biological effectiveness of X-ray and gamma radiation are equal, and independent of the energy of the photon, a comparative spectroscopic study of the effect of the two radiation types in cells may not provide new insights.

Michael Hermes opened the discussion of the paper by Achim Kohler: Is the discussed algorithm restricted to spherical particles like cells or can it also be used to compensate for the scattering effects occurring in more complicated structures like tissue sections? What effects can be expected on those?

Achim Kohler replied: The EMSC model employs a so-called meta model.[1] The meta model takes into account a whole range of refractive indices and sizes of spheres. It allows us to linearly superimpose this range of parameters (a range of spheres with different sizes and refractive indices). This is the reason for the flexibility of the model that allows for compensating these factors and also for scattering occurring in more complicated structures.

1 A. Kohler, J. Sulé-Suso, G. D. Sockalingum, M. Tobin, F. Bahrami, Y. Yang, J. Pijanka, P. Dumas, M. Cotte, H. Martens, Estimating and correcting Mie scattering in synchrotron based microscopic FTIR spectra by extended multiplicative signal correction (EMSC), *Appl. Spectrosc.*, 2008, **62**, 259–66.

Hugh J. Byrne commented: The approximation of spherical scatterers is embedded in the original Mie theory. Where do the finer ripples on the spectra come from?

Achim Kohler responded: The Mie theory describes the scattering of light at a dielectric sphere exactly.

The formula used by us in our Faraday discussions paper, is an approximation formula published in ref. 1. It models the broad Mie oscillations (wiggles) and it employs a complex refractive index that takes into account the fluctuations of the real part of the refractive index due to absorption resonances. It does not describe ripples as they appear in the exact Mie theory. The ripples in the Mie absorbance spectra correspond to partial-wave resonances in the Mie coefficients. They have been observed and described recently for the first time in the FTIR spectra of PMMA spheres.[2] Ripples are connected to so-called whispering gallery modes that appear in perfect spheres. Since cells have shapes that differ considerably from spheres, ripples cannot usually be observed in the spectra of human cells.

1 H. C. Van De Hulst, *Light scattering by small particles*, Courier Corporation, 1957.
2 R. Lukacs, R. Blümel, B. Zimmerman, M. Bagcioglu and A. Kohler, Recovery of absorbance spectra of micrometer-sized biological and inanimate particles, *Analyst*, 2015, **140**(9), 3273–3284.

Bayden Wood asked: Does Mie-scattering or light dispersion still manifest in the second derivative and would it improve classification to use the second derivative as the model spectrum in the RMie-EMSC? Would you recommend performing the RMie-EMSC on the second derivative?

Achim Kohler answered: The broad Mie oscillations (wiggles) are suppressed in the second derivative spectra. The effect of the fluctuations of the real refractive index due to strong absorption resonances may even be increased in the second derivative. The RMie-EMSC algorithm is developed for raw spectra. If one wants to apply it to second derivative spectra, the code has to be changed and second derivative spectra have to be used for building the meta-model used in the EMSC. That is in principle possible and an interesting idea that may be followed up.

Peter Gardner said: First I have a comment and then a question. The comment is that I wanted to clarify to those who are unfamiliar with these algorithms that the cell is not approximated as one single scattering sphere. The complexity of the cell shape can be approximated by using combinations of many different spheres of differing refractive index.

The question to Achim is whether he sees the main value of the new algorithm over the 2010 version, as being increased speed (approximately 100 fold) or delivering a more accurate correction? For single cells, speed is not such an issue since the number of spectra are usually low but for correcting large tissue images it is essential.

Achim Kohler replied: The most important practical achievement in my opinion is the increased speed. As with the algorithm from 2010, the new algorithm suffers from the problem that the corrected spectra tend towards the reference spectrum and it depends on the number of iterations used in the algorithm. Therefore, it is difficult to say currently, how big the effect of the use of a more precise model is. Further testing and refinements of the algorithm will answer this question. Notwithstanding, if a more precise model is available (i.e. a model that is closer to exact theory) and if this model can be used without any other costs such as computation time, it is appropriate to modify the existing algorithm with the new model.

Gianfelice Cinque addressed Achim Kohler, Rohit Bhargava and Hugh Byrne: In Mie scattering the refractive index real and imaginary parts are involved, and these are related by the Kramers-Kronig transformation. In the similar case of the standing wave artefacts in transflectance measurements, 3 years ago our IR group at Diamond published a paper in ABC about the use of the pure reflectance measurements on cells to retrieve the refractive index since the KK relations are easier to obtain from reflectance data:[1] have you considered this as a possible approach to validate your theory?

1 K. Wehbe, J. Filik, M. D. Frogley and G. Cinque, *Anal. Bioanal. Chem.*, 2013, **405**, 1311–1324.

Achim Kohler answered: That is an interesting idea, which may be followed up.

Rohit Bhargava replied: Yes, indeed. It is indeed very appropriate to test theories but we have not specifically applied it to the nice data you have provided. The general framework we have developed should work and we did test the predictions from the theory with polymer films on a reflective substrate as validation. Thus far, we have applied our rigorous algorithms[1] to rather simple cases, e.g. films,[2] cylinders,[3] spheres,[4] and periodic structures,[5] as summarized in ref. 6. An important difference in our approach and others is that we model both the morphology and the spectral data concurrently. We do not assume the object to consist of spheres or other building blocks to fit the spectra. I also want to emphasize that it is not just the sample but also the optics that influence recorded data, providing extensive opportunities for improving the information content of IR imaging.[7,8]

1 B. J. Davis, P. S. Carney and R. Bhargava, Theory of mid-infrared absorption microspectroscopy. II. Heterogeneous samples, *Anal. Chem.*, 2010, **82**, 3487–3499.
2 B. J. Davis, P. S. Carney and R. Bhargava, Theory of mid-infrared absorption microspectroscopy. I. Homogeneous samples, *Anal. Chem.*, 2010, **82**, 3474–3486.
3 B. J. Davis, P. S. Carney and R. Bhargava, Theory of infrared microspectroscopy for intact fibers, *Anal. Chem.*, 2011, **83**, 525–532.
4 T. van Dijk, D. Mayerich, P. S. Carney and R. Bhargava, Recovery of absorption spectra from Fourier transform infrared (FT-IR) microspectroscopic measurements of intact spheres, *Appl. Spectrosc.*, 2013, **67**, 546–552.
5 R. K. Reddy, M. J. Walsh, M. V. Schulmerich, P. S. Carney and R. Bhargava, High-definition infrared spectroscopic imaging, *Appl. Spectrosc.*, 2013, **67**, 93–105.

6 R. Bhargava, Infrared spectroscopic imaging: The next generation, *Appl. Spectrosc.*, 2012, **66**, 1091–1120.
7 T. van Dijk, D. Mayerich, R. Bhargava and P. S. Carney, Rapid spectral-domain localization, *Opt. Express*, 2013, **21**, 12822–12830.
8 D. Mayerich, T. van Dijk, M. J. Walsh, M. V. Schulmerich, P. S. Carney and R. Bhargava, On the importance of image formation optics in the design of infrared spectroscopic imaging systems, *Analyst*, 2014, **139**, 4031–4036.

Hugh J. Byrne added: In fact, the original exploration of the resonant contributions of the refractive index to the "spectral artefact" was done through comparison of transflection, transmission and reflection measurements, and it was demonstrated that reflection measurements on a transparent substrate give relevant biological data, as the "penetration depth" at the reflecting surface is of the order of the wavelength.[1] Pure reflection measurement could therefore be used for analysis, although the low signal requires long accumulation times for good quality spectra.

1 P. Bassan, H. J. Byrne, J. Lee, F. Bonnier, C. Clarke, P. Dumas, E. Gazi, M. D. Brown, N. W. Clarke and P. Gardner, Reflection contributions to the dispersion artefact in FTIR spectra of single biological cells, *Analyst*, 2009, **134**, 1171–1175.

Wolfgang Petrich commented: We keep referring to the real and imaginary part of the refractive index as if we were measuring one or the other solely. In any typical tissue sample however, we always measure both components at the same time and the task you are addressing is how to disentangle the "mixed" observation (which we call the hyperspectral image) into the two components. As was pointed out already the concept behind this disentanglement is the quest for minimizing the errors in the things we want to find out (*e.g.* the imaginary part of the refractive index) and to shift the error budget towards those things which are of no apparent interest to us. However, what I do not really understand is why we would thus infer an error onto the size of our scatterers. Let me take a simple example of taking mid-IR images of polystyrene spheres in solution. Personally I would guess that the size (distribution) of these spheres is among the things which are known best in the whole setting. So why would you still use a broad size distribution for your calculation? Why not take the particle size as a given - at least if the example is as simple as this one?

Achim Kohler replied: First, I would like to answer the comment related to the real and imaginary part of the refractive index. It is correct that both the real and the imaginary part of the refractive index are needed to estimate the extinction efficiency. They cannot be estimated independently. The EMSC algorithm in our paper estimates iteratively both the real and the imaginary part of the refractive index by minimizing an error. The error is defined by the part of the extinction efficiency, which is not described by the Mie model employing the estimates of the real and the imaginary part of the refractive index. Further, the EMSC algorithm takes into account a relevant range of the constant part of the real refractive index and a relevant range of the size of scatterer. It is correct that this is not strictly necessary for the example of polystyrene spheres, where the size is known and the constant part of the real refractive index is approximately known (it should be noted that the constant part of the real refractive index of polystyrene spheres is not very well known for the whole infrared range).

Estimations of the refractive index for polystyrene spheres with an EMSC model with a pre-defined refractive index are done in ref. 1. However, when working with highly scattering cells, we cannot obtain the size of the cells for every single spectrum experimentally. The refractive indices are unknown and the size of the cells deviate considerably from a sphere. Thus, we developed an EMSC model based on a meta-model covering a whole range of the constant part of the real refractive index and a whole range of sizes.

1 R. Lukacs, R. Blumel, B. Zimmerman, M. Bagcioglu, A. Kohler, Recovery of absorbance spectra of micrometer-sized biological and inanimate particles, *Analyst*, 2015, **140**(9), 3273–3284.

Klaus Gerwert opened a general discussion of the paper by Luca Quaroni: You showed assignments of bands in the IR spectra. How did you assign the bands? Did you use isotopic labeling to make a clear cut assignment? Or have you used other modifications like mutagenesis?

Luca Quaroni answered: Most of the assignments are based on the analysis of the correlation properties of the bands in the difference spectra, following a protocol that we called CSM, described in a previous publication.[1] The idea is based on the use of 2D-COS to identify clusters of bands that evolve with exact synchrony in difference spectra as a function of time (or of some other changing parameter). These band clusters are assigned either to the same molecule or to more molecules the concentration of which changes exactly in phase (*e.g.* two products of the same reaction). This approach allows us to simplify the problem by reducing the system to different groups of bands that cluster together. We then apply a form of fingerprinting: we look up the spectra of relevant molecules and see if they match any of the clusters of bands that we have identified. This is much more reliable than making assignments based on single bands, which is often meaningless in a complex system like a cell. In this particular experiment we are facilitated by the fact that, under the experimental conditions in use, we know that glycolysis dominates the metabolism of the cells, so we restrict our analysis mostly to the bands of metabolites involved in this pathway, plus other known abundant cellular components. We have also integrated this approach with isotopic substitution, which is one of the most solid methods for band assignment in complex systems. We showed in ref. 1 that we can confirm the assignment of the starting compound (glucose) in this metabolic pathway *via* isotopic substitution. In this case the CSM method is also very helpful because it allows us to cluster together the bands of the isotopically labelled glucose, thus helping its identification. Without a CSM analysis it would be difficult to identify the isotopically shifted bands in the complexity of the cellular difference spectra.

Isotopic labelling could also be used to identify other metabolites. However, an effective way to introduce an isotopic label into a metabolite *in vivo* must be found. This must be devised case by case. Simply labelling glucose and allowing the label to "move along the metabolic pathway", as is done for metabolic tracing experiments, was not useful in our specific experiment: several downstream metabolites end up being only partially labelled or labelled in different positions, in part because of interplay with other pathways, thus making the spectra too complex to interpret. However, this very same approach may well provide complete labelling of some intermediates in other cases and contribute to the assignment.

1 L. Quaroni and T. Zlateva, Real-Time Metabolic Analysis of Living Cells with Correlated Cellular Infrared Microscopy, *Anal. Chem.*, 2014, **86**(14), 6887–6895.

David Perez-Guaita said: Table 1 of the paper establishes the minimum concentration of small molecules and metabolites that can be detected by IR. According to the table, molecules at low ppm values can be detected but the IR spectra also have proteins at percentage level. Can you detect analytes with concentrations ranging 6 orders of magnitude in the same spectrum? Can you detect at such levels? Regarding the ongoing discussion about the assignment of the bands, I think that they cannot be made to molecules, but to functional groups. Some authors are sceptic of any assignment and others assign multiple metabolites, even if they are not detectable (that is the case in this paper). In my opinion we can meet in the middle, since the spectra are affected by a combination of well-known major components and the secondary conformation of proteins. I believe we need to investigate and establish a consensus on what we can and cannot assign.

Luca Quaroni responded: The question addresses several important issues and will receive separate answers.

1 – Concerning the content of Table 1, I would like to stress that the table is not meant to give absolute detection limits for FTIR experiments. It rather compares the concentrations reported for some molecules in cells with one of the lowest concentrations detected in an FTIR measurement. This is meant to convey the message that there are plenty of small molecules that are potentially detectable in an FTIR measurement and that can provide a measurable contribution to the FTIR spectrum, although their identity and detection limits should be addressed in each separate experiment.

2 – Measuring over six orders of magnitude of absorbance values is possible with FTIR instrumentation, so it is also possible to cover six orders of magnitude of concentration, although when working with biological samples it is more common to work over a range of four to five orders of magnitude of absorbance. The accessible concentration range depends on the instrument and the experimental design. Working with a monolayer of cells in an aqueous environment the main limitation is water absorption. In particular the region around 1650 cm^{-1}, where both polypeptides and water absorb, is the most difficult one to address. Even if sample thickness is adjusted so that saturation does not occur, light throughput is so reduced that the signal to noise ratio in this region is always poor, making it hard to detect absorbance changes smaller than 10^{-1} or 10^{-2} au. This is why this region has been excluded from the present study. In contrast, we can appreciate absorbance changes of 10^{-4} au in other regions of the spectrum.

3 – Difference spectroscopy is traditionally used to highlight small variations of absorbance in the presence of strong absorption from various sample components. Of course this requires a dynamic system, which evolves over time or as a function of a perturbation. This approach has been used by many for decades to study the fine details of protein dynamics by FTIR spectroscopy, including some delegates at this conference. In these experiments variations of 10^{-4} absorbance

units, sometimes less, are measured over a total sample absorption of 10^0–10^{-1} absorbance units. This is a typical example of an experiment in which we measure small absorbance values in the presence of a large absorption background from the sample. We have also used this approach in our experiments on cells to highlight the small changes due to metabolite turnover.

4 – The statement "others assign multiple metabolites, even if they are not detectable (that is the case in this paper)" is unclear. The bands discussed in the publication are detected: they are seen in the 2D-COS spectra (they are also seen in the differential absorbance spectra but there most of them cannot be separated). The question is which molecule they are assigned to.

5 – The idea behind this work (this paper and its closely related predecessor[1]) is not actually in opposition to the statement that bands in cellular spectra should not be assigned. Quite the contrary, its conception is based on that concern and builds on it.

I believe that assigning bands in cellular and tissue spectra by relying on tables of band frequencies, as it is currently often done in many publications, is completely unreliable given the complexity of the sample. I must point out that I am not innocent of it myself in past publications. Part of the aim of Table 1 in this paper is to show how many contributions can be expected in cellular spectra from molecules that are typically completely neglected in these sorts of assignments. The rest of the work (this paper and ref. 1) is aimed at improving this situation and provides a methodology to improve the reliability of assignments beyond what is currently considered as the state-of-the-art. The work uses a combination of correlation spectroscopy and isotopic substitution to perform assignments. The need for reliable band assignments is not new to vibrational spectroscopy. It is actually one of its defining and recurring issues and has been carefully addressed since the early days of vibrational spectroscopy. However this seems to have been forgotten in the current application of vibrational spectroscopy to complex biological samples. As early as thirty years ago, band assignments in the spectra of something as simple as a single purified protein were considered unreliable unless they were substantiated by isotopic labelling or calculations or both. Functional group tables were considered to be useful for general guidance and for assignments in simple, low molecular weight, molecules. A "simple" protein was already considered too complex for this approach. Major journals would turn down articles that drew major conclusions from spectral assignments without a convincing assignment. Assignments based on tables of functional group frequencies were not considered reliable in this context. I find it very worrying that nowadays, decades later, as a community we routinely accept assignments in the spectra of samples such as cells, which are orders of magnitude more complex than single proteins, without any hard experiments supporting the assignment. I think that we have severely and unacceptably lowered the bar. The result is a drop in the reliability of our conclusions. This can only have negative consequences on the reputation of our community.

We have to define new rigorous standards that make assignments acceptable. In the absence of these, I agree with Klaus Gerwert that we should refrain from providing molecular interpretations to cellular and tissue bands. As he rightly pointed out, assignments are not necessary in spectral cytopathology applications. For the latter we just need to rely on differences in spectral traces, for classification purposes, whatever the bands arise from. Everything risks being just words, if not substantiated. If we decide to take on the challenge, we will have to establish as a community a clear standard for the acceptance of band assignments in cellular samples.

1 L. Quaroni and T. Zlateva, Real-Time Metabolic Analysis of Living Cells with Correlated Cellular Infrared Microscopy, *Anal. Chem.*, 2014, **86**(14), 6887–6895.

Sergei Kazarian commented: You may need to provide additional support of your statement about Table 1 in the paper and its presentation, particularly for the position of its solid vertical line since the detection limit is affected by the chemical structure of the drug (different to metabolites) and a number of optical factors.

Luca Quaroni replied: Indeed the vertical line of Table 1 is not meant to represent an absolute detection limit for the technique. It is rather a way to provide an approximate indication of the order of magnitude of the lowest concentrations that have been detected to date in an optimal situation, using an ATR cell for aqueous measurements and an analyte which is relatively easy to discriminate from cellular components. The exact detection limit depends on the molecular properties, instrumentation and experiment design and must be determined for each molecule and for each optical configuration. The detector, light source and instrument transfer function all play a part in determining the detection limit of an experiment, as well as the measurement parameters such as interferometer scan rate, resolution, and the use of optical and electronic filters. Sample thickness, sampling geometry and choice of optical substrate are also factors that must be considered as affecting detection limits. In the specific case of spectromicroscopy experiments, which are pertinent to this paper, detection limits are expected to be much higher than in the case of a macroscopic ATR experiment. Values in the range 0.1 mM to 1 mM are more commonly observed for specific molecules in microscopy experiments. For microscopy measurements, the size of the confocal aperture is an important factor in determining detection limits, and a trade off is typically necessary between spatial resolution (when above diffraction limited values) and detection limit in a specific experiment.

Christopher Phillips remarked: We have developed a technique, so-called "scattering-scanning-nearfield-optical-microscopy" (s-SNOM), that can now perform this type of IR spectroscopic imaging at a spatial resolution down to $\sim$10 nm, i.e. good enough to look inside a single cell for the first time (see DOI: 10.1039/c5fd00150a and references therein). Can you suggest an aspect of the cell metabolism problem that would be interesting to study with this new technique?

Luca Quaroni responded: If we want to take advantage of the high spatial resolution of s-SNOM, we should try to study processes that occur at the single organelle level. If we restrict ourselves to metabolic processes in eukaryotic cells, the obvious choice would be to study oxidative phosphorylation in single mitochondria, which in many cases are 0.5 – 1 μm in size. The resolution of s-SNOM is such that, at least in theory, processes could be selectively monitored within the intermembrane space of these organelles. I say "in theory" because the direct study of metabolic processes, as well as other biochemical processes occurring in real time, requires the capability to perform time resolved measurements on live cells with a sufficient signal to noise level to detect and resolve the spectral contribution from multiple species at the detection limit of the technique. This landmark performance has not been achieved yet and may be many years away for whole cell measurements. I think that a more realistic milestone for the next few years would be to aim at using s-SNOM to study metabolic processes in single isolated organelles or in single prokaryotic cells.

Bayden Wood opened the discussion of the Maria Paula Marques's paper: You stated that inoculating the cells with the drug caused the DNA within the cells to adopt the A-DNA conformation. However, when drying down or fixing the cells the DNA adopts the A-DNA conformation regardless. So how can you deduce that the changes observed in the spectra are consistent with the formation of A-DNA when the cells are formalin fixed and the DNA is in the A-DNA conformation irrespective of the drug treatment?

Maria Paula Marques replied: In the present study the effect of three anti-tumour agents on cell metabolites, including DNA, is assessed. Although it is a known fact that formalin cell fixation and air-drying may induce conformational rearrangements in DNA (e.g. B- to A-DNA, ref. 1), incubation with the tested cisplatin-like agents prior to the cell fixation procedure is also expected to have a significant impact on DNA (in view of the recognised mode of action of the DNA-binding drug cisplatin). Therefore, the spectral data finally measured, through Raman and IR, specifically for DNA should be the result of both effects: the impact of the metal-based compounds plus the fixation/drying process (accepting that the latter does not induce a complete B- to A-DNA rearrangement). This justifies the hypothesis that the new Pt_2Spm and Pd_2Spm agents disrupt the DNA native conformation (*via* covalent binding predominantly to the purine bases, similarly to cisplatin).[2] This is indeed corroborated by the fact that the spectral profile assigned to A-DNA (e.g. by the ν(OPO) backbone band at 805 cm^{-1}) is not identical for all three complexes probed (cisplatin, Pt_2Spm and Pd_2Spm), which would be the case if we were detecting solely the effect of cell fixation/dehydration.

Furthermore, the impact of the antineoplastic agents presently studied is based on comparative measurements – treated versus untreated-cells, as well as cisplatin-exposed versus cells incubated with the new potential drugs (Pt_2Spm and Pd_2Spm). Hence, reliable results are anticipated, provided these compounds are administered to the cells under exactly the same conditions (e.g. concentration, incubation time), and the samples are then handled in the same way before spectroscopic analysis. Any drug-prompted impact on the cellular metabolic profile occurs prior to the fixation/dehydration procedure, which is not expected to completely mask the drug-induced impact on DNA.

1 B. R. Wood, *Chem. Soc. Rev.*, 2016, **45**, 1980.
2 M. P. M. Marques, D. Gianolio, G. Cibin, J. Tomkinson, S. F. Parker, R. Valero, R. P. Lopes and L. A. E Batista de Carvalho, *Phys. Chem. Chem. Phys.*, 2015, **17**, 5155.

Hugh J. Byrne continued the discussion of the paper by Luca Quaroni: As was discussed, in the context of diagnostics based on tissue and other complex matrices, it is difficult and/or impossible to assign specific bands to specific biomolecules or groups of biomolecules. However, in *in vitro* studies, in which external variables such as time or dose, in for example drug or radiation exposure studies, can be precisely controlled, we can develop and explore the sensitivities and specificities of data mining methods to extract spectroscopic variations which are correlated with the external perturbation or variable. These data mining methods are best validated using simulated models of the spectral responses. Potential applications are for *in vitro* screening of novel drug efficacies and mechanisms of interaction, reducing reliance on animal models, and even patient specific screening, towards so-called "Companion Diagnostic" applications. The emergence of full spectrum nonlinear Raman techniques will enable videorate visualisation of, for example, drug uptake, receptor binding, and cellular response – label free, high content spectroscopic screening.

Bayden Wood commented: I would just like to add that it is possible to assign some maker bands in tissues and cells without the use of isotopes. For instance specific DNA conformational marker bands can be assigned by hydrating and dehydrating unfixed cells/tissues. In this way it is possible to distinguish RNA from DNA by monitoring the position of the asymmetric phosphodiester stretching band which will appear at $\sim$1220 cm^{-1} for B-DNA and $\sim$1240 cm^{-1} for RNA when the cells/tissues are hydrated. Other marker bands for B-DNA appear at $\sim$1715 cm^{-1} that are assigned to a base stacking mode, which shifts to $\sim$1710 cm^{-1} when the DNA is in the A-form following dehydration. It is also possible to determine whether there is an increase in unsaturated *versus* saturated fatty acids by looking at the intensity of the methyne stretching mode ($\sim$3100–3000 cm^{-1}). If this increases relative to the CH_3 –CH_2 stretching bands then one can conclude there has been an increase in unsaturated versus saturated fatty acids. Other approaches to assist in band assignments include changing the pH, which can assist in assigning the carboxylate groups associated with proteins and temperature studies which can help in identifying the marker bands for protein denaturation. Spiking samples with cells or infectious agents can also give insight into the spectral bands associated with the pathological condition and help to determine whether or not primary markers or secondary inflammation markers are responsible for the classification. DNA, RNA and glycogen digestion studies as performed by Professor Diem's group can also be invaluable in the identification of bands.[1] Definitive markers for collagen can also be readily identified especially the collagen triplet between 1300–1200 cm^{-1} observed in the FTIR spectra of connective tissue. In the case of malaria parasites versus control uninfected red blood cells there are a number of very distinct marker bands for hemozoin, lipids and nucleic acids. I am of the opinion that is vitally important that we as spectroscopists make every attempt to assign the bands in our spectra where possible as this takes away the black box perception of the technology, which will lead to more respectability in the clinical setting. Most diagnostic approaches in the market have known markers that the technology is built around, whether these are antibodies or specific nucleic acid sequences or molecular fragments as in the case of MALDI. More work on establishing the biochemical species that are responsible for the classification will provide more robust modelling and ultimately more respect from the clinical community for spectroscopic based diagnostics *i.e.* a bottom up approach instead of the top down approach.

1 L. Chiriboga, H. Yee and M. Diem, Infrared Spectroscopy of Human Cells and Tissue. Part VII: FT-IR Microspectroscopy of DNase- and RNase-Treated Normal, Cirrhotic, and Neoplastic Liver Tissue, *Appl. Spectrosc.*, 2000, **54**, 480–485.

Klaus Gerwert added: I fully agree with Hugh Byrne, especially Raman and maybe CARS in the future are excellent techniques for companion diagnostics. This will become very important in the future, when novel drugs will enter into the clinical application.

I fully agree with Bayden that in rare exceptions one can assign bands by hand. However, if this approach will be used in the clinics in an automated approach, we need software which identifies the marker bands, classifies the cancer and provides a label-free automated differential diagnosis. Such classification in the clinic cannot be done on individual band assignments by a spectroscopist. For individual specific questions such an approach might work in rare examples, but not in a general approach in the clinic.

Luca Quaroni responded: I agree that validating techniques for data mining is indeed indispensable and this should be done first *in vitro*, on models of increasing complexity. There is a need to define parameters that allow us to quantify how reliable our assignments are. However, there will always be the caveat that *in vitro* models will never exhaustively reproduce *in vivo* systems. One of the challenges in this respect will be developing data mining methods that can handle unknown variables. Overall, if we want to develop the capability to make reliable band assignments *in vivo*, I expect that we will still need to use additional experimental methods, such as isotopic labelling and mutagenesis, to confirm many of the assignments. The fact that a band assignment is reliably carried out *in vitro* by itself does not guarantee that the same band in a system *in vivo* will have the same assignment, at the very least because of the possibility of overlapping contributions for other molecules. I think that the assignments to DNA bands that are discussed in another response to this comment are case in point. These assignments are very convincingly done for purified DNA. However, it cannot be ruled out that *in vivo* at least part of the spectral contribution in the region where DNA bands absorb is due to other molecules. Plenty of phosphate and alcohol derivatives, plus polypeptides absorb in these regions, and these contributions would also be subject to changes from dehydration/rehydration, thus limiting the accuracy of the assignment.

Klaus Gerwert continued the discussion of the paper by Maria Paula Marques: We have taken a FTIR spectrum of *Halobacterium salinarium*. In the absorbance spectra we cannot observe any of the bacteriorhodospin bands, even though about 50% of the surface contains only the protein bacteriorhodospin. There is already an unusually large accumulation of a single protein. Only by performing light-induced FTIR difference spectra, by which only bacteriorhodopsin is specifically activated, do the individual protein bands show up. They are about three orders of magnitude smaller than the bands seen in the absorbance spectra of the whole bacterium. Therefore I do not think that individual bands of protein groups can be assigned in the absorbance spectra of tissue or cells.

Samir El-Mashtoly asked: How did you monitor the physiological cell response to the drug? Did you perform real-time cell analysis?

Maria Paula Marques replied: No real-time analysis was performed in this particular study. Instead, the drug effect was probed for a definite incubation time (48 h) with each of the anticancer agents tested, at distinct concentrations, with these conditions having been chosen in the light of previous work on the same systems (e.g. by biological assays[1] and NMR metabolomic experiments[2,3]). The results presently reported reflect the direct effect of the tested agents on the cellular biochemical profile, as well as the physiological response from the cell to the chemotherapeutic insult. These were not always clearly distinguished, variations in the bands assigned to lipids and proteins being the combined outcome of direct drug interaction and the cellular reaction to the drugs – namely through the formation of membrane vesicles for exocytosis or biochemical events following apoptotic stimuli such as protein degradation.

In order to attain a better description of the whole extent of the drug impact on the cell metabolic profile, and distinguish the drug impact from the physiological cellular response, a multidisciplinary approach is needed. This is being applied for the systems presently reported, by coupling several vibrational spectroscopic techniques (Raman, FTIR and neutron scattering) to dedicated theoretical methods, and applying EXAFS and XANES to the elucidation of the structural characteristics and stability of the drug adducts with their main pharmacological targets (e.g. DNA), as well as NMR metabolomic experiments to complement the vibrational data acquired for the same cellular systems and drugs. In particular, the effect of an anticancer drug on the structural and dynamical profiles of intracellular water (known to play a pivotal role in cellular biochemical processes) is under way by *quasi*-elastic neutron spectroscopy in human cancer cells, which is an innovative way of tackling a drug's pharmacodynamics leading to an improved understanding of its *in vivo* mode of action. Such a multidisciplinary approach will allow us to gather information on the distinct cellular features prone to be affected by a chemotherapeutic agent, from the metabolic profile and chemical content to the intracellular *millieu*, at different spatial- and timescales.

1 S. M. Fiuza, J. Holy, L. A. E Batista de Carvalho and M. P. M. Marques, *Chem. Biol. Drug Des.*, 2011, 77, 477.
2 I. F. Duarte, I. Lamego, J. Marques, M. P. M. Marques B. J. Blaise and A. M. Gil, *J. Proteome Res.*, 2010, **9**, 5877.
3 I. Lamego, I. F. Duarte, M. P. M. Marques and A. M. Gil, *J. Proteome Res.*, 2014, **13**, 6033.

William Tipping communicated: I would like to enquire if there has been any efforts in understanding the local intracellular concentration of the cisplatin-like complexes, and whether it would be possible to incorporate a bioorthogonal Raman functional group, for example a CO ligand into the design of the drugs to enable their intracellular detection?

Maria Paula Marques responded: The overall intracellular concentration of the reported complexes has been determined in preliminary experiments through mass spectrometry of drug-treated cell pellets. Nevertheless, the drug biodistribution within the cell has not been assessed to this date, although this is a relevant goal for upcoming studies. There are several ways to tackle this issue: (i) indirectly, by Raman and/or IR microspectroscopy imaging at the subcellular level, identifying the drug effect on particular cellular metabolites (*e.g.* nucleus, nucleolus, cytoplasmic proteins, membrane), that can also be isolated from the pre-treated cells in order to attain a more accurate assignment; (ii) through detection of the drug itself by mass spectrometry of the isolated cellular constituents (*e.g.* proteins, DNA, lipids) after drug exposure (for distinct concentrations and incubation times); (iii) using the recently reported correlated cellular IR spectromicroscopy (CSM) (see L. Quaroni's paper, DOI: 10.1039/c5fd00156k) that allows time- and spatially-resolved measurements of cells to identify small molecules, which would be ideal for detecting chemotherapeutic agents in the cellular environment if (or when) the technique is extended to Raman microspectroscopy and imaging.

Regarding the incorporation of a Raman marker group (such as a bioorthogonal CO) into the complexes, this is chemically possible. I can understand your point, aiming at a better drug localisation which is certainly very important. However, we would obtain a different agent displaying an altered reactivity relative to the parent cisplatin-like complexes presently reported – both the intracellular hydrolysis and interaction with the drug targets (*e.g.* binding to DNA) would certainly be changed, and hence its biological activity (namely its cytotoxicity) could be completely different. This is, in fact, the procedure followed to optimise a lead drug in order to obtain a distinct agent with improved effectiveness and decreased deleterious effects (toxicity or acquired resistance).

Hugh J. Byrne addressed Maria Paula Marques, Luca Quaroni and Achim Kohler: In the development of *in vitro* screening methods, it is important to build confidence in the data mining methods and the interpretation of the spectroscopic fingerprints which are revealed. The possibility of correlation with external variables is an advantage, and the use of simulated datasets in which systematic perturbations have been introduced can help in this regard. PLSR and cross correlation techniques are data mining possibilities. Does the panel have any feeling for how they compare? Which techniques are the best?

Maria Paula Marques answered: In terms of the work we perform, PLSR could potentially be used for data mining, with a view to determine the impact of different drugs on the cells. The scores and loadings plots obtained could then be used to identify what aspects of the spectra differ for each system probed – different drugs (namely mono- *vs.* polynuclear or Pt *vs.* Pd), concentrations, exposure times and even distinct cancer cell lines (*e.g.* ref. 1). This is a procedure we aim to apply in the near future, namely on recently acquired FTIR imaging data for cancer cells in the presence of several cisplatin-like agents (distinct drug types and concentrations). A requirement for this procedure is to synchronise the cells regarding cell cycle, since the biochemical changes associated with cell cycle progression may be greater than those due to the drug impact which would mask the metabolic variations we aim to detect. Also, for infrared data the spectra must be Mie corrected, to ensure that we are identifying variations in the cell biochemical profile and not changes in morphology (due to differing amounts of associated scattering).

1 M. Jimenez-Hernandez, M. D. Drown, C. Hughes, N. W. Clarke and P. Gardner, *Analyst*, 2015, **140**, 4453.

Luca Quaroni replied: I am not aware of a systematic and quantitative comparison of PLSR and 2D-COS, at least for the type of problems addressed in my work. A qualitative comparison will be made here, addressing a few points. Other delegates might want to add more. One reason why I personally have so far focused on the use of 2D-COS for the analysis of cellular spectra is the fact that the asynchronous plots in 2D-COS (*i.e.* the imaginary part of the 2D Fourier Transform of the correlation function between spectra) contain information about the sequence of events in a reaction mixture. These properties, although not discussed in the paper presented for this meeting, can help to identify unknown processes or cause and effect relationships in the cellular environment, without having to build accurate kinetic spectral traces. This analysis can be carried out without any previous knowledge of the specific molecules involved, although some information about the general form or the evolution (*e.g.* sinusoidal versus monotonic) is required.

Another advantage for using 2D-COS for work on a cellular system is that the analysis is not affected by incomplete information on some variables. Even if I do not know anything about the system under study, I can still calculate 2D-COS plots and identify band multiplets for specific molecules or sets of molecules. These can then be assigned to specific molecules if sufficient bands are available, or remain as unidentified molecules but without affecting the rest of the analysis. In contrast, for PLSR it is important to know at least some if not most of the X-variables, to build a reliable model. Too many unknowns make the analysis unreliable. This can be a real challenge when working with cellular samples, where identifying *a priori* all the contributions to the spectrum can be a challenge.

An advantage of PLSR is that it provides, as a direct result, a plot of spectral contributions as a function of time or of a perturbation. This can provide quantitative information on reaction rates in the system under study.

One feature which is shared by both PLSR and 2D-COS is the capability to handle noisy data sets. Both of them can help identify weak components in noisy cellular spectra, something which is critical if we want to analyse metabolic processes. I am not aware of a quantitative comparison between the two approaches. It would be a very useful project. Finally I would like to conclude that it may be worth looking into the sequential use of 2D-COS analysis and PLSR to study cellular processes. 2D-COS can rapidly identify all band multiplets that contribute to spectral changes, as function of time or of a perturbation. The outcome of this analysis can then be used to build a complete set of X-variables for a PLSR model. PLSR can then be used to extract reaction rates for all the identified components.

Achim Kohler responded: Cross-validation is a validation technique that can be used for any multivariate prediction or classification method. It is often used in the context of PLSR, but is not restricted to PLSR. In cross-validation, segments of the sample set are alternately kept out as test sets. The advantage is that the complete data set can be used for both model building and testing. Cross-validation has to be applied carefully using the following two requirements: (1) One has to be sure that all the segments that are kept out are independent. For example if independent experiments were performed to obtain independent test sets, then every experiment has to be taken out as one segment in a cross-validation procedure. (2) The second comment concerns the fact that cross-validation is often used to optimize parameters in the calibration or classification model. In this case, it is important to be aware that cross-validation does not represent an independent testing, since the cross-validation segments are used to adjust parameters. The degree of the potential overfitting depends on the flexibility of the calibration or classification model. Therefore, it is in general recommended, to use an independent test set after parameter estimation by cross-validation. This can be done by cross-model validation as for example shown for sparse PLSR in ref. 1.

1 I. Karaman, N. P. Nørskov, C. C. Yde, M. S. Hedemann, K. E. B. Knudsen and A. Kohler, Sparse multi-block PLSR for biomarker discovery when integrating data from LC-MS and NMR metabolomics, 2015, *Metabolomics*, **11**(2), 367–379.

David Perez-Guaita continued the discussion of Achim Kohler's paper: You mention that we can only do quantifications with one strain of bacteria. I disagree with that because that depends on the complexity of the system. Go to simplest model, serum. It has been studied by ATR during the last 30 years, it has been established so that you can predict the main components but anything below 0.1% cannot be predicted.

Achim Kohler replied: The comment which I made and which you refer to is the following: Calibration models for microbial cellular compounds have to be validated very carefully with respect to the application range of the obtained models. We have established calibration models for fatty acids in yeasts and moulds which we use to optimize biorefinery processes (see for example ref. 1 and 2). In these studies, we predict PUFA, MUFA and SAT. We observe that when these models are validated on completely new strains that are grown under similar conditions, the prediction models fail. Still, these models are useful for the development of biorefinery processes, since the amount of strains used is often very limited and the number of different conditions for example harvesting time points *etc.* is high. Thus, we can afford to include all relevant strains in our calibration model for the applications we consider. We believe that the reason why the models fail when applied to new strains is that new strains introduce completely new variability through a whole range of new metabolites. If one wants to deal with this variability, one has to increase the size of the dataset used for establishing the model considerably. This is in general possible, but then the question is if it makes sense to replace conventional wet-chemical analysis by vibrational spectroscopy. These are the same problems we do not account for when working with more homogenous materials such as milk. In this case the estimation of SAT, PUFA and MUFA works very well. See for example ref. 3.

1 V. Shapaval, N. K. Afseth and A. Kohler, Fourier transform infrared spectroscopy for the prediction of fatty acid profiles in Mucor fungi grown in media with different carbon sources, *Microb. Cell Fact.*, 2014, **13**, 86
2 A. Kohler, U. Böcker, V. Shapaval, A. Forsmark, M. Andersson, J. Warringer, H. Martens, S. W. Omholt and A. Blomberg, High-Throughput Biochemical Fingerprinting of Saccharomyces cerevisiae by Fourier Transform Infrared Spectroscopy, *PLoS ONE*, 2015, **10**(2), e0118052.

3 N. K. Afseth, H. Martens, L. Giskehaug, B. Narum, K. Jørgensen, S. Lien, A. Haug, A. Kohler, Predicting fatty acid composition of milk – A comparison of two FTIR sampling techniques, *Appl. Spectrosc.*, 2010, **64**(7), 700–707.

Achim Kohler continued the discussion of the paper by Luca Quaroni: Is the reason for using second derivatives that you remove fringes in your spectra?

Luca Quaroni responded: The main reason for the use of second derivatives is the need to separate the spectral contribution due to absorption changes from the contribution due to changes in the baseline. Baseline changes are common in IR absorption measurements of live cells. Cells move and change shape during the measurement and the resulting changes in the refraction and scattering of the beam can cause major changes in the spectral baseline. However, most of the latter changes are approximated by polynomial functions. When calculating the second derivative of the spectrum, the relative contribution of polynomial and sinusoidal functions is drastically decreased or cancelled relative to that of the Lorentzian/Gaussian functions that describe absorption bands. Therefore most baseline changes practically disappear in the second derivative spectrum. Failure to remove baseline variations would make the 2D-COS spectra difficult or impossible to analyze, since these variations would give broad correlation bands with all other spectral changes. We did not observe appreciable sinusoidal fringes in our spectra, mostly because we use CaF_2 windows, the refractive index of which is close to that of the aqueous medium and of the cells. However, if they were present, their contribution would also be reduced by the use of second derivative spectra.

Gavin Lloyd asked Achim Kohler: Is the real power of the EMSC algorithm to not to completely remove an artefact, but to neutralise its effect? A true reference (e.g. mean spectrum of sample) may not be artefact free, but EMSC can still effectively neutralise its effect and reduce the impact on subsequent analysis. Is it therefore necessary/appropriate to completely remove an artefact? What do we gain by correcting spectra *e.g.* to a reference that is from a phantom, other than peak assignments that are tentative at best anyway?

Achim Kohler replied: An estimation of scatter contributions and their removal in the infrared spectroscopy of cells has two aims. (1) To neutralize its effects for the subsequent multivariate data analysis. (2) To improve spectral interpretation. It is correct that for the first purpose it is not strictly necessary to use a scatter free spectrum as reference spectrum in EMSC. If a not scatter free spectrum is used as a reference spectrum, the EMSC correction adjusts every measured spectrum to the reference spectrum such that it contains approximately the same scatter contribution as the reference spectrum. When spectra are mean-centred after EMSC correction for example for subsequent PCA and PLSR analysis, the use of a not scatter free reference spectrum has no impact on the subsequent analysis. This is demonstrated in ref. 1. However, if peak heights and peak ratios are investigated, the use of a not scatter free spectrum as a reference spectrum for the EMSC analysis might have an effect.

1 N. K. Afseth and A. Kohler, Extended multiplicative signal correction in vibrational spectroscopy, a tutorial, *Chemom. Intell. Lab. Syst.*, 2012, **117**, 92–99.

Peter Gardner addressed everyone: As has already been said the scattering contribution is not an artefact. However it must be separated from the absorption spectrum because the morphological effects on the spectrum are large compared with the more subtle biochemical effects. If you do not remove the scattering the vast majority of the variance in your data will be due to the difference in size and shape of the cells. This does not mean that this information is not useful. We have shown for example that you can classify cells in different phases of the cell cycle reasonably well just using the scattering part of the spectrum that is usually discarded.[1]

1 M. Jimenez-Hernandez, C. Hughes, P. Bassan, F. Ball, M. D. Brown, N. W. Clarke, P. Gardner, Exploring the spectroscopic differences of Caki-2 cells progressing through the cell cycle while proliferating in-vitro, *Analyst*, 2013, **138**(14), 3957–3966.

Hugh J. Byrne responded: It should be noted that, even in a perfectly homogeneous, chemically and physically, thin film, there are contributions of the resonantly enhanced real refractive index to the surface reflection.

This was demonstrated in ref. 1. These are very apparent as transflection for strongly absorbing samples, but will contribute in all cases, also to transmission measurement.

These come from the same physical phenomenon as the scattering, but I presume in the case of films are best described by a simple reflection theory, whereas with a rougher the surface, or a more inhomogeneous sample, they are better described by scattering theory. In, for example paraffin embedded samples, this contribution may be more dominant than the scattering.

1 P. Bassan, H. J. Byrne, J. Lee, F. Bonnier, C. Clarke, P. Dumas, E. Gazi, M. D. Brown, N. W. Clarke and P. Gardner, Reflection contributions to the dispersion artefact in FTIR spectra of single biological cells, *Analyst*, 2009, **134**, 1171–1175

Max Diem commented: The goal of the correction should be to reverse as much of the R-Mie scattering induced frequency shifts of the bands, since they affect classification algorithms profoundly.

Peter Gardner responded: I agree completely. The intensity of a band is not so much of a problem but band shifts are and they have to be removed.

Bayden Wood asked: The size and morphology of the nucleus will have its own physical infrared signature and potentially be diagnostically useful for cancer detection as an example. By not correcting the spectra for scattering are we potentially throwing out information that could improve the diagnostic capability?

What are your thoughts on this?

Achim Kohler replied: Morphological parameters carry often important information and are thus useful for diagnosis and discrimination. While chemical variation in first approximation introduces linear (Lambert Beer type) responses in the spectral data, the morphological information contributes to non-Lambert Beer type responses in the spectral data. Therefore, the separation of morphological parameters from chemical information may improve classification models and regression models considerably. EMSC allows the separation of morphological parameters to use them either separately or in combination with the pure absorbance spectra. For examples see ref. 1 and 2.

EMSC parameters may also be added as additional columns to the pure absorbance spectra prior to building the regression or classification model.

1 A. Kohler, C. Kirschner, A. Oust and H. Martens, Extended multiplicative signal correction as a tool for separation and characterisation of physical and chemical information in Fourier transform infrared microscopy images of cryo-sections of beef loin, *Appl. Spectrosc.*, 2005, **59**(6), 707–716.

2 B. Zimmermann, Z. Tkalčec, A. Mešić and A. Kohler, Characterizing Aeroallergens by Infrared Spectroscopy of Fungal Spores and Pollen, *PloS One*, 2015, **10**(4), e0124240.

Max Diem added: The correction for dispersive effects affects the entire spectrum. All peaks are shifted due to dispersive components, and the easiest way to correct for all these shifts is by a phase correction method that unmixes the real and imaginary spectral components.

Samir El-Mashtoly commented: Peter mentioned that the morphological changes in cells are huge while the change in chemistry is subtle upon for example treating cells with a drug.

Morphological changes are huge and the changes in chemistry are also huge but the main problem is the detection limit of vibrational spectroscopy. This is can be seen clearly when we compare our vibrational spectroscopic results with the results of other techniques for instance western blot or fluorescence where large changes are shown.

Peter Gardner replied: Samir is right that both the morphological changes and chemical changes can be large when cells are treated with a drug. The more subtle differences are in the disease states of tissue. The chemical changes are small and the resultant spectral changes may be even smaller. As Klaus Gerwert mentioned earlier in the meeting you may have several hundred proteins up regulated and several hundred down regulated and you only see the net difference in the spectra.

Faraday Discussions

PAPER

Non-invasive chemically specific measurement of subsurface temperature in biological tissues using surface-enhanced spatially offset Raman spectroscopy

Benjamin Gardner,[a] Nicholas Stone*[a] and Pavel Matousek*[b]

Received 5th November 2015, Accepted 2nd December 2015
DOI: 10.1039/c5fd00154d

Here we demonstrate for the first time the viability of characterising non-invasively the subsurface temperature of SERS nanoparticles embedded within biological tissues using spatially offset Raman spectroscopy (SORS). The proposed analytical method (T-SESORS) is applicable in general to diffusely scattering (turbid) media and features high sensitivity and high chemical selectivity. The method relies on monitoring the Stokes and anti-Stokes bands of SERS nanoparticles in depth using SORS. The approach has been conceptually demonstrated using a SORS variant, transmission Raman spectroscopy (TRS), by measuring subsurface temperatures within a slab of porcine tissue (5 mm thick). Root-mean-square errors (RMSEs) of 0.20 °C were achieved when measuring temperatures over ranges between 25 and 44 °C. This unique capability complements the array of existing, predominantly surface-based, temperature monitoring techniques. It expands on a previously demonstrated SORS temperature monitoring capability by adding extra sensitivity stemming from SERS to low concentration analytes. The technique paves the way for a wide range of applications including subsurface, chemical-specific, non-invasive temperature analysis within turbid translucent media including: the human body, subsurface monitoring of chemical (*e.g.* catalytic) processes in manufacture quality and process control and research. Additionally, the method opens prospects for control of thermal treatment of cancer *in vivo* with direct non-invasive feedback on the temperature of mediating plasmonic nanoparticles.

Introduction

The measurement of subsurface temperature in turbid media is a topical area in analytical sciences with potential applications ranging from monitoring

[a]*Biomedical Physics, School of Physics, College of Engineering, Mathematics and Physical Sciences, University of Exeter, Exeter, UK. E-mail: N.Stone@exeter.ac.uk*

[b]*Central Laser Facility, Research Complex at Harwell, STFC Rutherford Appleton Laboratory, Harwell, Oxford, OX11 0QX, UK. E-mail: pavel.matousek@stfc.ac.uk*

subsurface temperatures in the human body *in vivo* during hyperthermia treatments for cancer, to monitoring chemical and materials processes during production and storage. Current mainstream methods, such as contact thermometers and infrared probes, are confined to measuring the surface temperature in non-IR transparent or diffusely scattering media.[1] There are only a handful of methods capable of measuring temperature at depth in turbid media. Microwave-based temperature sensing can penetrate deep inside turbid media; although these sensors have limited spatial resolution due to the long wavelength of radiation used, and do not provide high selectivity to chemical subcomponents.[2] Another approach to deep temperature measurement is Magnetic Resonance Imaging (MRI) thermometry.[3] However, as this is reliant upon temperature-dependant water proton resonance frequency shifts, this mechanism is not always applicable. There are other temperature-dependent properties that can be probed with MRI, such as T_1 (spin–lattice) relaxation time and T_2 (spin–spin) relaxation time of water molecules. Both of these approaches are complex to accurately calibrate, and are limited to specific tissue types.[3] In addition, MRI methods are often prohibitively costly for many practical applications.

Recently we have demonstrated a new approach[4] combining the conventional anti-Stokes to Stokes temperature monitoring capability of Raman spectroscopy with the subsurface capability of Spatially Offset Raman Spectroscopy (SORS)[5,6] – Temperature SORS (T-SORS). The concept is based on the SORS capability to non-invasively analyse the chemical composition of stratified turbid media. It overcomes the limitation of conventional Raman spectroscopy near surfaces within diffusely scattering samples.

In general, SORS itself has paved the way for a host of new applications including the scanning of liquids in sealed bottles in aviation security, the monitoring of final products and incoming raw materials in pharmaceutical quality control, and also for non-invasive breast cancer and bone disease diagnosis.[7,8] The principle of T-SORS relies on the measurement of both the Stokes and anti-Stokes scattering spectrum of subsurface components using SORS. The temperature is derived from the ratio of the intensities of counterpart Stokes and anti-Stokes lines.[9,10] The ratio is temperature-dependent due to the fact that anti-Stokes Raman bands are solely due to vibrationally excited molecules, whereas the Stokes lines derive their intensity from molecules both in the ground as well as vibrationally excited states.[4] In the first proof-of-concept study,[4] the temperature of a turbid sublayer of polytetrafluoroethylene (PTFE) was determined through a highly diffusely scattering overlayer of polyoxymethylene POM (3 mm thick) using the intensity ratios of the anti-Stokes and Stokes bands of sublayer (PTFE) recovered non-invasively using SORS measurements. Using calibration data set obtained *a priori*, it was possible to predict the sublayer temperature with an accuracy of +/−0.16 °C over a clinically relevant range of 36–39 °C. The laser excitation wavelength was 830 nm and a standard Raman detection system was used.

Here we expand on the capability of the concept into another related area by demonstrating its viability in monitoring embedded SERS nanoparticles using SORS in a modality termed – T-SESORS. In a similar manner to its earlier counterpart, the T-SESORS approach relies on measuring the anti-Stokes to Stokes line of molecules attached or adjacent to SERS nanoparticles (Fig. 1) dispersed in a sample. This also provides a unique sensing capability enabling probing of the

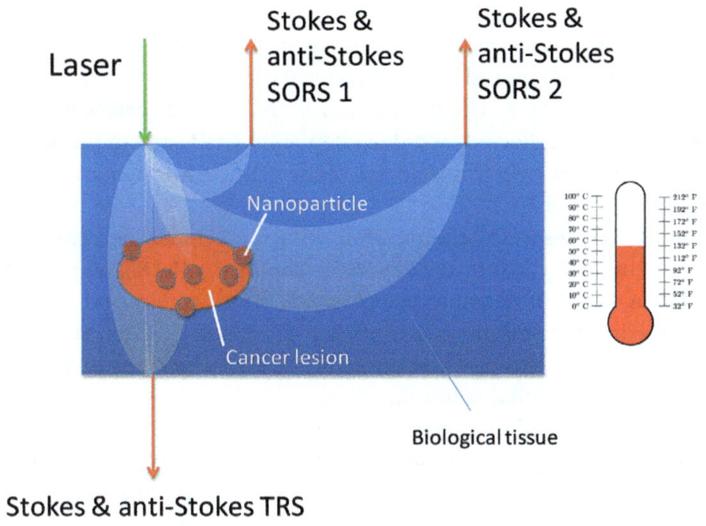

Fig. 1 Schematic diagram of the T-SESORS thermometry concept. Both the SORS and TRS geometries permit subsurface monitoring of sample temperature mediated by SERS nanoparticles embedded in turbid medium. The temperature of nanoparticles can also be elevated above background levels and monitored using laser radiation in plasmonic photothermal therapy for the most optimum treatment.

local environment of these particles rather than the bulk subsurface environment derivable from conventional T-SORS. This feature can be beneficial, for example, in monitoring the temperature of cells adjacent to SERS nanoparticles with high sensitivity stemming from the large enhancement of the SERS effect one typically encounters. As such this approach could also be combined with plasmonic photothermal therapy,[11] where nanoparticles labelled with antibodies are selectively attached to cancer cells. These then could be used both to absorb laser radiation to elevate their local temperature and destroy surrounding cancer cells and at the same time to report back on their local environment temperature *via* the T-SESORS mechanism. This would permit the optimisation of treatment for the most effective cancer eradication whilst minimising collateral damage to healthy cells further away.

Experimental

The experiments were carried out using a home-built instrument based around a transmission Raman geometry.[12,13] Transmission Raman spectroscopy represents a variant of Spatially Offset Raman Spectroscopy (SORS), with the spatial offset present in TRS being at the extreme – *i.e.* the illumination zone is on the other side of the sample to the collection zone. Both the geometries achieve subsurface monitoring capability in a comparable manner; with TRS the sensing is biased towards the centre of sample by default, whereas with SORS the exact choice of offset defines the penetration depth. With either of the approaches, irrespective of their exact geometry, the detection of distinct subsurface chemical signatures within the sample is the key prerequisite and common feature enabling the unique identification of the temperature of a subsurface entity. The basic setup was described previously.[14] The excitation wavelength was 830 nm and was delivered by a spectrum-stabilised laser (Innovative Photonic Solutions: I0830MM0350MF-EM) passed through two 830 nm band pass filters (Semrock) to clean up the beam profile. The laser power at the sample surface was ~400 mW. Raman spectra were collected on a deep depletion CCD camera (Andor iDus-420) coupled to a Kaiser spectrometer (Holospec 1.8i). All Raman spectra were collected for a total of 5 s.

SERS nanoparticles were produced following a previously described method.[15] 1 mL of 0.05 mg mL^{-1} nanoparticles (nanoComposix) (60 & 100 nm diameters) and 100 μL of 1 mM 4-Mercaptobenzoic acid (MBA) were mixed for 5 minutes. Following mixing the sample was centrifuged at 3000 rpm for 10 minutes, the supernatant was removed and the nanoparticles resuspended in 1 mL of phosphate-buffered saline (PBS) solution.

The sample consisted of a quartz cell containing the SERS nanoparticles surrounded by a layer of porcine tissue. The total thickness of the sample was ~10 mm. The sample was kept in a large quartz cell filled with water, which was thermostatically controlled by a Thermo Scientific (Haake C10 water bath) coupled to a copper water block.

The experiments were carried out between 25 and 42 °C, with the temperature increased in increments of ~3.5 °C. The temperature was allowed to equilibrate for 40 minutes prior to spectral acquisitions. At each temperature the sample was measured for 40 minutes, generating 480 spectra.

The temperature was measured from the sample in the quartz cell by two type K thermocouples connected to a Pico – TC-08 Thermocouple Data Logger. The temperature was measured every second during measurements and the average temperature over the two thermocouples was used.

The data was analysed using Matlab 2014a, both by fitting a Gaussian curve to either of the chosen SERS bands at 1085 and 1580 cm^{-1} and when using the partial least-squares regression (PLS) approach. The PLS models were created from every other data point, and validated with the second half (test set), except when otherwise stated.

Results and discussion

In Surface Enhanced Spatially Offset Raman Spectroscopy (SESORS) measurements[16,17] Raman spectra are collected from the sample surface at locations that are separated from the laser illumination point (see Fig. 1). The larger the separation (spatial offset Δs), the larger the contrast between the sublayers and surface layers. Two or more spectra would typically be collected and processed using scaled subtraction, or a larger number acquired and processed using multivariate analysis to recover the estimates of pure spectra of individual layers. In the transmission configuration, bulk information within the probe volume can be obtained and the subsurface temperature from any subsurface regions with distinct Raman signatures recovered in a single measurement. Fig. 2 presents a schematic diagram of the experimental setup for the temperature measurements used here, in which the sample is probed in transmission geometry.

The general viability of using SERS as a temperature reporter has been established earlier.[18] In a first set of experiments we have confirmed that the nanoparticles selected for non-invasive studies here are also viable for monitoring

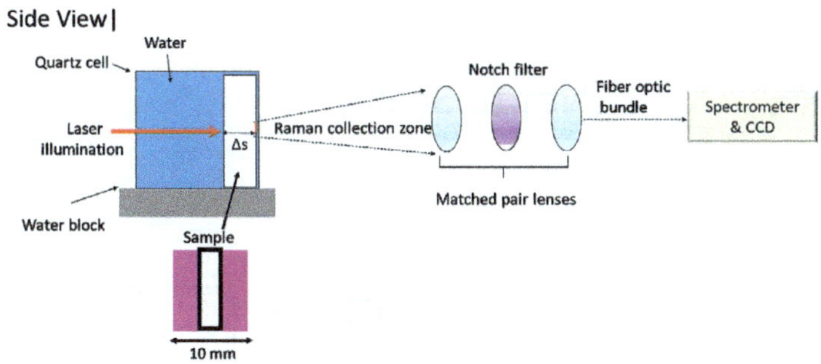

Fig. 2 Schematic of the experimental setup. The 830 nm laser beam is passed through the sample in transmission geometry. The sample consists of two layers of porcine meat, one on either side (wrapped around) of a quartz cell containing SERS nanoparticles. The Raman photons are collected by a 50 mm diameter $f = 60$ mm lens, filtered by a Kaiser notch filter (830 nm), and focussed with a 50 mm diameter $f = 60$ mm lens onto a fibre bundle (Ceramoptec). The circular cross-section fibre bundle is rearranged into a line of fibres matching entry into a Kaiser Holospec 1.8i spectrometer with a low dispersion transmission grating and an Andor iDus 420 deep depletion CCD.

temperature using the Stokes and anti-Stokes signals (Fig. 3). A solution with dispersed nanoparticles was placed in a plastic tube and measured directly in transmission across a temperature range of 25–42 °C. It should be noted that the ratio of anti-Stokes to Stokes Raman bands, in general, is not a linear function of temperature as the dependence involves an exponential function.[4] However, across a relatively narrow range of temperatures as used here the linear approximation of the dependence is entirely adequate, as demonstrated. (If a wider range of temperatures was monitored where nonlinear dependence was revealed in the calibration stage, then a non-linear univariate or multivariate analysis could be invoked and used to fit the calibration data and then predict unknown temperatures).

Due to differences in the SERS enhancement as a function of nanoparticle diameter (as expected) there are also differences in the anti-Stokes/Stokes ratio of the chosen Raman band at 1580 cm^{-1} (Fig. 3B) at the same temperature. Through using the univariate analysis approach, *i.e.* monitoring the ratio between the anti-Stokes/Stokes ratio, it was possible to create a model and predict the temperature with the test set. As is shown in Fig. 3C, with the 100 nm nanoparticles an RMSE of 0.43 °C was achieved, and a similar value was also achieved with the 60 nm particles (data no shown).

To further explore the limits of the technique and reach higher accuracies we deployed a multivariate analysis using the entire spectral range measured in both the Stokes and anti-Stokes spectral regions. A PLS model has been constructed to predict temperature from a calibration data set (Fig. 3D). The model yielded an RMSE of 0.21 °C, about twice as accurate as the basic univariate approach above.

The concept of T-SESORS has been demonstrated initially on a quartz vial containing a solution with nanoparticles dispersed within it, concealed behind a total of 5 mm of porcine tissue (~2.5 mm on each side). The Raman spectra of the nanoparticles and the signal when buried in porcine tissue are shown in Fig. 4.

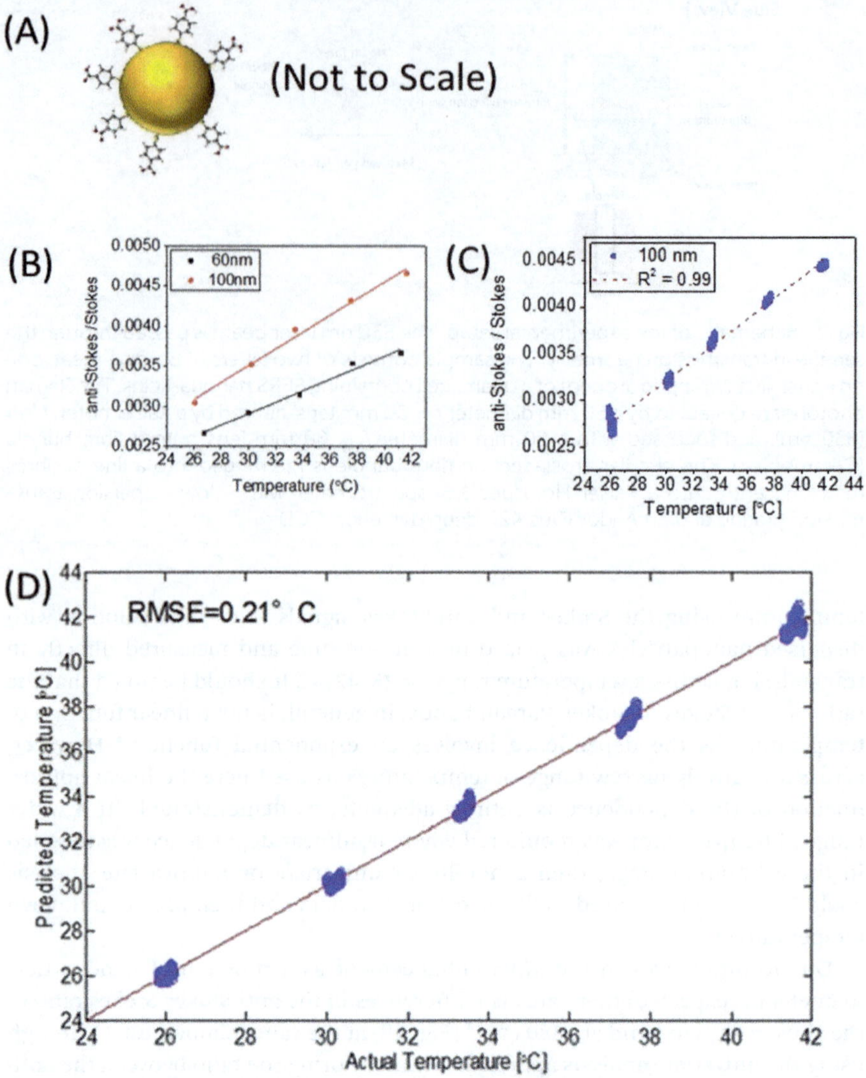

Fig. 3 Testing of the T-SERS concept using non-buried SERS nanoparticles. (A) Illustration of the thiol attachment of the chosen SERS reporter to the gold nanoparticles. (B) Anti-Stokes/Stokes ratio of a chosen Raman band (1580 cm^{-1}) for two nanoparticle sizes (60 & 100 nm). (C) Univariate model for predicting temperature for 100 nm SERS nanoparticles based upon the anti-Stokes/Stokes ratio of the 1580 cm^{-1} peak. (D) Performance of PLS model for predicting the temperature of SERS nanoparticles.

The spectra were measured at different temperatures in a clinically relevant range of 25–42 °C. The data were split into calibration and prediction data sets; both contained half the measured data.

Again a univariate analysis was performed on a SERS band (~1085 cm^{-1}), the average of which is shown for the anti-Stokes and Stokes intensities respectively for each measured temperature (Fig. 5a). A calibration curve was produced (Fig. 5b), which achieved a prediction error (RMSE) of 1.4 °C (Fig. 5c). An unusual

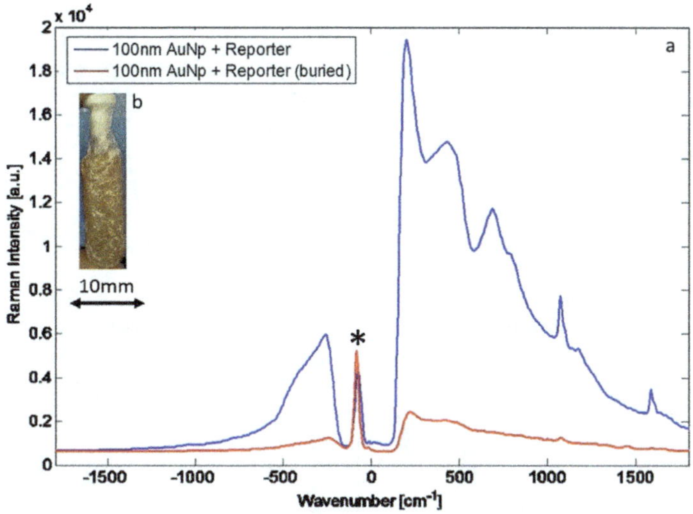

Fig. 4 Anti-Stokes and Stokes Raman spectra of gold nanoparticles alone + buried in porcine tissue. (a) The Raman spectra of the SERS signal when measured in a quartz cell (blue), and when the cell is buried in porcine tissue (red). (b) The quartz cell wrapped with porcine tissue. An artefact due to the elastically scattered and unfiltered laser line is labelled with an asterisk.

phenomenon was observed at the Stokes Raman band (Fig. 5a – right panel), which decreased in intensity with increasing temperature. This phenomenon was not observed for SERS nanoparticles alone, suggesting that the absorption or scattering properties of surrounding tissue change with temperature and cause this trend (it should be noted that this had no impact on the accuracy or performance in general of both the univariate and multivariate methods, as the effect is accounted for in the calibration stages with both the approaches).

As above, we have also applied a multivariate analysis to the problem to derive a higher predictive accuracy. In constructing a PLS model it is crucial to select the optimal number of components; too few and the model could under-fit, while too many can lead to over-fitting, either of which can lead to a poor predictive ability with future samples. The full spectral range was used in a PLS model to create a model and predict unknown temperatures, the results of which are shown in Fig. 6. Here a leave-every-other-replicate-out PLS model was constructed from the experimental data. The number of PLS components *versus* the percentage of explained variance (Fig. 6a) shows that most of the variance is explained within the first 5 components. The model constructed using PLS (Fig. 6b) showed an improved R^2 in comparison to that achieved through the univariate approach (0.99 *vs.* 0.92). More importantly though, there was an improvement in the prediction power with a decrease in the RMSE from 1.4 °C to 0.20 °C.

To further explore the limits of T-SESORS, a thicker layer of porcine tissue was introduced; 10 mm thickness in total, with the cell buried in the middle. By doubling the thickness of the tissue layer there was a small increase in the level of error (Fig. 7), with a small increase in the RMSE from 0.20 to 0.23 °C in the PLS model.

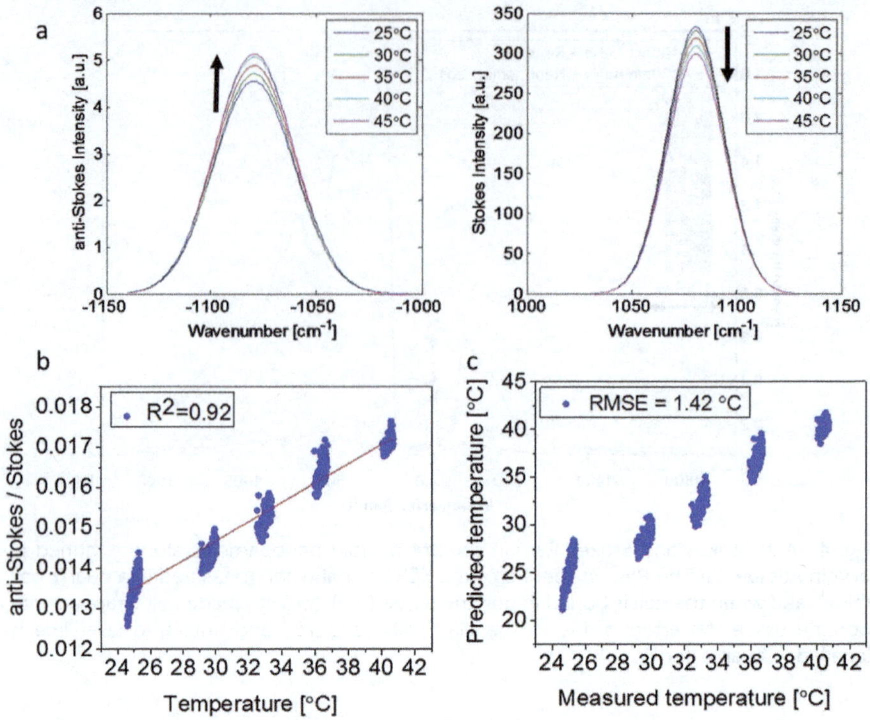

Fig. 5 Univariate temperature monitoring of buried gold nanoparticles. (a) The average anti-Stokes & Stokes bands of the SERS Raman reporter (~1080 cm^{-1}) at each temperature point. (b) Calibration curve of the SERS reporter. (c) Correlation between the measured temperature and the predicted temperature on a prediction (test) sample.

It should be noted that further optimisation of the technique is envisaged by improving data pre-processing and experimental apparatus. The results presented here represent only the first conceptual demonstration of T-SESORS.

An important practical aspect of these measurements is how to derive appropriate calibration T-SESORS data at different temperatures to enable the prediction of unknown temperatures from T-SESORS spectra. In situations where invasive access to the sample can be facilitated, *e.g.* in some industrial process control situations, this is relatively straightforward and the subsurface calibration temperature can easily be ascertained. In many situations, however, this may not be possible. In such cases an alternative option exists consisting of acquiring calibration spectra from a sample that is in full thermal equilibrium with its environment, *i.e.* ensuring any thermal gradients are absent during the calibration measurements. In such a situation the subsurface temperature can be assumed to be the same as that of the sample surface and simply acquired from a surface measurement. By altering the temperature of the ambient environment and letting the sample equilibrate fully again, additional T-SORS spectra at different temperatures can be further acquired if desired, or necessary. In most situations where it is known how a specific anti-Stokes Raman spectral region responds to temperature (response curve slope), only one calibration data point (temperature) may be required to compensate for any absorption within the

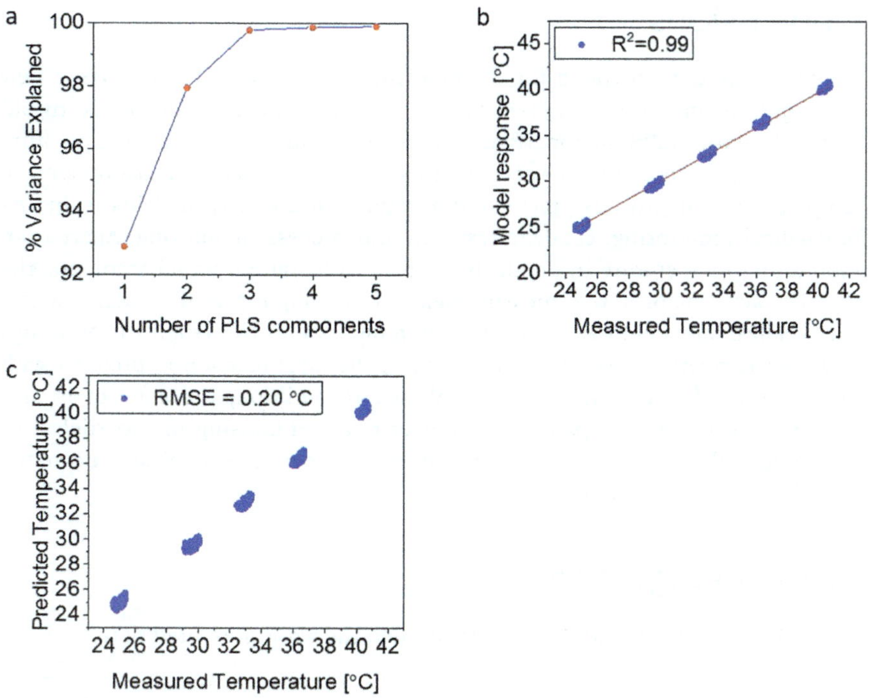

Fig. 6 PLS model of heating buried nanoparticles (25–42 °C). (a) The percentage of variance explained for the model data, as a function of the number of components used in the PLS model of nanoparticles being heated in a layer of porcine tissue. (b) The fit response *vs.* observed response for the PLS model (calibration data set). (c) Predicted temperatures *vs.* the measured temperature for nanoparticles in porcine tissue for the final test set of data (prediction data set).

sample that could distort true anti-Stokes to Stokes band ratios in a specific sample and experimental geometry. In the biomedical field this could be the normal (equilibrated) body temperature.

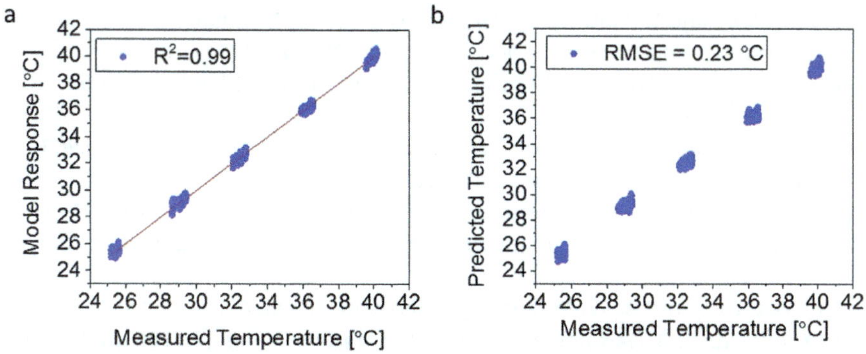

Fig. 7 PLS model of heating nanoparticles buried in 10 mm porcine tissue (25–42 °C). (a) The fit of the model response and the measured temperature. (b) Predicted temperatures *vs.* the measured temperature of the gold nanoparticles.

Conclusions

Here we have demonstrated for the first time the viability of T-SESORS, a new technique for monitoring non-invasively the temperature of subsurface turbid materials with SERS nanoparticles embedded within them as temperature reporters. The development of a method for non-invasive measurement of temperature at depth has practical applications in a variety of fields including biomedical monitoring, catalytic research and process monitoring. Apart from monitoring temperature in depth in samples with static thermal gradients, the method also enables the measurement of non-equilibrated systems where chemically distinct entities are at different temperatures from their surroundings and their temperature evolves in time (*e.g.* a chemical reaction in progress with direct monitoring of reagents or catalysts). In addition, the concept is also potentially useable in hyperthermia cancer treatments using nanoparticles, by providing a direct means of monitoring the temperature of plasmonic nanoparticles by non-invasive means.

Acknowledgements

An EPSRC grant (EP/K020374/1) partly funded the work presented here.

References

1 P. R. N. Childs, J. R. Greenwood and C. A. Long, *Rev. Sci. Instrum.*, 2000, **71**, 2959–2978.
2 A. Levick, D. Land and J. Hand, *Meas. Sci. Technol.*, 2011, **22**, 065801.
3 K. Hynynen, N. McDannold, R. V. Mulkern and F. A. Jolesz, *Magn. Reson. Med.*, 2000, **43**, 901–904.
4 B. Gardner, P. Matousek and N. Stone, *Anal. Chem.*, 2016, **88**, 832–837.
5 P. Matousek, I. P. Clark, E. R. C. Draper, M. D. Morris, A. E. Goodship, N. Everall, M. Towrie, W. F. Finney and A. W. Parker, *Appl. Spectrosc.*, 2005, **59**, 393–400.
6 P. Matousek, M. D. Morris, N. Everall, I. P. Clark, M. Towrie, E. Draper, A. Goodship and A. W. Parker, *Appl. Spectrosc.*, 2005, **59**, 1485–1492.
7 K. Buckley and P. Matousek, *Analyst*, 2011, **136**, 3039–3050.
8 P. Matousek and N. Stone, *Chem. Soc. Rev.*, 2016, **45**, 1794–1802.
9 H. Baokun, T. Yanjie, L. Zuowei, G. Shuquin and L. Zhaokai, *Instrum. Exp. Tech.*, 2007, **50**, 282–285.
10 E. Smith and G. Dent, *Modern Raman Spectroscopy – A Practical Approach*, John Wiley & Sons Ltd, Chichester, 2005, ch. 6.
11 X. Huanga and M. A. El-Sayed, *Alexandria Journal of Medicine*, 2011, **47**, 1–9.
12 B. Schrader and G. Bergmann, *Fresenius' Zeitschrift für Analytische Chemie*, 1967, **225**, 230.
13 P. Matousek and A. W. Parker, *Appl. Spectrosc.*, 2006, **60**, 1353–1357.
14 M. Z. Vardaki, B. Gardner, N. Stone and P. Matousek, *Analyst*, 2015, **140**, 5112–5119.
15 A. Jaworska, L. E. Jamieson, K. Malek, C. J. Campbell, J. Choo, S. Chlopicki and M. Baranska, *Analyst*, 2015, **140**, 2321–2329.

16 N. Stone, K. Faulds, D. Graham and P. Matousek, *Anal. Chem.*, 2010, **82**, 3969–3973.

17 N. Stone, M. Kerssens, G. R. Lloyd, K. Faulds, D. Graham and P. Matousek, *Chem. Sci.*, 2011, **2**, 776–780.

18 E. A. Pozzi, A. B. Zrimsek, C. M. Lethiec, G. C. Schatz, M. C. Hersam and R. P. van Duyne, *J. Phys. Chem. C*, 2015, **119**, 21116–21124.

Faraday Discussions

PAPER

High resolution FTIR imaging provides automated discrimination and detection of single malaria parasite infected erythrocytes on glass†

David Perez-Guaita,[a] Dean Andrew,[b] Philip Heraud,[a] James Beeson,[bcd] David Anderson,[b] Jack Richards[bcd] and Bayden R. Wood*[a]

Received 18th November 2015, Accepted 22nd December 2015

DOI: 10.1039/c5fd00181a

New highly sensitive tools for malaria diagnostics are urgently needed to enable the detection of infection in asymptomatic carriers and patients with low parasitemia. In pursuit of a highly sensitive diagnostic tool that can identify parasite infections at the single cell level, we have been exploring Fourier transform infrared (FTIR) microscopy using a Focal Plane Array (FPA) imaging detector. Here we report for the first time the application of a new optic configuration developed by Agilent that incorporates 25× condenser and objective Cassegrain optics with a high numerical aperture (NA = 0.81) along with additional high magnification optics within the microscope to provide 0.66 micron pixel resolution (total IR system magnification of 61×) to diagnose malaria parasites at the single cell level on a conventional glass microscope slide. The high quality images clearly resolve the parasite's digestive vacuole demonstrating sub-cellular resolution using this approach. Moreover, we have developed an algorithm that first detects the cells in the infrared image, and secondly extracts the average spectrum. The average spectrum is then run through a model based on Partial Least Squares-Discriminant Analysis (PLS-DA), which diagnoses unequivocally the infected from normal cells. The high quality images, and the fact this measurement can be achieved without a synchrotron source on a conventional glass slide, shows promise as a potential gold standard for malaria detection at the single cell level.

1. Introduction

Malaria remains one of the most devastating diseases to afflict humanity. In 2015, there were approximately 214 million reported cases of malaria resulting in

[a]Centre for Biospectroscopy, Monash University, Clayton, 3800, Victoria, Australia. E-mail: bayden.wood@monash.edu

[b]Centre for Biomedical Research, Burnet Institute, Melbourne, 3004, Victoria, Australia

[c]Department of Microbiology, Monash University, Clayton, 3800, Victoria, Australia

[d]Department of Medicine, University of Melbourne, Parkville, 3050, Victoria, Australia

† Electronic supplementary information (ESI) available. See DOI: 10.1039/c5fd00181a

438 000 deaths.[1] Most deaths occur in children under 5 years of age with approximately 90% of cases occurring in Sub-Saharan Africa.[1] Much progress has been made in recent years in decreasing this burden and this has led to a strong global commitment to malaria elimination. Achieving this objective will require highly sensitive diagnostics that detect low-density parasitemia in asymptomatic carriers (*i.e.* people who have contracted the infection but do not express the clinical symptoms of chills and fevers). These individuals currently provide a vast submicroscopic reservoir of infection that is transmissible to mosquitoes. Identifying and treating these individuals will require diagnostics that can be used for rapid mass screening at very low cost. Routine tests for malaria screening include optical microscopy,[2] Rapid Diagnostic Tests (RDTs) based on antigen capture[3] and nucleic acid amplification like the Polymerase Chain Reaction (PCR).[4,5] Optical microscopy on thick blood films is a relatively inexpensive technique but requires an experienced microscopist to make the diagnosis and has a reported sensitivity of 5–10 parasites per μL.[6] RDTs are used worldwide but are often limited in their sensitivity and ability to identify multiple species of *Plasmodium*. RDTs are very cheap, ~USD $2, but take 15–25 minutes to obtain a result and provide no quantitative information on the level of parasitemia.[3] PCR amplification is the current gold standard with a detection limit of 1 parasite per μL but needs multiple primers to speciate, is costly and requires considerable technical expertise. In pursuit of a new diagnostic tool we have trialled a number of spectroscopic approaches including Raman and Fourier transform infrared (FTIR) spectroscopy. These techniques rely on the detection of vibrating functional groups characteristic of the parasite.

Raman spectroscopy can be applied to diagnose parasites in single erythrocytes.[7–17] The technique principally relies on the detection of the haemozoin chromophore, which forms as a by-product from the digestion of haemoglobin by the parasite. The technique can detect early blood-stage parasites, providing they have haemozoin crystal clusters on the order of 300 nm. Consequently it is not good at detecting early stage rings and gametocytes, which are the dominant forms of the parasite found in peripheral blood in *Plasmodium falciparum*. Surface Enhanced Raman Spectroscopy (SERS)[18–20] and Tip Enhanced Raman Spectroscopy (TERS)[21] have the potential to detect smaller haemozoin inclusions but the expense and complexity of the measurements make it difficult to implement in a hospital. The other consideration is the time taken to analyse a field of cells with this technique, which can take several hours. Raman acoustic levitation spectroscopy (RALS) has shown potential to diagnose haemozoin from a lysed pellet of erythrocytes[13] but the levitated blood drop can become unstable and explode making it unsafe to the health worker.

In pursuit of a more routine spectroscopic approach that does not rely solely on the presence of haemozoin, we have investigated the potential of Fourier transform infrared (FTIR) spectroscopy.[22–24] Unlike Raman spectroscopy, which is a scattering based technique; FTIR relies on the detection of functional groups absorbing infrared photons. The technique is very suited to detecting functional groups that have a strong change in their dipole moment as the nuclei oscillate. Consequently, proteins, nucleic acids, lipids and carbohydrates are all good infrared absorbers and provide distinct markers for malaria parasites, especially when the technique is used on isolated erythrocytes, which contain haemoglobin but no nucleus or other large organelles to confound the diagnosis. By applying synchrotron FTIR transmission spectroscopy with a neural network we were able to discern rings, trophozoites, and schizonts with very high specificity on a thin film of cells fixed in methanol on CaF_2 windows.[23] Using Attenuated Total Reflection (ATR-FTIR) spectroscopy we have demonstrated a detection limit of 0.00001% parasitemia (0.1 parasites per μL) and a quantification limit of 0.001% parasitemia (10 parasites per μL) using spiked isolated erythrocytes fixed in methanol.[22] More recently we utilised a FTIR microscope equipped with a Focal Plane Array (FPA) detector coupled to a synchrotron light source at the IRENI beamline located in the former Synchrotron Radiation Centre, University of Wisconsin-Madison, to image and diagnose single cells infected with the malaria parasite. The microscope was fitted with a 74× Cassegrain objective (Ealing Inc., Rocklin, CA, USA) with a Numerical Aperture (NA = 0.65) resulting in a pixel resolution of 0.54 μm. While the images showed that single cells could be resolved, it did not resolve subcellular detail and individual parasites could not be identified within single cells. The substrate used in this case was BaF_2, which is very expensive for routine measurements. More recently Bassan *et al.*[25] showed the potential of using glass-based substrates for FTIR imaging of tissue biopsies. The strong absorbance from the Si–O bonds in glass resulted in only the 3800–2500 cm^{-1} spectral range being diagnostically useful, however, the authors demonstrated that tissue could be classified into four basic tissue cell types and that using just the epithelial cells, reasonable discrimination of normal and malignant breast tissue could be found.

Recently Agilent developed a new 25× Cassegrain objective with a high NA = 0.81 for their Agilent Cary 620 microscope. The high NA of this objective, coupled with the high magnification optics within the microscope, results in a pixel resolution of 0.66 μm giving a magnification of 61× and a spatial resolution of 1.4 μm at 3750 cm^{-1}. Recently Reddy *et al.*[26] demonstrated that a pixel resolution of 0.6 μm provides the ideal degree of spatial oversampling for a NA = 0.81 system, for truly diffraction limited imaging. Pixel resolution smaller (or magnification higher) than this results in over-sampling and no improvement in the image quality. Over-sampling and no improvement in image quality could be achieved above this NA. In this paper, we show unprecedented image quality of single malaria infected erythrocytes recorded on a glass substrate and demonstrate how FTIR imaging can be used to diagnose infected cells from the extracted spectra using Partial Least Squares-Discriminant Analysis (PLS-DA).[27] The extremely high resolution on cheaper glass substrates paves the way forward for single cell diagnosis with an affordable substrate. Moreover, the extremely high spatial resolution enables distinct visualisation of the parasite's digestive vacuole within the fixed cell, demonstrating subcellular imaging on a blood cell for the first time.

2. Experimental

2.1 Sample preparation

The experiments were performed in compliance with the relevant laws and institutional guidelines. Ethics approval for the use of human blood samples in this study was obtained from the Alfred Human Research and Ethics Committee. Written informed consent was obtained from blood donors. Blood was collected from Melbourne control patients into lithium heparin vacutainers (BD) and centrifuged at 1500 rpm for 5 minutes to remove plasma. Blood was cultured at 3%

haematocrit in RPMI media supplemented with 10% AlbuMAX II (Gibco) and infected with *Plasmodium falciparum*, laboratory strain D10. Thin blood smears of malaria parasite infected erythrocytes were air dried for 20 minutes, then fixed in 100% methanol for 5 minutes.

2.2 Infrared microscopy and image acquisition

Infrared images were obtained using an Agilent (Santa Clara, USA) Cary 620 microscope equipped with a 25× Cassegrain objective upgraded with high magnification optics within the microscope providing a 5× magnification boost over the "standard magnification" mode. Hyperspectral images were recorded using transmission geometry at a resolution of 4 cm^{-1} in the glass transparent region (3600–2500 cm^{-1}) by co-adding 400 interferograms. The applied zero filling factor of 2 provided 572 points per spectrum. The same configuration was used for recording a background image from a clean region on the glass slide. The 128 × 128 pixels of the FPA detector of the microscope covered a region of 102.2 × 102.2 μm of the slide and mosaics of 3 × 2 tiles were obtained. After image acquisition the slides were stained with 10% Giemsa stain for 5 minutes, and visual images were captured for identification of the infected cells.

2.3 Data analysis

Hyperspectral images of the mosaics were imported to Matlab v2014b from Mathworks (Natik, USA). The data processing was carried out using in-house Matlab scripts and functions available in the PLStoolbox v7.0 package from Engeinvector (Manson, USA). For single erythrocyte spectra (results section 3.1), the function *improfile* was used for obtaining the spatial profile of the images. The images were resized to five times the original size using the imresize function prior to circle identification. The function *imfindcircles* was applied to detect circles with a radius ranging between 23 to 28 pixels (3.7–4.5 μm). For the intracellular resolution (results Section 3.2), noise reduction was applied using the first 20 PCs prior to the PLS-DA. The PLS-DA was performed over the 3400–2800 cm^{-1} region using a 25 point Savitzky–Golay smoothing filter, a baseline correction procedure (based on an automatic weighted least squares routine) and mean centring. For the detection of infected erythrocytes from their average spectra (results Section 3.3), PLS-DA was carried out between 3400–2800 cm^{-1}, using a 21 points Savitzky–Golay smoothing filter and standard normal variate normalization. The noise in the spectra necessitated excessive smoothing. The tuning parameters for the PLS-DA (*e.g.* number of latent variables regions and pre-processing methods) were selected according to visual inspection of the spectra and the cross-validation error, with smaller cross-validation errors for the more robust models.

3. Results

3.1 Single erythrocyte spectroscopy

The high magnification imaging configuration was able to provide sufficient spatial resolution to ensure that a group of spectra corresponded to a single cell. By applying chemometric analysis a snapshot of the metabolic composition of each cell could be obtained. Single cell resolution using conventional IR microscopy with standard 15× Cassegrain optics is problematic due to the small

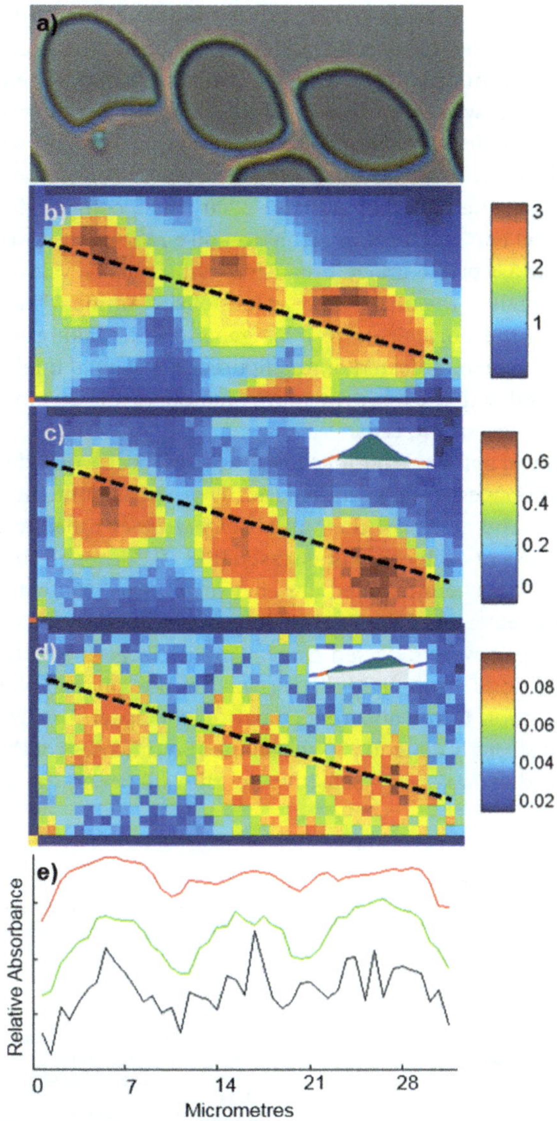

Fig. 1 High resolution infrared images of three adjacent erythrocytes. (a) Visible image, (b) integrated area between 3600–2500 cm^{-1}, (c) 3400–3200 cm^{-1}, (d) 2990–2850 cm^{-1}, (e) profile of the false-colour images along the three cells indicated by the dashed line on (b) (red), (c) (green) and (d) (black).

size of the erythrocytes (~7 μm in diameter), which is at the limit of the spatial resolution of IR light.[28] In order to investigate the absolute spatial resolution that could be achieved using this optic configuration we performed oversampling of the image. Fig. 1 depicts different maps extracted from the hyperspectral image of three cells. The visible image (Fig. 1a) highlights cells adjacent to each other and exemplifies the capability of the methodology for distinguishing between contiguous erythrocytes. The following figures show the false-colour map representing the integrated area between 3600–2500 cm^{-1} of the absorbance spectra

(Fig. 1b); along with the integrated area of the Amide A band, which is assigned principally to the N–H stretching modes of proteins (Fig. 1c); and the C–H stretching region from lipids (Fig. 1d). A visual inspection of the maps indicates that each erythrocyte was comprised of 75–100 pixels, depending on the diameter of the cell, and was clearly distinguishable from adjacent cells, with a gap of 1.4 μm (*i.e.* two pixels). The visual discrimination is wavelength dependent and is therefore poorer in the C–H region (3100–2800 cm^{-1}) compared to the Amide A region (3600–3200 cm^{-1}). The poor signal-to-noise ratio of the C–H bands further reduces the contrast of the image because the absorbance values are in general three times lower than the absorbance values of the Amide A band.

Fig. 1e shows the normalised profiles generated by taking a line of spectra through the three cells (indicated on the figures as a dashed line). The three cells are clearly resolved with the nodes at approximately 10 and 20 μm, corresponding to the adjacent edges of the cells.

The capability of software routines to extract the spectra of individual cells was assessed so that images could be automatically processed mitigating the need for human analysis. The ability of the Matlab *imfindcircles* function for identifying and selecting pixels of each cell was assessed (Fig. 2). Although the cells were well defined, when using the integrated area between 3600–2500 cm^{-1}, the software was only able to detect 26 of the 32 erythrocytes observed in the visible image. In addition, some cells were detected twice. This is likely due to the spectra containing contributions from light dispersion in the phase component of the spectrum in addition to the absorbance part of the spectrum. The extent of the light dispersion strongly depends on the shape and thickness of the cells. The cell detection algorithm was very sensitive to the shape of the cells, with the

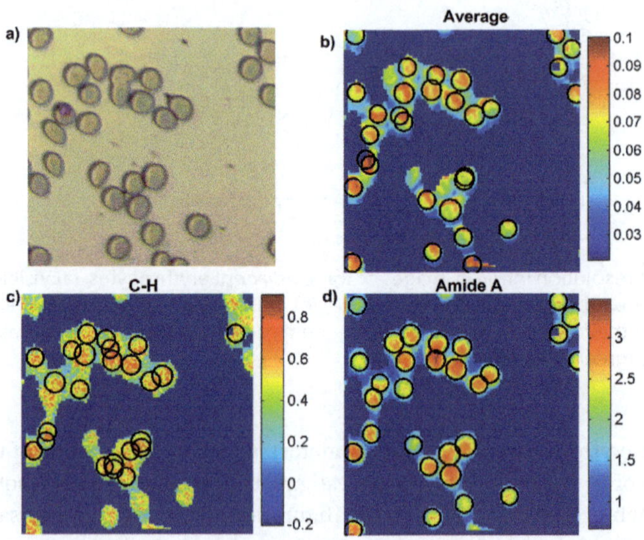

Fig. 2 The automatic identification of erythrocytes in hyperspectral infrared images. (a) Visible image showing a group of infected and non-infected cells. (b) Integrated area between 3600–2500 cm^{-1}. (c) 2990–2850 cm^{-1}, (d) 3400–3200 cm^{-1}. The cells identified by the *imfindcircles* function for the integrated area between 3600–2500 cm^{-1}, C–H and Amide A maps are circled in (b–d), respectively.

diagnostic potential of the technique compromised when cells were not spherical. The infrared images obtained from the integration of the C–H stretching region did not provide enough discrimination for the identification of individual cells due to the wavelength dependence of the resolution and the low absorbance in this region. However, in the case of the Amide A images, all the cells were detected using the algorithm, demonstrating that the images had the required spatial resolution for automatic detection of individual cells.

3.2 Intracellular resolution

Mature trophozoites were investigated to determine if subcellular structures like the digestive vacuole could be identified. An example of a cell infected with a trophozoite is shown in Fig. 3a. Plots of the integrated area (3600–2500 cm^{-1}, *z*-spatial direction) *versus* the spatial position (x, y) of the trophozoite shows the digestive vacuole as a "hole", which correlates to a reduction in optical density. Images of six infected erythrocytes were used to generate a false-colour map based on the integrated area between 3600–2500 cm^{-1} of the raw absorbance spectrum for each pixel (Fig. 3b). These data therefore contain both absorption and scattering effects. Warm colours indicate a high quantity of organic matter, while cold colours represent an absence of it. It can be seen that the parasite shows a significant decrease in organic matter in the region of the vacuole, which can be detected as a circular-shaped hole in each of the infrared images.

Fig. 4 shows maps of 9 infected erythrocytes generated by integrating the area between 3600–2500 cm^{-1}. The digestive vacuole of each trophozoite is clearly resolved in the images and circled in red, while the erythrocyte cytoplasm is circled in blue. The spectral information extracted from these circled areas was poor and no significant chemical differences were observed between the parasite and the cytoplasm of the cell. Maps based on the integration of bands and other unsupervised multivariate procedures were also assessed for finding chemical differences between the regions but were also unsuccessful, in part due to the low

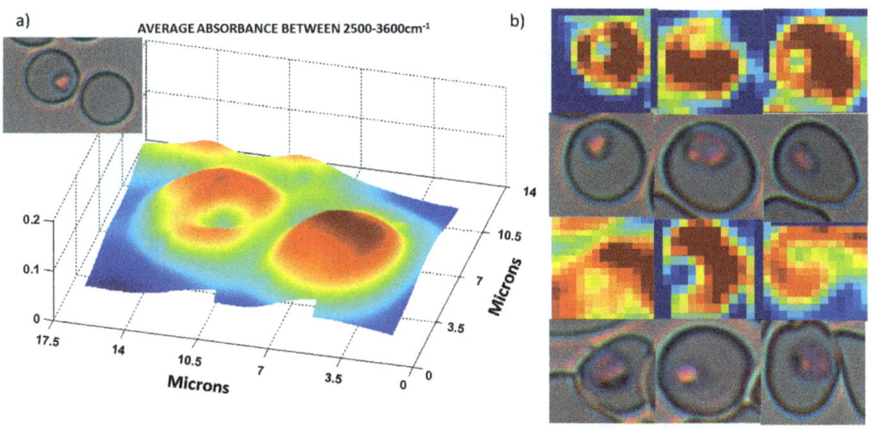

Fig. 3 Infrared images of trophozoites inside infected erythrocytes. (a) 3D representation of an infected and an uninfected cell. (b) False-colour images of 6 erythrocytes infected with trophozoites and their visible images. Colour scale corresponding to the integration area underneath each spectrum (pixel).

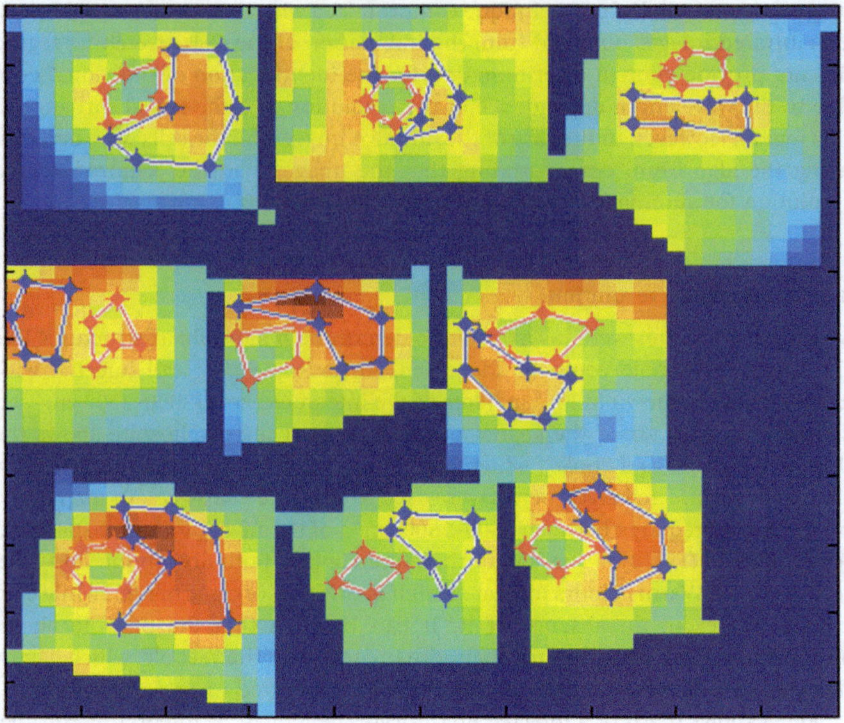

Fig. 4 Regions selected as trophozoites (red) and erythrocyte cytoplasm (blue) for the PLS-DA performed in Section 3.2.

signal-to-noise ratio of the pixels in the digestive vacuole region. To enhance the probability of detecting chemical differences between infected and uninfected cells a supervised multivariate modelling approach was employed.

Partial least squares-discriminant analysis (PLS-DA) is a supervised approach where the loadings are weighted to enhance the discrimination. PLS-DA was

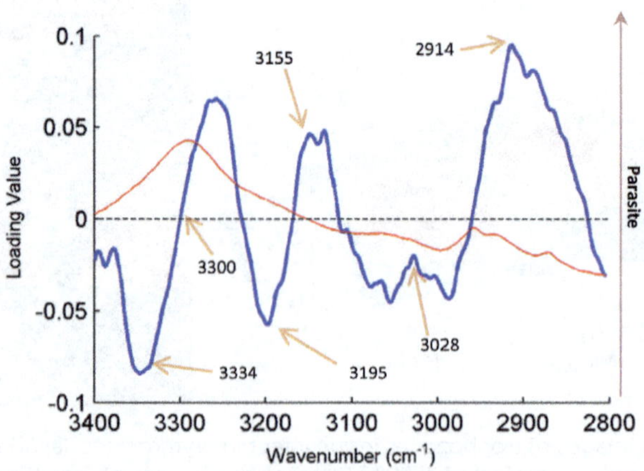

Fig. 5 Regression vector for the PLS-DA model described in Section 3.2. The red spectrum denotes the mean spectrum of the dataset.

performed using 9 infected erythrocytes. The important spectral features assigned to normal erythrocyte cytoplasm (mainly haemoglobin) and the trophozoite's digestive vacuole are shown in Fig. 5. Seventy-three pixels were assigned to the trophozoite vacuole and 200 pixels to the regular erythrocyte cytoplasm. A cross validation PLS-DA model was built using 5 latent variables and showed 81% specificity and 83% sensitivity. In this study, sensitivity and specificity were defined as the true positive and true negative rate obtained by comparing the values predicted by the model to the actual values obtained using a visual inspection of the stained cells. A permutation test, developed by our group,[29] found the model to be significant ($p < 0.005$). Fig. 5 shows the regression vector for the classification along with the mean spectrum of the samples. The bands responsible for the classification are: (a) the C–H bands at 2914 cm^{-1}, 3150 cm^{-1} and 3028 cm^{-1}; (b) a shift in the Amide A band centred at 3300 cm^{-1}. Strong loadings located at 2914 cm^{-1} and 3150 cm^{-1} and the weaker loading at 3028 cm^{-1} are attributed to a higher concentration of unsaturated fatty acids in the region infected by the parasite, as evinced by the broad band at 2914 cm^{-1} (assigned to the CH_2 stretching vibration) along with bands at 3150 cm^{-1} and 3028 cm^{-1} (assigned to the C=C–H stretching vibration), which are typically found in triglycerides and unsaturated fatty acids. The redshift of the Amide A band in the region of the parasite compared to the corresponding band in the uninfected part of the cell is possibly due to changes in the conformation of the proteins in the presence of the parasite.

3.3 Detection of infected erythrocytes from their average spectrum

The spectra from the photomicrographs of 14 infected and 34 uninfected cells were manually selected (ESI Fig. S1†). The mean spectra of the uninfected cells (denoted by black numbers) were extracted from the area contained by the blue circles and the mean spectra of the infected cells (denoted as blue numbers) were extracted from the area contained by the pink circles. The average spectra extracted from the infected cells had a better signal-to-noise ratio (average of 60 to 80 spectra) and were considered to be representative of the entire cell macromolecular chemistry.

The leave-one-out PLS-DA cross validation model showed an 87% classification rate (88% specificity and 87% sensitivity) and the permutation test performed ($n = 200$) indicated that this value was statistically significant ($p < 0.005$). The receiver operating curve (ROC) for the cross validation is shown (Fig. 6). This curve represents the different true positive and true negative rates obtained when a threshold (in our case the y-cross validation limit value) is varied. These results demonstrate the capability of the imaging methodology for differentiating between infected and non-infected cells. The regression vector depicts the bands responsible for the separation (Fig. 7). The bands at 3300, 2952 and 2870 cm^{-1} are characteristic of uninfected cells. These bands can be assigned to the Amide A and the CH_3 symmetric and asymmetric stretching vibrations, respectively. Other bands correlated with the presence of the parasite include those at 2922 cm^{-1} and 2850 cm^{-1}, which are assigned to the symmetric and antisymmetric stretching modes of the CH_2 group. As stated above, the CH_3 and amide bands are from proteins while the presence of CH_2 suggests the formation of long chain fatty acids and triglycerides. These results indicate a change in the lipid to protein ratio, caused by the consumption of haemoglobin by the parasite.

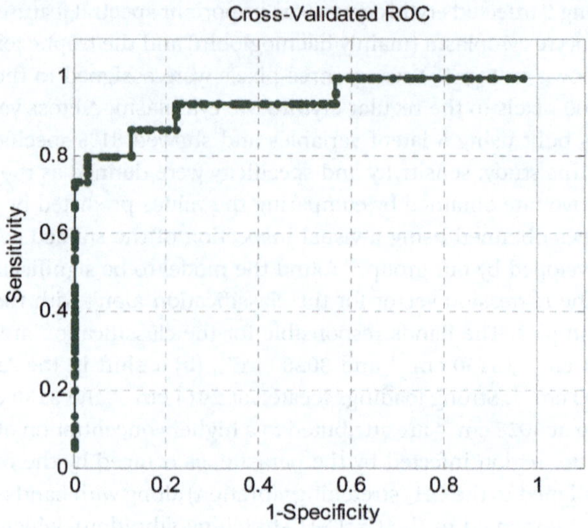

Fig. 6 Receiver operating curve of the cross validation of the PLS-DA model performed in Section 3.3.

It should be noted that there are differences between the regression vector for the first model where the classification was performed on the visible parasite subcellular structure against the rest of the cytoplasm of the red blood cell, and the second classification model performed on the average spectrum of the infected erythrocytes against the average spectrum of the uninfected erythrocytes. In regards to the position of the Amide A mode, in the first classification model there was a redshift in the case of the parasite pixels, while in the second classification there was a general increase in absorption of the Amide A mode in the uninfected cells. In addition, the band assigned to the C=C–H stretching mode noted in the first model disappears for the classification performed between cells in the second model. This may be explained by the fact that the differences are more subtle when comparing different regions within the cell. In contrast, when comparing the average spectrum from whole cells the differences are dominated

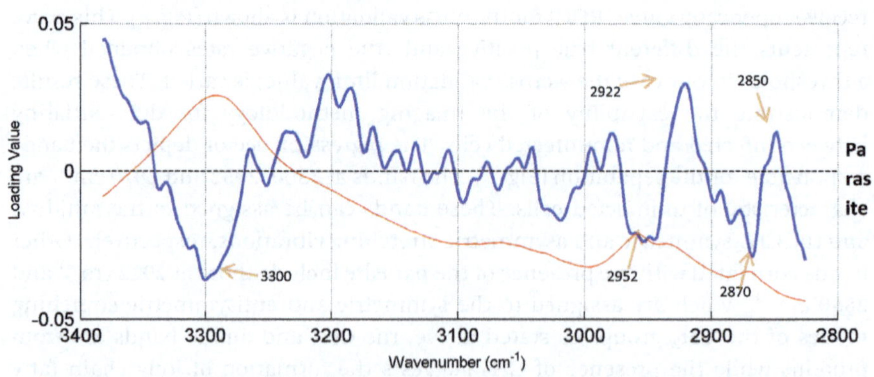

Fig. 7 Regression vector for the PLS-DA model described in Section 3.3. The red spectrum denotes the mean spectrum of the dataset.

by the major metabolic and phenotypic changes of the entire parasite–host system rather than changes in the haemoglobin to lipid ratio within a single cell.

4. Conclusions

We have demonstrated that high-resolution infrared images of single cells on a glass slide can be obtained using a FPA with a conventional globar IR source by incorporating a 25× high numerical aperture (NA = 0.81) Cassegrain objective system. The use of an inexpensive glass slide, as opposed to expensive IR transmission slides, offers a number of advantages in terms of cost and following current clinical practises for thick and thin film blood smears. The use of an inexpensive glass slide, as opposed to expensive transmission windows, offers a number of advantages in terms of cost and following current clinical practises for thick and thin film blood smears. The high quality images could resolve the digestive vacuole of the trophozoite demonstrating sub-cellular resolution. Moreover, the high-resolution images of individual cells enabled the successful application of an algorithm to auto-detect individual cells from absorbance maps based on wavenumber values greater than 3500 cm^{-1}. The spectra could then be extracted from all of the pixels associated with that cell, averaged and modelled using PLS-DA, which achieved an 87% classification rate for a leave-out-one cross-validation. The main differences identified related to: (i) changes in the Amide A band, possibly due to protein conformational changes caused by the parasite and (ii) more unsaturated fatty acids in the infected cells. More work is required to ascertain if the approach can be used to identify early ring stage parasites and the more elongated gametocyte forms found in peripheral blood but these initial studies demonstrate the potential of high resolution FPA imaging for single cell diagnosis on a blood film.

Acknowledgements

Funding to the authors was provided by the Australian Research Council (Future Fellowship FT120100926 to BRW), National Health and Medical Research Council of Australia (Program grant and Senior Research Fellowship to JGB; Early Career Fellowships JSR; Infrastructure for Research Institutes Support Scheme Grant to the Burnet Institute); Victorian State Government Operational Infrastructure Support Grant to the Burnet Institute. We acknowledge Mr Finlay Shanks for instrumental support and Dr Mustafa Kansiz (Agilent Technologies) for loaning us the 25× Cassegrain objectives used for the measurements.

References

1 WHO, 10 facts on malaria, http://www.who.int/features/factfiles/malaria/en/.
2 B. E. Barber, T. William, M. J. Grigg, T. W. Yeo and N. M. Anstey, *Malar. J.*, 2013, **12**, 8.
3 B. E. Barber, T. William, M. J. Grigg, K. Piera, T. W. Yeo and N. M. Anstey, *J. Clin. Microbiol.*, 2013, **51**, 1118–1123.
4 H. A. Dawoud, H. M. Ageely and A. A. Heiba, *J. Egypt. Soc. Parasitol.*, 2008, **38**, 339–350.

5 H. A. Dawoud, H. M. Ageely and A. A. Heiba, *J. Egypt. Soc. Parasitol.*, 2008, **38**, 329–338.

6 N. Tangpukdee, C. Duangdee, P. Wilairatana and S. Krudsood, *Korean J. Parasitol.*, 2009, **47**, 93–102.

7 B. R. Wood, A. Hermelink, P. Lasch, K. R. Bambery, G. T. Webster, M. A. Khiavi, B. M. Cooke, S. Deed, D. Naumann and D. McNaughton, *Analyst*, 2009, **134**, 1119–1125.

8 B. R. Wood and D. McNaughton, *Expert Rev. Proteomics*, 2006, **3**, 525–544.

9 M. Bilal, M. Saleem, S. T. Amanat, H. A. Shakoor, R. Rashid, A. Mahmood and M. Ahmed, *J. Biomed. Opt.*, 2015, **20**, 017002.

10 M. Bruckner, K. Becker, J. Popp and T. Frosch, *Anal. Chim. Acta*, 2015, **894**, 76–84.

11 C. W. Ong, Z. X. Shen, K. K. H. Ang, U. A. K. Kara and S. H. Tang, *Appl. Spectrosc.*, 1999, **53**, 1097–1101.

12 C. W. Ong, Z. X. Shen, K. K. H. Ang, U. A. K. Kara and S. H. Tang, *Appl. Spectrosc.*, 2002, **56**, 1126–1131.

13 L. Puskar, R. Tuckermann, T. Frosch, J. Popp, V. Ly, D. McNaughton and B. R. Wood, *Lab Chip*, 2007, **7**, 1125–1131.

14 A. Schulte, F. Andrade, C. Schwarz, L. Ayong and D. Chakrabarti, *Biophys. J.*, 2007, 383A.

15 B. R. Wood, S. J. Langford, B. M. Cooke, F. K. Glenister, J. Lim and D. McNaughton, *FEBS Lett.*, 2003, **554**, 247–252.

16 K. M. Marzec, D. Perez-Guaita, M. de Veij, D. McNaughton, M. Baranska, M. W. Dixon, L. Tilley and B. R. Wood, *ChemPhysChem*, 2014, **15**, 3963–3968.

17 M. Kozicki, J. Czepiel, G. Biesiada, P. Nowak, A. Garlicki and A. Weselucha-Birczynska, *Analyst*, 2015, **140**, 8007–8016.

18 N. L. Garrett, R. Sekine, M. W. Dixon, L. Tilley, K. R. Bambery and B. R. Wood, *Phys. Chem. Chem. Phys.*, 2015, **17**, 21164–21168.

19 C. Yuen and Q. Liu, *J. Biomed. Opt.*, 2012, **17**, 017005.

20 C. Yuen and Q. Liu, *Analyst*, 2013, **138**, 6494–6500.

21 B. R. Wood, E. Bailo, M. A. Khiavi, L. Tilley, S. Deed, T. Deckert-Gaudig, D. McNaughton and V. Deckert, *Nano Lett.*, 2011, **11**, 1868–1873.

22 A. Khoshmanesh, M. W. A. Dixon, S. Kenny, L. Tilley, D. McNaughton and B. R. Wood, *Anal. Chem.*, 2014, **86**, 4379–4386.

23 G. T. Webster, K. A. De Villiers, T. J. Egan, S. Deed, L. Tilley, M. J. Tobin, K. R. Bambery, D. McNaughton and B. R. Wood, *Anal. Chem.*, 2009, **81**, 2516–2524.

24 B. R. Wood, K. R. Bambery, M. W. A. Dixon, L. Tilley, M. J. Nasse, E. Mattson and C. J. Hirschmugl, *Analyst*, 2014, **139**, 4769–4774.

25 P. Bassan, J. Mellor, J. Shapiro, K. J. Williams, M. P. Lisanti and P. Gardner, *Anal. Chem.*, 2014, **86**, 1648–1653.

26 R. K. Reddy, M. J. Walsh, M. V. Schulmerich, P. S. Carney and R. Bhargava, *Appl. Spectrosc.*, 2013, **67**, 93–105.

27 D. Ballabio and V. Consonnia, *Anal. Methods*, 2013, **5**, 3790–3798.

28 P. Lasch and D. Naumann, *Biochim. Biophys. Acta*, 2006, **1758**, 814–829.

29 D. Perez-Guaita, J. Kuligowski, S. Garrigues, G. Quintas and B. R. Wood, *Analyst*, 2015, **140**, 2422–2427.

Faraday Discussions

PAPER

Rapid characterisation of *Klebsiella oxytoca* isolates from contaminated liquid hand soap using mass spectrometry, FTIR and Raman spectroscopy†

Ralf Dieckmann,*[a] Jens Andre Hammerl,[a] Hartmut Hahmann,[d] Amal Wicke,[d] Sylvia Kleta,[a] Piotr Wojciech Dabrowski,[e] Andreas Nitsche,[c] Maren Stämmler,[b] Sascha Al Dahouk[a] and Peter Lasch[b]

Received 13th November 2015, Accepted 7th December 2015

DOI: 10.1039/c5fd00165j

Microbiological monitoring of consumer products and the efficiency of early warning systems and outbreak investigations depend on the rapid identification and strain characterisation of pathogens posing risks to the health and safety of consumers. This study evaluates the potential of three rapid analytical techniques for identification and subtyping of bacterial isolates obtained from a liquid hand soap product, which has been recalled and reported through the EU RAPEX system due to its severe bacterial contamination. Ten isolates recovered from two bottles of the product were identified as *Klebsiella oxytoca* and subtyped using matrix-assisted laser desorption/ionisation time-of-flight mass spectrometry (MALDI TOF MS), near-infrared Fourier transform (NIR FT) Raman spectroscopy and Fourier transform infrared (FTIR) spectroscopy. Comparison of the classification results obtained by these phenotype-based techniques with outcomes of the DNA-based methods pulsed-field gel electrophoresis (PFGE), multi-locus sequence typing (MLST) and single nucleotide polymorphism (SNP) analysis of whole-genome sequencing (WGS) data revealed a high level of concordance. In conclusion, a set of analytical techniques might be useful for rapid, reliable and cost-effective microbial typing to ensure safe consumer products and allow source tracking.

[a]*Federal Institute for Risk Assessment, Department of Biological Safety, Max-Dohrn-Str. 8-10, D-10589 Berlin, Germany. E-mail: ralf.dieckmann@bfr.bund.de; Fax: +49 30 18412-2791; Tel: +49 30 18412-2182*

[b]*Robert Koch-Institut, ZBS 6 – Proteomics and Spectroscopy, Seestraße 10, D-13353 Berlin, Germany*

[c]*Robert Koch-Institut, ZBS 1 – Highly Pathogenic Viruses, Seestraße 10, D-13353 Berlin, Germany*

[d]*Landesamt für Verbraucherschutz Sachsen-Anhalt, Fachbereich Lebensmittelsicherheit, Freiimfelder Str. 68, D-06112 Halle, Germany*

[e]*Robert Koch-Institut, NG4 – Bioinformatics, Seestraße 10, D-13353 Berlin, Germany*

† Electronic supplementary information (ESI) available. See DOI: 10.1039/c5fd00165j

Introduction

Bacteria belonging to the genus *Klebsiella* are a frequent cause of human nosocomial and community-acquired infections leading to a wide range of diseases such as pneumonia, urinary tract infections, antibiotic-associated haemorrhagic colitis (AAHC), septicaemia, and skin and soft tissue infections. Klebsiellae are nonmotile, rod-shaped, Gram-negative bacteria with a prominent capsular polysaccharide. In particular, *Klebsiella* (*K.*) *pneumoniae* and *K. oxytoca* are associated with nosocomial infections often affecting immunocompromised patients and those requiring intensive care. They are both opportunistic pathogens found in the environment and in mammalian mucosal surfaces. Hand sanitizer and liquid soap dispensers have frequently been the source of hospital-acquired infections[1–7] and multiple nosocomial outbreaks have been reported as a result of inadequate antisepsis or disinfection.[8] Personal care products with a high water content such as hand soaps or shampoos as well as disinfectants used in private or public areas are at risk of being contaminated by bacteria, leading to product deterioration and health risks to consumers.[9–12] Intrinsic (occurring during manufacture) as well as extrinsic (occurring during use) contamination of personal care products or disinfectant solutions have been implicated in human infections.[8]

Bacterial typing is necessary to identify sources and routes of product contamination and is a prerequisite for targeted control measures. The term subtyping refers to characterisation beyond the species or subspecies level, allowing the determination of clonal relationships and phylogenetic relatedness of bacterial strains. Nowadays, classical typing methods such as lysotyping or serotyping are more and more replaced by molecular biological approaches. Pulsed-field gel electrophoresis (PFGE) is the molecular typing method of choice for several bacterial agents because of its high discriminatory power, epidemiological relevance and standardisation.[13–15] It is based upon the variable migration of large DNA restriction fragments in an electrical field of alternating polarity and has been widely applied to characterise numerous bacterial species including *K. oxytoca*,[16–19] however, it is technically demanding, time-consuming (approx. 1 week) and costly. Multi-locus sequence typing (MLST) is based on the allelic differences between housekeeping genes and provides unambiguous data that are suitable for global epidemiology and evolutionary studies,[20] and databases for many bacterial species are publicly available (http://www.pubmlst.org/databases/). Recently, a MLST method was developed for *K. oxytoca*, and analysis of nosocomial isolates showed that MLST can discriminate between isolates from patients suffering from nosocomial pneumonia and from antibiotic-associated haemorrhagic colitis.[21] In recent years, whole-genome sequencing (WGS) has developed rapidly and has become a feasible typing method for microbiology laboratories in recent years. It provides ultimate discriminatory power and the potential to gather additional information such as the presence of antibiotic resistance genes, virulence genes or extrachromosomal elements.[22,23] For single nucleotide polymorphism (SNP) analysis, whole-genome sequence data can be mapped against a reference genome and proved to be suitable for epidemiological typing and outbreak investigations.[24–28]

Various analytical techniques often referred to as whole-organism fingerprinting methods are currently under investigation for rapid, objective and reliable bacterial identification at the species level and for bacterial typing. Such techniques not only reduce costs and efforts *via* automation and high-throughput solutions, but also generate results more rapidly. Particularly, matrix-assisted laser desorption/ionisation time-of-flight (MALDI TOF) mass spectrometry (MS), Fourier transform infrared (FTIR) spectroscopy and Raman spectroscopy are promising tools because these methods are rapid, simple and cost-effective. Furthermore, the requirements for sample preparation are modest. In contrast to PFGE, MLST, or WGS, which target the bacterial genome, these "phenotypic" techniques are based on the detection of large numbers of spectral features originating from various molecular building blocks (proteins, lipids, polysaccharides and nucleic acids). Since different microorganisms differ in their overall molecular composition, the spectra generated can be interpreted as highly complex taxon-specific fingerprints. MALDI TOF mass spectrometry mainly targets highly abundant ribosomal and nucleic acid binding proteins.[29,30] Vibrational spectroscopic techniques, *i.e.* infrared (IR) and Raman spectroscopy, are based on the specific vibrational motions of small functional groups: while IR spectroscopy measures the absorption of infrared light, Raman spectroscopy uses the samples' inelastic scattering properties following excitation by a monochromatic light source (laser). All three phenotypic methods allow for label-free analysis of microbial samples and therefore might be suitable for monitoring purposes and outbreak investigations without prior knowledge of the bacterial targets. Advantageously, they do not require prior amplification of nucleic acids like PCR-based methods, but detect biomolecules that are constitutively expressed by each individual cell.

In the present study, ten bacterial isolates of *K. oxytoca* were analysed along with a selection of clinical and reference strains by a panel of phenotypic (MALDI TOF MS, FTIR and Raman spectroscopy) and genotypic (PFGE, MLST, WGS) characterisation techniques. Bacterial isolates were obtained from two lots of an intrinsically contaminated liquid hand soap product. This sanitary product was reported in 2013 through the EU RAPEX system (Rapid Alert System) for non-food dangerous products and recalled because it posed a microbiological risk due to severe contamination with *K. oxytoca*, with an average concentration of 2.5×10^4 cfu ml^{-1} (cfu: colony forming unit).

The main objectives of the study were (i) to examine the phylogenetic relatedness of the *K. oxytoca* hand soap isolates, (ii) to determine the potential utility of three phenotyping techniques *i.e.* MALDI TOF MS, FTIR and NIR FT Raman spectroscopy for subtyping *Klebsiella* spp. in terms of reproducibility, accuracy and taxonomic resolution and (iii) to assess the degree of correlation between the results provided by the phenotype-based techniques and those attained by DNA-based approaches. In summary, a combined analysis of classification data obtained by distinct but complementary microbial characterisation techniques should be established to ensure product hygiene measures and allow tracing back to a contamination site.

Materials and methods

Bacterial strains and cultivation

Ten isolates originating from two bottles (lot I and lot II) of an intrinsically contaminated liquid hand soap produced in Greece, four clinical isolates, among them three from different infants with enteritis in 1995, and three reference strains were included in this study (Table 1). The liquid soap strains were isolated within routine product quality monitoring using standardised microbiological methods and a standard protocol for germ count. *K. oxytoca* was confirmed by standard biochemical methods and a commercial test kit (BBL CRYSTAL E/NF, Becton Dickinson, Sparks, USA). The product was recalled because it posed a microbiological risk due to a high concentration of *K. oxytoca* and was reported through the EU Rapid Alert System for non-food dangerous products (EU RAPEX, A12/1308/13). The average concentration of bacteria detected in the soap was 2.5×10^4 cfu ml^{-1}. Bacterial samples were prepared by growing each strain in three independent batches by two passages under aerobic conditions. Cells were cultured for nucleic acid-based characterisation on tryptic soy agar with neutralising agents histidine, lecithin, Tween-80 (Caso-HLT-Agar, Merck, Darmstadt, Germany) for 72 h at 30 °C. For phenotypic characterisation cultivation was carried out on caso agar plates (Merck, Darmstadt, Germany) for 24 h at 37 °C.

Multi-locus sequence typing (MLST)

Genomic DNA (gDNA) of the isolates was prepared using the QIAamp DNA mini kit (Qiagen, Hilden, Germany) as recommended by the manufacturers. All isolates were analysed by MLST of seven housekeeping genes (*gapA*, *infB*, *mdh*, *pgi*, *phoE*, *rpoB*, *tonB*) as described by Herzog *et al.*[21] PCR products were purified with the QIAquick PCR purification kit (Qiagen) and sequenced by MWG Eurofins (Ebersberg, Germany). Sequence analysis was performed using the software packages of DNAstar (Lasergene) and Accelrys DS Gene. The sequences of the genetic loci investigated were submitted to the PubMLST database (http://www.pubmlst.org/koxytoca/) to assign sequence types (ST). For confirmation of MLST data, sequences of the housekeeping genes were extracted from WGS data

Table 1 Overview of strains of *Klebsiella* spp. isolates included in this study

Strain	Country of origin and year of isolation	Source
K. oxytoca PHS-890	Greece, 2013	Hand soap, lot I
K. oxytoca PHS-891	Greece, 2013	Hand soap, lot I
K. oxytoca PHS-892	Greece, 2013	Hand soap, lot I
K. oxytoca PHS-893	Greece, 2013	Hand soap, lot I
K. oxytoca PHS-894	Greece, 2013	Hand soap, lot I
K. oxytoca PHS-895	Greece, 2013	Hand soap, lot II
K. oxytoca PHS-896	Greece, 2013	Hand soap, lot II
K. oxytoca PHS-897	Greece, 2013	Hand soap, lot II
K. oxytoca PHS-898	Greece, 2013	Hand soap, lot II
K. oxytoca PHS-899	Greece, 2013	Hand soap, lot II
K. oxytoca CB4063	Germany, 1995	Child, enteritis
K. oxytoca CB4074	Germany, 1995	Child, enteritis
K. oxytoca CB4072	Germany, 1995	Child, enteritis
K. oxytoca CCUG 15788	Sweden	Environmental
K. oxytoca Oman 61	Oman, 2011	Clinical isolate
K. oxytoca ATCC 13182	USA	Pharyngeal tonsil
K. pneumoniae subsp. ozaenae ATCC 25926	Belgium	Human blood

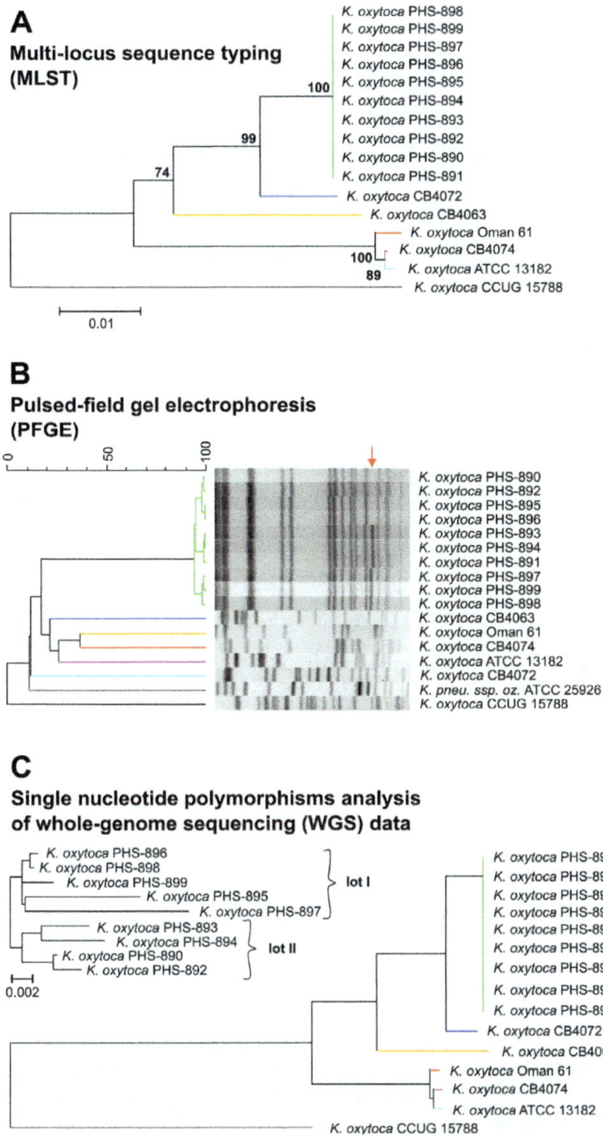

Fig. 1 Genotyping results of *K. oxytoca* hand soap isolates, clinical isolates and reference strains. (A) Dendrogram produced by multi-locus sequence typing (MLST) as described by Herzog *et al.*[21] For MLST seven housekeeping genes (*gapA*, *infB*, *mdh*, *pgi*, *phoE*, *rpoB*, *tonB*) were used. (B) Analysis of pulsed-field gel electrophoresis (PFGE) patterns (XbaI restriction) of *K. oxytoca* and *K. pneumoniae*. (C) SNP analysis of whole-genome sequencing (WGS) data of *K. oxytoca*. Inset: detailed view of the clade containing only the data of the hand soap isolates.

of the *Klebsiella* spp. isolates using MLST 1.8 at https://www.cge.cbs.dtu.dk/services/MLST/.[31] Phylogenetic analysis of MLST sequences was performed using the Mega 6.0 software.[32] Neighbour-joining (NJ) trees for concatenated sequences were constructed using the Kimura-2 parameter model to estimate the genetic distances.[33,34]

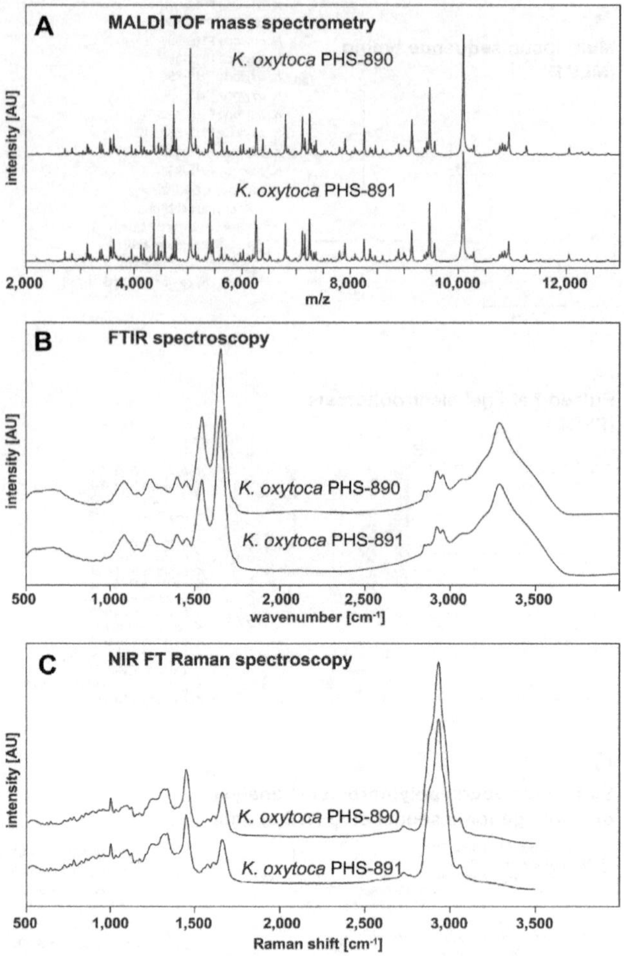

Fig. 2 Spectra acquired by means of phenotyping methods (MALDI TOF MS, FTIR and FT Raman spectroscopy) from cultures of *K. oxytoca* isolates PHS-890 and PHS-891 (hand soap isolates). Each spectral trace was generated by averaging the available spectra from three biological replicates. For more clarity, spectra were min–max normalised and shifted along the *y* axis. (A) MALDI TOF mass spectra of the *K. oxytoca* isolates PHS-890 (upper trace) and PHS-891 (lower trace). (B) FTIR spectra in the mid-IR spectral region ranging from 500 to 4000 cm^{-1}. (C) NIR FT Raman spectra of *K. oxytoca* isolates. The wavelength λ of the NIR Raman laser was 1064 nm.

Pulsed-field gel electrophoresis (PFGE)

Whole DNA from each *Klebsiella* spp. isolate was analysed by macrorestriction profiling using XbaI, and resolved by PFGE analysis in accordance with the CDC (Centers for Disease Control and Prevention) Pulse Net protocol (http://www.cdc.gov/pulse.net) as described by Hunter *et al.*,[35] using a CHEF-DRIII SYS220/240 (Bio-Rad Laboratories GmbH, Munich, Germany) system. PFGE patterns were analysed using the BioNumerics software package (version 6.6, Applied Maths, Sint-Martens-Latem, Belgium). Similarity analysis was performed

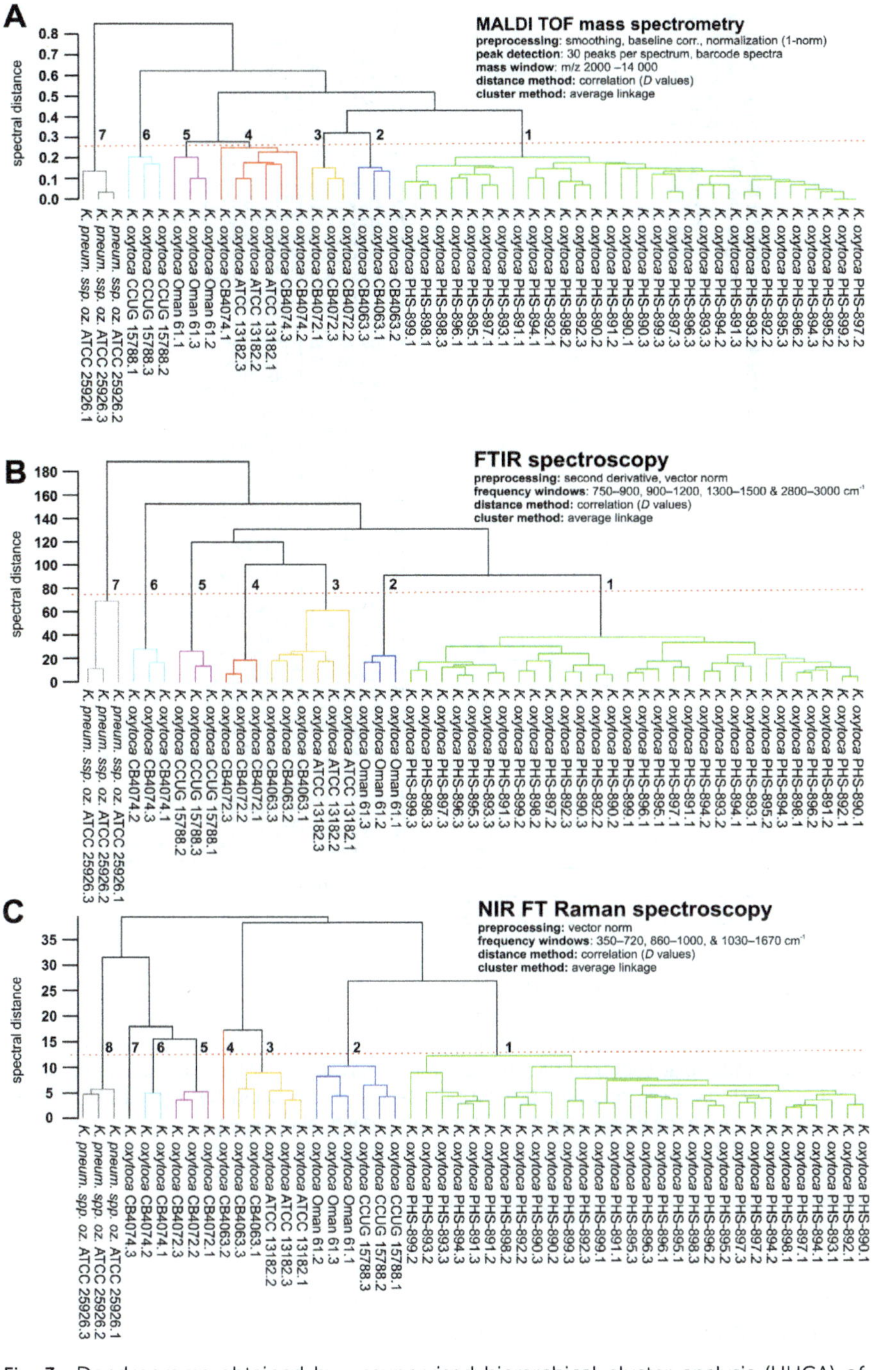

Fig. 3 Dendrograms obtained by unsupervised hierarchical cluster analysis (UHCA) of MALDI TOF mass, FTIR and NIR FT Raman spectra from *Klebsiella* spp. strains. Each microbial strain was characterised in triplicate. Horizontal lines (red, dotted) indicate the similarity cutoff value used arbitrarily for group definition and differentiation of *Klebsiella* spp. strains below the species level. (A) Dendrogram produced from MALDI TOF mass spectra of *Klebsiella* spp. strains. (B) Dendrogram obtained by clustering FTIR spectra. (C) Dendrogram obtained by clustering NIR FT Raman spectra.

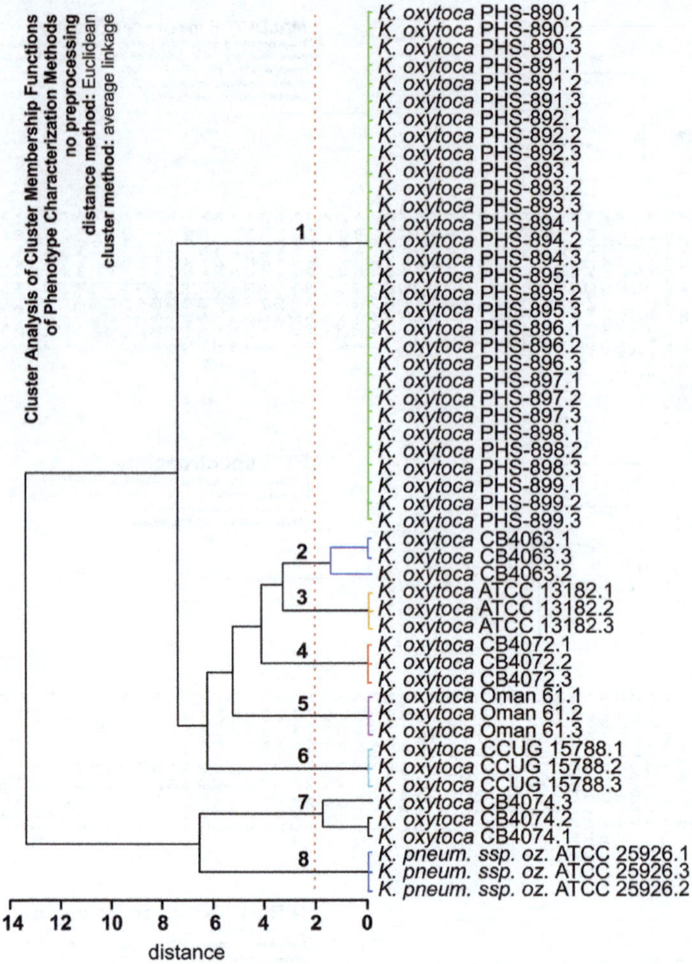

Fig. 4 Dendrogram of a combined analysis of microbial phenotypes. Spectral data from *Klebsiella* spp. cultures were obtained by MALDI TOF mass spectrometry, FTIR and FT Raman spectroscopy (triplicate measurements of individual microbial cultures by each technique). Microbial spectra collected by the different phenotyping techniques were subsequently individually analysed by unsupervised hierarchical clustering (UHCA). Cluster membership functions were established for each individual culture by setting the number of clusters to 7 (MALDI, FTIR), or 8 (Raman). The membership functions served as inputs for a top-level UHCA. The dendrogram of the combined analysis suggests that all *K. oxytoca* hand soap isolates are indistinguishable (see text for details).

using Dice coefficients with a band matching tolerance of 0.5% and 1% optimisation. Ward's method was used as the clustering method.

Analysis of single nucleotide polymorphisms (SNP) of whole-genome sequencing (WGS) data

Isolation of genomic DNA (gDNA) was conducted as described previously[36] using the QIAamp DNA mini kit (Qiagen, Hilden, Germany). A total of 1 ng of gDNA was subjected to library preparation using the Illumina Nextera® XT DNA Sample

Preparation Kit. Samples were tagged, pooled, and paired-end 2 × 300 bp sequencing of the DNA was performed using an Illumina MiSeq, which yielded 47.2 million reads with a total of 10.5 billion bases. SNPs differentiating the strains from each other were identified by mapping the reads to the sequence of *K. oxytoca* strain M1 (CP008841.1) using bowtie2.[37] A total of 38 214 discriminating SNPs were identified using an in-house pipeline based on GATK 3.4.[38] Tree calculation was conducted with PhyML 3.0 under the GTR + γ model.[39]

FTIR spectroscopy

For IR absorbance measurements, bacterial cells were harvested and prepared as previously described.[40–42] Briefly, small amounts of late-exponential-phase cells (~10 to 60 μg dry weight) were carefully removed with a platinum loop from regions of confluent colony growth and suspended in 80 μl of distilled and sterilised water by vortexing. An aliquot of 35 μl of the suspension was transferred to a ZnSe optical plate in a multisampling cuvette and dried in a desiccator over a drying agent (Sicapent®, Merck) with the application of a moderate vacuum (approx. 8 kPa) to form a transparent film suitable for FTIR measurements. Prior to spectral measurements the sample holder was sealed with a KBr cover plate to control the humidity and to protect the instrument from contamination. Spectra were recorded in the region between 500 and 4000 cm^{-1} using an IFS 28/B FTIR spectrometer (Bruker Optics, Ettlingen, Germany). The instrument was specially designed for the measurement of microorganisms and was equipped with a deuterated triglycine sulphate (DTGS) detector. For each FTIR spectrum 64 scans were co-added and averaged. Spectra were recorded at a nominal spectral resolution of 6 cm^{-1}. Fourier transformation was done using a Blackmann–Harris 3-term apodisation function and a zerofilling factor of 4, giving a point spacing of approximately 1 cm^{-1}. The spectrometer was continuously purged by dry air to reduce interference with atmospheric water vapour and CO_2. IR spectra were acquired from three independent cultivations (biological replicates) of each individual bacterial strain.

Near-infrared (NIR) FT Raman spectroscopy

Approximately 5–10 wire loops corresponding to 4–8 mg dry weight of bacterial material were transferred into cavities of custom-made steel cups produced by Körber & Körber (Birkenwerder, Germany). These steel cups have a diameter of 10 mm and a height of 4 mm. A centred hole of 2 mm in diameter and a depth of 0.5 mm allowed deposition of the bacterial biomass. Samples were dried in a desiccator over Sicapent® (Merck) with the application of a moderate vacuum (8–10 kPa). The cups were covered with custom-designed CaF_2 plates (Korth Kristalle, Altenholz, Kiel, Germany) to prevent contamination of the instrument. FT Raman measurements were performed using a MultiRam NIR FT Raman spectrometer (Bruker Optics) equipped with a Nd:YAG laser operating at 1064 nm and a liquid nitrogen cooled germanium (Ge) detector. The laser power was adjusted to 300–550 mW. FT Raman spectra were obtained with a spectral resolution of 6 cm^{-1}. Fourier transformation was done using a Blackmann–Harris 3-term apodisation function and a zero filling factor of 4. For each spectrum 250 individual scans were co-added. To further increase the signal-to-noise-ratio (SNR), three technical replicate spectra were consecutively collected from each of the three biological replicates (individual cultures) of each strain. For further analyses (cluster analysis), spectra from technical replicates were averaged.

MALDI TOF mass spectrometry

For MS sample preparation the modified trifluoroacetic acid (TFA) inactivation protocol for highly pathogenic microorganisms was used.[43] Briefly, 3–5 loops from bacterial cultures (growth conditions as described above) were suspended in 20 µl HPLC grade water (Mallinckrodt Baker B.V., Deventer, Netherlands) before 80 µl of pure TFA (Uvasol®, Merck) were added. After an incubation period of 30 min, the solutions were diluted ten-fold with HPLC grade water (Mallinckrodt Baker B.V.). A volume of 2 µl of the microbial dilutions was mixed with an equal volume of a 12 mg ml^{-1} α-cyano-4-hydroxycinnamic acid (HCCA) solution (Bruker Daltonics, Bremen, Germany). The HCCA solution was prepared by dissolving HCCA in a 2 : 1 (v/v) mixture of 100% acetonitrile and 0.3% TFA. 2 µl of each mixture was spotted onto ground steel sample targets from Bruker Daltonics. Microbial mass spectra were acquired using an Autoflex I mass spectrometer (Bruker Daltonics) controlled by the FlexControl 3.0 data acquisition software. The mass spectrometer was equipped with a nitrogen laser (λ = 337 nm) operating at a pulse rate of 10 Hz (manual mode of spectra acquisition). The pulse ion extraction time was 200 ns and the sampling rate was set to 0.5 GHz. Mass spectral measurements were carried out in the linear mode using an acceleration voltage of 20.00 (ion source 1) and 18.25 (ion source 2) kV. The lens voltage was 6.50 kV. Spectra were generated in the range between m/z 2000 and 20 000. For calibration we used *E. coli* K12 strain DSM 3871 as an external standard. At least 300 individual laser shots were co-added for each MALDI TOF mass spectrum. Two technical replicate spectra were collected from each of the three biological replicates of each strain. For further analyses, spectra from technical replicates were averaged.

Analysis of FTIR, Raman and MALDI mass spectra

Recording of spectra, data storage and analysis of FTIR and NIR FT Raman spectra were performed by means of the OPUS software package provided by Bruker Optics. Spectra were pre-processed by applying a second-derivative Savitzky–Golay smoothing/derivative filter with nine smoothing points (FTIR only) and vector-normalisation (FTIR, FT Raman). Pre-processed spectra were subjected to unsupervised agglomerative hierarchical cluster analysis (UHCA) using the information contained in the spectral windows 750–900 cm^{-1}, 900–1200 cm^{-1}, 1300–1500 cm^{-1} and 2800–3000 cm^{-1} for FTIR or 350–720 cm^{-1}, 860–1000 cm^{-1} and 1030–1670 cm^{-1} for FT Raman, respectively. Dendrograms were obtained by means of the OPUS options *scaling to first range* (=D values as the distance method)[44] and *average linkage* (linkage method). For processing microbial mass spectra, including calculation of mean spectra and UHCA, the MicrobeMS software[45] was utilised. For UHCA, mass spectra were smoothed, baseline corrected and normalised.[46,47] Peak detection was carried out by setting the number of peaks for each individual spectrum to 30. These peak tables were converted to barcode spectra, in which the peak amplitude information (intensity) has been omitted. Barcode spectra serving as inputs for UHCA and dendrograms were constructed with the same distance and linkage method used for FTIR and FT Raman spectra (see above).

Combined cluster analysis of phenotype data

Cluster membership functions obtained on the basis of UHCA of microbial MALDI TOF mass, FTIR and FT Raman spectra served as inputs for a combined phenotypic characterisation method. To this end, vectors containing the individual cluster membership values were compiled for each spectrum of a biological replicate after fusion steps 1–6 (MALDI and FTIR), or 1–7 (Raman). These vectors were, in the following, subjected to a top-level unsupervised hierarchical cluster analysis carried out by means of the *pdist* (Euclidean distance) and *linkage* (average linkage) functions of Matlab's statistics toolbox (Matlab version R2014a, The Mathworks, Natick, MA, USA). The dendrogram was generated using the *dendrogram* function of Matlab.

Results & discussion

Genotyping

The genetic relatedness of the isolates was determined by MLST, PFGE and SNP analysis of WGS data. Allelic profiles at seven loci based on the MLST scheme established by Herzog *et al.*[21] were assigned according to the PubMLST database (see ESI Table S1†). The ten hand soap isolates of *K. oxytoca* shared a single MLST profile (3;5;21;13;20 (^{62}C → T);6;12). The most similar sequence type found in the PubMLST database was ST151 (3;5;21;13;20;6;12), which displayed only a single nucleotide polymorphism. The corresponding strain (no. M936) was isolated from human faeces in France, 2009 (see ESI, Fig. S1†). Sequence data were concatenated in the order of loci used to define the allelic profile, leading to a single sequence of 3003 bp for each strain. The resulting neighbour-joining phylogeny is depicted in Fig. 1A. In addition, a combined analysis with all *K. oxytoca* isolates present in the PubMLST database was undertaken (see Fig. S1†). According to the nomenclature defined by Herzog *et al.*,[21] the ten hand soap isolates were found in a branch of cluster B1 together with strains that were mainly isolated from respiratory cultures. Within our study, *K. oxytoca* CB 4072 was most closely related to the hand soap isolates and clustered in a subbranch of cluster B1. Isolate CB 4063 was found in a branch of cluster B2, while all other isolates were found in cluster A except strain *K. oxytoca* CCUG 15788. According to MLST, the assignment of this reference strain to the species *K. oxytoca* seems questionable. As pointed out by Herzog *et al.*, the majority of AAHC (antibiotic-associated haemorrhagic colitis) isolates of *K. oxytoca* belong to cluster A, while respiratory isolates are almost exclusively found in subcluster B1, suggesting a pathogenic potential of the soap isolates that is associated with pneumonia rather than gastrointestinal tract infections.[21]

All isolates included in this study were examined by PFGE using XbaI macrorestriction enzyme. Between 10 and 20 bands per isolate were obtained. Their restriction patterns and a dendrogram created using Dice similarity coefficient and Ward's method is shown in Fig. 1B. In congruency to the MLST phylogeny, the hand soap isolates of *K. oxytoca* formed a separate PFGE cluster and belonged to a single pattern. Six profiles contained an additional band at about 72 kb, which was missing in the remaining four patterns and presumably represented a plasmid fragment (see red arrow in Fig. 1B). The other analysed clinical isolates and reference strains had unique genotypes and exhibited no genetic relatedness based on PFGE data.

Whole-genome based SNP analysis was used to infer the relationships of the isolates, and a phylogenetic tree was reconstructed using SNPs differentiating these genomes (Fig. 1C). A SNP was defined as a position in the genome sequence at which at least two strains displayed different bases. Two outlier strains (*K. pneumoniae* ssp. *ozaenae* ATCC 25926 and *K. oxytoca* CCUG15788) and one hand soap isolate (PHS-891) were omitted from this analysis due to the low coverage of their genome sequences. The SNP tree was computed from 38 214 concatenated qualified SNPs using the complete genome sequence of strain *K. oxytoca* M1 (Acc. No. CP008841) as the reference. Whole-genome based SNP analysis clustered all hand soap isolates into one single cluster, differentiating them accurately from the reference strains and other clinical isolates. The inset of Fig. 1C shows a detailed view of the hand soap isolate clade, differentiating isolates from the two hand soap bottles, lot I and lot II, based on 134 of the 38 214 SNPs.

Phenotyping

Similarity trees (dendrograms) produced by the different phenotype-based analytical techniques, MALDI TOF mass spectrometry, FTIR and NIR FT Raman spectroscopy, were compared with phylogenetic trees obtained by the different genotype-based methods (MLST, PFGE and WGS). Phenotype-based analytical techniques are known to produce taxon-specific spectral fingerprints suitable for bacterial identification at the species or subspecies level (MALDI), or even at the strain level (FTIR, FT Raman). Fig. 2A–C presents typical examples of MALDI TOF mass, FTIR and NIR FT Raman spectra from two exemplarily chosen isolates of *K. oxytoca*. Since in phenotype-based analyses some variation is common between spectra of technical and/or biological replicates, each of the *Klebsiella* spp. isolates or strains given in Table 1 was characterised from three individual microbial cultures.

MALDI TOF mass spectrometry produced about 80 to 100 mass peaks in the m/z range from 2000 to 13 000 that mainly originate from highly abundant ribosomal proteins and from nucleic acid binding proteins.[29,30] Lipids, carbohydrates and nucleic acids are usually not detectable by MALDI TOF mass spectrometry using common experimental protocols for microbial identification. Due to the high level of standardisation and the manual mode of MS measurements, the SNR and the spectral reproducibility were particularly high (*cf.* spectra of *K. oxytoca* hand soap isolates in Fig. 2A).

In contrast, the two vibrational spectroscopic methods, FTIR and NIR FT Raman spectroscopy, can be used to record signatures carrying signals from all main building blocks of microbial cells (*e.g.* proteins, lipids, nucleic acids, carbohydrates). Both methods are known to be suitable for differentiation, identification and classification of diverse microbial species down to the strain level.[44,48–51] In IR and Raman spectra of microbial samples, the useful information is generally distributed over broad spectral regions; in combination with various methods of spectral pre-processing the spectral information can be extracted from the typically broad and complex spectral contours by multivariate pattern recognition methods such as cluster analysis, or analysis by neural networks.[44,52–55] As shown by Fig. 2B and C, the respective raw FTIR, or NIR FT Raman spectra of the *K. oxytoca* isolates PHS-890 and PHS-891 exhibit a remarkable degree of similarity. The latter aspect is quite typical: in vibrational spectroscopy, the usually very small, but highly characteristic, differences between spectra of closely related microbial strains can often be visualised only by application of advanced methods of resolution enhancement (derivation, or Fourier self-deconvolution) and normalisation.[52]

Whole-cell MALDI TOF MS was initially used to confirm species identification by the MALDI Biotyper system from Bruker Daltonics.[56] All isolates were identified at the species level with log score values above 2.2 (data not shown). Unsupervised hierarchical cluster analysis (UHCA) of the mass peak tables revealed some degree of intraspecies diversity and the dendrogram (see Fig. 3A) demonstrated a good correlation with phylogenetic data based on molecular typing methods, indicating the discriminatory power of the method. Spectra from the liquid soap isolates formed a homogeneous subcluster clearly distinct from all other *Klebsiella* spp. isolates (*cf.* cluster 1 in the dendrogram of Fig. 3A). The degree of heterogeneity within this cluster is comparable with the heterogeneity found in the remaining clusters 2–7. In the dendrograms of Fig. 3 the horizontal red lines are arbitrarily set cutoff values used for group definition and differentiation of the *Klebsiella* spp. strains investigated.

Sets of discriminating biomarker peaks were tentatively assigned to protein identities by database searches using experimental peak positions of the MALDI TOF mass spectra. Major biomarker ions discriminating the hand soap isolates from the reference and control strains included the 30S ribosomal subunit protein S15 (MW of the $[M\text{-}Met + H]^+$ ion of 10 079 and 10 095 Da corresponding to an A → S amino acid exchange), integration host factor subunit alpha (m/z $[M\text{-}Met + H]^+$ of 11 237 and 11 251 Da, D → E) and 50S ribosomal protein L25 (m/z $[M + H]^+$ of 10 650 and 10 677 (L → V; N → T; L → I) and m/z $[M + H]^+$ 10 637 (unknown sequence)).

Dendrograms obtained by UHCA of FTIR and NIR FT Raman spectra from the *Klebsiella* spp. data set are given in Fig. 3B and C, respectively. The dendrogram of the FTIR data (Fig. 3B) demonstrates an acceptable level of separation between spectra from *K. pneumoniae* and *K. oxytoca* and – with the exception of *K. oxytoca* strains ATCC 13182 and CB4063, which form a mixed cluster – the presence of strain-specific clusters. Outliers are not included in this dendrogram, and there is a high degree of similarity among spectra of the hand soap isolates: these isolates form a large and homogenous cluster, cluster 1, which interestingly does not show any signs of internal substructure, neither in terms of subclusters composed by replicate spectra of individual isolates, nor in terms of isolates from the different sample lots (see Table 1). Analysis of the NIR FT Raman spectra by UHCA indicates a somewhat reduced level of separation: although all spectra of the hand soap isolates also form one large cluster, we observe outliers (*K. oxytoca* CB4074.3 and CB4063.2) and the formation of two mixed clusters, cluster 2, which is formed by spectra from *K. oxytoca* strains Oman 61 and CCUG 15788, and cluster 3 composed by spectra from *K. oxytoca* strains CB4063 and ATCC 13182. Furthermore, the dendrogram of Fig. 3C was obtained by omitting the spectral information contained in the CH-stretching region (2800–3050 cm^{-1}). This region is known to be dominated by contributions of lipids. Cluster analysis attempts of FT

Raman spectra that included the CH-stretching region revealed a tendency towards a less clear separation between the individual strains and more outliers (not shown). Although the main structure of the dendrogram of Fig. 3C is similar to the dendrogram structures of Fig. 3A and B, this example suggests that FT Raman spectroscopy provides less taxonomic information than the two other phenotypic characterisation methods. Whether this observation represents a fundamental limitation of NIR FT Raman, or whether this is merely only a consequence of a reduced SNR – UHCA with derivative spectra does not improve the results – cannot be decided on the basis of the relatively small spectral data set investigated in this study.

Fig. 4 shows the results of a combined cluster analysis of data acquired by the different microbial characterisation techniques MALDI TOF mass spectrometry, FTIR and FT Raman spectroscopy. Spectra obtained from the different biological replicates of the hand soap isolates fall in one single cluster (cluster 1). The dendrogram of Fig. 4 indicates that replicates of each strain compose strain-specific clusters with a maximum intra-strain distance that is significantly lower than any distance between strains. This and the absence of outliers suggest that all *K. oxytoca* hand soap isolates are phenotypically indistinguishable.

Comparison of techniques

Rapid and reliable bacterial typing systems are key tools to identify routes of transmission and infection, so that targeted control measures can be initiated in a timely manner. The potential of rapid phenotypic techniques for discriminating *K. oxytoca* isolates below the species level has been examined in this work and compared with the respective performance of DNA-based methods. Clustering of spectra from the *Klebsiella* isolate collection, which included ten *K. oxytoca* strains from two different lots of a contaminated liquid hand soap product, revealed a high level of concordance for all six microbial typing techniques used. MALDI TOF mass spectrometry and the vibrational spectroscopic methods of FTIR and NIR FT Raman spectroscopy are all physico-chemical techniques capable of rapid and non-subjective fingerprinting of bacteria. All three phenotype-based characterisation techniques have in common that spectral analysis is based on pattern recognition, in which spectral fingerprints of unknowns are compared against libraries of bacterial reference spectra. Phenotypic characterisation techniques can be universally applied without prior knowledge of genus or species identity, and are easily adaptable to routine analysis. Measurements can be automated under high throughput conditions providing results within minutes. The techniques have the advantages of low costs and of low requirements in terms of sample amount and sample preparation efforts. For the above reasons, phenotypic methods have been suggested in the past as valuable alternatives for classifying microorganisms below the species level and for epidemiological typing (FTIR;[57–63] Raman;[64–67] MALDI TOF MS).[68–73]

The depth of taxonomic resolution attainable by the different phenotypic characterisation techniques has been the subject of intense research. While MALDI TOF mass spectrometry is particularly robust for diagnostics at the genus and species level with some potential for strain level identification,[71] FTIR spectroscopy suffers from the fact that a high level of intraspecies diversity of the IR spectral fingerprints may lead to reduced accuracy of species identification.[48] On the other hand, the discriminative potential of FTIR spectroscopy for microbial typing is well documented.[49,50,74,75] This discrepancy and the proposed higher strain typing capability of FTIR spectroscopy have been explained by the particular information content of the IR spectra carrying information from several classes of biomolecules, including lipids, nucleic acids and carbohydrates (lipopolysaccharides, LPS). The sensitivity of IR spectroscopy towards carbohydrates and LPS present in outer membrane structures of Gram-negative bacteria, including *Klebsiella*, may constitute the compositional basis for its ability to discriminate bacterial strains. In contrast, MALDI TOF MS detects mainly conserved ribosomal and to a lesser extend nucleic acid binding proteins that exhibit only limited strain specificity. Whether the aforementioned considerations are also valid for the complementary vibrational spectroscopic method of NIR FT Raman spectroscopy, is difficult to judge on the basis of the available data. The dendrograms produced from the experimental IR and Raman spectra suggest, however, a reduced taxonomic resolution of NIR FT Raman compared to FTIR spectroscopy (see outliers in Fig. 3C). This and the finding of a generally reduced separability of the individual microbial strains in the NIR FT Raman setup are likely a result of an only moderate SNR. It must be stated however, that the number of strains used in this study is rather limited and does not allow us to draw general conclusions. Furthermore, the taxonomic resolution of all three phenotypic characterisation techniques is certainly taxon-dependent.[76]

Vibrational spectral fingerprints are dependent on the metabolic state of the cells under study. Growth and nutrient conditions may exert a notable influence on the spectral patterns. Therefore, microbial characterisation by FTIR and NIR FT Raman spectroscopy requires a high level of standardisation to assure adequate spectral reproducibility. Interestingly, the standardisation requirements are less strict in a MALDI-TOF MS-based workflow (see above), whereas DNA-based techniques are considered to be invariant to the cell's metabolic state. The presence of spectral databases suitable for exchange between different laboratories is a further requirement of a phenotype-based microbial identification methodology. While such libraries are not available for IR or Raman-based applications, MALDI TOF MS has been already implemented in the routine practice of microbial identification, mainly because of the fact that commercial spectral databases are today readily available.[77] It was certainly the higher robustness of the mass spectral fingerprints and – compared to IR and Raman spectroscopy – the lower requirements in terms of standardisation (growth conditions, type of sample preparation) that have enforced the development of commercial MALDI TOF MS databases. This is to our opinion the main reason why MS and not IR or Raman typing has found its way into the microbiological routine.

Among DNA-based typing techniques, PFGE is currently the gold standard for epidemiological studies and outbreak detection.[78] PFGE has a very high discriminative power but is labour-intensive and offers only low sample throughput. Using PFGE for source-tracking purposes, it should be kept in mind that differences in PFGE profiles of isogenic strains may be attributed to differences in the contents of mobile genetic elements (*e.g.* plasmids or prophages).[79] In the highly related liquid hand soap isolates, PFGE revealed the presence of additional single bands in some isolates, which are likely due to the carriage of large plasmids. MLST also has a high discriminatory power and has the advantage of providing highly standardised data readily transportable between laboratories. However, because genetic mutations accumulate slowly in housekeeping genes, classical MLST schemes may be more suited to study long-term global epidemiology rather than local epidemiology. Like PFGE, MLST is known as a labour-intensive, time-consuming and costly technique. Higher resolution methods utilizing whole-genome sequencing (WGS) will clearly improve surveillance and tracing pathogens and are likely to replace other typing methods in the near future. WGS has already been used for back-tracking outbreaks including *K. pneumoniae*.[80] SNP analysis using WGS shows great promise for use as an epidemiological typing tool. In our study it provided the highest discriminatory power and allowed differentiation even of highly related isogenic strains. In fact, WGS was the only method that was able to separate the liquid hand soap isolates into distinct lot-specific clades (see inset of Fig. 1C). WGS is often perceived as a relatively complex and time-consuming technique. However, this is not always true, as time-consuming cultivation can be sometimes avoided. In cases where a sufficient bacterial load is available and DNA from other sources is absent, sequencing can be performed directly from the isolated sample DNA. In such cases, WGS has a major speed advantage over methods that require cultivation. It is important to emphasise that – as in the present study – WGS analysis from isolated sample DNA is not always feasible.

In conclusion, the present study suggests that phenotype-based screening could assist in dereplication of *K. oxytoca* isolates with different profiles. Phenotype methods may be applied as rapid alternatives to PFGE and MLST for initial classification, or source-tracking of bacterial pathogens before results from DNA-based techniques are available. While MALDI-TOF MS will be the method of choice due to its availability in many microbiological laboratories, FTIR spectroscopy appears to be particularly well-suited among the phenotype-based techniques because of its high discriminatory power, simplicity and the advantage of low costs. The results of NIR FT Raman measurements suggest that under the given experimental conditions (NIR excitation at $\lambda = 1064$ nm, Ge-detector, FT setup) this method is not optimally suitable for microbial typing.

Whole genome sequencing is certainly the most powerful technique that still could benefit from combination with rapid phenotyping methods. Indeed, techniques like MALDI-TOF MS or FTIR spectroscopy could assist and guide the time-consuming data analysis procedures of WGS investigations. The combination of genotyping and phenotyping techniques would be helpful to overcome the inherent slowness of a purely genotype-based identification system. For further pathogen characterisation, *e.g.* regarding the presence of virulence or antibiotic resistance genes and extrachromosomal elements, and unambiguous confirmation of the typing results obtained by phenotypic analyses, WGS will certainly play a key role in the future.

Abbreviations

AAHC	Antibiotic-associated haemorrhagic colitis
CFU	Colony forming unit
DTGS	Deuterated triglycine sulphate detector
FT	Fourier transform

FTIR	Fourier transform infrared
HCCA	α-Cyano-4-hydroxycinnamic acid
IR	Infrared
MALDI TOF	Matrix-assisted laser desorption/ionization time-of-flight
MLST	Multi-locus sequence typing
MS	Mass spectrometry
NIR	Near-infrared
NJ	Neighbour-joining
PFGE	Pulsed-field gel electrophoresis
SNP	Single nucleotide polymorphism
SNR	Signal-to-noise ratio
ST	Sequence type
TFA	Trifluoroacetic acid
UHCA	Unsupervised hierarchical cluster analysis
WGS	Whole genome sequencing

Acknowledgements

The authors are thankful for the excellent technical assistance of Marett Splett, Jane Kowall and Anna Louisa Hauffe at the BfR Berlin and of Andy Schneider and Stefanie Herfort from the Proteomics and Spectroscopy Unit (ZBS6) at the RKI Berlin. Furthermore, the authors are grateful to Dr Lothar Beutin (BfR Berlin) and Dr Daniela Jacob (RKI Berlin, Highly Pathogenic Microorganisms/ZBS2) for providing selected strains of *K. oxytoca*. The authors acknowledge intramural funding at the German Federal Institute for Risk Assessment (BfR), Grants 47-001 and 1322-621.

References

1 S. D. Eiref, I. M. Leitman and W. Riley, Hand sanitizer dispensers and associated hospital-acquired infections: friend or fomite?, *Surg. Infect.*, 2012, **13**(3), 137–140.

2 S. E. Brooks, M. A. Walczak, R. Hameed and P. Coonan, Chlorhexidine resistance in antibiotic-resistant bacteria isolated from the surfaces of dispensers of soap containing chlorhexidine, *Infect. Control Hosp. Epidemiol.*, 2002, **23**(11), 692–695.

3 S. E. Brooks, M. A. Walczak, S. Malcolm and R. Hameed, Intrinsic *Klebsiella pneumoniae* contamination of liquid germicidal hand soap containing chlorhexidine, *Infect. Control Hosp. Epidemiol.*, 2004, **25**(10), 883–885.

4 M. Drancourt, Nosocomial *Serratia marcescens* infections associated with extrinsic contamination of a liquid nonmedicated soap, *Infect. Control Hosp. Epidemiol.*, 2000, **21**(3), 196–199.

5 S. Buffet-Bataillon, V. Rabier, P. Bétrémieux, A. Beuchée, M. Bauer, P. Pladys, *et al.* Outbreak of *Serratia marcescens* in a neonatal intensive care unit: contaminated unmedicated liquid soap and risk factors, *J. Hosp. Infect.*, 2009, **72**(1), 17–22.

6 S. Lanini, S. D'Arezzo, V. Puro, L. Martini, F. Imperi, P. Piselli, *et al.* Molecular epidemiology of a *Pseudomonas aeruginosa* hospital outbreak driven by a contaminated disinfectant-soap dispenser, *PLoS One*, 2011, **6**(2), e17064.

7 E. Aktaş, E. Taşpinar, D. Alay, E. D. K. Ögedey, C. Külah and F. Cömert, Extrinsic contamination of liquid soap with various Gram-negative bacteria in a hospital in Turkey, *Infect. Control Hosp. Epidemiol.*, 2010, **31**(11), 1199–1201.

8 D. J. Weber, W. A. Rutala and E. E. Sickbert-Bennett, Outbreaks associated with contaminated antiseptics and disinfectants, *Antimicrob. Agents Chemother.*, 2007, **51**(12), 4217–4224.

9 L. A. Lorenz, B. D. Ramsay, D. M. Goeres, M. W. Fields, C. A. Zapka and D. R. Macinga, Evaluation and remediation of bulk soap dispensers for biofilm, *Biofouling*, 2012, **28**(1), 99–109.

10 M. Chattman, S. L. Maxwell and C. P. Gerba, Occurrence of heterotrophic and coliform bacteria in liquid hand soaps from bulk refillable dispensers in public facilities, *J. Environ. Health*, 2011, **73**(7), 26–29.

11 C. A. Zapka, E. J. Campbell, S. L. Maxwell, C. P. Gerba, M. J. Dolan, J. W. Arbogast, *et al.* Bacterial hand contamination and transfer after use of contaminated bulk-soap-refillable dispensers, *Appl. Environ. Microbiol.*, 2011, **77**(9), 2898–2904.

12 T. A. Madani, S. Alsaedi, L. James, B. S. Eldeek, A. A. Jiman-Fatani, M. M. Alawi, *et al. Serratia marcescens*-contaminated baby shampoo causing an outbreak among newborns at King Abdulaziz University Hospital, *J. Hosp. Infect.*, 2011, **78**(1), 16–19.

13 B. Swaminathan, T. J. Barrett, S. B. Hunter and R. V. Tauxe, PulseNet: the molecular subtyping network for foodborne bacterial disease surveillance, United States, *Emerging Infect. Dis.*, 2001, **7**(3), 382–389.

14 B. Swaminathan, P. Gerner-Smidt, L. K. Ng, S. Lukinmaa, K. M. Kam, S. Rolando, *et al.* Building PulseNet International: an interconnected system of laboratory networks to facilitate timely public health recognition and response to foodborne disease outbreaks and emerging foodborne diseases, *Foodborne Pathog. Dis.*, 2006, **3**(1), 36–50.

15 R. D. Arbeit, M. Arthur, R. Dunn, C. Kim, R. K. Selander and R. Goldstein, Resolution of recent evolutionary divergence among *Escherichia coli* from related lineages: the application of pulsed field electrophoresis to molecular epidemiology, *J. Infect. Dis.*, 1990, **161**(2), 230–235.

16 D. S. Hansen, R. Skov, J. V. Benedí, V. Sperling and H. J. Kolmos, *Klebsiella* typing: pulsed-field gel electrophoresis (PFGE) in comparison with O:K-serotyping, *Clin. Microbiol. Infect.*, 2002, **8**(7), 397–404.

17 D. Decré, B. Burghoffer, V. Gautier, J. C. Petit and G. Arlet, Outbreak of multi-resistant *Klebsiella oxytoca* involving strains with extended-spectrum β-lactamases and strains with extended-spectrum activity of the chromosomal β-lactamase, *J. Antimicrob. Chemother.*, 2004, **54**(5), 881–888.

18 V. C. C. Cheng, W. C. Yam, L. L. Tsang, M. C. Y. Yau, G. K. H. Siu, S. C. Y. Wong, *et al.* Epidemiology of *Klebsiella oxytoca*-associated diarrhea detected by simmons citrate agar supplemented with inositol, tryptophan, and bile salts, *J. Clin. Microbiol.*, 2012, **50**(5), 1571–1579.

19 A. Tsakris, A. Poulou, F. Markou, V. Pitiriga, E. T. Piperaki, I. Kristo, *et al.* Dissemination of clinical isolates of *Klebsiella oxytoca* harboring CMY-31, VIM-1, and a new OXY-2-type variant in the community, *Antimicrob. Agents Chemother.*, 2011, **55**(7), 3164–3168.

20 M. C. J. Maiden, J. A. Bygraves, E. Feil, G. Morelli, J. E. Russell, R. Urwin, *et al.* Multilocus sequence typing: a portable approach to the identification of clones within populations of pathogenic microorganisms, *Proc. Natl. Acad. Sci. U. S. A.*, 1998, **95**(6), 3140–3145.

21 K. A. T. Herzog, G. Schneditz, E. Leitner, G. Feierl, K. M. Hoffmann, I. Zollner-Schwetz, *et al.* Genotypes of *Klebsiella oxytoca* isolates from patients with nosocomial pneumonia are distinct from those of isolates from patients with antibiotic-associated hemorrhagic colitis, *J. Clin. Microbiol.*, 2014, **52**(5), 1607–1616.

22 S. R. Leopold, R. V. Goering, A. Witten, D. Harmsen and A. Mellmann, Bacterial whole-genome sequencing revisited: portable, scalable, and standardized analysis for typing and detection of virulence and antibiotic resistance genes, *J. Clin. Microbiol.*, 2014, **52**(7), 2365–2370.

23 E. Zankari, H. Hasman, R. S. Kaas, A. M. Seyfarth, Y. Agersø, O. Lund, *et al.* Genotyping using whole-genome sequencing is a realistic alternative to surveillance based on phenotypic antimicrobial susceptibility testing, *J. Antimicrob. Chemother.*, 2013, **68**(4), 771–777.

24 A. M. Hammerum, F. Hansen, M. N. Skov, M. Stegger, P. S. Andersen, A. Holm, *et al.* Investigation of a possible outbreak of carbapenem-resistant *Acinetobacter baumannii* in Odense, Denmark using PFGE, MLST and whole-genome-based SNPs, *J. Antimicrob. Chemother.*, 2014, **70**(7), 1965–1968.

25 P. Leekitcharoenphon, R. S. Kaas, M. C. Thomsen, C. Friis, S. Rasmussen and F. M. Aarestrup, snpTree – a web-server to identify and construct SNP trees from whole genome sequence data, *BMC Genomics*, 2012, **13**(Suppl 7), S6.

26 P. Leekitcharoenphon, E. M. Nielsen, R. S. Kaas, O. Lund and F. M. Aarestrup, Evaluation of whole genome sequencing for outbreak detection of *Salmonella enterica*, *PLoS One*, 2014, **9**(2), e87991.

27 D. Stucki, M. Ballif, T. Bodmer, M. Coscolla, A. M. Maurer, S. Droz, *et al.* Tracking a tuberculosis outbreak over 21 years: strain-specific single-nucleotide polymorphism typing combined with targeted whole-genome sequencing, *J. Infect. Dis.*, 2015, **211**(8), 1306–1316.

28 A. J. Taylor, V. Lappi, W. J. Wolfgang, P. Lapierre, M. J. Palumbo, C. Medus, *et al.* Characterization of foodborne outbreaks of *Salmonella enterica* serovar enteritidis with whole-genome sequencing single nucleotide polymorphism-based analysis for surveillance and outbreak detection, *J. Clin. Microbiol.*, 2015, **53**(10), 3334–3340.

29 V. Ryzhov and C. Fenselau, Characterization of the protein subset desorbed by MALDI from whole bacterial cells, *Anal. Chem.*, 2001, **73**(4), 746–750.

30 F. J. Pineda, M. D. Antoine, P. A. Demirev, A. B. Feldman, J. Jackman, M. Longenecker, *et al.* Microorganism identification by matrix-assisted laser/dissorption ionization mass spectrometry and model-derived ribosomal protein biomarkers, *Anal. Chem.*, 2003, **75**(15), 3817–3822.

31 M. V. Larsen, S. Cosentino, S. Rasmussen, C. Friis, H. Hasman, R. L. Marvig, *et al.* Multilocus sequence typing of total-genome-sequenced bacteria, *J. Clin. Microbiol.*, 2012, **50**(4), 1355–1361.

32 K. Tamura, G. Stecher, D. Peterson, A. Filipski and S. Kumar, MEGA6: molecular evolutionary genetics analysis version 6.0, *Mol. Biol. Evol.*, 2013, **30**(12), 2725–2729.

33 M. Kimura, A simple method for estimating evolutionary rates of base substitutions through comparative studies of nucleotide sequences, *J. Mol. Evol.*, 1980, **16**(2), 111–120.

34 N. Saitou and M. Nei, The neighbor-joining method: a new method for reconstructing phylogenetic trees, *Mol. Biol. Evol.*, 1987, **4**(4), 406–425.

35 S. B. Hunter, P. Vauterin, M. A. Lambert-Fair, M. S. Van Duyne, K. Kubota, L. Graves, *et al.* Establishment of a universal size standard strain for use with the pulsenet standardized pulsed-field gel electrophoresis protocols: converting the national databases to the new size standard, *J. Clin. Microbiol.*, 2005, **43**(3), 1045–1050.

36 J. A. Hammerl, P. Lasch, A. Nitsche, P. W. Dabrowski, H. Hahmann, A. Wicke, *et al.* Draft Genome Sequences of *Klebsiella oxytoca* Isolates Originating from a Highly Contaminated Liquid Hand Soap Product, *Genome Announc.*, 2015, **3**(4), e00820-15.

37 B. Langmead and S. L. Salzberg, Fast gapped-read alignment with Bowtie 2, *Nat. Methods*, 2012, **9**(4), 357–359.

38 A. McKenna, M. Hanna, E. Banks, A. Sivachenko, K. Cibulskis, A. Kernytsky, *et al.* The Genome Analysis Toolkit: a MapReduce framework for analyzing next-generation DNA sequencing data, *Genome Res.*, 2010, **20**(9), 1297–1303.

39 S. Guindon, J. F. Dufayard, V. Lefort, M. Anisimova, W. Hordijk and O. Gascuel, New algorithms and methods to estimate maximum-likelihood phylogenies: assessing the performance of PhyML 3.0, *Syst. Biol.*, 2010, **59**(3), 307–321.

40 D. Helm, H. Labischinski and D. Naumann, Elaboration of a procedure for identification of bacteria using Fourier-Transform IR spectral libraries: a stepwise correlation approach, *J. Microbiol. Methods*, 1991, **14**(2), 127–142.

41 D. Helm, H. Labischinski, G. Schallehn and D. Naumann, Classification and identification of bacteria by Fourier-transform infrared spectroscopy, *J. Gen. Microbiol.*, 1991, **137**(1), 69–79.

42 D. Naumann, D. Helm and H. Labischinski, Microbiological characterizations by FT-IR spectroscopy, *Nature*, 1991, **351**(6321), 81–82.

43 P. Lasch, H. Nattermann, M. Erhard, M. Stämmler, R. Grunow, N. Bannert, *et al.* MALDI-TOF mass spectrometry compatible inactivation method for highly pathogenic microbial cells and spores, *Anal. Chem.*, 2008, **80**(6), 2026–2034.

44 P. Lasch and D. Naumann, Infrared Spectroscopy in Microbiology, *Encyclopedia of Analytical Chemistry*, 2015.

45 P. Lasch, *MicrobeMS: A Matlab Toolbox for Analysis of Microbial MALDI-TOF Mass Spectra*, 2015. http://www.microbe-ms.com.

46 P. Lasch, T. Wahab, S. Weil, B. Palyi, H. Tomaso, S. Zange, *et al.* Identification of Highly Pathogenic Microorganisms by Matrix-Assisted Laser Desorption Ionization-Time of Flight Mass Spectrometry: Results of an Interlaboratory Ring Trial, *J. Clin. Microbiol.*, 2015, **53**(8), 2632–2640.

47 P. Lasch, C. Fleige, M. Stämmler, F. Layer, U. Nübel, W. Witte, *et al.* Insufficient discriminatory power of MALDI-TOF mass spectrometry for typing of *Enterococcus faecium* and *Staphylococcus aureus* isolates, *J. Microbiol. Methods*, 2014, **100**(1), 58–69.

48 M. Wenning, F. Breitenwieser, R. Konrad, I. Huber, U. Busch and S. Scherer, Identification and differentiation of food-related bacteria: a comparison of

FTIR spectroscopy and MALDI-TOF mass spectrometry, *J. Microbiol. Methods*, 2014, **103**, 44–52.

49 L. Beutin, Q. Wang, D. Naumann, W. Han, G. Krause, L. Leomil, *et al.* Relationship between O-antigen subtypes, bacterial surface structures and O-antigen gene clusters in *Escherichia coli* O123 strains carrying genes for Shiga toxins and intimin, *J. Med. Microbiol.*, 2007, **56**(2), 177–184.

50 T. Grunert, M. Wenning, M. S. Barbagelata, M. Fricker, D. O. Sordelli, F. R. Buzzola, *et al.* Rapid and reliable identification of *Staphylococcus aureus* capsular serotypes by means of artificial neural network-assisted Fourier transform infrared spectroscopy, *J. Clin. Microbiol.*, 2013, **51**(7), 2261–2266.

51 C. A. Rebuffo-Scheer, C. Kirschner, M. Staemmler and D. Naumann, Rapid species and strain differentiation of non-tubercoulous mycobacteria by Fourier-Transform Infrared microspectroscopy, *J. Microbiol. Methods*, 2007, **68**(2), 282–290.

52 P. Lasch, Spectral pre-processing for biomedical vibrational spectroscopy and microspectroscopic imaging, *Chemom. Intell. Lab. Syst.*, 2012, **117**, 100–114.

53 D. Helm, H. Labischinski, G. Schallehn and D. Naumann, Classification and identification of bacteria by Fourier-transform infrared spectroscopy, *J. Gen. Microbiol.*, 1991, **137**(1), 69–79.

54 J. Schmitt, T. Udelhoven, T. Löchte, H. C. Flemming and D. Naumann, Artificial Neural Network applied to FTIR and FT-Raman spectra in biomedical applications, *AIP Conf. Proc.*, 1998, **430**, 260–263.

55 D. Naumann, FT-IR spectroscopy of microorganisms at the Robert Koch-Institute: experiences gained during a successful project, *Proc. SPIE*, 2008, 6853.

56 T. Maier, S. Klepel, Z. Renner and M. Kostrzewa, Fast and reliable MALDI-TOF MS-based microorganism identification, *Nat. Methods*, 2006, **3**, 324–334.

57 C. A. Rebuffo-Scheer, J. Schmitt and S. Scherer, Differentiation of *Listeria monocytogenes* serovars by using artificial neural network analysis of fourier-transformed infrared spectra, *Appl. Environ. Microbiol.*, 2007, **73**(3), 1036–1040.

58 S. Kim, B. L. Reuhs and L. J. Mauer, Use of Fourier transform infrared spectra of crude bacterial lipopolysaccharides and chemometrics for differentiation of *Salmonella enterica* serotypes, *J. Appl. Microbiol.*, 2005, **99**(2), 411–417.

59 A. E. Kuhm, D. Suter, R. Felleisen and J. Rau, Identification of *Yersinia enterocolitica* at the species and subspecies levels by fourier transform infrared spectroscopy, *Appl. Environ. Microbiol.*, 2009, **75**(18), 5809–5813.

60 C. Patrícia Coutinho, I. Sá-Correia and J. Almeida Lopes, Use of Fourier transform infrared spectroscopy and chemometrics to discriminate clinical isolates of bacteria of the *Burkholderia cepacia* complex from different species and ribopatterns, *Anal. Bioanal. Chem.*, 2009, **394**(8), 2161–2171.

61 M. Vaz, L. Meirinhos-Soares, C. C. S. Sousa, M. Ramirez, J. Melo-Cristino and J. A. Lopes, Serotype discrimination of encapsulated *Streptococcus pneumoniae* strains by Fourier-transform infrared spectroscopy and chemometrics, *J. Microbiol. Methods*, 2013, **93**(2), 102–107.

62 R. Davis and L. J. Mauer, Subtyping of *Listeria monocytogenes* at the haplotype level by Fourier transform infrared (FT-IR) spectroscopy and multivariate statistical analysis, *Int. J. Food Microbiol.*, 2011, **150**(2–3), 140–149.

63 O. Preisner, R. Guiomar, J. Machado, J. C. Menezes and J. A. Lopes, Application of fourier transform infrared spectroscopy and chemometrics for differentiation of *Salmonella enterica* serovar enteritidis phage types, *Appl. Environ. Microbiol.*, 2010, **76**(11), 3538–3544.

64 K. Guyot, V. Biran, C. Doit, D. Moissenet, T. Guillard, L. Brasme, *et al.* Raman spectroscopic analysis of the clonal and horizontal spread of CTX-M-15-producing *Klebsiella pneumoniae* in a neonatal intensive care unit, *Eur. J. Clin. Microbiol. Infect. Dis.*, 2012, **31**(10), 2827–2834.

65 D. Willemse-Erix, T. Bakker-Schut, F. Slagboom-Bax, J. W. Jachtenberg, N. Lemmens-Den Toom, C. C. Papagiannitsis, *et al.* Rapid typing of extended-spectrum β-lactamase- and carbapenemase-producing *Escherichia coli* and *Klebsiella pneumoniae* isolates by use of spectracell RA, *J. Clin. Microbiol.*, 2012, **50**(4), 1370–1375.

66 D. F. M. Willemse-Erix, M. J. Scholtes-Timmerman, J. W. Jachtenberg, W. B. Van Leeuwen, D. Horst-Kreft, T. C. B. Schut, *et al.* Optical fingerprinting in bacterial epidemiology: Raman spectroscopy as a real-time typing method, *J. Clin. Microbiol.*, 2009, **47**(3), 652–659.

67 K. Maquelin, L. Dijkshoorn, T. J. K. Van Der Reijden and G. J. Puppels, Rapid epidemiological analysis of *Acinetobacter* strains by Raman spectroscopy, *J. Microbiol. Methods*, 2006, **64**(1), 126–131.

68 M. Treviño, D. Navarro, G. Barbeito, C. García-Riestra, C. Crespo and B. J. Regueiro, Molecular and epidemiological analysis of nosocomial carbapenem-resistant *Klebsiella* spp. using repetitive extragenic palindromic-polymerase chain reaction and matrix-assisted laser desorption/ionization-time of flight, *Microb. Drug Resist.*, 2011, **17**(3), 433–442.

69 M. Berrazeg, S. M. Diene, M. Drissi, M. Kempf, H. Richet, L. Landraud, *et al.* Biotyping of Multidrug-Resistant *Klebsiella pneumoniae* Clinical Isolates from France and Algeria Using MALDI-TOF MS, *PLoS One*, 2013, **8**(4), e61428.

70 A. Mencacci, C. Monari, C. Leli, L. Merlini, E. De Carolis, A. Vella, *et al.* Typing of nosocomial outbreaks of *Acinetobacter baumannii* by use of matrix-assisted laser desorption ionization-time of flight mass spectrometry, *J. Clin. Microbiol.*, 2013, **51**(2), 603–606.

71 T. R. Sandrin, J. E. Goldstein and S. Schumaker, MALDI TOF MS profiling of bacteria at the strain level: a review, *Mass Spectrom. Rev.*, 2013, **32**(3), 188–217.

72 R. Dieckmann and B. Malorny, Rapid screening of epidemiologically important *Salmonella enterica* subsp. *enterica* serovars by whole-cell matrix-assisted laser desorption ionization-time of flight mass spectrometry, *Appl. Environ. Microbiol.*, 2011, **77**(12), 4136–4146.

73 R. Dieckmann, R. Helmuth, M. Erhard and B. Malorny, Rapid classification and identification of *Salmonellae* at the species and subspecies levels by whole-cell matrix-assisted laser desorption ionization-time of flight mass spectrometry, *Appl. Environ. Microbiol.*, 2008, **74**(24), 7767–7778.

74 G. D. Sockalingum, W. Bouhedja, P. Pina, P. Allouch, C. Mandray, R. Labia, *et al.* ATR-FTIR spectroscopic investigation of imipenem-susceptible and -resistant *Pseudomonas aeruginosa* isogenic strains, *Biochem. Biophys. Res. Commun.*, 1997, **232**(1), 240–246.

75 M. Wenning, N. R. Buchl and S. Scherer, Species and strain identification of lactic acid bacteria using FTIR spectroscopy and artificial neural networks, *J. Biophotonics*, 2010, **3**(8–9), 493–505.

76 J. Ghyselinck, K. Van Hoorde, B. Hoste, K. Heylen and P. De Vos, Evaluation of MALDI-TOF MS as a tool for high-throughput dereplication, *J. Microbiol. Methods*, 2011, **86**(3), 327–336.

77 M. L. Faron, B. W. Buchan, J. Hyke, N. Madisen, J. L. Lillie, P. A. Granato, *et al.* Multicenter Evaluation of the Bruker MALDI Biotyper CA System for the Identification of Clinical Aerobic Gram-Negative Bacterial Isolates, *PLoS One*, 2015, **10**(11), e0141350.

78 A. J. Sabat, A. Budimir, D. Nashev, R. Sá-Leão, J. van Dijl, F. Laurent, *et al.* Overview of molecular typing methods for outbreak detection and epidemiological surveillance, *Euro Surveill.*, 2013, **18**(4), 20380.

79 S. M. Soto, I. Rodríguez, M. R. Rodicio, J. Vila and M. C. Mendoza, Detection of virulence determinants in clinical strains of *Salmonella enterica* serovar enteritidis and mapping on macrorestriction profiles, *J. Med. Microbiol.*, 2006, **55**(4), 365–373.

80 E. S. Snitkin, A. M. Zelazny, P. J. Thomas, F. Stock, D. K. Henderson, T. N. Palmore, *et al.* Tracking a hospital outbreak of carbapenem-resistant *Klebsiella pneumoniae* with whole-genome sequencing, *Sci. Transl. Med.*, 2012, **4**(148), 148ra116.

Faraday Discussions

PAPER

Fiber-optic Raman spectroscopy for *in vivo* diagnosis of gastric dysplasia

Jianfeng Wang,†[a] Kan Lin,†[a] Wei Zheng,†[a] Khek Yu Ho,[b] Ming Teh,[c] Khay Guan Yeoh[b] and Zhiwei Huang*[a]

Received 2nd November 2015, Accepted 20th November 2015

DOI: 10.1039/c5fd00151j

This study aims to assess the clinical utility of a rapid fiber-optic Raman spectroscopy technique developed for enhancing *in vivo* diagnosis of gastric precancer during endoscopic examination. We have developed a real-time fiber-optic Raman spectroscopy system capable of simultaneously acquiring both fingerprint (FP) (*i.e.*, 800–1800 cm^{-1}) and high-wavenumber (HW) (*i.e.*, 2800–3600 cm^{-1}) Raman spectra from gastric tissue *in vivo* at endoscopy. A total of 5792 high-quality *in vivo* FP/HW Raman spectra (normal ($n = 5160$); dysplasia ($n = 155$), and adenocarcinoma ($n = 477$)) were acquired in real-time from 441 tissue sites (normal ($n = 396$); dysplasia ($n = 11$), and adenocarcinoma ($n = 34$)) of 191 gastric patients (normal ($n = 172$); dysplasia ($n = 6$), and adenocarcinoma ($n = 13$)) undergoing routine endoscopic examinations. Partial least squares discriminant analysis (PLS-DA) together with leave-one-patient-out cross validation (LOPCV) were implemented to develop robust spectral diagnostic models. The FP/HW Raman spectra differ significantly between normal, dysplasia and adenocarcinoma of the stomach, which can be attributed to changes in proteins, lipids, nucleic acids, and the bound water content. PLS-DA and LOPCV show that the fiber-optic FP/HW Raman spectroscopy provides diagnostic sensitivities of 96.0%, 81.8% and 88.2%, and specificities of 86.7%, 95.3% and 95.6%, respectively, for the classification of normal, dysplastic and cancerous gastric tissue, superior to either the FP or HW Raman techniques alone. Further dichotomous PLS-DA analysis yields a sensitivity of 90.9% (10/11) and specificity of 95.9% (380/396) for the detection of gastric dysplasia using FP/HW Raman spectroscopy, substantiating its clinical advantages over white light reflectance endoscopy (sensitivity: 90.9% (10/11), and specificity: 51.0% (202/396)). This work demonstrates that the fiber-optic FP/HW Raman spectroscopy technique has great

[a]*Optical Bioimaging Laboratory, Department of Biomedical Engineering, Faculty of Engineering, National University of Singapore, 9 Engineering Drive 1, Singapore 117576. E-mail: biehzw@nus.edu.sg; Fax: +65 6872 3069; Tel: +65 6516 8856*

[b]*Department of Medicine, Yong Loo Lin School of Medicine, National University of Singapore, National University Health System, Singapore 119260*

[c]*Department of Pathology, Yong Loo Lin School of Medicine, National University of Singapore, National University Health System, Singapore 119074*

† Wang, Lin and Zheng contributed equally.

promise for enhancing *in vivo* diagnosis of gastric precancer during routine endoscopic examination.

Introduction

Gastric cancer is the fifth most commonly diagnosed malignancy and also the third leading cause of cancer death worldwide, with an estimated 951 600 new cases and 723 100 deaths occurring in 2012, accounting for 6.8%, and 8.8%, respectively of all the estimated new cancer cases (14 090 100) and deaths (8 201 600).[1] The incidence rates of gastric cancer are highest in Eastern Asia (*e.g.*, Japan, China, South Korea, Mongolia) with an incidence rate of 35.4% in males and 13.8% in females.[1] Early detection and localization with immediate removal and treatment (*e.g.*, endoscopic submucosal dissection (ESD)) of premalignant lesions (*e.g.*, high-grade dysplasia [HGD]) is crucial to improving the survival of gastric patients who underwent endoscopic screening of the stomach.[2–5] Current routine diagnosis of gastric dysplasia is clinically challenging, as the conventional white-light reflectance (WLR) endoscopy used relies heavily on visual identification of gross morphologic changes. While in 60% of the gastric patients, gastric dysplasia was shown as flat mucosa in the absence of specific endoscopic alterations, resulting in limited diagnostic accuracy using WLR endoscopy.[3–5] Well-performed but randomized endoscopic biopsies followed by conventional hematoxylin and eosin (H&E) staining remains the current gold standard for gastric cancer screening and diagnosis.[3–5] However, random biopsies associated with sampling errors make it labor-intensive and difficult for screening of high-risk gastric patients with multiple suspicious lesions.[3–5] Further, the diagnosis is also compromised with intra- and/or inter-operator variations.[3–5] There is profound clinical need to develop more advanced diagnostic techniques, providing rapid, objective and enhanced diagnosis of gastric precancer (dysplasia) with molecular specificity.

Optical spectroscopic techniques (*e.g.*, diffuse reflectance, autofluorescence, and Raman spectroscopy) cast new light on either uncovering tissue optical properties (*e.g.*, absorption and scattering coefficients), endogenous fluorophore distribution, or the constituent-specific (*i.e.*, DNA, proteins, lipids, and water *etc.*) biochemical/biomolecular information, which offer objective insights and therefore have attracted a great deal of interest for precancer and cancer diagnosis in internal organs (*e.g.*, stomach, esophagus, colon, bladder, and lung, *etc.*).[6–28] Raman spectroscopy is a label-free optical vibrational technique capable of probing biochemical/biomolecular changes of tissue associated with neoplastic transformation.[6–27,29] Currently, the exploration of biomedical Raman spectroscopy for gastric cancer and precancer diagnosis is mostly centered on the fingerprint (FP) range (*i.e.*, 800–1800 cm^{-1}), which uncovers a wealth of biochemical information (*i.e.*, proteins, lipids and nucleic acids) in cells and tissue.[8–20,27,28] However, FP Raman spectroscopy can suffer from relatively weak Raman intensity but overwhelmed tissue autofluorescence background in certain organ sites (*e.g.*, stomach, lung, liver).[12,20] High-wavenumber (HW) (*e.g.*, 2800–3600 cm^{-1}) Raman spectroscopy, on the other hand, can provide complementary biochemical information that is not achievable with FP Raman spectroscopy (*e.g.*, tissue water content), while it has stronger tissue Raman signals with less autofluorescence interference.[21–24] Several rationales exist by combining FP and HW Raman spectroscopy for gastric precancer and early cancer diagnosis at endoscopy: for tissues (*e.g.*, stomach) exhibiting intense autofluorescence overwhelming the tissue FP Raman signals, the HW Raman spectroscopy could still contain reasonably intense tissue HW Raman peaks associated with gastric tissue neoplasia; further, the FP and HW Raman spectroscopy are complementary, and the integrated FP/HW Raman spectroscopy could have advantages for improving tissue characterization and diagnosis.[25,26] Very recently, we have successfully developed an endoscope-based fiber-optic Raman spectroscopy system capable of simultaneously acquiring both fingerprint (FP) (*i.e.*, 800–1800 cm^{-1}) and high-wavenumber (HW) (*i.e.*, 2800–3600 cm^{-1}) tissue Raman spectra *in vivo* at endoscopy.[25,26] In this work, we investigate the clinical utility of the integrated FP/HW Raman spectroscopy technique developed to enhance real-time *in vivo* diagnosis of gastric dysplasia during clinical endoscopic examinations. Partial least squares-discriminant analysis (PLS-DA) and leave-one-patient-out cross validation (LOPCV) are implemented to develop robust spectral diagnostic algorithms for differentiation between dysplasia and normal gastric tissue. The diagnostic performances of the FP/HW Raman spectroscopy technique are also compared with the routine WLR endoscopic examination.

Materials and methods

Raman instrumentation and data processing

We have developed a rapid fiber-optic FP/HW Raman endoscopic system[25,26] that consists of a 785 nm diode laser (maximum output: 300 mW, B&W TEK Inc.), a customized beveled fiber-optic Raman probe (1.8 mm in outer diameter),[27] and a high-throughput reflective imaging spectrograph (Acton LS-785 f/2, Princeton Instruments Inc.) equipped with a customized high-diffraction efficiency (>85% within in range 800–3600 cm^{-1}) gold-coated grating (830 gr mm^{-1}) and a thermoelectric-cooled, NIR-optimized deep-depletion charge-coupled device (CCD) camera (PIXIS 400BR eXcelon, Princeton Instruments Inc.) for tissue Raman signals. The beveled fiber-optic Raman endoscopic probe (1.8 mm in outer diameter) compatible with most medical endoscopes was specially designed and fabricated for optimizing epithelial tissue Raman measurements:[27] *i.e.*, ~85% of the FP/HW Raman signal collected by the beveled Raman endoscopic probe arises from the top ~200 μm epithelium layer; whereas ~15% of the Raman collection is from below ~(200–800) μm of the gastric tissue, facilitating the *in vivo* diagnosis of gastric precancer.[27] The 785 nm laser power incident on the tissue surface is ~12 mW (equivalent to ~1.5 W cm^{-2} within the spot size of ~500 μm) permissible by the American National Standards Institute (ANSI) standard.[30] To further evaluate the photothermal effect of the laser power used, thermal modeling was conducted on the stomach,[31–34] showing a non-significant temperature rise (<0.07 °C) without causing photothermal damage to the gastric tissue within 1 minute of laser irradiation time.[35] The system was wavelength-calibrated using a mercury/argon calibration lamp (HG-1 and AR-1, Ocean Optics Inc., Dunedin, FL). The spectral response correction for the wavelength-calibrated system was conducted using a standard lamp (RS-10, EG&G Gamma Scientific, San Diego, CA).[28]

The measured gastric tissue spectra represented a combination of weak tissue Raman scattering, intense AF and the noise. Therefore, the spectra were preprocessed by a third-order Savitzky–Golay smoothing filter (a window width of 3 pixels) to reduce the noise. A fifth-order polynomial was found to be optimal for fitting the AF background in the noise-smoothed spectrum over the range of 800–1800 cm^{-1}, and this polynomial was then subtracted from the measured FP spectrum to yield the FP tissue Raman spectrum alone. While in the HW range (2800–3600 cm^{-1}), the first-order polynomial fit was used for removing the HW AF background. The tissue FP/HW Raman spectrum was finally normalized to the integrated area of the entire FP and HW range to compare the differences in spectral shapes and relative Raman band intensities between dysplastic and normal gastric tissue. One notes that there are no Raman peaks observed in the wavenumber range of 1800–2800 cm^{-1}, and this silent range is therefore excluded for data analysis. All the preprocessing is completed within 30 ms, and the processed Raman spectra and diagnostic outcomes can be displayed on the computer screen in real-time during clinical endoscopic examinations.[28]

Clinical trial protocol

The ethical protocol of the present study was approved by the Institutional Review Board (IRB) of the National Healthcare Group (NHG) of Singapore. Prior to Raman measurements, all patients (21 to 80 years old) signed an informed consent permitting the *in vivo* Raman spectroscopic measurements during gastric endoscope examinations. The trial was registered (registration number: ISRCTN15587241) on June 23, 2015 into the ISRCTN registry (http://www.isrctn.com/), which is a publicly accessible primary register that participates in the World Health Organization (WHO) International Clinical Trial Registry Platform. Also, the trial was performed in accordance with the International Conference on Harmonization (ICH) for Good Clinical Practice (GCP) guidelines, Declaration of Helsinki (2000). A total of 191 gastric patients (115 men and 76 women with mean ages of 55; normal ($n = 172$), dysplasia ($n = 6$), and adenocarcinoma ($n = 13$)), were enrolled in the gastrointestinal Raman endoscopic examination programme. The Raman probe passes down to the instrument channel of medical endoscopes (GIF FQ260Z; Olympus Medical Systems, Tokyo, Japan) under the guidance of WLR endoscopic imaging.[13] The Raman probe tip was visible approximately 0.5 cm in front of the endoscope camera. During endoscopic examination of suspicious lesions, the Raman probe was placed in gentle contact with the gastric mucosal surface, and the positioning of the Raman probe against the tissue sites was verified on the computer monitor by the endoscopist in charge at endoscopy. Multiple spectra (~5–15) for each tissue site were measured with scanning times of 0.1 to 0.5 s, which permits a rapid survey of large tissue areas. Raman spectra acquired in non-contact with gastric tissues were automatically discarded online using Raman diagnostic software developed by our group.[28] Immediately after the tissue Raman acquisitions, each tissue site measured was biopsied and sent for histopathological examination by three senior gastrointestinal pathologists who were blinded to the Raman scans. The consented histopathology assessments on the biopsied tissues serve as the gold standard to determine the diagnostic ability of the FP/HW fiber-optic Raman technique for identifying gastric dysplasia from normal gastric tissue.

Statistical analysis

Cohen's κ statistics were calculated to examine the agreement among the three pathologists.[36] One-way analysis of variance (ANOVA) with a Fisher *post hoc* least significant difference (LSD) test was used to evaluate the Raman spectral differences among the gastric tissue subtypes (*i.e.*, normal, dysplasia and adenocarcinoma).[28] Partial least squares (PLS)-discriminant analysis (DA) was applied for developing spectral diagnosis models.[26] Leave-one-patient-out cross-validation was further used to assess and optimize the PLS-DA model complexity, while reducing the risk of over-fitting. The overall discriminatory accuracy of the FP, HW, and the integrated FP/HW diagnostic models were evaluated through the use of the receiver operating characteristic (ROC) analysis.[37] A DeLong test was implemented on the ROC curves of the three (*i.e.*, FP, HW, and the integrated FP/HW) diagnostic models to evaluate whether the areas under their curves (AUC) were statistically significant.[38] The majority voting strategy was applied for final classification of each tissue site measured. The above multivariate statistical analysis was performed using in-house written scripts in the Matlab programming environment (Mathworks. Inc., Natick, MA).

Results

A total of 5792 FP/HW gastric tissue Raman spectra (normal (n = 5160), HGD (high-grade dysplasia, $n = 155$), and adenocarcinoma ($n = 477$)) were successfully acquired in real-time *in vivo* from 441 tissue sites (normal ($n = 396$), HGD ($n = 11$), and adenocarcinoma ($n = 34$)) of the 191 gastric patients recruited (normal (n = 172), HGD ($n = 6$), and adenocarcinoma ($n = 13$)). Fig. 1a shows the mean *in vivo* FP/HW Raman spectra ±1 standard deviation (SD) (shaded area) of normal, dysplasia and adenocarcinoma gastric tissues. Prominent gastric tissue Raman peaks with tentative assignments can be observed in the FP range,[8–20] *i.e.*, 853 (ν(C–C) proteins), 1004 (ν_s(C–C) ring breathing of phenylalanine), 1078 (ν(C–C) of lipids), 1265 (amide III ν(C–N) and δ(N–H) of proteins), 1302 (CH_2 twisting and wagging of lipids), 1335 (CH_3CH_2 twisting of proteins and nucleic acids), 1445 ($\delta(CH_2)$ deformation of proteins and lipids), 1618 (ν(C=C) of porphyrins), 1655 (amide I ν(C=O) of proteins) and 1745 cm^{-1} (ν(C=O) of phospholipids). Intense Raman peaks with assignments are also observed in the HW range,[21–26] *i.e.*, 2850 and 2885 cm^{-1} (symmetric and asymmetric CH_2 stretching of lipids), 2940 cm^{-1} (CH_3 stretching of proteins), 3009 cm^{-1} (asymmetric =CH stretching of lipid), ~3300 cm^{-1} (amide A (NH stretching of proteins)) and the broad Raman band of water (OH stretching vibrations peaking at ~3250 and ~3400 cm^{-1}) that are related to the local conformation and interactions of OH bonds in the intracellular and extracellular space of gastric tissue. One notes that the intense broad Raman band of water above 3000 cm^{-1} has also been observed in other soft tissues (*e.g.*, brain and oral cavity).[39,40] Fig. 1b shows the difference Raman spectra (*i.e.*, normal–dysplasia, normal–adenocarcinoma, dysplasia–adenocarcinoma) $\pm$1SD (shaded area), reflecting the different Raman-active component contribution for different gastric lesions. The significant difference ($p = 1.2 \times 10^{-9}$, one-way analysis of variance (ANOVA) with Fisher *post hoc* least significant difference (LSD)) in Raman spectra of normal and neoplastic gastric tissue discerned (Fig. 1b) demonstrates the potential of the integrated FP/HW fiber-optic Raman

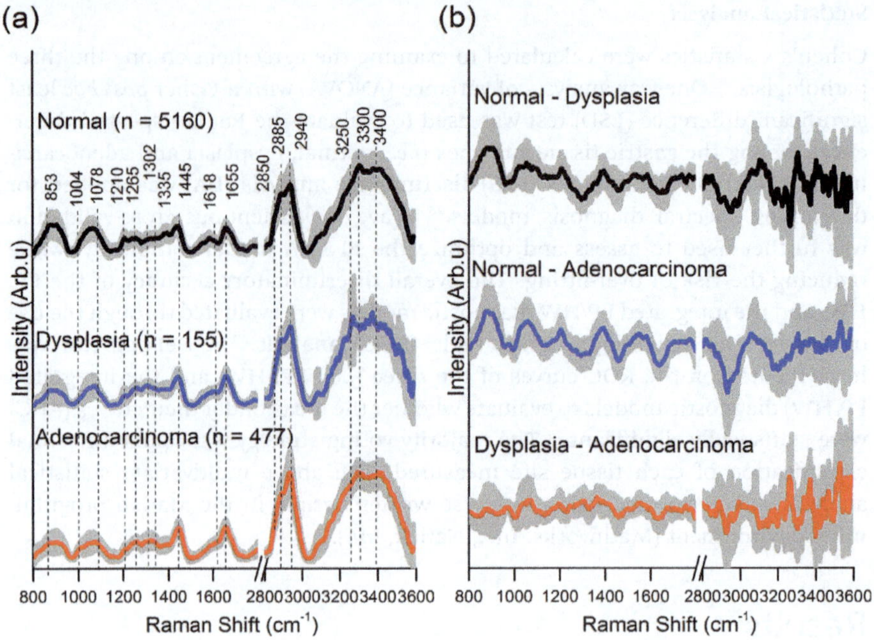

Fig. 1 (a) The mean *in vivo* FP/HW Raman spectra ±1 standard deviations (SD) of normal ($n = 5160$), dysplasia ($n = 155$) and adenocarcinoma ($n = 477$) acquired from 441 tissue sites of 191 gastric patients during clinical endoscopy. (b) Difference spectra (*i.e.*, normal–dysplasia, normal–adenocarcinoma, dysplasia–adenocarcinoma) ±1 SD resolving the distinct spectral features of gastric dysplasia.

spectroscopy for real-time *in vivo* diagnosis of gastric dysplasia and adenocarcinoma.

To elucidate the diagnostically important Raman-active components, Fig. 2a shows a logarithmic plot of the calculated *p*-values (one-way analysis of variance (ANOVA) with Fisher *post hoc* least significant difference (LSD) test at the 0.05 level.) for each of the Raman intensities in the entire spectral range (*i.e.*, 800–1800 cm^{-1} and 2800–3600 cm^{-1} with a set of 779 intensities). In particular, nine spectral sub-regions with statistically significant differences ($p < 1 \times 10^{-9}$) among normal gastric mucosa, dysplasia and gastric adenocarcinoma were found: *i.e.*, 830–940 cm^{-1}, 1000–1100 cm^{-1}, 1176–1220 cm^{-1}, 1256–1337 cm^{-1}, 1385–1410 cm^{-1}, 1440–1480 cm^{-1}, 1600–1670 cm^{-1}, and 2860–3000 cm^{-1}, related to proteins, lipids and nucleic acids. Significant spectral differences were also observed in bound water in the ranges of 3160–3260 cm^{-1}. Fig. 2b displays a histogram of the most statistically different Raman peak intensities (mean ± 1 SD) from both FP and HW ranges at 853 cm^{-1}, 1004 cm^{-1}, 1078 cm^{-1}, 1210 cm^{-1}, 1265 cm^{-1}, 1302 cm^{-1}, 1335 cm^{-1}, 1445 cm^{-1}, 1618 cm^{-1}, 1655 cm^{-1}, 2850 cm^{-1}, 2885 cm^{-1}, 2940 cm^{-1}, 3250 cm^{-1}, and 3300 cm^{-1}, confirming the capability of the rapid FP/HW fiber-optic Raman spectroscopy to reveal the biochemical/biomolecular constituents (*e.g.*, proteins, lipids, DNA and water *et al.*) associated with gastric carcinogenesis *in vivo*. Besides, we have also correlated the *in vivo* Raman spectra with representative histopathology slides of the three different pathology categories (Fig. 3) including (a) normal gastric mucosa; (b) high-grade

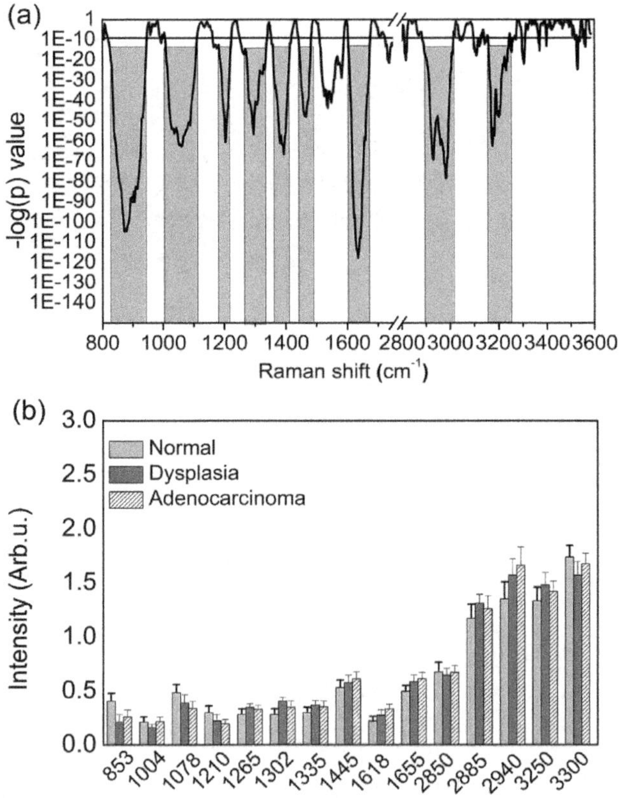

Fig. 2 (a) ANOVA of the three tissue categories over the entire spectral range (*i.e.*, 800–1800 cm^{-1}, and 2800–3600 cm^{-1}). (b) Histogram ±1 SD of the most diagnostically significant Raman peaks for the three tissue category classifications (853 cm^{-1}, 1004 cm^{-1}, 1078 cm^{-1}, 1210 cm^{-1}, 1265 cm^{-1}, 1302 cm^{-1}, 1335 cm^{-1}, 1445 cm^{-1}, 1618 cm^{-1}, 1655 cm^{-1}, 2850 cm^{-1}, 2885 cm^{-1}, 2940 cm^{-1}, 3250 cm^{-1}, and 3300 cm^{-1}).

dysplasia showing abrupt transition of dysplastic epithelium, and (c) gastric adenocarcinoma. While the histopathology identifies prominent cellular and architectural anomalies associated with gastric carcinogenesis (Fig. 3), FP/HW Raman spectra uncovers the molecular specificity inherent with the gastric dysplasia and adenocarcinoma (Fig. 1).

To further develop robust multivariate spectral diagnostic algorithms and concurrently gain deeper insights into the underlying molecular biomarkers accompanied with the gastric neoplasia development, PLS-DA together with LOPCV was applied on the *in vivo* tissue Raman spectra acquired (Fig. 1). Essentially, the PLS-DA approach assigns orthogonal latent variables (LVs) on the basis of the spectral variance that correlate with the gold standard of histopathology (*i.e.*, normal, dysplasia, and adenocarcinoma Fig. 3). The molecular features associated with gastric carcinogenesis can be further extracted and visualized through loadings (Fig. 4). Loading on LV1 largely captured specific FP Raman peaks arising from saturated and unsaturated lipid moieties (*i.e.*, 1078 cm^{-1} (ν(C–C)), 1302 cm^{-1} (CH_2 twisting and wagging), 1445 cm^{-1} ($\delta(CH_2)$ deformation), 1655 cm^{-1} (ν(C=C)), 1745 cm^{-1} (ν(C=O)), and HW Raman peaks

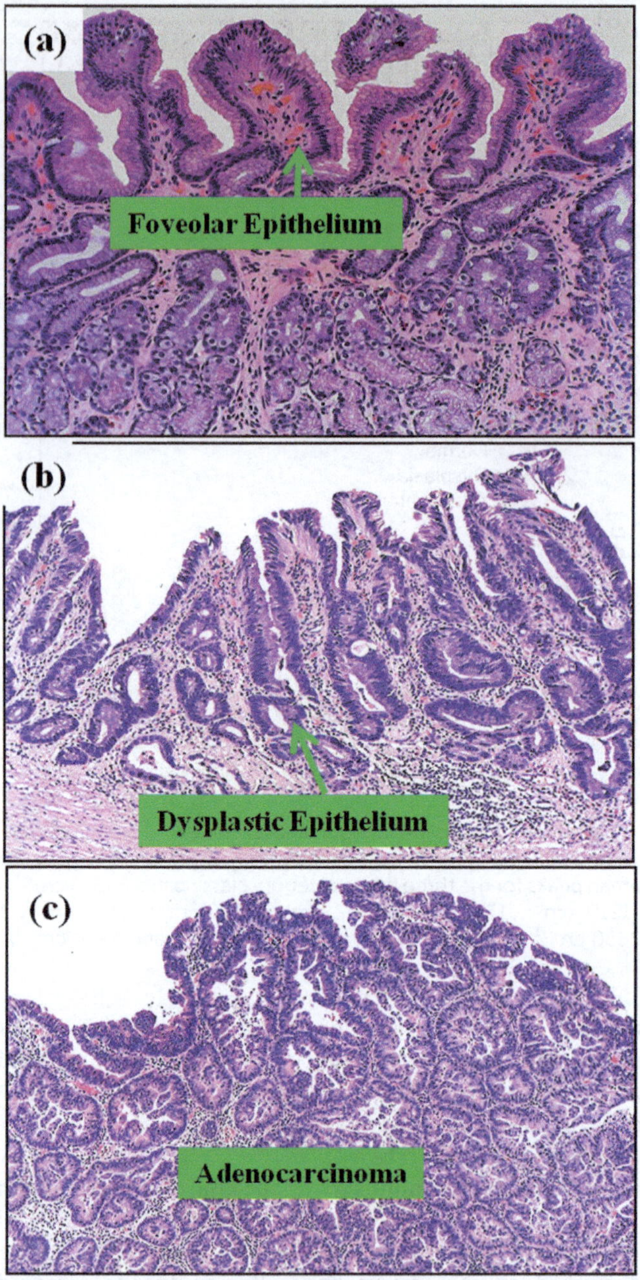

Fig. 3 Representative hematoxylin and eosin (H&E)-stained histopathological slides (original magnification, ×200) corresponding to different gastric tissue types measured. (a) Normal foveolar epithelium; (b) dysplastic epithelium, and (c) adenocarcinoma.

at 2850 and 2885 cm^{-1} (symmetric and asymmetric CH_2 stretching), 3009 cm^{-1} (asymmetric =CH stretching of lipids) as well as the symmetric and asymmetric OH vibrations in the HW range.[21–26] On the other hand, loading on LV2 generally captured Raman peaks assigned to proteins and nucleic acids (*i.e.*, 1004 cm^{-1}

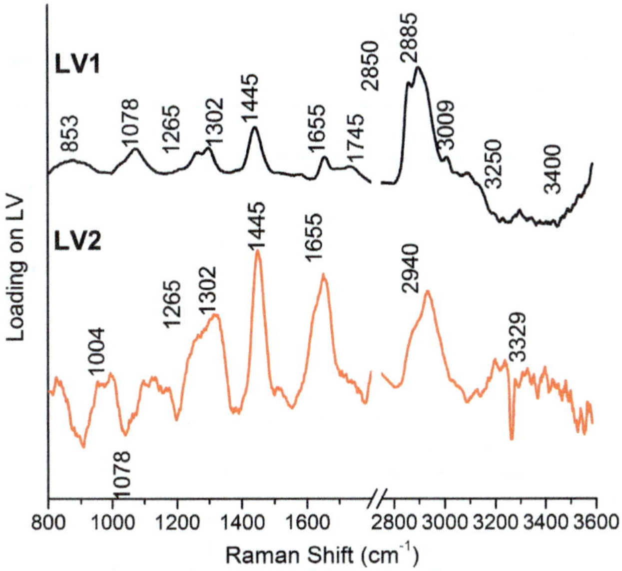

Fig. 4 The first two diagnostically significant LVs of Raman spectral features for gastric tissue classification (*i.e.*, normal, dysplasia and adenocarcinoma) algorithm development using partial least squares (PLS)-discriminant analysis (DA) and leave-one-patient-out cross-validation (LOPCV).

(ν_s(C–C)), 1265 cm^{-1} (amide III ν(C–N) and δ(N–H)), 1335 cm^{-1} (nucleic acids), 1445 cm^{-1} (δ(CH_2) deformation), 1655 cm^{-1} (amide I ν(C=O)), 2940 cm^{-1} (CH_3 stretching), and ~3329 cm^{-1} (amide A (N–H stretching)).

Fig. 5 shows the 2-dimensional ternary plot of the posterior probabilities of each Raman prediction using PLS-DA-LOPCV and majority voting for (a) FP, (b) HW, and (c) integrated FP/HW Raman spectroscopy, respectively. A good consistency for the gastric tissue groupings (*i.e.*, normal, dysplasia and adenocarcinoma) was obtained among the three independent pathologists as indicated by the Cohen's kappa of 0.89.[36] The prediction results are also summarized in Tables 1–3. We found that normal mucosa could be detected with an accuracy of

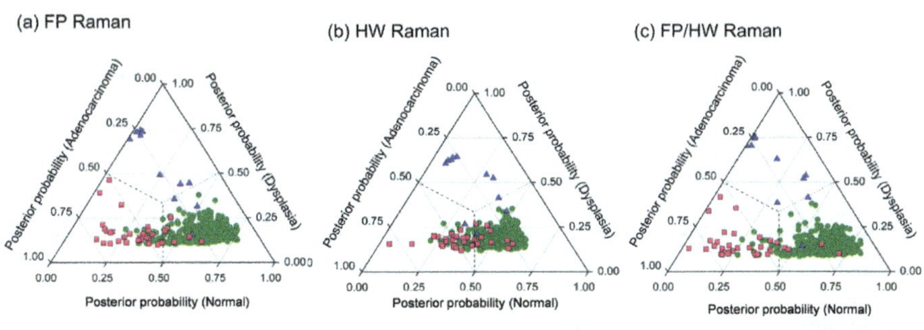

Fig. 5 The posterior probabilities of 191 gastric patients belonging to normal stomach (n = 396 sites), gastric dysplasia (n = 11 sites) and adenocarcinoma (n = 34 sites) based on the (a) FP, (b) HW and (c) the integrated FP/HW Raman techniques, respectively. (●) normal; (▲) dysplasia; and (■) adenocarcinoma.

Table 1 Confusion matrix and tissue site-based diagnostics using FP/HW Raman spectroscopy together with PLS-DA and majority voting

FP/HW Raman prediction			
Histopathology	Normal	Dysplasia	Adenocarcinoma
Normal	380	0	16
Dysplasia	2	9	0
Adenocarcinoma	4	0	30
Sensitivity (%)	96.0	81.8	88.2
Specificity (%)	86.7	100	96.0
Accuracy (%)	95.0	99.5	95.5

Table 2 Confusion matrix and tissue site based diagnostics using FP Raman spectroscopy together with PLS-DA and majority voting

FP Raman prediction			
Histopathology	Normal	Dysplasia	Adenocarcinoma
Normal	373	0	23
Dysplasia	3	8	0
Adenocarcinoma	6	0	28
Sensitivity (%)	94.2	72.7	82.4
Specificity (%)	80.0	100	94.3
Accuracy (%)	92.7	99.3	93.4

95.0% (sensitivity: 96.0%, specificity: 86.7%), dysplastic tissues was identified with an accuracy of 99.5% (sensitivity: 81.8%, specificity: 100%), whereas adenocarcinoma was identified with an accuracy of 95.5% (sensitivity: 88.2%, specificity: 96.0%) by using integrated FP/HW Raman spectroscopy (Table 1), superior to using either FP (Table 2) or HW (Table 3) Raman techniques alone.

We further conducted a dichotomous PLS-DA analysis to examine whether the FP and HW Raman modalities were complementary for distinguishing gastric dysplasia from normal gastric tissue. The integrated FP/HW Raman spectroscopy yielded a diagnostic accuracy of 95.7% [sensitivity of 88.4% (137/155) and a specificity of 95.9% (4950/5160)], higher than those using either FP (accuracy of

Table 3 Confusion matrix and tissue site based diagnostics using HW Raman spectroscopy together with PLS-DA and majority voting

HW Raman prediction			
Histopathology	Normal	Dysplasia	Adenocarcinoma
Normal	363	0	33
Dysplasia	2	8	1
Adenocarcinoma	10	0	24
Sensitivity (%)	91.7	72.7	70.6
Specificity (%)	73.3	100	91.6
Accuracy (%)	89.8	99.3	90.0

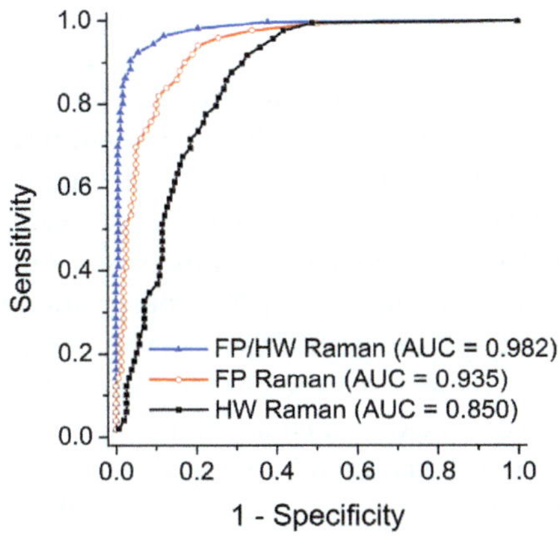

Fig. 6 Receiver operating characteristic (ROC) curves for separating gastric dysplasia from normal gastric tissue. The areas under the ROC curves (AUC) are 0.935, 0.850 and 0.982 for the FP, HW, and the integrated FP/HW Raman techniques, respectively.

92.8%; sensitivity of 79.3% (123/155), and a specificity of 93.2% (4809/5160)) or HW (accuracy of 93.3%; sensitivity of 63.2% (98/155), and a specificity of 94.2% (4861/5160)) Raman technique alone. The receiver operating characteristic (ROC) curves for the differentiation between dysplastic and normal gastric tissue are also generated (Fig. 6), with the integration areas under the ROC curves being 0.935, 0.850 and 0.982 for the FP, HW, and the integrated FP/HW Raman techniques, respectively. Further DeLong test on the ROC–AUCs shows that the AUC of the integrated FP/HW Raman technique is significantly higher than that of either the FP ($p = 1.4 \times 10^{-3}$, DeLong test) or HW ($p = 1.2 \times 10^{-6}$, DeLong test) Raman techniques,[38] illustrating the diagnostic advantages of the integrated FP/HW Raman technique over either the FP or HW Raman technique. Overall, the above results reaffirm that the integrated FP/HW Raman technique provides the best diagnostic performance for the *in vivo* detection of gastric dysplasia as compared to the FP or HW Raman spectroscopy techniques.

Discussion

We have applied an integrated FP/HW fiber-optic Raman spectroscopy system to acquire high-quality *in vivo* Raman spectra of gastric tissue in real time during clinical endoscopic examinations. One main advantage of this system over prior Raman techniques is its ability to acquire both the FP Raman spectra and its complementary HW Raman spectra simultaneously at endoscopy. While histopathology (Fig. 3) identifies the abrupt transition of the dysplastic epithelium, the integrated FP/HW Raman spectroscopy reveals the biochemical and biomolecular changes occurring in the epithelium. Significant Raman spectral differences were observed between different pathological subtypes (Fig. 1b, and 2), confirming the diagnostic potential of the integrated FP/HW Raman spectroscopy for *in vivo* diagnosis of preneoplastic and neoplastic gastric lesions. FP Raman spectra reveal significant biochemical/biomolecular differences associated with gastric carcinogenesis. For instance, the increased thickness of the epithelial layer of gastric dysplasia attenuates the laser power delivered and also obscures the Raman signals from deeper subcutaneous fat layers,[41] leading to a significantly reduced Raman peak at 1078 cm^{-1} attributed to phospholipids. Further, a decreased percentage of phenylalanine is indicated by the Raman peak at 1004 cm^{-1}. This finding is consistent with the previous study that established a link between essential amino acid deficiency (*e.g.* phenylalanine) and carcinogenesis.[42] We also found that gastric dysplasia is characterized by increased Raman peaks assigned to proteins (*e.g.*, 1265, 1445, and 1655 cm^{-1}). This observation is in agreement with cytometric studies that dysplastic gastric tissues are accompanied with enhanced metabolic activities and increased hyperchromatism.[43–45] Specifically, dysplastic hyperchromatism induced an elevated concentration of histones, the main protein component that makes up the chromatin for dysplasia tissue, as indicated by the stronger Raman intensity at 1445 cm^{-1} (ref. 4 and 46). The increased histone concentration is also supported by the relatively increased Raman-active components of amide III band (1265 cm^{-1}) and amide I band (1655 cm^{-1}), which are indicative of α-helix structures[47] (Fig. 1). Further comparison of the difference Raman spectra (Fig. 1b) *in vivo* brings the aggregated β-pleated sheet to light, which is indicative of enhanced biochemical interactions between the proteins and the cell microenvironment related to the increase of mitotic activity, one of the cellular alteration characteristics of gastric dysplasia.[48] Besides, the Raman band at 1335 cm^{-1} (CH_3CH_2 twisting of proteins and nucleic acids) is higher for gastric dysplasia, signifying a dramatically increased DNA contribution to the development of gastric dysplasia. This is probably due to the fact that the dysplastic epithelium contains enlarged, hyperchromatic and crowded nuclei.[45] Besides, the increased DNA observed are also consistent with DNA changes associated with tissue precancer and cancer transformation interrogated using the Raman spectroscopy technique.[8–20,25,26] Meanwhile the FP Raman spectra show specific but relatively weak Raman peaks that are predominantly related to backbone structures of proteins, lipids and DNA.[8–20] The HW Raman spectra, on the other hand, offer exclusive new insights into the CH_2 and CH_3 stretching information and structure as well as the interactions of intra- and inter-cellular water *in situ*. The overall intensities of Raman signals of gastric dysplasia in the 2800–3020 cm^{-1} range involving proteins (*e.g.*, 2940 cm^{-1}) and lipids (*e.g.*, 2850 and 2885 cm^{-1}) are significantly higher than those in normal gastric tissue. The protein to lipid Raman peak ratio (*i.e.*, $I_{2940}/I_{2850} \pm$ SD) reveals a significant increase ($p < 0.001$) from normal (1.996 $\pm$ 0.009) to dysplastic (2.434 $\pm$ 0.001) gastric tissue, a well-known indicator of increased protein components, while simultaneously reduced lipid content associated with gastric carcinogenesis.[4,5,11,19] Moreover, the HW Raman spectrum resolves the changes of the subtle =CH stretch mode at 3009 cm^{-1} of gastric dysplasia that provides unequivocal evidence as a marker for saturated fatty acids (Fig. 2).[49] Such insights are crucial to deciphering the increasingly appreciated role of lipid metabolism in cancerous tissues.[49] With the integrated FP/HW fiber-optic Raman spectroscopy, it is now possible to further strengthen the link between lipid metabolism and carcinogenic onset and progression *in situ*. In particular, prominently increased water content was also observed for the dysplastic and cancerous stomach as indicated by the significantly increased Raman peaking at 3250 cm^{-1} (Fig. 2b), reflecting the rearrangement of hydrogen-bonded networks in the epithelial layer caused by local interaction with macromolecules such as collagen.[50] Hence, the integrated FP/HW Raman endoscopy technique is not only capable of harvesting a wealth of information on proteins, lipids, and nucleic acids, but also provides information on the bound water content in gastric tissue for improving early diagnosis of gastric precancer and early cancer at endoscopy.

Further PLS-DA analysis demonstrates the diagnostic advantages of the integrated FP/HW Raman spectroscopy for early detection of gastric dysplasia. The complementary properties of the FP and HW Raman spectral modalities for enhancing tissue diagnosis can partially be explained by back-tracing the incorrect predictions of each Raman modality. The FP and HW Raman modalities made incorrect predictions on 32 (normal: 23, dysplasia: 3, and adenocarcinoma: 6) and 46 (normal: 33, dysplasia: 3, and adenocarcinoma: 10) tissue sites, respectively, of the total 191 patients recruited. Integrated FP/HW Raman spectroscopy reduced the number of incorrect predictions to 22 (normal: 16, dysplasia: 2, and adenocarcinoma: 4). Further investigations show that the Raman signals of misclassified FP Raman spectra are extremely weak but with higher AF background and its resultant low FP spectral signal-to-noise ratios (SNR) $\sim 0.0118 \pm 0.0054$ (SNR: defined as the Raman intensity of the normal gastric tissue Raman peak at 1445 cm^{-1} divided by its AF background); while the SNR (defined as the mean values of Raman intensity of the normal gastric tissue Raman peak at 1445 cm^{-1} divided by its AF background, and the Raman intensity of the normal gastric tissue Raman peak at 2940 cm^{-1} divided by its AF background) of the integrated FP/HW Raman spectroscopy is 0.0777 ± 0.0287, which shows significant enhancement through its addition of the HW Raman modality, therefore improving the performance of the tissue classification with integrated FP/HW Raman spectroscopy. In addition, the enhanced diagnostic performances of the FP/HW Raman spectroscopy were confirmed by its comparisons against the routinely used WLR endoscopy technique (Table 4). On a per-tissue site basis, the FP/HW Raman spectroscopy provides a diagnostic accuracy of 99.3% (sensitivity: 90.9% (10/11); specificity: 99.9% (394/396)), which are higher than that of the WLR endoscopy (accuracy: 52.1%; sensitivity: 90.9% (10/11); specificity: 51.0% (202/396)) (Table 4). Besides, a Dutch study[51] reported a sensitivity of 51% and a specificity of 67% for the diagnosis of premalignant lesions including gastric dysplasia using WLR endoscopy, which are similar to the performances of WLR endoscopy in our study (Table 4), reconfirming the unrivaled advantages of the FP/HW Raman spectroscopy system developed for the diagnosis of gastric dysplasia. The real-time FP/HW fiber-optic Raman spectroscopy system developed is now routinely used by both experienced and novice endoscopists without difficulty at the Endoscope Centre at the National University Health System

Table 4 Comparisons of diagnostic performance for the differentiation between normal and dysplastic gastric tissues on a tissue site basis using real-time fiber-optic FP/HW Raman spectroscopy and conventional WLR endoscopy

Diagnostic performance comparisons			
	Sensitivity	Specificity	Accuracy
WLR endoscopy	90.9% (10/11)	51.0% (202/396)	52.1% (212/407)
Raman spectroscopy	90.9% (10/11)	99.5% (394/396)	99.3% (404/407)

(NUHS), Singapore, as an adjunct for screening and surveillance of gastric patients and targeted biopsies during clinical endoscopic examination. Some limitations of the current study should be pointed out. One notes that the current *in vivo* FP/HW tissue Raman datasets acquired are unbalanced with a relatively larger number of true negatives but a smaller amount of pathologic tissue (*e.g.* dysplasia), which may give rise to a biased overall accuracy within the PLS-DA models. The patients recruited are not randomized and might have selection bias (*i.e.*, non-respondent bias). Large prospectively randomized multicenter clinical trials are underway for further assessing the true clinical diagnostic value of the real-time FP/HW Raman spectroscopy in gastrointestinal examinations.

In summary, we demonstrate that high-quality *in vivo* gastric tissue Raman spectra can be acquired from gastric patients in real-time by using a simultaneous FP/HW fiber-optic Raman spectroscopy technique. Significant FP/HW Raman spectral differences are observed among normal, dysplasia and adenocarcinoma of the stomach. The use of complementary biochemical/biomolecular information harvested through the integrated FP/HW Raman spectroscopy significantly enhances the diagnosis of gastric dysplasia as compared to either the FP or HW Raman techniques alone. Further comparison with the routinely used WLR endoscopy reconfirms the advantages of the integrated FP/HW Raman spectroscopy for gastric dysplasia diagnosis. This unique biomolecular endoscopic approach based on the simultaneous FP/HW fiber-optic Raman spectroscopy opens a new avenue for enhancing real-time, *in vivo* diagnosis of gastric precancer and early cancer at the molecular level during clinical endoscopy.

Acknowledgements

This project was supported by the National Medical Research Council (NMRC) (CIRG/1331/2012; BnB/0012c/2014; BnB13dec037), and the Academic Research Fund (AcRF)-Tier 2 from the Ministry of Education (MOE) (MOE2014-T2-1-010), Singapore.

References

1 L. A. Torre, F. Bray, R. L. Siegel, J. Ferlay, J. Lortet-Tieulent and A. Jemal, *Ca-Cancer J. Clin.*, 2015, **65**, 87–108.

2 T. Gotoda, H. Yamamoto and R. M. Soetikno, *J. Gastroenterol.*, 2006, **41**, 929–942.

3 M. Rugge, F. Farinati, R. Baffa, F. Sonego, F. Di Mario, G. Leandro, F. Valiante and D. Interdisciplinary Group on Gastric Epithelial, *Gastroenterology*, 1994, **107**, 1288–1296.

4 G. Y. Lauwers and R. H. Riddell, *Gut*, 1999, **45**, 784.

5 M. Lansdown, P. Quirke, M. F. Dixon, A. T. Axon and D. Johnston, *Gut*, 1990, **31**, 977–983.

6 B. Mayinger, M. Jordan, T. Horbach, P. Horner, C. Gerlach, S. Mueller, W. Hohenberger and E. G. Hahn, *Gastrointest. Endosc.*, 2004, **59**, 191–198.

7 G. Zonios, L. T. Perelman, V. Backman, R. Manoharan, M. Fitzmaurice, J. van Dam and M. S. Feld, *Appl. Opt.*, 1999, **38**, 6628–6637.

8 L. M. Almond, J. Hutchings, N. Shepherd, H. Barr, N. Stone and C. Kendall, *J. Biophotonics*, 2011, **4**, 685–695.

9 J. C. C. Day, R. Bennett, B. Smith, C. Kendall, J. Hutchings, G. M. Meaden, C. Born, S. Yu and N. Stone, *Phys. Med. Biol.*, 2009, **54**, 7077.

10 R. O. P. Draga, M. C. M. Grimbergen, P. L. M. Vijverberg, C. F. P. V. Swol, T. G. N. Jonges, J. A. Kummer and J. L. H. Ruud Bosch, *Anal. Chem.*, 2010, **82**, 5993–5999.

11 Z. Huang, M. S. Bergholt, W. Zheng, K. Lin, K. Y. Ho, M. Teh and K. G. Yeoh, *J. Biomed. Opt.*, 2010, **15**, 037017.

12 Z. Huang, A. McWilliams, H. Lui, D. I. McLean, S. Lam and H. Zeng, *Int. J. Cancer*, 2003, **107**, 1047–1052.

13 Z. Huang, S. K. Teh, W. Zheng, J. Mo, K. Lin, X. Shao, K. Y. Ho, M. Teh and K. G. Yeoh, *Opt. Lett.*, 2009, **34**, 758–760.

14 Y. Komachi, H. Sato, K. Aizawa and H. Tashiro, *Appl. Opt.*, 2005, **44**, 4722–4732.

15 J. T. Motz, M. Hunter, L. H. Galindo, J. A. Gardecki, J. R. Kramer, R. R. Dasari and M. S. Feld, *Appl. Opt.*, 2004, **43**, 542–554.

16 M. G. Shim, B. C. Wilson, E. Marple and M. Wach, *Appl. Spectrosc.*, 1999, **53**, 619–627.

17 M. A. Short, S. Lam, A. McWilliams, J. Zhao, H. Lui and H. Zeng, *Opt. Lett.*, 2008, **33**, 711–713.

18 N. Stone, C. Kendall, J. Smith, P. Crow and H. Barr, *Faraday Discuss.*, 2004, **126**, 141–157.

19 S. K. Teh, W. Zheng, K. Y. Ho, M. Teh, K. G. Yeoh and Z. Huang, *Br. J. Cancer*, 2008, **98**, 457–465.

20 C. A. Patil, I. J. Pence, C. A. Lieber and A. Mahadevan-Jansen, *Opt. Lett.*, 2014, **39**, 303–306.

21 S. Koljenović, T. C. Bakker Schut, R. Wolthuis, B. de Jong, L. Santos, P. J. Caspers, J. M. Kros and G. J. Puppels, *J. Biomed. Opt.*, 2005, **10**, 031116.

22 K. Lin, D. L. P. Cheng and Z. Huang, *Biosens. Bioelectron.*, 2012, **35**, 213–217.

23 J. Mo, W. Zheng, J. J. H. Low, J. Ng, A. Ilancheran and Z. Huang, *Anal. Chem.*, 2009, **81**, 8908–8915.

24 L. F. Santos, R. Wolthuis, S. Koljenović, R. M. Almeida and G. J. Puppels, *Anal. Chem.*, 2005, **77**, 6747–6752.

25 M. S. Bergholt, K. Lin, J. Wang, W. Zheng, H. Xu, Q. Huang, J.-l. Ren, K. Y. Ho, M. Teh, S. Srivastava, B. Wong, K. G. Yeoh and Z. Huang, *J. Biophotonics*, 2015, **201**, 1–10.

26 J. Wang, K. Lin, W. Zheng, K. Y. Ho, M. Teh, K. G. Yeoh and Z. Huang, *Sci. Rep.*, 2015, **5**, 12957.

27 J. Wang, M. S. Bergholt, W. Zheng and Z. Huang, *Opt. Lett.*, 2013, **38**, 2321–2323.

28 S. Duraipandian, M. Sylvest Bergholt, W. Zheng, K. Yu Ho, M. Teh, K. Guan Yeoh, J. Bok Yan So, A. Shabbir and Z. Huang, *J. Biomed. Opt.*, 2012, **17**, 081418.

29 C. V. K. Raman and K. S. Krishnan, *Nature*, 1928, **121**, 501–502.

30 American National Standards Institute, *American National Standard for the Safe Use of Lasers, ANSI Standard 2136.1-1986*, Washington, D.C., 1986.

31 J. H. Torres, M. Motamedi, J. A. Pearce and A. J. Welch, *Appl. Opt.*, 1993, **32**, 597–606.

32 F. P. Incropera, *Fundamentals of heat and mass transfer*, John Wiley & Sons, 2011.

33 M. D. Keller, R. H. Wilson, M.-A. Mycek and A. Mahadevan-Jansen, *Appl. Spectrosc.*, 2010, **64**, 607–614.

34 C. Holmer, K. S. Lehmann, J. Wanken, C. Reissfelder, A. Roggan, G. Mueller, H. J. Buhr and J.-P. Ritz, *J. Biomed. Opt.*, 2007, **12**, 014025–014028.
35 S. Thomsen, *Photochem. Photobiol.*, 1991, **53**, 825–835.
36 J. Carletta, *Comput. Lang.*, 1996, **22**, 249–254.
37 J. A. Hanley and B. J. McNeil, *Radiology*, 1982, **143**, 29–36.
38 E. R. DeLong, D. M. DeLong and D. L. Clarke-Pearson, *Biometrics*, 1988, **44**, 837–845.
39 R. Wolthuis, M. van Aken, K. Fountas, J. S. Robinson Jr, H. A. Bruining and G. J. Puppels, *Anal. Chem.*, 2001, **73**, 3915–3920.
40 E. M. Barroso, R. W. H. Smits, T. C. Bakker Schut, I. ten Hove, J. A. Hardillo, E. B. Wolvius, R. J. Baatenburg de Jong, S. Koljenović and G. J. Puppels, *Anal. Chem.*, 2015, **87**, 2419–2426.
41 K. Badizadegan, V. Backman, C. W. Boone, C. P. Crum, R. R. Dasari, I. Georgakoudi, K. Keefe, K. Munger, S. M. Shapshay and E. E. Sheets, *Faraday Discuss.*, 2004, **126**, 265–279.
42 B. S. Worthington, J. A. Syrotuck and S. I. Ahmed, *J. Nutr.*, 1978, **108**, 1402–1411.
43 J. H. Hughes, C. J. Leigh, S. S. Raab, S. Y. Hook, M. B. Cohen and M. J. Suhrland, *Cancer Cytopathol.*, 1998, **84**, 289–294.
44 I. Georgakoudi, B. C. Jacobson, M. G. Müller, E. E. Sheets, K. Badizadegan, D. L. Carr-Locke, C. P. Crum, C. W. Boone, R. R. Dasari, J. van Dam and M. S. Feld, *Cancer Res.*, 2002, **62**, 682–687.
45 P. Correa and M. B. Piazuelo, *J. Dig. Dis.*, 2012, **13**, 2–9.
46 G. J. Thomas, B. Prescott and D. E. Olins, *Science*, 1977, **197**, 385–388.
47 O. H. Olsen, M. R. Samuelsen, S. B. Petersen and L. Nrskov, *Phys. Rev. A*, 1989, **39**, 3130.
48 P. Correa, *Cancer Res.*, 1988, **48**, 3554–3560.
49 T. Pohle, S. Brändlein, N. Ruoff, H. K. Müller-Hermelink and H. P. Vollmers, *Cancer Res.*, 2004, **64**, 3900–3906.
50 S. Leikin, V. A. Parsegian, W. Yang and G. E. Walrafen, *Proc. Natl. Acad. Sci. U. S. A.*, 1997, **94**, 11312–11317.
51 L. G. Capelle, J. Haringsma, A. C. de Vries, E. W. Steyerberg, K. Biermann, H. van Dekken and E. J. Kuipers, *Dig. Dis. Sci.*, 2010, **55**, 3442–3448.

Faraday Discussions

PAPER

Investigation of intervertebral disc degeneration using multivariate FTIR spectroscopic imaging†

Kerstin T. Mader,*[a] Mirte Peeters,[bc] Suzanne E. L. Detiger,[bc] Marco N. Helder,[bc] Theo H. Smit,[bc] Christine L. Le Maitre[d] and Chris Sammon[a]

Received 11th November 2015, Accepted 14th January 2016
DOI: 10.1039/c5fd00160a

Traditionally tissue samples are analysed using protein or enzyme specific stains on serial sections to build up a picture of the distribution of components contained within them. In this study we investigated the potential of multivariate curve resolution-alternating least squares (MCR-ALS) to deconvolute 2nd derivative spectra of Fourier transform infrared (FTIR) microscopic images measured in transflectance mode of goat and human paraffin embedded intervertebral disc (IVD) tissue sections, to see if this methodology can provide analogous information to that provided by immunohistochemical stains and bioassays but from a single section. MCR-ALS analysis of non-degenerate and enzymatically *in vivo* degenerated goat IVDs reveals five matrix components displaying distribution maps matching histological stains for collagen, elastin and proteoglycan (PG), as well as immunohistochemical stains for collagen type I and II. Interestingly, two components exhibiting characteristic spectral and distribution profiles of proteoglycans were found, and relative component/tissue maps of these components (labelled PG1 and PG2) showed distinct distributions in non-degenerate *versus* mildly degenerate goat samples. MCR-ALS analysis of human IVD sections resulted in comparable spectral profiles to those observed in the goat samples, highlighting the inter species transferability of the presented methodology. Multivariate FTIR image analysis of a set of 43 goat IVD sections allowed the extraction of semi-quantitative information from component/tissue gradients taken across the IVD width of collagen type I, collagen type II, PG1 and PG2. Regional component/tissue parameters were calculated and significant correlations were found between histological grades of degeneration and PG parameters (PG1: $p = 0.0003$, PG2: $p < 0.0001$); glycosaminoglycan (GAG) content and

[a]*Sheffield Hallam University, Materials and Engineering Research Institute, Sheffield, S1 1WB, UK. E-mail: K.Mader@shu.ac.uk*

[b]*Department of Orthopaedic Surgery, VU University Medical Center, Amsterdam, The Netherlands*

[c]*Skeletal Tissue Engineering Group Amsterdam (STEGA) and MOVE Research Institute, Amsterdam, The Netherlands*

[d]*Sheffield Hallam University, Biomolecular Science Research Centre, Sheffield, S1 1WB, UK*

† Electronic supplementary information (ESI) available. See DOI: 10.1039/c5fd00160a

PGs (PG1: $p = 0.0055$, PG2: $p = 0.0001$); and MRI T2* measurements and PGs (PG1: $p = 0.0021$, PG2: $p < 0.0001$). Additionally, component/tissue parameters for collagen type I and II showed significant correlations with total collagen content ($p = 0.0204$, $p = 0.0127$). In conclusion, the presented findings illustrate, that the described multivariate FTIR imaging approach affords the necessary chemical specificity to be considered an important tool in the study of IVD degeneration in goat and human IVDs.

Introduction

Low back pain (LBP) affects millions of people worldwide, and has been linked to degenerative changes of the intervertebral disc (IVD).[1,2] The IVD is a structurally and chemically complex cartilaginous tissue consisting of distinct regions; a softer gelatinous inner core known as the nucleus pulposus (NP) and a highly organised fibrous outer region known as the annulus fibrosus (AF).[3] The proteoglycan and collagen type II concentrations gradually decrease from the NP out towards the AF, whilst the collagen type I concentration increases.[4,5] Additionally, the presence and importance of elastic fibres was studied and an intricate network of elastin fibres, microfibrils and collagen fibres has been found.[6] During aging and degeneration, IVD cells produce abnormal amounts of matrix components and matrix-degrading enzymes.[7–11] Eventually the IVD matrix composition changes, for example, the degradation and loss of proteoglycans, a change in the distribution and composition of collagens, the denaturation of collagen type II and increased decorin and fibronectin concentrations have been reported.[12,13] On a macroscopic level these changes in matrix composition lead to changes in the structure of the IVD, the boundary between AF and NP becomes less distinct, annular lamella become irregular, bifurcating and interdigitating, and cleft and fissure formation occurs, which leads to a loss in the functionality of the IVD.[12] In recent years a wide range of potential therapeutic strategies have been developed allowing more sophisticated designs and enhancing the success of regenerative therapies.[14–16] It has therefore become of great interest to develop comprehensive analytical methodologies to objectively define critical characteristics of target matrices as well as to assess repair and regeneration efficacy.

Fourier transform infrared (FTIR) spectroscopy provides information about the chemical species in a sample based on the frequency of the vibrations of its covalent bonds.[17] Additionally, using a FTIR imaging set-up it is possible to measure spatially resolved quantitative biochemical information. This together with technological developments in this field,[18] which now allow a typical experiment to be completed and analysed within hours, make this technique a viable tool for rapid disease screening and diagnosis in pathologies where tissue biopsies are collected as part of routine diagnosis.[18] FTIR imaging has been used for the analysis of biological samples such as cells, bone and cartilaginous tissue in its native, repaired and regenerated state.[19–25] Different components in the extracellular matrix of connective tissue *e.g.* collagens and proteoglycans, show discriminatory spectroscopic characteristics and information about multiple species can be derived from the measurement of a single section without the need for lengthy immunohistochemical staining from multiple successive sections.[24,25]

However, despite its great potential FTIR imaging of biological samples is not without drawbacks; significant overlap of infrared spectral profiles of different tissue components (Fig. 1) complicates the extraction of chemically specific parameters.[26] The analysis of FTIR spectroscopic imaging data can be further hindered by spectral artefacts related to sample preparation[27,28] and collection mode.[29,30] This is particularly true of spectra collected in transflectance mode, which has some advantages over transmission mode, for example the use of low cost substrates, ease of sample preparation and a great potential for automation, which has been reported to be affected by spectral distortions due to reflective and optical phenomena such as Mie scattering and electric field standing wave effects.[31–38] For FTIR spectroscopy to become a diagnostically useful tool the challenge lies in the development of analytical strategies which enable the development of robust methods with high chemical specificity. Current methodologies utilise 2nd derivative infrared spectra and multivariate analysis methods such as principal component analysis (PCA), partial least squares (PLS) and cluster analysis to enhance the data interpretation of the FTIR imaging data of cartilaginous tissue.[39–43] Only recently the potential of multivariate curve resolution-alternating least squares (MCR-ALS) methods for the analysis of biological samples was highlighted.[44] Like PCA, MCR methods are able to describe data matrices of multi-component measurements, without any prior knowledge of the nature and composition of the mixtures. The only assumption is a bilinear structure of the data matrices. In spectroscopic data, a bilinear structure arises through obedience of the Beer–Lambert law which implies a linear relationship between the absorbance and concentration of a component and the additivity of these absorbencies in a mixture.[45,46] However, while PCA analysis of infrared images results in abstract factors, MCR analysis transforms abstract results into noise filtered, physically and/or chemically meaningful spectra and distribution maps of single components through the application of constraints related to the physicochemical or mathematical properties of the mixture components.[45–48]

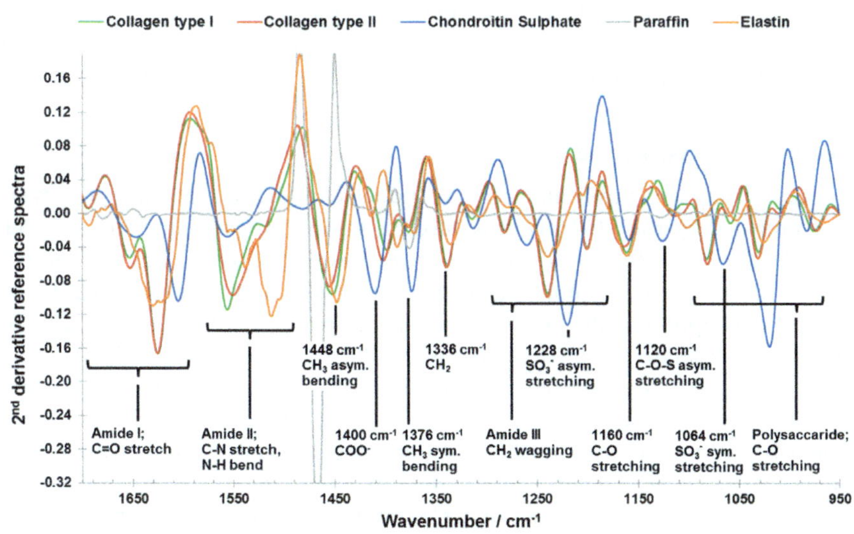

Fig. 1 Vector normalised 2nd derivative ATR-FTIR reference spectra of collage type I (green), collagen type II (red), chondroitin sulphate (blue), paraffin (grey) and elastin (orange) components (see data collection in the materials and methods section: reference materials), including peak assignments.[43]

Iterative MCR methods such as MCR-ALS have been shown to work especially well for the resolution of vibrational spectroscopic images of biological samples.[49–53] Furthermore, MCR-ALS results can then be used as input information into segmentation analysis for histopathological classification.[44]

To date, there have been limited studies of IVDs using FTIR spectroscopy and multivariate data analysis.[54,55] In this work, we examine the potential of MCR-ALS analysis to extract spatially resolved biochemical information of non-degenerated and enzymatically *in vivo* degenerated goat IVDs from 2nd derivative FTIR microscopic imaging data collected in transflectance mode. Additionally, the transferability of the described multivariate imaging approach from goat to human IVD sections is investigated. As a second objective, the derived methodology is applied to a second set of goat IVDs discs ($n = 43$) obtained from an *in vivo* goat study investigating the effect of two different growth factors (BMP2 and BMP2/7) in combination with a fibrin hyaluronic acid hydrogel on the regeneration of the intervertebral disc. Extracted multivariate spectroscopic parameters are correlated with parameters derived from histological grading, biochemical and MRI T2* measurements.

Experimental

Reference materials

Collagen type II (collagen from bovine tracheal cartilage), collagen type I (collagen from bovine achilles tendon), chondroitin sulphate A sodium salt (chondroitin sulphate A sodium salt from bovine trachea) and elastin (elastin from bovine neck ligament) were purchased from Sigma Aldrich (Gillingham, Dorset, UK). Attenuated total reflectance (ATR) FTIR spectra of reference material were collected at a spectral resolution of 4 cm^{-1} using a Nexus FTIR spectrometer (Thermo Scientific, UK) fitted with a diamond Golden Gate ATR accessory (Specac Limited, Orpington, Kent, UK). ATR-FTIR spectra were vector normalised (950–1800 cm^{-1}, ISys 5.0.0.14) and second derivatives of the spectra were generated using the Savitzky–Golay transformation (filter order 3, filter length 15, ISys 5.0.0.14).

Goat IVD samples; set one

Non-degenerated (control) and enzyme induced mildly degenerated goat IVDs were provided by the VU University Medical Centre, Amsterdam. The samples originate from a larger set of samples used to study the biomechanical and NP viscoelastic properties of non-degenerate and mildly degenerated goat IVDs. A detailed description of the experimental set-up is outlined in Detiger *et al.*[56] In short, eight skeletally mature female Dutch milk goats were obtained from a local farmer. During a surgical procedure three out of six lumbar IVDs in each goat were injected with chondroitinase ABC (CABC), an enzyme which cleaves proteoglycans and more specifically glycosaminoglycans (GAGs), which thus induces the onset of degeneration.[57–60] The research protocol was approved by the Scientific Board and the Animal Ethics Committee of VUmc. To confirm that mild degeneration had occurred, radiographic and MRI analyses were conducted. After biomechanical and viscoelastic testing the samples were stored at −20 °C until further analysis. After thawing, IVDs were cut from the endplates with a surgical knife, formalin fixed (10% v/v, overnight at 4 °C) and paraffin embedded (sagittal directions). Consecutive 4 μm sections of each sample were mounted on glass slides (Lavender X-Tra Slides, Leica, Milton Keynes, UK) for histological and immunohistochemical staining, and on custom made reflective 316 stainless steel slides for FTIR imaging analysis.

Histology and immunohistochemistry. Sections were deparaffinised, rehydrated and subjected to H&E, Alcian blue (pH 2.5) and Masson Trichrome (Masson Trichrome with aniline blue, Bio-Optica, Milan, Italy) staining using standard protocols. Additionally, immunohistochemistry was used to visualise the distribution of collagen type I and collagen type II. For the immunohistochemical staining paraffin sections were deparaffinised, rehydrated and endogenous peroxidase blocked using hydrogen peroxide. After washing in distilled water, sections for collagen type II staining were treated with a hyaluronidase-protease enzyme antigen retrieval solution for 1 h at 37 °C (0.2% w/v protease and 2% w/v hyaluronidase in TBS; Sigma-Aldrich, Gillingham, Dorset, UK). Following washing, non-specific binding sites were blocked at room temperature for 45 minutes in 25% w/v rabbit serum (Abcam, Cambridge, UK) in 1% w/v BSA/TBS (Sigma, Gillingham, Dorset, UK). Sections were incubated overnight at 4 °C with type II collagen monoclonal primary antibody (1 : 100 dilution; Developmental Studies Hybridoma Bank, The University of Iowa, Iowa City, USA). Additional sections were treated with mouse IgG1 isotype controls (NCG01) (Abcam, Cambridge, UK) and used as negative controls. To determine collagen type I content within a section, a heat/citric acid antigen retrieval method (0.05 M Tris–HCL, pH = 9.5, microwave 5 min at 40% power followed by 5 min at 20% power (Sanyo Microwave, 900 W)) was used and non-specific binding sites were blocked at room temperature for 45 minutes with 25% w/v donkey serum in 1% w/v BSA/TBS (Sigma, Gillingham, Dorset, UK). Sections were incubated overnight at 4 °C with rabbit polyclonal to collagen type I (1 : 100 dilution; Abcam, Cambridge, UK) and rabbit polyclonal IgG isotype controls (Abcam, Cambridge, UK) were used for negative controls. On the following day all sections were incubated with secondary antibodies for 30 min at room temperature (collagen type II: 1 : 400 dilution of rabbit polyclonal secondary antibody to mouse IgG H&L (Biotin) and collagen type I: 1 : 200 dilution of Donkey anti-rabbit IgG H&L (Biotin) (Abcam, Cambridge, UK)). Following washes in TBS sections were incubated in ABC solution (R.T.U. Vectastain® Universal Elite® ABC kit, Vector Laboratories, Inc., CA, USA) for 30 minutes at room temperature. Following TBS washes, antibody staining was disclosed by incubation with 3,3′-diaminobenzidine tetrahydrochloride solution (Sigma, Gillingham, Dorset, UK) for 20 min at room temperature and counterstained with Mayers Haematoxylin (Raymond A Lamb, Eastbourne, East Sussex, UK). Sections were dehydrated and mounted in Pertex (Leica, Milton Keynes, UK).

FTIR imaging and data analysis. Mid-infrared microscopic images were collected using an Agilent 680-IR FTIR spectrometer coupled to an Agilent 620-IR FTIR imaging microscope. The microscope was fitted with a liquid nitrogen cooled 64 × 64 mercury–cadmium–telluride focal plane array detector (FPA) and an automated sampling stage. FTIR mosaic images (23 × 57 images, pixel aggregation 256, image pixel dimensions: 92 × 228) covering an area of approximately 8.05 × 19.95 mm of the tissue sections were collected in transflectance mode at a spectral resolution of 4 cm^{-1}. Infrared data matrices of the control and injected samples were collated resulting in a 184 × 228 pixels data matrix. Data was pre-processed by performing a second derivative transformation on the spectra (Savitzky–Golay: filter order 3, filter length 15; ISys 5.0.0.14 software, Malvern Instruments Limited, Malvern, Worcestershire, UK). Pixels within the imaging data set where spectra indicated only the presence of paraffin or substrate were identified and masked using spectral statistics (histogram at 1554 cm^{-1}, masked at 0; ISys 5.0.0.14). The data sets were analysed using a MCR-ALS algorithm described in detail by Wang *et al.*[61] The basic procedures of the MCR-ALS algorithm comprised the transformation of abstract factors deduced from non-linear iterative partial least squares (NIPALS) decomposition into chemical and concentration information by applying a modified iterative alternating least-square optimisation.[61] MCR-ALS was carried out using the MCR-ALSv1.6 software with the following settings; decomposition method: NIPALS; maximum number of iterations: 500; soft constraints: MALS-2D (MCRv1.6 Copyright© 2003–2004 Unilever, UK). Two spectral regions 950–1300 and 950–1600 cm^{-1} were analysed using a range of four to six components in the MCR-ALS model. No spectral correction or exclusion of the spectral region ~1400–1500 cm^{-1}, which is dominated by absorbance peaks of the embedding medium paraffin, was carried out. Both paraffin and potential spectral distortions are considered as additional signal contributions during MCR-MLS resolution.[62]

Human IVD samples

Two human cadaveric discs were obtained within 72 h of death from the Leeds tissue bank (ethical approval: H1306/98). Samples were obtained from a 45 year old male L4/5 disc and 74 year old female L1/2 disc.

Histology. A wedge of IVD tissue encompassing AF and NP was formalin fixed (10% v/v) and paraffin embedded. Consecutive 4 μm sections of each disc were mounted on glass slides (Lavender X-Tra Slides, Leica, Milton Keynes, UK) for histological analysis and on custom made reflective 316 stainless steel slides for FTIR imaging analysis. Sections were histologically graded between 0 and 12 based on the presence of cell clusters, fissures, loss of demarcation and haematoxophilia (indicating reduced proteoglycan content) with each component scored out of 3. A score of 0 to 3 indicates a histologically non-degenerate IVD and a grade of ≥4 indicates evidence of degeneration, as described previously by Le Maitre *et.al.*[63] Discs were graded as: male L4/5 disc: grade 2 (non-degenerate); female L1/2 disc: grade 8 (degenerate).

FTIR imaging and data analysis. FTIR mosaic images of human sections were collected covering an area of approximately 9.8 × 21.7 mm (pixel aggregation 256; image pixel dimensions: 112 × 248 pixels) and the same data analysis as described for the goat samples was applied.

Goat IVD samples; set two

A second set of 43 IVDs were obtained from an *in vivo* goat study investigating the effect of two different growth factors (BMP2 and BMP2/7) in combination with a fibrin hyaluronic acid hydrogel on the regeneration of the intervertebral disc as reported in Peeters *et al.*[64] The research protocol was approved by both a Scientific Board as well as the Animal Ethics Committee of the VU University Medical Center. Seven skeletally mature female Dutch milk goats were used for this study. In the first surgical procedure, mild intervertebral disc degeneration was induced by injecting ±200 μl chondroitinase ABC (CABC) 0.25 Units per ml in the lumbar IVDs using a 29 G needle of 1 cm length. Another disc (T13-L1) was left as a non-degenerate control.[57–60] During the second surgery, twelve weeks later, IVDs were injected with hydrogel in combination with either the BMP-2 or BMP-2/7 growth factor and one level was injected with hydrogel only. Detailed discussions about the safety and efficacy of the nanobiopolymeric fibrin–hyaluronic acid (FBG–HA) conjugated hydrogels, alone or in combination with BMP-2 or BMP2/7 growth factors are reported in Peeters *et al.*[64] In short it was found, that although mild degeneration was induced after injection with CABC, which was indicated by a significant disc height loss of injected IVDs, no significant differences in histological grades, MRI T2* mapping, GAG and total collagen content were found between the control and treated IVDs. However, it was found that MRI T2* mapping showed strong and significant correlations with biochemistry and histology in the NP.

MRI T2* relaxation time mapping. MRI scans were acquired from all lumbar spines within 2–3 hours after autopsy using a 1.5 T MR scanner (Magnetom Symphony, Syngo MR VA30; Siemens Healthcare). Sagittal scans were performed using a T2-weighted turbo spin echo sequence, followed by a multi-echo gradient echo sequence for T2* mapping (echo times 5.7, 10.9, 16.05, 21.2 and 26.4 ms).[57] Each IVD was divided into 5 different regions of interest (ROI) covering the IVD from anterior to posterior where ROI 1 and 3 covered 27.5% of the total disc diameter whereas ROI 2,4 and 5 each covered 15% (Centricity RA 600, Radworks, USA). This way ROI 1, 3 and 5 represent the anterior annulus fibrosus (aAF), NP and posterior annulus fibrosus (pAF) respectively. ROI 2 and 4 represent the transition zones between AF and NP and were not used for analysis. Mean T2* relaxation times were calculated for each ROI fitting the signal intensities of the five echo times by a linear-log least-squares method using Microsoft Excel (Microsoft Office 2010).

Histology. After obtaining MRI scans, IVDs including endplate were dissected from the spine using an oscillating band saw. Mid-sagittal slices of 4 mm were cut, fixed in 4% formalin, decalcified in Kristense fluid and subsequently embedded in paraffin. Consecutive 3 and 4 μm sections of each sample were mounted on glass slides for histological staining, and on custom made reflective 316 stainless steel slides for FTIR imaging analysis respectively. Sections were stained with Alcian blue (Alcian Blue-Periodic Acid Schiff). Stained sections were scored for disc degeneration by two independent researchers using a histological grading scale adapted for goats as described by Hoogendoorn *et al.*[59] Differences in scoring were resolved by consensus, resulting in a final histological scoring ranging between 0 (non-degenerate) and 6 (degenerate).

FTIR imaging and data analysis. FTIR mosaic images (~5.3 × 18.2 mm; pixel aggregation 256; image pixel dimensions: 60 × 208) of the tissue sections were collected. For the correlation of spectroscopic parameters with results from biochemical, histological and magnetic resonance imaging only areas of interest (AoI: 10 × 208 pixels) across each section (Fig. 2A) were collated for all investigated samples resulting in a 420 × 208 pixel data matrix. Data pre-treatment and multivariate analysis was performed as described in the previous FTIR imaging and data analysis section. Extracted MCR-ALS spectral profiles were assigned to collagen type I, collagen type II, and the proteoglycan parameters PG1 and PG2 according to visual matches with reference spectra and the findings as described

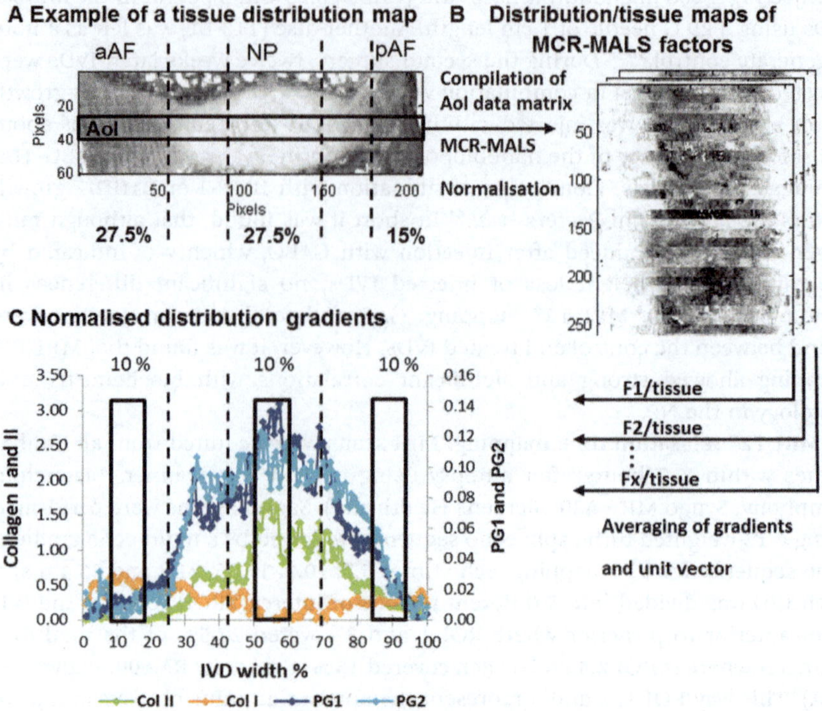

Fig. 2 Schematic representation of the extraction of multivariate FTIR parameters. Showing an example of a tissue distribution map (peak area of the 2nd derivative of the amide III peak; 1186–1297 cm^{-1}) of a paraffin embedded goat IVD section. Black lines indicate IVD regions where an IVD is divided into zones across the width of a sagittal section (described in the section MRI T2* relaxation time mapping) (A). An area of interest (AoI) across the middle of each disc spanning the whole width of each image (10 × 208 pixels) is selected and a new AoI data matrix is generated. MCR-ALS is carried out and distribution maps of selected factors are divided by tissue distribution maps to generate relative component distribution maps (B). To be able to compare IR parameters with MRI and biochemical parameters, the relative distribution maps are further processed. Firstly, the average estimated intensity across each section is taken and secondly, the actual IVD width is determined for each section and set to unity (100%). Thirdly, areas under the curve equivalent to 10% IVD width located in the aAF (10–20%), NP (51–61%) and pAF (85–95%) regions are integrated (C).

in the results and discussion section. Relative MCR-ALS scores maps were calculated as component/tissue maps. These relative component distribution maps were further processed: firstly, the average estimated intensity across each section was taken and secondly, the actual IVD width was determined for each section and set to unity (100%) (Fig. 2B). Thirdly, areas under the curve equivalent to 10% IVD width located in the anterior AF (aAF) (10–20%), NP (51–61%) and posterior AF (pAF) (85–100%; 85–95%) regions, comparable with ROI 1,3 and 5 used for MRI T2* analysis, were integrated (ISys) (Fig. 2C).

Bioassays for GAG and total collagen content. Tissue samples for biochemical analysis were obtained from consecutive 4 mm slices used for histological, immunohistochemical and FTIR imaging analyses. Tissue samples were harvested from the anterior AF, NP and posterior AF analogous to ROI 1, 3 and 5 respectively, as used for MRI T2* mapping. The samples were freeze dried (speed vac) and digested in 1.5 ml papain digestion solution containing 0.1 M sodium acetate, 0.01 M L-cysteine, 0.01 M EDTA and 0.33% papain (w/v) (all Merck Millipore, USA), the pH of the papain solution was titrated to 6.6 using 1 M NaOH. Samples were digested overnight at 65 °C in a continuously shaking warm water bath. GAG content was measured using a colorimetric 1,9 dimethyl-methylene blue assay according to the manufacturer's protocol (Biolcolor Ltd., Carrickfergus, UK). Hydroxyproline (HYP) content, representing total collagen content of the tissue, was quantified using a dimethylamino-benzaldehyde assay adapted from Paul *et al.*[65] Both GAG and HYP were normalized by tissue dry weight (μg (mg DW)$^{-1}$).

Statistical analysis. Non-parametric linear regression analysis was performed between extracted MCR-ALS parameters and histological grades, biochemical parameters and MRI T2* measurements (StatsDirect 3.0126). P values ≤ 0.05 were considered statistically significant and scatter diagrams were plotted using the GraphPad Prism® software (Version 6.05).

Results and discussion

Sagittal sections of paraffin embedded goat IVDs were measured and an average intensity map (950–1800 cm^{-1}) of a non-degenerated IVD section was generated, and the posterior AF (pAF, blue), anterior AF (aAF, green) and NP (red) regions of an IVD were highlighted (Fig. 3A). Overall the average intensity is similar across sample regions, however local higher average intensities can be observed, accentuating artificial structures such as folds formed during sample preparation. Raw spectra show considerable baseline variations (data not shown) and are off-set corrected by setting the minimum value of each spectrum (950–1800 cm^{-1}) to zero (Fig. 3B). To compare spectra from different regions average spectra were calculated of the areas indicated with squares (Fig. 3A). Representative, average transflectance spectra of the aAF, pAF and NP regions show characteristic bands of collagens and PGs (amide I, amide II, amide III/sulphate and polysaccharide region) as well as characteristic bands of the embedding medium paraffin (1473, 1464 and 1378 cm^{-1}). While paraffin peaks at 1473 and 1464 cm^{-1} have similar intensities across the sample regions; the amide I, amide II and amide III peaks show varying intensities. The band intensities of amide I, II and III in the aAF spectrum are approximately double the band intensities exhibited in the NP spectrum. The amide I peak of the aAF spectrum shows a peak shift to a higher wavenumber and asymmetry. The convoluted peak associated with polysaccharide species (900–1100 cm^{-1}) shows a comparatively higher intensity in relation to the amide bands in the pAF and NP spectra, indicating a higher concentration of PGs in the pAF and NP regions. Performing a second derivative of FTIR spectra can enhance the discrimination of broad overlapping bands,[29,36] and more detailed differences between the aAF, pAF and NP regions, especially in the polysaccharide region are revealed. Average second derivative spectra are shown in Fig. 3C and an expansion of the polysaccharide region shows a clear discrimination between the spectra, based on the negative intensity at ~1064 cm^{-1}. The second derivative amide II bands in the pAF and aAF spectra show a subtle shift to higher wavenumbers compared to the NP spectrum. The amide I

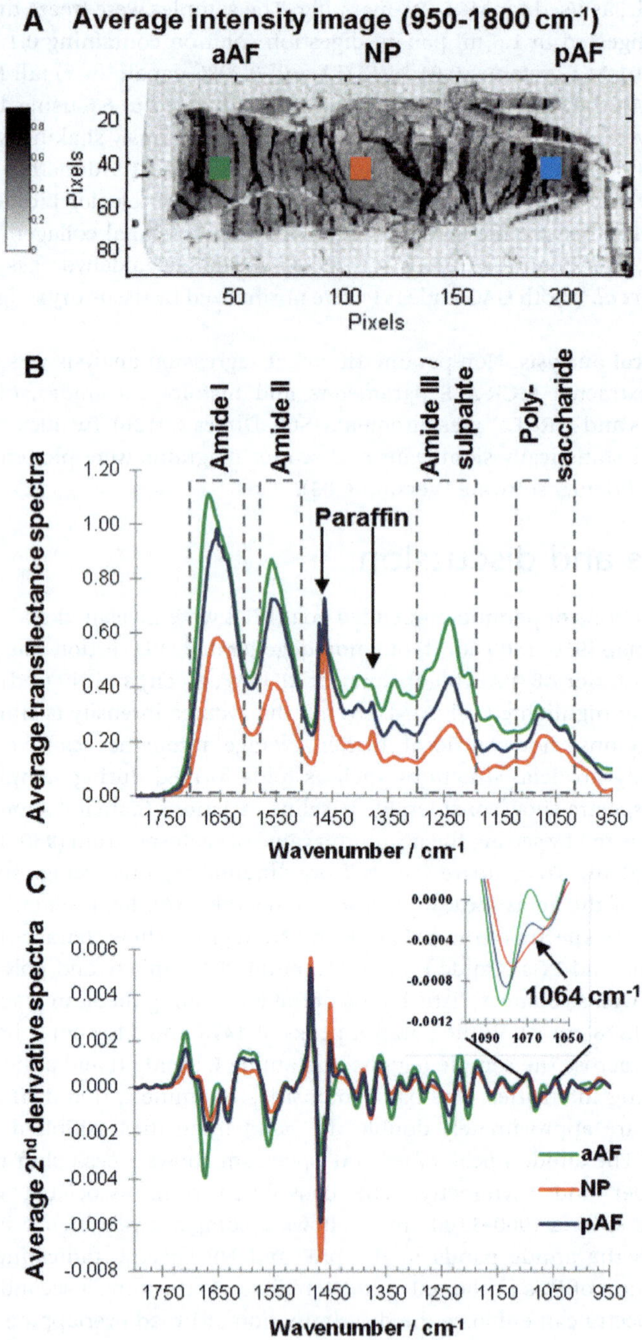

Fig. 3 Average intensity (950–1800 cm^{-1}) image of a non-degenerated paraffin embedded goat IVD section, highlighting anterior annulus fibrosus (aAF, green), posterior annulus fibrosus (pAF, blue) and nucleus pulposus (NP, red) regions (A). Off-set corrected average transflectance FTIR spectra (B) and 2nd derivative spectra (C) of aAF (green), pAF (blue) and NP(red) regions.

band of the aAF spectrum shows a different peak shape and a shift to higher wavenumbers. Paraffin embedded goat IVD sections measured in transflectance mode show region dependent changes in spectral intensity leading to a saturation of the signal in some sample areas. Based on these findings the amide I spectral region traditionally used as a biomarker for collagens[25] was excluded from further analysis and MCR-ALS deconvolution was carried out in the spectral region 950–1600 cm^{-1}.

MCR-ALS analysis

The number of components, or factors, used for the MCR-ALS resolution should not be a critical parameter, studying the evolution of extracted spectral profiles and corresponding distribution maps over a range of factors is good practice and can give additional information about the data set and help with the data interpretation.[47] Details of all MCR-ALS results derived from the spectral regions 950–1600 cm^{-1} and 950–1300 cm^{-1} modelled using 4–6 factors can be found in ESI (Fig. S1 and S2†). While a detailed discussion of the MCR-ALS results is beyond the scope of this manuscript, the major findings are summarised.

Theoretically, MCR-ALS analysis yields spectral profiles of single/pure components and in order to assist with the assignment of the biochemical identity of the extracted factors, estimated spectral profiles are compared to reference spectra of single components. However within this work the extracted spectral profiles may deviate from reference spectra for a number of reasons: (I) reference spectra were measured with ATR-FTIR, in comparison IVD sections were measured using FTIR transflectance microscopy meaning that the relative peak heights and peak positions[66] would differ due to the non-linear effective sampling depth of the ATR measurement. (II) Intermolecular interactions between matrix components can lead to peak shifts in the spectral profiles.[17] (III) Iterative MCR-ALS deconvolution leads to a set of feasible solution and the final results depend on the initial estimates and chosen optimisation processes.[47] In this work, the MCR-ALS deconvolution is carried out without any input of *a priori* information such as the spectral or distribution characteristics of an IVD. Initial estimates are based on abstract factors derived from *eigenanalysis*. The four most commonly used mathematical methods to perform *eigenanalysis* are singular value decomposition (SVD), NIPALS, the power method and the Jacobi method.[46] Although these techniques use different approaches to extract abstract factors the results should be the same.[46] Initial investigations of loadings and scores derived from SVD and NIPALS deconvolution of the spectral regions 950–1600 cm^{-1} and 950–1300 cm^{-1} modelled using 5 factors, however showed that results can vary (S3 and S4). Nevertheless, it was found that after iterative partial least squares optimisation both sets of abstract factors lead to almost identical MCR-ALS solutions. Further inspections of the extracted MCR-ALS spectral profiles showed deviations from the reference spectra, including increased/decreased relative peak heights. One possible explanation for the observed differences could be that the used combination of abstract factors as initial estimates together with the modified iterative alternating least-square optimisation[61] leads to spectral profiles, which emphasise the biggest spectral change in the data set rather than pure spectral profiles. For example, this could explain the observed differences between the extracted spectral profiles for PGs and the reference spectrum of chondroitin

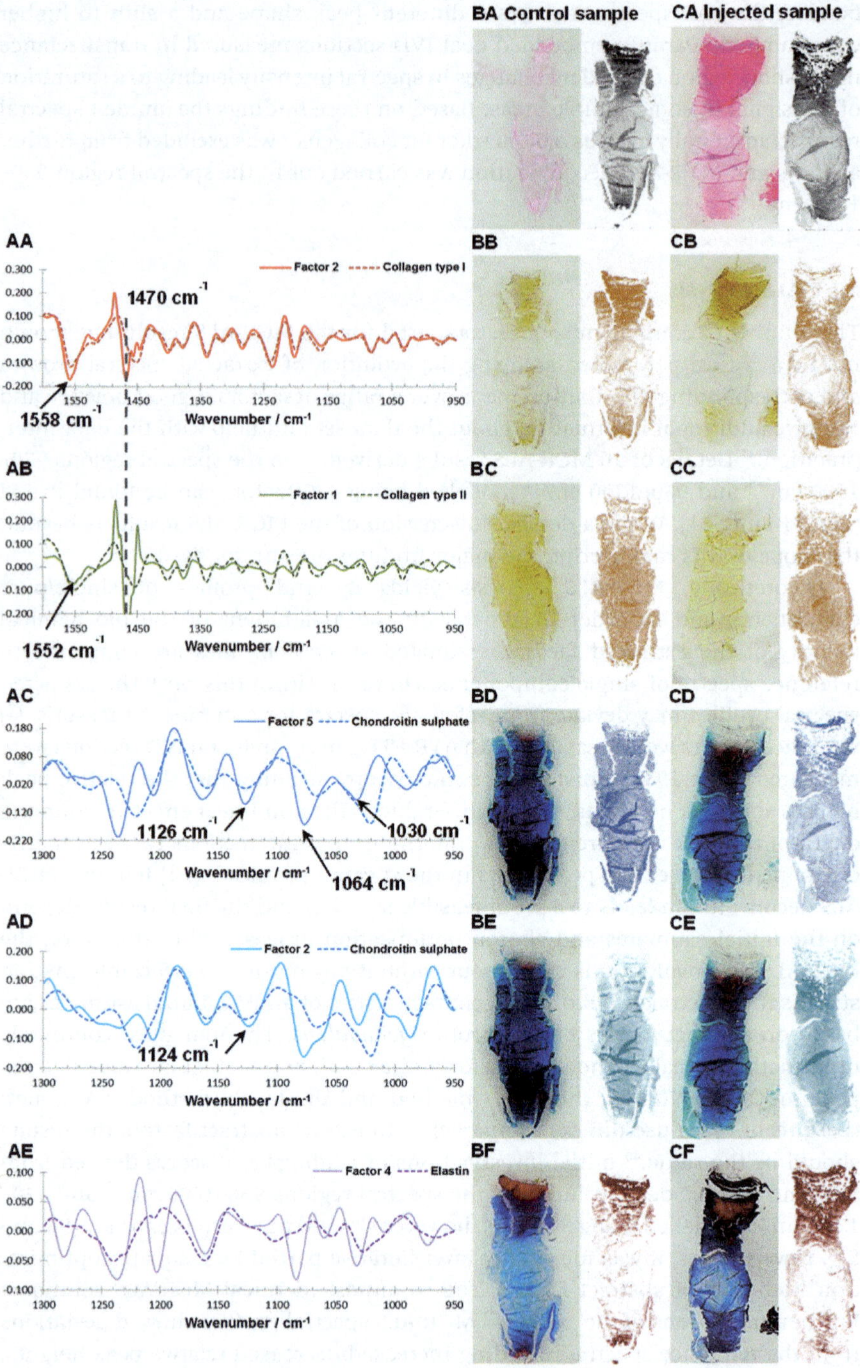

Fig. 4 Comparison of reference spectra and extracted spectral profiles after MCR-ALS decomposition of Factor 2 (950–1600 cm^{-1}) and collagen type I (AA), Factor 1 (950–1600 cm^{-1}) and collagen type II (AB), Factor 5 (950–1300 cm^{-1}) and chondroitin sulphate (AC), Factor 2 (950–1300 cm^{-1}) and chondroitin sulphate (AD) and Factor 4 (950–1300 cm^{-1}) and elastin (AE). Consecutive sagittal sections of a non-degenerated and a mildly degenerated goat IVD stained with H&E (general structure: cytoplasm and extracellular

sulphate (Fig. 4AC and AD, 6BE and BF). Both natural degeneration and enzymatic degeneration induced by CABC causes a breakdown and loss of chondroitin sulphate.[12,13,59,60] Therefore a linear decrease of all the characteristic peaks of chondroitin sulphate (*e.g.* 1376, 1228, 1120, 1064 and 1030 cm^{-1}) would be expected.[43] However the extracted spectral profiles accentuate the peak at 1064 cm^{-1} previously found to show the biggest relative changes in the investigated wavenumber region before and after enzymatic removal of chondroitin sulphate.[43] (IV) In spectroscopic data matrices linear or near-linear relationships within a set of variables, also referred to as co-linearity, can affect the deconvolution power and stability of the MCR-ALS model. Examples of incomplete deconvolution can be seen in the spectral profiles of collagen type I and II, which show contributions of a strong peak at ~1470 cm^{-1} characteristic of paraffin (Fig. 4AA and AB). Although, extracted spectral profiles of collagen type I and II show contributions of paraffin, corresponding distribution maps show markedly different distributions in comparison to the paraffin distribution. Furthermore, it was found that the MCR-ALS results are stable when modelled over a range of four to six factors (Fig. S1†). (V) Finally, in transflectance mode the lateral structure in heterogeneous samples leads to significant distortions of the measured spectra[34] arising from a combination of reflective and optical factors such as a coupling between wavelengths, sample geometry, optical properties within the sample, the presence of interfaces, and the optical setup.[34] While artefacts caused by Mie scattering can be effectively removed through the application of correction algorithms,[35,36] spectral artefacts caused by electric field standing waves have been reported to have a detrimental effect on the sensitivity of transflectance measurements to detect biochemical differences and can limit the transferability of spectral biomarkers from one measurement mode to another.[29]

Results of multivariate FTIR imaging of goat IVD sections (set one): comparison with histological and immunohistological stains

As discussed in the previous section, because of differences between the extracted spectral profiles and collected reference spectra, the biochemical identity of each MCR-ALS factor was established based on a visual assessment, firstly on the correlation between the reference spectra and extracted spectral profiles (Fig. 4A), and secondly by a comparison between the histological and immunohistochemical stains and the estimated distribution maps (Fig. 4B and C). The best MCR-ALS results were selected from both investigated spectral regions, and overall five MCR-ALS factors showing distinct distribution maps and spectral profiles were chosen for further discussion (Table 1).

Fig. 4BA to BF and CA to CF show consecutive sections of a non-degenerated and a mildly degenerated goat sample stained with H&E, collagen type I, collagen

matrix: pink, cell nuclei: purple) (BA, CA), collagen type I (BB, CB), collagen type II (BC, CC), Alican blue (PGs: blue stain) (BD, CD and BE, CE) and Masson Trichrome (collagens: blue, other connective tissues *e.g.* keratin and elastin: red) (BF, CF) are shown at the left. Matching tissue distribution (BA, CA) and MCR-ALS distribution maps are shown at the right as follows: Factor 2 (950–1600 cm^{-1}) (BB, CB), Factor 1 (950–1600 cm^{-1}) (BC, CC), Factor 5 (950–1300 cm^{-1}) (BD, CD), Factor 2 (950–1300 cm^{-1}) (BE, CE) and Factor 4 (950–1300 cm^{-1}) (CF, CF).

Table 1 MCR-ALS resolution of the data was carried out in two spectral regions 950–1300 cm^{-1} and 950–1600 cm^{-1} using a range of four to six components. Best matches based on the visual assessment of the correlation between histological and immunohistochemical stains and estimated distribution maps as well as reference spectra and extracted spectral profiles were selected as follows

Component	Wavenumber region [cm^{-1}]	No. of factors used for the model	Factor index after MCR
Collagen I	950–1600	5	2
Collagen II	950–1600	5	1
Elastic fibres	950–1300	5	4
PG 1	950–1300	5	5
PG 2	950–1300	5	2

type II, Alcian blue and Masson Trichrome. Please note that a pixel by pixel comparison of the distribution of stained biomolecules between stains as well as stains and MCR-ALS distribution maps is not possible as the consecutive sections have slightly different shapes and sizes due to sample preparation and different staining protocols. Nevertheless, the macroscopic comparison of stains and extracted distributions is a good initial validation method. Second derivative peak integration maps of the amide III spectral region (1186–1297 cm^{-1}) are proposed as a measure of overall extracellular matrix or tissue structure and match H&E stained tissue sections. Like the H&E stained sections, tissue distribution maps highlight the overall structure of an IVD (Fig. 4BA and CA). Sections show a densely packed aAF region at the top, a smaller less dense pAF region at the bottom and an only lightly stained NP region in the middle. The distribution map of Factor 2 extracted from the spectral region 950–1600 cm^{-1} shows a gradual increase of intensity from the NP towards the AF typical for collagen type I distribution in IVDs and matches well with the immunohistochemical stained section (Fig. 4BB and CB). The distribution map of Factor 1 (950–1600 cm^{-1}) shows a decrease in intensity from the NP towards the AF typical for collagen type II (Fig. 4BC and CC). As discussed in the previous section, both collagen type I and type II spectral profiles show contributions of a paraffin peak at ~1470 cm^{-1}. Additionally, a peak shift of the amide II band from 1558 cm^{-1} in collagen type I to 1552 cm^{-1} in collagen type II can be observed. Distribution maps of Factor 5 (950–1300 cm^{-1}) and Factor 2 (950–1300 cm^{-1}) show a decrease in intensity from the NP region towards the AF region and are in good agreement with the Alcian blue stain (Fig. 4BD, CD and BE, CE). Both spectral profiles show characteristic PG peaks at ~1030, 1064, 1160 and 1124 cm^{-1}.[43] See also the results and discussion section: MCR-ALS analysis for further comments concerning the comparison between reference spectra and extracted PG signatures. The distribution map of Factor 4 (950–1300 cm^{-1}) shows a high intensity in the AF regions and matches well with the red areas of the Masson Trichrome stained sections. The calculated spectral profile of this factor resembles the reference spectrum of elastin (Fig. 4AE). Yet, while there is a good agreement between extracted distribution maps and staining, further validation of the biochemical identity of this factor against more molecular specific stains for elastic fibres is necessary.

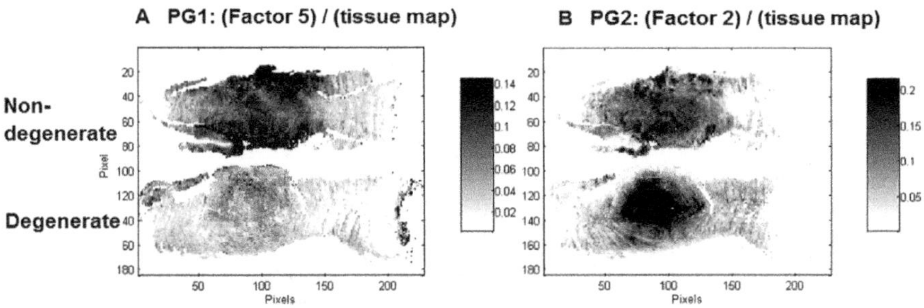

Fig. 5 PG/(tissue map) ratios are proposed as a measure of relative PG content and distribution maps of PG1: (Factor 5)/(tissue map) (A) and PG2: (Factor 2)/(tissue map) (B) extracted from the spectral region 950–1300 cm^{-1} modelled with five factors are shown for the control (top) and injected (bottom) goat IVD samples.

Interestingly, two factors were observed (PG1: Factor 5 (950–1300 cm^{-1}) and PG2: Factor 2 (950–1300 cm^{-1})), which show characteristic spectral and distribution maps of PGs. Information about PGs and more specifically aggrecan and GAG content in the NP, is of particular interest as it can be directly related to disc degeneration.[67] PG/amide I ratios have previously been used as a measure of relative PG content in cartilage.[21] In this work PG distributions are normalised with second derivative peak integration maps of the amide III spectral region (1186–1297 cm^{-1}) (further referred to as tissue map) and relative PG distribution maps reveal distinct PG distributions in non-degenerate *versus* degenerate samples (Fig. 5A and B). While the relative component distribution map based on PG1 is less intense in the degenerate sample in comparison to the non-degenerate sample, the relative component distribution map of PG2 is more intense in the degenerate sample in comparison to the non-degenerate sample. The described MCR-ALS signatures for the two PG factors were found in all three data sets presented in this work. Initial investigations into the nature of the two PG factors have shown reproducible differences in the extracted spectral MCR-ALS signatures of PG1 *versus* PG2, including a peak shift of the C–O–S peak to a lower wavenumber from PG1 (Factor 5, 1126 cm^{-1}) to PG2 (Factor 2, 1124 cm^{-1}). A peak shift of the C–O–S peak has previously been observed by Rieppo *et al.*,[43] where the authors noted a peak shift to lower wavenumbers in samples treated with the CABC enzyme, which facilitates the breakdown of GAG chains. Based on these findings PG1 would represent the "original" PG/GAG population which is decreased in degenerated samples, while PG2 could represent an altered PG/GAG population which is increased in degenerated discs. As the goat degeneration model is mimicking the onset of degeneration, much of the long term remodelling has not yet taken place and the altered PG/GAG molecules are localised in the NP.

Results of multivariate FTIR imaging of human IVD sections

Extracted spectral profiles were compared with reference spectra and spectral profiles derived from the MCR-ALS analysis of goat IVDs. It was found that MCR-ALS analysis of human samples results in comparable spectral profiles as extracted from the goat samples and five factors showing similar spectral characteristics to the spectral profiles extracted for goat IVDs can be seen

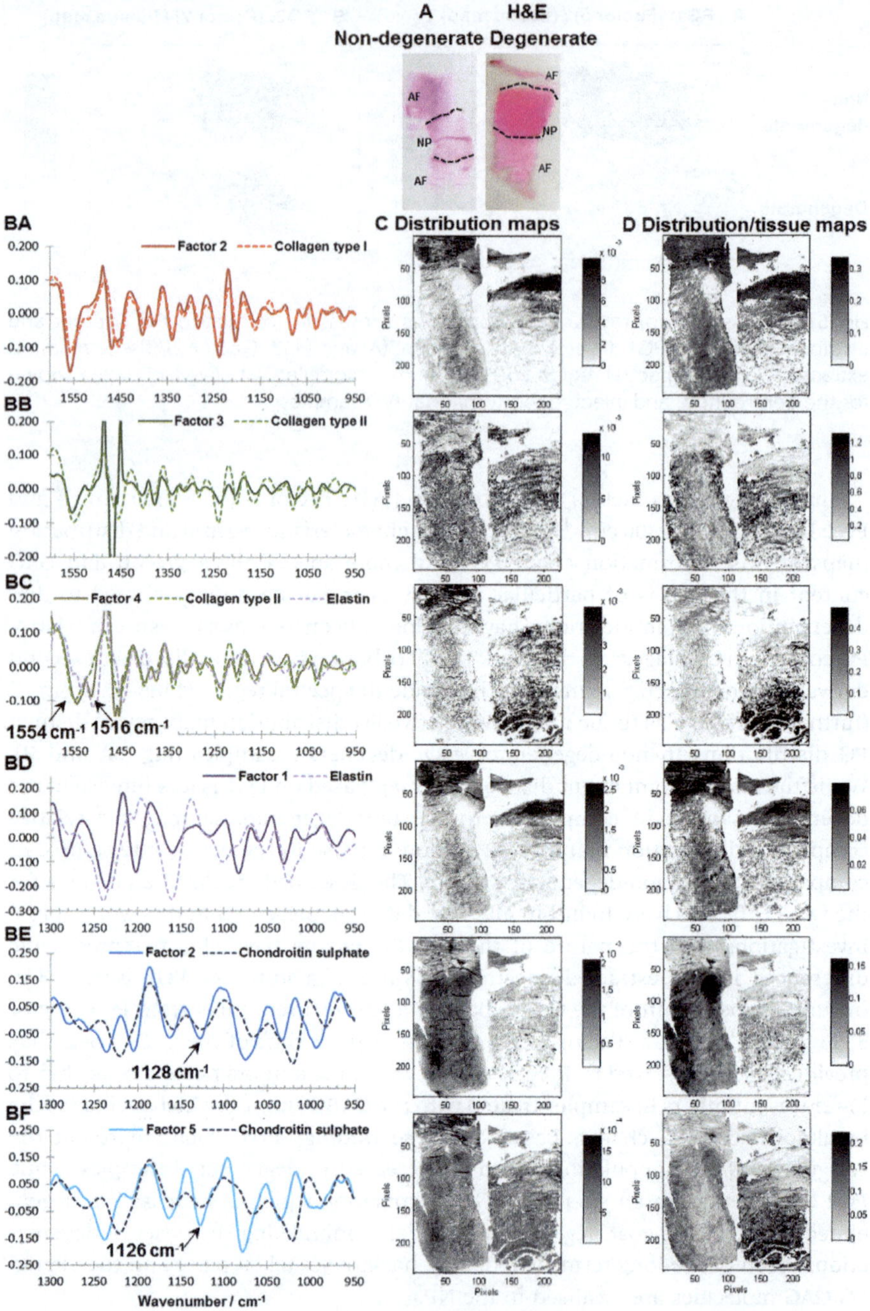

Fig. 6 H&E of non-degenerate (left) and degenerate (right) human IVDs (A). MCR-ALS calculated spectral profiles *vs.* reference spectral profiles of Factor 2 (950–1600 cm^{-1}, modelled with five factors) *vs.* collagen type I (BA), Factor 3 (950–1600 cm^{-1}, modelled with five factors) *vs.* collagen type II (BB), Factor 4 (950–1600 cm^{-1}, modelled with five factors) *vs.* collagen type II and elastin (BC), Factor 1 (950–1300 cm^{-1}, modelled with five factors) *vs.* elastin (BD), Factor 2 (950–1300 cm^{-1}, modelled with five factors) *vs.* chondroitin sulphate (BE) and Factor 5 (950–1300 cm^{-1}, modelled with five factors) *vs.* chondroitin sulphate (BF) as well as corresponding distribution maps (C) and distribution/tissue maps (D) of non-degenerate (left) and degenerate (right) human IVD sections.

(Fig. 6BA–BF). H&E stains, corresponding distribution maps and distribution/tissue maps of non-degenerate (left) and degenerate (right) human discs are shown in Fig. 6A, C and D. Although the degenerate human sample is at a later stage of degeneration in comparison to the goat samples, two MCR-ALS PG signatures can be observed following the same trend; where the distribution/tissue maps of PG1 (Factor 2, 1128 cm^{-1}) show a lower presence in the degenerated disc. Additionally, a localisation of PG1 within cells in the posterior AF region can be seen (Fig. 6BE). Furthermore, PG2 (Factor 5, 1126 cm^{-1}) shows the highest intensity in the posterior AF region of the degenerated disc (Fig. 6BF). This region has been identified as a major location for hernias to occur and transport of PGs/GAGs into the AF region was also observed in Paul *et al.*[65] One additional spectral profile extracted from the spectral region 950–1600 cm^{-1} not detected in the goat sections was observed, showing two peaks at ~1554 and 1516 cm^{-1} matching spectral characteristics of the reference spectra of both collagen and elastin (Fig. 6BC). Estimated distribution/tissue maps of this factor show a higher component content in the degenerated disc, suggesting this could be a degradation product produced during native disc degeneration in humans, and could be an additional marker for degeneration at a later stage.

Results of multivariate FTIR imaging of goat IVD sections (set two): correlation with biochemical and MRI T2* parameters

Relative distribution gradients for MCR-ALS extracted factors assigned to collagen type I, collagen type II, PG1 and PG2 have been compared for the lowest histological grade 0 and the two highest histological grades 5 and 6 (Fig. 7). The figure shows increasing PG and collagen type II gradients towards the NP. Both PG1 and PG2 content is markedly decreased in IVDs with a high histological grading. In

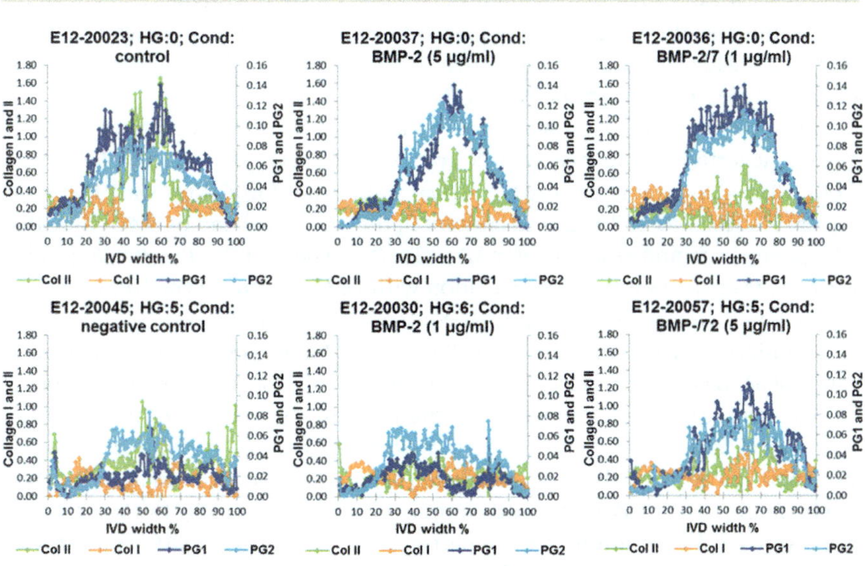

Fig. 7 Examples of relative distribution gradients of MCR-ALS extracted factors assigned to collagen type I (Col I, orange), collagen type II (Col II, green), PG1 (dark blue) and PG2 (light blue) of samples with the lowest histological grade 0 and the two highest histological grades 5 and 6.

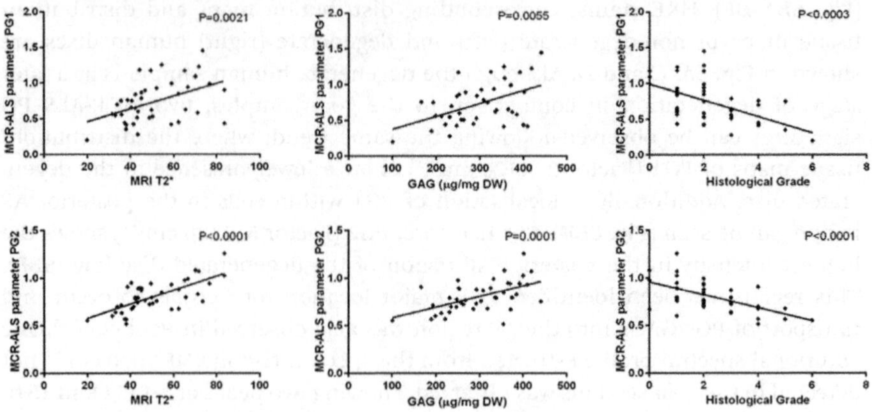

Fig. 8 Correlation graphs between PG1 and MRI T2*, GAG content (µg (mg DW)$^{-1}$) and histological grade (top) and PG2 and MRI T2*, GAG content (µg (mg DW)$^{-1}$) and histological grade (bottom).

Table 2 Non-parametric linear regression between histological grades, GAG content (µg (mg DW)$^{-1}$), total collagen content (HYP (µg (mg DW)$^{-1}$)), MRI T2* and extracted MCR-ALS matrix parameters related to proteoglycans in the IVD matrix for NP

NP	Histology	GAG (µg (mg DW)$^{-1}$)	MRI T2*	HYP (µg (mg DW)$^{-1}$)
Collagen type II	—	$P = 0.0001$	$P = 0.0017$	$P = 0.0127$
Collagen type I	—	$P = 0.0022$	$P = 0.0067$	$P = 0.0204$

the IVDs with low histological grades PG1 shows a slightly higher content in the aAF, while a slightly higher content of PG2 can be observed in discs E12-20045 and E12-20030 with histological grades 5 and 6. Additionally, a slight decrease in collagen type I content in the NP can be observed.

MCR-ALS extracted parameters for PG1 and PG2 were correlated with MRI T2*, GAG concentration and histological grades in the aAF, NP and pAF IVD areas. Both MCR-ALS parameters for PGs show significant correlation with the histological grades, MRI T2* and GAG content in the NP (Fig. 8) and between the MCR-ALS parameters and GAG in the pAF (PG1: $P = 0.0017$; PG2: $P = 0.0039$). Interestingly, PG2 shows a higher correlation with GAG content in the NP, while PG1 shows a higher correlation with GAG content in the pAF.

Significant correlation between collagen type I and type II and GAG content, total collagen content and MRI T2* was observed only for the NP region (Table 2).

No significant correlations were found between the MCR-ALS extracted parameters and GAG or total collagen content for the aAF region. One possible explanation might be that volumes taken for the biochemical analysis are relatively small (~10% IVD width) and cannot be taken from exactly the same part of the IVDs used for sectioning. This is particularly relevant in the pAF, as this region is relatively large 1–27% IVD width and in comparison to the NP structurally more organised (spatially heterogeneous). This difference in sample size and position used for the different techniques may reduce the correlation between the investigated parameters. Furthermore, in this study MCR-ALS parameters for collagen type I and type II were correlated with total collagen and further studies using more specific biochemical assays are recommended.

Conclusions

In this study, we have examined the potential of MCR-ALS analysis to extract spatially resolved biochemical information of non-degenerated and degenerated goat IVDs from 2nd derivative FTIR microscopic imaging data collected in transflectance mode. The presented results show that MCR-ALS allows the resolution of 2nd derivative FTIR microscopic imaging spectra of paraffin embedded IVD sections and important biochemical matrix components can be detected and their distribution visualised. Furthermore, matrix components specifically related to healthy and degenerate states were found. An analytical strategy was presented for the extraction of semi-quantitative biochemical information from IVD sections. Multivariate FTIR image analysis reveals the relative distribution gradients for collagen type I, collagen type II and two PG components. MCR-ALS parameters extracted from the NP region of the IVDs of both PG components show significant correlations with histology ($p < 0.001$), GAG content ($p < 0.01$) and MRI T2* measurements ($p < 0.01$). The FTIR derived parameters for collagen type I and II show significant correlations with the total collagen content ($p = 0.02$, $p = 0.01$).

Exploratory MCR-ALS results highlight the ability of multivariate curve resolution techniques to deconvolute the overall signal into single component contributions, but also the importance of understanding the molecular information contained in spectral profiles extracted using MCR-ALS analysis. Collinearity was observed in the 950–1600 cm^{-1} data set, however distinct factors for collagen type I and II can still be resolved, and good matches between the extracted spectral profiles and distribution maps with immunohistochemical stains were found.

While more molecular specific comparative experiments are recommended to further investigate the chemical nature of the extracted PG parameters and to validate correlations between the extracted collagen type I and type II parameters with collagen type I and II content, the presented results show the potential to study IVD degeneration and enable the characterisation of extracellular matrix components and their distribution in a single slice of IVD tissue, therefore allowing extensive investigations to be performed on limited tissue samples. Additionally, the described multivariate FTIR imaging method was shown to be easily transferable to human sections. The facile application across species provides a major advantage of traditional techniques such as immunohistochemistry where selective antibodies for each species are required, which involve extensive optimisation. In addition the ability of FTIR imaging to screen for variants of the extracellular matrix molecules enables identification of novel components, modifications or degradation products which is not possible *via* directed immunohistochemical analysis.

Acknowledgements

Research funding was provided by the European Union (NPMIMETIC ref. 246351) for 7th Framework Programme funding. The authors acknowledge, that the type

II collagen antibody developed by Thomas F. Linsenmayer was obtained from the Development Studies Hybrdioma Bank developed under the auspices of the NICHD and maintained by the University of Iowa, Department of Biology, Iowa City, IA 52242.

References

1 D. G. Hoy, E. Smith, M. Cross, L. Sanchez-Riera, F. M. Blyth, R. Buchbinder, *et al.*, *Ann. Rheum. Dis.*, 2014, **74**(1), 4–7.

2 T. Driscoll, G. Jacklyn, J. Orchard, E. Passmore, T. Vos, G. Freedman, *et al.*, *Ann. Rheum. Dis.*, 2014, **73**, 975–981.

3 H. Shankar, J. A. Scarlett and S. E. Abram, *Tech. Reg. Anesth. Pain Manag.*, 2009, **13**, 67–75.

4 H. K. Beard, R. Ryvar, R. Brown and H. Muir, *Immunology*, 1980, **41**, 491–501.

5 D. R. Eyre and H. Muir, *Biochem. J.*, 1976, **157**, 67–270.

6 J. Yu, U. Tirlapur, J. Fairbank, P. Handford, S. Roberts and C. P. Winlove, *J. Anat.*, 2007, **21**, 460–471.

7 K. S. Emanuel, P. P. A. Vergroesen, M. Peeters, R. M. Holewijn, I. Kingma and T. H. Smit, *Eur. Cells Mater.*, 2015, **29**, 330–341.

8 M. A. Adams, *Surgery*, 2012, **30**(7), 347–350.

9 A. Pockert, S. Richardson, C. L. Le Maitre, M. Lyon, J. Deakin, D. J. Buttle, A. J. Freemont and J. A. Hoyland, *Arthritis Rheum.*, 2009, **60**(2), 482–491.

10 C. L. Le Maitre, A. Pockert, D. J. Buttle, A. J. Freemont and J. A. Hoyland, *Biochem. Soc. Trans.*, 2007, **35**, 652–655.

11 C. L. Le Maitre, A. J. Freemont and J. A. Hoyland, *J. Pathol.*, 2004, **204**, 47–54.

12 J. P. G. Urban and S. Roberts, *Arthritis Res. Ther.*, 2003, **5**(3), 120–130.

13 J. P. G. Urban, S. Roberts and J. R. Ralphs, *Am. Zool.*, 2000, **40**, 53–61.

14 C. L. Le Maitre, A. L. Binch, A. A. Thorpe and S. P. Hughes, *J. Neurosurg. Sci.*, 2015, **59**(1), 47–61.

15 D. Sakai and S. Grad, *Adv. Drug Delivery Rev.*, 2015, **84**, 159–171.

16 G. Fontana, E. See and A. Pandit, *Adv. Drug Delivery Rev.*, 2015, **84**, 146–158.

17 N. B. Colthup, L. H. Daly and S. E. Wiberley, *Introduction to Infrared and Raman Spectroscopy*, Academic Press, London, 3rd edn, 1990.

18 M. J. Baker, J. Trevisan, P. Bassan, R. Bhargava, H. J. Butler, K. M. Dorling, *et al.*, *Nat. Protoc.*, 2014, **9**(8), 1771–1791.

19 R. Bhargava, *Appl. Spectrosc.*, 2012, **66**(10), 1091–1120.

20 M. Diem, M. Miljković, B. Bird, T. Chernenko, J. Schubert, E. Marcsisin, *et al.*, *Spectroscopy*, 2012, **27**(5–6), 463–496.

21 S. Gamsjaeger, R. Mendelsohn, A. L. Boskey, S. Gourion-Arsiquaud, K. Klaushofer and E. P. Paschalis, *Curr. Osteoporos. Rep.*, 2014, **12**, 454–464.

22 J. P. Spalazzi, A. L. Boskey, N. Pleshko and H. H. Lu, *PLoS One*, 2013, **8**(9), 1–16.

23 A. Hanifi, J. B. Richardson, J. H. Kuiper, S. Roberts and P. Pleshko, *Osteoarthritis Cartilage*, 2012, **20**, 988–996.

24 A. Boskey and N. P. Camacho, *Biomaterials*, 2007, **28**, 2465–2478.

25 M. Kim, X. Bi, W. E. Horton, R. G. Spencer and N. P. Camacho, *J. Biomed. Opt.*, 2005, **10**, 031105.

26 L. Rieppo, S. Saarakkala, T. NärhI, J. Holopainen, M. Lammi and S. Helm, *Microsc. Res. Tech.*, 2010, **73**, 503–512.

27 V. Zohdi, D. R. Whelan, B. R. Wood, J. T. Pearson, K. R. Bambery and M. J. Black, *PLoS One*, 2015, **10**(2), e0116491.

28 F. Lyng, E. Gazi and P. Gardner, in *Biomedical Applications of Synchrotron Infrared Microspectroscopy*, ed. D. Moss, RSC Analytical Spectroscopy Monographs No. 11, 2011, vol. 185, ch. 5, pp. 145–189.

29 M. J. Pilling, P. Bassan and P. Gardner, *Analyst*, 2015, **140**, 2383–2392.

30 A. Hanifi, C. McGoverin, Y.-T. Ou, F. Safadi, R. G. Spencer and N. Pleshko, *Anal. Chim. Acta*, 2013, **779**, 41–49.

31 P. Bassan, J. Lee, A. Sachdeva, J. Pissardini, K. M. Dorling, J. S. Fletcher, A. Henderson and P. Gardner, *Analyst*, 2013, **138**, 144–157.

32 J. Filik, M. D. Frogley, J. K. Pijanka, K. Wehbe and G. Cinque, *Analyst*, 2012, **137**, 853–861.

33 B. J. Davis, P. S. Carney and R. Bhargava, *Anal. Chem.*, 2012, **82**, 3487–3499.

34 B. J. Davis, P. S. Carney and R. Bhargava, *Anal. Chem.*, 2012, **82**, 3474–3486.

35 K. R. Bambery, B. R. Wood and D. McNaughton, *Analyst*, 2012, **137**, 126–132.

36 P. Bassan, A. Sachdeva, A. Kohler, C. Hughes, A. Henderson, J. Boyle, J. H. Shanks, M. Brown, N. W. Clarke and P. Gardner, *Analyst*, 2012, **137**, 1270–1377.

37 P. Bassan, H. J. Byrne, F. Bonnier, J. Lee, P. Dumas and P. Gardner, *Analyst*, 2009, **134**, 1586–1593.

38 M. Romeo and M. Diem, *Vib. Spectrosc.*, 2005, **38**, 129–132.

39 L. Rieppo, S. Saarakkala, J. S. Jurvelin and J. Rieppo, *J. Biomed. Opt.*, 2014, **19**, 027003.

40 L. Rieppo, T. Närhi, H. J. Helminen, J. S. Jurvelin, S. Saarakkala and J. Rieppo, *J. Biomed. Opt.*, 2013, **18**, 097006.

41 A. Hanifi, H. McCarthy, S. Roberts and N. Pleshko, *PLoS One*, 2013, **8**(5), 1–16.

42 Y. Kobrina, L. Rieppo, S. Saarakkala, H. J. Pulkkinen, V. Tiitu, P. Valonen, I. Kiviranta, J. S. Jurvelin and H. Isaksson, *Osteoarthritis Cartilage*, 2013, **21**, 462–469.

43 L. Rieppo, S. Saarakkala, T. Närhi, H. J. Helminen, J. S. Jurvelin and J. Rieppo, *Osteoarthritis Cartilage*, 2012, **20**, 451–459.

44 J. Felten, H. Hall, R. Tauler, A. de Juan and A. Gorzsas, *Nat. Protoc.*, 2015, **10**(2), 217–240.

45 S. C. Rutan, in *Introduction to Multivariate Curve Resolution*, ed. S. Brown, R. Tauler and B. Walczak, Elsevier, Amsterdam, 2009, ch. 2.15, pp. 249–259.

46 E. R. Malinowski, *Factor Analysis in Chemistry*, John Wiley and Sons, Inc., New York, 2002.

47 A. de Juan, S. C. Rutan and R. Tauler, in *Two-Way Data Analysis: Multivariate Curve Resolution – Iterative Resolution Methods*, ed. S. Brown, R. Tauler and B. Walczak, Elsevier, Amsterdam, 2009, ch. 2.19, pp. 325–344.

48 R. Tauler, M. Maeder and A. de Juan, in *Multiset Data Analysis: Extended Multivariate Curve Resolution*, ed. S. Brown, R. Tauler and B. Walczak, Elsevier, Amsterdam, 2009, ch. 2.24, pp. 473–505.

49 S. Piqueras, L. Duponchel, M. Offroy, F. Jamme, R. Tauler and A. de Juan, *Anal. Chem.*, 2013, **85**, 6303–6311.

50 Y. Pu, W. Wang, Y. Yang and R. R. Alfano, *Appl. Opt.*, 2013, **52**, 1293–1301.

51 M. Miljković, T. Chernenko, M. Romeo, B. Bird, C. Matthäus and M. Diem, *Analyst*, 2010, **135**, 2002–2013.

52 S. Piqueras, C. Kraft, C. Beleites, K. Egodage, F. von Eggeling, O. Guntinas-Lichius, *et al.*, *Anal. Chim. Acta*, 2015, **881**, 24–36.
53 T. M. Hancewicz and J.-H. Wang, *Chemom. Intell. Lab. Syst.*, 2005, 77, 18–31.
54 I. Hadjab, L. Epure, J. Antoniou, L. Yahia and F. Mwale, *Global Spine J.*, 2014, **04**, 064.
55 D. Southern, G. Lutz, A. Bracilovic, P. West and M. Spevak, *HSS J.*, 2006, **2**, 49–54.
56 S. E. L. Detiger, R. J. W. Hoogendoorn, A. J. van der Veen, B. J. van Royen, M. N. Helder, G. H. Koenderink, *et al.*, *J. Orthop. Res.*, 2013, **31**(5), 703–709.
57 S. E. L. Detiger, R. M. Holewijn, R. J. W. Hoogendoorn, *et al.*, *Eur. Spine J.*, 2015, **24**, 1935–1943.
58 S. E. L. Detiger, M. N. Helder, T. H. Smit and R. J. W. Hoogendoorn, *Eur. Spine J.*, 2015, **24**(9), 1992–2000.
59 R. J. W. Hoogendoorn, M. N. Helder, R. J. Kroeze, R. A. Bank, T. H. Smit and P. I. Wuisman, *Spine*, 2008, **33**(9), 949–954.
60 R. J. W. Hoogendoorn, P. I. J. M. Wuisman, T. H. Smit, V. Everts and M. N. Helder, *Spine*, 2007, **32**(17), 1816–1825.
61 J.-H. Wang, P. K. Hopke, T. M. Hancewicz and S. L. Zhang, *Anal. Chim. Acta*, 2003, **476**, 93–109.
62 S. Piqueras, L. Duponchel, R. Tauler and A. de Juan, *Anal. Chim. Acta*, 2011, **705**, 182–192.
63 C. L. Le Maitre, A. Freemont and J. A. Hoyland, *Arthritis Res. Ther.*, 2005, **7**, R732–R745.
64 M. Peeters, S. E. L. Detiger, L. S. Karfeld-Sulzer, K. Mevorat Kaplan, T. H. Smit, A. Yayon, F. E. Weber and M. N. Helder, *BioRes. Open Access*, 2015, **4.1**, 398–406.
65 C. P. L. Paul, T. School, H. A. Zuiderbaan, *et al.*, *PLoS One*, 2013, 8e62411.
66 R. T. Graf, J. L. Koenig and H. Ishida, *Polymer Science and Technology Series*, 1999, **36**, 385–395.
67 S. S. Sivan, E. Wachtel and P. Roughley, *Biochim. Biophys. Acta*, 2014, **1840**(10), 3181–3189.

Faraday Discussions

PAPER

Assessing corneal biomechanics with Brillouin spectro-microscopy

Guillaume Lepert,[a] Ricardo M. Gouveia,[b] Che J. Connon[b] and Carl Paterson*[a]

Received 2nd November 2015, Accepted 7th December 2015

DOI: 10.1039/c5fd00152h

A new Brillouin spectro-microscope was designed and built to investigate the mechanical properties of bovine and human corneas. This instrument integrates a single-stage virtually imaged phased array spectrometer with a novel adaptive-optics interferometric filter to achieve unprecedented rejection of the elastic background signal. As a result, highly-resolved, reproducible data from both thin and thick collagen-based materials were obtained. In particular, this technique is capable of rigorously measuring the relative stiffness of different areas of human corneas, thus providing a true non-contact method to characterise the fundamental mechanical features of both live and fixed biological tissue samples.

Brillouin and Raman scattering both constitute useful techniques to study material properties, although through very different regimes. Raman spectroscopy probes intra-molecular vibrational bonds (optical phonons) in the THz range (100 s to 1000 s of cm^{-1}) and reveals information about chemical composition.[1] By contrast Brillouin scattering probes bulk acoustic waves (acoustic phonons) at GHz frequencies ($\sim$0.1–1 cm^{-1}) and is linked to viscoelastic properties.[2] Brillouin scattering is based on the principle that photons scattered by travelling density fluctuations (spontaneous thermal phonons) acquire a Doppler shift. Measuring this shift is measuring the speed of sound in a given material, which in turn is related to the mechanical properties and moduli of the material.[3] Thus, although at a fundamental level the underlying physics is similar, the two techniques address very different needs.

Brillouin spectroscopy is a useful tool in solid mechanics, and has long been used to characterise the mechanical properties of transparent fluids and crystalline solids.[4] However, only recently has this technique been applied to the study of organic matter, thanks primarily to the introduction of virtually imaged phased arrays (VIPAs)[5] *in lieu* of multi-pass Fabry–Pérot spectrometers,[6] and to the

[a]*Imperial College London, Blackett Laboratory, London, SW7 2BW, UK. E-mail: guillaume.lepert07@imperial.ac.uk*

[b]*Institute of Genetic Medicine, Newcastle University, International Centre for Life, Central Parkway, Newcastle upon Tyne, NE1 3BZ, UK*

realisation that the geometric broadening of the Brillouin spectra arising from focused illumination does not preclude high-NA Brillouin microscopy.[7] The invention and widespread adoption of VIPAs in recent years has triggered something of a renaissance for Brillouin scattering by cutting the acquisition times to seconds or less and enabling Brillouin imaging.[5] Compared with mechanical methods such as atomic force microscopy (AFM), where measurements require direct interactions between the sample and a probe, Brillouin imaging offers the advantage of a truly local non-contact analysis at microscopic resolution. For example, several recent studies have used Brillouin scattering to characterise the mechanical properties of biological tissues, including live cells,[8] components of the extracellular matrix,[9] and other natural or synthetic biomaterials.[10] Although these developments have yet to be translated to a clinical settings, they show great promise for the diagnosis and treatment of arteriosclerosis plaque[11] and keratoconus.[12]

In the present work we review the fundamentals of Brillouin scattering and introduce a new Brillouin spectro-microscope, dedicated to soft biomaterials, that enables significant advances in suppressing parasitic elastic scattering in aqueous media. This instrument was used to evaluate the mechanical properties of dense collagen type-I gels produced after plastic compression, as well as of corneal tissue of bovine and human origin. The results reported here provide important information about the structure and organisation of the biomaterials analysed. We also address the challenges inherent to this method, notably in relating the measured Brillouin shift to the biomechanical properties evaluated using standard rheological and nano-indentation techniques.

1 Principles and apparatus for Brillouin spectro-microscopy

1.1 Brillouin scattering

In Brillouin scattering, an incoming photon (angular frequency ω_i, wave vector $\vec{k}_i$) is inelastically scattered off a travelling thermal phonon (angular frequency ω_p, wave vector $\vec{k}_p$) at an angle θ, with a new frequency ω_s and wave vector $\vec{k}_s$, as depicted in Fig. 1. In a quantum description, the photon will have either gained or lost a phonon quantum of energy through annihilation or creation of a phonon, respectively referred to as Stokes and anti-Stokes processes. The scattering angle θ is set by the experiment's illumination/collection geometry, which also constrains the phonon wavelength *via* the conservation of energy and momentum:

$$k_p = 2k_i \sin(\theta/2) = \frac{4\pi n \sin(\theta/2)}{\lambda}. \quad (1)$$

The phonon wavelength is comparable to that of the photons, while the frequency is in general of the order of 10 GHz. In the classical description, the phonon is an acoustic wave, a periodic modulation of density and therefore of refractive index, which can be regarded as a reflective Bragg mirror moving at the speed of sound. Then the above equation is just the Bragg condition; the Stokes/anti-Stokes processes correspond to waves travelling in opposite directions, with the Brillouin shift corresponding to a Doppler shift. The phase velocity v_p of the acoustic wave is

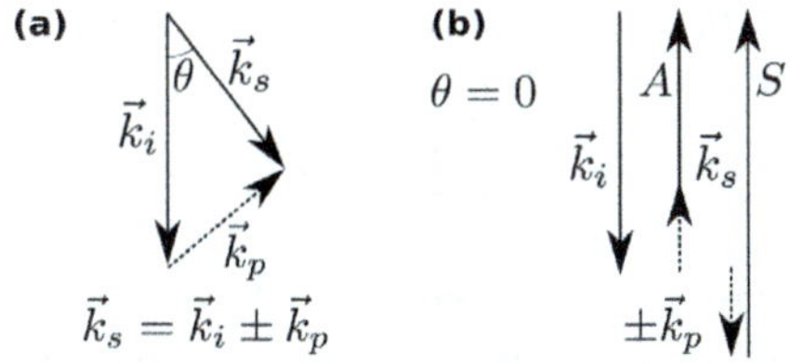

Fig. 1 Momentum conservation in Brillouin scattering. (a) The general case; i, and s denote the incident and scattered photons, and p the phonons. (b) The back-scattering geometry ($\theta = 0$) used in this paper, showing the generation of the Stokes (*S*) and anti-Stokes (*A*) scattered photons.

$$\nu_\mathrm{p} = \frac{\omega_\mathrm{p}}{k_\mathrm{p}} = \frac{\omega_\mathrm{s} \pm \omega_\mathrm{i}}{k_\mathrm{p}} = \Delta\nu_\mathrm{B}\frac{\lambda}{2n\sin(\theta/2)} \tag{2}$$

where $\Delta\nu_\mathrm{B} = (\omega_\mathrm{s} \pm \omega_\mathrm{i})/2\pi$ is the experimentally measured Brillouin shift.

In general, transverse and longitudinal acoustic waves propagate at different velocity, and will give rise to separate peaks in the Brillouin spectrum. However in the back-scattering geometry ($\theta = 180°$) used in this work only longitudinal waves participate in the scattering processes,[13] and the spectrum consists of a single Stokes/anti-Stokes doublet.

1.2 Relation of the Brillouin shift to mechanical moduli

The acoustic velocity is correlated to the mechanical moduli of the material. Since we are probing longitudinal waves only, and because the phonon wavelength is much smaller then the material's extent, the relevant moduli is the so-called p-wave or longitudinal modulus *M*, which is related to the velocity by $\nu_\mathrm{p} = \sqrt{M/\rho}$ (ρ is the density of the medium). Rheological or nano-indentation techniques measure Young's modulus *E*, and it would therefore be desirable to be able to relate these two quantities. For a number of reasons this is not straightforward; indeed preliminary comparative experiments[14] have shown that the Brillouin modulus is two or three orders of magnitude larger than Young's modulus. First, the acoustic isotropy of highly ordered tissues such as corneal collagen is questionable. Second, *M* is measured at GHz acoustic frequencies whereas *E* is a quasi-static measurement. Acoustic dispersion could therefore

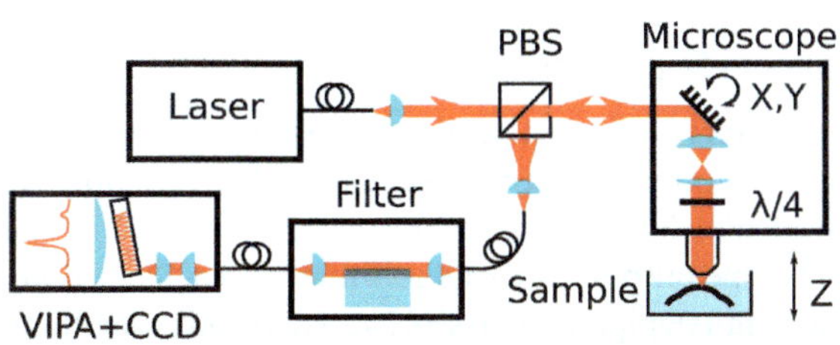

Fig. 2 Simplified schematics of the experimental set-up, showing the confocal microscope, elastic scattering filter, VIPA spectrometer, all connected by single-mode fibres, and the sample in its immersion medium.

significantly increase the moduli in the hypersonic regime.[15] Thirdly, one may be tempted to use the usual relation[3]

$$E = M\frac{(1+\nu)(1-2\nu)}{1-\nu} \tag{3}$$

where ν is the Poisson ratio. For biomaterials of interest, rich in water and therefore nearly incompressible, the Poisson ratio is close to 0.5. Examination of eqn (3) shows that it diverges at that point, and the conversion is therefore prone to very large errors. Finally, in heterogenous materials acoustic confinement effects (due to the presence of various acoustically reflective structures and interfaces) may alter the phonon density of states, causing a distortion of the Brillouin lineshape and affecting the extraction of the Brillouin shift. Such effects have been reported on the Brillouin spectrum of water in the presence of planar and cylindrical acoustic resonators.[16]

At present, none of these issues have been properly investigated, much less resolved. In what follows we shall therefore not attempt to convert the measured Brillouin shifts into mechanical moduli and use the Brillouin shift $\Delta\nu_B$ as the metric.

1.3 Epifluorescence confocal microscope set-up

The Brillouin spectro-microscope is schematically depicted in Fig. 2. A frequency-doubled Nd:YVO_4 diode-pumped solid-state laser (CNI, MSL-FN-671) operating at $\lambda_{ex} = 671$ nm, coupled into a single-mode polarisation-maintaining fibre (Corning SM600), provided single mode illumination (transverse and longitudinal) with a spectral linewidth smaller than 1 MHz. The fibre output was collimated to a 3 mm diameter beam, reflected off a pair of scanning mirrors (Thorlabs GVS002), and expanded to an 8 mm diameter by two achromatic doublets (focal lengths 75 and 180 mm) to fill the back focal plane of a 20× water dipping microscope objective (Olympus XLUMPLFLN20xW, numerical aperture 0.5). The doublets served as the scan relay optics by imaging the mirrors onto the objective's back focal plane. In addition, a quarter-wave plate inserted just before the objective was used to transform the laser linear polarisation into a circular polarisation. Scattered light was collected by the same objective, descanned by the same mirrors, separated from the excitation light by a polarising beam-splitter, and coupled into a single-mode fibre which also acted as the confocal pinhole. The final imaging resolution was calculated to be 1 × 1 × 5 μm. For depth sectioning, the sample was mounted on a vertical translation stage (Newport MFA-CC).

1.4 Suppression of elastic scattering

The elastically-scattered light usually represents a much stronger signal compared to the Brillouin scattering itself. In order to suppress the former, an original wavefront division interferometer was introduced in the system. Light exiting the microscope output fibre was collimated to a 4 mm diameter beam and intersected by a parallelepipedic glass prism such that exactly half of the beam travels above the prism, and half inside. The path difference δ between the two halves depends on the angle θ_i between the prism and the incident beam. It was calculated as

$$\delta = d(n-1)\left(1+\frac{\theta_i^{\,2}}{2n}\right) \tag{4}$$

and then adjusted by tuning the prism angle to achieve destructive interference at the laser wavelength when recoupling the beam into a single-mode fibre, while allowing the passage of inelastically scattered (Brillouin) light. The filter transmission is a periodic function of the prism angle and of the frequency, with a free spectral range (FSR) c/δ (c is the speed of light in a vacuum). Using a 30 mm long BK7 prism of refractive index $n = 1.51$, the FSR is about 20 GHz. Assuming the amplitude of the two paths is exactly matched (through translation of the prism), the maximum extinction $S_{\max}$ is limited only by the laser linewidth σ and calculated as:

$$S_{\max} = \frac{\pi\sigma}{\mathrm{FSR}}. \tag{5}$$

1.5 VIPA spectrometer

The VIPA (LightMachinery) has a free spectral range of 30 GHz. Light exiting the filter output fibre was collimated to a 0.8 mm diameter and focused on the VIPA entrance window by a 100 mm cylindrical lens. A 400 mm cylindrical lens imaged the VIPA's angular spectrum onto a CCD camera (Andor iXon DU888, operated in non-electron-multiplying mode) with the help of a 50 mm cylindrical lens, orientated orthogonal to the others, to provide focussing along the non-dispersed axis.

1.6 Data acquisition and processing

The spectra imaged on the CCD camera were acquired with a typical exposure time of 100 to 1000 ms, increasing with the penetration depth in the sample. Vertical binning was used on the sensor so that one-dimensional spectra were produced without further processing. The beam scanning, sample vertical motion, and interferometer feedback were controlled using a Python-based, custom-made software specifically developed for this purpose (MicroscoPy, pypi.python.org/pypi/MicroscoPy). Raw spectra were stored as a multidimensional array in HDF5 files, alongside position information and other metadata. Data processing was performed using Mathematica. Briefly, the spectral axis was calibrated against reference spectra of water or solid polystyrene, with frequency expressed as a quadratic function of pixel coordinates, and a linear least-square fit to the positions of 3 to 4 Rayleigh peaks and their associated Brillouin peaks. Both the Stokes and anti-Stokes peaks of the brightest interference order were then fitted with a Lorentzian function, and the Brillouin shift was calculated as the average of both the Stokes and anti-Stokes frequencies, making measurements insensitive to laser drift and temperature fluctuations (which can cause a global spectral drift of the order of 1 GHz h^{-1}).

1.7 Filter feedback

Unlike the measurement of Brillouin frequency, the interferometric filter is sensitive to frequency drifts (on a timescale of minutes) and requires continuous adjustments to maintain its extinction to the required level. For this purpose, a simple feedback mechanism was incorporated by the addition of a piezoelectric actuator to periodically modulate the interferometer angle to produce a ~10%

modulation in the (post-filtering) elastic amplitude. The latter is measured directly on the acquired spectrum by integrating the pixel intensities around the elastic peak. A digital PID (proportional–integral–differential) control loop was used to keep the prism angle close to its optimum position (corresponding to a minimum in the elastic amplitude), with a time constant of the order of minutes to average out the large fluctuations in the amplitude of the elastic scattering while scanning through the sample.

1.8 Preparation of compressed collagen gels

Compressed collagen gels were prepared as described elsewhere.[17] Briefly, ice-cold rat tail collagen type-I (2 g L^{-1} in 0.6% acetic acid; First Link Ltd, UK) was mixed with 10× Modified Essential Medium (MEM; Life Technologies, CA, USA) and neutralised with 1 M NaOH in an 8 : 1 : 1 volume ratio, respectively, and allowed to gel at 37 °C for 30 min. Polymerised gels were subjected to plastic compression under a 134 g load for 5 min at room temperature by placing the gels between layers of gauze and on top of 10 sheets of filter paper. Plastically-compressed collagen gels were then transferred to a Petri dish onto 3- or 4-gel stacks and washed at room temperature for at least 30 min in distilled water before being analysed by Brillouin microscopy. Chemically-stiffened collagen gels were analysed after fixing the compressed collagen gels in 4% (w/v) paraformaldehyde, with extensive washing in distilled water before imaging. Experiments were performed in triplicate.

1.9 Preparation of corneal tissue

Bovine corneal tissue (depicted in Fig. 3) was isolated from fresh cow eyes obtained from an abattoir (Linden Food Burradon Ltd, UK). The tissue corresponding to the cornea and the surrounding sclera was excised from whole cow eyes, kept on ice during the isolation, incubated with a penicillin–streptomycin solution (1000 U mL^{-1}; Thermo Scientific, MA, USA) for 30 min at room temperature to inhibit bacterial contamination, washed in an excess of phosphate buffered saline (PBS), and kept at 4 °C in either PBS supplemented with 0.025%

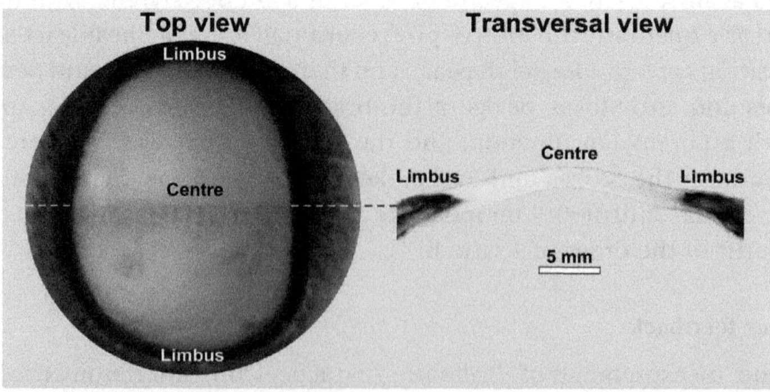

Fig. 3 Whole bovine corneal tissue isolated from fresh cow eyes. The left panel shows a top-down view of the bovine cornea, with its translucent ellipsoidal centre, and the opaque, pigmented limbus that surrounds it. The right panel shows a transversal section of the cornea to illustrate the tissue thickness. Scale bar = 5 mm.

sodium azide preservative or at room temperature in Carry-C tissue preservation medium (Alchimia, Italy) until further analysis was performed. Human corneal tissues were kindly provided by Dr Francisco Figueiredo, FRCOphth (Royal Victoria Infirmary Newcastle, UK). Tissues remaining from corneal transplant procedures were kept up to 30 days after isolation from cadaveric donors (ages ranging 58–72 years old; male–female donor ratio of 1 : 1; no prior history of corneal diseases or ocular trauma), in accordance with the Newcastle University and Newcastle-upon-Tyne Hospital Trust Research Ethics Committees' guidelines. Isolated tissues were then extensively washed in PBS, dried of excess adsorbed liquid, and then maintained at room temperature in Carry-C tissue preservation medium until further analysis was performed. Tissues were imaged in Petri dishes while completely submerged in PBS or Carry-C medium.

1.10 Statistical analysis

The Brillouin shift was measured at eight different z-positions from five distinct collagen gels ($n = 5$), with differences between average shifts analysed using an unpaired two-tailed Student's t-test, with $p = 0.0001$ and $R^2 = 0.91$.

2 Results and discussion

2.1 Brillouin spectro-microscope assembly and specifications

The new Brillouin spectro-microscope was specifically dedicated to study the mechanical properties of soft biological tissues. These systems are characterised by low scattering signals, high elastic backgrounds, and the requirement of sub-GHz resolutions. For this particular instrument we used a $\lambda_{ex} = 671$ nm DPSS laser with high spectral purity and sub-MHz linewidth (Fig. 2). At this wavelength, the Stokes/

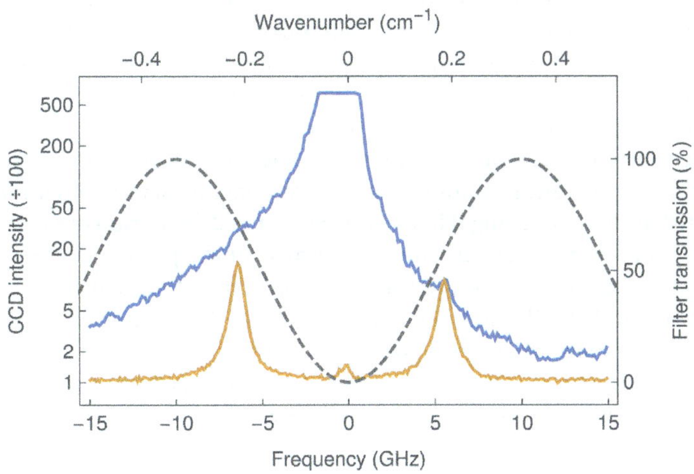

Fig. 4 Brillouin spectrum of water, measured with the filter tuned to maximize (blue curve) and minimise (orange) the transmission at the laser frequency. In the first case, the Brillouin peaks are essentially invisible, overwhelmed by the stronger elastic scattering. In the latter, the Brillouin peaks are easily resolved and the elastic peak has been suppressed by nearly 50 dB. The grey line shows the sinusoidal transmission of the filter, whose free spectral range is 20 GHz.

anti-Stokes components of Brillouin scattering are typically shifted by 5–7 GHz in aqueous media.[18] Because such small shifts are inaccessible to grating spectrometers, Brillouin spectrometry has traditionally relied on multi-pass scanning Fabry–Pérot interferometers to achieve both the necessary resolution and fringe contrast; the acquisition time for a single spectrum was measured in minutes or even hours.[19] This limitation is overcome by VIPAs, which combine the large dispersive power of the etalon with the spatial dispersion of gratings. They are essentially Fabry–Pérot etalons, dispersing focused light into fringes of equal inclination, with the important difference that the input light is focused by a cylindrical lens in only one direction and enters the etalon through a narrow antireflection-coated window. All of the input light is therefore transmitted, forming a one-dimensional spectrum instead of circular fringes, and resulting in a more efficient device. However, VIPAs have limited finesse (100 at most), and therefore poor fringe contrast (<40 dB), insufficient to resolve (anti-)Stokes Brillouin peaks amidst a strong elastic background, such as that inevitably arising from epi-fluorescence confocal microscopy. This is commonly addressed by cascading two VIPAs in a cross-axis configuration,[20] at the cost of reduced throughput and considerable complexity. Another approach involves the absorption of the elastic scattering in hot vapour cells, although this often distorts the Brillouin peaks as well.[21]

In the present work, we used an original adaptive-optics wavefront division interferometer, described above, that enables the interferometric filtering of the elastic scattering, as demonstrated in Fig. 3. With the elastic suppression turned off (interferometer detuned), the signal was overwhelmed by a strong elastic background (blue line) as shown in Fig. 4. By tuning the interferometer to minimise the amplitude of the central elastic peak, the Brillouin Stokes and anti-Stokes components became clearly resolved (orange line). The Brillouin shift was detected at 5.86 ± 0.01 GHz (or 0.195 cm^{-1}), corresponding to a sound velocity of 1475 ± 10 m s^{-1}, as expected for distilled water at 20 °C.[22] Furthermore, the elastic suppression ratio exceeded 46 dB, a significantly better outcome compared to other methods currently in use.[20,21,23] Consequently, the strong extinction made possible by the interferometric filter allows the implementation of a single-stage VIPA configuration instead, which constitutes, to the extent of our knowledge, a novel development in Brillouin microscopy.

Furthermore, the use of an immersion objective allows the samples to be observed in physiological conditions, as well as enabling several other important benefits for Brillouin imaging. Firstly, it prevents the dehydration of the biological tissues, a parameter that greatly affects their mechanical properties.[24] Secondly, it reduces specular reflection at the sample surface, therefore reducing the elastic background. And finally, it limits the loss of resolution normally caused by spherical aberration when focusing deep into thick samples. This issue is particularly relevant in imaging cornea specimens, particularly with bovine origin, where the tissue can be up to 2 mm thick (see below).

2.2 Mechanical properties of high-density collagen gels

In order to explore the potential uses and limitations of our instrument, we started by evaluating dense collagen gels produced by a plastic compression method. These thin biomaterials have been extensively used as versatile models of connective tissue such as cornea[17,24] and skin.[25] Furthermore, their mechanical properties (*e.g.*, elastic modulus) have been shown to be highly-defined and dependent on the compression load and duration.[24,26,27] As such, collagen gels are relevant as models for the determination of the biomechanical properties of soft materials. The Brillouin shift observed for compressed collagen gels was shown to be very consistent for all gels analysed, both from those in the same stack (4 gels per stack) or between different replicates, with an average shift of 6.06 ± 0.04 GHz (Fig. 5a), and despite the strong elastic background scatter due to the proximity of the interface. In addition, the shifts detected at the surface and within the gels were similar, and their variation smaller than that observed between gels (Fig. 5a). This suggested that the mechanical properties of the compressed collagen gels were uniform between gels and throughout their thickness. Moreover, stiffened collagen gels (crosslinked with paraformaldehyde) showed significantly higher Brillouin shifts of 6.35 ± 0.07 GHz (Fig. 5b), a clear indication that alterations in the stiffness of the collagen gels results in significantly different shifts in Brillouin scatter. Interestingly, the Brillouin profiles also allowed us to determine gross sample depth, with the stiffened collagen gels showing an approximately 50% reduction in thickness compared to the untreated compressed collagen gels. This shrinkage constitutes a common effect of conventional cross-linking methods,[28] and may partially account for the increased gel stiffness.

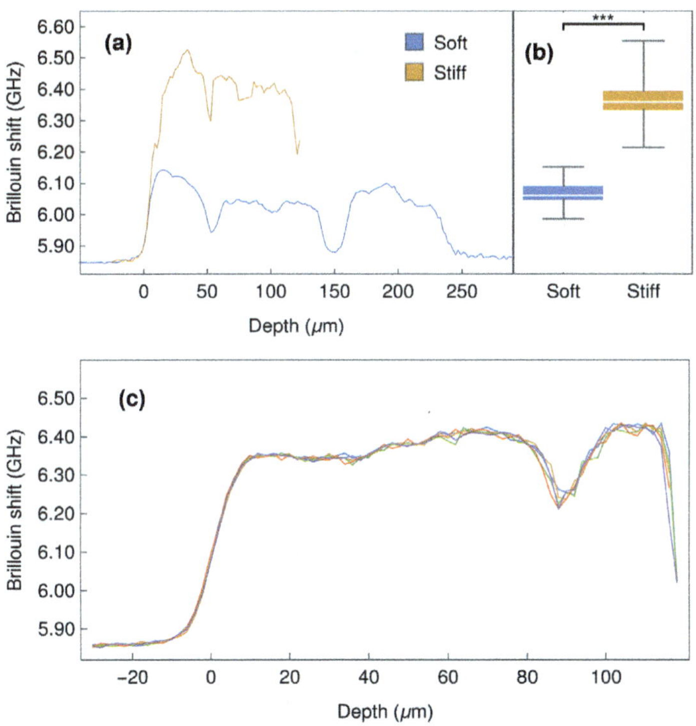

Fig. 5 Brillouin imaging of compressed collagen gels in water. (a) Representative depth profile of Brillouin shifts of a stack of 4 collagen gels, before (blue) and after stiffening treatment with PFA (orange). (b) Box plot of average shift distribution from five independent control and stiffened collagen gels ($n = 5$; *** corresponds to $p = 0.0001$). (c) Repeated Brillouin z-scans of a stack of stiffened collagen gels, illustrating the high stability and reproducibility of Brillouin imaging. Distance between measurements in $z = 2$ μm.

2.3 Mechanical properties of corneal tissue

Having determined the Brillouin shift of thin collagen samples, we then evaluated the limitations of the technique for the analysis of thick biological samples. Bovine corneas were used for these experiments, as these tissues are relatively transparent despite the 1.2–1.5 mm thickness of their centre region (Fig. 3). In addition, as the correct evaluation of mechanical properties in biological tissues requires a constant and physiologically-relevant level of tissue hydration, Brillouin spectro-microscopy measurements should be performed in such conditions. Corneas are naturally hyperosmotic, slightly dehydrated tissues *in vivo*.[29] To

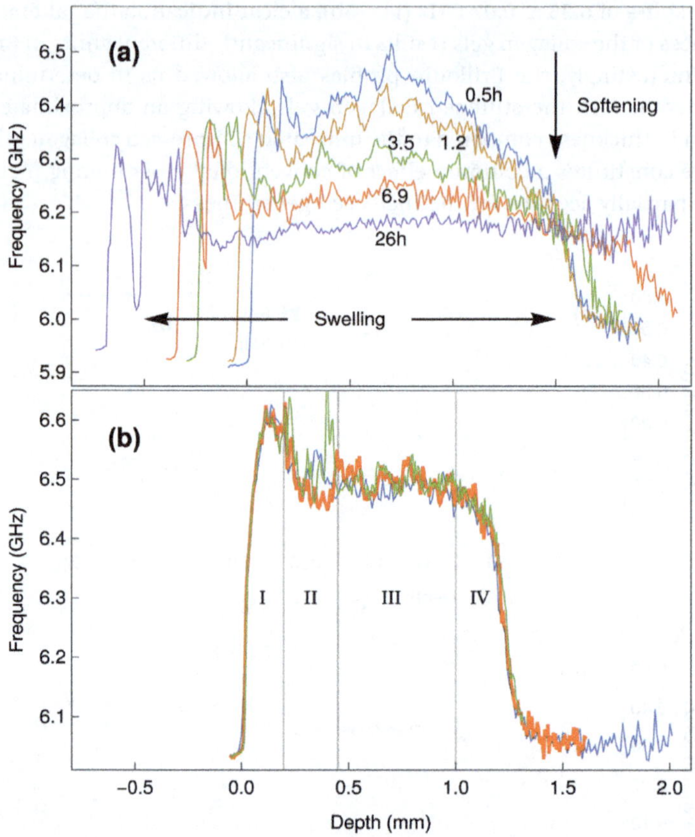

Fig. 6 (a) Time-lapse Brillouin imaging of bovine corneal tissue in PBS. Representative depth profiles of Brillouin shifts from the centre region of the bovine corneas are shown at different times from 0.5 to 26 h after immersion. Progressive swelling was illustrated by the increased depth of the scans (horizontal arrows), whereas tissue softening was represented by the lower shift values (vertical arrow). Disruption of structural integrity of the swollen tissue was illustrated by abrupt changes in shift pattern after 6.9 h in PBS (detachment of anterior layer). Distance between measurements in $z = 10$ μm. (b) Brillouin imaging of bovine corneal tissue in Carry-C. At the time of measurement, the samples had been immersed in the medium for 3 days, yet showed no sign of swelling. The different anatomical–morphological subdivisions of the tissue can be identified: corneal epithelium (I), anterior stroma (II), posterior stroma (III), corneal endothelium (IV). Distance between measurements in $z = 10$ μm.

replicate a dehydration state, whole bovine corneas were immersed in glycerol immediately after isolation, and then washed and kept in PBS to remove the glycerol and evaluate the effects of re-hydration through time on tissue stiffness and thickness (Fig. 6a). The scan series performed throughout the entire depth of the tissues showed that corneas swell progressively, with their thickness almost doubling after 26 h in PBS. This constituted a clear indication of increased hydration well beyond normal *in vivo* levels.[30] Concomitantly, the Brillouin shift measured at all depths of the corneal tissue decreased with time, indicating that the swollen corneas became softer. Specifically, at $t = 0.5$ h, the corneas showed shifts of 6.3–6.5 GHz at their anterior and 6.5–6.2 GHz at their posterior regions, respectively, whereas at $t = 26$ h, shifts varied between 6.0 and 6.3 GHz. Furthermore, the pattern of the shifts measured in the swollen corneas indicated a loss of structural integrity within the tissue (Fig. 6a), a common consequence of excessive corneal hydration and oedema.[31,32]

Overall, these results highlighted the importance of performing the measurements in appropriate conditions (*i.e.*, physiological-like) for a given tissue or organ. When measuring the in-depth Brillouin scattering of bovine corneas maintained in Carry-C, a medium specifically developed to preserve corneas in their physiological hydration state, the profiles were remarkably similar to those previously reported for the central regions of human corneas (Fig. 6b).[33] In particular, these profiles were divided into four regions (Fig. 6b): an anterior-most peak with 6.5–6.6 GHz shifts corresponding to the corneal epithelium (I), a slope ranging between 6.55 and 6.5 GHz corresponding to the anterior stroma (II), a plateau of shifts of approximately 6.5 GHz corresponding to the posterior stroma (III), and a steep slope corresponding to the posterior-most region of the cornea (IV).[34] Furthermore, this pattern was maintained unaltered, along with the natural thickness of the bovine corneal tissue, even after keeping it for more than a week in Carry-C medium.

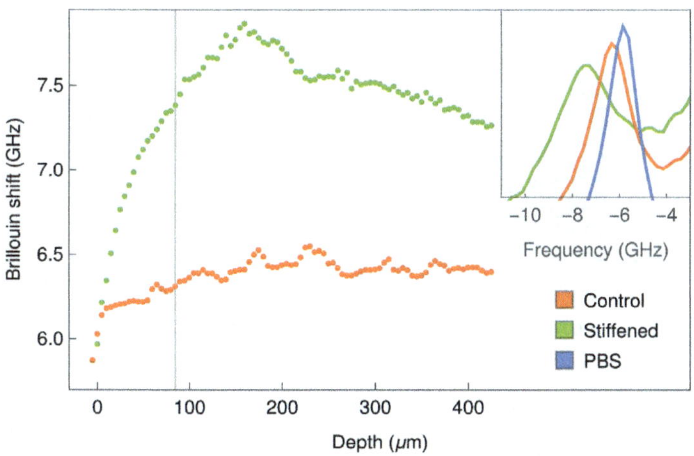

Fig. 7 Brillouin imaging of stiffened bovine corneas. Representative depth profiles of Brillouin shifts of the anterior centre region of corneas before (red line) and after stiffening treatment with PFA (green line). Distance between measurements in $z = 5$ μm. Example spectra (Stokes peak) at $z = 85$ μm are shown in the inset, together with the spectrum of PBS for comparison (blue line).

The Brillouin images were also capable of distinguishing differences in stiffness from normal and stiffened-treated corneal tissue. The anterior-most region of the whole bovine corneas fixed with paraformaldehyde showed significantly higher shifts compared to the same region from the untreated tissues (Fig. 7). Moreover, this increase in shift was observed not only at the surface, where the tissue was treated, but also in-depth (Fig. 7, insets), providing information about the range of penetration for such treatment.

These results also highlight the non-contact nature of Brillouin spectromicroscopy for investigating the mechanical properties in the interior of biological tissues. However, this ability was shown to be limited by the transparency of the material. For example, imaging more opaque areas of the bovine cornea (such as the limbus, Fig. 3) resulted in a considerable reduction in the signal-to-noise ratio at deeper levels of the tissue (data not shown). In contrast, this limitation was not evident in the human cornea, where the (species specific) improved

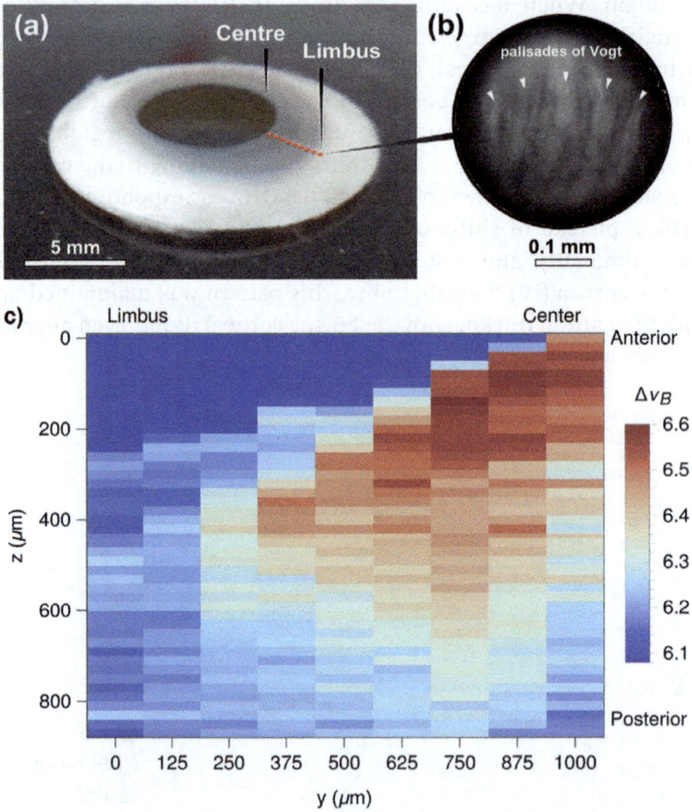

Fig. 8 (a) Human corneal ring tissue used to analyse the stiffness in corneal centre and limbus regions. (b) Micrograph of the limbus area, showing the palisades of Vogt. (c) Brillouin imaging of the human corneal tissue in Carry-C. *z*-scans were taken in 125 μm steps from the centre to the limbus, at the locations approximately indicated by red dots in (a). The vertical step size was 20 μm. The heat map represents Brillouin shift values, with areas in dark blue corresponding to regions filled with medium, whereas corneal tissue was shown as light blue (soft) to dark red pixels (stiffer tissue). Scale bars, (a) and (b) = 5 and 0.1 mm respectively.

transparency of the limbus (Fig. 8a) still allowed highly-resolved spectra to be obtained. In particular, human corneal tissues maintained in the Carry-C medium were able to be imaged in a sequence of in-depth measurements, thus allowing the assembly of *z*-scans into 2D heat maps (Fig. 8c). Remarkably, these sequences not only provided a notion of the tissue's morphology, but also the finer differences in stiffness between different regions of the tissue (Fig. 8c: anterior (upper) *vs.* posterior (lower region); limbus (left) *vs.* central cornea (right)). Specifically, the values for the Brillouin shift observed in the limbus were lower than those in the central cornea, particularly in the anterior-most region of the tissue. This data was supported by previous, although much less detailed, studies using AFM[35] or stress–strain analysis of the cornea.[36] Taken together, these results represent a more complete overall view of the stiffness of biological tissues, providing information about the mechanical properties of human corneas of unparalleled accuracy and resolution.

3 Conclusions

In summary, we have introduced a new Brillouin microscope consisting of a fibre-coupled confocal scanning microscope, an interferometric filter to suppress elastic scattering, and a single-stage VIPA spectrometer. The microscope's ability to acquire depth-resolved Brillouin spectra in very thin as well as very thick samples (*e.g.* in 50μm-thick collagen gels or in bovine corneas up to 2 mm) has been demonstrated, thanks to the interferometric filter that allows sufficient rejection of specular reflections at the sample surfaces.

This technique was used to monitor the response of compressed collagen gels and corneal tissue to stiffening (cross-linking) treatments. Brillouin imaging of the limbus in human corneal rings also revealed a clear contrast between the stiffer inner stroma of the centre and the softer limbus. These results validate Brillouin spectro-microscopy as a potent tool for assessing the mechanical properties of soft biomaterials, particularly transparent ones. We thus envisage many possible applications for this technique, including fundamental characterisation studies and as a tool to diagnose the rigidity of many tissues, namely the cornea.

However, the quantification of the absolute stiffness of soft tissues by Brillouin spectroscopy (*i.e.*, using Brillouin shift values to calculate the elastic modulus) constitutes a challenge that remains unresolved.

References

1 A. Smekal, *Naturwissenschaften*, 1923, **11**, 873–875.
2 L. Brillouin, *Ann. Phys.*, 1922, **17**, 88–122.
3 C. Kittel, *Introduction to Solid State Physics*, Wiley, 8th edn, 2005.
4 G. Benedek and K. Fritsch, *Phys. Rev.*, 1966, **149**, 647–662.
5 G. Scarcelli and S. H. Yun, *Nat. Photonics*, 2007, **2**, 39–43.
6 R. Mock, B. Hillebrands and J. R. Sandercock, *J. Phys. E: Sci. Instrum.*, 1987, **20**, 656–659.
7 G. Antonacci, M. R. Foreman, C. Paterson and P. Török, *Appl. Phys. Lett.*, 2013, **103**, 221105.

8 G. Scarcelli, W. J. Polacheck, H. T. Nia, K. Patel, A. J. Grodzinsky, R. D. Kamm and S. H. Yun, *Nat. Methods*, 2015, **12**, 1132.

9 F. Palombo, C. P. Winlove, R. S. Edginton, E. Green, N. Stone, S. Caponi, M. Madami and D. Fioretto, *J. R. Soc., Interface*, 2014, **11**, 20140739.

10 A. J. Traverso, J. V. Thompson, Z. A. Steelman, Z. Meng, M. O. Scully and V. V. Yakovlev, *Anal. Chem.*, 2015, **87**, 7519–7523.

11 G. Antonacci, R. M. Pedrigi, A. Kondiboyina, V. V. Mehta, R. de Silva, C. Paterson, R. Krams and P. Török, *J. R. Soc., Interface*, 2015, **12**, 20150843.

12 G. Scarcelli, S. Besner, R. Pineda and S. H. Yun, *Invest. Ophthalmol. Visual Sci.*, 2014, **55**, 4490–4495.

13 R. Vacher and L. Boyer, *Phys. Rev. B: Solid State*, 1972, **6**, 639–673.

14 G. Scarcelli, P. Kim and S. H. Yun, *Biophys. J.*, 2011, **101**, 1539–1545.

15 K. F. Herzfeld and T. A. Litovitz, *Absorption and Dispersion of Ultrasonic Waves*, Academic Press, New York and London, 1959.

16 K. Hattori, K. Sakai and K. Takagi, *Jpn. J. Appl. Phys., Part 1*, 1999, **38**, 4932–4935.

17 R. M. Gouveia, R. R. Jones, I. W. Hamley and C. J. Connon, *Biomater. Sci.*, 2014, **2**, 1222–1229.

18 M. Damzen, *Stimulated Brillouin Scattering: Fundamentals and Applications*, Institute of Physics, 2003.

19 J. R. Sandercock, *Phys. Rev. Lett.*, 1972, **29**, 1735–1738.

20 G. Scarcelli and S. H. Yun, *Opt. Express*, 2011, **19**, 10913–10922.

21 Z. Meng, A. J. Traverso and V. V. Yakovlev, *Opt. Express*, 2014, **22**, 5410–5415.

22 W. Marczak, *J. Acoust. Soc. Am.*, 1997, **102**, 2776.

23 G. Antonacci, G. Lepert, C. Paterson, P. Török and P. Torok, *Appl. Phys. Lett.*, 2015, **107**, 061102.

24 J. W. Foster, R. R. Jones, C. A. Bippes, R. M. Gouveia and C. J. Connon, *Exp. Eye Res.*, 2014, **127**, 37–41.

25 M. E. Smithmyer, L. A. Sawicki and A. M. Kloxin, *Biomater. Sci.*, 2014, **2**, 634–650.

26 R. A. Brown, M. Wiseman, C. B. Chuo, U. Cheema and S. N. Nazhat, *Adv. Funct. Mater.*, 2005, **15**, 1762–1770.

27 R. R. Jones, I. W. Hamley and C. J. Connon, *Stem Cell Res.*, 2012, **8**, 403–409.

28 W. J. Howat and B. A. Wilson, *Methods*, 2014, **70**, 12–19.

29 D. W. Leonard and K. M. Meek, *Biophys. J.*, 1997, **72**, 1382–1387.

30 Y. L. Kim, J. T. Walsh Jr, T. K. Goldstick and M. R. Glucksberg, *Phys. Med. Biol.*, 2004, **49**, 859–868.

31 K. M. Meek, D. W. Leonard, C. J. Connon, S. Dennis and S. Khan, *Eye*, 2003, **17**, 927–936.

32 C. M. Hsueh, W. Lo, W. L. Chen, V. A. Hovhannisyan, G. Y. Liu, S. S. Wang, H. Y. Tan and C. Y. Dong, *Biophys. J.*, 2009, **97**, 1198–1205.

33 G. Scarcelli and S. H. Yun, *Opt. Express*, 2012, **20**, 9197–9202.

34 G. Scarcelli, R. Pineda and S. H. Yun, *Invest. Ophthalmol. Visual Sci.*, 2012, **53**, 185–190.

35 J. A. Last, S. M. Thomasy, C. R. Croasdale, P. Russell and C. J. Murphy, *Micron*, 2012, **43**, 1293–1298.

36 J. O. Hjortdal, *J. Biomech.*, 1996, **29**, 931–942.

Faraday Discussions

DISCUSSIONS

Clinical Spectroscopy: general discussion

Royston Goodacre, Valter Sergo, Hugh Barr, Christopher Sammon, Zachary D. Schultz, Matthew J. Baker, Duncan Graham, Maria Paula Marques, Josep Sulé-Suso, James Livermore, Karen Faulds, Faris Sinjab, Pavel Matousek, Colin J. Campbell, Richard Dluhy, Peter Gardner, Christopher Phillips, Max Diem, Bayden Wood, Alexander Apolonskiy, Sergei Kazarian, Leanne Fullwood, Kathleen Gough, Wolfgang Petrich, Gavin Lloyd, Ola Ibrahim, Gianfelice Cinque, Ganesh Dhruvananda Sockalingum, Nick Stone, Catherine Kendall, Sarah McAughtrie, David Perez-Guaita, Louise Clark, Klaus Gerwert, Alois Bonifacio, Ioan Notingher, Peter Lasch, Rohit Bhargava, Guillaume Lepert, Kerstin Mader and Carl Paterson

DOI: 10.1039/C6FD90013E

Duncan Graham opened a general discussion of the paper by Pavel Matousek: Regarding the temperature measurements, how did you do them? Stokes/anti-Stokes measurement? The values are not what I'd expect and seem lower. What was the wavelength and power of the laser and did you calculate what you thought the temperature measurement should be based on that? How far does the temperature gradient reach? The scattering : absorption ratio changes with nanoparticle size and composition. Have you thought about using more highly absorbing nanoparticles such as hollow gold nanoshells?

Pavel Matousek responded: The anti-Stokes to Stokes intensity ratio depends on multiple factors including the position of the bands with respect to the plasmon resonance of the SERS nanoparticles. Additional factors affecting the intensity ratio include the instrumental spectral response (the presented data are not corrected for this response). However all these effects are accounted for and eliminated in the calibration step. The laser wavelength was 830 nm and the laser power was 400 mW. In these experiments we purposely did not induce temperature gradients by the probe laser. The laser was merely used to "read" the nanoprobe SERS signals and not to induce any temperature changes itself. This was ensured by using a largely defocused illumination spot size (~1–2 mm) and by further intensity dilution in the sample due to photon diffusion. The intentional increase of temperature by a laser beam (*e.g.*, in photothermal therapy) will be the subject of our future studies. Further optimisation of the SERS enhancement effect would indeed be highly beneficial in enhancing the sensitivity and penetration depth of T-SESORS and will also be the subject of our future research.

Zachary D. Schultz said: The enhancement observed in SERS intensities is associated with the Raman shift wavelength relative to the localized surface plasmon resonance, how does this affect the Stokes and anti-Stokes intensity measurements? Does this affect the reproducibility of the sensors, such that calibration of every batch of probes is necessary?

Pavel Matousek replied: Indeed, each type of SERS nanoprobes would need to be calibrated individually as the SERS enhancement can be considerably different for the Stokes and anti-Stokes regions.

Karen Faulds said: Are you measuring the localised heating at the nanoparticle surface or the temperature of the surrounding environment? The reporter molecules are directly attached to the nanoparticles and it is the changes in the SERS response of these reporter molecules that is being measured, and it is known that nanoparticles can experience large degrees of localised heating under laser interrogation. The temperature changes at the surface of the nanoparticle should therefore have a greater effect on the Stokes/anti-Stokes ratio than the environment?

Pavel Matousek responded: In Temperature Spatially Offset Raman Spectroscopy (T-SESORS) experiments one indeed measures the local heating around the nanoparticles. Since the temperature was equilibrated in the nanoparticle medium, in our experiments, this also probed the surrounding environment. However in a non-equilibrated thermal situation nanoparticles would indeed be reporting on the local temperature, *i.e.*, in the immediate vicinity of the nanoparticle, and not on the bulk environment temperature. If the nanoparticles were purposely heated (*e.g.*, by another (pump) laser) as, for instance, in photothermal therapy, these nanoparticles would be exclusively reporting on the local temperature. In such experiments one could envisage using another set of nanoparticles not tuned to absorb the radiation from the pump laser and these could be then used to report on the bulk temperature of the environment further away from the nanoparticles receiving heat (assuming these were randomly distributed in the sample and not co-located with the nanoparticles receiving pump laser radiation).

Faris Sinjab addressed Pavel Matousek and Nick Stone: For Stokes/anti-Stokes shifts of 1080 cm^{-1} excited at 830 nm, the absolute separation of the photon wavelengths will be roughly 150 nm (912 nm for anti-Stokes photons, 762 nm for Stokes photons). Due to this relatively large separation (compared to peaks usually used for SORS in the Stokes spectrum), would the difference of the mean free path of absorbed/scattered photons introduce an appreciable systematic error into the intensity ratio used for temperature measurement? If so, could this be calibrated for to reduce such an error?

Nick Stone answered: The tissue attenuation profile will have an appreciable affect on the relative Stokes/anti-Stokes signals measured. This can be measured directly or the effect of it can be measured by using a Raman calibration sample at the surface of the tissue. Please also see Pavel's response.

Pavel Matousek replied: The difference in the scattering properties of the sample matrix can indeed be significant but these effects can be indeed calibrated out. Similar effects can be induced by sample absorption if present but this effect could also be calibrated out.

Sergei Kazarian asked: This is an elegant approach for measurements of the distribution of subsurface temperature within a sample. Is it possible to achieve similar information in terms of depth profiling using infrared emission detection systems or approaches that detect the distribution of temperature within the sample?

Pavel Matousek answered: Generally any black body radiation-based systems (passive) would be strongly biased towards the surface of the sample. One could potentially envisage an active NIR absorption system deployed in a spatially offset configuration reporting on subsurface temperature through (high lying) vibrational combination and overtone bands (observable in deep tissue NIR absorption spectroscopy within the tissue transparency window of around 650–950 nm) if these were changing bandwidth or band position with temperature. This would be a rather sample-specific technique.

Sarah McAughtrie asked: The paper (DOI: 10.1039/C5FD00154D) states that the temperature can be predicted with an accuracy of +/− 0.16 °C, but in terms of clinical relevance how sensitive would the temperature sensing need to be for implementation in the clinic?

Pavel Matousek replied: It would depend on the specific application. For example, in photothermal therapy where the elevation of temperature to around 40–50 °C is envisaged, such accuracy would be more than adequate.

Nick Stone commented: It may be necessary when undertaking hypothermia treatment to raise the temperature of abnormal tissues, by only a few degrees, for a period of minutes. Direct monitoring, non-invasive monitoring of the raised temperature would allow careful light-based mediation of the resulting treatment.

Valter Sergo asked: The bandwidth carries information on the distribution of energy states and, hence, on temperature distribution in the sampled volume. Did you check for trends of the Stokes/anti-Stokes bandwidths with temperature?

Pavel Matousek responded: Raman bandwidths vary with temperature and this may indeed convey useful information. However, in our basic study we did not investigate this dependence.

Colin J. Campbell opened a general discussion of the paper by Bayden R. Wood: What is the pathway for translation of the FTIR imaging technology to the developing world where instrumentation may not be as sophisticated?

Bayden Wood answered: At this stage there is no pathway for FTIR imaging in the developing world but there is a very clear pathway for portable Attenuated Total Reflection (ATR) spectroscopy, which we are currently trialling in Thailand. Besides portability the ATR approach is inexpensive, easy to use, can be powered by battery and offers the potential for quality control at the point of spectral collection and cloud-based diagnosis for real-time monitoring.

FTIR imaging on glass substrates would be more suitable for the modern clinical laboratory and offers a way to diagnose malaria at the single cell level, providing a gold standard for malaria diagnosis.

Sergei Kazarian addressed Bayden Wood and Peter Lasch: I do not think you are actually resolving single cells chemically; based on the data presented we can see evidence of absorbance contrast. The spatial resolution is not increased simply because there are now more pixels per unit area if the sampling was already above the diffraction limit. In your case of a 0.8 numerical aperture (NA) objective (DOI: 10.1039/C5FD00181A), the maximum spatial resolution is about 2.3 micrometres using the Rayleigh criterion at a 3 micrometre wavelength. Due to discrete sampling, at least 2× (2.5× would be better to avoid aliasing) would be needed to fully resolve the features. You were careful to mention in your paper that too much oversampling does not result in increased image quality. Some oversampling is necessary to obtain features on the very edge of the diffraction limit (assuming other optical phenomena do not degrade the image on the way to the detector). In this case, oversampling is about 3.5× the diffraction limit (at NA = 0.8 in your case) so it would appear that it is excessive. However, maybe such oversampling is needed given other optical imperfections within the system?

Note that the resolution targets or so-called "pixel resolution" significantly overestimates the "chemical" spatial resolution achieved, which is significantly worse if you apply more stringent tests such as measuring a sharp interface between two different polymers. See how the achieved spatial resolution was assessed for imaging in transmission, for example by Chan and Kazarian,[1] and also our paper in this Discussion (DOI: 10.1039/C5FD00158G). The spatial resolution in FTIR imaging has also been discussed in a dedicated section in a very recent review.[2]

1 K. L. A. Chan and S. G. Kazarian, *Anal. Chem.*, 2013, **85**, 1029–1036.
2 K. L. A. Chan and S. G. Kazarian, *Chem. Soc. Rev.*, 2016, **45**, 1850–1864.

Peter Lasch answered: I fully agree, see also my comment to the next question from Wolfgang Petrich.

Bayden Wood responded: I would agree, but the system when in "std mag" mode is not entirely "diffraction limited" but it is to some extent "pixel size limited", especially at the high wavenumber end, the end that we are using for our analysis. Increasing the magnification (*via* the built-in high mag optics) reduced the pixel size to a level to ensure a fully diffraction limited system.

You are correct in stating that the degree of oversampling is higher than absolutely necessary but such optics are fixed, resulting in a slightly smaller field of view (FOV). We recognize that as this sample is a 20 Å thick Cr layer on 2mm glass, it presents an ideal sample for determining the spatial resolution, as it has an extremely high chemical contrast and the Cr layer is so thin, and the sample thickness can be considered almost negligible. The USAF target is convenient in that it is a widely available target that is well characterized and is useful in comparing instruments or optical setups, but that its extrapolation to "real world" samples should not be done literally.

We have addressed the point regarding air force resolution targets above but take the point that the polymer blend would be a more suitable target to prove the spatial resolution. However, the paper was focused on diagnosing parasites inside cells using a glass substrate and not on spatial resolution. We have unequivocally proven that we can resolve parasites inside infected cells statistically using multivariate methods where we have extracted spectra from the parasite and the surrounding haemoglobin and have obtained distinct chemical differences. We have used permutation and combination testing to ensure that the classification is robust. Moreover, the regression vector from the Partial Least Squares-Discriminant Analysis (PLS-DA) shows that the chemical changes we are seeing in both the amide A and CH stretching regions are consistent with the spectral changes we observe using ATR and synchrotron focal plane array imaging on malaria infected cells. We have also shown that we can resolve two cells that are 1.4 μm apart and can use software to select cells that are separated by this distance. We have then selected the individual cells using a Matlab routine and extracted all of the spectra from the cell. We have then averaged these spectra from each cell and taken the second derivative. We then use PLS-DA to discriminate the infected from non-infected cells and the regression vector shows distinct chemical differences once again in the amide A band and CH stretching vibrations from lipids. Hence we are demonstrating both intercellular and intracellular resolution based on chemical changes within the individual cells.

Wolfgang Petrich addressed Bayden Wood and Peter Lasch: Sorry for taking your presentation as an example but you are the first one in this discussion meeting who clearly states a number and refers to it as "spatial resolution" (DOI: 10.1039/C5FD00181A). I think there is some hygiene to be done in terms of nomenclature. The term "spatial resolution" is well-defined in the context of microscopy. Maybe you wish to specify which resolution criterion is being used in detail, but resolution always refers to the actual capability to spatially resolve two identical particles, which in all the microscopes we use is by no means sub-diffraction limited. The term "pixel resolution" does not make the obtainable spatial resolution any better. I would thus appreciate if we call "spatial resolution" what has been defined in physics and what is governed by the fundamental principles pointed out by Abbe. If we wish to refer to other numbers, we should be strict in that we do NOT call these "spatial resolution". Personally, I don't like the word "pixel resolution" too much since I hear that some people perceive this number as "spatial resolution".

Admittedly, spatial resolution refers to identical particles which can not be distinguished based on their intrinsic properties. If we have two non-identical, *i.e.* distinguishable particles next to each other, we may be able to resolve them. This is true for including the spectral analysis into the spatial analysis (in analogy to the localisation microscopy based on two adjacent but different fluorophores). However, such an improvement in the localization capabilities among distinguishable particles on the basis of their (for that matter: fluorescence, IR, Raman, or any other) spectra is not a unique property of the microscope but strongly depends on the sample being investigated. Localizing two distinguishable, different particles is not what we should refer to as spatial resolution. Hence, my "credo" is that a) when we talk about "spatial resolution" I'd like to stick to strictly using the term spatial resolution as it is defined by Abbe (or the modulation transfer function if you wish), and b) when talking about (highly sample-specific) localising of distinguishable particles, say "localising" and tell me about your model system.

Peter Lasch answered: To be honest I do not really understand why we have to do some hygiene. It is my impression that the term "spatial resolution" is well-defined and used accurately by most of us. I agree that the term "pixel resolution" is incorrect; "pixel size" would be more appropriate. In optics spatial resolution is always defined for a single wavelength. Often the Rayleigh criterion is used: two points can be separated when the central maximum of the first Airy disk is just one radius away of the other one. Rayleigh's criterion of separation corresponds under ideal conditions to a distinct contrast value, namely of 26.4%.[1] If the IR and perhaps also the Raman imaging community could agree to this definition (which was my impression in the past) then possible misunderstandings could be avoided for the future.

1 P. Lasch and D. Naumann, *Biochim. Biophys. Acta*, 2006, **1758**, 814–829.

Bayden Wood replied: Fair points. The spatial resolution as presented in slide 5 from Agilent are based on the Rayleigh criterion spatial resolution measurements using a "hi res" version of the standard chrome-on-glass USAF resolution target with the calculation following the method of Lasch and Naumann.[1] We recognize that as this sample is a 20 Å thick Cr layer on 2mm glass, it presents an ideal sample for determining the spatial resolution, as it has an extremely high chemical contrast and the Cr layer is so thin, and the sample thickness can be considered almost negligible (see response to the question above). We agree that "pixel resolution" on its own does not necessarily make the obtainable spatial resolution better. However, it can improve spatial resolution where the system is in a "pixel size limited" state, as often is the case with the Agilent system, particularly at the lower wavenumber end. On the flip side, this particular "feature" is what gives the system such a large field of view (FOV), which is often very advantageous for our line of work. The use of the "high mag" optics which provides an additional 5× magnification over standard mag optics, brings the system into a truly diffraction limited state across the entire wavelength, by ensuring adequate oversampling. Professor Rohit Bhargava calculated the ideal "pixel resolution/size" for a given numerical aperture of the objective, demonstrating that for 0.8 NA, the pixel size should be ideally ~0.8 μm (at 4000 cm^{-1}), matching well with the 0.66 μm provided by the high mag optics and the 25× obj.[2] It is worth stressing here we recognize that the primary "driver" for spatial resolution is NA, and pixel size/resolution (which is really just the total system magnification) can either aid in achieving the diffraction limited performance by appropriate spatial oversampling, or it can hinder it if the pixel size is too large, though in that instance, it is often a trade-off between maximum FOV and spatial resolution.

We agree furthermore that all too often, we see "pixel resolution" being confused for "spatial resolution", the former being a fixed value dependent only on the total system magnification, and the latter being dependent on the objective NA and at what wavenumber/wavelength the spatial resolution has been calculated.

1 P. Lasch and D. Naumann, *Biochim. Biophys. Acta*, 2006, **1758**, 814–829.
2 R. K. Reddy, M. J. Walsh, M. V. Schulmerich, P. S. Carney and R. Bhargava, *Appl. Spectrosc.*, 2013, **67**, 93–105.

Rohit Bhargava added: The spatial resolution is defined for a wavelength and for identical point objects to be separated. The classical physics definition is well-agreed upon and works well. The highest image quality can be achieved by appropriate optical design in which we find the sizes to be smaller than the wavelength and, of course, resolution. This "pixel size" should not be termed oversampling unless the limits described by theory[1] are exceeded and should not be confused with resolution.

1 R. K. Reddy, M. J. Walsh, M. V. Schulmerich, P. S. Carney and R. Bhargava, *Appl. Spectrosc.*, 2013, **67**, 93–105.

Sergei Kazarian commented: I would like to use this opportunity to discuss this issue further. We define "spatial resolution" properly in our paper (DOI: 10.1039/C5FD00158G), using the Rayleigh criterion which differs only from the Abbe criterion through the use of the coefficient of 0.5 rather than 0.61. Within the context of optical physics and microscopy then I agree; the visible contrast between two points can be used as a measure of spatial resolution. We used this definition in our paper, where we measured an ideal sample (1951 USAF target) and assessed the achieved spatial resolution using the Rayleigh criterion for a contrast of 26.4%. The emphasis though, is that achieved spatial resolution is lower than the theoretical value based on numerical aperture, and moreover, the USAF target contains no spectral features, it only provides contrast. For real samples studied with spectroscopic imaging, resolving the changes in spectral features within the sample is the main objective. Therefore spatial resolution based on Rayleigh/Abbe criteria significantly overestimates the achieved resolution, and when using the value obtained from a USAF target, one needs to be aware of its implications within a spectroscopic context as it too is an overestimation of the achieved resolution for resolving chemical features.[1] We show that, for FTIR spectroscopic imaging, which is inherently a chemical imaging technique, the criterion for "resolved" needs to be stricter, as if two adjacent particles or materials which are chemically different can be considered resolved by the Abbe criterion, "spectral contamination" between both would render chemical analysis infeasible as there would be no way to determine the true chemical signature of each. This is a problem for spectroscopic studies of heterogeneous materials and as an example, applies to biological cells. They may appear resolved with respect to Rayleigh/Abbe criteria, but there would be significant "spectral contamination" between each cell, hindering the proper chemical analysis of each individual cell. Indeed, the spatial resolution based on Rayleigh or Abbe criteria was often used in publications to give a false impression of the spatial resolution obtained in spectroscopic imaging, while the actual spatial resolution achieved was significantly worse (2 to 4 times lower) than the theoretical value because of optical aberrations. Thus, the spatial resolution may be defined by Rayleigh/Abbe criteria but it cannot be applied to a real system without its validation, and we show the achieved spatial resolution in our work not only with a USAF target (and Rayleigh criterion of 26.4% contrast), but also with a polymer laminate sharp interface, assessing how characteristic spectroscopic bands change across it. Given this, it would be appropriate to use the spectral change across an interface between 5% and 95% of the maximum absorbance of a spectral band (edge profile) to define the achieved chemical spatial resolution. For an example system of two particles, only 5% of the absorbance of a characteristic band of one particle would be present in the other if they are considered resolved by this criterion, which would be an acceptable amount of "spectral cross-contamination". The use of a polymer laminate interface, and the change in the spectrum across this interface, is an appropriate test in defining the achieved spatial resolution in the context of spectroscopic imaging, and the use of an edge function as a way to assess spatial resolution has been done by us in this paper and previous publications,[1,2] and also by others in infrared and Raman spectroscopy who likewise make evident its applicability.[3–5]

1 K. L. A. Chan and S. G. Kazarian, *Chem. Soc. Rev.*, 2016, **45**, 1850–1864.
2 K. L. A. Chan and S. G. Kazarian, *Appl. Spectrosc.*, 2003, **57**, 381–389.
3 A. J. Sommer, L. G. Tisinger, C. Marcott and G. M. Story, *Appl. Spectrosc.*, 2001, **55**, 252–256.
4 D. J. Gardiner, M. Bowden, P. R. Graves and R. B. Thompson, *Philos. Trans. R. Soc., A*, 1986, **320**, 295–306.
5 G. L. Carr, *Rev. Sci. Instrum.*, 2001, 72, 1613–1619.

Gianfelice Cinque commented: Our experience at Diamond both in scanning IR microscopy and focal-plane array (FPA) imaging in full field is that oversampling by a factor of 3 with respect to the diffraction limit spatial resolution is generally beneficial for IR maps; *vice versa* grouping pixels for the region of interest (ROI) is useful to enhance the signal-to-noise ratio for successive spectroscopy data analysis.

Klaus Gerwert resumed the general discussion of the paper by Bayden R. Wood: Can you describe how you performed the difference spectra you showed in your presentation?

Bayden Wood answered: Difference spectra were calculated directly from vector-normalized raw spectra in the absorbance mode and then converted to the second derivatives. The converted spectra enable easier visualisation of bands within a broad background of multicomponent absorption bands and minimised the effects of Mie scattering.

Max Diem asked: How does the relatively poor S/N ratio of the measurement affect the sensitivity, which you stated is one out of 100 000 infected cells?

Bayden Wood responded: I was referring to the poor signal-to-noise ratio for the FTIR imaging approach. The poor S/N ratio really affects the CH stretching region but not the amide A band. The intense amide A band virtually vanishes if the cell is infected with a parasite because the majority of the haemoglobin has been catabolized by the parasite into haemozoin. The dramatic reduction of the amide A mode is a very good marker band to determine infected cells in FTIR images. The sensitivity of 1 in 100 000 cells was in reference to the ATR approach where we get an excellent S/N ratio and is currently the technique we are using in our Thai trial.

Josep Sulé-Suso asked: Can you differentiate the different stages of the parasite life cycle in infected cells?

Bayden Wood answered: Using the ATR approach we have unequivocally identified and quantified different stages of the parasite life cycle including the gametocyte, ring, schizont and trophozoite stages.[1] When sampling peripheral blood it is important to be able to detect gametocyte and ring stage parasites as trophozoites and schizonts are only found in small numbers in peripheral blood. ATR has the required sensitivity to detect these cell types at very low parastemia (1 parasite per microlitre of isolated red blood cells).

1 A. Khoshmanesh, M. W. A. Dixon, S. Kenny, L. Tilley, D. McNaughton and B. R. Wood, *Anal. Chem.*, 2014, **86**, 4379–4386.

Sergei Kazarian queried: Is it just contrast that is being measured? Resolution targets are not suitable for the assessment of the chemical spatial resolution you achieve because those targets are idealised samples without spectral features. The absorbance profiles across the sharp interface between two spectrally dissimilar polymers is a more realistic system for the assessment of spatial resolution, and the distance between 90–10% or 95–5% of the maximum absorbance is the true useful spatial resolution you could achieve. Such a test provides the real spatial resolution that you achieve while the 5% in this test accounts for spectral noise. It is the chemical information that you are trying to achieve in studies of live cells, not just the observation of features due to the increased contrast. We used this test for assessing spatial resolution in both ATR and transmission modes.

1 K. L. A. Chan, S. G. Kazarian, A. Mavraki and D. R. Williams, *Appl. Spectrosc.*, 2005, **59**, 149–155.
2 K. L. A. Chan and S. G. Kazarian, *Appl. Spectrosc.*, 2003, **57**, 381–389.

Bayden Wood replied: I have addressed these points in the answer above but thank you for the references, I will read them with interest.

Royston Goodacre opened a general discussion of the paper by Peter Lasch: Peter's study is a very nice systemic analysis using MS, FTIR and Raman for bacterial differentiation. If I could buy only a single instrument why would I buy an IR or Raman instrument when there are now thousands of MALDI-ToF MS instruments in clinical microbiology laboratories for routine bacterial identification?

Peter Lasch responded: In our opinion, the decision of which instrument to buy will depend on the application. Mass spectrometry (specifically MALDI-TOF MS) offers a number of advantages if the application is microbial species identification (*e.g.*, in clinical microbiology). MALDI-TOF MS for microbial identification at the species level is today widely accepted as an accurate, robust and cost-effective method and spectral databases are commercially available. The question, however, of whether this technique is suitable for strain differentiation is still intensively discussed.

The results of our and many other studies suggest that vibrational spectroscopy offers a taxonomic resolution at the strain level. Therefore, in cases where strain identification/differentiation is required we would recommend buying an IR or Raman spectrometer.

Royston Goodacre added: Many thanks for your nice reply Peter. Readers might like to look at the results of a small poll on twitter asking which analytical method people think is best for bacterial identification (Fig. 1).

Bayden Wood asked: Can MALDI discriminate between two different strains of bacteria in the same micro-colony? What about FTIR?

Ganesh Dhruvananda Sockalingum also asked: In your study, you are analysing different types of data coming from different techniques and using pure isolates. Can you comment on the advantages of vibrational spectroscopy over MALDI if you have mixed isolates (real samples)?

Peter Lasch responded: To answer the first question, routine MALDI-TOF MS cannot be used to resolve two different strains present in a mixed culture. Admittedly, in the literature there are some attempts to unmix MALDI-TOF mass spectra obtained from cultures with two or more microbial strains.[1] However, to my knowledge none of these routines are implemented in actual versions of software for MALDI-TOF MS-based identification.

The challenges of unmixing MALDI spectra from mixed cultures arise mainly from the fact that the spectrum of the mixed culture is not a simple superposition of the spectra from the pure components. The key words here are competitive ionization, ion suppression, relative abundance, *etc.*

Contrary to this, FTIR spectroscopy has the advantage that it can be coupled with microscopy. Using FTIR microspectroscopy, it is possible to characterize microcolonies which usually contain cells from only one strain. We have developed a stamping technique that can be used to reproducibly transfer the upper cell layers from microcolonies from solid agar plates onto IR-transparent substrates.[2] In this way, FTIR microspectra from microcolonies are helpful to identify bacteria from mixed cultures.

1 L. Zhang, S. Smart and T. R. Sandrin, *Sci. Rep.*, 2015, 5, 15834.
2 P. Lasch and D. Naumann, Infrared Spectroscopy in Microbiology, in *Encyclopedia of Analytical Chemistry*, John Wiley & Sons, 2015, DOI: 10.1002/9780470027318.a0117.pub2.

Colin J. Campbell said: The correlation between the genotype and phenotype is really interesting. Is there a role for this type of spectroscopic phenotyping in the study of disorders associated with the gut microbiome, perhaps using endoscopic Raman?

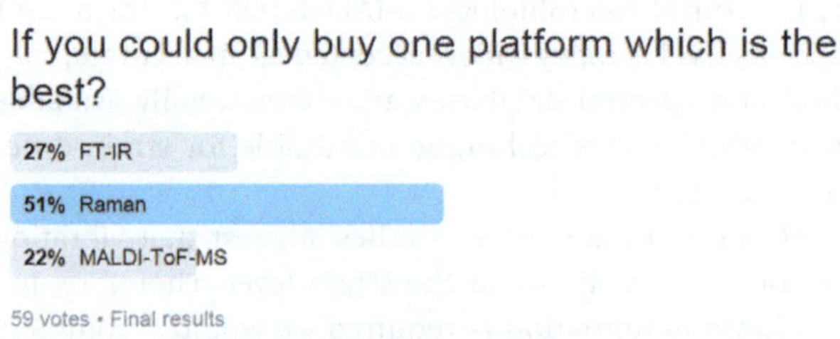

Fig. 1 The results of a twitter poll on which analytical platform people think is best for bacterial identification. The poll posted by @roygoodacre started on 22 March 2016 and lasted for 24 h.

Peter Lasch answered: I do not believe that the sensitivity of vibrational spectroscopy will be sufficient at the present stage to obtain high quality IR or Raman spectra suitable for microbial identification. My skepticism is based also on the problems associated with the need for spectral databases which ideally should contain spectral entries acquired under standardized conditions from well-characterized microbial strains. Both problems pose considerable challenges for an *in vivo* spectroscopic phenotyping technique in the gut.

Gianfelice Cinque asked: Could you explain how you did the cluster analysis of the phenotype data and combined different techniques ?

Peter Lasch replied: Cluster membership functions obtained on the basis of Unsupervised Agglomerative Hierarchical Cluster Analysis (UHCA) of microbial MALDI-TOF mass, FTIR and FT Raman spectra served as inputs for a combined phenotypic characterisation method. To this end, vectors containing the individual cluster membership values were compiled for each spectrum of a biological replicate after fusion steps 1–6 (MALDI and FTIR), or 1–7 (Raman). These vectors were then subjected to a top-level unsupervised hierarchical cluster analysis carried out by means of the pdist (Euclidean distance) and linkage (average linkage) functions of Matlab's statistics toolbox (Matlab version R2014a, The Mathworks, Natick, MA, USA). The dendrogram was generated using the dendrogram function of Matlab (see the Materials and Methods section, DOI: 10.1039/C5FD00165J). Please note also Fig. 2, which illustrates the cluster membership input functions used for the combined phenotype analysis.

Louise Clark resumed the general discussion of the paper by Bayden R. Wood: With malaria, the amount of platelets within the blood deteriorates rapidly in the late stages. Are you able to quantify the amount of platelets in the blood?

Bayden Wood responded: This is an interesting point that we have not addressed. The ability to determine platelet number using FTIR would not only be

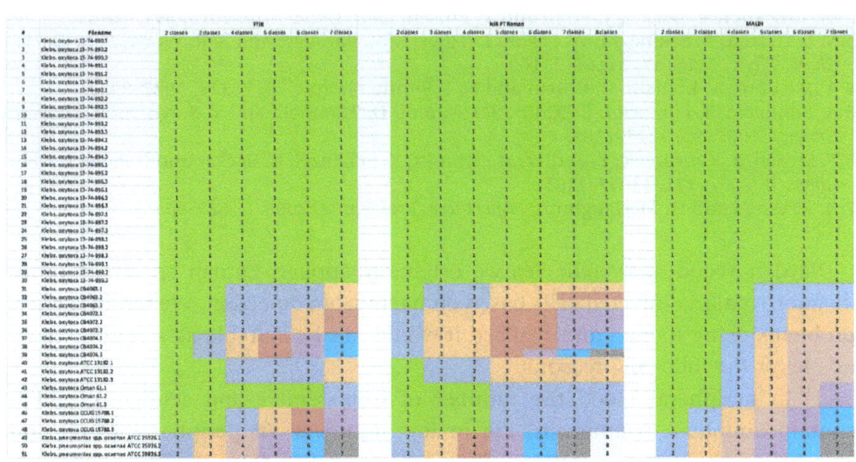

Fig. 2 Combined phenotype analysis.

beneficial to malaria diagnosis but also for a range of other blood disorders. I imagine that it would be relatively easy to set up a series of plasma samples spiked with different numbers of platelets to test this hypothesis. If you perform a light buffy coat separation the platelets remain mainly in the serum. If we are able to count these and get a number we could then perform serial dilutions with serum and using PLS regression determine if we can count them.

David Perez-Guaita returned to the general discussion of the paper by Peter Lasch: Cluster functions are useful to avoid the heterogeneity of data when you have portions of data from different sources (*i.e.* the need to process them independently to make them equivalent). However, in my opinion you pay the price of losing molecular information. Have you looked at correlation between the three methodologies?

Peter Lasch replied: I agree, one will lose molecular information when using the combined cluster analysis approach. However, the main goal of the study was to test the taxonomic resolution of the phenotyping methods (FTIR spectroscopy, NIR FT-Raman spectroscopy and MALDI-TOF mass spectrometry) and to compare the results with the outcomes of nucleic acid-based methods (MLST, PFGE, SNP-WGS, DOI: 10.1039/C5FD00165J). Therefore, the loss of structural and compositional information was acceptable in this particular study. Correlations between the three methods were not studied.

Alois Bonifacio resumed the general discussion of the paper by Bayden R. Wood: There are publications, many of which you co-authored, about the use of Raman spectroscopy for haemozoin characterization, imaging and detection, even within infected erythrocytes.[1–7] Using visible excitation, haemozoin detection with Raman benefits the electronic resonance. In spite of these publications about Raman spectroscopy and malaria, in the end you decided to use FTIR as the tool for malaria diagnosis. Could you please state the main reasons in favor of FTIR as a diagnostic tool for plasmodium detection in blood smears?

1 M. Kozicki, J. Czepiel, G. Biesiada, P. Nowak, A. Garlicki and A. Wesełucha-Birczyńska, *Analyst*, 2015, **140**, 8007–8016.

2 M. Brückner, K. Becker, J. Popp and T. Frosch, *Anal. Chim. Acta*, 2015, **894**, 76–84.

3 N. L. Garrett, R. Sekine, M. W. Dixon, L. Tilley, K. R. Bambery and B. R. Wood, *Phys. Chem. Chem. Phys.*, 2015, **17**, 21164–21168.

4 A. J. Hobro, A. Konishi, C. Coban and N. I. Smith, *Analyst*, 2013, **138**, 3927–3933.

5 A. Bonifacio, S. Finaurini, C. Krafft, S. Parapini, D. Taramelli and V. Sergo, *Anal. Bioanal. Chem.*, 2008, **392**, 1277–1282.

6 T. Frosch, S. Koncarevic, L. Zedler, M. Schmitt, K. Schenzel, K. Becker and J. Popp, *J. Phys. Chem. B*, 2007, **111**, 11047–11056.

7 B. R. Wood and D. McNaughton, *Expert Rev. Proteomics*, 2006, **3**, 525–544.

Bayden Wood answered: Yes we originally applied Raman spectroscopy to diagnose malaria-infected cells based on haemozoin detection. We ran into a few problems using Raman and realized it would not be suitable as a routine diagnostic for the following reasons:

1) In peripheral blood the number of trophozoites and schizonts, which contain abundant haemozoin, are in low numbers. In peripheral blood you mainly find ring stage parasites and gametocytes that contain negligible amounts

of haemozoin, and detecting these cells with Raman is rather difficult, especially when performing measurements on bulk samples. However, infected cells do contain lipids, DNA, RNA and carbohydrates, which are all strong IR absorbers, that do not appear in uninfected red blood cells and hence are easily detected using ATR and FTIR imaging.

2) In general the time taken to image a blood smear using Raman spectroscopy is on the order of hours. FPA imaging is much quicker and ATR is even faster.

3) Portable ATR spectrometers are less expensive than portable Raman spectrometers or Raman microscopes and much easier to use, which is important when training health workers.

Josep Sulé-Suso asked: Do haemoglobin diseases affect the spectra of cells? Can changes in their shape affect also their spectra?

Bayden Wood responded: It is unlikely that there would be many haemoglobin-based diseases that can obscure the diagnosis because the markers for the parasites are very distinct for both ATR and FTIR imaging. However, that being said we have had a few false positives in our Thailand ATR trial where the spectrum differed from the control regression vector. These false positives appeared right on the edge of the confidence limit for normal and infected red blood cells. The patients were subsequently re-tested and it was found that one had thalassaemia, one had haemorrhagic dengue and the other two had undiagnosed infections. Increasing the data base with other haem-related diseases will further reduce the false positive rate and improve the specificity of the modelling.

Ganesh Dhruvananda Sockalingum remarked: In your single cell measurements, you show circular shapes drawn around the cells (DOI: 10.1039/C5FD00181A). Does this mean you only measured those? What happened to the other shapes? Were they discarded? Have you other ways to select non-circular cells?

Bayden Wood replied: At this stage we can only use the *imfindcircles* routine in Matlab to find circular shapes. This is also the reason why our sensitivity and specificity are on the low side (81% specificity and 83% sensitivity) compared to our ATR pilot study in Thailand where we are achieving above 95% specificity and 93% sensitivity. We will need to modify the routine to pick up different pixel dimensions, which will enable the auto-selection of different shaped cells and improve the sensitivity and specificity.

Hugh Barr remarked: There is increasing resistance to anti-malarial drugs particularly in South East Asia where your trial is occurring. Artemisinin is also losing its effect. Do you have any insights as to how spectroscopy can detect resistance?

Bayden Wood answered: We plan to address this question in a new trial to be undertaken in Papua New Guinea. In this trial we will collect spectra from patients on a daily basis following treatment with artemisinin and determine if we can observe the kill-off point, *i.e.* when the parasitemia rapidly decreases in response to the drug. The ability to quantify parasitemia using the ATR approach offers this possibility. Moreover, because the data is being uploaded directly from the sites where the patients attend the clinic through the "cloud" interface it will be possible to observe outbreaks in resistance at the site of blood collection.

Maria Paula Marques communicated: When referring to future studies on antimalarial drug response in this type of sample, which spectral biomarkers do you intend to monitor?

Bayden Wood communicated in reply: We plan to monitor the same markers, *i.e.* nucleic acids, lipids and haemozoin, that we use to detect and quantify the parasites. The markers are very unique for infected red blood cells and stand out from uninfected red blood cells which are essentially bags of haemoglobin with only very small amounts of lipids and nucleic acids and no haemozoin. The evidence that the drug is working will be the decrease of these marker bands in response to the drug.

Catherine Kendall opened a general discussion of the paper by Zhiwei Huang: On the video shown in the presentation we saw a flag saying "no contact". How do you determine whether there is good contact with the tissue for a measurement to be taken? Do you also assess the quality of the measured spectra? If so, how do you do this and what is your rejection rate?

Zhiwei Huang replied: To assess the quality of the measured Raman spectra in real-time, we have built a large good-quality Raman database acquired under proper probe-tissue contacts. A principal component analysis (PCA) model was developed for contact and Raman spectra quality verification and validation. For each new Raman spectrum measured, Hotelling's T2 and Q-residual analysis were implemented with a 95% confidence interval (CI) verifying whether the new Raman spectra are within the common tissue variations of the PCA model for acceptance. Those Raman spectra located outside 95% CI will be discarded.[1] Currently, the fingerprint high-wavenumber (FP/HW) Raman spectroscopy system developed is routinely used by both experienced and novice endoscopists without any difficulty at the Endoscope Center at National University Hospital, Singapore for the screening and surveillance of GI patients undergoing an endoscopy. The Raman spectra rejection rate could vary from ~5% to ~30% depending on the endoscopists' operations and the motions of organs during clinical Raman measurements.

1 S Duraipandian, M. S. Bergholt, W. Zheng, K. Y. Ho, M. Teh, K. G. Yeoh, J. B. Yan So, A. Shabbir and Z. Huang, *J. Biomed. Opt.*, 2012, **17**, 081418.

Ganesh Dhruvananda Sockalingum asked: Very nice progress in clinical *in vivo* Raman during surgery. Recently it was shown that one can differentiate between tumours and non-tumours by looking at the water band? Is your calculation in the high wavenumbers also taking into account the water band and/or other peaks?

Zhiwei Huang replied: Thanks for the comments. Yes, our clinical Raman studies on gastrointestinal tracts indicate that high wavenumber (HW) (*i.e.*, 2800–3600 cm^{-1}) Raman spectroscopy could be used for differentiation between cancer and non-cancer tissues. In this work, we have taken HW Raman signals (*e.g.*, lipids [2850 and 2885 cm^{-1}: symmetric and asymmetric CH_2 stretching of lipids; 3009 cm^{-1}: asymmetric =CH stretching of lipid], proteins [2940 cm^{-1}: CH_3 stretching of proteins], ~3300 cm^{-1} [amide A (NH stretching of proteins)] and the broad Raman band of water [OH stretching vibrations peaking at ~3250 and ~3400 cm^{-1}]) into account for gastric dysplasia identification. The diagnostic accuracy for identifying gastric dysplasia from normal gastric tissue is limited (only ~63.2%) based on the HW Raman spectroscopy technique alone. In contrast, by using the integrated FP/HW Raman spectroscopy technique, the diagnostic accuracy for separating gastric dysplasia from normal gastric tissue can increase to up to 88.4%, suggesting that the enhanced diagnostic accuracy can be achieved by using the integrated FP/HW fiber-optic Raman spectroscopy technique. This is probably due to the fact that FP (e.g., 800–1800 cm^{-1}) Raman spectroscopy can uncover a wealth of biochemical information (*i.e.*, proteins, lipids and nucleic acid) in cells and tissue. On the other hand, HW (*e.g.*, 2800–3600 cm^{-1}) Raman spectroscopy can provide complementary biochemical information (*e.g.*, tissue water content, proteins and lipids) with stronger tissue Raman signals but less autofluorescence interference that is not achievable with FP Raman spectroscopy. Therefore, the integrated FP/HW Raman technique that uncovers more comprehensive tissue diagnostic information has great potential to improve precancer and early cancer diagnosis by endoscopy.

Hugh Barr enquired: Your excellent video in the presentation showed that after narrow band imaging there was a morphological change. In the oesophagus we are trying to target with our probe the stem cell niche where cancer starts (crypt dysplasia): the very earliest completely invisible biochemical changes. How confocal is your probe and can it image the crypts where the stem cells reside approximately two thirds the way down?

Zhiwei Huang answered: Our Monte Carlo simulations[1] show that with the use of the fabricated beveled fiber-optic Raman probe, ~85% of the total Raman signal collected arises from the top ~200 μm of the epithelial layer, whereas only ~15% of the Raman collection is from tissue depths of ~200–800 μm, facilitating the *in vivo* detection of epithelial precancer and cancer. We have already demonstrated that the beveled fiber-optic Raman probe can be used to detect dysplastic Barrett's esophagus (BE) *in vivo*.[2]

1 J. Wang, M. S. Bergholt, W. Zheng and Z. Huang, *Opt. Lett.*, 2013, **38**, 2321–2323.
2 M. S. Bergholt, W. Zheng, K. Y. Ho, M. Teh, K. G. Yeoh, J. B. Yan So, A. Shabbir and Z. Huang, *Gastroenterology*, 2014, **146**, 27–32.

Alexander Apolonskiy asked: The sensitivity of your technique – what you have shown in your presentation – is below 80%. Is there a way to improve this parameter up to more than 90%?

Zhiwei Huang replied: Our study shows that if using the fingerprint (FP) or high-wavenumber (HW) Raman spectroscopy techniques alone, the sensitivity for differentiating gastric dysplasia from normal tissue and adenocarcinoma is 72.7%; the sensitivity for differentiating gastric adenocarcinoma from normal tissue and gastric dysplasia using the HW spectroscopy technique is 70.6%; the sensitivities for identifying gastric dysplasia from normal tissue are 79.3% and 63.2%, respectively, using the FP and HW Raman spectroscopy techniques. On the other hand, by using the integrated FP/HW fiber-optic Raman spectroscopy technique, the sensitivity for identifying gastric dysplasia from normal tissue and adenocarcinoma can increase up to 81.8%, and the diagnostic accuracy for separating gastric dysplasia from normal gastric tissue is 88.4% using the integrated FP/HW Raman technique. This study suggests that the enhanced diagnostic accuracy can be achieved by using the integrated FP/HW fiber-optic Raman spectroscopy technique. This is probably due to the fact that FP (*e.g.*, 800–1800 cm^{-1}) Raman spectroscopy can uncover a wealth of biochemical information (*i.e.*, proteins, lipids and nucleic acids) in cells and tissue. In contrast, HW (*e.g.*, 2800–3600 cm^{-1}) Raman spectroscopy can provide complementary biochemical information (*e.g.*, tissue water content, proteins and lipids) with stronger tissue Raman signals but less autofluorescence interference that is not achievable with FP Raman spectroscopy. Therefore, more comprehensive tissue diagnostic information can be acquired using the integrated FP/HW Raman technique, proving its potential for improving gastric precancer diagnosis by endoscopy. Currently, further clinical FP/HW Raman measurements on a larger series of gastric patients are in progress to strengthen the *in vivo* tissue Raman database for Raman diagnostic software enhancement. With the increase of gastric precancer cases, we will be able to develop more robust diagnostic algorithms to couple with the FP/HW fiber-optic Raman technique for further improving the diagnostic accuracy of *in vivo* gastric precancer diagnosis by endoscopy.

Richard Dluhy asked: Your use of ANOVA with the Fisher *post hoc* least significant difference test seems to work quite well for spectral discrimination. Have you tried other methods for feature selection and how do they compare?

Zhiwei Huang replied: Besides the ANOVA with Fisher's LSD test, we also applied the Tukey Honest Significant Difference (HSD) method for the Type I error rate across multiple comparisons, and the results are practically identical to using the ANOVA–LSD method. We have also tried many other feature selection methods,[1] such as Bayesian feature selection,[2] logistic regression feature selection,[3] random forests algorithm for feature selection,[4] principal component analysis,[5] support vector machines,[6] *etc.* We found that different feature selection methods exhibit equally good feature selection abilities for tissue classification. Specifically, for Raman spectroscopy to detect gastric dysplasia, we found that the most statistically significant Raman peak intensities exist not only in the FP regime (853 cm^{-1}, 1004 cm^{-1}, 1078 cm^{-1}, 1210 cm^{-1}, 1265 cm^{-1}, 1302 cm^{-1}, 1335 cm^{-1}, 1445 cm^{-1}, 1618 cm^{-1}, 1655 cm^{-1}), but also in the HW regime (2850 cm^{-1}, 2885 cm^{-1}, 2940 cm^{-1}, 3250 cm^{-1}, 3300 cm^{-1}). The results reaffirm the capability of Raman spectroscopy to reveal the specific biochemical/biomolecualr changes associated with tissue neoplasia transformation at a molecular level. We should point out that besides the diagnostic robustness, another consideration in adopting specific feature selections for discrimination is the computation efficiency suited for real-time *in vivo* Raman diagnosis in clinical settings.

1 Y. Saeys, I. Inza and P. Larrañaga, *Bioinformatics*, 2007, **23**, 2507–2517.

2 K. E. Lee, N. Sha, E. R. Dougherty, M. Vannucci and B. K. Mallick, *Bioinformatics*, 2003, **19**, 90–97.
3 J. G. Thomas, J. M. Olson, S. J. Tapscott and L. P. Zhao, *Genome Res.*, 2001, **11**, 1227–1236.
4 S. K. Teh, W. Zheng, D. P. Lau and Z. Huang, *Analyst*, 2009, **134**, 1232–1239.
5 V. G. Tusher, R. Tibshirani and G. Chu, *Proc. Natl. Acad. Sci. U. S. A.*, 2001, **98**, 5116–5121.
6 E. Widjaja, W. Zheng and Z. Huang, *Int. J. Oncol.*, 2008, 32, 653–662.

(505:[505]505) **Nick Stone** asked: Your patient case mix is weighted heavily towards low risk/normal patient groups. Have you considered Bayesian discrimination models enabling you to take this into account using prior probabilities?

Zhiwei Huang responded: Thanks for the comments and suggestions related to the current unbalanced Raman database and discrimination models. Our preliminary analysis using the Bayesian discrimination model on the gastric dysplasia shows a diagnostic accuracy of 95.6% (sensitivity of 89.0% (138/155) and a specificity of 95.8% (4945/5160)) by using the integrated FP/HW fiber-optic Raman spectroscopy technique developed, which is similar to the diagnostic performance based on partial least squares discriminant analysis (PLS-DA) together with leave-one-patient-out cross validation (LOPCV). With the implementation of Bayesian discrimination models, the integrated FP/HW Raman spectroscopy technique is confirmed to be able to provide a much higher diagnostic accuracy as compared to either FP (accuracy of 93.3%; sensitivity of 78.1% (121/155), and a specificity of 93.8% (4839/5160)) or HW (accuracy of 93.4%; sensitivity of 63.2% (98/155), and a specificity of 94.3% (4866/5160)). Apparently, both the discriminant techniques provide equally effective classification of dysplasia from normal gastric tissue based on the current gastric Raman dataset. With the increase of gastric precancer cases under ongoing Raman clinical measurements on a larger series of gastric patients, we will further compare their diagnostic performance for gastric precancer classifications among the Bayesian model discrimination and other multivariate techniques.

Christopher Phillips commented: This technique is being used on humans in a clinical setting. Can you clarify how much of the clinical decision is being effectively based on your technique, and how much remains with the clinician? What sort of ethical approval was required before the work could proceed?

Zhiwei Huang replied: One of the objectives to develop the rapid fiber-optic FP/HW Raman technique in this work is to assist the clinicians in targeted biopsies instead of random biopsies during endoscopic screening of high-risk GI patients with multiple suspicious lesions. In the current Raman study on the gastric patients recruited, we have shown the capability of the *in vivo* FP/HW Raman spectroscopy technique to much reduce the unnecessary negative biopsies compared to white-light reflectance (WLR) endoscopy. That is, the biopsy number taken for gastric dysplasia diagnosis is 204 as determined by the clinicians with WLR, and only 10 of the 204 biopsies were positively indicated as dysplasia, providing a diagnostic yield of ~5% (10/204). In comparison, the biopsy number suggested by FP/HW Raman spectroscopy is 12, with a much enhanced diagnostic yield of 83% (10/12). It should be pointed out that in the current study, we are not stopping the clinicians from taking the biopsies during endoscopic procedures. The clinicians in the endoscope center at the National University Hospital,

Singapore, still take biopsies as is the norm during endoscopic examinations for histopathology to serve as the gold standard to determine the diagnostic performance of the FP/HW fiber-optic Raman technique for *in vivo* GI tissue diagnosis and characterization. Before the start of this clinical Raman project, the ethical protocol of the study was approved by the Institutional Review Board (IRB) of the National Healthcare Group (NHG) of Singapore. Prior to the *in vivo* tissue Raman measurements, all patients signed informed consent permitting *in vivo* Raman spectroscopic measurements during endoscope examinations. Further, the trial has also been registered (registration number: ISRCTN15587241) into the ISRCTN registry (http://www.isrctn.com), which is a publicly accessible primary register that participates in the World Health Organization (WHO) International Clinical Trial Registry Platform.

Ioan Notingher asked: In terms of the gold standard, when we look at tissues we struggle to match the locations where the Raman spectroscopy was measured to the histopathology. How does this work in your case? Every time you take a Raman measurement, do you pick up a biopsy from that location – is the tissue homogeneous? While this can be addressed in the training stage (*i.e.* you don't include spectra where you are unsure about the histopathology), what about when you validate? If you only select samples large enough to be homogeneous, I assume this leads to bias?

Zhiwei Huang responded: During endoscopic measurements, the tissue sites measured by Raman spectroscopy were suction-marked with a cap to ensure that the biopsied tissues were taken exactly from the tissue sites measured for histopathology examination. One notes that during the building of the *in vivo* tissue Raman spectrum database, we only choose the Raman spectra acquired from homogeneous tissues with consensus with the pathologic results as the training dataset for diagnostic model development. We understand that living tissue is complex and seldom uniform, with possible overlap of different cell types. Therefore, we adopt the majority voting scheme (*i.e.*, majority wins) in Raman diagnostic algorithms to perform *in vivo* tissue Raman diagnosis.

Nick Stone asked: You use 3 pathologists to provide you with sample grading. How do you incorporate this information into your models and their testing? Do you use a majority or consensus vote for training/testing? Is this for each specific Raman measurement point?

Zhiwei Huang replied: In the current study, we only choose the Raman spectra acquired from homogeneous tissues with consensus histopathologic tissue types to build an *in vivo* tissue Raman database serving as the training dataset for diagnostic model development. Cohen's κ statistics were also calculated to examine the agreement among the three pathologists who were blind to the Raman results. A good consistency for the gastric tissue groupings (*i.e.*, normal, dysplasia and adenocarcinoma) was obtained among the three independent pathologists as indicated by the Cohen's κ of 0.89. For testing and prospective prediction of each tissue Raman measurement, we utilize the majority voting scheme (*i.e.*, majority wins) in Raman diagnostic algorithms to perform *in vivo* tissue Raman diagnosis at the tissue sites tested.

Leanne Fullwood asked: What laser powers did you use and how did you decide they were acceptable for clinical use?

Zhiwei Huang answered: To our knowledge, there are still no clear guidelines available for defining the maximum permissible laser irradiation in internal organs (*e.g.*, esophagus, stomach, colon, lung, bladder, *etc.*). In this work, the 785 nm laser power incident on the tissue surface is ~10–12 mW (equivalent to ~1.5 W cm^{-2} with a light beam size of ~500 μm), which is within the maximum permissible power as set out by the American National Standards Institute (ANSI) standard for skin irradiation.[1] We have evaluated the photothermal effect of the 10–12 mW laser power used on gastric tissue. The photothermal modeling of the gastric tissue shows that there is only a tissue temperature rise of <0.07 °C (without consideration of body fluids flowing, convection, *etc.*) within the one minute of 12 mW 785 nm laser irradiation, which does not cause any obvious photothermal damage to tissue during the clinical Raman measurements (each tissue spectrum is acquired in <1 sec in our study). Hence, the current laser power used is relatively safe for clinical Raman measurements in GI tracts.

1 ANSI Standard 2136.1-1986, American National Standard for the Safe Use of Lasers, American National Standards Institute, Washington, D.C., 1986.

Wolfgang Petrich commented: There is a lot of work on using tissue autofluorescence for the diagnosis of tissue, especially for endoscopy. In your Raman system, you will get autofluorescesnce for free (whether you want or not). It may, admittedly, be quite low given the 785 nm laser wavelength but you may still notice some fluorescence background. Did you discard this fluorescence background signal or did you consider this signal as an additional aid for your measurements?

Zhiwei Huang replied: Yes, the measured tissue spectra under 785 nm laser excitation represent a combination of weak tissue Raman scattering superimposed on the intense tissue autofluorescence (AF) background. We found that the tissue autofluorescence intensity could vary up to 100 times depending on different tissue sites in the stomach, which is unsuitable for reliable tissue diagnosis model development. Hence, we discard the AF background by using polynomial fittings (*i.e.*, the fifth-order polynomial was used for removing the AF background over the fingerprint (FP) range of 800–1800 cm^{-1}, whereas the first-order polynomial fit was used for removing the AF in the high-wavenumber (HW) (2800–3600 cm^{-1}) range). We only utilize tissue FP/HW Raman signals extracted for gastric precancer diagnosis and characterization.

Ola Ibrahim asked: Are you going to follow up with the patients you took samples from? It would be interesting to see if the Raman diagnoses any early dysplasia that the pathologists missed.

Zhiwei Huang responded: One of the main objectives to develop the rapid fiber-optic FP/HW Raman technique in this work is to assist the clinicians in targeted biopsies instead of random biopsies to reduce unnecessary negative biopsies during endoscopic screening of high-risk GI patients with multiple suspicious lesions. We also apply the rapid fiber-optic FP/HW Raman technique developed for the follow-up monitoring of the post-treatment efficacy of the patients who received the treatments. We agree that it would be interesting to perform Raman scans for the follow-up patients after revising our IRB protocol for approval.

Hugh Barr stated: The big problem remains the various shades of change. Your presentation showed high risk and normal patients (high risk patients have a greater than 10–20% chance of cancer developing within 5 years; most patients (90%) will not develop cancer within their lifetime). Particularly in the oesophagus there is "indefinite for dysplasis". We have shown that the Raman "fingerprint" can predict ultimate degeneration over a period of years. This identifying the "Bad Barretts". This is important because we can increase the surveillance in these patients and reassure the majority of patients by decreasing the level of surveillance. We must concentrate on those at greatest risk.

Zhiwei Huang replied: We fully agree with Prof. Barr's comments. It would be interesting to follow up with the patients at greatest risk, and evaluate how the integrated FP/HW Raman spectroscopy performs in the prediction of "indefinite for dysplasis" degeneration over time during endoscopic examination.

Nick Stone opened a general discussion of the paper by Kerstin T. Mader: We all know the issues with transflection approaches, you explained them well in your paper (DOI: 10.1039/C5FD00160A), if you designed your study again would you reconsider using this approach or have you overcome these issues? Have you explored the difference between your stainless steel slides and low-E transflection slides?

Kerstin Mader responded: One of the main considerations when designing this approach was for it to be a screening tool. To achieve a high throughput, low cost approach the transflectance mode was chosen. Also, the sample preparation is the same as for histology and immunohistological analysis and therefore easy integration into already existing workflows is possible. In the presented approach we have explored multivariate curve resolution-alternating least squares (MCR-ALS) as a means of discerning the underlying individual spectral contributions, not only to improve the deconvolution of highly overlapping sample component signals (*e.g.*, collagen type I *vs.* collagen type II) but also the deconvolution of biochemical information from physical/structural information. The results show that the described approach delivers the desired biochemical information at the same level as immunohistochemistry. Also, the extracted multivariate IR parameters for PGs show a better correlation with histological grading than the GAG/mg DW values derived from a bioassay, therefore I would not design the approach differently. However, the results also show that the deconvolution is not complete, and further experiments are planned to investigate the prediction accuracy of the MCR-ALS models and if necessary to further optimise the models *via* the introduction of refined constraints.

We have not looked at different substrates yet. In this article we have concentrated on the chemical information, however it would be very interesting to further explore the physical information contained in these measurements and to study the variance between different transflectance substrates but also transmission measurements.

Peter Gardner commented: I just want to clarify my issue with transflection. Transflection is a perfectly good method for obtaining infrared spectra of a wide range of samples. If you are not looking for subtle differences in spectra it is fine. I urge caution, however, when the differences you are interested in are very subtle, for example when looking at two slightly different disease states or if you have cells that are different sizes.

Bayden Wood added: Are there any examples in the literature to date where the sensitivity and specificity has been improved by performing the measurement in the transmission mode as opposed to the transflection mode? We have tested this with liver cancer and malaria infection and have achieved the same sensitivity and specificity irrespective of the substrate.

Rohit Bhargava replied: Even perfectly homogeneous films in the transmission mode will show effects arising from focusing (angular diversity of light) at the sample.[1,2]

1 B. J. Davis, P. S. Carney and R. Bhargava, *Anal. Chem.*, 2010, **82**, 3474–3486.
2 B. J. Davis, P. S. Carney and R. Bhargava, *Anal. Chem.*, 2010, **82**, 3487–3499.

Matthew J. Baker asked: Can you comment on the effect of sample preparation and spectral processing on the quantitative measure delivered by the IR methodology?

Kerstin Mader replied: In the presented methodology, quantification is based on the Beer–Lambert law and therefore the experimental set-up and data processing methods will influence quantification. The potential advantage of the presented methodology over other multivariate calibration methods such as for example PLS with respect to the extraction of quantitative results, is the ability of iterative MCR-ALS algorithms to build a calibration model for only one "pure" component irrespective of the mixture composition.[1] This simplifies the necessary calibration matrix to a great extent and enhances the transferability of the model, for example, between species or from *in vitro* to *in vivo* systems.

1 R. Tauler and A. de Juan, Data Handling in Science and Technology, ed. A. M. de la Peña, H. C. Goicoechea, G. M. Escandar and A. C. Olivieri, Elsevier Ltd., 2015, vol. 29, ch. 5, pp. 247–292.

Kathleen Gough addressed Kerstin Mader: I am very interested in the vibrational spectroscopic analysis of collagen, partly because it can be distinguished easily, and partly because its inherent organizational architecture makes it an excellent subject for study under polarized light. You will have seen this topic addressed in our paper (DOI: 10.1039/C5FD00168D). While there will be overlap with spectral profiles of other tissue components, the dramatic alteration in the collagen FTIR spectral profile under polarized light can be exploited in some cases. Within the disc structure that you are studying, is there architectural order to the collagen and would it change in degenerated discs? Have you done polarization-dependent IR imaging and would this be useful as part of the proposed FTIR analysis?

Kerstin Mader responded: The main interest in this study was to the characterisation of disc degeneration at a very early stage/mild disc degeneration where no major restructuring has taken place yet (highlighted in Fig. 4, DOI: 10.1039/C5FD00160A). The earliest changes in disc matrix composition caused by disc degeneration have been reported as a loss of proteoglycan molecules in the nucleus pulposus region[1] where the collagen network has been shown to be randomly arranged.[2] We have therefore focused more on the glycosaminoglycan content. However, as degeneration progress restructuring of the highly structured annulus fibrosus region is taking place, polarization-dependent IR imaging together with compositional information would be very useful for example to build composition–structure and structure–function relationships. We have not done polarization-dependent IR imaging in this study, but having recently acquired polarisers that sit in our IR microscope we would hope to do so in the future.

1 M. A. Adams, *Surgery*, 2012, **30**, 347–350.
2 J. P. G. Urban and S. Roberts, *Arthritis Res. Ther.*, 2003, **5**, 120–130.

Nick Stone commented: A three part question:

1. You have explored the relationship between pure substances measured with the ATR-IR approach and formalin-fixed paraffin-embedded tissue samples. Have you corrected for the wavelength dependence of the signals with ATR? Is this important for the MCR-ALS approach?
2. You also plotted your spectral results against T2*-weighted MR images. Can you explain what you are plotting from the MR data?
3. How relevant is the spinal goat model in relation to human spines?

Kerstin Mader responded: 1. In this study we have not performed an ATR correction. In the presented approach the pure substances are only used to validate the MCR-ALS results after analysis to distinguish between major spectral characteristics, for example proteoglycans *vs.* collagens. However if the reference profile were to be used as a constraint to guide the iterative optimisation process, great care needs to be taken to match the reference profile with the actual target component profile as a mismatch can distort the outcome. This can be very challenging as considerations should not only include experimental conditions such as sampling techniques, spectral and spatial resolution, but also "molecular resolution", for example, will the measured spectrum represent 1) an average population of a range of molecular forms of a generic family in a specific environment, 2) an average population of a range of molecular forms of the same macromolecular structure in a specific environment, 3) one specific form of a macromolecule in a specific environment, or indeed 4) one molecule in a specific environment.

2. The spectral results are plotted against the mean T2* relaxation times in ms for the nucleus pulposus region of interest (RoI) (Fig. 2A and experimental section: MRI T2* relaxation time mapping, DOI: 10.1039/C5FD00160A).

3. An animal model is a living, non-human model used to investigate human disease for the purpose of understanding its cause and evaluating its treatment. Intervertebral discs containing a nucleus pulposus are only found in mammals. We chose the disc as our model for various reasons: 1. The intervertebral disc is a non-vascularised structure. This implies that the transport of nutrition and waste products relies on diffusion and convection of fluids. Large discs thus are physically different from small discs, which makes large animals (like goats) a more relevant model for the human condition than small animals (like rabbits or mice). 2. In many species like rabbits and mice, the nucleus pulposus contains notochordal cells, which have an early embryonic origin and function as stem cells of the disc. It is well established that notochordal cells contribute to the generation and regeneration of the matrix of the nucleus pulposus. Humans older than 10 years no longer have notochordal cells in their discs. For the relevance of the disease model (the degenerating intervertebral disc), the animal model also should be without notochordal cells, which is the case in goats.[1] 3. A common misunderstanding about quadrupeds is that their spine is loaded differently than in humans. However, axial compression is by far the largest load component in both animal and human spines. Considering the animal as a bridge, the torso experiences bending which is counterbalanced by tension in the ventral structures (*linea alba*, thorax, musculature) and by compression in the spine. This is confirmed by the architecture of the vertebral trabecular bone.[2] 4. Our group has established and validated models for disc degeneration in goats. Degeneration was induced by the injection of chondroitinase ABC, which cuts the glycosaminoglycans from their hyaluronan backbone. The degeneration observed closely resembled the disc degeneration as described in humans from the radiological, mechanical and biochemical perspective. While there still are differences between goat and human discs, the most relevant aspects of disc physiology and degeneration seem to be covered by the goat disc.

1 R. J. Hoogendoorn, Studies on the degeneration and regeneration of the intervertebral disc, PhD thesis, Vrije Universiteit Amsterdam, 2009.
2 T. H. Smit, *Eur. Spine J.*, 2002, **11**, 0940-6719.

Louise Clark opened a general discussion of the paper by Guillaume Lepert: Is the hydration of the sample accurate to what it would be in the patient?

Guillaume Lepert replied: Hydration control is indeed essential. Fig. 3 of our paper (DOI: 10.1039/C5FD00152H) shows the consequence of over-hydration of the bovine cornea immersed in PBS over 24 h: swelling and decrease of the Brillouin shift. All experiments must be carried out with samples immersed in a medium that relevantly emulates the physiological state of the tissue under analysis. In the present case we used the Carry-C preservation medium (Alchemia, Italy), which contains a de-swelling agent (dextran) that maintains corneal tissue in its natural physiological hydration state. See also Laggner *et al.*[1]

1 M. Laggner, A. Pollreisz, G. Schmidinger, R. A. Byrne, C. Scheinecker, U. Schmidt-Erfurth and Y. T. Chen, *Invest. Ophthalmol. Visual Sci.*, 2015, **56**, 3584–3592.

Christopher Sammon asked: What are the challenges that must be overcome in taking this approach away from a microscopic tool towards a macroscopic tool that could be used to assess for example, *in vitro* tissue culture, such that you could use a non-contact mechanical measure to determine when an implant was ready for implantation? Is the approach limited to optically transparent samples?

Guillaume Lepert responded: This technique can be fully and immediately applied to *in vitro* tissue culture assays, such as the analysis of cellular responses to the mechanical stimuli induced by different surfaces. Moreover, as this is a non-contact confocal microscopy technique, it can be applied to thicker samples, being particularly useful to analyse cells encapsulated within a scaffold, as well as the scaffold itself. We have examined collagen scaffolds of 100 μm thickness, however the exact limitations on scaffold thickness and translucency have yet to be fully defined. Tissue absorption and scattering will limit the penetration depth through scaffolds but a useful signal may still be acquired near the surface. Loss of signal due to scaffold heterogeneity (other than absorption) may be corrected by adaptive optics. However, a temperature, humidity, and CO_2-controlled environment should be assembled around the microscope stage to ensure optimal growth conditions for cells and tissues are maintained during the experiments.

Alexander Apolonskiy asked: Brillouin scattering gives information about elastic properties. On the other hand, laser light in a bio-sample doesn't penetrate deep. What about photoacoustics? MHz acoustic waves penetrate very deep and give information about elastic properties, with the spatial resolution of the order of up to 50 micrometers. Can you comment on this?

Guillaume Lepert answered: Photoacoustics is primarily an optical absorption imaging technique. The incident laser light must be absorbed to generate acoustic waves, and therefore the technique is not suitable for use in transparent tissues. Moreover, the resolution is limited to ~50 μm, two orders of magnitude lower than that offered by confocal microscopy. Contact with the ultrasound probe is required, whereas confocal microscopy is non-contact.

As an elastography modality, photoacoustics is used to image deformation in a stress–strain measurement. Photoacoustics and Brillouin microscopy are therefore not interchangeable and address different needs.

Nick Stone asked: Can you comment on how the Brillouin scattering cross-section changes with wavelength?

Guillaume Lepert responded: Just like Raman scattering, the Brillouin cross-section scales as λ^{-4}. This is a general property of quasi-elastic scattering from small scattering elements.[1]

1 L. D. Landau and E. M. Lifshitz, Electrodynamics of Continuous Media, Pergamon, New York, USA, 1958.

Ganesh Dhruvananda Sockalingum said: Can Brilloin spectromicroscopy differentiate between different types of collagen? Collagen is involved in various pathophysiological processes. Is the technique capable of differentiating between types of cross-linking?

Guillaume Lepert replied: Fig. 8 of our paper (DOI: 10.1039/C5FD00152H) shows a clear contrast between the highly organised collagen of the stroma and the disorganised collagen of the sclera (both are type I). Different types of collagen will be differentiated under Brillouin microscopy if they are mechanically distinct; this depends on other factors than just the collagen type. Similarly with cross-linking, if different methods elicit the same mechanical response, they will be undistinguishable.

Nick Stone asked: Can you explain whether there is a direct correlation between the Young's modulus of the sample and the Brillouin-measured high frequency modulus? Is it possible to validate your findings using other physical techniques?

Guillaume Lepert replied: As discussed in Section 1.2 of our paper (DOI: 10.1039/C5FD00152H), the measured Brillouin shift is proportional to the ratio between the speed of sound and speed of light. The usual linear elastic relations may in principle be used to relate the speed of sound to Young's modulus; in practice this is problematic for a number of reasons including: 1) imprecise knowledge of the optical refractive index and density on a microscopic scale; 2) the potential effects of acoustic dispersion; 3) a critical dependence on the Poisson ratio in nearly incompressible materials; 4) the fact that biological tissues are not linearly elastic.

Valter Sergo remarked: The acquisition time of your experiments can be up to one second. Can you comment on the viscoelastic behaviour of the system under observation during that time frame?

Guillaume Lepert answered: Although this is not something we have investigated, we do not expect this to be much of a problem since tissue is essentially unaffected by the measurement. However, for longer experiment times (in the range of hours or days), measures were taken to reduce the evaporation of the liquid medium used for sample immersion, as this can affect the hydration of biological tissues and consequently their viscoelastic properties (see the question from Louise Clark).

Nick Stone remarked: You focus on the cornea in this study. Is this there a clear clinical need, or is it because it is a transparent tissue you can use?

Guillaume Lepert responded: Our current work is driven by a real biological question. Previously we hypothesised that the mechanism of human corneal epithelial homeostasis depends on the different mechanical properties between the limbus and the central region of this organ.[1] In addition, within the same paper, we also hypothesised that the stem cell properties of limbal epithelial stem cells are modulated by the mechanical properties of their surrounding matrix, and that the success of stem cell transplantation therapy can be improved by modulating the stiffness of the limbal niche.[1] However, a detailed characterisation of these corneal tissues was not feasible until now. Our microscope allows direct, non-contact measurement of the mechanical properties of the cornea, thus helping us investigate these hypotheses. Brillouin spectroscopy can also be used in relation to other cornea-related studies, notably keratoconus and cross-linking.[2]

1 J. W. Foster, R. R. Jones, C. A. Bippes, R. M. Gouveia and C. J. Connon, *Exp. Eye Res.*, 2014, **127**, 37–41.
2 G. Scarcelli, S. Kling, E. Quijano, R. Pineda, S. Marcos and S. H. Yun, *Invest. Ophthalmol. Visual Sci.*, 2013, **54**, 1418–1425.

Nick Stone asked: In terms of your system, you have highlighted that the interference filter makes a significant difference. What are the disadvantages? Stability, bandwidth? What kind of temperature stability do you need in your lab in order to make this work?

Guillaume Lepert replied: The filter simplifies the setup by being more efficient (the throughput is only limited by a fibre coupler) and allowing the use of a single-stage virtually imaged phased array (VIPA) spectrometer.

The filter requires active feedback to maintain a good suppression ratio. Better temperature and laser frequency stability make this feedback easier, but are not essential. In the current set-up, daily temperature variations of 3 °C are common and unproblematic.

Since the filter response is a sinusoidal function of frequency, the 3 dB bandwidth is half the free spectral range, or about 10 GHz. The Brillouin shifts of interest for soft aqueous tissues lie in a 3 GHz-wide window. Therefore there is no limitation associated with the bandwidth.

Pavel Matousek asked: Is there an equivalent of SERS enhancement in Brillouin scattering?

Guillaume Lepert replied: Surface-enhanced Brillouin scattering was investigated back in the 1980s, see for example Moretti *et al.*[1] For a more recent take, see Meng *et al.*[2]

1 A. L. Moretti, W. M. Robertson, B. Fisher and R. Bray, *Phys. Rev. B*, 1985, **31**, 3361.
2 Z. Meng, V. V. Yakovlev and Z. Utegulov, Proc. SPIE 9340, Plasmonics in Biology and Medicine XII, 2015, 93400Z.

Nick Stone asked Guillaume Lepert, Zhiwei Huang and Kerstin Mader: Can the three of you give us a vision of where these things will be in 5 years' time?

Guillaume Lepert responded: It is not unrealistic to imagine that a clinically viable Brillouin microscope can be developed in that time frame. First, however, more work needs to be done at a fundamental level to understand better the relationship between the Brillouin shift and the mechanical properties (see earlier related question).

Moreover, the application of this technique to *in vitro* tissue culture systems (see Christopher Sammon's question) can greatly improve our understanding of cell behaviour in response to mechanical stimuli.

Zhiwei Huang replied: In the past two decades, tremendous advancements in high-throughput Raman spectrograph, sensitive charge-coupled device camera, near-infrared diode laser, and fiber-optic Raman probe design and fabrication have been made, laying a foundation of developing fiber-optic Raman spectroscopy techniques for real-time biomedical diagnostic applications. We have shown that rapid fiber-optic Raman spectroscopy coupled with discriminant models can be used to detect gastric dysplasia with good diagnostic accuracy, illustrating the potential of applying fiber-optic Raman techniques for non-invasive or minimally invasive detection and diagnosis of precancer and early cancer *in vivo* during clinical endoscopic examination. However, to fully bring the fiber-optic Raman technology into routine endoscopic diagnostic practices, several critical challenges should be tackled: (i) The cost-effective and miniaturized fiber-optic Raman probes (*i.e.*, 1–2 mm in diameter) with high collection efficiency but with minimal interference of the silica fiber background should be developed to comply with medical endoscopes for access to different internal organs. (ii) The excitation laser power should be kept very low to prevent photothermal damage to tissue. Hence, light dosimetry for defining the maximum permissible laser irradiation in internal organs should be established, which is currently falling far behind the fiber-optic Raman technology advancement. (iii) Robust chemometric methods together with turbid tissue modeling related to tissue optical properties (*e.g.*, scattering and absorption coefficients, anisotropy) should be built for extracting biochemical and morphological diagnostic information for accurate tissue Raman diagnosis and characterization. In addition, combining rapid fiber-optic Raman spectroscopy with other optical imaging techniques (*e.g.*, white-light reflectance imaging, narrow-band imaging, hyperspectral imaging, diffuse optical tomography, optical coherence tomography, confocal endomicroscopy imaging, and multiphoton imaging) and/or biomedical imaging (*e.g.*, ultrasound, computed tomography, and magnetic resonance imaging) could also be the direction to explore for label-free functional imaging in biomedical systems. The above challenges can only be solved through close collaborations between bioengineers, biophysicists, biochemists, biospectroscopists, biostatisticians, radiologists, pathologists, and clinicians. We anticipate that in the next five years, many of the above challenges can be resolved. The fiber-optic Raman spectroscopy technique will be emerging as a new routine endoscopic tool, enabling non-invasive real-time disease detection at early and manageable stages, monitoring therapeutic effects and treatment outcomes, and guiding interventions, thereby providing the medical community with critical diagnostic information for improving patient care and management.

Kerstin Mader replied: Vibrational spectroscopy is a fantastic technique for qualitative and quantitative measurements. A lot has been learned in the past 12 years, and this knowledge can now be utilised to further enhance the selectivity and sensitivity of the results and to build comprehensive analytical strategies. This would include advanced soft modelling multivariate data analysis methods such as iterative MCR-ALS to allow the extraction of quantitative results.

Hugh Barr resumed the general discussion of the paper by Zhiwei Huang: The solution to the huge burden of pathology consuming many staff hours is technological. The National Health Service (NHS) has stated this in no uncertain terms.

Zhiwei Huang replied: I agree with the comments. We believe that using the rapid fiber-optic FP/HW Raman technique developed, the endoscopic medical practice could undergo a revolution by eliminating the subjectivity in endoscopic procedures. The optical biopsy at the biomolecular level in this work may change the rules of clinical diagnostics by improving patient care, that is, delivering more accurate and earlier diagnoses and surgery, enabling real-time decision making in order to avoid unnecessary procedures, and providing better options for patients at reduced costs.

Christopher Phillips returned to the general discussion of the paper by Kerstin T. Mader: An appeal of this approach is that the chemical concentrations can, in principle, be measured quantitatively and reproducibly. Can you comment on the effects of using derivative spectra in this regard? For example, do the conditions under which the derivatives are taken (smoothing, spectral resolution *etc.*) affect which of the spectral features are chosen by the statistical analysis software?

Kerstin Mader responded: As highlighted by Max Diem,[1] using second derivatives has both advantages and disadvantages. Using second derivatives decreases the S/N ratio and can introduce an intensity distortion related to the curvature of the raw spectral peak. To account for this, it is important to have a common protocol for all measurements as well as to match these conditions in the calibration matrix (please see also Matthew J. Baker's question). On the other hand, using second derivatives eliminates sloping baselines and increases spectral resolution, which in turn increases the selectivity during MCR-ALS analysis.

1 Max Diem, in *Modern Vibrational Spectroscopy and Micro-Spectroscopy: Theory, Instrumentation and Biomedical Applications*, Wiley, 1st edn, 2015, ch. 12, pp. 251–282.

Gavin Lloyd resumed the general discussion of the paper by Zhiwei Huang: How do you calibrate your Raman probe and have you done any tests to see if you can transfer your discriminant model between different versions of your probe and spectrometer?

Zhiwei Huang replied: As for the inter-Raman system transferability with the same Raman probe, we have conducted a large scale of Raman experiments on GI tissues with a few same-model Acton Raman spectrometers (LS-785). The good transferability is achievable among them as long as both the wavelength and intensity response of the entire system are properly calibrated before the Raman measurements. As for the inter-Raman probe transferability with the same Raman system, we have also conducted a very large scale of Raman experiments on GI tissues with at least 10 customized Raman probes. The good transferability with discriminant models is achievable but not so straightforward, as compared to the inter-Raman system transferability. The wavelength and spectral resolution and the intensity response of the entire system always need to be calibrated prior to the Raman measurements. However the inter-Raman probes' characterization should be performed as one of the input parameters in the Raman discriminant models. For instance, the fluorescence background and silica peaks of the fiber Raman probes with respect to laser excitation power could vary among different Raman probes, which should be documented, extracted and removed with proper algorithms. With the use of multivariate techniques, *e.g.*, partial least-squares (PLS) regression modeling on the extracted but characteristic internal reference

Raman signals (*e.g.*, shoulder of the prominent fused silica boson peak (~130 cm^{-1}); distinct sapphire ball-lens peaks (380, 417, 646, and 751 cm^{-1})) from the ball-lens fiber-optic Raman probe, we could monitor and correct the variations of the laser excitation power and fiber coupling efficiency in situ for standardizing the tissue Raman intensity among different Raman probes.[1]

1 M. S. Bergholt, S. Duraipandian, W. Zheng and Z. Huang, *Anal. Chem.*, 2013, **85**, 11297–11303.

Matthew J. Baker addressed Zhiwei Huang: During your design/validation phase you mentioned that a lot of work has gone into the development of the correct pressure on the contact to use inside the stomach, and overcoming issues with fluorescence. Due to this have you considered developing an ATR-FTIR probe-based analysis?

Zhiwei Huang answered: In our clinical Raman endoscopic measurements, the fiber-optic Raman probe is placed in gentle contact with the tissue sites measured to avoid possible tissue damage arising from the probe–tissue contact pressure rendered on the tissue surface. We have investigated the effect of the probe–tissue contact pressures varying from gentle, mild to firm in the gastric tissue Raman spectra acquired. We found that except for dramatic intensity fluctuations in the tissue autofluorescence background, no significant Raman spectral changes are observed under different probe–tissue contact pressures on the tissue surface (results to be published elsewhere). Therefore, currently, we would still continue focusing on rapid fiber-optic Raman spectroscopy coupled with multivariate techniques for *in vivo* tissue diagnosis and characterization by endoscopy. For the long term, we may consider incorporating the complementary ATR-FTIR technique into Raman endoscopic procedures to assess the feasibility for further enhancing the *in vivo* diagnosis of precancer and early cancer in gastrointestinal tracts.

Christopher Phillips addressed Zhiwei Huang and Hugh Barr: We use IR imaging to augment the grading of standard "H+E" *ex vivo* tissue biopsies, but I am often asked if we could develop an *in vivo* alternative. The answer is, for the moment at least, no, but actually we see this as an advantage rather than a drawback. The point is that the profession will continue to take the biopsies in any case, so the *ex vivo* method introduces no extra patient procedures. This makes it much easier to trial, and to get accepted than, say, trying to get the profession to add an extra endoscopic examination to the patient plan.

Hugh Barr responded: One of the major concerns in the NHS is the amount of normal or non-concerning biopsies performed to exclude disease. This consumes substantial health care resources. Thus an *in vivo* system to reduce this burden is now seen as vital. In particular the USA health system is burdened by non-neoplastic polyps removed for biopsy. There are systems being discussed to leave some polyps if there is no worry of neoplasia. *In vivo* analysis would be enormously reassuring.

Zhiwei Huang replied: Currently, the white-light endoscopy with biopsies followed by pathological examination on H&E-stained sectioned tissue slides remains the gold standard for cancer screening and diagnosis. However, random biopsies associated with sampling errors make the procedures labor-intensive, and difficult for screening high-risk patients with multiple suspicious lesions, particularly for early cancer and precancer with limited diagnostic accuracy. The diagnosis accuracy can also be compromised due to inter-pathologists' variations. Indeed, there is a profound clinical need to develop a rapid fiber-optic Raman diagnostic technique for objective, *in vivo* tissue diagnosis with biochemical/biomolecular specificity. As shown in our study, the advantages of using rapid fiber-optic FP/HW Raman spectroscopy developed for *in vivo* gastric tissue diagnosis are multifold. For instance, the fiber-optic Raman technique with biochemical specificity could be used to assist endoscopists in targeted tissue biopsies, instead of random biopsies, to significantly reduce unnecessary negative biopsy numbers, thereby reducing the pain and medical costs of patients. The patients with a higher risk of developing neoplasia could also be detected and diagnosed early by fiber-optic Raman spectroscopy for determining the lesions' margin with immediate treatment (*e.g.*, ESD) during the endoscopic examination.

James Livermore commented: It has been suggested in this discussion that the application of clinical spectroscopy should be limited to enhancing the pathological diagnosis of biopsy samples, and that vibrational spectroscopy does not have a place in pre-biopsy tissue analysis. It was also suggested that the spectroscopist is not going to stop the clinician from undertaking biopsies as this practice is too ingrained into the diagnostic workflow. In my field of neurosurgery I would argue that there is significant need for such an application. We routinely undertake needle biopsies of brain and spine lesions when they are not amenable to formal surgical resection. The ability to differentiate tumour subtypes without the need for taking the actual biopsy would decrease the risk of haemorrhage associated with the removal of biopsy tissue.[1] This is particularly the case when undertaking biopsies *via* an endoscope where significant bleeding from a biopsy site could result in complete loss of vision of the surgical field with possible catastrophic results. Furthermore, the biopsy of necrotic tissue often results in a non-diagnostic sample which may require re-biopsy, subjecting the patient to further surgery. Raman spectroscopy has been shown to be able to differentiate necrosis from tumour and normal brain tissue,[2,3] and may, therefore, have a role in decreasing the risk of a negative biopsy if it were to be used to analyse tissue pre-biopsy. Possibly the most clinically important application for vibrational spectroscopy in neurosurgery is for the intra-operative assessment of tumour boundaries. If spectroscopy can be used intra-operatively to analyse brain tissue *in situ* and give the surgeon a method to demarcate tumour boundaries,[4] we can increase the possibility of complete surgical resection which has been shown to increase patient survival.

1 L. J. Livermore, R. Ma, S. Bojanic and E. A. C. Pereira, *Br. J. Neurosurg.*, 2014, **28**, 637–644.
2 J. Desroches, M. Jermyn, K. Mok, C. Lemieux-Leduc, J. Mercier, K. St-Arnaud, K. Urmey, M.-C. Guiot, E. Marple, K. Petrecca and F. Leblond, *Biomed. Opt. Express*, 2015, **6**, 2380–2397.
3 R. Kast, G. Auner, S. Yurgelevic, B. Boradbent, A. Raghunathan, L. M. Poisson, T. Mikkelsen, M. L. Rosenblum and S. N. Kalkanis, *J. Neurooncol.*, 2015, **125**, 287–295.
4 M. Jermyn, K. Mok, J. Mercier, J. Desroches, J. Pichette, K. St-Arnaud, L. Bernstein, M.-C. Guiot, K. Petrecca and F. Leblond, *Sci. Transl. Med.*, 2015, 7, 274ra19.

Hugh Barr replied: I agree it is very important to target biopsy if necessary since biopsy is destructive. If bleeding occurs then other areas of abnormality can be obscured. In addition we can reduce the damage to normal healthy tissue which need not be removed. Immediate diagnosis followed by therapy can facilitate immediate treatment and cure (see Fig. 3).[1,2]

1 O. J. Old, L. M. Fullwood, R. Scott, G. R. Lloyd, L. M. Almond, N. Shepherd, N. Stone, H. Barr and C. Kendall, *Anal. Methods*, 2014, **6**, 3901–3917.
2 L. M. Almond, J. Hutchings, G. Lloyd, M. Wadley, N. Shepherd, S. Sanders, J. Day, O. Stevens, N. Stone, C. Kendall and H. Barr., *Gastrointest. Endosc.*, 2014, **79**, 37–45.

Christopher Phillips answered: We can see that the brain may be a special case, and also, we appreciate the general "unmet need" for vibrational spectroscopies in defining intra-operative margins. That said though, it is still true that the majority of clinicians we interact with are adamant that the biopsy system is here to stay, at least for the forseeable future, and there is very little appetite for "automating" the decision making process. The consensus is that the factors being taken into account by a trained clinician are so subtle and sophisticated, and the decision so important, that no-one would trust it to a computer algorithm. Certain aspects might be welcomed, *e.g.*, automated measures of mitotic

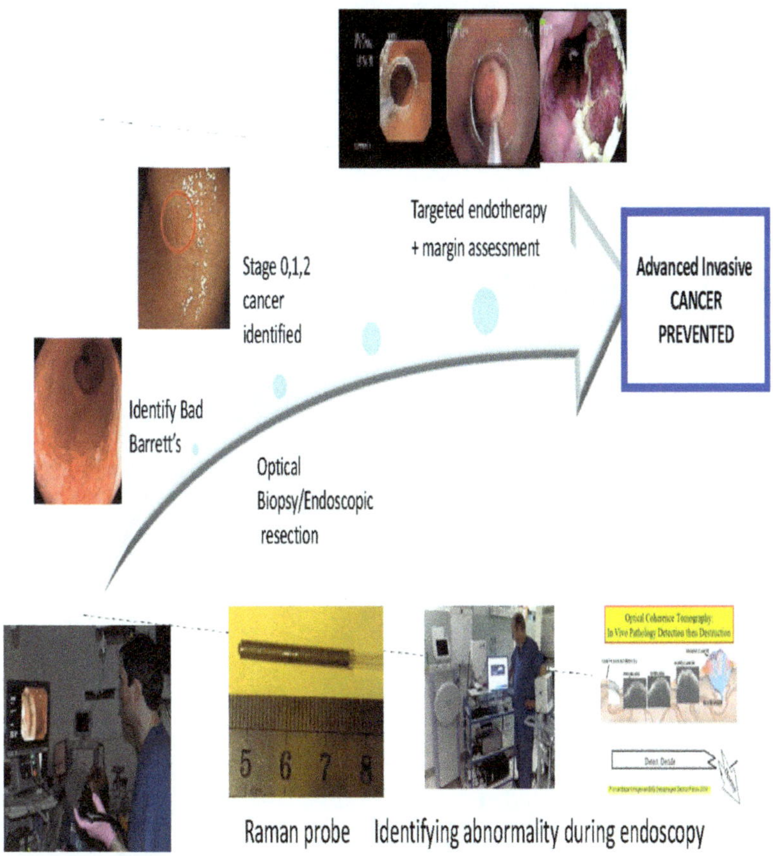

Fig. 3 Illustrative pathway for endoscopic real-time diagnosis and therapy.

activity and screening biopsies for tumour presence would be welcomed because they could free pathologists of tedious and time-consuming jobs.

Zhiwei Huang responded: We fully agree with the comments on the importance of *in vivo* clinical Raman applications in real-time medical diagnosis and tumor boundaries assessment. The rapid fiber-optic FP/HW Raman technique with depth-resolved ability developed in this work has great potential to be a clinically powerful tool for improving patient care by delivering accurate and earlier diagnosis and surgery, enabling real-time decision making to avoid unnecessary procedures, *etc.* We would like to explore the applications of FP/HW fiber-optic Raman spectroscopy for brain tumor staging as well as brain tumor demarcation in the near future.

Faraday Discussions

PAPER

From synthetic DNA to PCR product: detection of fungal infections using SERS†

Samuel Mabbott,[a] David Thompson,[a] Narayana Sirimuthu,[a] Graeme McNay,[b] Karen Faulds[a] and Duncan Graham*[a]

Received 15th November 2015, Accepted 24th December 2015

DOI: 10.1039/c5fd00167f

We report the use of silver hydroxylamine nanoparticles functionalised with single stranded monothiolated DNA for the detection of fungal infections. The four different species of fungi that were targeted were *Candida albicans*, *Candida glabrata*, *Candida krusei* and *Aspergillus fumigatus*. Rational design of synthetic targets and probes was carried out by carefully analysing the 2-D folding of the DNA and then by global alignment of the sequences to ensure specificity. The effects of varying the concentrations of the DNA and dye surrounding the nanoparticles on the resultant surface enhanced Raman scattering (SERS) signal were also investigated to ensure compatibility of the probes in a multiplexed environment. Using principal components analysis (PCA) it was possible to detect the individual presence of each target and group them accordingly. The move to detect the *C. krusei* single stranded PCR product (ssPCR) was significant to demonstrate that the methodology could be employed for the detection and diagnosis of invasive fungal infections (IFDs) within a clinical setting. Initially the PCR product was subjected to an alkali shock method in order to separate the strands ready for detection using the nanoparticle probes system. This time 18 base probes were employed to enhance hybridisation efficiency and dextran sulfate was found to have a vital role in ensuring that detection of the *C. krusei* target was achieved. This demonstrated the use of DNA functionalised silver nanoparticle for the detection of clinically relevant DNA relating to a specific fungal infection and offers significant promise for future diagnostic applications.

Introduction

Candida and *Aspergillus* fungi are ubiquitous within the environment yet are amongst the most well known pathogens responsible for causing fungal infections.[1] The pathogenic fungi from the *Aspergillus* genus is most often found

[a]*Centre for Molecular Nanometrology, Department of Pure and Applied Chemistry, University of Strathclyde, Technology and Innovation Building, 99 George Street, Glasgow, G1 1RD, UK. E-mail: duncan.graham@strath.ac.uk*

[b]*Renishaw Diagnostics Limited, Nova Business Park, 5 Robroyston Oval, Glasgow, G33 1AP, UK*

† Electronic supplementary information (ESI) available. See DOI: 10.1039/c5fd00167f

growing on decomposing organic material, whilst *Candida* is a genus of yeasts, hence its presence in humans being diagnosed as a yeast infection.

Both types of fungi present little risk to healthy individuals, however invasive fungal infections (IFDs) have been identified as important causes of morbidity and death in chronically immunosuppressed patients whose weakened immune systems are unable to fight the fungal infection.[2] Patients at high risk of infection from *Aspergillus* species include those with acute myelogenus leukemia or myelodysplastic syndrome during remission induction chemotherapy, patients undergoing allogenic haemotopoietic stem cell transplantation, recipients of solid organ transplants and those with other conditions of severe and prolonged immunosuppression.[3] *Candida* also poses a risk to the immunocompromised, but defining patients who are at high risk for invasive candidiasis (IC) is much more difficult. Patients at the highest risk from IC include those who have a central venous catheter, people being treated in intensive care units, those with kidney failure, people being treated with broad spectrum antibiotics, recipients of gastrointestinal surgery and diabetics.[4]

Candida infections account for 70% to 90% of all IFDs.[5] Invasive aspergillosis (IA) is less prevalent, however it is estimated to affect 5–13% of patients who undergo bone marrow transplants, 5–25% of solid organ transplant recipients and 10–20% of people with leukaemia undergoing intensive chemotherapy. Mortality rates are also particularly high in people with IA where rates can rise as high as 90% under certain circumstances.[6]

Diagnosis of both types of invasive infections is difficult mainly because the patients are already ill from other medical conditions. The most common symptoms are fever and chills, but it is often not until unsuccessful treatment with antibiotics that other treatments are pursued. Antifungal drugs for the treatment of the pathogens can generally be placed into three categories; polyenes,[7,8] azoles[7–9] and echinocandins.[8,10–12]

Some types of *Candida*, most prominently *C. glabrata* (studied here) have been shown to be increasingly resistant to antifungal treatments such as fluconazole (7% resistance) and echinocandins (1%). *Candida* is not alone in developing resistance,[13] *Aspergillus* has also been shown globally to be resistant to 3–6% of marketed azole drugs.[14] Alternative treatment options are available but these are uneconomical and have toxic side effects, which are undesirable for patients who are already ill. It is becoming increasingly evident that for effective treatment and reduced resistance discrimination of fungi at a species level is needed to ensure targeted treatment and to ensure that resistance to the existing drugs does not reach the current levels of antibiotic resistance in bacteria which is now a major global issue.

One of the most readily used methods of infection diagnosis alongside blood testing is culturing.[15,16] Culturing is the most common means of identifying fungi to allow selection of the most appropriate drug for treatment, however it is not without flaws. The main issue is that diagnosis can take anywhere from days to weeks and another issue is the susceptibility of the method to false positives. False diagnosis can be down to a number of factors including human collection error, specimen delay in reaching the laboratory and harmless colonisation due to the ubiquitous nature of the fungi.[17] Other non-invasive diagnostic tools involve searching for serum biomarkers such as galactomannan[18–20] and β-D-glucan,[20,21] but these tests are not species specific therefore neither is the treatment given to the patient.

Currently there is a lack of information on the true potential for dual infection as none of the current techniques look for both *Candida* and *Aspergillus* infections simultaneously (*e.g.* real-time PCR for *Aspergillus* or *Candida* species, Roche Septifast – *Candida*, Galactomannan – *Aspergillus*). However, recent examples of dual infection have been seen in clinical evaluations carried out by Renishaw Diagnostics Limited (unpublished). Given the low prevalence of disease, the true benefit of a higher order multiplex is confidently saying that someone doesn't have one of the disease targets so as to not give treatment unnecessarily (the negative predictive value). The other benefit is that when someone does have a disease target present, it is possible to diagnose and target the treatment appropriately, which is the reason we purposefully detect *C. glabrata*, *C. krusei* and *A. terreus* as these are treated with the use of different drugs. Diagnosis made on the basis of DNA extraction and PCR amplification can be made specific, targeting at a species level and when coupled with surface enhance Raman scattering (SERS) it offers excellent analytical sensitivity and multiplex capabilities as discussed in this paper.

SERS is a vibrational spectroscopy that uses plasmonic materials to enhance Raman scattering by several orders of magnitude yielding an ultra-sensitive and selective analytical technique.[22,23] In this paper we make use of silver nanoparticles and use a biologically specific event to exploit the fact that aggregated nanoparticles give rise to an increase in SERS. In our original demonstration of this we used DNA functionalised silver nanoparticles with a Raman reporter to detect a specific DNA sequence.[24] Follow up papers investigated the different geometries and a variety of Raman tags[25] but to date biologically relevant, PCR product DNA has not been investigated using this approach. In addition, the multiplexing capability of the system has not been pushed beyond two model sequences.[24]

Here, the detection of single stranded DNA (ssDNA) specific to the fungal infection through the aggregation of nanoparticle probes induced by sequence specific hybridisation is reported. We initially start with synthetic targets and demonstrate a multiplex of four before demonstrating the ability to detect a specific fungal DNA sequence from the PCR product.

Experimental

Synthesis and characterisation of silver nanoparticles (AgNPs)

Silver nanoparticles were synthesised according to a protocol developed by Leopold and Lendl.[26] Initially all glassware was soaked in fresh *aqua regia* to remove contaminants, the clean glassware was then washed with copious amounts of deionised water and left to dry before synthesising the nanoparticles. Hydroxylamine hydrochloride (1.67×10^{-3} M) and sodium hydroxide (3.33×10^{-3} M) were added to 450 mL of deionised water. This solution was then stirred vigorously whilst silver nitrate (50 mL, 1×10^{-2} M) was added immediately, turning the solution a dark brown colour. The sol was left to stir for a further 15 min until a colour change to golden brown signified the successful formation of AgNPs. A λ_{max} of 410 nm and NP concentration of ~480 pM (extinction coefficient = 2.87×10^{-10} M) were determined using spectrophotometry, and SEM images verified the spherical nature and size (53 ± 5 nm) of the AgNPs. Prior to DNA functionalisation, the concentration of the particles was diluted to 200 pM using deionised water and the SERS activity/signal reproducibility was assessed using a range of isothiocyanate dyes to ensure their validity over a three month time period.

Functionalisation of AgNPs with single-stranded monothiolated DNA and dye

All thiolated DNA was purchased from ATDbio (Southampton, UK), treated with DTT then HPLC purified to remove di-thiols and other potential contaminants prior to use. Whilst the site of thiol functionalisation could be differed depending on the hybridisation orientation being achieved (discussed in the Results section), the modified oligonucleotide always consisted of the same basic moieties; DNA sequence for specific detection, three hexaethylene glycol (HEG) units to give the oligonucleotides flexibility away from the NP surface and a thiol for tethering to the NPs. For fast attachment of the thiolated-DNA to the NP surface a pH-assisted methodology was employed.[27] Spectrophotometry measurements of the functionalised nanoparticles were carried out to elucidate their concentration. The average yield of the particles was around 72% (~144 pM). Once re-suspended in PBS (phosphate buffered saline solution), the dye was added. The amount added differed from dye to dye but isothiocyanates generally varied from 1000 to 5000 molecules per NP. After the dye addition the nanoparticles were left to equilibrate for 18 h, then centrifuged and re-suspended in 0.1 M PBS. Yields of the NP probes after the addition of dye were around 64% (~128 pM).

Hybridisation to synthetic DNA target

In a general hybridisation experiment, 10 pM of each probe conjugate (P1 and P2) was added to a plastic PCR tube along with 10 nM of the synthetic target (final concentration). The volume was then made up to 60 μL using 0.3 M PBS and left to hybridise for 30 min. 6 samples were created for each probe set; 3 consisted of a complementary synthetic target, whilst the other 3 contained a non-complementary synthetic target (control). Successful hybridisation of the nanoparticle probes to the target was signified by the loss of yellow colour (although this can often be difficult to observe). After 30 min, the solutions were transferred to a 96 well plate and a further 90 μL of 0.3 M PBS was added to each replicate.

SERS analysis

SERS analysis was carried out on a Ren-DX Raman plate reader (Renishaw Diagnostics Ltd) equipped with a 532 nm laser with a maximum laser power of 50 mW, and a 20× objective. Samples were deposited in a black 96 well plate for investigation. Unless stated, all sample volumes were 150 μL and spectra were collected using 5% laser power at the sample (~2.5 mW) and a 1 s accumulation. The system was calibrated using silicon (520 cm^{-1}) and the distance between the sample and objective (*z*-axis) was optimised using ethanol (150 μL) before sample investigation.

Results and discussion

The basic principle of the assay is based on the specific nanoparticle assembly of silver nanoparticles by DNA hybridisation. This changes the intensity of the SERS signal from a Raman reporter attached to the nanoparticle indicating which DNA sequence has been involved in the hybridisation. Multiple combinations of DNA functionalised nanoparticles can be added together and the exact sequence match determined by examining the change in intensity of the Raman reporter coding for that sequence. The basic components of this approach are shown in Fig. 1. The head-to-tail conformation was used for the 12 base probes because it enabled sufficiently quick hybridisation to the target DNA whilst maintaining an acceptable distance between the nanoparticles for sufficient SERS enhancement of the reporter (dye). Head-to-tail (B) is presumed to be a compromise between all three orientations available. Whilst tail-to-tail (A) hybridises the quickest to the target, the larger nanoparticle separation reduces the plasmonic coupling so the SERS signal tends to be lower than with the other orientations. Head-to-head (C) produces a large SERS response, yet the reduced distance between the nanoparticles causes steric hindrance reducing their ability to bind effectively to the target. When moving to longer overall DNA sequences steric influences become more prevalent therefore we investigated the use of both 12 and 18 base probes as well as examining the different geometries. The majority of the synthetic probes were examined in a head-to-tail arrangement (B) unless otherwise stated and tail-to-tail was used for the PCR product detection.

Multiplexed DNA sequence detection

Sequences for each of the species were acquired from the basic local alignment search tool (BLAST) and unique sections of the fungal DNA were selected for

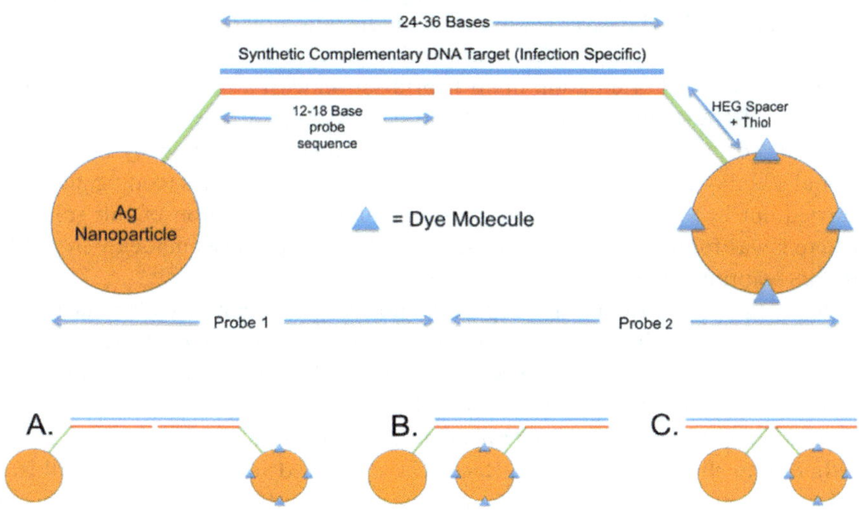

Fig. 1 The top scheme shows the hybridisation of nanoparticle probes to complementary single stranded DNA. The length of the probe sequence can be varied to increase discrimination and stability. The Raman reporter (dye) is added to the nanoparticle surface along with the DNA probe sequence to create a unique code for this particular DNA sequence. Schematics (A–C) represent the three different orientations that the probes can take when attaching to a complementary single strand of DNA due to the stereochemistry induced by the deoxyribose sugar on DNA. (A) Tail-to-tail, (B) head-to-tail, (C) head-to-head.

specific targeting. Careful selection of the DNA target and hence the probes used to functionalise the nanoparticles are of paramount importance for target specificity and to achieve multiplexed detection. However, choosing the DNA with which to modify the nanoparticles is not trivial. Single stranded DNA probes are not only capable of hybridising extremely well to their targets, but also to themselves or other probes within a multiplex, rather than the target if a degree of complementarity is present. Hybridisation of probes to themselves is termed homodimerisation whilst probes which hybridise to each other are called heterodimers. In an attempt to overcome this problem, 2-D folding models generated from amplified sequences were analysed at 25 °C and 0.3 M (basic hybridisation conditions) to identify regions that were not involved in the formation of loops or base pairing, thus allowing for more direct targeting using the probe sequences. Alongside the 2-D structure formation, global alignment of each of the species amplified regions was carried out to highlight unique bases for targeting. After carrying out a thorough assessment of the sequences and highlighting target regions, twelve base probes were developed. The main goal of this DNA assessment was to ensure that the probes used to detect synthetic DNA could be easily translated for the detection of the single stranded PCR product (ssPCR).

Nanoparticle functionalisation. Modification of the oligonucleotide sequences with a thiol (for attachment of the probe to the silver surface) and HEG groups (allowing extension of the DNA away from the nanoparticle) can take place on either the 5′ or 3′ ends, but changing the modification site changes the conformation with which the nanoparticle probes bind to the target. Fig. 1A–C highlights the different orientations of the probes that can be achieved, depending on the site of thiol modification. The 12 base probes were all modified at the 5′ termini therefore they hybridised to the target DNA in a head to tail conformation. The 18 base probes however feature thiol modification on opposite ends. Probe 1 (P1) has a thiol attached to the 3′ end whilst probe 2 is modified on the 5′ site, giving a tail to tail orientation upon hybridisation with a complementary single strand of DNA. Functionalisation of the hydroxylamine reduced nanoparticles was carried out using the same pH reduction method reported by Liu *et al.*[27] Initial stability testing of the DNA functionalised nanoparticles (conjugates) was carried out by subjecting the particles to a high concentration of salt solution where it was found that the DNA coating on the nanoparticles made them stable in PBS solutions up to 0.7 M.

The selection of the Raman reporters (normally resonant dyes for maximum intensity) is also an important factor in realising a SERS multiplex. The DNA alone does not have a sufficiently large Raman cross-section making it inappropriate to use as a Raman reporter for the identification of the target DNA. Fortunately, there are a range of fluorescent dyes commercially available that are capable of tethering to the nanoparticles surface. The dyes used in this work were all isothiocyanates (ITC) that attach to the nanoparticles *via* the sulphur. The dyes used for developing the multiplex were substituted rhodamine isothiocyanate (XRITC), tetramethylrhodamine isothiocyanate (TRITC), rhodamine B isothiocyanate (RBITC) and BODIPY-ITC (structures shown in Fig. S1†).

Each of the probe sets were tested separately by adding the complementary sequence to ensure that they showed sufficient discrimination between the presence of the complementary and non-complementary target (Fig. 2A).

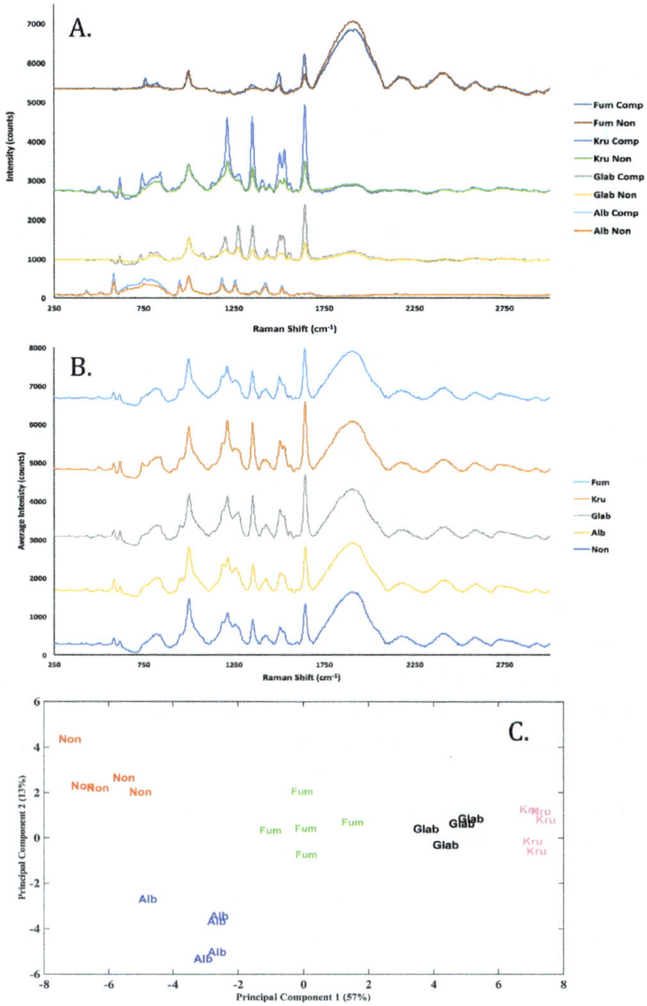

Fig. 2 SERS data from the nanoparticle probes with the target alone and then as a multiplex. Spectra have been baseline corrected and offset for clarity. The four probes *A. fumigatus*, *C. krusei*, *C. glabrata* and *C. albicans* have been abbreviated to Fum, Kru, Glab and Alb respectively in all plots whilst the nonsense is represented by Non. (A) Shows the discrimination achieved by each of the fungal probes when hybridised alone in the presence of complementary and then non-complementary single stranded synthetic DNA. (B) Shows the spectral signal arising from a multiplex, consisting of all the fungal probes together with the individual targets. Discrimination of these spectra is difficult by eye. (C) Discrimination of the multiplexed targets is possible when PCA is applied to the dataset.

All of the DNA probe sets used with the isothiocyanate dyes showed discrimination between the complementary and non-complementary single stranded DNA targets. The next step was to employ the conjugates in a multiplexed environment. To set up the multiplex 10 pM of each probe set was added to a PCR tube along with 10 nM of the fungal target. Each fungal target along with a non-complementary sequence was added in turn to the mixture of probes to see whether discrimination was possible. The average spectra from each probe set in response to the complementary target presence is displayed in Fig. 2B. Spectral changes caused by the presence of the different complementary fungal sequences are very difficult to differentiate by eye therefore chemometric methods had to be applied. Principal components analysis (PCA) was used to evaluate the data and showed that the probes could be effectively separated depending on which target was present in the multiplexed environment (Fig. 2C). Prior to evaluation using PCA, all data from 1700 cm^{-1} to 3000 cm^{-1} was removed as background, the remaining data were then baseline corrected and scaled. This is a significant result demonstrating that the DNA specific hybridisation of the appropriate probe set was capable of displaying a discernible change in the SERS in a background of the other combinations. However, this data were acquired using synthetic target sequences that were the same length as the combined probe sequences and to translate this into a clinically useful measurement, longer, biologically relevant sequences must be used. Therefore, the measurement of a fungal target sequence of DNA generated from a plasmid using PCR was attempted, however due to the target now being a much longer sequence (200–400 bases) this required significant further development and optimisation.

Progression towards PCR target discrimination

We have demonstrated that short, synthetic DNA targets can be identified, however this represents an idealised system, whereby the probes (12 bases each) form a perfect overlap with the 24 base target derived from the fungal BLAST sequence. In reality the amplified sequences are much longer, consisting of between ~200 and 400 bases. This presents challenges as the hybridisation of the probes to the longer sequences is much more difficult. The longer sequences are prone to forming 2-D folded structures when subjected to physiological buffer conditions at 25 °C. In order to demonstrate the diminished hybridisation of the 12 base probes (*C. krusei* with MGITC functionalised nanoparticles), longer synthetic sequences consisting of 48, 96 and 144 bases were used. The results shown in Fig. S2† demonstrate how the SERS signal decreases as the length of the target increases. The reduced 'off to on' ratio in comparison with the non-complementary target signal therefore presents an issue in the discrimination of the PCR product sequences.

In an attempt to address the reduction in signal an exclusion buffer (dextran sulfate) was employed.[28] The main action of the dextran sulfate has been reported to enhance hybridisation by denying volume to the DNA-NPs and targets, thereby minimising the effects of diffusion. The effects of the buffer addition on the SERS signal when using a 144 base target can be seen in Fig. S3.†

Increasing the w/v percentage of dextran sulfate present in the hybridisation buffer results in an increase in the SERS intensity for longer targets. Without dextran the longer targets show only a small increase in SERS intensity compared with the non-complementary control. The SERS signal increases until a plateau is reached at 3% final concentration. However, it should be noted that the addition of the exclusion buffer increases the background observed but it is not significant enough to affect the ability to discriminate between complementary and non-complementary targets.

18 Base probes and ssPCR detection. Initial attempts to detect the ssPCR product was carried out using 12 base probes, however this was not successful with a number of possible reasons postulated; the long length of the ssPCR target impedes hybridisation, the potentially low concentration of the ssPCR and that the alkali shock method employed did not produce viable DNA. By incubating the double stranded DNA in a NaOH solution at room temperature, the production of single stranded DNA was hoped to be quicker than subjecting the DNA to urea during multiple heat/cool cycles to denature the double strand. To potentially circumvent these problems longer oligonucleotide probes were examined. By extending the sequence conjugated to the nanoparticle from 12 to 18 this would address the length and concentration concerns. Using DNA-NPs specific for *C. krusei* the SERS response of the 18 base DNA-NPs were tested with an exact length complement synthetic target (36 bases) and the ssPCR target (Fig. 3). Fig. 3 shows that the 18 base targets gave a similar off-to-on SERS increase when using the synthetic complement, but no change in intensity was observed when using ssPCR (*C. krusei* = 380 bases). To further push the conjugates to hybridise to the ssPCR the concentration of dextran sulfate used was increased from 3% to 10% in the assay. This is close to the maximum concentration of dextran sulfate possible as the solution becomes saturated at 15–20%.

Fig. 4 shows the results when 18 base DNA-NPs were used in conjunction with a high concentration of dextran sulfate and the ssPCR. The SERS spectra show a clear increase in intensity when the complement (*C. krusei*) is used compared to a non-complementary target (*A. fumigatus*). This successfully demonstrated the potential application of the SERS methodology described here in the investigation of clinical samples based on PCR products.

After the successful detection of the ssPCR target the focus shifted to ascertaining the concentration dependence. After treatment of the double stranded DNA with NaOH ~125 μL of the ssPCR product is generated from 10 μL of 1000 input copy number dsPCR (double stranded PCR) product. This approximately

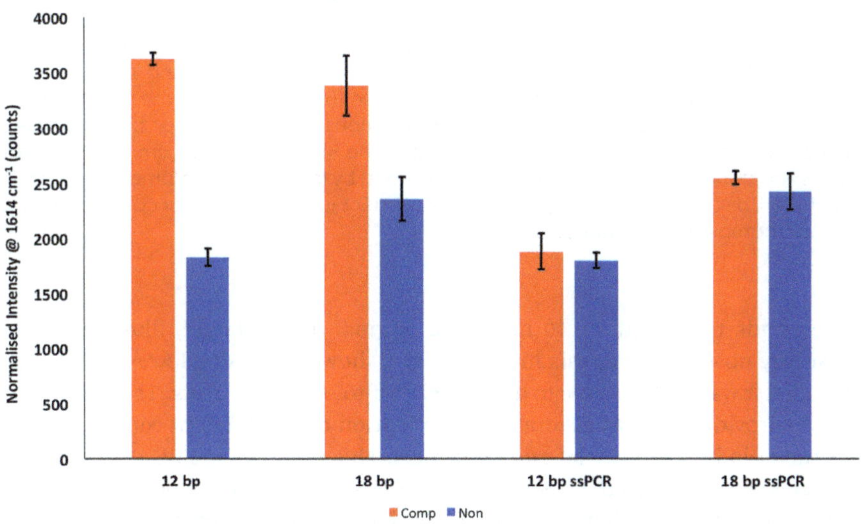

Fig. 3 The 12 and 18 base probes (12 bp and 18 bp respectively) show comparable performance in complementary and non-complementary discrimination when hybridised to exact complements (24 bases and 36 bases). However, the levels of discrimination between the complementary and non-complementary ssPCR product is much lower.

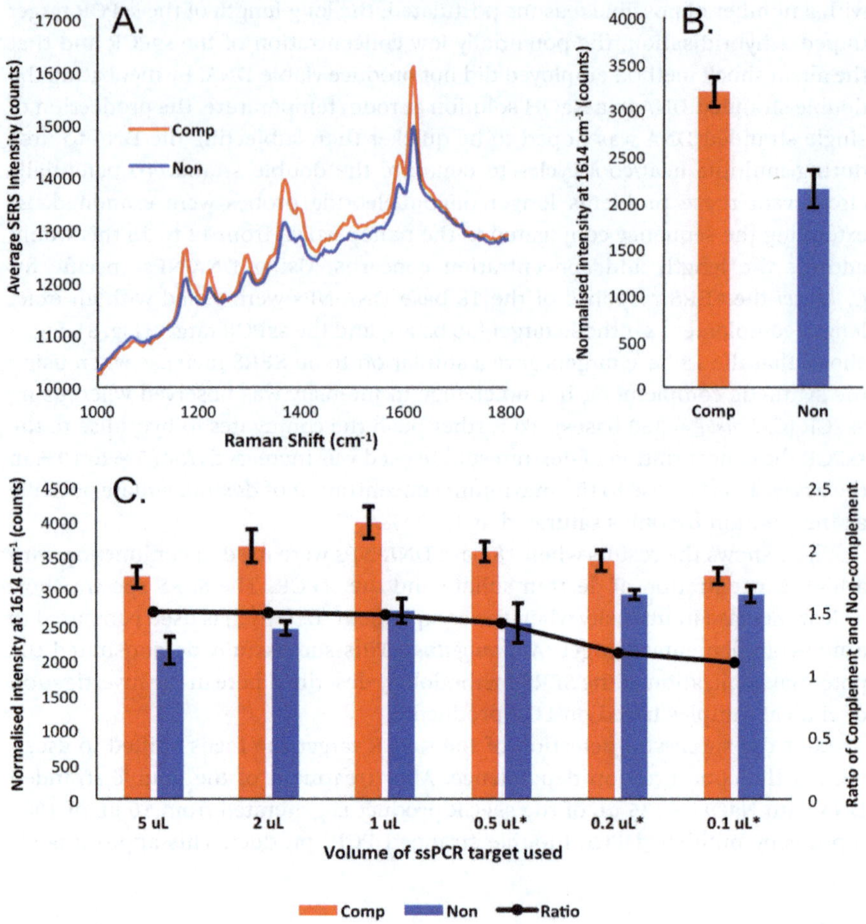

Fig. 4 This figure shows the discrimination of *C. krusei* using the ssPCR target. The plot in (A) shows the spectral discrimination between the complementary and non-complementary (*A. fumigatus*) targets, whilst the bar chart in (B) demonstrates the mean level of discrimination and accompanying standard deviations (4 replicates for both complementary and non-complementary targets) using the intensity at 1614 cm^{-1}. The mean signals and standard deviations in chart (C) show that by lowering the volume of the ssPCR target used, discrimination is still possible to around 0.1 μL (4 replicates for complementary and non-complementary targets).

corresponds to 0.04 μL or 40 nL of the original dsPCR target, illustrating the sensitivity possible using this method. Fig. 4 shows that using lower volumes of the target it was still possible to detect clear off-to-on SERS results. The parameter used to provide the most accurate discrimination of the different concentrations was the ratio between the off and on signals. This is shown in Fig. 4B and C and indicates that although the absolute intensities vary, the ratio of 'off to on' follows a concentration dependence. Therefore, we have clearly demonstrated that these nanoparticle conjugates can be used to detect a specific PCR product relating to a fungal infection and shows promise for future development into a multiplexed based measurement for multiple infections.

Conclusion

Successful detection of both synthetic and ssPCR targets has been demonstrated using DNA functionalised nanoparticles and SERS. Careful design of the probes at the initial stages was of paramount importance, not only to allow specific detection of the fungal targets but to ensure that unwanted hybridisations between the probes themselves was avoided. The rational design also allowed sequences to be easily extended for use in detection of longer single strands as seen with the ssPCR target. Steric issues with the longer PCR based target were overcome by using 18 base probes and dextran sulfate in the buffer system. The concentration dependence of the SERS signal was observed and this preliminary set of experiments indicates promise for this approach in combination with end point PCR measurement in a multiplexed manner. The work carried out here lays important foundations for a move towards specific detection of fungal species within a clinical setting using this nanoparticle assembly approach.

Acknowledgements

The authors acknowledge support from Renishaw Diagnostics Ltd and the TSB/EPSRC (EP/K502303/1) *via* the TSB call in nanosensors for healthcare. DG acknowledges support from the Royal Society in the form of a Wolfson Research Merit Award.

References

1 R. Saral, *Rev. Infect. Dis.*, 1991, **13**, 487–492.
2 B. Tutt, *OncoLog*, May 2013, Vol. 58, No. 5.
3 M. J. G. T. Ruping, J. J. Vehreschild and O. A. Cornely, *Drugs*, 2008, **68**, 1941–1962.
4 M. Pfaller and D. Diekema, *Clin Microbiol Rev*, 2007, **20**, 133–163.
5 J. Delaloye and T. Calandra, *Virulence*, 2014, **5**, 161–169.
6 A. A. Brakhage, *Curr. Drug Targets*, 2005, **6**, 875–886.
7 M. A. Ghannoum and L. B. Rice, *Clin. Microbiol. Rev.*, 1999, **12**, 501–517.
8 T. T. Liu, R. E. B. Lee, K. S. Barker, R. E. Lee, L. Wei, R. Homayouni and P. D. Rogers, *Antimicrob. Agents Chemother.*, 2005, **49**, 2226–2236.
9 D. J. Sheehan, C. A. Hitchcock and C. M. Sibley, *Clin. Microbiol. Rev.*, 1999, **12**, 40–79.
10 S. C. A. Chen, M. A. Slavin and T. C. Sorrell, *Drugs*, 2011, **71**, 11–41.
11 D. W. Denning, *Lancet*, 2003, **362**, 1142–1151.
12 D. R. Hospenthal, *Infect. Med.*, 2004, **21**, 476–478.
13 A. A. Cleveland, M. M. Farley, L. H. Harrison, B. Stein, R. Hollick, S. R. Lockhart, S. S. Magill, G. Derado, B. J. Park and T. M. Chiller, *Clin. Infect. Dis.*, 2012, **55**, 1352–1361.
14 M. C. Arendrup, *Clin. Microbiol. Infect.*, 2014, **20**, 42–48.
15 D. A. Stevens, V. L. Kan, M. A. Judson, V. A. Morrison, S. Dummer, D. W. Denning, J. E. Bennett, T. J. Walsh, T. F. Patterson and G. A. Pankey, *Clin. Infect. Dis.*, 2000, **30**, 696–709.

16 J. L. Vincent, J. Rello, J. Marshall, E. Silva, A. Anzueto, C. D. Martin, R. Moreno, J. Lipman, C. Gomersall, Y. Sakr, K. Reinhart and E. I. G. Investigators, *JAMA, J. Am. Med. Assoc.*, 2009, **302**, 2323–2329.
17 J. P. Latge, *Clin. Microbiol. Rev.*, 1999, **12**, 310–350.
18 K. A. Marr, S. A. Balajee, L. McLaughlin, M. Tabouret, C. Bentsen and T. J. Walsh, *J. Infect. Dis.*, 2004, **190**, 641–649.
19 M. A. S. H. Mennink-Kersten, J. P. Donnelly and P. E. Verweij, *Lancet Infect. Dis.*, 2004, **4**, 349–357.
20 W. W. Hope, T. J. Walsh and D. W. Denning, *Lancet Infect. Dis.*, 2005, **5**, 609–622.
21 J. W. Pickering, H. W. Sant, C. A. P. Bowles, W. L. Roberts and G. L. Woods, *J. Clin. Microbiol.*, 2005, **43**, 5957–5962.
22 K. Hering, D. Cialla, K. Ackermann, T. Dorfer, R. Moller, H. Schneidewind, R. Mattheis, W. Fritzsche, P. Rosch and J. Popp, *Anal. Bioanal. Chem.*, 2008, **390**, 113–124.
23 A. Campion and P. Kambhampati, *Chem. Soc. Rev.*, 1998, **27**, 241–250.
24 D. Graham, D. G. Thompson, W. E. Smith and K. Faulds, *Nat. Nanotechnol.*, 2008, **3**, 548–551.
25 F. McKenzie and D. Graham, *Chem. Commun.*, 2009, 5757–5759.
26 N. Leopold and B. Lendl, *J. Phys. Chem. B*, 2003, **107**, 5723–5727.
27 X. Zhang, M. R. Servos and J. W. Liu, *J. Am. Chem. Soc.*, 2012, **134**, 7266–7269.
28 G. M. Wahl, M. Stern and G. R. Stark, *Proc. Natl. Acad. Sci. U. S. A.*, 1979, **76**, 3683–3687.

Faraday Discussions

PAPER

Sheath flow SERS for chemical profiling in urine

Colleen M. Riordan, Kevin T. Jacobs, Pierre Negri and Zachary D. Schultz*

Received 6th November 2015, Accepted 11th January 2016
DOI: 10.1039/c5fd00155b

The molecular specificity and sensitivity of surface enhanced Raman scattering (SERS) makes it an attractive method for biomedical diagnostics. Here we present results demonstrating the utility and complications for SERS characterization in urine. The chemical fingerprint characteristics of Raman spectra suggest its use as a label free diagnostic; however, the complex composition of biological fluids presents a tremendous challenge. In particular, the limited number of surface sites and competing absorption tend to mask the presence of analytes in solution, particularly when the solution contains multiple analytes. To address these problems and characterize biological fluids we have demonstrated a sheath-flow interface for SERS detection. This sheath-flow SERS interface uses hydrodynamic focusing to confine analyte molecules eluting out of a column onto a planar SERS substrate where the molecules are detected by their intrinsic SERS signal. In this report we compare the direct detection of benzoylecgonine in urine using DSERS with chemical profiling by capillary zone electrophoresis and sheath-flow SERS detection. The SERS spectrum from the observed migration peaks can identify benzoylecgonine and other distinct spectra are also observed, suggesting improved chemical diagnostics in urine. With over 2000 reported compounds in urine, identification of each of the detected species is an enormous task. Nonetheless, these samples provide a benchmark to establish the potential clinical utility of sheath-flow SERS detection.

Introduction

Biomedical diagnostics increasingly rely on multiplex indicators to identify health and disease in patients. While measuring cholesterol provides one data point for monitoring cardiovascular disease, it is more common to include additional risk factors to assess a patient's condition. Progressively, systems biology approaches are being investigated to evaluate how changes in multiple indicators correlate to the overall changes in different systems.[1] The field of metabolomics, for example, uses differences in more than 40 000 known

Department of Chemistry and Biochemistry, University of Notre Dame, Notre Dame, IN, 46556, USA. E-mail: Schultz.41@nd.edu

metabolites to assess changes in biochemical pathways relevant to health and disease.[2] While a large number of metabolites have been reported, typical assays only identify about 10–20% of the known metabolites in a given sample. In a recent comprehensive study examining urine, which is reported to contain 2651 metabolites, only 445 metabolites were identified.[4] The challenge of metabolite identification is tremendous, as it was recently noted that less than 2% of the recorded mass spectra in metabolomics experiments can be attributed to a specific metabolite.[3] Improved diagnostics with increased identification power could have tremendous impact.

Currently, the technology available for metabolomics largely consists of nuclear magnetic resonance (NMR) and mass spectrometry (MS).[5–8] These methods take advantage of distinctly different chemical properties to identify metabolites. NMR offers universal identification arising from the deconvolution of NMR peaks associated with the nuclear spins of individual metabolites. The downside to NMR is that samples can require millimolar concentrations and milliliter volumes, making detection of many biological samples difficult.[6] The increased sensitivity of MS has made it an attractive detection technology, as it is capable of characterizing nanomolar concentrations and nanoliter volumes.[1] Even yoctomolar (10^{-24} M) metabolite detection, comparable to single molecule fluorescence, was demonstrated by nanostructure initiated MS.[9,10] While LC-MS has proven quite capable of detecting low concentrations of analyte, there are certain classes of molecules that present challenges, such as structural isomers and poorly ionizing molecules. Some of these molecules are identifiable using chemical ionization or complex sample derivatization and gas chromatography (GC).[1] Commercial metabolomics services utilize NMR and MS for assays. A study comparing the 3 commercial metabolite panels used to analyze blood detected a total of 482 metabolites.[11] The NMR based panel identified 24 metabolites (including 4 unique) and MS methods identified the rest. Though NMR spectra can be deconvoluted without using a separation technique, simplifying sample analysis, the sensitivity of MS generally provides increased identification. Orthogonal and chemically specific detection methods with sensitivity comparable to LC-MS would provide additional diagnostic information.

Raman spectroscopy, particularly surface enhanced Raman scattering (SERS), is an attractive option for biomedical diagnostics.[12] Similar to NMR, the signals observed in SERS arise from the structure of the molecule and are thus chemically specific. Also, very low concentrations of molecules can be detected by SERS at levels comparable to MS detection. The sensitivity and chemical specificity of SERS suggest an appealing approach for biochemical profiling. In fact, many classes of biologically relevant molecules that are difficult to discern by MS are readily discriminated by their Raman spectrum. Examples of SERS based metabolomics include the work that identified and quantified nicotine and two of its major metabolites,[13] that separated and identified pyrimidine and purine bases following liquid chromatography,[14] and that detected methamphetamines in urine using Au nanorods and a portable Raman detector.[15]

There are also a number of challenges associated with using SERS for biochemical profiling. Mass transport of analytes to the enhancing nanostructures is one challenge.[16] We recently demonstrated that, in solution, only molecules that at least temporarily adsorbed to the nanostructures were detected by SERS.[17] This result is consistent with the fact that most high sensitivity SERS assays either use nanoparticles in solution, where mixing is more effective, or deposition and evaporation onto a substrate, which confines the molecules to the surface. In both of these approaches, spot to spot variation in the nanostructures can alter the observed SERS signal, complicating identification. Because SERS requires analytes to interact with the nanostructures, the number of available surface sites can also limit detection. In these cases, molecules at significantly higher concentrations can inhibit the detection of low concentration species of interest. In many ways, this challenge is similar to MS detection and matrix ion suppression.

In SERS, the nature and organization of the nanostructures also affects the detected signal. It has been shown that the nature of the plasmonic hot spot can have a profound impact on the observed SERS signal.[18] Ideally, the same plasmonic construct is used for the detection of all molecules. This can be challenging with colloidal nanoparticles, but is more straightforward for planar SERS substrates. However, the transport issues discussed above have to be addressed, which has promoted the use of colloids thus far. Even in a microliter drop of solution on a substrate, it requires time for molecules to diffuse to the surface and the molecules can also diffuse away. In general, this transport problem has limited the utility of substrate based sensing assays.

Recently, our laboratory demonstrated the use of hydrodynamic focusing to improve transport and the efficiency of SERS detection in solution on planar substrates.[19] By confining analyte molecules near the SERS substrate, sheath-flow SERS shows more than 1000× improvement in the limit of detection. Additionally, the directed sample delivery minimized the amount of sample necessary to less than 0.1 microliters. An unexpected but useful aspect of the sheath-flow SERS detector is that low concentrations of analyte do not foul the surface, which enables repeated detection. The advantages realized with sheath-flow SERS make substrate based sensing competitive with colloidal assays. The sheath-flow SERS detector has been successfully coupled with capillary zone electrophoresis for the detection of amino acid[20] and peptide mixtures,[21] as well as dye molecules[22] and lipid vesicles.[23]

In this report, we examine the ability of sheath-flow SERS to identify metabolites in urine. The chemical complexity of urine introduces substantial matrix effects and represents an important test case for establishing the clinical relevance of sheath-flow SERS for metabolomics. Specifically, we examine the ability to identify benzoylecgonine (BZE), the primary metabolite of cocaine that can be detected in urine following drug use. We demonstrate two approaches, direct detection of BZE in urine as well as identifying BZE in a CZE separation of a urine sample. In both cases, BZE can be identified without the use of sophisticated chemometrics.

Experimental

Materials and reagents

Certified drug free urine was obtained from UTAK Laboratories. The urine was diluted in borate buffer at a pH of 9.4 to a 2 : 1 dilution of buffer to urine. Reference benzoylecgonine (BZE) dissolved at 1 mg mL^{-1} in methanol was obtained from Sigma Aldrich.

Sheath-flow SERS detector

A previously described sheath-flow SERS detector was used to focus the analyte to the surface.[19,22] Borate buffer was used as the sheath fluid for all studies. The continuous flow studies used a sheath flow of 150 μL min^{-1} and when coupling with CZE a sheath flow of 15 μL min^{-1} was used. These flow rates were previously determined to be most successful for the sample capillary flow rates used. In the continuous sample flow experiments, a pressure driven flow rate of approximately 4 μL min^{-1} was used. The electrokinetic flow rate in the CZE experiments was measured to be 0.2–0.5 μL min^{-1}. The capillary used had an inner diameter of 72 μm and an outer diameter of 143 μm. The capillary was 30 cm long for the steady state experiments and 50 cm long for the CZE experiments. The capillary and SERS substrate were cleaned by washing with 0.01 M NaOH, nanopure water, and borate buffer between studies. All SERS spectra were obtained using sheath-flow unless otherwise specified.

Raman measurements

The home-built Raman microscope,[23] equipped with a 660 nm diode laser and a cylindrical lens for line focus excitation, and the Ag SERS surfaces used have both been previously described.[24] The line focus is diffraction limited in width at approximately 1 μm, and the width is estimated to be approximately 60 μm,[23] similar to the inner diameter of the sample capillary. The excitation laser was focused onto the SERS flow cell using a 40× water-immersion objective. The excitation power measured at the output of the objective was 3–4 mW. Spectra were recorded in a kinetic series with acquisition times of 200–250 ms and a varying number of spectral acquisitions.

Capillary zone electrophoresis

A Spellman CZE1000R power supply provided a constant voltage of 15 kV to a Pt electrode in the injection block, where it sat in the solution of borate buffer along with one end of the sample capillary. A two second pressure injection (*ca.* 130 nL) was used to deliver the analyte into the capillary preceding the start of the CZE run. The system was grounded from the outlet port of the flow cell.

Analysis

Continuous sample flow experiments were analyzed using the DSERS method, as described by Scott and Carron, in order to separate the SERS signals from background Raman signals.[25] In DSERS, the fluctuations associated with SERS detection can be extracted from the steady-state background, recognizing that the SERS signals contribute more substantially to the standard deviation than does the invariant signal response. All data analysis was performed in MatLab.

Results and discussion

Fig. 1 shows the time-dependent response observed from the sheath-flow detection of urine spiked with 1 μg mL^{-1} of BZE continuously pumped over the SERS substrate. The observed SERS spectrum shows a high degree of continuity. Small

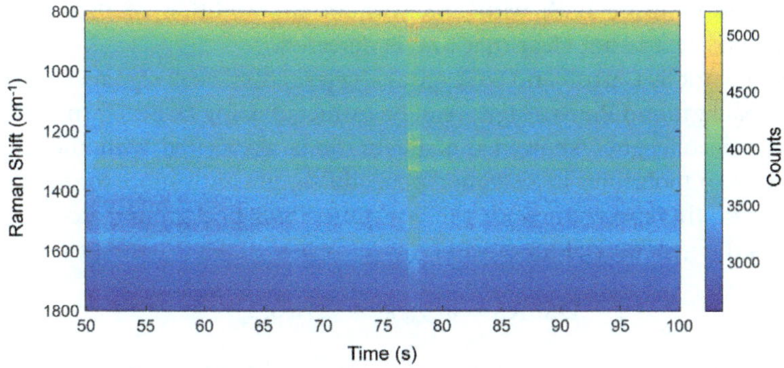

Fig. 1 Heatmap of the SERS signal of urine spiked with 1 μg mL^{-1} BZE. Under sheath-flow confinement, the sample was measured with 250 ms acquisitions over 100 s. The acquisitions between 50 and 100 s when the analyte eluted are shown. The signal is largely invariant over the 200 spectra plotted.

variations in intensity are observed, but no clear fluctuations in the Raman shift of the acquired signals are observed. Experiments using unspiked urine show the same overall features. Fig. 2A plots the average spectrum observed from 4000

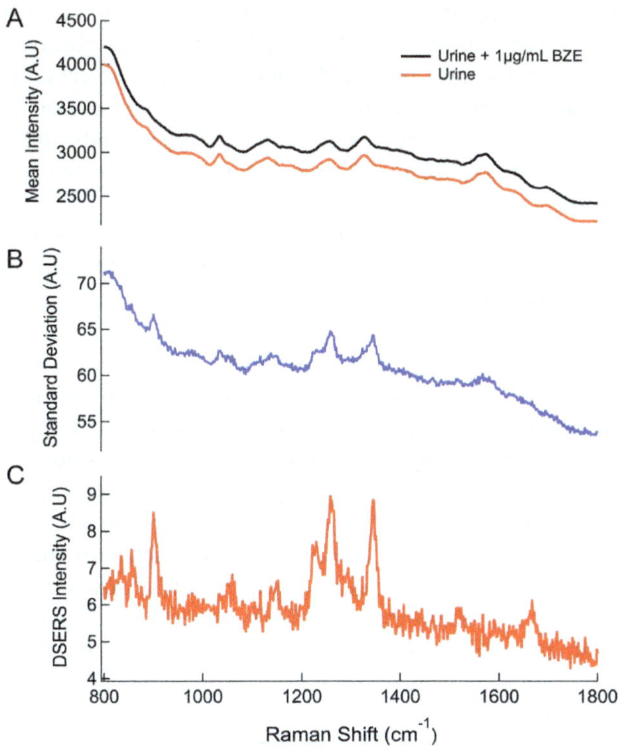

Fig. 2 The DSERS analysis of urine spiked with 1 μg mL^{-1} of BZE is shown. (A) The mean SERS spectra obtained from 200 separate acquisitions of urine (red) and urine spiked with BZE (black) are shown. (B) The standard deviation of the signal of urine spiked with BZE shows deviations attributable to BZE peaks more than the average urine background. (C) The DSERS spectrum obtained from (A) and (B) shows peaks characteristic of BZE.

acquisitions of urine both with and without the addition of BZE. From the observed signal it is not clear that BZE is detected.

Recent work by Carron and colleagues suggests that weak signals observed on a strong background Raman signal can be extracted using DSERS.[25] In their work, Carron and colleagues showed that fluctuations associated with the transient association of molecules in SERS hotspots add significantly more to the standard deviation of the spectrum than the constant SERS background does. Eqn (1) describes the general DSERS process:

$$\mathrm{DSERS} = \sigma_{\mathrm{total}} - A \times I_{\mathrm{average}} \tag{1}$$

where σ_{total} is the standard deviation of the spectra, A is a scaling coefficient and I_{average} is the mean intensity over multiple acquisitions. The difference in the standard deviation spectrum and the mean spectrum can be used to pull out SERS signals as weak as 1 part in 1000.

Fig. 2 uses the DSERS approach to examine the data plotted in Fig. 1. In Fig. 2A, it is difficult to discern the differences in the observed spectra in the absence and presence of BZE. The observed peaks are similar to those observed from uric acid in controlled SERS experiments. Uric acid is found at high concentrations in urine and is known to complicate the detection of trace species. This suggests that uric acid is preferentially adsorbed to the Ag substrate, preventing the BZE from contributing significantly. When the BZE spectra are averaged over long periods without the BZE on the surface, the BZE signal is difficult to observe. This is also true in the DSERS analysis. We previously showed a similar effect for the transient detection of lipid vesicles, where the interaction with the surface lasted on the order of 100 ms and signal acquisitions of several seconds did not provide detectable signals.[26] In Fig. 2C, we observe that a clear spectrum can be obtained from DSERS analysis.

Fig. 3A shows the reference spectrum from BZE on our SERS substrate. In this experiment, the BZE in methanol was deposited onto the SERS substrate and the methanol was allowed to evaporate. The BZE reference spectrum was background subtracted, using a difference over sum model, where the background was measured by taking spectra from a portion of the SERS substrate where the analyte was not present. Fig. 3B shows the DSERS result from both the BZE spiked and unspiked urine. The DSERS spectrum of blank urine shows a number of small peaks attributable to any of the 2000+ components found in urine. Interestingly, the DSERS spectrum of the BZE spiked sample shows a spectrum with a high degree of similarity to the reference spectrum in Fig. 3A.

Fig. 3 clearly demonstrates that DSERS in combination with sheath-flow SERS can detect the cocaine metabolite in urine. This result is consistent with previous results showing that methamphetamines can be detected in urine using DSERS.[15] It is quite amazing that the spectrum is so clearly detected. This may correspond to the concentration (1 μg mL^{-1}) of BZE added; however, urine is known to naturally contain mg mL^{-1} concentrations of other compounds such as uric acid. The strong BZE signal may be associated with the carboxyl and amine functional groups, which would be expected to interact strongly with a positively charged silver surface. The role of sheath-flow in this experiment is not obvious, but it enables ready comparison of blank and analyte-containing urine on the same substrate. This may enable further quantitative analysis.

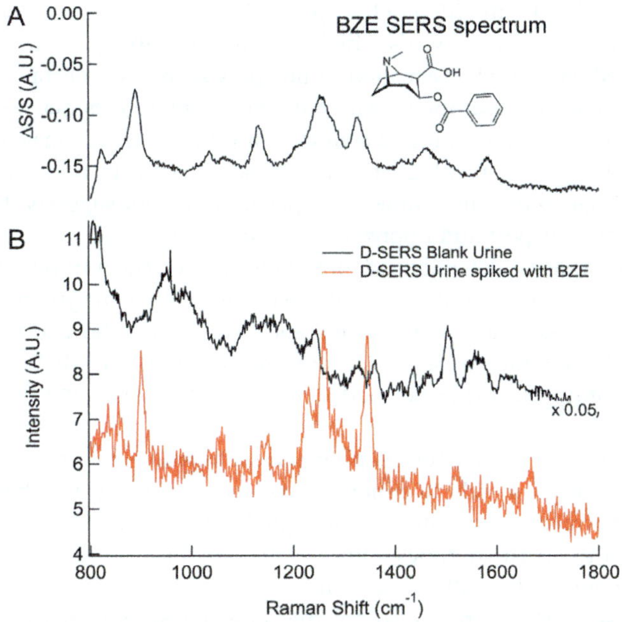

Fig. 3 (A) The reference spectrum of BZE deposited on our SERS substrate is shown. The spectrum was obtained by depositing a drop of BZE in methanol onto the surface and allowing the methanol to dry. (B) The DSERS results from blank and BZE spiked urine are shown. The spectra in (B) are offset for clarity.

Fig. 4 shows DSERS results obtained from 30 ng mL^{-1} of BZE spiked into a urine sample and sampled using the sheath-flow SERS flow cell. The peaks observed in Fig. 4A are near the limit of detection, but show agreement with the more concentrated sample in Fig. 3. Fig. 4B shows several peaks with increased intensity. The variation observed at low concentration may be attributed to the

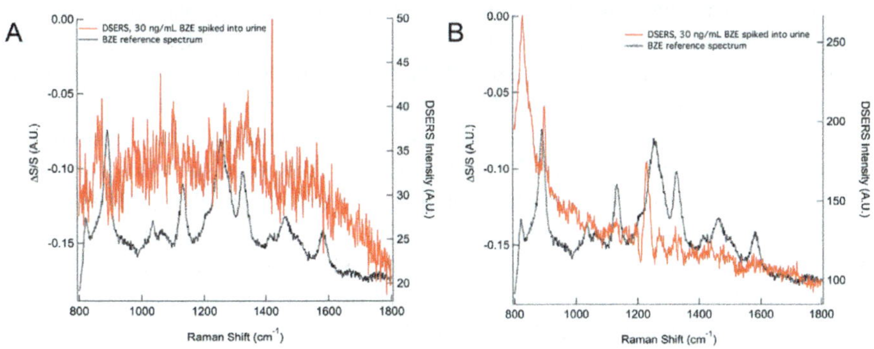

Fig. 4 The DSERS analysis of urine spiked with 30 ng mL^{-1} (100 nM) of BZE are shown in (A) and (B). The two panels represent replicate measurements. The intensity of the observed features varies, but the frequencies are consistent with the reference spectra shown in black. It should be noted that peaks below 1000 cm^{-1} Raman shift can be obscured by the presence of oxide on the Ag SERS substrate.

detection of molecules in different hot-spots, a known challenge in SERS analysis. The differences in intensity also highlight reproducibility issues known in quantitative SERS analysis. Detection in hotspots at low concentrations is often indicative of stochastic sampling, where individual molecules may exhibit spectral differences. Use of an internal standard or other approach may address the problem of quantification at higher concentrations. The detection of 30 ng mL^{-1} of BZE is similar to the environmental exposure level and suggests that sheath-flow facilitates transport and improves detection capabilities.

While the DSERS detection of BZE presents a straightforward, targeted metabolite diagnostic, it would be beneficial to detect other analytes for metabolic profiling that would enable pathway analysis. As discussed above, this is typically done using LC-MS methodology. To test the utility of sheath flow SERS for metabolic profiling in urine, we performed CZE-SERS on a BZE spiked urine sample.

Fig. 5A shows the mean detected photons as a function of migration time. The CZE run lasts approximately 30 minutes and consists of a large number of spikes, which each correlate to specific Raman signals. In Fig. 5B, the Raman spectra acquired at the specified times indicate that distinct compounds are being detected throughout the CZE separation.

In contrast to MS characterization, the observed SERS signal is difficult to attribute to specific molecules. Equally confounding, co-migrating analytes may exhibit a composite SERS signal provided that both species can equally adsorb to the SERS substrate. As discussed above, the analyte with the strongest interaction to the surface will be preferentially adsorbed and detected. This is consistent with our previous results published regarding a mixture of amino acids methionine, serine, and arginine, where the observed SERS spectrum was dominated by the signal from methionine.[23]

A significant advantage of sheath-flow SERS is that the detected compounds do not readily foul the surface. Thus, sequentially eluting compounds can be investigated at the same spot on the SERS substrate. Furthermore, it has been shown that, in some cases, closely migrating compounds can be discerned from each other. In a CZE separation of 8 peptides, the peaks attributed to each analyte were baseline resolved. Performing an equivalent separation with MS detection showed the peptides co-eluted.[21] The peak narrowing in CZE was attributed to Langmuir effects associated with SERS detection; however, it also suggests that narrow features detected in the urine SERS electropherogram may be associated with low concentration species. In our studies of amino acid and protein mixtures, a spot focus ($\sim$1 μm diameter) was used for detection. This limited the number of hot spots, or molecules, detected and likely corresponds to very low numbers (possibly 1) of molecules giving rise to the observed signal. The line focus used in the present work increases the number of molecules sampled, which should increase signal to noise, but also increases the possibility of different molecules being detected in each spectrum. Clearly good separations are still necessary for identifying complex mixtures.

To definitively assign the observed spectra to the compounds of interest, statistical algorithms are needed. Fig. 6 uses the Pearson correlation coefficient determined between every spectrum in the CZE-SERS trace against the reference spectrum for BZE. Pearson's correlation coefficient determines the degree of correlation between two spectra on a scale from −1 to 1. A coefficient of 1

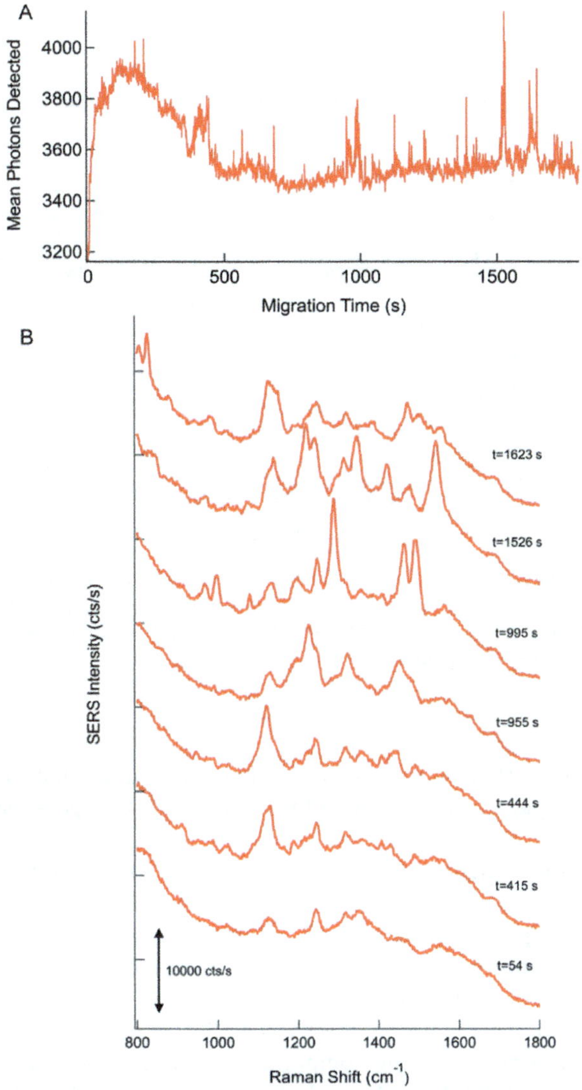

Fig. 5 (A) The CZE-SERS electropherogram is plotted by averaging the total number of photons detected during each 200 ms acquisition at each time in the CZE separation. (B) The SERS spectra acquired at each time point indicated are shown. The SERS spectra are the average of the 5 spectra at the indicated time ±0.5 s.

indicates a strong positive correlation, while a value of −1 means the signals anti-correlate. A score of 0 means there is no correlation. The background correlation appears to result for the background associated with the SERS substrate. The highest correlation in our experiment is observed at the migration time of 955 s. Fig. 6B shows the spectrum at $t = 955$ s with the reference BZE spectrum (Fig. 3A) superimposed. There is some variation in the peak at 1275 cm^{-1}, which we attribute to solvo-chromatic effects observed in the SERS reference spectra. There is a change in pH between the reference and the CZE run associated with the basic

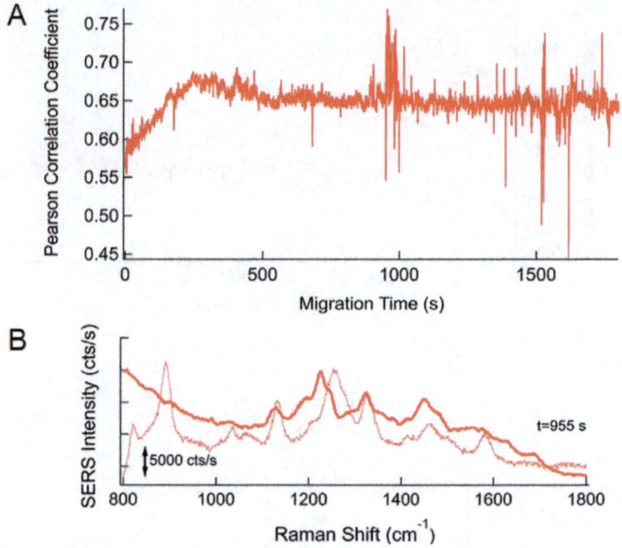

Fig. 6 (A) The Pearson correlation coefficient between each spectrum in the CZE-SERS run and the reference SERS spectrum of BZE is plotted. (B) The spectrum showing the highest correlation coefficient is plotted (solid) along with the reference spectrum of BZE (dotted), showing the strong agreement.

running buffer, or there may be co-migrating compounds that affect the spectrum. The strong agreement suggests this is the migration peak associated with BZE.

There is some evidence of uric acid throughout the separation, which may arise from adsorption in sequential runs. In the CZE-SERS experiments, it appeared that improved detection was achieved after the first run, suggesting some form of conditioning may be necessary for detection on the surface. This effect requires further study into the mechanism of sheath-flow SERS.

This approach of using library spectra is similar to methods used with GC-MS for the identification of volatile compounds based on fragmentation patterns. Still in its infancy, SERS has all the challenges of MS for identification, and as mentioned above only ~2% of MS spectra can be attributed to a specific metabolite. The importance of sheath-flow SERS detection is that different analytes are clearly observed during the CE-separation and that these analytes show agreement with the reference spectra of the compound.

The sheath-flow SERS detector has the potential for in-line use prior to, or concurrently with, MS detection. Again, the amount of sample needed for SERS is quite small. This may enable SERS elucidation of features detected by MS. One direct application would be the discrimination of structural isomers, which typically requires sample derivatization in MS analysis. The fact that SERS samples require no pretreatment and can also be analyzed by MS may enable improved coverage of the metabolome in future experiments.

One common challenge to SERS detection is the irreproducibility of the signals. Indeed, the SERS spectrum of BZE taken from aggregated colloids and our planar substrate show some spectral differences. This is typically attributed to the heterogeneous nature of the plasmonic environment associated with the different SERS assays. This presents a challenge to the development of spectral libraries and is something that must be taken into consideration going forward.

Conclusions

Sheath-flow SERS is capable of detecting and identifying specific metabolites in complex biological samples, such as urine. We have successfully detected BZE in urine at 30 ng mL^{-1}, which is relevant to clinical diagnostics. The importance of adsorption in the SERS process is both facilitated and remediated using this sheath-flow SERS detector. The sheath-flow promotes transport and adsorption to the surface for detection, but also promotes desorption and subsequent detection of other compounds in the CZE experiment. The development of library spectra will be important for the identification of multiple diverse analytes. Signals observed in CZE-SERS suggest that complementary characterization may enable a more complete coverage of the metabolome and will facilitate systems biology approaches to clinical diagnostics.

Acknowledgements

The work was supported by the National Science Foundation award DBI-1455445 and National Institutes of Health award R21 GM107893. The authors thank Neal Taub for helpful discussion and suggesting we examine urinalysis, as well as Keith Carron for discussions on DSERS.

References

1 C. H. Johnson and F. J. Gonzalez, *J. Cell. Physiol.*, 2012, **227**, 2975–2981.
2 D. S. Wishart, T. Jewison, A. C. Guo, M. Wilson, C. Knox, Y. Liu, Y. Djoumbou, R. Mandal, F. Aziat, E. Dong, S. Bouatra, I. Sinelnikov, D. Arndt, J. Xia, P. Liu, F. Yallou, T. Bjorndahl, R. Perez-Pineiro, R. Eisner, F. Allen, V. Neveu, R. Greiner and A. Scalbert, *Nucleic Acids Res.*, 2013, **41**, D801–D807.
3 R. R. da Silva, P. C. Dorrestein and R. A. Quinn, *Proc. Natl. Acad. Sci. U. S. A.*, 2015, **112**, 12549–12550.
4 S. Bouatra, F. Aziat, R. Mandal, A. C. Guo, M. R. Wilson, C. Knox, T. C. Bjorndahl, R. Krishnamurthy, F. Saleem, P. Liu, Z. T. Dame, J. Poelzer, J. Huynh, F. S. Yallou, N. Psychogios, E. Dong, R. Bogumil, C. Roehring and D. S. Wishart, *PLoS One*, 2013, **8**, e73076.
5 G. Theodoridis, H. G. Gika and I. D. Wilson, *TrAC, Trends Anal. Chem.*, 2008, **27**, 251–260.
6 O. Beckonert, H. C. Keun, T. M. D. Ebbels, J. Bundy, E. Holmes, J. C. Lindon and J. K. Nicholson, *Nat. Protoc.*, 2007, **2**, 2692–2703.
7 C. H. Johnson and F. J. Gonzalez, *J. Cell. Physiol.*, 2012, **227**, 2975–2981.
8 K. Suhre, *J. Endocrinol.*, 2014, **221**, R75–R85.
9 M. P. Greving, G. J. Patti and G. Siuzdak, *Anal. Chem.*, 2011, **83**, 2–7.
10 T. R. Northen, J.-C. Lee, L. Hoang, J. Raymond, D.-R. Hwang, S. M. Yannone, C.-H. Wong and G. Siuzdak, *Proc. Natl. Acad. Sci. U. S. A.*, 2008, **105**, 3678–3683.
11 K. Suhre, C. Meisinger, A. Doering, E. Altmaier, P. Belcredi, C. Gieger, D. Chang, M. V. Milburn, W. E. Gall, K. M. Weinberger, H.-W. Mewes,

M. H. de Angelis, H. E. Wichmann, F. Kronenberg, J. Adamski and T. Illig, *PLoS One*, 2010, **5**, e13953.
12 D. I. Ellis and R. Goodacre, *Analyst*, 2006, **131**, 875–885.
13 O. Alharbi, Y. Xu and R. Goodacre, *Analyst*, 2014, **139**, 4820–4827.
14 C. Carrillo-Carrión, S. Armenta, B. M. Simonet, M. Valcárcel and B. Lendl, *Anal. Chem.*, 2011, **83**, 9391–9398.
15 R. Dong, S. Weng, L. Yang and J. Liu, *Anal. Chem.*, 2015, **87**, 2937–2944.
16 I. M. White, S. H. Yazdi and W. W. Yu, *Microfluid. Nanofluid.*, 2012, **13**, 205–216.
17 S. M. Asiala and Z. D. Schultz, *Anal. Chem.*, 2014, **86**, 2625–2632.
18 J. Ye, F. F. Wen, H. Sobhani, J. B. Lassiter, P. Van Dorpe, P. Nordlander and N. J. Halas, *Nano Lett.*, 2012, **12**, 1660–1667.
19 P. Negri, K. T. Jacobs, O. O. Dada and Z. D. Schultz, *Anal. Chem.*, 2013, **85**, 10159–10166.
20 P. Negri and Z. D. Schultz, *Analyst*, 2014, **139**, 5989–5998.
21 P. Negri, S. A. Sarver, N. M. Schiavone, N. J. Dovichi and Z. D. Schultz, *Analyst*, 2015, **140**, 1516–1522.
22 P. Negri, R. J. Flaherty, O. O. Dada and Z. D. Schultz, *Chem. Commun.*, 2014, **50**, 2707–2710.
23 K. T. Jacobs and Z. D. Schultz, *Anal. Chem.*, 2015, **87**, 8090–8095.
24 S. M. Asiala and Z. D. Schultz, *Analyst*, 2011, **136**, 4472–4479.
25 B. L. Scott and K. T. Carron, *Anal. Chem.*, 2012, **84**, 8448–8451.
26 S. M. Asiala and Z. D. Schultz, *Chem. Commun.*, 2013, **49**, 4340–4342.

Faraday Discussions

PAPER

Toward SERS-based point-of-care approaches for therapeutic drug monitoring: the case of methotrexate†

Stefano Fornasaro,‡[a] Silvia Dalla Marta,‡[a] Marco Rabusin,[b] Alois Bonifacio*[a] and Valter Sergo*[a]

Received 16th November 2015, Accepted 8th January 2016

DOI: 10.1039/c5fd00173k

To date, in spite of their toxicity, the plasmatic concentration of most chemotherapeutic drugs is difficult to monitor in oncological patients, because their quantitative determination is expensive and time consuming. This contribution reports a first attempt for the direct quantitative determination of a chemotherapeutic drug in human serum samples by means of Surface Enhanced Raman Spectroscopy (SERS). In this study, SERS substrates constituted by Au nanoparticles deposited on paper by a simple dipping method have been used for rapid (few minutes) analysis of diluted human serum spiked with different concentrations of methotrexate, MTX. The drug concentrations were chosen in a range designed to cover typical therapeutic plasmatic values (from nanomolar to millimolar) in oncological patients, and the pertinent calibration was obtained by Partial Least-Squares Regression (PLSR). Stability selection was employed to evaluate the capability of the PLSR model to accurately predict and extract spectral variations correlated to MTX concentration. Such a quantitative determination is crucial for frequent, and hence adherent, therapeutic drug monitoring, TDM, of chemiotherapic drugs, given their heavy side effects. Its low cost, rapid response and the possibility of obtaining spectra with simple and compact instruments, make SERS particularly apt for implementing effective TDM. The promising results obtained in the analytical validation indicate which steps are to be taken on the way toward a clinical validation with real samples from oncological patients, for MTX as well as for other chemotherapeutic drugs.

Introduction

Therapeutic drug monitoring (TDM) is highly recommended in clinical settings to provide individualised patient treatment, optimising the

[a]*Department of Engineering and Architecture, University of Trieste, Trieste, Italy. E-mail: sergo@units.it; abonifacio@units.it*

[b]*Institute for Maternal and Child Health (I.R.C.C.S.) Burlo Garofolo, UO Pediatric Hemato-Oncology, Trieste, Italy*

† Electronic supplementary information (ESI) available. See DOI: 10.1039/c5fd00173k

‡ These authors contributed equally.

efficacy of drugs with a narrow therapeutic window while minimising side effects.[1]

Methotrexate (MTX, 2,4-diamine-*N*,10-methylpteroyl glutamic acid) is a folate antagonist included in anti-neoplastic and anti-rheumatic drugs. It is one of the most widely used anti-cancer agents, with indications and established protocols for a range of children's and adult cancers.[2]

Unlike other chemotherapeutic agents, MTX is used in a wide variety of doses. Intermediate- and lower-dose MTX regimens (*e.g.*, 20 mg m^{-2}) are used in maintenance chemotherapy and in the treatment of benign conditions such as psoriasis and rheumatoid arthritis,[3] while much higher dose regimens (*e.g.*, 1000 mg m^{-2} to 33 000 mg m^{-2}) *via* prolonged intravenous (IV) infusion are used for the treatment of some leukaemias and osteosarcomas.[4] The latter dose range is referred to as high-dose MTX (HDMTX).

The effective concentration range of MTX is limited to a relatively narrow therapeutic window, and dose adjustments during MTX therapy are routinely undertaken to address the high inter- and intra-patient variability in MTX pharmacokinetics.[5] Serum MTX concentrations can vary over 5 orders of magnitude (10 nM to 1 mM) from one patient to another using a single fixed dose and in the same patient during treatment.[4] Pharmacokinetic variability is generally greater in paediatric patients than in adults. For HDMTX therapy, it is usually desirable to reach initial serum concentrations between 10 and 100 μM, maintained for prolonged periods (12 to 36 hours), which must drop to less than 200 nM after 72 hours. The plasma MTX concentration at 48 hours after the start of HDMTX infusion should be <1 μM, as high-risk toxic, adverse effects are associated with concentrations >10 μM.[6]

After HDMTX, monitoring MTX serum levels is thus crucial to reverse its side effects, such as myelosuppression, nephrotoxicity, hepatotoxicity, and death.[6,7]

MTX concentrations are routinely monitored, and rescue therapy is regulated on the basis of these measurements.[8]

Several analytical methods have been reported for the TDM of MTX in biological fluids. Existing MTX assays include the Enzyme-Multiplied Immunoassay Technique (EMIT),[9] enzyme inhibition assays,[10] the fluorescence polarisation immunoassay (FPIA),[11] radioimmunoassay,[12] capillary zone electrophoresis,[13] and liquid chromatography coupled with tandem mass spectrometry (HPLC-MS/MS).[14–18] Very recently, an approach for indirect MTX determination, which is based on SPR and has an excellent performance, was presented by Zhao *et al.*[19]

HPLC-MS/MS is considered the reference standard technique for MTX analyses in biological fluids, though it is less commonly employed than immunoassays in clinical settings, since it is time consuming, expensive and requires specialised operators. Moreover, many of these routine methods lack precision, sensitivity, and a broad dynamic range; they still require protein precipitation or clean-up steps by passage through a pre-treatment column.

Because of its remarkable advantages such as rapidity, simplicity, and relatively low costs, recent data on buffered aqueous MTX solutions suggested that the combination of Surface-Enhanced Raman Scattering (SERS) spectroscopy with multivariate statistical analysis could be a promising analytical tool for therapeutic drug monitoring of MTX.[20,21] However, SERS approaches quantifying MTX directly in serum or other complex biofluids have not been reported yet.

In this work, we report promising preliminary results related to the application of SERS combined with chemometrics for rapid MTX quantification in a complex, real-life sample such as diluted human serum. To obtain a predictive model as stable and as realistic as possible, a stability selection approach is employed to evaluate the capability of the model to accurately predict and extract spectral variations correlated to MTX concentration.

Experimental

Chemicals and reagents

All chemicals and solvents were of analytical reagent grade; they were purchased from Sigma-Aldrich (Milano, Italy) and used as received. Filter paper with 2 μm pore size (qualitative filter paper, 410) was purchased from VWR International (Milano, Italy). Phosphate buffered saline solution (PBS) (pH 7.4) was prepared by dissolving a PBS tablet (Sigma-Aldrich) in Milli-Q water (200 mL). Serum samples used for method construction and validation were obtained from a healthy volunteer. The volunteer signed written informed consent to participate to the study, according to the recommendations of the declaration of Helsinki and the Italian law DL n° 675 of the 31-12-1996. The experimental protocols and methods were carried out in accordance with the approved guidelines. All glassware used for gold nanoparticle (AuNP) preparation was carefully cleaned with aqua regia and thoroughly rinsed with MilliQ water. For all cleaning procedures and preparation of solutions, MilliQ water was used.

SERS substrates preparation and characterisation

In-house built, solid SERS substrates were used in this study. Citrate-reduced gold nanoparticles (NPs) were synthesised according to a protocol described by Turkevich *et al.*, with slight modifications.[22] 10.6 mg of $NaAuCl_4$ were dissolved in 25 mL of Milli-Q water and heated to boiling. 750 μL of 1% sodium citrate was then rapidly added under vigorous magnetic stirring. The solution was kept boiling under stirring for 20 min. The AuNP colloidal solutions were stored in the dark at room temperature (RT) and were stable at least for eight months.[23] During that time, no aggregation was visible in the colloidal solutions, nor any changing of their UV-visible spectra.

The colloids were characterised by UV-visible absorption spectroscopy after each preparation using a Lambda 20 bio UV-Vis spectrometer (Perkin-Elmer, Monza, Italy). AuNPs feature a surface plasmon band at 540 ± 2 nm. The colloids were also characterised with a Transmission Electron Microscope (TEM, Philips EM 208). Both UV-Vis extinction spectra and TEM images of AuNPs for different preparations can be found in the ESI.†

The solid SERS substrates were obtained using the dip-coating method,[24–27] loading the NPs on the filter paper. A piece of 1 cm^2 was placed on the bottom of a cylindrical glass vial (total capacity of 10 mL) containing 3 mL of Au colloidal solution, to which sodium citrate was added up to a final concentration of 20 mM. The presence of the citrate leads to a colour change of the colloidal solution from red to grey-blue, indicating NP aggregation. The vials containing the filter paper were then stocked in the dark at room temperature for one week; after this time, all the NPs were deposited on bottom of the vial, covering the paper. The supernatant was extracted with a plastic syringe, paying attention not to touch or move the paper substrate. The substrates were then dried in the vials in air at room temperature and then stocked in Milli-Q water to avoid any loss of NP plasmonic activity. The repeatability of SERS measurements among different substrates has been checked using different analytes and the relative standard deviation over four different independent substrates was ≤15%.

Sample preparation and SERS measurements

A MTX stock standard solution was prepared by dissolving 5.0 mg of MTX hydrate (>98% HPLC, Sigma-Aldrich) powder in 0.2 mL of 0.1 M sodium hydroxide and diluting to 10 mL in a volumetric flask with PBS solution. The MTX stock standard solution was then diluted to 100 μM in PBS, aliquoted for single use and stored at −20 °C. Further dilutions were performed in 4% Bovine Serum Albumin (BSA)–PBS, a surrogate matrix frequently used to simulate plasma and serum, because its pH (7.4) and ionic strength (150 mM) are similar to those of the two biofluids. BSA was also added at a concentration of 40 g L^{-1} to simulate the protein content of serum. Such dilutions were performed at room temperature before analysis and in a dark room, to avoid MTX photodegradation. A set of eleven calibration standards was prepared in the concentration range of 0.1–20 μM for MTX. For the validation of the analytical method in a real-life matrix, calibration standards were prepared by spiking drug-free human serum with appropriate amounts of MTX stock solutions, keeping a constant 1 : 4, v/v ratio between serum and PBS. The final calibration range for MTX in diluted serum was 0.1–300 μM. The 5-fold dilution of serum standards upon spiking was chosen in analogy with the dilution procedures required for commercially available MTX immunoassays often used in clinical settings.

Small drops of 5 μL of both MTX solutions and spiked serum samples were deposited on the nanostructured paper substrates and allowed to dry (15 min).

Instrumentation

Normal Raman and SERS spectra were measured at room temperature (22 ± 0.5 °C) with an InVia Raman spectrometer (Renishaw plc, Wotton-under-Edge, UK) equipped with a Leica DMLM microscope (Leica Biosystems, Wetzlar, Germany) in a backscattering configuration. The excitation source was a high-power (360 mW) diode laser (Toptica Photonics AG, Germany) with an emission at 785 nm; the laser light was focused onto the sample *via* a ×10 objective lens (NA 0.25). The spectrograph was equipped with a 1200 lines per mm grating and a thermoelectrically cooled charge coupled device (CCD) camera. The calibration of the instrument was checked prior to each measurement using the 520 cm^{-1} band of a silicon reference sample.

To facilitate handling, each 8 mm × 8 mm paper substrate was adhered to a standard microscope slide (25 mm × 75 mm) immobilized onto the microscope stage.

SERS spectra were recorded over a wavenumber range of 300–1900 cm^{-1}, using the "extended" acquisition mode of the InVia Raman microscope. All spectra were recorded using an accumulation of three scans (10 s exposure each, for a total of 30 s) at 6 random locations on the substrate surface, with 18 mW of laser power delivered to the sample. Instrument settings were optimised to maximise signal and minimise saturation or sample degradation arising from laser excitation. In particular, the laser power density at the sample was decreased upon increasing the diameter of the laser spot using the "defocusing" option (90%) of the InVia Raman microscope. The system was controlled and the data were collected with WiRE 3.2 software (Renishaw).

Data pre-processing

Special attention was paid to spectral data pre-processing [*e.g.*, offset-Baseline Correction (BC), Area Normalisation (AN), and Multiplicative Scatter Correction (MSC)] for the purpose of outlier rejection, dimensionality reduction, and improvement in the robustness and accuracy of subsequent calibration models. First, the spectral region between 350 and 1700 cm^{-1} was selected for analysis. Then, the pre-processing workflow consisted of five steps: (i) removal of cosmic rays; (ii) BC; (iii) smoothing interpolation to increase the signal to noise ratio and to reduce the number of data points per spectrum (reduction of the dimensionality of the data); (iv) AN or MSC; and (v) outlier detection/removal by Principal Component Analysis (PCA).

All spectra pre-processing was performed within the R software environment for statistical computing and graphics.[28] In particular, data import and export, pre-processing, and visualisation were performed with the *hyperSpec* package[29] for R; for BC, a fourth-order polynomial baseline was fit automatically to the whole spectral range and subtracted from each spectrum of the dataset using the modpolyfit function from the package *baseline*;[30] peak picking was performed using detectPeaks from the *MALDIquant*[31] package. Anomalous spectra were automatically discarded by the pcout function from the package *mvoutlier*.[32] After the elimination of the outliers, the procedure was continually repeated until no more outliers were identified, in a self-consistent process. Eventually, a total of 6 spectra (over 99) were discarded as outliers (Fig. S4 and S5†).

Data analysis

All data analysis was performed within the R software environment for statistical computing and graphics[28] building on the *pls*[33] and *glmnet*[34] packages.

Partial least squares regression

For multivariate calibration, we employed Partial Least Squares Regression (PLSR) models to establish a relationship between the MTX levels and spectral data.[35] PLSR is a multivariate statistical method widely used in spectroscopy that aims at building a model that relates the variations of the spectral data (X matrix) to a series of relevant targets (Y matrix) according to the linear equation $Y = XB + E$, where B is a matrix of regression coefficients and E is a matrix of residuals. PLSR reduces the data to a small number of latent variables, maximising the covariance between the spectral data and the property to be modelled (in this case, MTX concentration).

The Repeated Double Cross-Validation (RDCV) strategy[36] was adopted for the optimisation of the complexity of PLSR models and to independently estimate the model performance for test set objects that have not been used in any step of model creation or optimisation. The optimum number of latent variables was selected by RDCV with 100 repetitions, applying the *one-standard-error-rule*.[37]

Model performance was judged on the basis of the Root Mean Square Error of Prediction (RMSEP).

Stability selection

Stability selection was exploited to assess the variability of the model coefficients, selecting those that remain consistently important after repeated subsampling of the data. Our settings were as follows: some uninformative variables were automatically eliminated by Elastic Net (EN) estimation ($\alpha = 0.5$) to subgroup the candidate variables and obtain a parsimonious model while retaining as much relevant information as possible. This procedure presents significant advantages for model interpretation. According to a strategy proposed for biomarker identification,[38] the perturbation of the data was obtained by leaving out 30% of the samples and considering only half of the variables at each iteration; the recursive PLSR calibration models were then built based on the strongly correlated variable obtained by the EN. Several PLSR models were calculated for the prediction of the known concentration of MTX from the spectral data, varying the pre-processing methods and EN setup.

Results and discussion

Comparison of SERS and Raman spectra of MTX samples

Fig. 1 shows the normal Raman spectrum of the MTX powder, along with the SERS spectra of 10 μM MTX in aqueous solutions and diluted human serum. For this preliminary study, we used diluted human serum instead of undiluted serum. This choice was made to limit the overall amount of biofluid needed, as well as because diluted samples were easier to work with than undiluted ones, which are rather viscous and tend to foam upon mixing. For these reasons, diluted serum is routinely used in analytical and clinical settings.[39] Moreover, we observed that dilution does not significantly influence SERS spectra as acquired using our substrates (Fig. S3A and B†).

The characteristic Raman bands of MTX were consistent with those previously reported for this drug.[20,40–42] Compared to the normal Raman, the SERS bands of the MTX solution in PBS had slightly different positions, relative intensities and widths, as is often observed for SERS spectra. Such differences are explained by invoking surface selection rules (for changes in relative intensities) and direct interaction of some molecule moieties with the metal surface (for shifts of band frequencies). Moreover, an underlying background due to traces of amorphous carbon in SERS spectra can be inferred from a generally increased intensity around 1360 and 1560 cm^{-1}, due to two broad bands (also called "cathedral bands") originating from the stretching modes of sp^2 carbons.[43–46] Such bands are often observed when employing solid SERS substrates, because of a minimal photodegradation as a consequence of laser illumination. However, amorphous carbon has a very high Raman cross-section, so that even small quantities will yield a detectable background. In our measurements, the presence of amorphous carbon is kept to a minimum by ensuring a low laser power density (see Methods), and thus it is not interfering with MTX detection. Also worth mentioning is the definite background signal given by the SERS substrates themselves, reported in light grey in Fig. 1 along with the SERS spectra. This background is largely due to

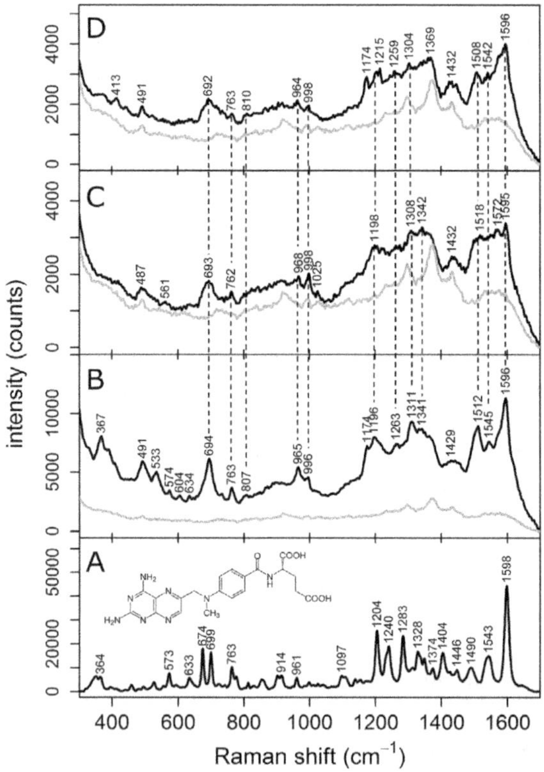

Fig. 1 (A) Normal Raman spectrum of MTX powder; (B) SERS spectrum of MTX 10 μM in PBS, (C) SERS spectrum of MTX in 4% BSA–PBS, and (D) SERS spectrum of MTX diluted in human serum (1 : 4, v/v). The background signal of the substrate (grey line) is reported for comparison together with SERS spectra. The chemical structure of MTX is reported for reference in (A). All spectra were collected using an excitation at 785 nm.

the SERS signal of citrate ions,[47] originally present as stabilizing adsorbates on the surface of colloidal Au NPs used to prepare the SERS substrates. Being constant, this substrate background can be assumed not to interfere with MTX detection.

Despite some differences, several bands in the SERS spectra were observed at Raman shifts very close to those of the normal Raman spectrum of MTX, so that in some cases a direct correlation between the Raman and SERS modes is straightforward. This is the case for the normal Raman bands at 699, 763, 961, 1204 and 1598 cm^{-1}, whose correspondence to their SERS counterparts is evident, for all three SERS spectra reported. A detailed and complete assignment of the SERS bands of MTX to vibrational modes, as well as a detailed description of the MTX–metal interaction, is out of the scope of this paper, and it would require carrying out an *ab initio* computational study at the level of Density Functional Theory (DFT). The reader interested in a tentative assignment of the SERS bands is referred to ref. 40–42. As this paper is concerned with MTX detection in diluted serum, it is noteworthy that the SERS bands appearing at 491, 694, 965, 996, 1512 and 1596 cm^{-1} in the spectrum of MTX in PBS solution (B), can be easily retrieved in both the SERS spectra of MTX in presence of albumin (C) and in diluted serum (D). In other words, MTX can be detected with SERS even in a complex medium such as diluted serum. It is interesting to note that the presence of albumin appears to affect the MTX SERS signal, both in terms of absolute intensity [(C) spectrum is less intense than (B)] as well as in terms of slight band shifts. This is not unexpected, since MTX is known to bind to albumin, so that only about 50% of the drug is found as a "free" molecule in serum, while the other 50% is tightly bound to the protein. Indeed, the absolute intensity of MTX in presence of albumin, as well as that of MTX in serum, is approximately 40% of the absolute intensity counts observed for the drug in PBS, *i.e.* without the protein. Considering that the enhancement due to SERS dramatically decreases with distance from the metal surface, so that only MTX molecules directly adsorbed onto the SERS substrate will significantly contribute to the signal, this decrease in SERS intensity can be readily explained assuming that albumin-bound MTX are prevented from adsorbing onto the metal, while free MTX must compete for the surface with albumins themselves.

In the case of human serum, the complexity and variability in the matrix composition pose an even harder challenge for calibration and analysis. Serum has several thousands of components, with more than 4000 metabolites,[48] some of which have a high affinity for gold surfaces, as proven by the intense SERS signal given by serum when using Au metal colloids.[49] To be observed, MTX must compete with all the serum constituents for adsorption. However, serum MTX concentration can be high, relative to most metabolites, and, while the presence of albumin appears to affect the overall intensity, MTX affinity for Au appears competitive enough to keep its SERS signal well detectable. This affinity can be partially explained by looking at its chemical structure: heterocyclic nitrogens, carboxylate groups as well as primary amine groups present in MTX are capable of a strong interaction with gold.[50] Such a strong MTX–Au interaction is responsible for the intense MTX SERS signal, and it makes the detection of this drug possible even in diluted serum.

Indeed, other authors have previously reported the quantification of aqueous solutions of MTX based on SERS spectra. Hidi *et al.* used the SERS band around 965 cm^{-1} for quantitative detection by relating the Lorentz fitted peak area directly to the MTX concentration in KOH solution.[20] However, because of the increased complexity and lower S/N ratio of our SERS spectra obtained in the presence of albumin or in diluted serum, the correlation of any individual band area with the MTX concentration was rather low and a univariate regression approach resulted in a poor prediction of the MTX content (Fig. S6†). Nonetheless, the spectral signature of MTX is specific, and many Raman shifts (to be considered as variables for chemometrics purposes) were correlated to the MTX concentration. These features made selective measurements of MTX possible by combining spectral information across a wide range of wavenumbers. As stated, in this study, we focused on the wide region between 350 and 1700 cm^{-1} to build a multivariate calibration model that incorporates much of the available spectral information.

Multivariate calibration

Assessment of the analytical performances. Multivariate calibration aims to establish a model that relates the variations of the spectral data to a series of known concentrations. PLSR modelling is particularly suited in multivariate calibration models for strongly correlated and multi-collinear spectral data, where the number of variables exceeds the sample size.[51] PLSR iteratively maximises the covariance between the spectral matrix and the reference data and produces highly interpretable Latent Variables (LVs) that describe diminishing quantities of spectral variance in the context of the reference data. The construction of an entirely accurate PLSR model required the careful selection of a number of latent variables (LVs). PLSR models were calculated using spectra with and without pre-processing with the application of different techniques. An overview of their influence on the model's ability to predict the MTX concentration from the SERS spectra is reported in Table 1. In general, the best models have as few components as possible and provide the lowest prediction error for the number of components used. Interestingly, pre-processing does not appear to dramatically affect model performance in terms of RMSEP.

Quantification of MTX levels in surrogate matrix

The EN-selected model shown in Fig. 2 provided a MTX concentration prediction of $R^2 = 0.9584$, RMSEP = 1.08. Only two LVs were needed for the model; this number is reasonable given the number of solution components and the spectral variance resulting from variations in the background spectra.

Stability selection. The EN-selected model for MTX in the surrogate matrix (Fig. 2A) achieved the lowest error using only 94 wavelengths and 2 LVs. It should be noted that the wavelength selection was not based on an explicit minimisation of prediction error, the criterion used in virtually all other variable selection methods. In fact, in cases with a low number of samples, this approach is usually not very reliable, increasing the risk of overfitting.

Stability selection, in contrast, using an approach inspired by random forests,[52] does not explicitly seek to minimise prediction error but only identifies

Table 1 Details of several models for the prediction of MTX levels in a surrogate matrix (4% BSA–PBS) and diluted human serum (HS), using different pre-processing methods. BC offset-baseline correction; AN area normalization; MSC multiplicative scatter correction; N number of wavelengths in the model; RMSEP units are in μM. LVs, latent variables in the double cross-validated models

		4% BSA–PBS ($n = 59$)				HS ($n = 93$)			
Pre-processing	Model	N	#LVs	R^2	RMSEP	N	#LVs	R^2	RMSEP
None	PLS	2701	4	0.939	1.49	2701	3	0.524	37.21
	EN-PLS	154	2	0.9474	1.35	138	2	0.5568	40.41
BC	PLS	2701	3	0.953	1.28	2701	2	0.5206	36.95
	EN-PLS	94	2	0.9584	1.08	145	2	0.6387	31.57
MSC	PLS	2701	3	0.9342	1.41	2701	1	0.5187	37.3
	EN-PLS	89	2	0.9428	1.39	110	1	0.6416	32.16
AN	PLS	2701	3	0.9475	1.25	2701	2	0.5277	36.74
	EN-PLS	591	2	0.957	1.15	1195	2	0.5492	34.93
BC + AN	PLS	2701	2	0.9575	1.18	2701	1	0.521	37.5
	EN-PLS	243	1	0.9613	1.19	597	1	0.6181	33.76
BC + MSC	PLS	2701	2	0.9596	1.11	2701	1	0.5187	37.38
	EN-PLS	88	1	0.9555	1.19	126	1	0.661	32.51

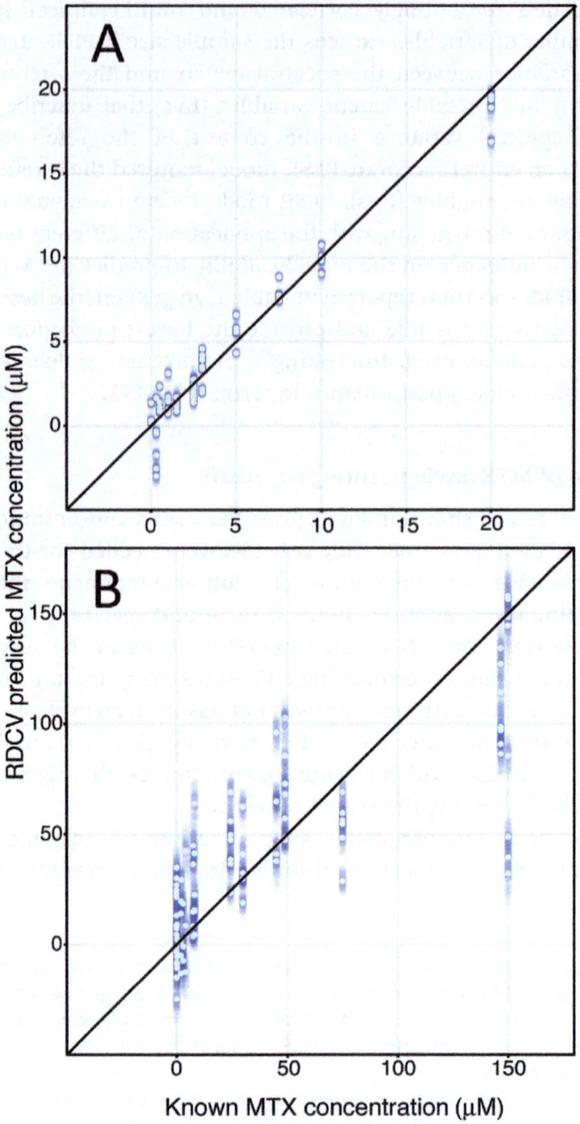

Fig. 2 Diagnostic plots from Repeated Double Cross Validation (RDCV) of SERS data collected in (A) 4% BSA–PBS, and (B) diluted human serum. The results from the 100 repetitions in RDCV are reported in blue (for 4% BSA–PBS they are very close together and therefore mostly hidden); the means of 93 predictions are reported in light blue.

the variables that are consistently important. The central idea is that real differences should be present consistently in the dataset and therefore should be found even under perturbation of the data by subsampling.[38,53] The EN approach is based on a relatively new version of penalised least squares. It employs a shrinkage with a penalty that is a compromise between the ridge regression penalty ($\alpha = 0$) and the Least Absolute Shrinkage and Selection Operator (LASSO) regression penalty ($\alpha = 1$).[54] Stability selection made the data more succinct and provided a good interpretation of the model, revealing an explicit relationship between the MTX levels and the spectral variables. As the Regression Coefficients (RCs) are descriptors of the spectral features that were used to build the model, we also aimed at assessing the accuracy with which the algorithm can faithfully use the known spectral profiles of MTX. In Fig. 3A, a direct comparison between the RC of the PLSR model and a MTX SERS spectrum is shown, highlighting the frequency with which different Raman shifts are chosen by the EN algorithm. This figure allows the data analyst to check that the variation of MTX levels, rather than the spectral variance of the matrix, plays a relevant role in the PLSR model construction. If so, both the EN-selected Raman shifts (coloured in red in Fig. 3) and the high values for the RC should coincide as much as possible with the SERS bands of MTX. This is clearly the case for the EN-PLS model for MTX in a surrogate matrix (Fig. 3A).

Quantification of MTX in human serum

The PLSR modelling process was repeated for the calibration of MTX concentration in diluted human serum, using the stability selection and pre-processing steps proposed in the assessment of the analytical performances section. Among blood sample types (whole blood, plasma, and serum), serum was selected in our study because (i) it retains most of the substances present in blood while not being subject to blood cell interference, (ii) it does not contain anticoagulants which might interfere with the spectroscopic signal due to the drug.[49] Actual measurements were performed on diluted serum samples, as detailed in the experimental section, for the reasons previously mentioned.

The EN-selected multivariate calibration achieved the lowest error, using only 145 wavelength and 2 LV, yielding a RMSEP of 31.57 μM, with an R^2 of 0.6387.

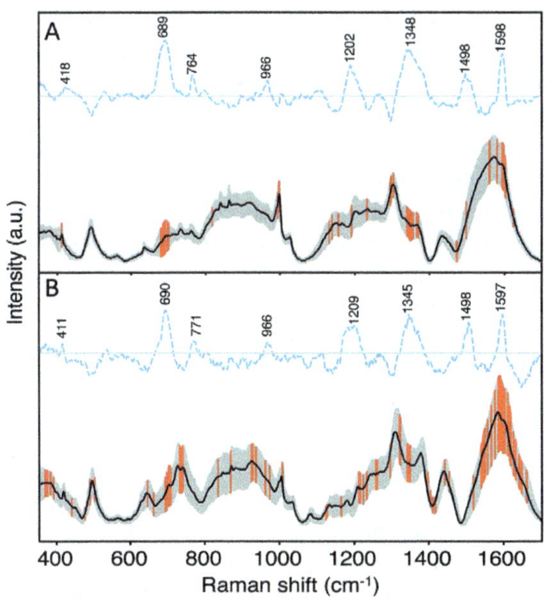

Fig. 3 Average spectra (black line) overlaid with EN selected. The blue dotted line shows a plot of the regression coefficient (RC) following PLSR. The RC has been offset for clarity.

Based on these values, it can be seen that the model produces a poorer prediction than the one built for the surrogate matrix. This is somewhat expected, considering the increased complexity of real serum with comparison to a buffered albumin solution. Considering that in this preliminary study we used serum samples from a single donor, inter-individual variations in serum composition can be ruled-out as a reason for the increased dispersion observed for data obtained from diluted serum with respect to those from buffered albumin solution. As serum itself does not yield intense SERS bands (see Fig. S3A and B†), intra-individual changes in composition, as well as the usual experimental errors involved in sample preparation and spiking, might be indirect sources of variation in the SERS data by modulating the interaction of MTX with the metal surface. Inspection of the MTX RC in Fig. 3B shows that the bands of the MTX SERS spectrum are faithfully reproduced and dominate the MTX RC. The correlation between EN-selected Raman shifts and MTX SERS bands is less straightforward. The different distribution of the EN-selected Raman shifts in the serum spectra is justified by the complexity of the biological matrix; the occurrence of so many components with the ensuing spectral complexity clearly affects the stability selection.[54]

Relevant implications

MTX requires careful diagnostic monitoring because of severe host side effects. Its concentration in plasma or serum is commonly screened with various techniques, and the dosage is regulated to the optimal therapeutic region by taking into account the patient characteristics.[5] Although there is an urgent demand for a faster and cheaper alternative to HPLC-MS/MS MTX assays, a sensitive and reliable SERS methodology for the TDM of MTX is still a challenge. To our knowledge, however, SERS-based MTX quantification, or even detection, in diluted or undiluted human serum has not yet been reported.

The methodology presented in this work clearly demonstrates that SERS in combination with multivariate data analysis has the capability for the quantitative detection of MTX in human serum. Furthermore, multivariate calibration in conjunction with stability selection, as demonstrated here, produces less complex and more accurate models because of the targeted inclusion of frequency bands specific to the analyte in question. As expected, in a comparison between the calibration and prediction errors listed in Table 1, 4% BSA–PBS spectra usually outperform the analogous models generated from diluted serum spectra. Scatter is noticeably greater, leading to a large standard deviation in response.

Quantification using SERS has been long recognized as a challenging task,[55] mainly because of poor repeatability and reproducibility, two flaws which were often reported for colloidal SERS substrates. The use of standard additions[56] or internal standards[57–59] has been proposed by some authors to tackle these problems, but it is evident that the availability of repeatable and reproducible inexpensive SERS substrates would be the main solution. The paper-based SERS substrates used in this study proved to be capable of quantification, as clearly shown in the case of MTX in a surrogate matrix.

When the method is applied to diluted serum measurements, the observed prediction error for MTX (31.78 $\mu g\ g^{-1}$) is still significantly higher than what is observed for established reference HPLC-MS/MS methods,[16] and thus an increase in the precision of the SERS method is still necessary. Considering the difficulty of achieving quantification in a complex matrix using SERS, however, the results reported in this study still represent a significant step forward.

Moreover, it is important to note that the SERS substrate used in this study was not specifically adapted to target MTX. A gold surface might be functionalised to promote a selective adsorption of the analyte of interest while maintaining the integrity and activity of the compound,[60] at the same time preventing the competition for the adsorption sites with the other serum components. Various strategies can be suggested to increase the affinity of the substrates for the drugs, including molecular recognition approaches such as MIP (molecularly imprinted polymers)[17] and artificial peptides. Such approaches are currently under investigation by our group, so that in the future the accuracy and precision of MTX quantification can be further improved. Furthermore, a functionalization of the SERS substrates would broaden the applicability of our SERS/EN-PLSR method, paving the way to its use with those chemotherapeutic drugs lacking the strong affinity for Au surfaces shown by MTX.

Conclusions

The present work clearly shows that, by using repeatable substrates and an adequate data analysis, the quantification of drugs by SERS in complex biological matrices, diluted serum in our case, is possible.

Also, our data provide an estimate of the difference in the quantification performance to be expected in real serum with respect to model solutions. The increased chemical complexity of serum with respect to model solutions, translates in a decrease of one order of magnitude in the precision (RSMEP) of the drug quantification.

Acknowledgements

SF acknowledges support from IRCCS Burlo Garofolo. This article is based upon work from COST Action BM1401 Raman-based applications for clinical diagnostics (Raman4Clinics), supported by COST (European Cooperation in Science and Technology).

References

1 E. Llorente Fernandez, L. Pares, I. Ajuria, F. Bandres, B. Castanyer, F. Campos, C. Farre, L. Pou, J. M. Queralto and J. To-Figueras, *Clin. Chem. Lab. Med.*, 2010, **48**, 437–446.

2 K. Barnhart, C. Coutifaris and M. Esposito, *Expert Opin. Pharmacother.*, 2001, **2**, 409–417.

3 J. Jolivet, K. H. Cowan, G. A. Curt, N. J. Clendeninn and B. A. Chabner, *N. Engl. J. Med.*, 1983, **309**, 1094–1104.

4 N. Graf, K. Winkler, M. Betlemovic, N. Fuchs and U. Bode, *J. Clin. Oncol.*, 1994, **12**, 1443–1451.

5 H. Seidel, A. Andersen, J. Terje Kvaløy, R. Nygaard, P. J. Moe, G. Jacobsen, B. Lindqvist and L. Slørdal, *Leuk. Res.*, 2000, **24**, 193–199.

6 Y. A. A. R. Ahmed and Y. Hasan, *J. Cancer Sci. Ther.*, 2013, **5**, 106–112.

7 B. C. Widemann and P. C. Adamson, *Oncologist*, 2006, **11**, 694–703.
8 A. Paci, G. Veal, C. Bardin, D. Leveque, N. Widmer, J. Beijnen, A. Astier and E. Chatelut, *Eur. J. Cancer*, 2014, **50**, 2010–2019.
9 M. P. Borgman, M. F. Hiemer, A. R. Molinelli, J. C. Ritchie and S. A. Jortani, *Ther. Drug Monit.*, 2012, **34**, 193–197.
10 B. C. Widemann, F. M. Balis and P. C. Adamson, *Clin. Chem.*, 1999, **45**, 223–228.
11 M. A. Pesce and S. H. Bodourian, *Ther. Drug Monit.*, 1986, **8**, 115–121.
12 J. J. Langone, *Methods Enzymol.*, 1982, **84**, 409–422.
13 C. Y. Kuo, H. L. Wu, H. S. Kou, S. S. Chiou, D. C. Wu and S. M. Wu, *J. Chromatogr. A*, 2003, **1014**, 93–101.
14 Y.-D. Li, Y. Li, N.-S. Liang, F. Yang and Z.-P. Kuang, *J. Chromatogr. B: Anal. Technol. Biomed. Life Sci.*, 2015, **1002**, 107–112.
15 R. C. Schofield, L. V. Ramanathan, K. Murata, M. Grace, M. Fleisher, M. S. Pessin and D. C. Carlow, *J. Chromatogr. B: Anal. Technol. Biomed. Life Sci.*, 2015, **1002**, 169–175.
16 E. Sonemoto, N. Kono, R. Ikeda, M. Wada, Y. Ueki and K. Nakashima, *Biomed. Chromatogr.*, 2012, **26**, 1297–1300.
17 S. Chen and Z. Zhang, *Spectrochim. Acta, Part A*, 2008, **70**, 36–41.
18 G. Rule, M. Chapple and J. Henion, *Anal. Chem.*, 2001, **73**, 439–443.
19 S. S. Zhao, N. Bukar, J. L. Toulouse, D. Pelechacz, R. Robitaille, J. N. Pelletier and J.-F. Masson, *Biosens. Bioelectron.*, 2015, **64**, 664–670.
20 I. J. Hidi, A. Mühlig, M. Jahn, F. Liebold, D. Cialla, K. Weber and J. Popp, *Anal. Methods*, 2014, **6**, 3943.
21 J. Yang, M. M.-C. Cheng, X. Tan and W.-C. Shih, *Biomed. Microdevices*, 2014, **16**, 673–679.
22 J. Turkevich, P. C. Stevenson and J. Hillier, *Discuss. Faraday Soc.*, 1951, **11**, 55–75.
23 J. Kimling, M. Maier, B. Okenve, V. Kotaidis, H. Ballot and A. Plech, *J. Phys. Chem. B*, 2006, **110**, 15700–15707.
24 L. Polavarapu and L. M. Liz-Marzan, *Phys. Chem. Chem. Phys.*, 2013, **15**, 5288–5300.
25 C. H. Lee, L. Tian and S. Singamaneni, *ACS Appl. Mater. Interfaces*, 2010, **2**, 3429–3435.
26 C. H. Lee, M. E. Hankus, L. Tian, P. M. Pellegrino and S. Singamaneni, *Anal. Chem.*, 2011, **83**, 8953–8958.
27 Y. H. Ngo, D. Li, G. P. Simon and G. Garnier, *Langmuir*, 2012, **28**, 8782–8790.
28 R Core Team, R: A language an environment for statistical computing, *R Foundation for Statistical Computing*, Vienna, Austria, 2016, https://www.R-project.org/
29 C. Beleites and V. Sergo, hyperSpec: a package to handle hyperspectral data sets in R, *R package version 0.98-20140523*, 2015, http://hyperspec.r-forge.r-project.org.
30 K. H. Liland and R. H. Mevik, baseline: Baseline Correction of Spectra, *R package version 1.2-1*, 2015, https://CRAN.R-project.org/package=baseline.
31 S. Gibb and K. Strimmer, *Bioinformatics*, 2012, **28**, 2270–2271.
32 P. Filzmoser and M. Gschwandtner, mvoutlier: Multivariate outlier detection based on robust methods, *R package version 2.0.6*, 2015, https://CRAN.R-project.org/package=mvoutlier.

33 B. H. Mevik and R. Wehrens, *J. Stat. Software*, 2007, **18**(2), 1–24.
34 J. Friedman and T. Hastie, *glmnet: Lasso and elastic-net regularized generalized linear models, R package*, 2009.
35 K. Varmuza and P. Filzmoser, *Introduction to multivariate statistical analysis in chemometrics*, 2009.
36 P. Filzmoser, B. Liebmann and K. Varmuza, *J. Chemom.*, 2009, **23**, 160–171.
37 T. Hastie, R. Tibshirani and J. H. Friedman, *The elements of statistical learning: data mining, inference, and prediction*, Springer, New York, NY, 2nd edn, 2013.
38 R. Wehrens, P. Franceschi, U. Vrhovsek and F. Mattivi, *Anal. Chim. Acta*, 2011, **705**, 15–23.
39 Q. A. Xu and T. L. Madden, *Analytical methods for therapeutic drug monitoring and toxicology*, Wiley-Blackwell, Oxford, 2011.
40 Y. Ozaki, R. W. King and P. R. Carey, *Biochemistry*, 1981, **20**, 3219–3225.
41 G. Seng, J. Bolard and L. Chinsky, *J. Raman Spectrosc.*, 1982, **13**(1), 100–102.
42 D. D. Saperstein, A. J. Rein and M. Poe, *J. Am. Chem. Soc.*, 1978, **100**, 4296–4300.
43 R. A. Alvarez-Puebla, *J. Phys. Chem. Lett.*, 2012, **3**, 857–866.
44 J. L. Castro, J. C. Otero and J. I. Marcos, *J. Raman Spectrosc.*, 1997, **28**, 765–769.
45 S. Sánchez-Cortés and J. V. García-Ramos, *J. Raman Spectrosc.*, 1998, **29**, 365–371.
46 J. S. Suh, M. Moskovits and J. Shakhesemampour, *J. Phys. Chem.*, 1993, **97**, 1678–1683.
47 C. H. Munro, W. E. Smith, M. Garner, J. Clarkson and P. C. White, *Langmuir*, 1995, **11**, 3712–3720.
48 N. Psychogios, D. D. Hau, J. Peng, A. C. Guo, R. Mandal, S. Bouatra, I. Sinelnikov, R. Krishnamurthy, R. Eisner, B. Gautam, N. Young, J. Xia, C. Knox, E. Dong, P. Huang, Z. Hollander, T. L. Pedersen, S. R. Smith, F. Bamforth, R. Greiner, B. McManus, J. W. Newman, T. Goodfriend and D. S. Wishart, *PLoS One*, 2011, **6**, e16957.
49 A. Bonifacio, S. D. Marta, R. Spizzo, S. Cervo, A. Steffan, A. Colombatti and V. Sergo, *Anal. Bioanal. Chem.*, 2014, **406**, 2355–2365.
50 L. Guerrini, Z. Jurasekova, C. Domingo, M. Pérez-Méndez, P. Leyton, M. Campos-Vallette, J. V. Garcia-Ramos and S. Sanchez-Cortes, *Plasmonics*, 2007, **2**, 147–156.
51 M. E. Keating, H. Nawaz, F. Bonnier and H. J. Byrne, *Analyst*, 2015, **140**, 2482–2492.
52 L. Breiman, *Machine learning*, 2001.
53 N. Meinshausen and P. Bühlmann, *J. Roy. Stat. Soc. B: Stat. Meth.*, 2010, **72**, 417–473.
54 G.-H. Fu, Q.-S. Xu, H.-D. Li, D.-S. Cao and Y.-Z. Liang, *Appl. Spectrosc.*, 2011, **65**, 402–408.
55 M. Sackmann and A. Materny, *J. Raman Spectrosc.*, 2006, **37**, 305–310.
56 E. Kammer, K. Olschewski, S. Stockel, P. Rosch, K. Weber, D. Cialla-May, T. Bocklitz and J. Popp, *Anal. Bioanal. Chem.*, 2015, **407**, 8925–8929.
57 W. Shen, X. Lin, C. Jiang, C. Li, H. Lin, J. Huang, S. Wang, G. Liu, X. Yan, Q. Zhong and B. Ren, *Angew. Chem.*, 2015, **54**, 7308–7312.
58 Y. Zhou, R. Ding, P. Joshi and P. Zhang, *Anal. Chim. Acta*, 2015, **874**, 49–53.
59 A. M. Fales and T. Vo-Dinh, *J. Mater. Chem. C*, 2015, **3**, 7319–7324.
60 R. A. Sperling and W. J. Parak, *Philos. Trans. R. Soc., A*, 2010, **368**, 1333–1383.

Faraday Discussions

PAPER

SERS as a tool for *in vitro* toxicology†

Kate M. Fisher,[a] Jennifer A. McLeish,[b] Lauren E. Jamieson,[a] Jing Jiang,[a] James R. Hopgood,[c] Stephen McLaughlin,[d] Ken Donaldson[b] and Colin J. Campbell*[a]

Received 9th December 2015, Accepted 22nd January 2016

DOI: 10.1039/c5fd00216h

Measuring markers of stress such as pH and redox potential are important when studying toxicology in *in vitro* models because they are markers of oxidative stress, apoptosis and viability. While surface enhanced Raman spectroscopy is ideally suited to the measurement of redox potential and pH in live cells, the time-intensive nature and perceived difficulty in signal analysis and interpretation can be a barrier to its broad uptake by the biological community. In this paper we detail the development of signal processing and analysis algorithms that allow SERS spectra to be automatically processed so that the output of the processing is a pH or redox potential value. By automating signal processing we were able to carry out a comparative evaluation of the toxicology of silver and zinc oxide nanoparticles and correlate our findings with qPCR analysis. The combination of these two analytical techniques sheds light on the differences in toxicology between these two materials from the perspective of oxidative stress.

Introduction

Redox potential is a function of the propensity of a chemical species to donate or accept electrons and the concentrations of the oxidised and reduced species. The Nernst equation allows the quantification of redox potentials under non-standard conditions:[1]

$$E = E^{\circ} + \frac{RT}{nF}\ln\frac{[\mathrm{Ox}]}{[\mathrm{Red}]}$$

where E is the redox potential (V), E° is the standard redox potential (V), R is the universal gas constant (J K^{-1} mol^{-1}), T is the temperature (K), n is the number of

[a]EaStCHEM, School of Chemistry, University of Edinburgh, EH9 3FJ, UK. E-mail: colin.campbell@ed.ac.uk

[b]MRC Centre for Inflammation Research, ELEGI Colt Laboratory, Queen's Medical Research Institute, University of Edinburgh, EH16 4TJ, UK

[c]Institute for Digital Communications, Joint Research Institute for Signal and Image Processing, School of Engineering, University of Edinburgh, EH9 3JL, UK

[d]School of Engineering and Physical Sciences, Heriot Watt University, Edinburgh, EH14 4AS, UK

† Electronic supplementary information (ESI) available. See DOI: 10.1039/c5fd00216h

electrons transferred, F is the Faraday constant (C mol^{-1}), [Ox] is the concentration of oxidised species, and [Red] is the concentration of reduced species.

Electron transfer drives energy transduction in biological cells. Electrons are transferred from reduced nicotinamide adenine dinucleotide (NADH) to molecular O_2 *via* a set of proteins that make up the electron transfer chain. As many functions of the cell are redox regulated (*e.g.* signalling; protein, DNA and RNA synthesis; and cell growth and death), the redox potential as defined by the Nernst equation clearly has biological significance as it both controls and reflects the biological activity of cells.[2–4]

The overall redox potential of a cell can be viewed as the balance between the generation of reactive oxygen/nitrogen (ROS/RNS) species and the antioxidants that degrade them. Most ROS/RNS generation is endogenous, as by-products of respiration, protein folding and NADPH oxidase activity.[5–7] Whilst some of these ROS/RNS species can cause harmful oxidation of biomolecules, some signalling pathways use these oxidised biomolecules to sense the redox status of the cell.[8–10] Antioxidant enzymes or small molecules transfer electrons from thiol/disulfide redox couples to reduce/eliminate ROS/RNS and oxidised biomolecules; these couples are oxidised in the process and then re-reduced by accepting electrons from other species.[11–13]

There are several thiol/disulfide redox couples in the cell. The main redox couple is glutathione/glutathione disulfide (GSH/GSSG) as it is mostly in the reduced form (GSH) and is the most abundant, with a concentration range of 1–11 mM.[2] The levels of GSH are therefore commonly used as a proxy for overall redox potential, however, it is important to note that there are many other redox couples in the cell that contribute to the overall intracellular redox potential. All of these redox couples are not necessarily in equilibrium with other each other, can vary independently of each other and are at different concentrations in different organelles.[4]

Dysregulation of redox potential occurs when antioxidant levels are overwhelmed by ROS/RNS levels, causing oxidative stress, which has been implicated in diseases such as chronic inflammation, cancer and neurodegeneration.[13–15] Oxidative stress can modify protein function through irreversible oxidative modification of protein residues, affecting cell function through inhibition of signalling pathways and resulting in apoptosis.

The gold standard for ratiometric redox potential measurement is the use of roGFPs, a green fluorescent protein modified with redox-active surface cysteine residues.[16] Each oxidation state of roGFP has a distinct excitation maximum, and a fluorescence measurement thus provides a ratio of oxidised to reduced species from which the redox potential can be calculated using the Nernst equation. However, roGFPs change oxidation state through interaction with glutaredoxins and this biases the measurement towards a measurement of GSH.[17] As stated above, overall redox potential is not a single analyte but is the result of many redox couples.

Our approach bypasses the problems involved in calculating overall redox potential from these individual redox couples, by using surface-enhanced Raman spectroscopy (SERS). We have designed redox-active reporter molecules based on quinones, which are active over a wider range of redox potentials than roGFPs.[18] These reporters change bond order depending on whether they are oxidised or reduced and these changes can be detected by Raman spectroscopy, thus allowing spectral discrimination between the oxidised and reduced forms of the reporter molecule. Covalent attachment of our reporter molecules to gold nanoshells amplifies the Raman signal by up to 10^8.[19] The use of lower energy infrared wavelengths as compared to visible wavelengths minimises phototoxicity. Furthermore cell components show autofluorescence at visible wavelengths but not at infrared wavelengths. We have shown that nanoshells coated with our reporter molecules are taken up into the cytoplasm and are non-toxic in various cell lines. We have used our nanosensors to show the correlation between redox potential and caspase activity during apoptosis.[18]

In this paper we detail the development of spectral processing and analysis algorithms that allow SERS spectra to be automatically processed to give an output of redox potential. In general we think that such automated analysis will allow greater translation of SERS to biological problems and in this particular case, the automated analysis has allowed us to carry out a comparative evaluation of the oxidative stress caused by silver and zinc oxide nanoparticles and correlate our findings with qPCR analysis.

Experimental

Functionalisation of nanoshells (NS)

The nanoshells have a peak absorbance at ~780 nm, with an absorbance of 1 corresponding to 2.2×10^9 NS per ml. Absorbances were measured with a Cary 50 UV-Vis spectrophotometer and spectra recorded from 200–900 nm.

NQ–NS. 1–2 mg of 1,8-diaza-4,5-dithia-1,8-di(2-chloro-[1,4]-naphthoquinone-3-yl)octane (NQ) was dissolved in 100% ethanol, then heated and filtered before 10-fold dilution with sterile filtered water (Sigma Aldrich). The solution was then added to 2.2×10^9 gold nanoshells (Nanospectra Biosciences Ltd) and incubated overnight at room temperature. The nanoshells (NQ–NS) were then washed 3 times with sterile filtered water and the absorbance measured.

MBA–NS. 1–2 mg of *para*-mercapto benzoic acid (MBA) was dissolved in 100% ethanol, filtered and diluted 10-fold with sterile filtered water (Sigma Aldrich). The solution was then added to 2.2×10^9 gold nanoshells (Nanospectra Biosciences Ltd) and incubated overnight at room temperature. The nanoshells (MBA–NS) were then washed 3 times with sterile filtered water and the absorbance measured.

Cell culture

A549 cells were grown as monolayers in Dulbecco's modified Eagle's medium (DMEM), supplemented with 10% heat-inactivated foetal bovine serum (FBS), 10 000 units per ml penicillin/streptomycin and 200 mM L-glutamine (all Life Technologies) (complete medium). NQ–NS incubation was carried out in serum-free medium, *i.e.* complete medium lacking FBS. Cells were grown in an incubator at 37 °C with a humidified 5% CO_2 atmosphere.

Surface-enhanced Raman spectroscopy

Calibration data. MBA: calibration data consisted of 3 independent datasets, each having 5 spectra per pH, making a total of 303 spectra, covering a range of

3.7 to 13.1 pH units. NQ: calibration data consisted of 5 spectra per redox potential, making a total of 40 spectra, covering a range of −460 to −250 mV.

Preparation of cells. A549 cells were plated at a density of 2×10^5 cells per dish on 35 mm diameter glass-bottomed imaging dishes (Greiner Bio-One). The following day the medium was replaced with a serum-free medium for 1.5 hours followed by overnight incubation with 200 fM NQ–NS. The cells were washed twice with PBS and 3 ml complete medium added. The cells were then either treated with 0.5 mg ml^{-1} silver nanoparticles (AgNP), 0.5 mg ml^{-1} zinc oxide nanoparticles (ZnONP) or 30 mM 2,2′-azobis(2-amidinopropane) dihydrochloride (AAPH; positive control) or were left untreated. The cells were incubated at 37 °C and 5% CO_2 until Raman spectroscopy was performed at 0, 1, 2, 3 and 4 hours after treatment.

Spectra acquisition. Raman spectroscopy was performed with a Renishaw InVia Reflex microRaman spectrometer with a 785 nm diode laser. Spectra were acquired between a Stokes Raman shift range of 1350–1800 cm^{-1}. Single spectra were acquired with a 50× Olympus super long working distance objective (NA = 0.45) to give a focal diameter of 2.1 mm using a point focus lens, and a power density at the sample of 66 mW mm^{-2}. The integration time was 30 s unless specified otherwise. SERS maps from cells were obtained with a 50× Olympus super long working distance objective (NA = 0.45) to give a line with dimensions 24.95 × 2.1 mm using a line focus lens. Raster scans were performed with a computer controlled *x*, *y*-stage, a step size of 5 mm, an integration time of 3 s and a power density at the sample of 330 mW mm^{-2}. The co-ordinates of signals from nanosensors were recorded and further Raster scans were performed centred on these co-ordinates with a step size of 1 mm, an integration time of 30 s and a power density at the sample of 66 mW mm^{-2}.

Data analysis

Individual spectra were processed as univariate signals with unknown parameters, the parameters of which were estimated using customised algorithms implemented in MATLAB® to determine the redox potential. Differences in redox potential between the treatment groups and the negative control were tested using ANOVA followed by the Holm-Šidák multiple comparison test. The MATLAB® code used is available upon request.

qPCR analysis

Total cellular RNA was isolated from A549 cells using the NucleoSpin RNA II kit (Machery-Nagel) according to the manufacturer's instructions. RNA concentrations were measured using the NanoDrop 2000c Spectrophotometer (Thermo Scientific). RNA was converted to cDNA using a high capacity cDNA Reverse Transcription Kit (Applied Biosystems). qRT-PCR was performed using SYBR i-Taq SYBR Green Supermix with ROX (Bio-rad), 0.5 μg cDNA and primers for either 18S (F: 5′-CATGGATTCAACGCAGAAG, R: 5′-GTAAAGTTGTGCGTCTCTGC) or HO-1 (F: 5′-CCAGCAACAAAGTGCAAG, R: 5′-CACATGGCATAAAGCCCT). Cycling parameters were 95 °C for 10 minutes, followed by 40 cycles of 95 °C for 15 seconds and 60 °C for 60 seconds. The specificity of the PCR product was confirmed through melting point analysis and the relative abundance of mRNA was calculated by a standard curve method, with 18S rRNA as the reference gene.

Results

SERS maps containing hundreds of spectra from single cells can be acquired in a short space of time (one map can be obtained in less than 5 minutes). The individual spectra require processing in order to determine redox potential, which is a time-consuming process if done manually. First, the user must select those spectra with a sufficiently intense SERS spectrum to warrant further attention. A smoothing algorithm is then applied to each chosen spectrum, an individual manually specified baseline is then subtracted and the peak heights obtained from the maximum value in the range of each peak. This manual processing takes approximately 3 minutes per spectrum and even though not every individual spectrum will require processing, a significant amount of time is required to be invested for a single map. In the development of sensors it is usual to perform manual processing of data, due to the small data sets generated. However, application of the sensors to biological data, where large data sets are common requires automation of data processing.

There are a number of drawbacks to manual processing in addition to the length of time required:

• Intra-individual bias and variance – because the baseline is manually specified, differences in its placement can occur due to human error, fatigue, mood *etc.* It is also possible that the baseline is subconsciously placed in order to achieve the expected results.

• Inter-individual bias and variance – different people will specify the baseline slightly differently which could lead to small but systematic differences in peak height ratios. Furthermore, each person may have differing criteria for deciding which spectra should be processed, leading to possible differences in the average redox potential determined through the exclusion or inclusion of spectra with a lower signal to noise ratio (SNR).

• The smoothing algorithm maintains peak height, thus also maintaining the noise, so the measurement of peak height is affected by noise. The presence of spikes of noise on one peak and not the other leads to over- or under-estimation of peak height ratios.

All of the above mean that while SERS is an attractive technique for making pH and redox measurements in live cells, the challenges of data analysis provide a barrier to its use for life-scientists unfamiliar with vibrational spectroscopic data. Automating the process of data analysis addresses these drawbacks in addition to being faster. A univariate peak fitting approach was used to ascertain peak height, which has the advantage of being less affected by noise than the manual method, as well as removing the bias and variance inherent in manual processing as the same criteria are applied to all spectra. Peak fitting also allows the investigation of other parameters such as peak width and area, which may also change with redox potential or pH, offering more options for ratiometric analysis. Therefore peak fitting can provide a more accurate method of estimating redox potential or pH from SERS spectra.

Peak fitting is a common method of analysis in spectroscopy and chromatography as the peaks have characteristic shapes that can be modelled by specific functions.[20,21] This allows a physical model to be tested, and data extracted from the model. The function parameters can provide information on the physical properties of the analyte under investigation. For example, in Raman, fluorescence and UV-visible spectroscopy the intensity of a peak is proportional to concentration.[22]

There are several examples of automated processing of Raman spectra in the literature, with most focusing on baseline subtraction to remove the large background signals from various sources, including autofluorescence from biological samples.[23–26] For many applications, SERS spectral analysis is performed in order to detect/classify components of a sample,[27] and accurate peak finding can be problematic when dealing with an unknown number of peaks, especially overlapping ones.[20] However, the peaks in our SERS spectra are due solely to our reporter molecule and, furthermore, we are not interested in fitting all of the peaks in a spectrum but only those needed for a ratiometric measurement. Therefore the parameters of the peak fitting were constrained to produce a more accurate physical model.

Our preliminary algorithm development was carried out using the pH-sensitive MBA reporter molecule as it has well-separated peaks with relatively flat background between them.

The pH-sensitive *para*-mercaptobenzoic acid (MBA; Fig. 1) molecule has been shown to be a good SERS-based reporter molecule for intracellular use.[28] Fig. 2 shows how peaks in the spectrum change with pH: the peak at ~1590 cm^{-1} is due to ring breathing;[29,30] it is the most intense peak in the spectrum in the region 1300–1800 cm^{-1} and is therefore a good candidate for a reference peak, being present at both high and low pH. The centre of this peak can be seen to shift to lower wavenumbers as pH increases. The peak at ~1400 cm^{-1} is barely distinguishable at low pH and increases as pH increases; it is due to COO^- stretching.[28–30] At pH > 8.5 this peak moves to higher wavenumbers (~1420 cm^{-1}) and becomes more intense. The peak at ~1700 cm^{-1} shows the opposite trend: it is highest at low pH and decreases with increasing pH; it is due to C=O stretching.[28,29] The opportunity therefore exists to use all three of these peak heights and areas in order to estimate pH, as well as the shift in the centre of the 1590 cm^{-1} peak.

Peak fitting can either operate on univariate data as in this paper, or on an ensemble of spectra which can be analysed using multivariate methods such as independent component analysis. The univariate case is simpler to deal with, and

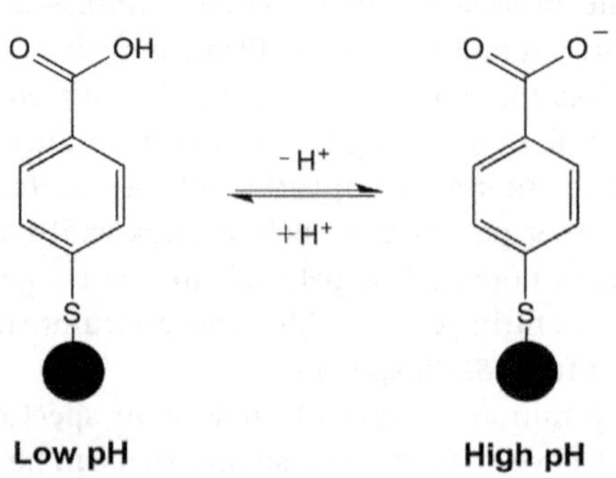

Fig. 1 Structure of the MBA reporter molecule conjugated to a gold nanoshell (black circle) *via* a gold–thiol bond. The protonated form (left) predominates at low pH, whereas the deprotonated form (right) predominates at high pH.

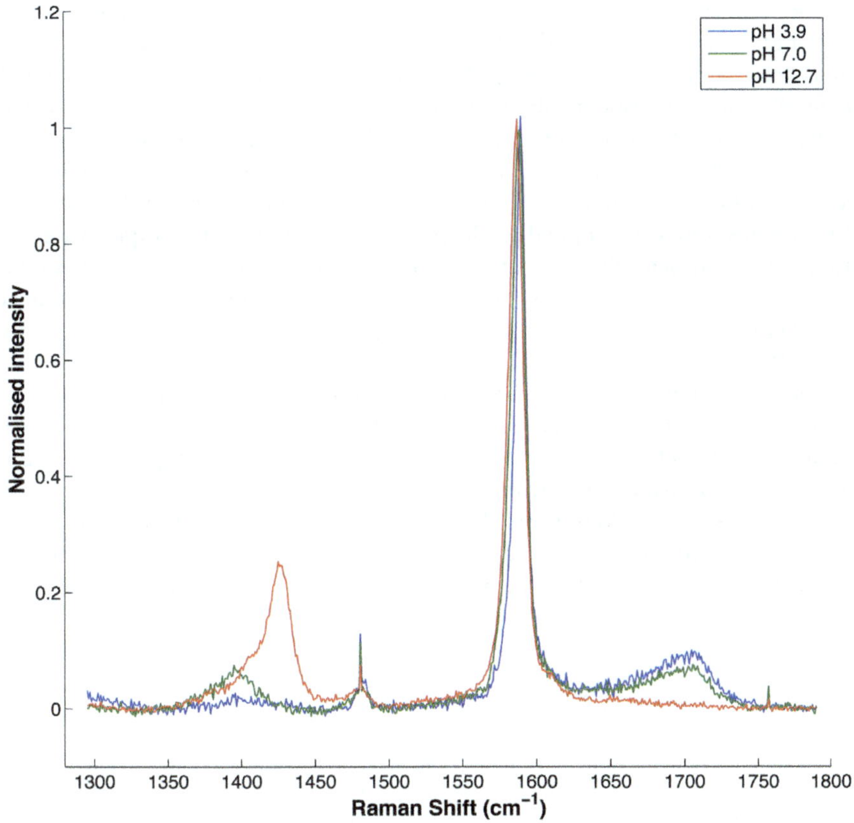

Fig. 2 Change in the position and intensity of three peaks in the MBA–NS Raman spectrum as a function of pH. Spectra have been background subtracted and normalised to the intensity of the peak at ~1590 cm^{-1}.

is an exercise in classical curve fitting using an error criterion, for example least squares. In the univariate model, the peak shape can be modelled using a variety of functions, including Gaussian, Voigt, pseudo-Voigt, and asymmetric variants.[21] However, the Lorentzian function is found to be the most parsimonious, requiring relatively few parameters, is simple to implement, and gives the least residual error.

Each of the three peaks in each spectrum is separately fitted around its peak centre using a Lorentzian peak model with an independent linear baseline:

$$y_n(x_c) = \frac{a_n}{1 + \left(\frac{x_c - b_n}{c_n}\right)^2} + m_n x_c + d_n$$

where $n \in \{1, \ldots, P\}$ is the index of the n-th underlying spectral peaks, $P = 3$ is the number of spectral peaks, x_c is the Raman shift associated to the n-th peak, $y_n(x_c)$ is the normalised intensity of the n-th peak, a_n is the (positive) amplitude of the n-th peak, b_n is the peak centre, c_n is the peak width, m is the linear baseline gradient and d is the baseline offset.

The peak centres b_n are known *a priori*, whereas the unknown parameters a_n, c_n, m and d can be determined by the standard nonlinear least squares algorithm, using for example the Levenberg–Marguardt algorithm.[31] This is implemented using the function "lsqcurvefit" in MATLAB. The Raman shift range over which each peak is fitted is determined experimentally as that which results in the least variation in the residuals over all the calibration spectra (excluding pH < 6 for the peak at ~1400 cm^{-1} and pH > 8.5 for the peak at around ~1700 cm^{-1}). The same three ranges were then applied to all spectra. Fig. 3 shows an example of a spectrum with each peak fitted separately. The peak width parameter, c_n, was constrained to be a minimum of 5 cm^{-1} and a maximum of 50 cm^{-1} (for the peak at ~1590 cm^{-1}) or 100 cm^{-1} (for the peaks at ~1400 and 1700 cm^{-1}) in case of very low SNR.

The calibration data were grouped by pH in order to investigate the relationships between various fit parameters and pH. Weighted means of the various parameters at each pH were calculated using inverse weighting to reduce the effect of values with a large error.

As expected, it was found that the centre of the peak at ~1590 cm^{-1} did vary with pH, the centre of the peak at ~1400 cm^{-1} shifted to higher wavenumbers at high pH, and the centre of the peak at ~1700 cm^{-1} did not change with pH. The

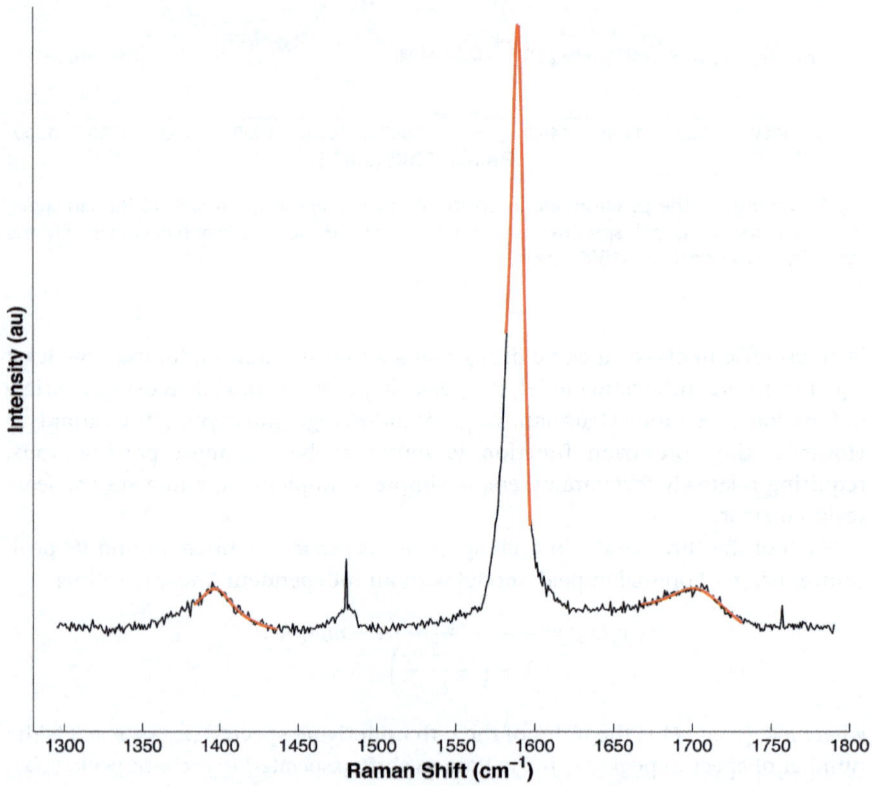

Fig. 3 Example of a calibration spectrum (black) with peak fits applied to three separate regions (red). The Raman shift range of each fit was determined experimentally as that which resulted in the least variation in the residuals over all calibration data (excluding pH < 6 for the peak at ~1400 cm^{-1} and pH > 8.5 for the peak at around ~1700 cm^{-1}).

following ratios of peak heights and areas were dependent on pH: 1590/1400, 1590/1700 and 1400/1700, indicating that a combination of seven parameters ($\sim$1590 cm^{-1} peak centre, three peak height ratios and three peak area ratios) could be used to estimate pH.

The three datasets were combined, a Boltzmann curve fitted and the 95% confidence bands calculated for each of the seven parameters. Fig. 4 shows the result of this fit for the change in the centre of the peak $\sim$1590 cm^{-1} (the other six parameters are shown in the ESI†).

Fig. 4 shows that the calibration is most accurate over a pH range of approximately 6–8; outside this range the estimated pH is subject to a large error. The aim of using this probe was to measure intracellular pH, which should be within this range, so acceptable ranges were therefore calculated for each parameter as the values of the Boltzmann fit at pH 6 and 8. The pH was then estimated from each of the seven calibration graphs, with an overall error in pH determined from both the error in the Boltzmann fit and the error in the peak centre, height ratio or area ratio. The estimated pH from each parameter was combined into a weighted mean to give a more accurate estimate of pH. Spectra were marked as 'pH < 6' or 'pH > 8' as appropriate if they met either of the following conditions:

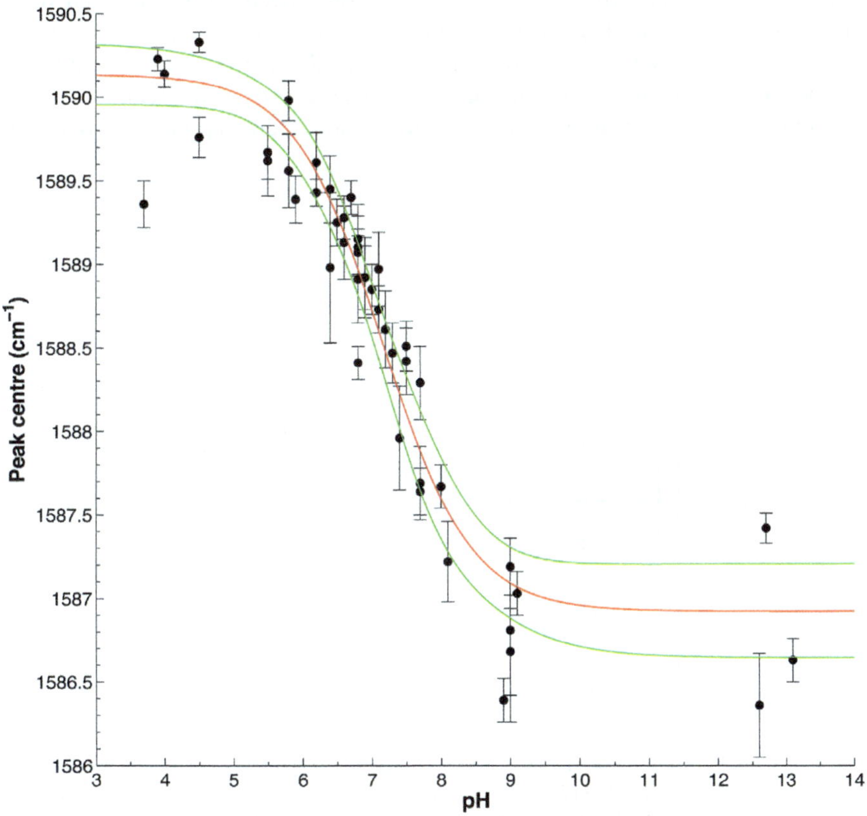

Fig. 4 Variation of the centre of the peak at $\sim$1590 cm^{-1} with pH. A Boltzmann curve (red line) has been fitted to points which are the weighted means from three independent datasets; error bars are the standard deviation of the weighted mean; green lines are the 95% non-simultaneous confidence bands.

- if 2 or more area ratios were outside their acceptable ranges.
- if 3 or more of the peak centre or height ratios were outside their acceptable ranges.

These conditions ensure that at least four of the seven parameters are used to provide an estimate of pH. Using fewer than four parameters can result in a less accurate estimated value; therefore spectra with fewer than four parameters are marked as being outwith the range of the sensor.

When used to analyse the calibration data, this method resulted in an average standard deviation of ±0.16 pH units. Individual 95% confidence intervals were calculated from the individual standard deviations of each spectrum, where the estimated pH was in the range 6–8; overall, 92.4% (very close to the expected value of 95%) of the calibration spectra were either correctly identified as being outwith the range of the sensor or the error range associated with the estimated pH contained the actual pH.

The automated algorithm was then compared to the manual processing method. The variability of spectra can be determined by acquiring consecutive spectra from the same aggregate of MBA-functionalised nanoshells dried onto a quartz coverslip. What is being tested is the variability between the manual and automatic processing methods, so it is important to obtain consecutive spectra that are very similar. To this end, 10 consecutive spectra were obtained from MBA–NS and processed by both methods. The manual process uses only one parameter to estimate pH: the ratio of the peak heights of the peaks at ~1400 and ~1590 cm^{-1}, whereas the automated process uses seven parameters. A comparison of the two methods is given in Fig. 5 and Table 1. Manual processing of the 10

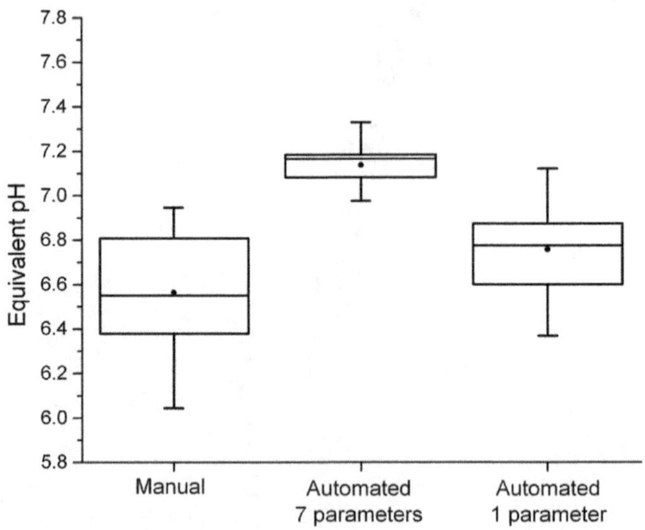

Fig. 5 Boxplots showing the median, 1st and 3rd quarter percentiles, range and mean (filled circle) for the estimation of pH from 10 consecutive spectra obtained from MBA–NS using both manual and automated processing methods. The manual method uses only one parameter to estimate pH: the height ratio of the peaks at ~1400 and ~1590 cm^{-1}; the pH estimated by the automated method using only this parameter is shown for comparison (automated 1 parameter). The automated method uses a total of seven parameters to estimate pH and results in a much smaller range than the manual method.

Table 1 Comparison of manual and automated processing methods in the estimation of pH from 10 consecutive spectra obtained from MBA–NS. The manual method does not provide an error associated with the estimated pH. The automated method results in much less variation in estimated pH from each of the 10 spectra, and the processing time per spectrum is reduced by nearly 200-fold

	Manual processing	Automated processing
Mean estimated pH	6.56	7.14
Standard deviation	0.29	0.09
Mean error	—	0.28
Time per spectrum	3 minutes	1 second

spectra resulted in a mean pH of 6.56 with a standard deviation of 0.29, whereas the automated processing resulted in a mean pH of 7.14 with a standard deviation of 0.09. The manual process does not provide an error associated with the estimated pH, whereas the automated process results in an associated mean error of 0.28 pH units. Furthermore, automated processing is nearly 200 times faster than manual processing and results in pH values spread evenly about the mean. The lower pH estimated by the manual method could be due to only using one parameter as the mean estimated pH of the same parameter by the automated method is 6.76 with a standard deviation of 0.21, lower than the mean derived from all seven parameters. These results show that not only is processing faster by two orders of magnitude, there is less variability in the automated processing method and the combination of seven parameters enables a more robust estimation of pH.

We used MBA–NS as a test-case for automated processing because the peaks used to measure pH are well resolved and easily identified. We then attempted to apply the same approach to automating redox potential measurement using the NQ–NS nanosensor developed in our lab (Fig. 6).[18] Fig. 7 shows how peaks in the spectrum change with redox potential: the peak at $\sim$1580 cm^{-1} is actually composed of two peaks at 1577 cm^{-1} and 1602 cm^{-1} (symmetric ring breathing and aryl ring stretching with N–H wagging, respectively). This peak is the most

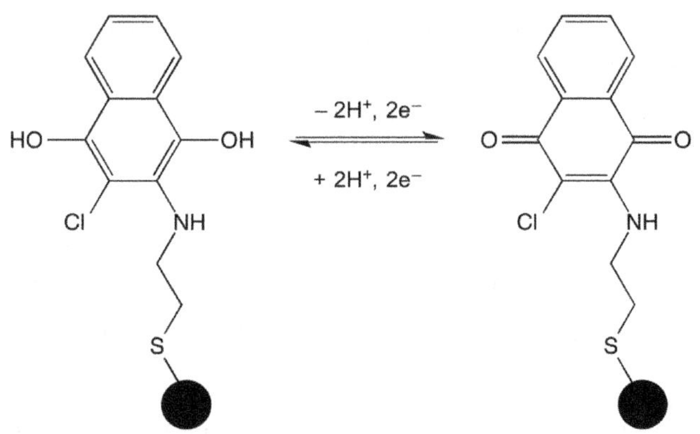

Fig. 6 Reduced (left) and oxidised (right) structures of the NQ reporter molecule conjugated to a gold nanoshell (black circle) *via* a gold–thiol bond.

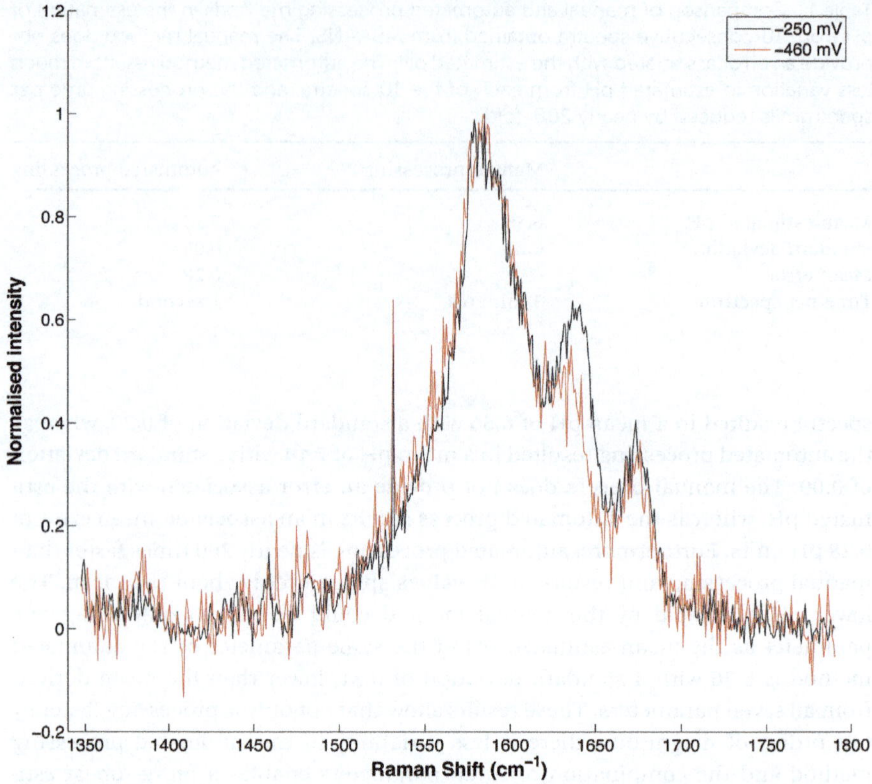

Fig. 7 Redox-dependent change in the intensity of the peak at ~1640 cm^{-1} in the NQ–NS Raman spectrum. Spectra have been background subtracted and normalised to the intensity of the peak at 1577 cm^{-1}.

intense peak in the region 1300–1800 cm^{-1}, so it is a good candidate for the reference peak, whereas the peak at ~1640 cm^{-1} is due to C=O stretching and can be used as a marker of increasing oxidation. For NQ, then, there is the possibility of using peak height and area ratios of a combination of the three peaks at 1577 cm^{-1}, 1602 cm^{-1} and 1640 cm^{-1}.

As the three peaks of interest overlapped, the data were modelled by the sum of three Lorentzian peaks:

$$y(x_c) = \sum_{n=1}^{P} \frac{a_n}{1 + \left(\frac{x_c - b_n}{c_n}\right)^2}$$

where again $n \in \{1, \ldots, P\}$ is the index of the n-th underlying spectral peaks, $P = 3$ is the number of spectral peaks, x_c is the Raman shift, a_n is the (positive) amplitude of the n-th peak, b_n is the peak centre, and c_n is the peak width at 1577, 1602 and 1640 cm^{-1}, respectively. As discussed earlier, the amplitude and peak widths are estimated using a nonlinear least squares algorithm and the peak centres are known *a priori*. The Raman shift range over which the fit is applied is 1560–1660 cm^{-1} as this is the range that results in the least variation in the residuals over all the calibration spectra. Constraints on peak widths were determined experimentally with a minimum of 40% and a maximum of 250% of the average peak width of the calibration spectra.

After fitting, the calibration data were grouped by redox potential in order to investigate the relationships between various fit parameters and redox potential. The weighted means and standard deviations of the various parameters at each redox potential were calculated using the same inverse weighting method as for the MBA–NS nanosensor. It was found that the ratio of the height of the peaks at 1577 and ~1640 cm^{-1} varied with redox potential, as did the equivalent area ratios. The centres of the peaks at 1577 and 1602 cm^{-1} did not change with redox potential, and the centre of the peak at ~1640 cm^{-1} only varied with redox potential at potentials more positive than −340 mV and thus could not be used to estimate redox potential. Therefore calibration curves were only constructed for the peak height and area ratios of the peaks at 1577 and ~1640 cm^{-1}.

As before, a Boltzmann curve was fitted and 95% confidence bands were calculated for both of these parameters (Fig. 8).

Acceptable ranges for the ratios were calculated as the values at −400 and −250 mV, as this is the region where the curves change most rapidly. The redox potential was then estimated from both calibration graphs, with the overall error in redox potential determined from both the error in the Boltzmann fit and the error in the peak height or area ratio. With only two parameters used to estimate redox potential, there is a possibility that one parameter with a large error can skew the estimated redox potential away from the actual potential. Therefore the estimated redox potential from each parameter was combined into a weighted mean to avoid this effect. Spectra were marked as 'reduced' or 'oxidised' if either or both ratios were outside their acceptable ranges, and thus outwith the range of the sensor.

When used to analyse the calibration data, this method resulted in an average standard deviation of 23.3 mV. Individual 95% confidence intervals were calculated from the individual standard deviations of each spectrum, where the estimated redox potential was in the range −400 to −250 mV. Overall, 85% of the calibration spectra were either correctly identified as being outwith the range of

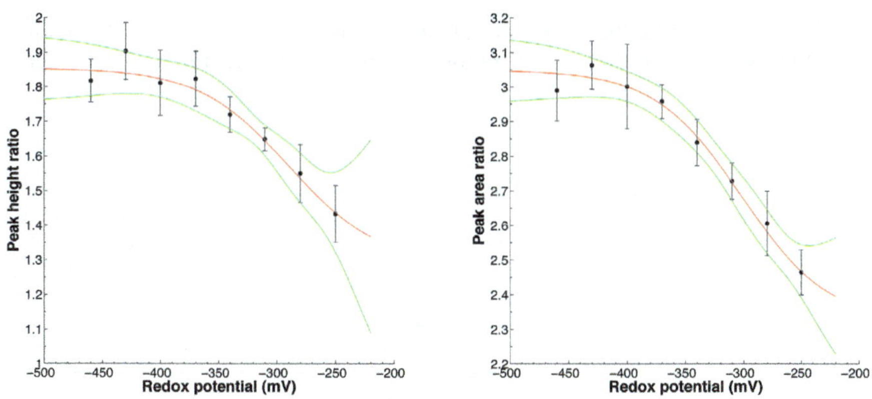

Fig. 8 Variation of the 1577/1645 cm^{-1} peak height (left) and peak area (right) ratios with redox potential. A Boltzmann curve (red line) has been fitted to points which are the weighted means of five data points; error bars are the standard deviation of the weighted mean; green lines are the 95% non-simultaneous confidence bands.

the sensor or the error range associated with the estimated redox potential contained the actual redox potential. Obtaining more data would improve the accuracy of the weighted mean, and also improve the error in the calibration curve. In addition, data were collected at equally spaced redox potentials; due to the sigmoid shape of the calibration curve, this could mean a further loss of accuracy in the region of −350 to −250 mV where the slope is steepest. The 95% confidence intervals are greatest at more reducing potentials. This is due to the larger error in the height of the peak at 1640 cm^{-1} as it decreases with decreasing redox potential. This peak becomes a shoulder on the combined peaks at 1577 and 1602 cm^{-1}, resulting in increased error in the fit in this region at more reduced potentials. This error could be minimised by improving the SNR of the detector, but the problem can also be overcome by designing reporters with more reductive standard potentials.[32,33]

The algorithm was then compared to the manual processing method. Ten consecutive spectra were obtained from NQ–NS dried onto an imaging dish and processed by both methods. The manual process uses only the peak height ratio, whereas the automated process uses both the peak height and area ratios. A comparison of the two methods is given in Table 2 and Fig. 9. Manual processing of the 10 spectra results in a mean redox potential of −257.4 mV with a standard deviation of 3.0 mV, whereas the automated processing results in a mean redox potential of −259.3 mV with a standard deviation of 2.8 mV. The manual process does not provide an error associated with the estimated redox potential, whereas the automated process results in an associated mean error of 20.9 mV.

The automated method was significantly faster than the manual method: nearly 200 times faster in processing a single spectrum, and 50–120 times faster in producing a colourmap of 147 spectra. The automated method was also more precise than the manual method for the MBA nanosensor and due to the lack of calibration spectra was no worse for the NQ nanosensor. However for the NQ nanosensor it was shown that the estimated values from the automated method were more evenly distributed about the mean than for the manual method, indicating that the automated method was less biased. All spectra were subject to the same criteria and intra- and inter-individual bias and variance were not present in the automated method.

Overall, the automated processing method is much faster and results in a less variable estimate of the pH or redox potential than the manual processing method.

Table 2 Comparison of manual and automated processing methods in the estimation of redox potential from 10 consecutive spectra obtained from NQ–NS. The manual processing method does not provide an error associated with estimated redox potential. The automated processing method reduces the processing time per spectrum by nearly 200-fold

	Manual processing	Automated processing
Mean estimated redox potential (mV)	−257.4	−259.3
Standard deviation (mV)	3.0	2.8
Mean error (mV)	—	20.9
Time per spectrum	3 minutes	1 second

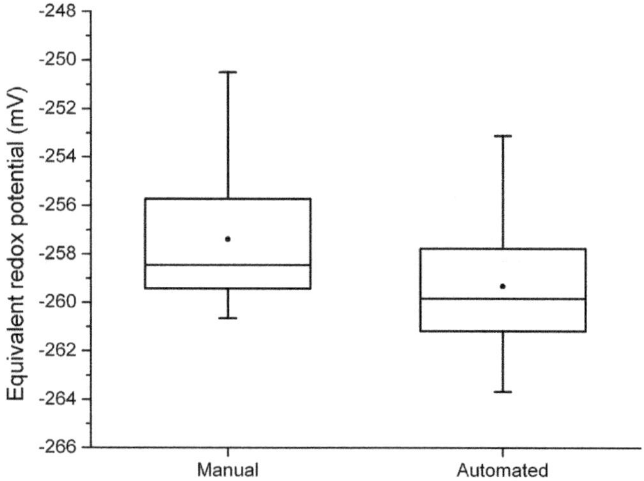

Fig. 9 Box plots showing the median, 1st and 3rd quarter percentiles, range and mean (filled circle) for the estimation of redox potential from 10 consecutive spectra using both manual and automated processing methods.

Effect of zinc oxide and silver nanoparticles on redox potential

The toxicological effects of inhaled particulate matter have been extensively investigated. Inhalation of ultrafine particles, such as in diesel exhaust, has been shown to cause inflammation in several cell types.[34] The toxicity of these species is thought to be related to their large surface area to volume ratio, and/or their ability to act as a carrier of transition metals into the lungs.[35] Nanoparticles have a smaller diameter and greater surface area to volume ratio compared to larger particles and are now manufactured in large quantities for industrial use (for example, in the manufacture of cosmetics, electronics, and paint[36]); therefore it is important to investigate the effects of nanoparticle inhalation during the manufacturing process. Engineered nanoparticles have been shown to have greater toxic effects than ultrafine particles (which are nanoscale ambient, as opposed to engineered, particles; for example air pollution particles), although the effects are strongly dependent on cell type.[34]

As well as being used as a model of lung cancer,[37,38] A549 cells are used in studies of inflammation[39] and nanoparticle toxicity.[35,40] Both ultrafine particles and metal nanoparticles cause production of ROS in A549 cells.[41] Elevated ROS levels cause upregulated expression of anti-oxidant genes through the Nrf2-KEAP pathway.[42] At low particle concentrations this upregulation is enough to degrade ROS, however, at higher concentrations, the antioxidant response is insufficient for the levels of ROS produced, and the resulting oxidative stress triggers apoptosis.[43]

This well-characterised response of A549 cells to nanoparticles makes them an ideal system in which to compare our measurements of intracellular redox potential with the traditional assays for oxidative stress.

A549 cells were incubated with NQ–NS overnight before either treatment with 30 mM 2,2′-azobis(2-amidinopropane) dihydrochloride (AAPH; positive control), sub-lethal treatment with 0.5 mg ml^{-1} silver nanoparticles (AgNP) or 0.5 mg ml^{-1}

zinc oxide nanoparticles (ZnONP) or were left untreated (see Experimental for full methods). SERS maps were acquired from untreated, AgNP-treated and ZnONP-treated cells over the following time periods after treatment: 0–1 hours; 1–2 hours; 2–3 hours; 3–4 hours and 4–5 hours. SERS spectra were acquired from AAPH-treated cells during the first hour after treatment only as there was extensive cell death after this period. Briefly, a cell or cells were rapidly scanned to locate any NQ–NS and then detailed scans covering an area of 6 × 25 mm were performed, centred on the location of each NQ–NS signal from the rapid scan.

The automated processing algorithm described above was used to process all SERS map spectra. After background subtraction, spectra were selected for further processing based on the following criteria:

- R^2 value ≥ 0.6.
- height of the peak at 1577 cm^{-1} ≥100 counts.

No spectra showed complete oxidation of NQ–NS. The automated processing method returned a redox potential with an associated error for each spectrum as described above. A weighted average and standard deviation were then calculated for each time period and each treatment, using the inverse weighting method. Fig. 10 shows the change in redox potential for each time period for all treatments. In the 0–1 hour time period only AAPH-treated cells showed a significant oxidative shift in redox potential compared to untreated cells ($p = 0.0014$). In the 1–2 hour time period, both AgNP- and ZnONP-treated cells showed a significantly more oxidative redox potential compared to untreated cells ($p < 0.0001$ and $p = 0.0015$, respectively). For the other time periods there was no significant difference in the redox potential between untreated cells and metal nanoparticle-treated cells.

From Fig. 10 it can be seen that both metal nanoparticle treatments showed the most oxidised redox potential during the 1–2 hour time period, with AgNP treatment resulting in a more oxidised potential than ZnONP treatment. All time periods showed significantly more positive redox potentials with respect to the potential measured during the 0–1 hour time period. In AgNP-treated cells, the redox potential during the 4–5 hour time period was significantly more reduced

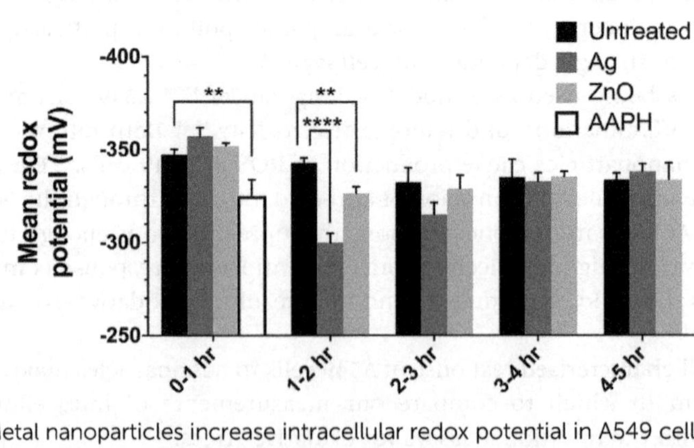

Fig. 10 Metal nanoparticles increase intracellular redox potential in A549 cells 1–2 hours after treatment. Cells containing NQ–NS were treated with one of 0.5 mg ml^{-1} AgNP, 0.5 mg ml^{-1} ZnONP or 30 mM AAPH (positive control), or were left untreated. Bars show the weighted mean of redox potentials within each time period and treatment; error bars represent the standard error of the weighted mean; **$p < 0.01$; ****$p < 0.0001$.

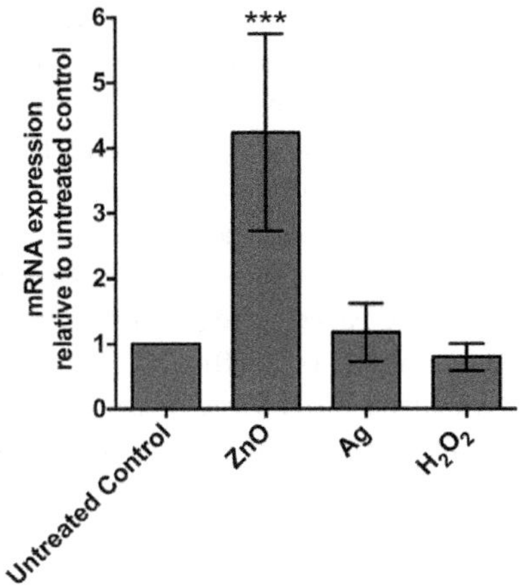

Fig. 11 HO-1 mRNA expression 3 hours after treatment with metal nanoparticles in A549 cells.

than the potential during the 1–2 hour or 2–3 hour time periods ($p = 0.0008$ and $p = 0.0208$, respectively), but it was still significantly more oxidised than during the 0–1 hour time period ($p = 0.0208$). ZnONP treatment showed the same trend but without reaching significance. The intracellular redox potential in untreated cells did not change significantly over the 5 hours.

These results were then compared to mRNA expression levels for selected antioxidant genes measured using qPCR. Nrf2 is a transcription factor that is responsive to changes in oxidative stress and an essential part of the cellular antioxidant response. Oxidative stress causes Nrf2 to translocate to the nucleus and upregulate expression of its target genes, many of which are antioxidants, including hemeoxygenase-1 (HO-1).[44] The levels of HO-1 mRNA transcripts were measured at three hours and showed that ZnONP, but not AgNP, treatment caused a significant upregulation of the HO-1 transcript (Fig. 11). However, the concentration of AgNP at which only 50% of cells remained viable (TD_{50}) was much lower than for ZnONP (data not shown). Together, these data indicate that the less oxidative effect of ZnONP treatment could be due to an increase in antioxidant response, which is not seen with AgNP treatment.

Discussion and conclusions

In this paper NQ–NS were used to measure the increase in intracellular redox potential in A549 cells treated with sub-lethal doses of engineered metal nanoparticles of Ag and ZnO, which have been shown to be toxic in A549 cells. The intracellular redox potential was shown to be most oxidised during the 1–2 hour time period for both metal NP treatments, with the AgNP-treated cells showing more oxidised potentials than the ZnONP-treated cells. The ZnONP-treated cells showed upregulation of antioxidant genes compared to the AgNP-treated cells and we speculate that the increased upregulation of antioxidant genes seen in the ZnONP-treated cells, compared to the AgNP-treated cells, would enable metal NP-induced ROS to be degraded more efficiently and thus could explain the smaller increase in redox potential.

For both metal NP treatments, the intracellular redox potential returned to a value which was not significantly different to the untreated control population indicating that the cellular antioxidant response had reversed the effects of metal NP treatment on this timescale, in the surviving cells. In the AgNP-treated cells, the redox potential became significantly more reductive after an initial oxidative shift, and eventually returned to a potential that is not significantly different to the untreated control cells. In the ZnONP-treated cells this trend did not reach significance, perhaps due to increased upregulation of anti-oxidant genes.

In summary, our SERS nanosensor method of measuring intracellular redox potential is capable of measuring differences in oxidative stress caused by nanoparticles in A549 cells, and is corroborated by traditional assays of oxidative stress.

References

1 A. C. Fisher, *Electrode Dynamics*, Oxford University Press, 1996.

2 F. Schafer and G. Buettner, Redox environment of the cell as viewed through the redox state of the glutathione disulfide/glutathione couple, *Free Radical Biol. Med.*, 2001, **30**(11), 1191–1212.

3 S. G. Menon and P. C. Goswami, A redox cycle within the cell cycle: ring in the old with the new, *Oncogene*, 2007, **26**(8), 1101–1109.

4 M. Kemp, Y.-M. Go and D. P. Jones, Nonequilibrium thermodynamics of thiol/disulfide redox systems: a perspective on redox systems biology, *Free Radical Biol. Med.*, 2008, **44**(6), 921–937.

5 J. M. Berg, J. L. Tymoczko and L. Stryer, *Biochemistry*, W. H. Freeman and Company, New York, 6th edn, 2007.

6 K. Bedard and K.-H. Krause, The NOX family of ROS-generating NADPH oxidases: physiology and pathophysiology, *Physiol. Rev.*, 2007, **87**(1), 245–313.

7 M. G. Pollard, K. J. Travers and J. S. Weissman, Ero1p: a novel and ubiquitous protein with an essential role in oxidative protein folding in the endoplasmic reticulum, *Mol. Cell*, 1998, **1**(2), 171–182.

8 Y. M. W. Janssen-Heininger, B. T. Mossman, N. H. Heintz, H. J. Forman, B. Kalyanaraman, T. Finkel, *et al.*, Redox-based regulation of signal transduction: principles, pitfalls, and promises, *Free Radical Biol. Med.*, 2008, **45**(1), 1–17.

9 C. C. Winterbourn, Reconciling the chemistry and biology of reactive oxygen species, *Nat. Chem. Biol.*, 2008, **4**(5), 278–286.

10 T. Finkel, Signal transduction by reactive oxygen species, *J. Cell Biol.*, 2011, **194**(1), 7–15.

11 J. Nordberg and E. S. Arnér, Reactive oxygen species, antioxidants, and the mammalian thioredoxin system, *Free Radical Biol. Med.*, 2001, **31**(11), 1287–1312.

12 J. Mates, Effects of antioxidant enzymes in the molecular control of reactive oxygen species toxicology, *Toxicology*, 2000, **153**(1), 83–104.

13 W. Dröge, Free radicals in the physiological control of cell function, *Physiol. Rev.*, 2002, **82**(1), 47–95.

14 J. K. Andersen, Oxidative stress in neurodegeneration: cause or consequence?, *Nat. Med.*, 2004, **10**, S18–S25.

15 M. Valko, D. Leibfritz, J. Moncol, M. T. D. Cronin, M. Mazur and J. Telser, Free radicals and antioxidants in normal physiological functions and human disease, *Int. J. Biochem. Cell Biol.*, 2007, **39**(1), 44–84.

16 C. T. Dooley, T. M. Dore, G. T. Hanson, W. C. Jackson, S. J. Remington and R. Y. Tsien, Imaging dynamic redox changes in mammalian cells with green fluorescent protein indicators, *J. Biol. Chem.*, 2004, **279**(21), 22284–22293.

17 A. J. Meyer, T. Brach, L. Marty, S. Kreye, N. Rouhier, J.-P. Jacquot, *et al.*, Redox-sensitive GFP in Arabidopsis thaliana is a quantitative biosensor for the redox potential of the cellular glutathione redox buffer, *Plant J.*, 2007, **52**(5), 973–986.

18 C. A. R. Auchinvole, P. Richardson, C. McGuinnes, V. Mallikarjun, K. Donaldson, H. McNab, *et al.*, Monitoring intracellular redox potential changes using SERS nanosensors, *ACS Nano*, 2012, **6**(1), 888–896.

19 K. C. Bantz, A. F. Meyer, N. J. Wittenberg, H. Im, Ö. Kurtuluş, S. H. Lee, *et al.*, Recent progress in SERS biosensing, *Phys. Chem. Chem. Phys.*, 2011, **13**(24), 11551–11567.

20 W. F. Maddams, The scope and limitations of curve fitting, *Appl. Spectrosc.*, 1980, **34**(3), 245–267.

21 V. B. Di Marco and G. G. Bombi, Mathematical functions for the representation of chromatographic peaks, *J. Chromatogr. A*, 2001, **931**(1–2), 1–30.

22 J. W. Robinson, E. M. S Frame and G. M. Frame II, *Undergraduate Instrumental Analysis*, 6th edn, Marcel Dekker, 2005.

23 J. Zhao, H. Lui, D. I. McLean and H. Zeng, Automated autofluorescence background subtraction algorithm for biomedical Raman spectroscopy, *Appl. Spectrosc.*, 2007, **61**(11), 1225–1232.

24 J. R. Beattie, Optimising reproducibility in low quality signals without smoothing; an alternative paradigm for signal processing, *J. Raman Spectrosc.*, 2011, **42**(6), 1419–1427.

25 H. G. Schulze, R. B. Foist, K. Okuda, A. Ivanov and R. F. B. Turner, A model-free, fully automated baseline-removal method for Raman spectra, *Appl. Spectrosc.*, 2011, **65**(1), 75–84.

26 S. K. Lau, P. Winlove, J. Moger, O. L. Champion, R. W. Titball, Z. H. Yang, *et al.*, A Bayesian Whittaker–Henderson smoother for general-purpose and sample-based spectral baseline estimation and peak extraction, *J. Raman Spectrosc.*, 2012, **43**(9), 1299–1305.

27 B. Sharma, R. R. Frontiera, A.-I. Henry, E. Ringe and R. P. van Duyne, SERS: materials, applications, and the future, *Mater. Today*, 2012, **15**(1), 16–25.

28 S. W. Bishnoi, C. J. Rozell, C. S. Levin, M. K. Gheith, B. R. Johnson, D. H. Johnson, *et al.*, All-optical nanoscale pH meter, *Nano Lett.*, 2006, **6**(8), 1687–1692.

29 J. Kneipp, H. Kneipp, B. Wittig and K. Kneipp, One- and two-photon excited optical ph probing for cells using surface-enhanced Raman and hyper-Raman nanosensors, *Nano Lett.*, 2007, **7**(9), 2819–2823.

30 M. A. Ochsenkuhn, P. R. T. Jess, H. Stoquert, K. Dholakia and C. J. Campbell, Nanoshells for Surface-Enhanced Raman Spectroscopy in Eukaryotic Cells: Cellular Response and Sensor Development, *ACS Nano*, 2009, **3**(11), 3613–3621.

31 D. W. Marquardt, An algorithm for least-squares estimation of nonlinear parameters, *J. Soc. Ind. Appl. Math.*, 1963, **11**, 431–441.

32 J. Jiang, C. Auchinvole, K. Fisher and C. J. Campbell, Quantitative measurement of redox potential in hypoxic cells using SERS nanosensors, *Nanoscale*, 2014, **6**(20), 12104–12110.

33 P. Thomson, V. L. Camus, Y. Hu and C. J. Campbell, Series of Quinone-Containing Nanosensors for Biologically Relevant Redox Potential Determination by Surface-Enhanced Raman Spectroscopy, *Anal. Chem.*, 2015, **87**, 4719–4725.

34 V. Stone, H. Johnston and M. Clift, Air pollution, ultrafine and nanoparticle toxicology: cellular and molecular interactions, *IEEE Transactions on NanoBioscience*, 2007, **6**(4), 331–340.

35 D. Brown, M. Wilson, W. MacNee, V. Stone and K. Donaldson, Size-dependent proinflammatory effects of ultrafine polystyrene particles: A role for surface area and oxidative stress in the enhanced activity of ultrafines, *Toxicol. Appl. Pharmacol.*, 2001, **175**(3), 191–199.

36 R. J. Vandebriel and W. H. De Jong, A review of mammalian toxicity of ZnO nanoparticles, *Nanotechnol., Sci. Appl.*, 2012, **5**, 61–71.

37 S.-M. Jeon, N. S. Chandel and N. Hay, AMPK regulates NADPH homeostasis to promote tumour cell survival during energy stress, *Nature*, 2012, **485**(7400), 661–665.

38 T. R. Kim, E. W. Cho, S. G. Paik and I. G. Kim, Hypoxia-induced SM22 alpha in A549 cells activates the IGF1R/PI3K/Akt pathway, conferring cellular resistance against chemo- and radiation therapy, *FEBS Lett.*, 2012, **586**(4), 303–309.

39 T. J. Standiford, S. L. Kunkel, M. A. Basha, S. W. Chensue, J. P. Lynch, G. B. Toews, *et al.*, Interleukin-8 gene expression by a pulmonary epithelial cell line. A model for cytokine networks in the lung, *J. Clin. Invest.*, 1990, **86**(6), 1945–1953.

40 A. Simon-Deckers, B. Gouget, M. Mayne-L'hermite, N. Herlin-Boime, C. Reynaud and M. Carrière, In vitro investigation of oxide nanoparticle and carbon nanotube toxicity and intracellular accumulation in A549 human pneumocytes, *Toxicology*, 2008, **253**(1–3), 137–146.

41 S. Lu, W. Zhang, R. Zhang, P. Liu, Q. Wang, Y. Shang, *et al.*, Comparison of cellular toxicity caused by ambient ultrafine particles and engineered metal oxide nanoparticles, *Part. Fibre Toxicol.*, 2015, **12**(1), 5.

42 X. Deng, F. Zhang, W. Rui, F. Long, L. Wang, Z. Feng, *et al.*, PM2.5-induced oxidative stress triggers autophagy in human lung epithelial A549 cells. Toxicol, *In Vitro*, 2013, **27**(6), 1762–1770.

43 C. Pisani, J.-C. Gaillard, V. Nouvel, M. Odorico, J. Armengaud and O. Prat, High-throughput, quantitative assessment of the effects of low-dose silica nanoparticles on lung cells: grasping complex toxicity with a great depth of field, *BMC Genomics*, 2015, **16**, 315.

44 R. K. Thimmulappa, K. H. Mai, S. Srisuma, T. W. Kensler, M. Yamamoto and S. Biswal, Identification of Nrf2-regulated genes induced by the chemopreventive agent sulforaphane by oligonucleotide microarray, *Cancer Res.*, 2002, **62**(18), 5196–5203.

Faraday Discussions

PAPER

Biofluid infrared spectro-diagnostics: pre-analytical considerations for clinical applications†

L. Lovergne,‡[abc] P. Bouzy,‡[d] V. Untereiner,‡[abe] R. Garnotel,[abf] M. J. Baker,§[c] G. Thiéfin§[abg] and G. D. Sockalingum§*[ab]

Received 19th November 2015, Accepted 10th December 2015
DOI: 10.1039/c5fd00184f

Several proof-of-concept studies on the vibrational spectroscopy of biofluids have demonstrated that the methodology has promising potential as a clinical diagnostic tool. However, these studies also show that there is a lack of a standardised protocol in sample handling and preparation prior to spectroscopic analysis. One of the most important sources of analytical errors is the pre-analytical phase. For the technique to be translated into clinics, it is clear that a very strict protocol needs to be established for such biological samples. This study focuses on some of the aspects of the pre-analytical phase in the development of the high-throughput Fourier Transform Infrared (FTIR) spectroscopy of some of the most common biofluids such as serum, plasma and bile. Pre-analytical considerations that can impact either the samples (solvents, anti-coagulants, freeze–thaw cycles…) and/or spectroscopic analysis (sample preparation such as drying, deposit methods, volumes, substrates, operators dependence…) and consequently the quality and the reproducibility of spectral data will be discussed in this report.

[a]Université de Reims Champagne-Ardenne, MéDIAN-Biophotonique et Technologies pour la Santé, UFR de Pharmacie, 51 rue Cognacq-Jay, 51095 Reims Cedex, France. E-mail: ganesh.sockalingum@univ-reims.fr

[b]CNRS UMR 7369, Matrice extracellulaire et Dynamique Cellulaire, MEDyC, 51096 Reims Cedex, France

[c]WESTChem, Department of Pure and Applied Chemistry, Technology and Innovation Centre, University of Strathclyde, Glasgow, G1 1RD, UK

[d]SATT NORD (Société d'Accélération du Transfert de Technologie), 4 bd de la Paix, 51100 Reims, France

[e]Plateforme en imagerie cellulaire et tissulaire (PICT), Université de Reims Champagne-Ardenne, 51 rue Cognacq-Jay, 51096 Reims Cedex, France

[f]Laboratoire de Biologie et Recherche Pédiatriques, CHU de Reims, 51092 Reims Cedex, France

[g]Service d'Hépato-Gastroentérologie, CHU de Reims, Hôpital Robert Debré, 51092 Reims Cedex, France

† Electronic supplementary information (ESI) available. See DOI: 10.1039/c5fd00184f

‡ Authors contributed equally to this work.

§ Authors contributed equally to this work and project supervision.

Introduction

During the last decade, a number of studies have highlighted the potential of vibrational spectroscopy applied to biofluids for screening, diagnostic and prognostic applications in the biomedical field.[1–9] These studies exploited the fact that biofluids exhibit spectral characteristics reflecting their biomolecular composition, which allows, through chemometric analysis, the identification of patterns reflecting sensitive and specific spectral biomarkers in various pathological conditions. Some biofluids such as serum and plasma appear as ideal media for routine clinical use as they are easily accessible, collectable by a minimally invasive method and repeatedly available for monitoring disease progression or therapeutic response. Accordingly, most clinical studies have been performed using blood components and the quest for spectral blood biomarkers has been widely explored in patients with malignant and non-malignant diseases, as reviewed recently by Baker *et al.*[10] Other biofluids such as bile or cerebrospinal fluid are less accessible but of particular interest as their proximity with diseased tissue may facilitate the identification of the spectral signatures of the disease.[11,12] Based on these data, biofluid vibrational spectroscopy appears as an innovative and promising facet of the concept of "liquid biopsy".

However, in spite of promising results, biofluid vibrational spectroscopy is still far from routine clinical application. A great majority of studies are proof-of-concept studies based on a small number of subjects with and without the disease of interest. The results of these pilot studies should be analysed by cross validation methods and interpreted cautiously until validated in clinical trials performed in independent cohorts with a large number of patients and using golden standard diagnostic methods or complementary techniques. Most importantly, at a time when multicenter studies are being considered, there is a crucial need to standardise and validate the modalities of sample collection, storage and spectral acquisition. Standardisation of sample-related factors is the first step of the pre-analytical validity of any technology intended for large scale and multicenter clinical applications. It is well-established that most errors in laboratory testing occur in the pre-analytical phase[13] and the need for the standardisation of the procedures has been demonstrated in other high-throughput technologies such as proteomics,[14,15] metabolomics[16,17] or genomics.[18]

The work presented here will address the issue of pre-analytical variability in the field of high-throughput biofluid infrared transmission spectroscopy with a particular focus on blood component preparation. Factors that may affect the IR spectral quality and profile of biofluids will be discussed including biofluid dilution, volume and deposition modalities, repetition of freeze–thaw cycles, drying conditions, types of blood collection tubes used, and intra- and inter-operator reproducibility.

Experimental

The studies were performed from a bank of serum and plasma samples stored at −80 °C, originally taken for a routine biochemical check-up at the Biochemistry

Laboratory of Reims University Hospital. Serum was obtained by centrifugation of freshly clotted blood and plasma by centrifugation of blood collected in tubes containing ethylenediaminetetraacetic acid (EDTA) or lithium heparin as anticoagulants. Bile was obtained during endoscopic retrograde cholangiopancreatography after selective biliary cannulation and before any injection of a contrast agent. After centrifugation, the cellular pellet was removed for routine cytological examination and the remaining supernatant was retrieved and stored at −80 °C at the Pathology Department of Reims University Hospital. An informed consent was obtained from all patients for performing a diagnostic and/or therapeutic endoscopic retrograde cholangiography.

1. Impact of biofluid volume and dilution on spectral quality test

The experimental procedures are detailed in Fig. 1. In brief, serum and plasma experiments were performed on pure samples and after 2-fold, 3-fold and 4-fold dilutions using normal saline (0.9% NaCl). As we have previously shown that bile can be analysed without dilution,[12] only pure bile samples were used. Samples were deposited onto silicon plates, either a 96-well plate with wells of 7 mm diameter or a 384-well plate with wells of 3 mm diameter. Sample volumes of 5 and 10 μL were studied using the 96-well plate and 3 and 5 μL using the 384-well plate. A deposit of 1 μL without dilution was also tested with the 384-well plate. The deposits were spread drops on the entire surface of the well or non-spread drops in the middle of the well. Ten replicates per sample were studied. After 1 h drying at room temperature, the different plates were inserted into a high-throughput module (HTS-XT) coupled to an FTIR-spectrometer (Tensor 27,

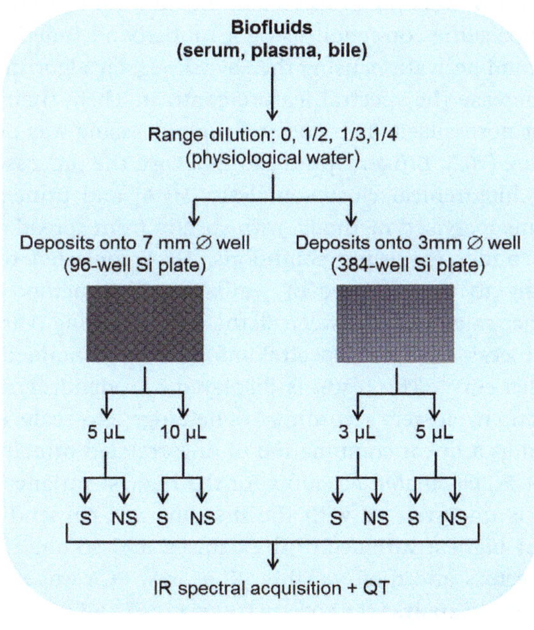

Fig. 1 Experimental protocol for studying the impact of dilution and volume on spectral quality test. S: spread dried drop, NS: non-spread dried drop, QT: quality test.

Bruker Optics GmbH, Ettlingen, Germany). Spectra were acquired in the transmission mode (OPUS v.6.5 software, Bruker Optics GmbH) in the 4000–400 cm^{-1} wavenumber range with a 4 cm^{-1} spectral resolution and 32 scans. A background reference spectrum was recorded before each sample analysis. The reference spectrum was automatically subtracted to obtain the final absorbance spectrum. A zero-filling factor of 2 and a Blackman-Harris-3-term function were applied for Fourier transformation.

The following quality tests (QT) were applied to each spectrum: (1) absorbance intensity: spectra were validated when the maximum absorbance (amide I band *c.a.* 1655 cm^{-1}) was below 1.8 arbitrary units; (2) signal-to-noise ratios were calculated according to the manufacturer's instructions for biological materials (Bruker Optics GmbH): the signals were calculated as the difference between the maximum and the minimum 1st derivative values in the range 1700–1600 cm^{-1} (signal 1 or S1) or as the difference between the maximum and the minimum 1st derivative values determined in the range 1260–1178 cm^{-1} (signal 2 or S2). Noise intensity (N) was calculated as the difference between the maximum and the minimum 1st derivative values in the range 2100–2000 cm^{-1}. Spectra met this quality test when the ratios S1/N and S2/N were above 50 and 20 respectively; (3) signal-to-water ratio: water signal (W) was evaluated using the difference between the maximum and the minimum 1st derivative values in the range 1842–1837 cm^{-1}. The ratio thresholds for spectral validation were S1/W and S2/W above 10 and 4 respectively.

Individual samples were considered as QT-validated for analysis when at least 8 replicate spectra out of 10 met the quality tests.

In order to analyse the impact of drop spreading, we performed experiments by successively using three-fold diluted serum and four-fold diluted plasma with spread and non-spread drops of 10 μL on 7 mm diameter wells. Spectra that passed the quality test were pre-processed in the wavenumber range from 1800 to 800 cm^{-1}. After baseline correction using a rubberband function, spectra were converted to second derivatives using the Savitsky–Golay algorithm, with 9-point smoothing to increase the spectral feature contrast. Then, their second derivatives were vector normalised. This spectral pre-processing was performed using the OPUS software (v6.5, Bruker Optics GmbH). Last, the processed spectra were classified using hierarchical cluster analysis (HCA) and principal component analysis (PCA) unsupervised methods, with spectra from spread and non-spread drops recorded under the same conditions. HCA consisted of grouping the spectra according to their degree of similarity. The method was based on Euclidean distance calculation between all the data sets using Ward's algorithm.[19] The merging process used the spectral information contained in the region between 1800–800 cm^{-1}. The result is displayed as a dendrogram showing the grouping of spectra in clusters according to a heterogeneity scale. In PCA, the data are reformulated as a linear combination of uncorrelated principal components (PCs).[20] The first PC calculated accounts for the highest variance in the dataset. The second PC is uncorrelated with the first one *i.e.*, perpendicular to it and explains the next highest variance of the dataset, and so on. The PCs are also called loading vectors and their weights PC scores. PCA was performed in the same spectral range to analyse the spectra from spread and non-spread drops and to determine if the two sets of spectra are separated.

2. Impact of the dilution solvent: distilled water *versus* physiological water

In this set of experiments, the impact of the solvent used for dilution, distilled water or physiological water, was analysed. Three-fold diluted serum and plasma were prepared with each solvent. Then, 5 μL of the samples were deposited onto a 384-well plate. After deposition, the drops were air-dried at room temperature for 1 h. Infrared spectra were acquired and pre-processed as described above. Finally, the pre-processed spectra were classified by HCA.

3. Impact of the type of anti-coagulant for plasma collection

Plasma samples were obtained by centrifugation of blood collected in tubes containing EDTA or lithium heparin as anticoagulants. After three-fold dilution, EDTA and lithium heparin plasma samples of 5 μL were deposited onto a 384-well plate and air-dried at room temperature for 1 h. Then, spectra were acquired and subjected to a quality test as described above. Spectra that passed the quality test were pre-processed and classified by HCA. A dendrogram was constructed to illustrate the level of similarity between the spectra of plasma samples from EDTA and lithium heparin collection tubes. In order to evaluate the impact of EDTA (MW: 292 Da) on the spectral profile, the FTIR spectra of the plasma samples were also analysed after dialysis with a 10 kDa membrane to filter out the EDTA molecules.

4. Spectral reproducibility

4.1 Impact of freeze–thaw cycles. The potential impact of repeated freeze–thaw cycles on the quality test and reproducibility was studied. Fresh serum and serum samples after up to five consecutive freeze–thaw cycles (1 h between each cycle) were studied. After thawing and three-fold dilution, serum samples of 5 μL were deposited onto a 384-well plate and dried at room temperature for 1 h. Then, spectra were acquired and subjected to a quality test as described above. Spectra that passed the quality test were pre-processed and classified by HCA to analyse the level of similarity between spectra according to the number of freeze–thaw cycles. In addition, a PCA was performed and a PCA score plot was constructed to determine if separation of the different sets of spectra could be obtained.

4.2 Inter-operator spectral reproducibility. Inter-operator reproducibility was studied by comparing the spectra obtained by three different operators analysing the same sample under the same experimental conditions. Three-fold diluted serum was used for this purpose with 5 μL deposits onto a 384-well plate. After 1 h drying at room temperature, spectra were acquired and those passing the quality test were pre-processed and analysed by HCA and PCA as described in the previous sections.

4.3 Intra- and inter-plate spectral reproducibility, day to day reproducibility. Reproducibility tests were performed using 5 μL of three-fold diluted serum spread by the same operator on a 384-well plate. After 1 h drying at room temperature, spectra were acquired and subjected to a QT as described above.

Intra-plate reproducibility was studied by depositing 360 replicates of the same sample under the same experimental conditions.

Inter-plate and day to day reproducibility was studied by comparing serum spectra obtained from three different patients using 3 different silicon plates on three different days.

Spectra that passed the quality test were pre-processed and classified by HCA to analyse the level of similarity between spectra.

5. Impact of drying modalities on spectral reproducibility

Since spectra from spread or non-spread drops are acquired after a drying period, the impact of drying duration on the QT and on the reproducibility was studied. Three-fold diluted serum and plasma as well as pure bile were used. Samples of 5 μL were deposited onto a 384-well plate. Then, they were dried for 45, 60, 120 min and 24 h before FTIR measurements. The experiments were performed at 25 °C in a close environment such as an incubator and at room temperature. Spectra were acquired and subjected to a QT as described above. Spectra that passed the QT were pre-processed and classified by HCA and PCA as described above.

Results and discussion

Due to the development of high-throughput molecular technologies and associated bioinformatics, the number of studies dedicated to the identification of new diagnostic biomarkers has increased exponentially over the last two decades. However, only a limited number of biomarkers have been validated for use in medical practice.[21] This is mainly explained by methodological flaws. Plebani and Carraro reported that most errors (68%) occurred in the pre-analytical phase and only 13% and 19% errors occurred in the intra-analytical and post-analytical phases respectively.[22] As for other high-throughput technologies, vibrational spectroscopic investigations are subject to the same problems although no statistics exist to date. The huge amount of data generated by spectroscopic analysis exposes this analysis to a significant risk of false positive findings. The risks should be minimised by rigorously controlling the sample and patient related factors in the exploratory phase and by standardising the conditions of spectral acquisition in the pre-analytical step. Thus, care should be taken to validate the sample-related factors and patient-related factors. Standardisation of specimen collection and storage is crucial to reach experimental reproducibility not only in an individual laboratory but also between different laboratories. In addition, investigators should be aware of the risks of contamination during sample handling.

In the context of developing a high-throughput FTIR (HT-FTIR) transmission method for clinical diagnostics, we have investigated some important aspects involved in biofluid sample handling and preparation, which can impact on the spectral data and consequently on the outcome, and that we believe should be considered in the pre-analytical step prior to clinical applications. The study is focused on serum and plasma, and to a lesser extent bile, as they are the most encountered biofluids in clinical diagnostics.

1. Impact of biofluid volume and dilution on spectral quality test

Absorption FTIR spectroscopy is directly related to the Beer–Lambert law and the amount of analysed molecules has an impact on the signal quality.

Table 1 Results of quality test on FTIR serum and plasma spectra from spread and non-spread dried drops at various dilutions and volumes

Si plate	7 mm diameter well (96-well plate)								3 mm diameter well (384-well plate)								
Drop volume (μL)	5				10				1	3				5			
Dilution	0	1/2	1/3	1/4	0	1/2	1/3	1/4	0	0	1/2	1/3	1/4	0	1/2	1/3	1/4
Serum spread drops																	
Sample QT validation[a]	● (red)	● (green)	● (green)	● (red)	● (red)	● (green)	● (green)	● (green)	● (red)	● (red)	● (red)	● (green)	● (green)	● (red)	● (red)	● (green)	● (green)
Validated spectra/sample[b]	0	9	10	0	0	10	10	9	7	0	5	10	10	0	1	10	10
Serum non-spread drops																	
Sample QT validation[a]	● (red)	● (red)	● (green)	● (green)	● (red)	● (red)	● (red)	● (green)	● (green)	● (red)	● (red)	● (green)	● (green)	● (red)	● (red)	● (green)	● (green)
Validated spectra/sample[b]	1	2	9	10	0	0	2	10	8	0	4	10	10	0	1	10	10
Plasma spread drops																	
Sample QT validation[a]	● (red)	● (green)	● (green)	● (green)	● (red)	● (green)	● (green)	● (green)	● (red)	● (red)	● (green)	● (green)	● (green)	● (red)	● (red)	● (green)	● (green)
Validated spectra/sample[b]	0	10	10	8	0	10	10	10	0	0	9	10	10	0	0	9	10
Plasma non-spread drops																	
Sample QT validation[a]	● (red)	● (red)	● (red)	● (green)	● (red)	● (red)	● (red)	● (green)	● (red)	● (red)	● (red)	● (green)	● (green)	● (red)	● (red)	● (red)	● (green)
Validated spectra/sample[b]	0	1	6	10	0	0	0	10	2	0	0	10	10	0	0	5	10

[a] The sample was validated (green dot) when at least 8 replicates out of 10 met the quality test (QT) and discarded when less than 8 replicates met the QT (red dot). [b] 10 replicates per sample were deposited.

The results are presented in Table 1. The dilution and volume of the samples as well as the size of the wells have a significant impact on the spectral QT. Spectra from the pure serum and pure plasma using a volume ≥ 3 μL did not meet the QT because the amide I band is saturated with a maximum absorbance intensity above 2 arbitrary units (Table 1). However, when the volume of pure serum was reduced to 1 μL, the majority of the spectra were validated for non-spread deposits (Table 1). This is a point of great interest for clinical application as the deposit of a biofluid without dilution and without spreading would be the ideal approach for a high-throughput technology since it minimizes the number of operator interventions and thus reduces the risk of variability in the results. However, pipetting such low volumes can introduce additional errors and impact on the reproducibility of the spectral data (ESI 1†). The results deserve to be confirmed on a larger number of samples. Furthermore, the approach has not been generalised to plasma samples (Table 1). The reason why the great majority of spectra from a pure plasma sample of 1 μL did not meet the QT is an open question but is probably related to the presence of coagulation proteins in plasma which are not in serum. When the plasma and serum samples were diluted, the amide I band absorbance decreased to the acceptable range due to the lower concentration of proteins (ESI 2†) in the sample and the proportion of QT-validated spectra increased. The two types of plates that are currently used in HT-FTIR spectroscopy are the 96-well and 384-well silicon substrates. Both types of plate have been tested in this study. Using the 96-well Si plate with 7 mm diameter wells, the optimal dilution for serum and plasma appears to be two-fold for spread drops and four-fold for non-spread drops (Table 1). Using the 384-well plate with 3 mm diameter wells, the optimal dilution was three-fold for serum and three- or four-fold for plasma depending on the volume (Table 1 and Fig. 2). Overall, higher dilution and lower volume appear necessary to meet the QT when using the 3 mm diameter wells compared with the 7 mm diameter wells. This underlines that the analyte amount in the deposit area is a limiting factor for obtaining QT validated spectra. As an example, in our experiments a serum volume of 10 μL 2-fold diluted and deposited on a 7 mm diameter well resulted in a QT-validated spectrum as did a volume of 3 μL 3-fold diluted and deposited on a 3 mm diameter well.

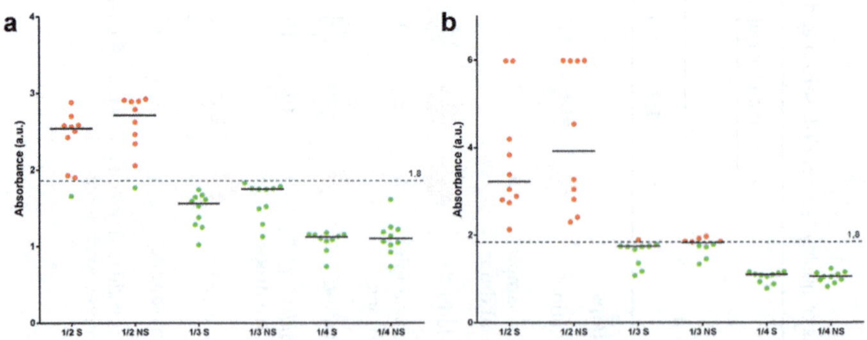

Fig. 2 Scatter plots based on the amide I absorbance value of serum (a) and plasma (b) replicate FTIR spectra from spread and non-spread dried drops (5 μL deposits onto 3 mm diameter well) at various dilutions. Validated and discarded spectra after the quality test are represented by green and red dots respectively. The median absorbance value is represented by the black bar. S: spread dried drop, NS: non-spread dried drop.

However, given the specificity of each instrument, it should be underlined that each set up necessitates a proper pre-analytical validation.

From HCA and PCA, it can be noted that the pre-processed spectra from 2-, 3- and 4-fold dilution spread drops on 7 mm diameter wells are separated in three different groups (ESI 3†) whereas the pre-processed spectra from 3- and 4-fold dilution spread drops on 3 mm diameter wells are not differentiated (ESI 4†). The above results indicate a higher heterogeneity with the 7 mm diameter wells compared to the 3 mm diameter wells although the spectra were from the same sample and were normalised. It should be mentioned that a 5 µL deposit onto 3 mm diameter wells covers the whole area of the well, which avoids drop spreading and represents an advantage in terms of pre-analytical validity. In this context and depending on the dilution factor, a bigger volume is needed for the 7 mm diameter wells, which will induce a longer drying time. This is illustrated in ESI 5† where photographs of different volumes of serum dried on the two types of plates are shown. The impact of serum drop spreading on the spectral profile was studied on both 7 and 3 mm diameter wells and the data were analysed by HCA and PCA. For the 7 mm diameter wells, the HCA plot showed two well-separated clusters of spectra corresponding to the spread and non-spread drops indicating that the modality of drop deposition on the surface of the plate had an influence on the spectral profile (ESI 6a†). These results were confirmed by PCA (ESI 6b†). In addition, PC1 and PC2 loadings tend to indicate modifications in the protein spectral profile (ESI 6c†). In contrast, for the 3 mm diameter wells, both HCA and PCA showed no distinction between the spread and non-spread dried drops spectra (ESI 7a–c†). This supports the above mentioned argument that the 5 µL spread or non-spread deposit onto a 3 mm diameter well covers the whole area of the well. Similar results were obtained for the plasma samples (data not shown). Thus, 5 µL of three-fold diluted serum or plasma appears to be the ideal volume for high-throughput analysis. Fig. 2a and b confirm these latter results respectively for serum and

Table 2 Results of quality test on FTIR bile spectra from spread and non-spread dried drops at various volumes

Silicon plate	7 mm diameter well (96-well plate)		3 mm diameter well (384-well plate)		
Drop volume (µL)	5	10	1	3	5
Dilution	0	0	0	0	0
Bile spread drops					
Sample QT validation[a]	🟢	🟢	🟠	🟢	🟢
Validated spectra/sample[b]	10	10	0	10	10
Bile non-spread drops					
Sample QT validation[a]	🟠	🟠	🟠	🟠	🟠
Validated spectra/sample[b]	6	6	0	6	7

[a] The sample was validated (green dot) when at least 8 replicates out of 10 met the quality test (QT) and discarded when less than 8 replicates met the QT (red dot). [b] 10 replicates per sample were deposited.

plasma *via* the scatter plots based on amide I absorbance by comparing the dilution factor (2-, 3- and 4-fold) as well as the spread and non-spread procedures.

Bile has been treated separately from serum and plasma because its composition is different. This is mainly due to lower concentrations of metabolites in bile compared to serum or plasma. Protein concentration in bile is about 5 g L^{-1} (ref. 23) as opposed to 60 to 80 g L^{-1} in serum. These specificities allow bile samples to be measured in their pure forms (ESI 8†). The low absorption, in particular the amide I band, required the implementation of an adapted QT. Our results show that all pure bile spectra (except for 1 μL) were QT-validated for spread dried drops. All spectra from non-spread dried drops were rejected (Table 2).

2. Impact of the dilution solvent: distilled water *versus* physiological water

Distilled water and physiological water are often used as solvents in many analytical assays. Since biofluids contain high protein concentrations, we wanted to investigate whether these solvents could have an effect on spectral analysis. The results are displayed in Fig. 3a where the dendrogram shows that spectra are clearly divided into two main clusters corresponding to each solvent used for dilution. This indicates that the solvent has a major impact on the spectral characteristics. It is known that distilled water induces protein precipitation (turbid solution), an effect which is not observed with physiological water (photograph inset, Fig. 3b). If sample dilution is necessary, it is recommended using physiological water rather than distilled water.

3. Plasma collection: impact of the type of anti-coagulant

For biochemical assays used routinely in clinics, blood samples are collected using either lithium heparin or EDTA tubes depending on the targeted analyses.

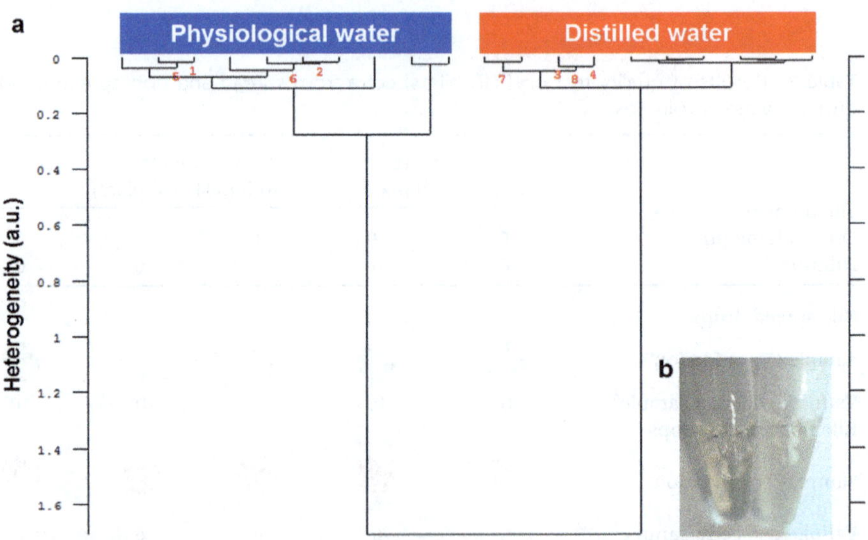

Fig. 3 (a) Hierarchical cluster analysis of FTIR pre-processed spectra from dried spread serum drops 3-fold diluted using either physiological or distilled water (deposits of 5 μL onto 3 mm diameter wells). (b) Photograph of serum diluted with physiological water (left) and distilled water (right).

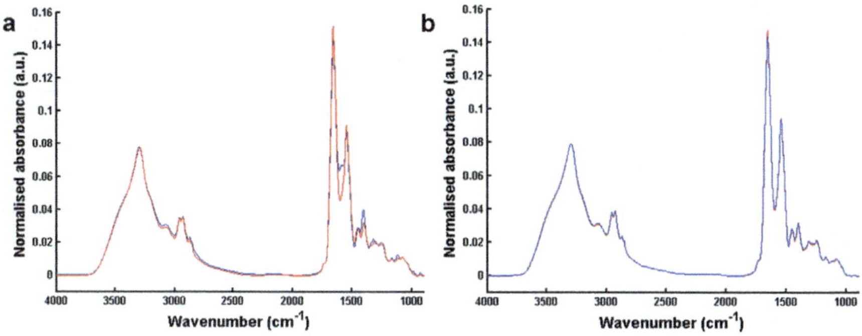

Fig. 4 Comparison of FTIR pre-processed mean spectra of plasma collected in tubes containing lithium heparin (red) or EDTA (blue) as anticoagulants without dialysis (a) or after dialysis with a 10 kDa membrane (b).

However, can these anti-coagulants have an effect on spectral data? To answer this question, spectra of plasmas from lithium heparin and EDTA tubes were acquired and compared in Fig. 4a. It can be observed that plasma from EDTA tubes exhibit additional spectral features compared to plasma from lithium heparin tubes, specifically in the fingerprint region (1800–800 cm^{-1}). After dialysis using a 10 kDa membrane, these artefactual peaks originating from small EDTA molecules can be removed (Fig. 4b). HCA classification displayed a complete separation between spectra from EDTA and lithium heparin plasma samples before dialysis. This clustering was no longer observed after dialysis (data not shown). Given these data, it appears more appropriate to use lithium heparin tubes for spectroscopic analysis of plasma. Unless there is a specific requirement for using plasma samples, serum appears to be the sample of choice for biofluid spectroscopy. The use of plasma is important in specific diseases (*e.g.*, hemophilia) where coagulating proteins are necessary for biochemical and hemostasis assays. In contrast with plasma, there is no need for using anticoagulants to collect serum rendering sample handling easier.

4. Spectral reproducibility

4.1 Impact of freeze–thaw cycles on spectral reproducibility. The impact of repeated freeze–thaw cycles before biofluid sample analysis is a matter of concern in biomedical technologies. Although they can be obtained non-invasively, serum and plasma are very valuable sample materials and they are stored at −80 °C where they are supposed to be relatively stable. However, for clinical studies, repeated freeze–thaw cycles are unavoidable when the sample volume is limited. Thus, inadequate conditions for sample storage can have an impact on the measured data as reported in a recent metabolomics study.[17] Taking into account this finding, we have studied the effect of freeze–thaw cycles on the spectral data. Fresh serum and serum samples after up to five consecutive freeze–thaw cycles were analysed and their spectral profiles compared. Fig. 5a displays the HCA results obtained in the 1800–800 cm^{-1} spectral range and clearly demonstrates a clear-cut delineation between fresh and freeze–thawed serum samples even after 1 cycle.

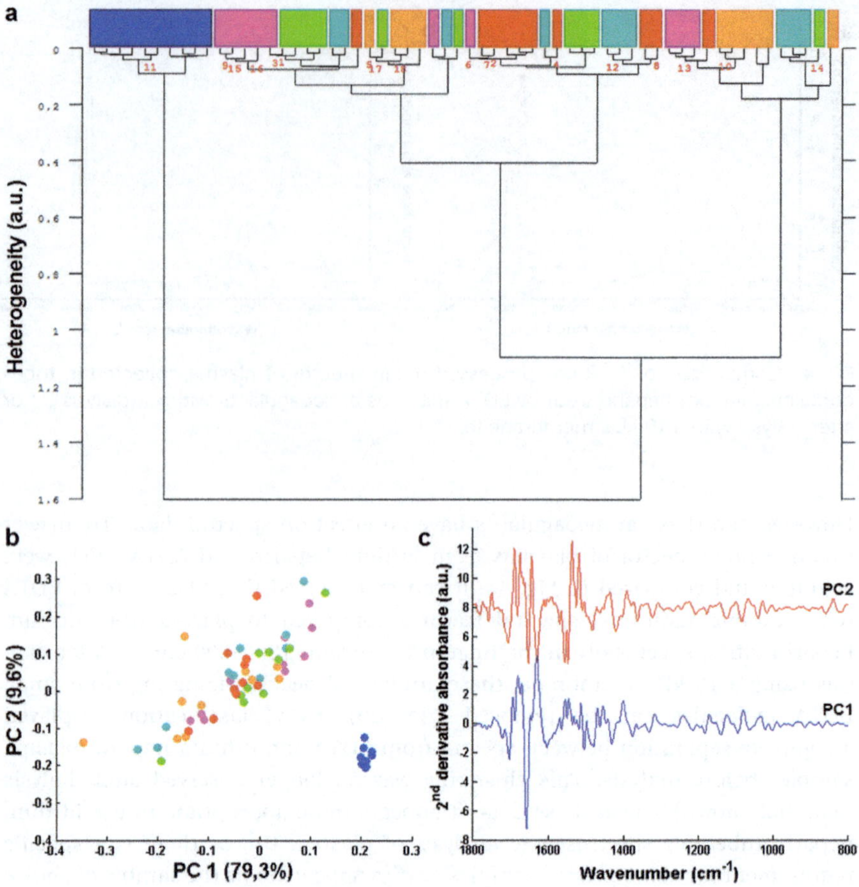

Fig. 5 (a) Hierarchical cluster analysis of FTIR pre-processed spectra from fresh serum and after up to 5 freeze–thaw cycles (deposits of 5 μL, 3-fold diluted onto 3 mm diameter wells). (b) PCA scatter plot of the spectral dataset from (a), (c) PC1 and PC2 loadings from (b). Fresh serum (dark blue), freeze–thaw cycle 1 (pink), 2 (light blue), 3 (orange), 4 (green), 5 (red).

The results were confirmed by PCA showing the fresh samples as a completely separate group (Fig. 5b). Both analyses show that all freeze-thawed data were mixed. Furthermore, the loadings of PC1 and PC2 tend to indicate that the most prominent modifications occur in the protein region (Fig. 5c). When the fresh samples were removed, data analysis did not reveal any structures in the spectral datasets of the five freeze–thaw cycles (data not shown). Similar results have been obtained when fresh plasma was compared with plasma samples after up to five consecutive freeze–thaw cycles (data not shown). These findings suggest that FTIR spectroscopy is sensitive enough to distinguish fresh from freeze-thawed serum and plasma samples but cannot differentiate between the five freeze–thaw cycles. On the other hand, metabolomics techniques being more sensitive have shown changes in metabolite ions[24] and macromolecules[25,26] after a few freeze–thaw cycles. Since sample freezing is unavoidable (large collection, multicentric studies, transportation…), it is therefore recommended to avoid multiple freeze–

thaw cycles of the same sample and favour multiple pure sample aliquots. However, our study has not taken into account the effect of freezing time on the spectral data, in particular for long term storage, and it would be interesting to investigate if FTIR spectroscopy could monitor any content deviations in metabolite ions and macromolecules like proteins, DNA and RNA.

4.2 Intra- and inter-plate spectral reproducibility, day to day reproducibility. In this study, intra- and inter-plate as well as day to day reproducibility has been investigated. For intra-plate reproducibility, 360 replicates of the same sample were deposited and measured by the same operator. The HCA did not show any specific clustering, all spectra were completely mixed and the replicates were distributed over the whole cluster (data not shown).

Day to day reproducibility was studied by analysing serum from three patients on three consecutive days. The dendrogram obtained after HCA (Fig. 6) shows that the clustering is only based on patient information. This is an important finding because the patient to patient variation is greater than day to day variation. This experiment has been reproduced on two other plates and no significant inter-plate spectral variability has been observed (data not shown).

4.3 Inter-operator spectral reproducibility. The inter-operator reproducibility was investigated by three different operators using the same instrumentation to analyse the same sample on the same plate and on the same day. This is to mimic a normal working day procedure in a clinical setup where the instrument could be used by several users. HCA and PCA have been used to visualise any structure in the dataset. Both the HCA and PCA results show that the spectral reproducibility is operator-independent (ESI 9†).

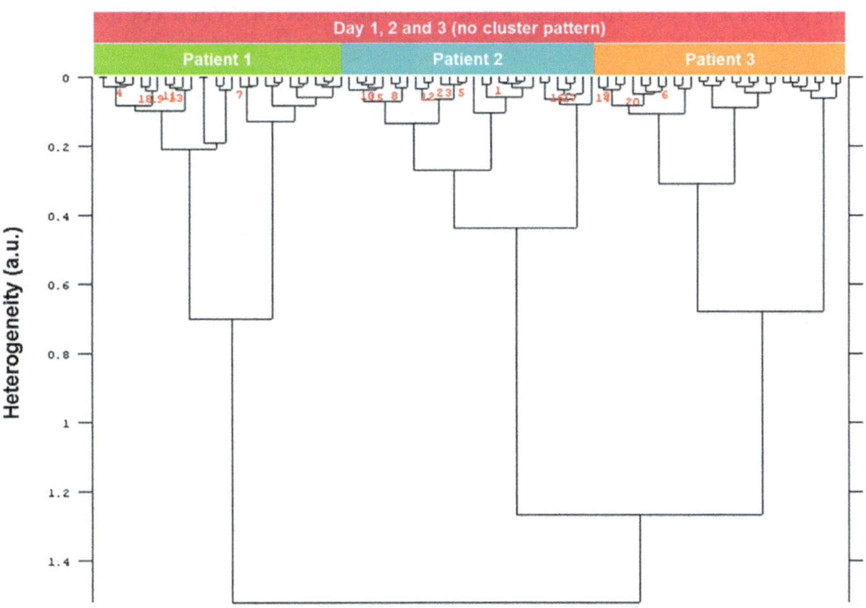

Fig. 6 Hierarchical cluster analysis of FTIR pre-processed spectra from the serum of three patient samples deposited on the same silicon plate on three different days (deposits of 5 μL, 3-fold diluted onto 3 mm diameter wells).

This is an important result for clinical implementation provided that the protocol is strictly followed.

The question is open whether this level of inter-operator reproducibility can be reached on different instruments of the same manufacturer or different manufacturers.

5. Impact of drying modalities on spectral reproducibility

High-throughput FTIR transmission spectroscopy of biofluids necessitates a drying phase for both spread and non-spread drops. This phase can impact on the QT and on the reproducibility of the spectral data. This effect has been studied on three-fold diluted serum and plasma as well as on pure bile. Five microlitre drops were dried at room temperature and at 25 °C for 45, 60, 120 min and 24 h before the FTIR measurements. Table 3 summarises the results obtained for the serum, plasma and bile samples and shows that a drying time of 45 min is not sufficient to meet the QT conditions (red dots). For all three biofluids, a minimum of one hour drying time is necessary at room temperature to pass the QT as opposed to two hours at 25 °C. However, it is important to note that it would be more appropriate to dry for a longer duration at room temperature to make sure that the drying process is complete. The spectra obtained after 2 h drying are differentiated by HCA from those recorded after 1 h and 24 h (ESI 10a†). Additionally, PCA shows a separation between 1 h and 2 h drying and between 2 h and 24 h drying at room temperature (ESI 10b and d†). The PC loadings clearly show that water absorption contributes to the spectral variance (ESI 10c and e†). The proximity of the clusters corresponding to data obtained after 1 h and 24 h drying suggests that sample rehydration may have occurred overnight. In a clinical application after the drying phase, it is therefore important to define the drying time and not to wait too long before recording spectra. When comparing the

Table 3 Quality test results on FTIR raw spectra of 3-fold diluted spread drops (5 μL) of serum (S), plasma (P) and pure spread bile (B) drops (5 μL) left dried at room temperature (RT) or 25 °C for 45, 60, 120 minutes and 24 h

	45 min			60 min			120 min			24 h		
	S	P	B	S	P	B	S	P	B	S	P	B
Drying at RT												
Sample QT validation[a]	Red	Red	Red	Green	Green	Green	Green	Green	Green	Green	Green	Green
Validated spectra/sample[b]	3	0	1	10	10	10	10	9	9	9	10	9
Drying at 25 °C												
Sample QT validation[a]	Red	Red	Red	Red	Red	Red	Green	Green	Green	Green	Green	Green
Validated spectra/sample[b]	0	0	0	0	0	0	10	10	10	10	10	10

[a] The sample was validated (green dot) when at least 8 replicates out of 10 met the quality test (QT) and discarded when less than 8 replicates met the QT (red dot). [b] 10 replicates per sample were deposited.

spectra of drops dried at room temperature and at 25 °C, we found that HCA shows no clustering according to these two conditions (data not shown).

This is an interesting aspect for clinical use because sample drying can be performed at room temperature without the need of any additional laboratory equipment for controlling drying conditions.

Conclusion

The development of any laboratory analytical technique, including vibrational spectroscopy, that has the potential of being translated to clinical applications requires an evaluation of the potential factors of the pre-, intra- and post-analytical phases that can impact on data quality and consequently on the result outcome. It has been reported that in the procedures described for the laboratory analysis of a patient's sample, two-thirds of the issues originate from the pre-analytical procedures. In this context, we have investigated using HT-FTIR spectroscopy some aspects of the pre-analytical procedures and their impact on the data quality and reproducibility. The results show that the type of biofluid to be analysed will condition the analytical procedures to be implemented. In a non-exhaustive manner, we have shown that the sample collection modality, the type of substrate, the dilution factor, the volume of sample deposited, the way the sample is deposited, the drying conditions, the inter-operator, and the day-to-day variabilities are some of the aspects that need to be investigated for error tracking in the pre-analytical phase. Indeed, more work is needed to evaluate the intra- and inter-centre variabilities as well as the effect of other factors involved in the intra- and post-analytical phases. The study has focused on the HT-FTIR methodology but it will be interesting to have similar studies on other vibrational spectroscopic modalities.

Acknowledgements

The Defence and Science Technology Laboratory (Dstl, UK) and the Direction Générale de l'Armement (DGA, France) as well as the Société d'Accélération du Transfert de Technologie (SATT) NORD are gratefully acknowledged for the research funding. We are thankful to the URCA PICT-IBiSA Technological Platform for providing the necessary instrumentation. MJB would like to thank EPSRC, Dstl, Rosemere Cancer Foundation, Brain Tumour North West, and the Sydney Driscoll Neuroscience Foundation for the funding. Dr C. Lacombe is acknowledged for her contribution.

References

1 F. Bonnier, F. Petitjean, M. J. Baker and H. J. Byrne, Improved protocols for vibrational spectroscopic analysis of body fluids, *J. Biophotonics*, 2014, **7**, 167–179.

2 J. L. González-Solís, J. C. Martínez-Espinosa, L. A. Torres-González, A. Aguilar-Lemarroy, L. F. Jave-Suárez and P. Palomares-Anda, Cervical cancer detection based on serum sample Raman spectroscopy, *Lasers Med. Sci.*, 2014, **29**, 979–985.

3 C. Hughes, M. Brown, G. Clemens, A. Henderson, G. Monjardez, N. W. Clarke and P. Gardner, Assessing the challenges of Fourier transform infrared spectroscopic analysis of blood serum, *J. Biophotonics*, 2014, 7, 180–188.

4 C. Lacombe, V. Untereiner, C. Gobinet, M. Zater, G. D. Sockalingum and R. Garnotel, Rapid screening of classic galactosemia patients: a proof-of-concept study using high-throughput FTIR analysis of plasma, *Analyst*, 2015, **140**, 2280–2286.

5 A. L. Mitchell, K. B. Gajjar, G. Theophilou, F. L. Martin and P. L. Martin-Hirsch, Vibrational spectroscopy of biofluids for disease screening or diagnosis: translation from the laboratory to a clinical setting, *J. Biophotonics*, 2014, 7, 153–165.

6 J. Ollesch, S. L. Drees, H. M. Heise, T. Behrens, T. Brüning and K. Gerwert, FTIR spectroscopy of biofluids revisited: an automated approach to spectral biomarker identification, *Analyst*, 2013, **138**, 4092–4102.

7 W. Petrich, K. Lewandrowski, J. Muhlestein, M. Hammond, J. Januzzi, E. Lewandrowski, R. Pearson, B. Dolenko, J. Früh and M. Haass, Potential of mid-infrared spectroscopy to aid the triage of patients with acute chest pain, *Analyst*, 2009, **134**, 1092–1098.

8 I. Taleb, G. Thiéfin, C. Gobinet, V. Untereiner, B. Bernard-Chabert, A. Heurgué, C. Truntzer, P. Hillon, M. Manfait and P. Ducoroy, Diagnosis of hepatocellular carcinoma in cirrhotic patients: a proof-of-concept study using serum micro-Raman spectroscopy, *Analyst*, 2013, **138**, 4006–4014.

9 X. Zhang, G. Thiéfin, C. Gobinet, V. Untereiner, I. Taleb, B. Bernard-Chabert, A. Heurgué, C. Truntzer, P. Ducoroy and P. Hillon, Profiling serologic biomarkers in cirrhotic patients *via* high-throughput Fourier transform infrared spectroscopy: toward a new diagnostic tool of hepatocellular carcinoma, *Transl. Res.*, 2013, **162**, 279–286.

10 M. J. Baker, S. R. Hussain, L. Lovergne, V. Untereiner, C. Hughes, R. A. Lukaszewski, G. Thiéfin and G. D. Sockalingum, Developing and understanding biofluid vibrational spectroscopy: a critical review, *Chem. Soc. Rev.*, 2016, **45**, 1803–1818.

11 M. Griebe, M. Daffertshofer, M. Stroick, M. Syren, P. Ahmad-Nejad, M. Neumaier, J. Backhaus, M. G. Hennerici and M. Fatar, Infrared spectroscopy: a new diagnostic tool in Alzheimer disease, *Neurosci. Lett.*, 2007, **420**, 29–33.

12 V. Untereiner, G. D. Sockalingum, R. Garnotel, C. Gobinet, F. Ramaholimihaso, F. Ehrhard, M. D. Diebold and G. Thiéfin, Bile analysis using high-throughput FTIR spectroscopy for the diagnosis of malignant biliary strictures: a pilot study in 57 patients, *J. Biophotonics*, 2014, 7, 241–253.

13 E. Schleicher, The clinical chemistry laboratory: current status, problems and diagnostic prospects, *Anal. Bioanal. Chem.*, 2006, **384**, 124–131.

14 A. Aguilar-Mahecha, M. A. Kuzyk, D. Domanski, C. H. Borchers and M. Basik, The effect of pre-analytical variability on the measurement of MRM-MS-based mid-to high-abundance plasma protein biomarkers and a panel of cytokines, *PLoS One*, 2012, 7, e38290.

15 J. A. Bons, D. de Boer, M. P. van Dieijen-Visser and W. K. Wodzig, Standardization of calibration and quality control using surface enhanced laser desorption ionization-time of flight-mass spectrometry, *Clin. Chim. Acta*, 2006, **366**, 249–256.

16 W. B. Dunn, D. Broadhurst, P. Begley, E. Zelena, S. Francis-McIntyre, N. Anderson, M. Brown, J. D. Knowles, A. Halsall and J. N. Haselden, Procedures for large-scale metabolic profiling of serum and plasma using gas chromatography and liquid chromatography coupled to mass spectrometry, *Nat. Protoc.*, 2011, **6**, 1060–1083.

17 P. Yin, R. Lehmann and G. Xu, Effects of pre-analytical processes on blood samples used in metabolomics studies, *Anal. Bioanal. Chem.*, 2015, **17**, 1–14.

18 Y. Zheng, T. Qing, Y. Song, J. Zhu, Y. Yu, W. Shi, L. Pusztai and L. Shi, Standardization efforts enabling next-generation sequencing and microarray based biomarkers for precision medicine, *Biomarkers Med.*, 2015, **9**, 1265–1272.

19 J. H. Ward, Hierarchical grouping to optimize an objective function, *J. Am. Stat. Assoc.*, 1963, **58**, 236–244.

20 H. Abdi and L. J. Williams, Principal component analysis, *Wiley Interdiscip. Rev. Comput. Stat.*, 2010, **2**, 433–459.

21 E. P. Diamandis, Cancer biomarkers: can we turn recent failures into success?, *J. Natl. Cancer Inst.*, 2010, **102**, 1462–1467.

22 M. Plebani and P. Carraro, Mistakes in a stat laboratory: types and frequency, *Clin. Chem.*, 1997, **43**, 1348–1351.

23 A. Reuben, Biliary proteins, *Hepatology*, 1984, **4**, 46S–50S.

24 P. Yin, A. Peter, H. Franken, X. Zhao, S. S. Neukamm, L. Rosenbaum, M. Lucio, A. Zell, H.-U. Häring and G. Xu, Preanalytical aspects and sample quality assessment in metabolomics studies of human blood, *Clin. Chem.*, 2013, **59**, 833–845.

25 B. L. Mitchell, Y. Yasui, C. I. Li, A. L. Fitzpatrick and P. D. Lampe, Impact of freeze–thaw cycles and storage time on plasma samples used in mass spectrometry based biomarker discovery projects, *Cancer Inf.*, 2005, **1**, 98–104.

26 A. J. Rai, P. M. Stemmer, Z. Zhang, B. L. Adam, W. T. Morgan, R. E. Caffrey, V. N. Podust, M. Patel, L. Ly and N. V. Shipulina, Analysis of Human Proteome Organization Plasma Proteome Project (HUPO PPP) reference specimens using surface enhanced laser desorption/ionization-time of flight (SELDI-TOF) mass spectrometry: multi-institution correlation of spectra and identification of biomarkers, *Proteomics*, 2005, **5**, 3467–3474.

Faraday Discussions

PAPER

New IR imaging modalities for cancer detection and for intra-cell chemical mapping with a sub-diffraction mid-IR s-SNOM

H. Amrania,[a] L. Drummond,[a] R. C. Coombes,[b] S. Shousha,[b] L. Woodley-Barker,[b] K. Weir,[a] W. Hart,[a] I. Carter[a] and C. C. Phillips*[a]

Received 2nd November 2015, Accepted 30th November 2015
DOI: 10.1039/c5fd00150a

We present two new modalities for generating chemical maps. Both are mid-IR based and aimed at the biomedical community, but they differ substantially in their technological readiness. The first, so-called "Digistain", is a technologically mature "locked down" way of acquiring diffraction-limited chemical images of human cancer biopsy tissue. Although it is less flexible than conventional methods of acquiring IR images, this is an intentional, and key, design feature. It allows it to be used, on a routine basis, by clinical personnel themselves. It is in the process of a full clinical evaluation and the philosophy behind the approach is discussed. The second modality is a very new, probe-based "s-SNOM", which we are developing in conjunction with a new family of tunable "Quantum Cascade Laser" (QCL) diode lasers. Although in its infancy, this instrument can already deliver ultra-detailed chemical images whose spatial resolutions beat the normal diffraction limit by a factor of ~1000. This is easily enough to generate chemical maps of the insides of single cells for the first time, and a range of new possible scientific applications are explored.

1. Introduction

The principles of vibrational spectroscopy, (VS), have been established over several decades, and, for much of this time, reliable off-the-shelf systems have been available that are capable of delivering high quality spectroscopic image (SI) datasets for use by the biomedical community. These datasets have potential utility over a wide range of biomedical disciplines that include histopathology, cytology, surgical targeting and drug discovery.[1]

Very recently, technical advances in array detectors and in tuneable "Quantum Cascade laser" (QCL) narrow line sources have opened up new lines of

[a]*Physics Dept., Imperial College, London, SW7 2AZ, UK. E-mail: Chris.phillips@imperial.ac.uk; Tel: +44 (0)207 594 7575*

[b]*Department of Cancer and Surgery, Faculty of Medicine, ICTEM, Room 145, Du Cane Road, London W12 0NN, UK*

experimentation, and most would agree that the mid-IR part ($\sim$3 μm < λ < $\sim$10 μm) of the spectrum is experiencing, in technology terms, a renaissance. On the other hand, the reality is that take-up of these technologies by the biomedical scientists themselves, particularly clinicians, is still modest.

This paper starts by examining possible reasons for this, (with the aim of stimulating discussion), to set a context within which IR SI implementations may be evaluated for their genuine biomedical potential. It concludes by using this context to examine two new VS-based imaging technologies that we have developed. They are at roughly opposite ends of the labyrinthine "translational pathway" that, ideally, can take a technology from the physical science research lab into the clinic.

The first example, the so-called "Digistain" imager, is a very simple, stripped-down, SI system that is currently at the clinical validation stage. It arose when, rather than asking the usual scientists/technologists question, *i.e.* "How can I apply my new research technology in a new and hopefully useful way?" we tried inverting the argument and, instead, asked the question "Can we use our research expertise to develop a system that really could end up being used by clinicians?" Over the course of a 10 year multidisciplinary development programme, we found that this second question demands a far more considered and complex answer than the first. Hopefully we will show how well such a simple incarnation of SI technology can perform in a genuine sharp-edged clinical trial environment.

The second example is a new, probe-based, SI system that is very expensive and, at the moment, very difficult to use. It is formed by marrying an existing high-end commercial s-SNOM (scattering-type Scanning Near-field Optical Microscope) machine to an array of newly available QCL diode lasers. It can deliver SI datasets down to a $\sim\lambda/1000$ spatial resolution of $\sim$10 nm, *i.e.* sufficient to generate chemical maps inside a single cell for the first time. It came into being in a more traditional way, *i.e.* by asking the first of the two questions discussed above. Although we believe it has the potential to deliver game-changing imaging performance across the biosciences, it is at a very early stage, and at the moment we can only give a selection of demonstration images that hint at its potential in the biomedical arena.

1.1 Issues affecting the translation of VS technologies into the biomedical arena

There are several issues that came into prominence while we were trying to answer the "Can we use our research expertise, to develop a system that really could end up being used by clinicians?" question, and we choose to list them here. Some may seem obvious, but we generally underestimated their importance, and often we chose to ignore them until it became clear that to do so was no longer tenable. In retrospect, we believe that, if one is genuinely interested in seeing VS technology applied in a biomedical setting, much effort and resource can be saved if one is prepared instead to face these issues as early and as honestly as possible.

1.1.1 Operator dependence and reproducibility. The central question here can be framed as "Can you genuinely envisage an incarnation of your technology that could be used by a clinician, and in a way that would give the same answer every time?" A more dramatic way to put it might be "Would you stake your life on the reproducibility of your technology?" This outwardly simple technical question is often the show-stopper because regulatory procedures will demand that you can prove that you can answer it in the affirmative. It is a purely technical question, and long before regulatory approval becomes an issue, you will have to convince investors that, at some stage, you will be able to deliver a positive answer to it. If they cannot be convinced, all commercialization prospects will wither, and no matter how powerful the technology is, it will simply never leave the developers lab.

Because this issue is so central, simplified and well-understood data acquisition and processing protocols become paramount. This is true even if, to technologists, they appear to be technically inferior to more sophisticated approaches. Only the former can be heuristically tested, and then tightly specified, and written into a so-called "Standard Operating Procedure".

This "SOP" document is critical and it has to be able to "lock-down" the procedures in a way that limits variability to acceptable levels when the measurements are performed with different machines, in different labs, and by different people. A successful trial should not be thought of as validating a technology, but rather it validates a technology paired with a SOP. For VS-based technologies, one of the biggest problems comes with so-called "unsupervised" SI data processing methods, because variations, *e.g.* in image uniformity or signal-to-noise (SNR) ratio, can have completely unpredictable effects on the data components that the software settles on in the course of its data analysis.[18]

1.1.2 "Occam's razor" and the quest for technology validation. Vibrational spectroscopies and SI technologies deliver huge and rich datasets, and this is often cited as one of their strengths by the scientific community. In most practical applications however, the overwhelming majority of this data will have to be discarded and the central question then becomes how this should be achieved.

In clinical settings the degree of data reduction becomes especially extreme. Medical practitioners are already factoring in huge quantities of "soft" information when they, say, reach a diagnosis, or decide on a patient's treatment plan. As a consequence, any new technologies that hope to augment the clinical process will likely have to boil an extremely complex informational dataset down to a simple yes/no decision, or, at most, a number on a scale of 1 to 10. If this is not done, the data is unlikely to be of realistic use to a clinician.

The problem with SI technologies is the huge, (effectively infinite) parameter space. There are so many potential ways to reduce the SI data that a form of data paralysis can take hold. A chicken-and-egg problem can set in when it becomes clear that it will be impossible to explore all the possible ways of reducing the dataset to find the best one.

For clinical applications this problem is compounded by the fact that the details of the SOP have to be completely "locked down", (and given adequate IP protection), before any clinical trials can be started. Even slight changes to the SOP will be deemed by the clinical profession to invalidate previously acquired trial data that were taken with the previous SOP. The result is that one has to "nail ones colours to the mast" as early as possible.

All these factors, coupled with the fact that resources are always limited in realistic development programmes, mean that the way forward is likely to involve a substantial degree of educated guesswork and/or some very pragmatic decision making. Ideally this will be guided by a knowledge of the physical/chemical/physiological mechanism that is underpinning the discriminating or diagnostic power of the SI dataset. Often though this understanding is either partial or completely absent. In this case, decisions can be just as usefully driven by technological considerations, such as the limitations of available IR sources and detectors.

1.1.3 Giving biomedical professionals VS-based information in a form they can use. The point has already been made that clinicians are already factoring large amounts of information into their clinical decisions. Most of this is qualitative, and is generated by the clinicians themselves, *via* their own senses. Very little of it is couched in the language of numbers favoured by scientists and engineers, and most clinicians reach their decisions almost entirely from data that reaches them in the form of images. This is unlikely to change in the foreseeable future.

Moreover, in most cases (*e.g.*, conventional X-ray images, biopsy microscopy, CAT scans, ultrasonic images, MRI *etc.*), the image data is evaluated almost entirely by eye. The clinician does not, (and does not need to) understand *e.g.* the contrast mechanisms, or spectroscopic details that generated the image, in order to be able to exploit its clinical value. Even if the images are available in a reliably-calibrated digitized format (and usually they are not), it is unusual for any numerical data to be extracted from them for diagnostic purposes; perhaps the measurement of a physical dimension, or some single-digit index figure.

Taken together, these facts argue strongly for the VS community to concentrate on ways of presenting their spectroscopic information to clinicians in a simple, clear, visual way, probably sweeping all the technical details into the background.

1.1.4 Technology innovation *vs.* "unmet need". At some point, perhaps halfway along the metaphorical "translational path", the language of the debate changes abruptly from issues that can be characterized by questions like "How good is this technology, say compared with the competition? What new things can it do?" that initially drives we technology developers, to a "What existing problem can it actually solve?" issue. Thenceforward, funders, investors and users all focus on the so-called "unmet need", *i.e.* the problem, and one that is agreed by the users as being an existing and important one, which the new technology can reasonably hope to address.

The key point is that the "unmet need" exists entirely independently of the technology. It was probably there even before the technology was developed. It is defined by consensus of the potential users of a technology, and not by our community of technology developers.

If such an "unmet need" can be identified early on it can have a dramatic effect on the trajectory of a technology and save much wasted effort. This may seem so obvious as not to need saying, but the reality is that many, if not most, VS-based technology development programmes, target an "application" that was chosen, sometimes even invented, by the technology developer themselves. Usually this is done without sufficient (or, sometimes, any) discussion with potential end users.

2. The "Digistain" IR imager, for cancer detection and diagnosis

2.1 Background and the "unmet need"

At the moment, 99% of cancers are detected and characterised by excising a small "biopsy" tissue sample from the patient, drying it, slicing it thinly (~μm), and staining it with two vegetable dyes that the histopathology community settled upon over 100 years ago. Haematoxylin stains the acidic parts of the specimen (*i.e.* the cell nuclei) blue, and eosin stains the cytoplasmic proteins pink. A trained histopathologist then "grades" this "H + E" specimen by eye, in a standard optical microscope. These "grades" often inform life-critical treatment decisions. Although the grading SOP protocols vary between different cancers, always the judgements are subjective, and often there is poor concordance between histopathology practitioners. Considered worldwide, this constitutes a very significant "unmet need".

2.2 The physical principles behind "Digistain"

Our Digistain imager uses the physics of mid-IR absorption in the same excised and dried tissue slice as a quantitative way of mapping out the ratio of phosphate moieties, found predominantly in cell-DNA,[2] to the proteins in the cell cytoplasm. The chemical concentrations are measured by taking a set of calibrated images in four mid-IR bands[2] (Fig. 1). The resulting chemical map is converted into a computer generated false colour image[3] and presented to the histopathologist in a way that allows it to be used to augment and inform the process of tissue grading.

The idea seems simple, but in execution it presents significant technical challenges. Reproducibility requires that the effects of non-uniformity and non-linearity in the detector array, together with variations arising from non-uniform sample illumination, and fluctuating thermal background signals, all have to be identified and engineered out.[4] Also the 4 images, which are taken at widely separated wavelengths, have to be accurately calibrated and registered for the ratioing process to work. Finally the optical throughput and image acquisition has to be refined to the point where adequate SNR images can be obtained with practical acquisition times of only a few minutes.

2.3 Factors affecting likelihood of successful clinical application

The choice of these two particular chemical moieties for Digistain was determined by a mix of the factors outlined in the previous section. In the first place, they give prominent features in the IR spectra that are within reach of a combination of comparatively basic (*i.e.* uncooled) mid-IR emitters and detectors. This added to their attractiveness as the features of the dataset most worthy of preserving in the face of the "Occam's razor" principle (Section 1.2.2).

Secondly, these features provided an image that could be thought of as being an approximate analogue of the existing H + E one. The concentration of the phosphate moiety correlates with the acidic components in the tissue that turn out blue in the H + E images, and the amide moiety correlates with eosinophilic components of the tissue that stain pink. This way we aimed to reduce the SI image data to a form that was "pre-digested" into a format that was best suited for the histopathologists to assimilate and use, in line with the considerations of Section 1.2.3.

Thirdly, we only settled on this subset of the possible SI data after having already undergone a science-driven research programme consisting of several years of technologically complex laser based spectroscopy.[5,6] Although this early study itself did little to establish a route to a viable clinical application, it had involved numerous consultations with histopathology professionals. These early

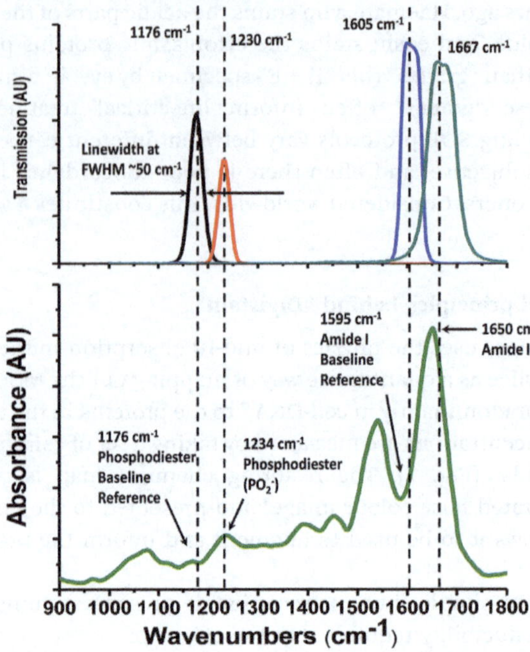

Fig. 1 Lower trace, a typical absorption spectrum of dried, unstained, human biopsy tissue, in this case, a 2.5 mm × 2.5 mm section of an oesophageal biopsy section, initially harvested using a standard "FFPE" (formalin fixed paraffin embedded), protocol, then subsequently deparaffinised. It is recorded at 4 cm^{-1} spectral resolution with an FTIR microscope. Upper trace, transmission spectra of the 4 bespoke mid-IR bandpass filters used in the "Digistain" imager. They work in pairs to acquire calibrated signal and baseline images at wavelengths corresponding to absorption by the amide I and phosphodiester chemical moieties.

consultations allowed us to identify and roughly target an important histopathology "unmet need" (Section 1.2.4), and steered the design principles behind the prototype imager of Fig. 2.

Fourthly, although there are of course many types of cancers it is well known that, in the majority of them, the cell reproduction cycle becomes significantly disrupted as healthy tissue becomes diseased. This is reflected in the fact that histopathologists frequently employ measures of both "pleomorphism" (changes in the relative size and shape of the cell nuclei) and "mitotic activity" (the rate of cell division) as key factors in their tissue grading SOPs. The "Digistain Index" (DI) that our imager measures is proportional to the ratio of concentrations of the phosphate to amide moieties, so we had grounds for hoping that a robust way could be found to correlate this data with tumour grade.[7–9] In other words we were using an educated guess (Section 1.2.3) to prejudge the data reduction process that would generate the most useful images, before committing to large-scale clinical trials.

Fifthly, with Digistain we were focussed on an optical measurement, *i.e.* mid-IR absorption in a smooth-sided tissue slice, which we could reasonably hope to be able to make with good robustness and reproducibility (Section 1.2.1). By taking an appropriately scaled and background corrected ratio[3] of the absorption

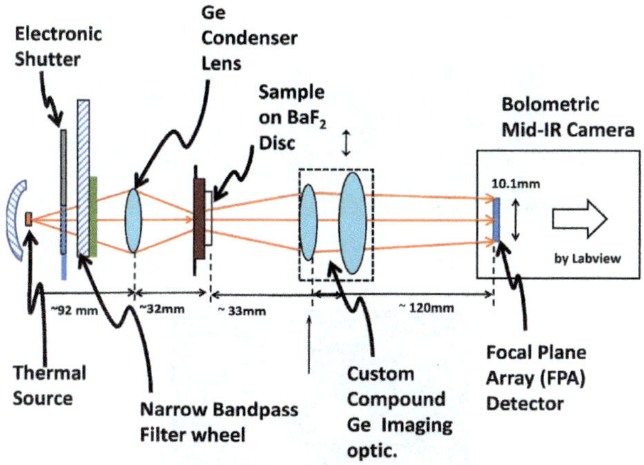

Fig. 2 Schematic of fully automated "Digistain" IR imager for histopathology. At the base (LHS) of the unit is a broadband thermal IR source with an electronically controllable shutter. The IR beam then passes through a set of bespoke, computer-selectable, narrowband IR filters, and then through a germanium condenser lens carefully designed to maximize the optical throughput. The biopsy sample is mounted on a 25 mm diameter BaF_2 disc and is imaged, through a custom designed low aberration imaging optic, onto a bolometric focal plane array with a magnification of $\times 3.1$ and at a diffraction-limited spatial resolution. The instrument has a 30×30 cm bench footprint, requires only a single 13 A socket, and can be operated, with a single mouse click, by untrained personnel.

of the moieties (Fig. 3) we could design a new SOP (3) that would be least sensitive to operator-dependent factors. The latter are mostly due to variations in the thickness of the tissue slice, but there is also the issue of fixing protocols, which can vary from site to site.

Sixthly, the Digistain approach presents the mid-IR spectroscopic information visually, and in a way that can readily be compared with existing H + E slides

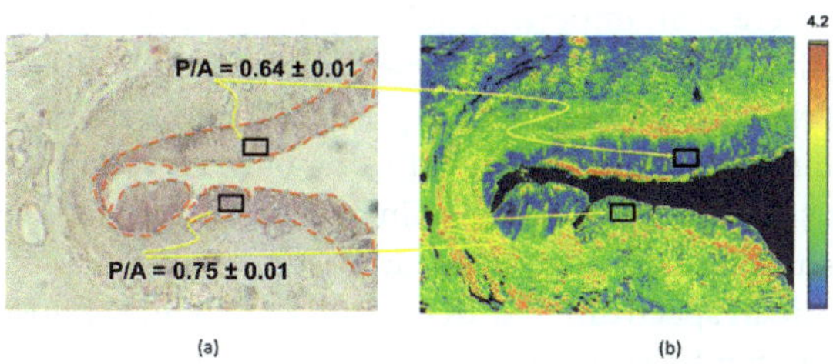

Fig. 3 Example of the discriminating power of a Digistain image, in this case of an oesophageal biopsy. Left: Conventional "H + E" stained biopsy section. The regions delineated with red dotted lines are epithelial tissue, and the upper one is healthy (as judged by a trained histopathologist), whereas the lower right one is dysplastic. Right: Corresponding "Digistain" image, taken from an unstained slice of tissue directly adjacent, on the FFPE tissue block, to the one used for the H + E stain. It is a false-colour image constructed from the measured "phosphide-to-amide" ratio, P/A, extracted from the mid-IR images, as described in Fig. 1 and 2. The healthy tissue shows up blue, and the dysplastic tissue green. Averaging this "P/A ratio" (which we call the "Digistain Index", DI), over the pixels in the two black rectangular areas gives numerical values for the two tissue types that differ at the $>33\sigma$ level.

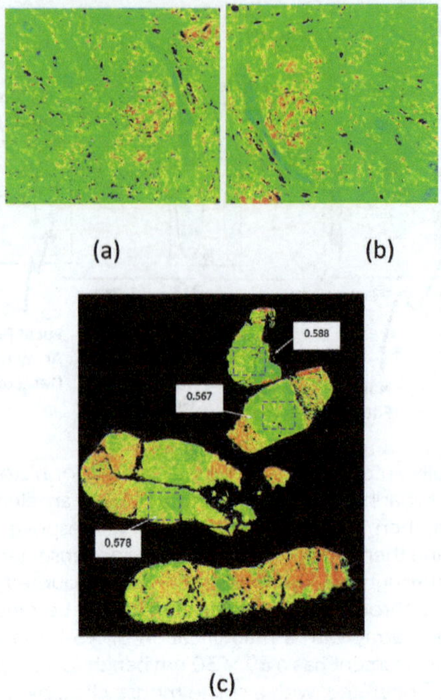

Fig. 4 Examples of Digistain reproducibility test. (a) and (b) are images of the same slice of breast biopsy tissue, but with the sample rotated and translated between scans. The circled areas are the same portion of tissue in each case, and return "Digistain Index" values of 1.0667 ± 0.009 and 1.061 ± 0.009 (3σ limits), implying a technical reproducibility better than one part in 100. (c) contains three different core biopsies taken simultaneously from the same histological "grade 2" breast tumour. The boxed green areas were identified as tumorous by a trained histopathologist on the basis of the corresponding H + E slides. The "Digistain Index" here is reproducible to DI $= 0.576 \pm 0.089$ across different parts of the tumour.

(Section 1.2.3). There is an important point here; because the H + E slides are the clinician's so-called "gold standard" they will continue to be used in any event. No clinician would risk running against the grain of the profession by attempting to diagnose a condition as serious as cancer using only a revolutionary and untested technology. Often we are asked if Digistain can be adapted for *in vivo* use and the answer, in the short term at least, is no, at least not without a major engineering programme on broad spectrum mid-IR compliant imaging endoscopes. However, in some ways this can be seen as an advantage rather than a shortcoming. If we had, say, produced a machine that required an extra *in vivo* endoscopy procedure on top of the established standard biopsy one, it would probably have been much harder to test it (it would have needed separate live patient trials) in order to get it adopted.

A critical factor in Digistain's favour is that after the H + E biopsies have been examined, the excised tissue samples are chemically preserved in wax (which can easily be dissolved away many years later), and archived, along with the original H + E slide, and a full medical history. There are millions of documented specimens of all types of cancer that one can go back to and study

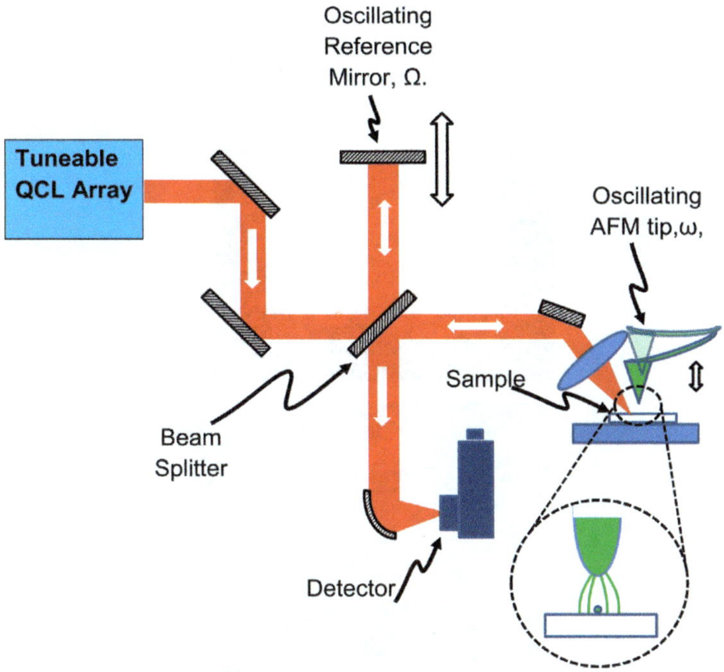

Fig. 5 Schematic of the mode of operation of a "scattering-type Scanning Near-field Optical Microscope" (s-SNOM). An AFM tip hovers close to the sample surface and vibrates, vertically, at a frequency ω. IR laser light is focused onto the tip region, and concentrated, into a spatial area roughly corresponding to the tip diameter (inset), by the "lightning rod" effect. By detecting the portion of the backscattered light fluctuating at 3ω (*i.e.* a harmonic of the tip vibration), the light backscattered from this small region can be isolated. An interferometer (with a reference mirror oscillating at $\Omega \ll \omega$) is used to measure both the amplitude and phase of the backscattered light, from which the optical properties of the very small sample area that is interacting with the AFM tip can be inferred. An image is assembled by rastering the AFM tip across the sample in the usual way.

with Digistain. This means it can be tested and validated with logistical and ethical hurdles that are far lower than if one had to secure access to cohorts of new live patients to trial it with. Of course the issues of reliability and reproducibility are just as key here; they require longitudinal trials to establish the degree of chemical stability, specifically, as judged by our new imaging method, afforded by the FFPE method (Fig. 4).

At present we are nearing the end of a clinical trial, involving upwards of 200 breast cancer patients. They were assessed according to the RECIST criteria (a set regulated assessment criteria that all breast cancer diagnostic methods must follow), and the biopsies were fixed using the RCP (Royal College of Pathologists) fixing protocol. To carry weight with the medical profession, these trials have to be performed, by a large team, under strict conditions of anonymity, double-blindness and with suitably experienced and accredited statistical analysis staff. Unfortunately, for this reason, we are not allowed to discuss the results at all until they have been formally released, but we hope to be able to discuss this in more detail at the meeting.

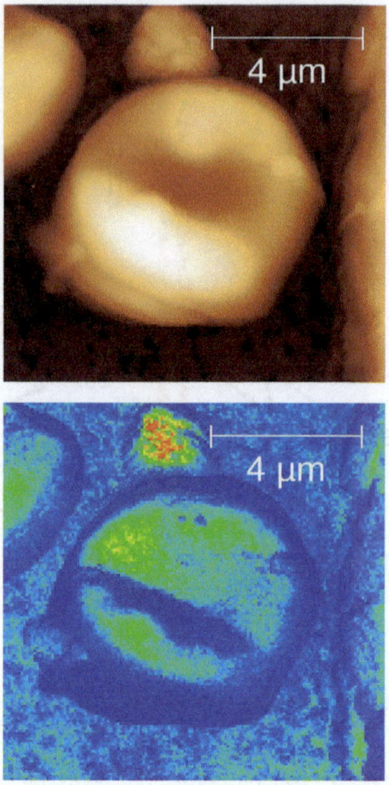

Fig. 6 Upper; AFM scan of the physical structure of a single red blood cell in a squamous epithelial oesophageal biopsy. Lower image, s-SNOM 3rd harmonic image, taken at $\lambda = 9.81$ μm, ~20 min scan time and 67 nm pixel resolution. The resolution here is ~140 times better than the diffraction limit at this wavelength and has allowed us to image chemical contrast within the cell for the first time.

3. A tuneable s-SNOM for intracellular chemical mapping

3.1 Introductory comments

Typical cells are 10s of μm across, and although ~μm-level diffraction-limited resolution mid-IR imaging is possible (5) SNR limitations mean that it is only practical with sophisticated bench-scale laser sources. These are very challenging to implement, and they still fail to deliver enough resolution to "look inside" a single cell. In principle, Raman methods can deliver sub-micron resolution, but the inefficiency of the Raman signal generation process means that getting useful signals from beams that are focussed so tightly involves power densities that are likely to compromise biological material.

The advent of a cell-level chemical analysis capability would have a wide ranging impact across the biosciences. At present, chemical analysis of changes to cell chemistry involve generating a sample of a large number of cells, (say by flow cytometry or cell breeding), breaking them up, and then analysing the mix with any one of a number of chemist's tools, including wet-chemistry assay, IR

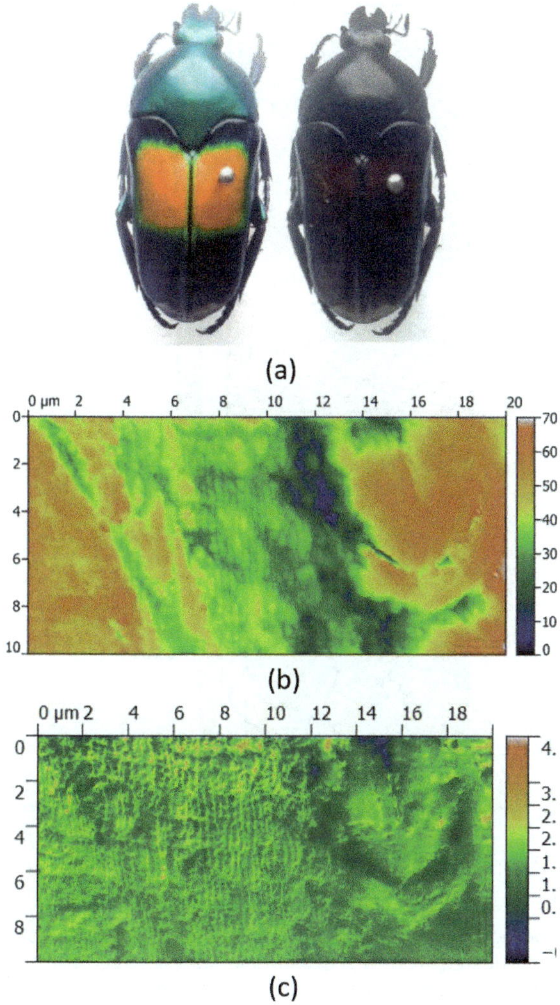

Fig. 7 (b) AFM topography and (c) $\lambda = 10.21$ μm s-SNOM scattered amplitude image of a "natural photonic crystal", *i.e.* the shell of an iridescent beetle (a) that has in-built periodic structure that gives it a chiral light scattering characteristic which means its shell markings only show up under circularly polarised light. The s-SNOM image has given the chemical contrast of the components that make up the multilayers in the insect shell. In order to generate the shell's interference features in the visible part of the spectrum, the concentrations of these components vary on a sub-micron length scale, and only the s-SNOM can deliver chemical images at this resolution.

spectroscopy, mass spectroscopy or labelling methods. Of course, all these "looking at the bits" approaches destroy the cell morphology information, and they necessarily average over any cell-to-cell variation in either chemistry or morphology.

Real cell-level imaging would have a major impact across the biosciences. The cell-to-cell variations within a population could give crucial insights into *e.g.* how cancer starts, (often with a single rogue cell), and, generally speaking cell-level pathology could be radically transformed. In pharmacology, the question of "How

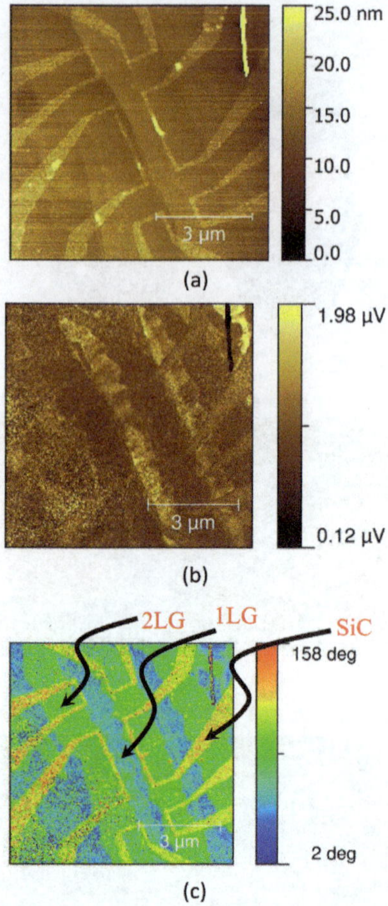

Fig. 8 (a) AFM topology, and $\lambda = 10.3$ µm s-SNOM amplitude (b) and phase (c) images of an FET transistor fabricated from a monolayer graphene (1LG) with occasional bilayer inclusions (2LG) sheet, etched away in places to reveal the SiC substrate.

does this drug act?" often boils down to "Where, and for how long, does this drug bind?" and being able to answer the latter, at a cell level, could have major implications for the drug discovery industry.

In this context, ultra-high-resolution imaging offers a unique "concentration advantage" feature. Fortunately, many pharmacologically active compounds have distinctive vibrational spectra, but most are active at mean concentrations in the 1 part in 10^9 region. Obviously this presents a major sensitivity challenge for VS-based technologies. If the drug binds at a particular site though, the local concentration can be dramatically elevated, and if the SI resolution is increased to the point where only a comparatively small number of molecules can be resolved, this effect can, in principle, lift the signal out of the noise.

So-called s-SNOMs promise to deliver this information.[10] They work by focussing mid-IR radiation onto the point where the probe tip of an atomic force microscope (AFM) touches a sample, and then analysing the degree to which the radiation is backscattered. The sharpness of the AFM tip concentrates the optical fields, *via* a "lightning rod" effect, into an area roughly commensurate with the AFM tip diameter. They offer spatial resolutions down to ~10 nm even with $\lambda \sim 10$ μm light, and they work with radiation densities and photon energies that are both low enough to be completely compatible with biological material.

However, the backscattered signal is small, and it requires a sophisticated interferometric signal recovery scheme[11] (Fig. 5) to pick it up. In turn this necessitates a coherent IR source. Usually this is a CO_2 laser that can only be tuned over a spectral range that is too narrow for useful IR spectroscopy. Until recently the only broad-band alternative source was a complex lab-scale laser system that used non-linear frequency mixing technology,[12] or an IR synchrotron,[13] but now the situation is rapidly evolving with the advent of commercially available "Quantum Cascade Lasers" (QCLs).

3.1.1 Brief synopsis of "Quantum Cascade Laser" technology. QCLs are a new type of laser diode[14] that use sophisticated semiconductor nano-structuring to produce a device that can be designed to lase across a wide wavelength range, from the mid-IR to the THz part of the spectrum. The technology is rapidly evolving, but at the moment, practical (*i.e.* operating at close to 300 K) commercially available devices roughly span the 5 μm ~ λ ~ 11 μm range.

However, each device typically tunes through 2 or 3 μm, so covering the normal mid-IR VS range is expensive because it needs 4 or 5 devices to be engineered into a single system. CW QCL operation is only available in parts of the spectrum, but they generate CW-equivalent powers in the mW range, with very narrow linewidths and close to diffraction-limited beam quality. The commercial QCL systems now offer turn-key operation, coupled to a convenient computer-controlled tuning system that makes them very easy to use.

Against this however, many of the QCL designs use external cavities for tuning the lasers, and these are extremely sensitive to small amounts of light being scattered back from the s-SNOM interferometer. This makes it difficult to get them to work stably enough to use with the s-SNOM. Also, the fact that some wavelengths are only available as pulse trains introduces technical complications to the signal chain[15] (as well as for imaging applications more generally) and in practice, when the QCL system switches between different laser devices as it is tuned, the s-SNOM optics usually have to be manually re-aligned.

3.1.2 Quantum Cascade Lasers applied to s-SNOM. Our s-SNOM system (NeaSNOM, Neaspec) uses probes with ~10 nm tip radius, oscillating at frequencies in the 50–300 kHz region. A range of QCLs are used, (Lasertune, Block Engineering, and MirCat, Daylight Solutions), both in CW and pulsed modes. When the lasers can only operate pulsed, steps have to be taken to synchronise the laser pulse rate to the sampling rate of the circuitry that digitises the detector signal, in order to avoid introducing large amounts of digitisation noise.[15]

Depending on the application, the backscattered light is polarised before being detected. In "plasmonic" experiments s-polarised incident light is used, because it can excite plasmonic modes in metal nanostructures without being sensitive to chemically based contrast mechanisms that arise between the different material components (usually a metal and a dielectric) that are used to make the nanostructure. If the detection channel is then p-polarised, (the so-called "cross polarisation scheme") it selects only the p-polarised field components that arise from the excitation of plasmons in the nano-photonic device, and is largely immune to chemical contrast effects.

One way to understand this is to notice that the incoming s-polarised light excites the resonant in-plane plasmon modes in the sample, without itself significantly interacting with the probe (which is a very poor antenna to anything but light polarised perpendicularly to the surface). For most biomedical experiments, it is the chemical contrast that is of interest, so both incoming and backscattered channels are p-polarised, as this is the polarisation that interacts with the tip and the sample most strongly (Fig. 5 inset). In all cases great care needs to be taken to avoid light backscattered from any of the optical components in the setup from reaching the QCL.

3.1.3 Demonstration images from the QCL s-SNOM. As well as the optical image, the s-SNOM can be operated as a standard AFM, and Fig. 6 shows the morphological information of a blood cell, taken from a deparaffinised FFPE biopsy slice, similar to those used in the Digistain experiments of Section 2. The lower trace is an early s-SNOM image, taken at a wavelength where stable QCL operation was possible. It's recorded at a 67 nm pixel resolution, and clearly records the cell structure at a $\sim\lambda/140$ resolution.

From the field of natural biophotonics, there are a number of iridescent beetle species that have evolved with periodic "photonic crystal" chiral shells, which give them distinctive markings that are visible only in circularly polarised light (Fig. 7). The question arises as to whether the chirality behind this is an accidental "self-organisation" effect arising from the natural chirality of the biomolecules that are forming the periodic structure in the shell, or whether the chirality has been generated by evolutionary forces. Fragments of the shell were polished to yield a smooth surfaced sample with minimal morphological contrast in the AFM image. The s-SNOM image however uses the chemical contrast to reveal the periodic structure of the chitin helices thought to be the origin of the effect.

Fig. 8 images a field effect transistor (FET) device fabricated for a graphene sheet grown epitaxially on a SiC substrate.[16,17] The AFM morphology image reveals the device structure. The s-SNOM phase image, in this example, approximately corresponds to the absorption in the graphene layer, and reveals regions of differing thickness, and the edge effects responsible for the device operation.

At the time of writing the QCL s-SNOM system is being reengineered to facilitate tuneable, full-spectrum spectroscopic imaging. The aim is to substantially increase the amount of scientific information that can be inferred from these images, and we hope to have new data to present at the meeting itself.

4. Conclusions

We have presented here two new ways of using mid-IR chemical imaging, and in both cases, as befits the topic of this discussion meeting, the focus has been on what they can offer the biomedical community. They are both at very different stages of technical readiness, and they face very different challenges in the future.

For Digistain the technical hurdles have been overcome, the clinical potential has essentially already been proven; the challenge now is to validate it using criteria and methods that will engage the support and interest of the clinicians. This is already underway, but we are well aware that it will have to be achieved on a wide enough scale to convince investors to produce them as tools to put in the medics' hands, because without that, the idea will come to nothing.

For the s-SNOM it is the beginning, and the challenges are still mostly technical. We believe it has a very bright future as a new scientific research tool for a wide range of biomedical problems, but, already, we are on the lookout for that elusive "unmet need".

Acknowledgements

Financial support from the UK Engineering and Physical Science Research council (grant numbers EP/K503733/1, EP/G031819/1 and EP/K 029398/1) is gratefully acknowledged. The Graphene FET samples were kindly supplied by Olga Kazakova and Vishal Panchal. We thank The Imperial Biomedical Research Centre and The Imperial Experimental Cancer Medicine Centre for technical assistance.

Notes and references

1 *Infrared and Raman Spectroscopic Imaging*, ed. R. Salzer and H. W. Siesler, 2009, Wiley-VCH, Weinheim, ISBN 978-3-527-31993-0.
2 H. Amrania, G. Antonacci, C.-H. Chan, L. Drummond, W. R. Otto, Nicholas A. Wright and C. Phillips, *Opt. Express*, 2012, **20**(7), 7290–7299.
3 C. C. Phillips and H. Amrania, *Infrared Imaging of Biological Material*, G B Patent, filed 29, April 2015, Application number PCT filed GB1507314.1.
4 L. Drummond, PhD thesis, Imperial College London, 2014.
5 H. Amrania, A. McCrow and C. Phillips, *Rev. Sci. Instrum.*, 2009, **80**, 123702.
6 H. Amrania, A. P. McCrow, M. R. Matthews, S. G. Kazarian, M. K. Kuimova and C. C. Phillips, *Chem. Sci.*, 2011, **2**(1), 107.
7 L. M. Merlo, L. S. Wang, J. W. Pepper, P. S. Rabinovitch and C. C. Maley, *Adv. Exp. Med. Biol.*, 2010, **676**, 1–13.
8 C. N. Battlehner, P. H. Saldiva, C. R. Carvalho, T. Y. Takagaki, G. S. Montes, R. N. Younes and V. L. Capelozzi, *Histopathology*, 1993, **22**(1), 31–34.
9 S. J. Jang, J. M. Gardner and J. Y. Ro, *Adv. Anat. Pathol.*, 2011, **18**(2), 165–172.
10 F. Keilmann and R. Hillenbrand, *Philos. Trans. R. Soc. London, Ser. A*, 2004, **362**, 787.
11 N. Ocelic, A. Huber and R. Hillenbrand, *Appl. Phys. Lett.*, 2006, **89**, 101124.
12 F. Huth, A. Govyadinov, S. Amarie, W. Nuansing, F. Keilmann and R. Hillenbrand, *Nano Lett.*, 2012, **12**(8), 3973–3978.
13 H. A. Bechtel, E. A. Muller, R. L. Olmon, M. C. Martin and M. B. Raschkeb, *Proc. Natl. Acad. Sci. U. S. A.*, 2014, **111**(20), 7191–7196.
14 J. Faist, F. Capasso, D. L. Sivco, C. Sirtori, A. L. Hutchinson and A. Y. Cho, *Science*, 1994, **264**, 553–556.
15 E. Yoxall, M. Navarro-C, M. Rahmani, S. A. Maier and C. C. Phillips, *Appl. Phys. Lett.*, 2013, **103**, 213110.
16 V. Panchal, R. Pearce, R. Yakimova, A. Tzalenchuk and O. Kazakova, *Sci. Rep.*, 2013, **3**, 2597.
17 V. Panchal, A. Lartsev, A. Manzin, R. Yakimova, A. Tzalenchuk and O. Kazakova, *Sci. Rep.*, 2014, **4**, 5881.
18 K. Atkinson, *Mid-IR Biological Imaging*, MSc thesis, Imperial College London, 2006.

Faraday Discussions

PAPER

High spatial resolution (1.1 μm and 20 nm) FTIR polarization contrast imaging reveals pre-rupture disorder in damaged tendon

Richard Wiens,[a] Catherine R. Findlay,[a] Samuel G. Baldwin,[b] Laurent Kreplak,[b] J. Michael Lee,[cd] Samuel P. Veres[ce] and Kathleen M. Gough*[a]

Received 15th November 2015, Accepted 11th January 2016

DOI: 10.1039/c5fd00168d

Collagen is a major constituent in many life forms; in mammals, collagen appears as a component of skin, bone, tendon and cartilage, where it performs critical functions. Vibrational spectroscopy methods are excellent for studying the structure and function of collagen-containing tissues, as they provide molecular insight into composition and organization. The latter is particularly important for collagenous materials, given that a key feature is their hierarchical, oriented structure, organized from molecular to macroscopic length scales. Here, we present the first results of high-resolution FTIR polarization contrast imaging, at 1.1 μm and 20 nm scales, on control and mechanically damaged tendon. The spectroscopic data are supported with parallel SEM and correlated AFM imaging. Our goal is to explore the changes induced in tendon after the application of damaging mechanical stress, and the consequences for the healing processes. The results and possibilities for the application of these high-spatial-resolution FTIR techniques in spectral pathology, and eventually in clinical applications, are discussed.

1. Introduction

We are using FTIR polarization contrast imaging (PCI) to examine mechanically damaged mammalian tendon. Characteristic features in the polarized FTIR spectra of collagen were recognized early in the development of infrared biospectroscopy,[1,2] even before the true triple helix molecular architecture had been discovered.[3–5] The polarization dependence results from the highly ordered arrangement of the basic units (tropocollagen) and their alignment within the

[a]*Department of Chemistry, University of Manitoba, Winnipeg, MB, R3T 2N2, Canada. E-mail: Kathleen.Gough@umanitoba.ca*

[b]*Department of Physics and Atmospheric Science, Dalhousie University, Halifax, NS, B3H 4R2, Canada*

[c]*School of Biomedical Engineering, Dalhousie University, Halifax, NS, B3H 3J5, Canada*

[d]*Department of Applied Oral Sciences, Dalhousie University, Halifax, NS, B3H 4R2, Canada*

[e]*Division of Engineering, Saint Mary's University, Halifax, NS, B3H 3C3, Canada*

aggregated structures of the microfibrils, fibrils and fibres.[6] The complex, hierarchical architecture of collagen facilitates a multitude of specialized, highly localized functions in tissues including tendon, skin, cartilage, ligament, scar tissue and bone. Under polarized infrared light, functional groups are vibrationally excited only when the alternating dipole change of the vibration is aligned with the polarization of the incident field. Vibrational spectroscopy has been an important technique for studying these tissues, providing molecular information on composition, orientation, post-translational modification and disease.[7–11]

Collagen fibrils are the fundamental load-bearing constituent of mammalian tissues. They are approximately 50–500 nm in diameter but can be mm in length.[12] Fibrils form as self-assembled linear aggregates of tropocollagen molecules (300 nm long $\times$ 1.5 nm in diameter); the latter consist of three helical chains wound together in a triple super-helix, stabilized by interchain hydrogen bonds.[13] Our knowledge of the molecular changes occurring within overload-damaged fibrils remains sparse. Using very high magnification SEM (100 000$\times$), Veres and Lee showed that both ruptured[14] and cyclically overloaded[15,16] bovine tendons contain collagen fibrils with "discrete plasticity": a characteristic damage morphology where discrete strain-induced kinks appear serially along the length of the nanoscale fibrils. Enzymatic probes and differential scanning calorimetry showed that some collagen molecules had undergone denaturation, *i.e.* uncoiling of the triple helix within fibrils that experienced discrete plasticity. This work led to the hypothesis that a molecular-based toughening mechanism is implicit in the structure of the mammalian collagen fibril. Our current hypothesis is that under discrete plasticity a designated subset of molecules within fibrils denature on overload, absorbing mechanical strain energy at many discrete locations along each fibril's length. Discrete plasticity in tendons may represent a naturally evolved, built-in, toughening design that is ultimately protective against actual tendon rupture.

Our goal is to obtain a molecular spectroscopic basis for understanding changes to the molecular structure of collagen fibrils, under mechanical overload injury. To that end, using a bovine tail tendon model in which overload-induced discrete plasticity damage has been well characterized,[14–16] we have studied matched-pair control and overloaded tendons with high-magnification FTIR PCI, confirming results with SEM and AFM. The highest spatial resolution far-field FTIR was attained at the IRENI beamline (Synchrotron Radiation Center, WI, now decommissioned) employing a Focal Plane Array (FPA) detector, to give 0.54 $\times$ 0.54 μm^2 pixels.[17] The Agilent high-magnification system used in the present work achieves the next best spatial resolution of 1.1 μm pixel dimension,[18] considered to be the practical limit for mid-IR with FPA.[19] This scale was sufficient to capture the orientation of bundled fibrils with fibres, including generalized sub-micron disorder, but could not provide sub-micron detail. To address the detection of finer details, single fibrils of the order of 300 nm in diameter were prepared from control tendon and examined with infrared near-field scattering scanning microscopy IR s-SNOM, at 20 nm spatial resolution. These experiments were performed with synchrotron source IR at the Advanced Light Source (ALS), Berkeley CA[20] and with a broadband mid-infrared nano-FTIR spectroscopy laser at Neaspec, GmbH, DE.[21–24] Additionally, near-field images were recorded at Neaspec using a mid-IR Quantum Cascade Laser (QCL). We show here that, regardless of source and system, PCI can be exploited in each case to reveal crucial aspects of organization in fibres and in fibrils. The potential for spectropathological and clinical applications will be discussed.

2. Methods

2.1 Tissue collection & mechanical damage procedure

Four tendons were dissected from the tails of two young adult steers killed for food at a local abattoir (Reid's Meats, Wolfville, NS, Canada). Once removed, tendons were immediately stored at 4 °C in phosphate-buffered saline (PBS) containing 1% antibiotic/antimycotic solution (10 000 units penicillin, 10 mg streptomycin, and 25 μg amphotericin B per mL; product A5955, Sigma-Aldrich Canada). Prior to the mechanical damage procedure, a 15 mm-long sample was removed from each tendon to serve as an unloaded control. Using a servo-hydraulic materials testing system, the remainder of each tendon was subjected to 5 cycles of subrupture overload at a strain rate of 1% per s, as described previously.[15] After the mechanical damage procedure, both the control and overloaded portions of each tendon were cut in half longitudinally yielding two sets of matched-pair samples, one set for SEM, and one set for FTIR. The tendon segments, approximately 10 mm long by 4 mm in diameter, were stored at room temperature for 24 hours in PBS containing 1% antibiotic/antimycotic solution and 1% protease inhibitor (2 mM 4-(2-aminoethyl)benzenesulfonyl fluoride, 14 μM E-64, 130 μM bestatin, 1 μM leupeptin, 0.3 μM aprotinin, and 1 mM EDTA; product S8820, Sigma-Aldrich Canada) prior to preparation for SEM and FTIR spectroscopy.

2.2 Scanning electron microscopy

In preparation for SEM, matched-pair control and overloaded tendon samples were fixed for one hour in 2.5% SEM-grade glutaraldehyde, rinsed in deionized water, dehydrated in graded ethanol, and critical point dried. The tendon samples were mounted on stubs using carbon tape with their exposed internal surfaces facing upward, and then sputter-coated with gold–palladium. Imaging was conducted with a Hitachi S-4700 scanning electron microscope operating at 3 kV, 15 μA.

2.3 FTIR sample preparation and imaging at normal (5.5 μm) and high (1.1 μm) magnification pixel size

The other matched-pair set, comprising 4 matched pairs of control and mechanically overloaded tendon, that is two pairs from each animal, were shipped overnight to the University of Manitoba, immediately removed from PBS medium, covered with OCT [Sakura Finetek Inc. USA], frozen in isopentane cooled in liquid N_2, and stored at −80 °C. Cryosections were cut to 5 or 8 μm thickness at −21 °C, mounted on BaF_2 windows and refrigerated overnight at 4 °C.[25]

Nine sections from the 4 control tendon samples and ten sections from the matching 4 overloaded tendon samples were imaged with an Agilent Cary 670 interferometer and 620 infrared microscope, equipped with a 64 × 64 FPA. Normal and high magnification optics (5.5 × 5.5 μm^2 and 1.1 × 1.1 μm^2 pixels, respectively)[18] were used to explore fibril organization. Sections were typically aligned on the stage such that the fibril direction was horizontal in the field of view. Spectral acquisition and data processing were accomplished with the

ResolutionsPro FTIR spectroscopy software (Agilent Inc. version 5.2.0). All spectra were acquired as sums of 128 scans ratioed against a background of 512 scans. Initial survey images with normal magnification were conducted on one matched pair, with no, 0° or 90° polarization, where 90° corresponds to the horizontal direction of the overall fiber orientation. The majority of the study was conducted with high magnification optics. Of these, 46, 40 and 124 individual FPA tiles were acquired from control tendon sections with no, 0° and 90° polarization, respectively; similarly, 36, 64 and 212 tiles were acquired from overload tendon sections. Some sections on BaF_2 windows were stored in a sealed container at 4 °C pending re-imaging with FTIR PCI or other analyses.

The amide I and II bands differ in intensity, as they arise from different vibrational modes that produce different intrinsic changes in molecular dipole. Moreover, they change with orientation relative to the polarization of the incident IR light. The FTIR PCI false colour images were created from the ratio of amide II to amide I integrated band areas, to compensate for differences in tissue thickness across a section. Integrated band areas for amide I were taken between 1675 and 1645 cm^{-1}, and for amide II between 1567 and 1537 cm^{-1}. The baseline was established from the average in the non-absorbing region between 1790 and 1760 cm^{-1} for both. Instrumental illumination differs under the different polarizations and a new background had to be recorded in each case, necessitating slightly different colour scales. Display limits were chosen to consistently illustrate variation in orientation: 0.1 to 1.3 for all 0° images; 0.6 to 2.0 for all 90° images, 0.6 to 1.2 for images acquired without the polarizer accessory. These scales were chosen such that a ratio of 1.0 represented spectra wherein the amide I and II bands also had approximately equal height, while the range of the false colours was primarily aqua-blue to yellow-green for normal tissue. Extremes are reached only when fibers within the field of view are strongly aligned with or against the polarization.

2.4 AFM imaging

Following the FTIR imaging experiments, two of the overload tendon sections were returned for AFM surface analysis at the Structural Nanomechanics Lab (Dalhousie University). Each section was imaged in air with a Bioscope Catalyst AFM (Bruker, USA) mounted on an Olympus IX71 inverted microscope (Olympus, USA). ScanAsyst Fluid probes (Bruker, USA) with a nominal cantilever spring constant of 0.7 N m^{-1} were used. The AFM was operated in peak force quantitative nanomechanical mapping mode at a scan rate of 0.5 to 1 Hz. The images were processed in Nanoscope Analysis (Bruker, USA) or SPIP (Image Metrology, Denmark). To ensure that areas being measured corresponded to those imaged with FTIR PCI, a stack of survey images were taken at 25 nm spatial resolution over 40 or 50 μm regions, and the morphology was matched to the IR and visible images. Higher resolution images (10 μm region, 10 nm spatial resolution) were obtained on selected regions.

2.5 Fibril preparation for AFM and nanoscale near-field IR

In order to conduct preliminary nanoscale near-field IR analysis of collagen fibrils, an additional tail tendon was acquired from a different young adult steer. The tendon was laid flat between sheets of gauze moistened with PBS, double bagged, and stored at −86 °C. After thawing at room temperature in its sealed bag, the tendon was rinsed in deionized water and dissected further using tweezers and a fine-tipped glass rod in 1 mL of deionized water until the mixture appeared cloudy. A 50 μL drop of the suspension was deposited on a 1 cm^2 gold-coated silicon wafer, allowed to settle for 10 min, rinsed with deionized water and dried with nitrogen. Visible images at 100× and 50× were used to identify single collagen fibrils prior to the first nano-FTIR experiments at the ALS, Berkeley CA.

ALS data collection. As per the standard protocol for mounting samples at Beam Line 5.4.1, ALS, the Au wafer was taped onto a magnetic mount, inserted within the chamber and manually adjusted to position the desired target into the field of view. The specially modified AFM (Innova, Bruker), and the source, detector and various optics were manually adjusted to acquire AFM images and Synchrotron Infrared NanoSpectroscopy (SINS) spectra.[20] Visible and AFM images revealed a multitude of residual fibres and fibrils; two fibrils with cross sections of the order of 300 nm were selected. Spectra were recorded with a modified commercial rapid-scan FTIR spectrometer (Nicolet 6700, Thermo-Scientific), with an asymmetric Michaelson interferometer. The second order amplitude ($s(\omega)$) and phase ($\varphi(\omega)$) spectra for the backscattered light were recorded at locations on and off the fibrils. For each background and sample, 512 scans were co-added, with a spectral data spacing of 8 cm^{-1}. Only 2 or 3 good quality spectra could be acquired per hour, since a background scan was required at least once every two hours, and 20 minutes was required to record individual spectra with sufficient signal-to-noise ratios. Post-processing custom programs with commercial fast Fourier transform analysis packages (Wavemetrics Igor) developed in-house at BL 5.4.1, ALS were used to convert the scattering data to a format that corresponds to an IR absorbance spectrum, see below.[20]

Neaspec data collection. The same two fibrils were examined with a scattering Scanning Near-Field Optical Microscope, sSNOM, with the broadband mid-IR nano-FTIR illumination unit from Neaspec.[21–24] In the first experiment, spectra were obtained by tuning the laser through three spectral ranges (800–1200 cm^{-1}, 1200–1800 cm^{-1} and 1800–2100 cm^{-1}), again recording the second order amplitude, $s(\omega)$, and phase, $\varphi(\omega)$, spectra for the backscattered light at locations on and off the fibril. Spectra were recorded with a full-optical travel range of 800 μm, which results in a real frequency resolution of 12.5 cm^{-1}. Zero-filling was added, reducing the frequency spacing of the data points to 3.125 cm^{-1}. Following standard post-processing, below,[21–24] similar to that performed on the ALS data, the mid-IR absorption spectra were assembled from the three spectral segments. In the second experiment, images were obtained across one of the fibrils by scanning the entire area at on- and off-resonant wavelengths, 1655 cm^{-1} and 1850 cm^{-1}, respectively, at 5 nm intervals, though with an anticipated 20 nm spatial resolution. Full spectra were recorded in about 3.5 minutes; single wavelength images were recorded at 20 ms per pixel over 100 × 200 pixels area, in less than 7 minutes per image. The amplitude and phase images showed that a minor shift in position occurred between the two acquisitions. After export to MatLab, the array from the 1655 cm^{-1} scan was shifted down one row and over three rows (5 nm and 15 nm) to achieve correct image registration of the single wavelength data files before creation of ratioed images.

Post-processing of near-field data. In each of the near-field experiments, amplitude, $s_2(\omega)$, and phase $\varphi_2(\omega)$ spectra were obtained from the imaginary function of the complex-valued second order scattering coefficient, (σ_2),[22,24] according to the relationship:

$$\sigma_2(\omega) = s_2(\omega)e^{i[\varphi(\omega)]} \tag{1}$$

The near-field nano-FTIR absorbance spectrum is obtained by normalizing the signal on a fibril against a signal from on a flat, non-absorbing surface, here, a nearby point on the gold-coated wafer:

$$\mathrm{Im}\{\sigma_2(\omega)\} = \mathrm{Im}\left\{\frac{s_{2\ (\mathrm{fibril},\omega)}}{s_{2\ (\mathrm{Au},\omega)}}\,e^{i[\varphi(\mathrm{fibril},\omega)-\varphi(\mathrm{Au},\omega)]}\right\} = \frac{s_{2\ (\mathrm{fibril},\omega)}}{s_{2\ (\mathrm{Au},\omega)}}\sin[\varphi(\mathrm{fibril}) - \varphi(\mathrm{Au})] \tag{2}$$

Single wavelength images recorded with the QCL source yielded amplitude and phase results that could be displayed in several formats: individual, normalized, or as the ratio of the near-field absorbance signals at 1655 and 1850 cm^{-1}.

3. Results

The infrared spectrum of normal, undamaged tendon (Fig. 1A) displays the numerous well-known spectral features that are unique to collagen assemblies,[26–28] and derive from the unusual molecular composition and superhelical structure. In each of the three α-chains that form a tropocollagen molecule, the repeat amino acid sequence is Gly–X–Y, where X and Y are often proline and hydroxyproline, respectively.[29] The alignment is such that the carbonyls are nearly perpendicular to the direction of the backbone; this orientation is retained to a significant extent in the assemblies. The intensity of the amide I band at 1660 cm^{-1} is enhanced when light is polarized parallel to carbonyl (0°); all other bands, particularly the amide II at 1545 cm^{-1}, are enhanced under light polarized

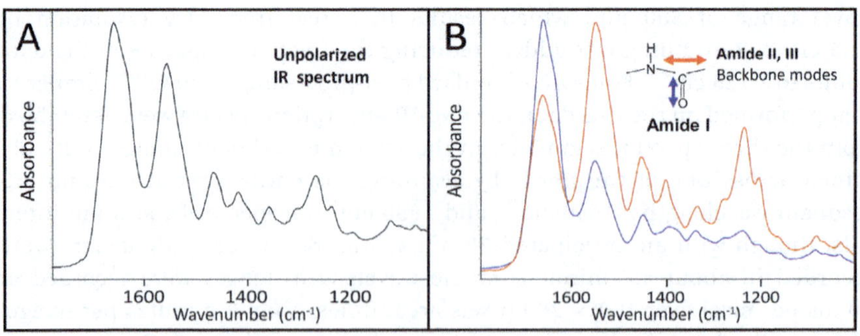

Fig. 1 Single-pixel FTIR spectra of control tendon extracted from FPA images acquired at normal magnification optics, 5.5 μm, illustrate the many potential spectral markers for fibril orientation. Spectra are shown (A) without polarizer (B) with polarizer oriented at 0° (blue, enhanced C=O stretch at 1660 cm^{-1}) and 90° (red, enhanced C–N stretch at 1545 cm^{-1} and backbone modes). Spectra are normalized to the stronger band in each case.

parallel to the fibril direction (90°), Fig. 1B. These spectral features can be easily exploited to reveal fibril orientation under polarized infrared light, *provided* that the internal organization has been retained at a molecular level. The contrast would be reduced by fibril disorganization and/or molecular denaturation.

The SEM images of collagen fibrils before and after tendon overload, and a typical series of stress–strain curves for tendon during sub-rupture overload, are shown in Fig. 2. Fibrils from control tendon exhibit the expected D-banding[5,14,30,31] that arises from the regular staggered arrangement of tropocollagen molecules within the fibrils, Fig. 2A. The interpretation of such stress–strain curves (Fig. 2B) is provided in detail elsewhere.[6,15,32] Briefly, under the application of low force to the tendon sample, out to ~0.1 strain in Fig. 2B, the 100 μm-scale collagen fibre crimp in the tissue straightens (see Fig. 3–5). If the extending force were removed, this crimp would reform as the tissue retracted. Beyond 0.1 strain, the collagen fibres are in direct tension, loading more or less elastically, and the stress–strain curve is linear. Sub-rupture overload is achieved by application of sufficient force to extend the tendon into the final non-linear region, beyond the yield stress point, (σ_y, Fig. 2B). In our sub-rupture experiments, the tendon sample was extended until the slope of the stress/strain curve became zero (~0.3 strain in the first overload cycle), and the extension was reversed to zero. This process was performed 5 times. Under cyclic overload, plastic damage accumulates in the tendon and greater final strain is achieved—with a smaller maximum force—in the four subsequent overload cycles.

SEM imaging of matched-pair control and overload samples confirmed that discrete plasticity damage was produced by repeated stretching beyond the tendon's yield point. As previously observed,[14,15] fibrils with discrete plasticity were characterized by the presence of kinks that repeated along the length of fibrils with nanoscale frequency, while some kinked fibrils also had a surface layer of denatured collagen, masking the fibrils' native D-banding (Fig. 2C).

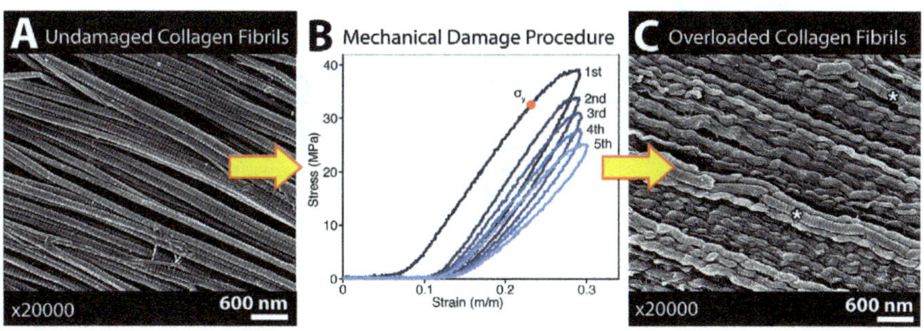

Fig. 2 SEM imaging confirmed that the subrupture overload procedure produced discrete plasticity damage in the tendon samples used for spectroscopic analysis. (A) SEM of a native, unloaded control tendon showing the normal straight, D-banded appearance of collagen fibrils. (B) Typical stress–strain curves showing 5 cycles of subrupture overload, where the tendon is repeatedly stretched beyond its yield point, σ_y. (C) SEM of fibrils with longitudinally repeating kinks called discrete plasticity damage. Some fibrils with discrete plasticity retain their D-banding (* in (C)), while in other kinked fibrils the D-banding is obscured by a surface layer of denatured collagen. Note that in the marked fibrils (* in (C)) distortion of the D-banding is an imaging artefact.

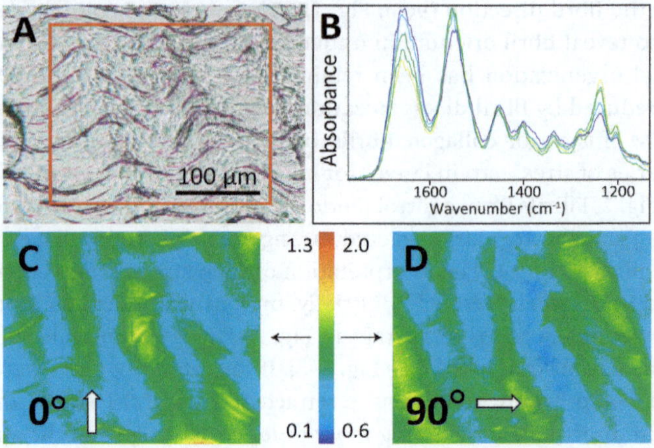

Fig. 3 The control tendon exhibits a normal crimp pattern under visible and polarized illumination. (A) White light 10× photoimage of a control tendon section on a BaF_2 window. The red box outlines the area examined with FTIR PCI (4 × 3 FPA tiles). (B) Spectra extracted from FTIR PCI mosaics to show variation observed under polarization at (C) 0° and (D) 90°. The false colour bar indicates the range of amide II to amide I band area values for (C) and (D), as described in Methods; the arrows indicate the image to which numerical scale applies.

The four control/overload sample pairs matching the set imaged with SEM were cryosectioned and imaged in FTIR transmission mode with normal and high magnification optics. The fibres of all control tendons exhibited the expected

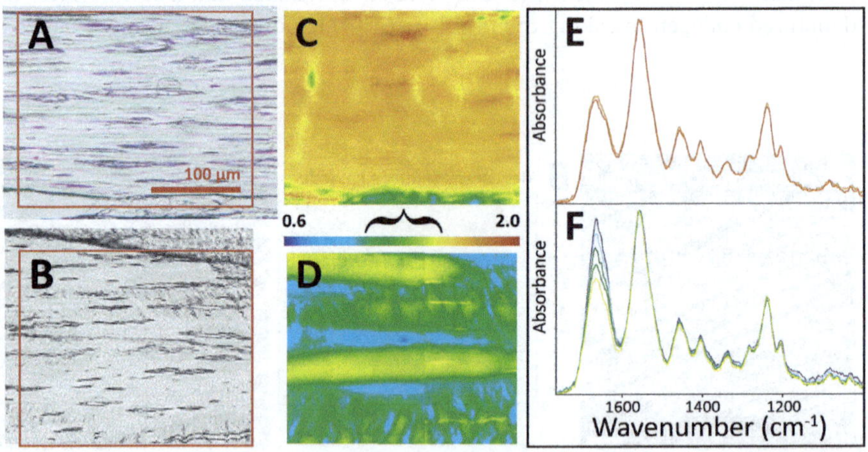

Fig. 4 FTIR PCI revealed sub-micron disorder in some stressed tendons. Visible images of tendon sections on BaF_2 windows show (A) loss of crimp structure on some sections and (B) residual crimp structure in other sections. FTIR PCI at 90° showed (C) a high degree of molecular alignment with fibrils in the case of uniform stretch and (D) alternating regions of alignment and crimp in the case of non-uniform stretch. Spectra extracted from FTIR images corresponding to the range of colours in FTIR PCI are shown in (E) and (F), normalized to amide II band. False colour scale bar as in Fig. 3 for 90°; the bracket indicates a range typical of normal crimped tendon. Spectra shown represent the full range of relative intensities observed.

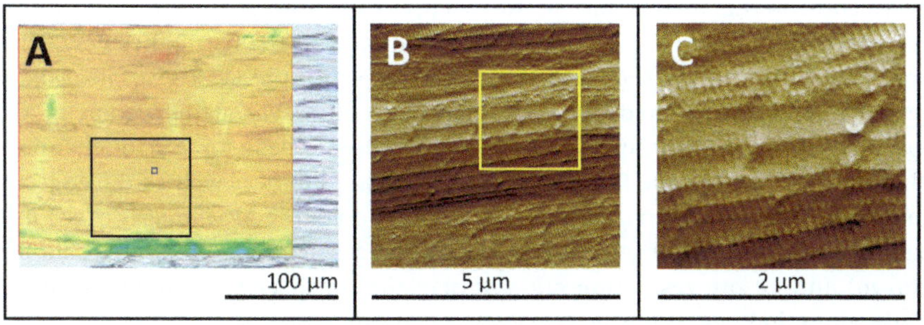

Fig. 5 Uniformly stressed tendon is lightly kinked. (A) The semi-transparent FTIR PCI false colour image from Fig. 4C is superimposed on a visible image. Black boxes outline the locations of 100 × 100 μm^2 (results not shown) and 5 × 5 μm^2 AFM images; (B) AFM 5 × 5 μm^2 image recorded at 10 nm resolution. (C) 2 × 2 μm^2 region as outlined by yellow box (B), enlarged to facilitate identification of low kink density in the otherwise normal, intact D-banded fibrils.

microscale crimps under white light (Fig. 3A), with an undulation repeat of about 100–150 μm. Note that this naturally occurring crimp pattern is 1000-fold greater than the nanokinks that are produced by mechanical damage within the individual fibrils that comprise the microscopic fibre bundles. Spectra were extracted from mosaic images, acquired as FTIR PCI with high magnification optics, and displayed at full scale to clarify the variation in the ratio of the integrated band areas of amide I and II with fibril orientation (Fig. 3B). The false colour scales in the processed images were adjusted separately for parallel and perpendicular polarizations to yield complementary images for perpendicular (0°) and parallel (90°) polarizations (Fig. 3C and D), as described in the Methods section. The high magnification optics images have sufficient spatial resolution to show that control tendon exhibits the expected polarization dependence, wherein the visible crimped appearance of the well-organized fibres is reflected in the FTIR PCI images. The colours in Fig. 3B match the colours in the processed images. For the example shown, the crimps only become parallel to the 90° light at the top of the wave, yellow spectrum and yellow regions in (Fig. 3B and C).

All tendons subjected to sub-rupture overload stress lost the characteristic undulating crimps, but not uniformly. In some regions, the fibres were visibly straightened under white light (Fig. 4A), but other sections appeared to have retained crimped regions (Fig. 4B). FTIR PCI at 90° mirrored the organization in the visible images in the first case (Fig. 4C and D). Spectra uniformly showed significantly enhanced amide II C–N stretch backbone modes across the section; the amide II to amide I ratio reached the high end of the false colour scale, yellow-red, as illustrated in the images (Fig. 4A, C and E). In the second example, where some crimp structure was retained, the ratios of the band areas were within normal bounds, translating as blue-green-yellow (Fig. 4B, D and F), implying that the tendon had undergone less, albeit uneven, stress during overload. However, closer analysis of the spectra from this FTIR PCI revealed differences strongly suggestive of disorder at a molecular scale: the enhancement of the amide II was much less than expected in regions that visibly appeared to be straight and uncrimped, along the deeper blue horizontal swaths (*vide infra*).

To confirm this interpretation, the two overload samples shown in Fig. 4 were returned for AFM imaging of the regions of interest in the tissue. Results of this correlated imaging are shown in Fig. 5 and 6. Part of the FTIR PCI false colour image has been rendered semi-transparent to allow visual inspection of the macroscopic axial alignment. All the AFM pictures are peak-force error channels, chosen to emphasize the finest details in the topography as well as the edges of individual fibrils. As seen in Fig. 5A, the orientation of the AFM images of the first sample matched that of the FTIR PCI. AFM images acquired at 25 nm (data not shown) and 10 nm resolution allowed detection of discrete plasticity, with the kinks along individual fibrils, spaced at ~800 nm intervals.

The second sample was not precisely aligned to match the FTIR PCI and AFM images, Fig. 6A, but this does not impact the analysis. A series of 40 μm survey AFM images were acquired at 25 nm spatial resolution inside the yellow rectangle (data not shown). Three of six higher-magnification AFM images are shown at two enlargement scales (Fig. 6B and C), showing aligned, moderately kinked fibrils in the top image and a progressively more disordered arrangement of kinked fibrils in the middle and lower images. The uppermost AFM image in Fig. 6C is similar to those in Fig. 5. The progressively disordered regions correspond to the regions revealed by FTIR PCI to contain submicron disorder, that is, a loss of axial alignment at the fibril or molecular scale (blue-green regions in false colour image, Fig. 6A).

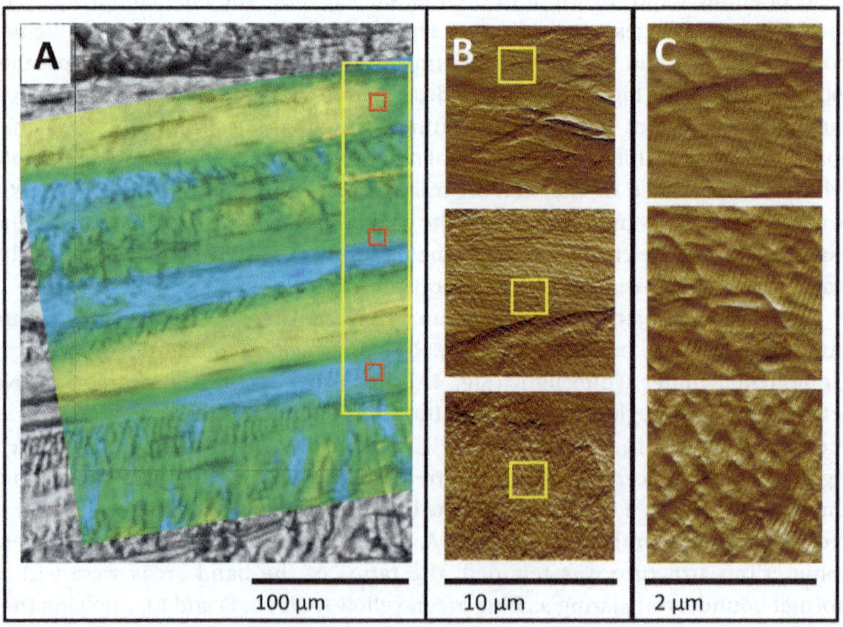

Fig. 6 Non-uniformly stressed tendon. (A) Visible image of tendon with semi-transparent FTIR PCI false colour image from Fig. 4C superimposed. The large yellow rectangle outlines the location of lower-resolution AFM images, taken as a series of 40 × 40 μm^2 areas (not shown). (B) Three 10 × 10 μm^2 AFM images, recorded at 10 nm spatial resolution at red squares in (A). (C) Excerpts from (B), outlined by yellow squares, enlarged to facilitate identification of kinking in the otherwise normal, intact D-banded fibrils.

The first Synchrotron Infrared NanoSpectroscopy (SINS) spectra from a control collagen fibril, recorded at ALS, are presented in Fig. 7A. The AFM image on the left was acquired rapidly in order to locate a useful target, and shows parts of two normal fibrils. Following Fourier transform of the second order scattered near-field light,[20] the nano-FTIR absorption spectra, Im $\sigma_2(\omega)$ are obtained. These match well with the far-field IR spectrum of collagen shown in Fig. 1. The proline ring mode ($\sim$1465 cm^{-1}) is just barely visible. Similar results were obtained on the same fibrils and locations from second order backscattering using the broadband nano-FTIR illumination source from Neaspec (Neaspec, GmbH, DE), shown beside the synchrotron spectra. The nano-FTIR absorption spectra are very similar to the results from ALS; the proline ring mode is clearly identifiable. The ALS spectra are noisier but should not be compared directly, since the Neaspec data were zero-filled.

Single wavelength images were acquired with the Neaspec QCL at the amide I maximum (1655 cm^{-1}), and at a non-absorbing wavelength (1850 cm^{-1}), Fig. 7B. The area was scanned in 5 nm steps, where the expected spatial resolution is 20 nm. All amplitude and phase images show the fibril features. The on- and off-resonance contrast in amplitude is much less than the contrast in phase, but both

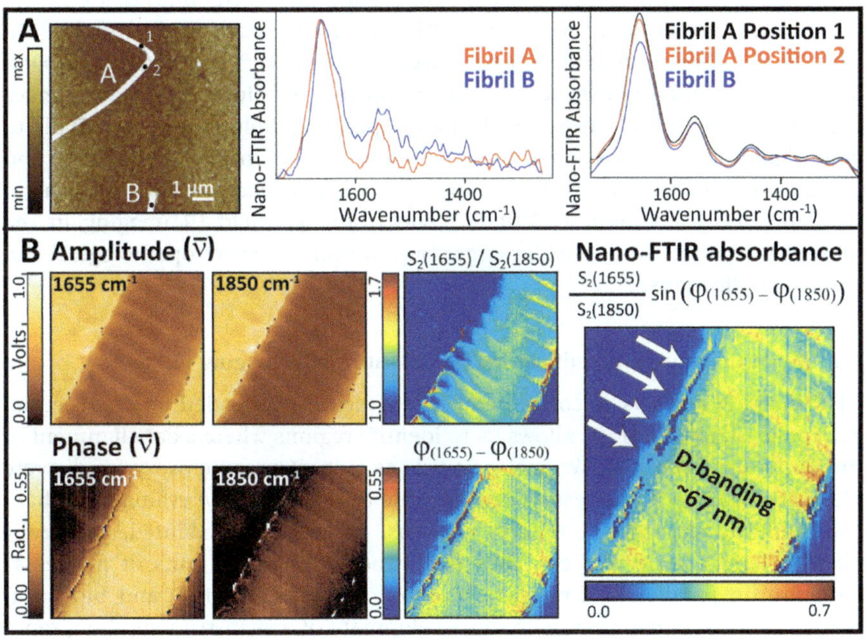

Fig. 7 Synchrotron Infrared NanoSpectroscopy (SINS) and infrared QCL s-SNOM of collagen fibrils demonstrate that fibril organization can be examined at nanoscale spatial resolution. (A) Mid-IR spectra from single points: (Left) AFM height image of the target fibrils (256 lines sampled at 512 points per line), at ALS, Berkeley USA; (Center) post-processed near-field nano-FTIR spectra (aka SINS). (Right) similar spectra acquired at points indicated by black dots, using the broadband nano-FTIR illumination source from Neaspec (Neaspec, GmbH, DE). (B) (Left) single wavelength QCL amplitude and phase images at 1655 and 1850 cm^{-1}. (Center) normalized amplitude and phase images. (Right) relative near field nano-FTIR absorbance intensity shows D-banding.

parts are required to construct the relative near-field FTIR absorbance image. The highest contrast seen at the edges of the fibril probably arises from artefacts in the edge scatter. The D-banding at ~67 nm spacing is well-resolved in all images, indicating a strong difference in optical response. The alteration in intensity arises from the gap-overlap regions caused by molecular stagger in the fibrils, just as would be seen in an SEM image.

4. Discussion

A little over a decade ago, a Faraday Discussion meeting was held to consider the state-of-the-art applications of spectroscopy in biomedicine. At that time, the state-of-the-art for FTIR spectroscopic imaging required a synchrotron source, and the highest spatial resolution possible was defined by the wavelength-dependent diffraction limit, from 10 to 2.5 μm.[33] Recent advances have made dramatic improvements on those limitations: benchtop FTIR instruments that can be used for spectropathology, with the necessary speed, sensitivity and high spatial resolution, are now commercially available. The diffraction limit to spatial resolution has been broken with near-field IR, allowing us to probe areas of the order of a few nanometers, in several different instrumental designs.[20,22] Here, we present results that employ all of these advances: (1) far-field diffraction-limited FTIR microscopy with a thermal bench source and improved optics that yield 1.1 μm pixels,[18] further enhanced with polarization contrast and (2) near-field FTIR with 20 nm resolution, with a synchrotron source FTIR[20] and with broadband nano-FTIR illumination source.[21–24] Our purpose is to illustrate the advantages these technological advances can bring to spectropathology, and eventually to clinical pathology in the context of our on-going study of mechanical damage and healing in tendons. We will first summarize what we have learned from high-magnification transmission FTIR PCI and from near-field IR imaging in the present study, and then consider larger implications for spectropathological and clinical applications.

4.1 What does high-magnification far-field infrared imaging tell us?

FTIR imaging analysis of control and mechanically overloaded tendon section, with polarization contrast, allows us to identify regions where axial alignment of the fibrils has been lost. We interpret this as a spectroscopic proxy for discrete plasticity kinking. These results demonstrate that we can now evaluate mechanically-induced damage in tendon sections, at an unprecedented, physically relevant length scale. We can thus study the changes in terms of molecular structure, in parallel with physico-structural (SEM, TEM, AFM) and biological methods. In the longer view, with the development of new instrumentation, rapid spectropathological screening and classification of tendon damage in a clinical setting can be envisioned.

Veres and Lee first demonstrated the existence of nanoscale serial kinking in collagen fibrils within ruptured and cyclically overloaded tendons,[14] co-articulating what we now term the discrete plasticity mechanism as a protective mechanism in tendon biomechanics. Recently, the overload-induced collagen kink morphology was shown to be a possible cue to cellular recognition of injury.[34] FTIR PCI is eminently suitable to monitor fibril organization, particularly with diffraction-limit optics that enable imaging at 1.1 μm pixel dimension, since this is a good match with natural fibril width. FTIR spectra provide molecular spectroscopic evidence of fibril orientation in control tendon, Fig. 1. Contrast is easily displayed, given the enormous difference in relative intensities of the amide I and II bands when the polarized infrared light is aligned either with the amide carbonyls (0°) or with the backbone (90°), as well as large differences in many of the other backbone modes.

In control tendon, fibrils are well organized; D-bands were clearly evident in SEM images (Fig. 2A), whereas mechanical overload introduced the expected serial nanoscale kinks characteristic of discrete plasticity (Fig. 2B and C). Some undulating crimps were easily seen in the visible images (Fig. 3A) and also in lower magnification SEM (not shown). Spectra extracted from a mosaic represent the typical variation that accompanies the undulating orientation of crimps in normal fibrils (Fig. 3B). Relative intensities are not directly comparable under different polarizations, but the FTIR PCI false colour scale in the processed images was adjusted to achieve the same, relatively narrow range of colour contrast, from yellow to blue (Fig. 3C and D) in control images. These ranges were optimized to allow for the full range observed in all tissues, including those that had been mechanically damaged.

In the mechanically damaged tendons, FTIR PCI revealed extreme alignment in uncrimped tendon fibrils in some cases, and an uneven loss of crimp, with concomitant molecular level disorder in others. The amide II band dominates in all spectra throughout the uncrimped region imaged in Fig. 4A, C and E. The calculated ratio for our chosen imaging parameters is consistently of the order of 1.7 to 2.0. Fibrils are visibly aligned; false colour FTIR images capture this alignment. The amide I band profile is well-known to be a good indicator of protein conformation. In normal self-assembled collagen, the typically observed profile is an asymmetric amide I band, with a maximum at around 1660 cm^{-1} and a lower energy shoulder at around 1635 cm^{-1}.[7–10] We note that spectral band shapes were unaltered from the norm and only the relative intensities had changed, indicating a high degree of physical order, but no evidence of denaturation.

Some of the spectral variations that were observed (Fig. 4B, D and F) suggested that despite their physical alignment, the fibrils were internally disorganized, *i.e.*, that nanoscale kinking had occurred. Undulations matching normal crimped tissue were evident in the lower part of these images. As before, the FTIR PCI results indicated fibre alignment without much nanoscale disorder in some regions where crimps were lost, seen here as yellow stripes in the false colour image Fig. 4D and yellow spectra in Fig. 4F. Interestingly, blue-green swaths were found between the yellow bands in Fig. 4D. In these regions, while the visible tissue image suggested that the crimps were lost and fibres were axially straightened, the amide II to amide I band area ratio was much lower than that expected for intact but well-ordered, aligned fibrils (Fig. 4F, blue and green spectra). In fact, the amide II band was frequently less intense than the amide I; the range of values for the amide II to amide I ratio was 0.9 to 1.2. Given the physical alignment of the fibre bundles with the polarization of the IR light, the most plausible explanation is that these regions underwent much greater, localized, tension, leading to significant kinking of fibrils within the fibre bundle.

Subsequent AFM surface imaging of these regions confirmed the existence of nanoscale kinking. AFM surface images of the well-aligned fibres in the visible image and FTIR PCI (Fig. 5A) show well-ordered fibrils with clear D-banding at the sub-micron scale, Fig. 5B and C. Light kinking is observed, corresponding to only a few kinks per 1.1 × 1.1 μm^2 pixel area sampled by a high magnification FTIR field of view. As noted, the spectra in the FTIR PCI are nearly uniform throughout (Fig. 4E). From the AFM results, we may conclude that the massive amide II enhancement in axially aligned tendon indicates nearly linear orientation of well-organized fibrils. FTIR PCI gives molecular evidence that the stress was sufficient to eliminate the microscale waveform crimp that fibrils normally follow, but, with only a few kinks per pixel, did not produce sufficient molecular disorder to cause a change in the PCI spectra.

In stark contrast, the AFM surface images in Fig. 6 exhibit alternating regions of uncrimped fibrils (yellow strips in FTIR image Fig. 6A; upper AFM images in Fig. 6B and C) and regions of disordered fibrils and with nanokinks in regions with low polarization contrast (blue-green strips in FTIR image Fig. 6A; middle and lower AFM images in Fig. 6B and C). Together, the results in Fig. 5 and 6 show that in some cases, the mechanical damage was heterogeneous, producing higher stress leading to regions of disorder immediately adjacent to regions that experienced relatively lower stress, such that the crimp structure was regained upon release of the mechanical force. This might be an artefact arising from an uneven grip in the experimental mechanical damage procedure. However, it has recently been shown that interfibrillar shear stresses occur under notch tension testing of rat tail tendon.[35] Both uneven tension and natural variations in the tendon architecture may contribute to this result, though the degree to which each is involved remains a question yet to be resolved.

4.2 What does nanoscale near-field infrared imaging give us?

With the FTIR imaging at the 1.1 micron scale, fibril damage at the nanometer scale may be suspected but cannot be proved. The SINS experiment at BL 5.4.1 is the first synchrotron site in the world to combine the high chemical specificity of FTIR with the nanoscale spatial resolution of AFM.[20] This facility was designed on the same basic principles as the broad band DFG laser s-SNOM, in that both employ an asymmetric Michelson interferometer with the sample located under the AFM tip in the extended fixed-mirror arm of the interferometer.[21–24] The brilliant broad band source has greater source intensity, but must be tuned through three spectral ranges to get the entire mid-IR spectrum. With the QCL, a fibril may be scanned at a single wavelength; the relative absorbance at different wavelengths can be used to create spectral contrast images.

The results shown in Fig. 7 demonstrate that near-field IR gives the required nanoscale resolution to image fibrils and even to identify differences in optical response across the D-banding structure. Several important features can now be exploited. First, comparison of the absorbance spectra with those in Fig. 1 illustrates the importance of polarization: spectra have an enhanced amide I band, relative to the amide II and the backbone modes. Here, however, it is the orientation of the electric field, which is polarized parallel to the tip and to some of the amide carbonyl groups, which are perpendicular to the fibril lying on the slide, that causes the spectra to be the same as that in the far-field 0° polarization spectrum.[23] While the far-field FTIR PCI experiments allowed us to evaluate polarization in the (x, y) plane of the tendon section, the near-field IR experiments allow us to evaluate polarization in the z-direction. The phenomenon is observed in both the synchrotron source and broad band source spectra, Fig. 7A. The absence of characteristic bands in the amide III region, especially those at 1337, 1284, 1240 and 1204 cm^{-1}, is understandable given that they are enhanced only for light polarized parallel to the fibril. From this, we can deduce that these other modes would become active if disorder caused the molecular backbone of fibrils to be aligned parallel to the tip.

The single wavelength images (Fig. 7B) present a dramatic example of the power of this technique for the study of oriented materials. The distinct D-band pattern is apparent in all of the images. The D-band defines a region that includes a gap between staggered tropocollagen molecules and the telopeptide sequences at the ends of the three individual chains in each triple helix.[15,32] Since the spectral contrast is based on the intensity of the carbonyl band, we can conclude that the amide group orientation is changing with local changes in molecular orientation down the length of the fibril. The D-band gap has been found to increase slightly under tension.[36] Although the contrast in the amplitude response is much lower than that of the phase response, the amplitude image is somewhat smoother. Since this is the first demonstration of a near-field nano-FTIR image of a single collagen fibril, and is taken only at two wavelengths, it is premature to speculate on the reasons for this difference in signal quality. Further work is planned to address these questions and to explore semi-quantitative analysis of the full spectral differences along the length of normal and damaged fibrils.

Band position is of interest, since different conformations might be anticipated in normal and in damaged tendon. Amide I and II band shapes and positions are sensitive to molecular conformation, thus, if disorder is at a molecular scale, *i.e.*, molecular denaturation, then we should also be able to see this in nearfield spectra. Subtle changes in molecular dynamics were detected with AFM-based nanomechanical mapping of single hydrated collagen fibrils at indentation speeds around 1 mm s^{-1}, at temperatures above 50 °C.[37] Using the same technique with overloaded fibrils, they have shown the presence of a poorly organized surface shell,[38] as seen in the many fibrils lacking D-banding in the SEM image presented here (Fig. 2C), and previously.[15] In this context, it will be critical to compare near-field IR results with the first surface and tip-enhanced Raman (SERS and TERS) analysis of the surface of collagen fibrils,[39] which showed the presence of different protein conformations at the surface of the fibrils. By permitting examination of polarization (orientation) dependence and band positions, near-field nano-FTIR thus opens the door into the spectroscopic evaluation of fibril structure at a molecular scale. This knowledge will be a unique window into what it means for soft tissues to be *damaged* in terms of structural change at the most fundamental level.

4.3 Potential and implications

With regard to the problem investigated here, the discrete plasticity response may be a fundamental feature of collagen's evolutionary design that helps to determine the remarkable toughness of connective tissues and the inflammatory and healing responses to injury. Spectropathological understanding of this key structural feature of soft tissue injuries can serve the development of rational strategies for modulation of injury recognition and optimized healing. This particular health issue is one example of damage occurring in an ordered system, damage that is amenable to analysis with FTIR PCI at the micron scale. The hierarchical, oriented structure is typical of many collagen-based materials, as is the fact that disease may often be equated with disorder. In the case of tendon, a spatial resolution of 1 μm is fortuitously a good match for fibre dimension. The polarization-dependent nature of the IR spectra of collagen fibres was recognized at the dawn of FTIR microscopy.[1] With the spatial resolution capability of current commercial FTIR microscopes, this approach can be applied to the study of disease-induced disorder in a broad range of tissues from extracellular matrix to bone, to monitor disease state and healing, and to evaluate synthetic collagen matrices and scaffolds for bioengineered tissue replacements.

FTIR PCI spectropathology as presented here has been performed in transmission, on tendon sections mounted on relatively expensive salt windows. This is perceived as a necessity, *versus* the much less expensive transflectance substrates, notably the low-emissivity Keveley MirrIR substrates, which cost about 1% that of a BaF_2 window, because of the standing wave artefacts associated with transflectance spectra of thin sections.[40,41] Despite these artefacts, valid biomarkers for transflectance FTIR PCI have been identified in cartilage sections, though mineralized tissues were more problematic.[42] Provided that reference spectra are understood and samples are prepared consistently, the less expensive transflectance substrates could be a cost-effective option for tissue studies.

Infrared fibre optic probes have been used in the clinical evaluation of connective tissue,[43,44] with spectral results that show evidence of cartilage degeneration in surgically excised human tissue. The probe contact surface was a ZnS ATR crystal, 1 mm in diameter, and tissue grading was achieved through PLS analysis of some 150 spectra taken at different points. The classification was based on changes in several spectral parameters, but general tissue orientation could only be assessed in tissue sections imaged with a standard bench microscope and IR polarizer, at 6.25 μm pixel resolution.[44] Nevertheless, instrumental capabilities are evolving and it may eventually be possible to include polarization and tissue orientation as part of a clinical probe.

Near-field nanoscale infrared imaging capabilities are evolving swiftly, raising the question not of whether but of how such tremendously detailed information can be exploited for spectropathology. As seems evident from this work, the fact that the near-field nano-FTIR spectra are *z*-polarized is an important aspect that can be exploited in the study of the molecular basis of disease, particularly those diseases involving the formation or destruction of protein aggregates. The results of the experiments presented here demonstrate that we have the capability of studying the spectra of collagen fibrils at the level of their molecular organization. In future work, we will study the changes that occur at the fibril and molecular levels in mechanically damaged tendon. With near-field nano-FTIR, we hope to obtain confirmation that collagen molecules are locally denatured, perhaps *via* specific mechanisms, at specific sites within overload-damaged fibrils. In future work, we plan to explore the question of when and how this mechanism appears in the human life cycle, from youth to old age.

5. Conclusions

Collagen is the critical load-bearing constituent of mammals and performs innumerable functions throughout the body, all of which are enabled by the highly ordered structures that scale from molecular tropocollagen molecules and fibrils, up to fibres, fasicles, tendons and the matrices of bones and teeth. In tendon, cartilage, and certain other tissues, disorder may often be equated with damage and disease. The first high-magnification FTIR PCI spectra of control and mechanically damaged tendons show spectral differences in the relative intensity of the conformationally sensitive amide I and II bands. Different degrees of mechanical damage and disorder within individual tendons can be detected spectroscopically at the fibre length scale of 1 μm. SEM images of parallel matched-pair tendons confirm the presence of mechanical damage, in the form of serial kinking in collagen fibrils. AFM surface images of the sections imaged with FTIR PCI confirm the presence of fibril disorder and the presence of kinks at the nanoscale. With near-field nano-FTIR spectroscopy, it has been possible to obtain the first ever spatially resolved infrared spectra and images of collagen fibrils, including remarkably distinct spectral variations with D-bands. These techniques offer significant advantages for fundamental research, spectropathology and clinical applications.

Acknowledgements

This research is funded by a grant from the Canadian Institutes for Health Research, MOP-133453. S. Baldwin acknowledges the support of an ASPIRE CREATE NSERC PhD Fellowship; C. Findlay acknowledges the support of a Research Manitoba Health Studentship (PhD). The authors thank Reid's Meats, NS, Canada, for their assistance with the collection of tendon samples and Pat Scallion, Institute for Research in Materials at Dalhousie University for assistance with scanning electron microscopy. We are grateful to Dr Hans Bechtel and Dr Michael C. Martin for assistance with data collection at Beam Line 5.4.1 at the Advanced Light Source, Berkeley CA, USA. The authors are very grateful to Neaspec GmbH DE, and in particular to Dr Max Eisele of Neaspec, for the nano-FTIR spectra and the phase and amplitude-resolved near-field images of collagen fibrils and for many helpful discussions.

References

1 R. D. B. Fraser, *Discuss. Faraday Soc.*, 1950, **9**, 379–383.
2 R. M. Badger and A. D. E. Pullin, *J. Chem. Phys.*, 1954, **22**, 1142.
3 J. T. Randall, R. D. B. Fraser, S. Jackson, A. V. W. Martin and A. C. T. North, *Nature*, 1952, **169**, 1029–1033.
4 F. O. Schmitt, *Proc. Am. Philos. Soc.*, 1956, **100**, 476–486.
5 A. J. Hodge and F. O. Schmitt, *Proc. Natl. Acad. Sci. U. S. A.*, 1960, **46**, 186–197.
6 P. Fratzl, K. Misof, I. Zizak, G. Rapp, H. Amenitsch and S. Bernstorff, *J. Struct. Biol.*, 1998, **122**, 119–122.
7 K. Esmonde-White, *Appl. Spectrosc.*, 2014, **68**, 1203–1218.

8 C. Eklouh-Molinier, T. Happillon, N. Bouland, C. Fichel, M.-D. Diébold, J.-F. Angiboust, M. Manfait, S. Brassart-Pasco and O. Piot, *Analyst*, 2015, **140**, 6260–6268.

9 A. Boskey and N. P. Camacho, *Biomaterials*, 2007, **28**, 2465–2478.

10 R. Wiens, M. Rak, N. Cox, S. Abraham, B. H. J. Juurlink, W. M. Kulyk and K. M. Gough, *Anal. Bioanal. Chem.*, 2007, **387**, 1679–1689.

11 A. Masic, L. Bertinetti, R. Schuetz, S.-W. Chang, T. H. Metzger, M. J. Buehler and P. Fratzl, *Nat. Commun.*, 2015, **6**, 5942.

12 N. S. Kalson, Y. Lu, S. H. Taylor, T. Starborg and D. F. Holmes, *eLife*, 2015, DOI: 10.7554/eLife.05958.001.

13 B. Brodsky and A. V. Persikov, *Adv. Protein Chem.*, 2005, **70**, 301–339.

14 S. P. Veres and J. M. Lee, *Biophys. J.*, 2012, **102**, 2876–2884.

15 S. P. Veres, J. M. Harrison and J. M. Lee, *J. Orthop. Res.*, 2013, **31**, 731–737.

16 S. P. Veres, J. M. Harrison and J. M. Lee, *Matrix Biol.*, 2014, **33**, 54–59.

17 M. J. Nasse, M. J. Walsh, E. C. Mattson, R. Reininger, A. Kajdacsy-Balla, V. Macias, R. Bhargava and C. J. Hirschmugl, *Nat. Methods*, 2011, **8**, 413–416.

18 C. R. Findlay, R. Wiens, M. Rak, J. Sedlmair, C. J. Hirschmugl, J. Morrison, C. J. Mundy, M. Kansiz and K. M. Gough, *Analyst*, 2015, **140**, 2493–2503.

19 R. K. Reddy, M. J. Walsh, M. V. Schulmerich, P. S. Carney and R. Bhargava, *Appl. Spectrosc.*, 2013, **67**, 93–105.

20 H. A. Bechtel, E. A. Muller, R. L. Olmon, M. C. Martina and M. B. Raschke, *Proc. Natl. Acad. Sci. U. S. A.*, 2014, **111**, 7191–7196.

21 T. Taubner, R. Hillenbrand and F. Keilmann, *Appl. Phys. Lett.*, 2004, **85**, 5064–5066.

22 F. Huth, A. Govyadinov, S. Amarie, W. Nuansing, F. Keilmann and R. Hillenbrand, *Nano Lett.*, 2012, **12**, 3973–3978.

23 I. Amenabar, S. Poly, W. Nuansing, E. H. Hubrich, A. A. Govyadinov, F. Huth, R. Krutokhvostov, L. Zhang, M. Knez, J. Heberle, A. M. Bittner and R. Hillenbrand, *Nat. Commun.*, 2013, **4**, 1–9.

24 S. Mastel, A. A. Govyadinov, T. V. A. G. de Oliveira, I. Amenabar and R. Hillenbrand, *Appl. Phys. Lett.*, 2015, **106**, 023113.

25 D. M. Stitt, M. Z. Kastyak-Ibrahim, C. R. Liao, J. Morrison, B. Albensi and K. M. Gough, *Vib. Spectrosc.*, 2012, **60**, 16–22.

26 W. H. Moore and S. Krimm, *Biopolymers*, 1976, **15**, 2439–2464.

27 G. S. Young, *Mater. Res. Soc. Symp. Proc.*, 1992, **267**, 859–867.

28 N. P. Camacho, P. West, P. A. Torzilli and R. Mendelsohn, *Biopolymers*, 2001, **62**, 1–8.

29 M. D. Shoulders and R. T. Raines, *Annu. Rev. Biochem.*, 2009, **78**, 929–958.

30 B. B. Doyle, D. W. L. Hukins, D. J. S. Hulmes, A. Miller, C. J. Rattew and J. Woodhead-Galloway, *Biochem. Biophys. Res. Commun.*, 1974, **60**, 858–864.

31 J. P. R. O. Orgel, T. C. Irving, A. Miller and T. J. Wess, *Proc. Natl. Acad. Sci. U. S. A.*, 2006, **103**, 9001–9005.

32 M. J. Buehler, *Proc. Natl. Acad. Sci. U. S. A.*, 2006, **103**, 12285–12290.

33 M. J. Tobin, M. A. Chesters, J. M. Chalmers, F. J. M. Rutten, S. E. Fisher, I. M. Symonds, A. Hitchcock, R. Allibone and S. Dias-Gunasekara, *Faraday Discuss.*, 2004, **126**, 27–39.

34 S. P. Veres, E. P. Brennan-Pierce and J. M. Lee, *J. Biomed. Mater. Res., Part A*, 2015, **103**, 397–408.

35 S. E. Szczesny, J. L. Caplan, P. Pedersen and D. M. Elliott, *Sci. Rep.*, 2015, **5**, 14649–14658.

36 M. P. E. Wenger and P. Mesquida, *Appl. Phys. Lett.*, 2011, **98**, 163707.

37 S. J. Baldwin, A. S. Quigley, C. Clegg and L. Kreplak, *Biophys. J.*, 2014, **107**, 1794–1801.

38 S. J. Baldwin, in prep.

39 C. Gullekson, L. Lucas, K. Hewitt and L. Kreplak, *Biophys. J.*, 2011, **100**, 1837–1845.

40 P. Bassan, A. Sachdeva, A. Kohler, C. Hughes, A. Henderson, J. Boyle, J. H. Shanks, M. Brown, N. W. Clarke and P. Gardner, *Analyst*, 2012, **137**, 1370–1377.

41 J. Filik, M. D. Frogley, J. K. Pijanka, K. Wehbe and G. Cinque, *Analyst*, 2012, **137**, 853–861.

42 A. Hanifi, C. McGoverin, Y.-T. Oua, F. Safadi, R. G. Spencer and N. Pleshko, *Anal. Chim. Acta*, 2013, **779**, 41–49.

43 A. Hanifi, H. McCarthy, S. Roberts and N. Pleshko, *PLoS One*, 2013, **8**, e64822.

44 A. Hanifi, X. Bi, B. Kavukcuoglu, P. C. Lin, E. DiCarlo, R. G. Spencer, M. P. G. Bostrom and N. Pleshko, *Am. J. Sports Med.*, 2012, **40**, 2853–2861.

Faraday Discussions

DISCUSSIONS

Biofluids and other techniques: general discussion

Royston Goodacre, Matthew J. Baker, Duncan Graham, Zachary D. Schultz, Max Diem, Maria Paula Marques, Gianfelice Cinque, Robbin Vernooij, Josep Sulé-Suso, Hugh J. Byrne, Karen Faulds, Michael Hermes, Holly Fleming, Alois Bonifacio, Richard Dluhy, Peter Gardner, Samir El-Mashtoly, Bayden Wood, Kathleen Gough, Stefano Fornasaro, Sergei Kazarian, Lauren Jamieson, Wolfgang Petrich, Ganesh Dhruvananda Sockalingum, Nick Stone, Catherine Kendall, Faris Sinjab, Parvez Haris, Abdu Subaihi, Stan Remiszewski, Petra Hellwig, Valter Sergo, Klaus Gerwert, Christopher Phillips and Colin J. Campbell

DOI: 10.1039/C6FD90014C

Richard Dluhy opened a general discussion of the paper by Duncan Graham: In your example of a heterogeneous solution-based assay for multicomponent analysis, what is the concentration of the target fungal ssPCR DNA that is used, and how do you manage the kinetics of the reaction such that the target reaches the probe in a time frame appropriate for a clinical assay?

Duncan Graham answered: This is an end point PCR measurement and we started with 1000 input copies. We typically use 40 cycles so the concentration of the DNA detected in the end is actually quite high, *e.g.* pico to nanomolar. We haven't had much of an issue with the kinetics so far but we do see an issue with the steric bulk of the longer sequences reducing the hybridisation efficiency.

Royston Goodacre commented: You mentioned that starting with blood and getting to a readout (diagnosis) typically takes 6 h. Could you be a bit more quantitative in terms of how long each step takes? In addition, how does this timeline compare with the current competition methods already used for diagnostics?

Duncan Graham responded: The 6 h refers to our commercially available assay from Renishaw Diagnostics Ltd for fungal infections in hospitals. This consists of polymerase chain reaction (PCR) which is around 2.5 h, surface enhanced Raman scattering (SERS) plate reading which is 15 min, and the rest is sample preparation and handling. The research in the discussion paper (DOI: 10.1039/C5FD00167F) is designed to be single tube and reduce the sample handling and preparation steps. The current gold standard for fungal testing is culture-based and takes a few days.

Nick Stone asked: You outlined the aggregation of lots of nanoparticles (NPs) needed to provide the "on" signal. Do you know if you need all of them or could a few dimers give you the signal? Have you investigated this? You outline the NP separation distance in terms of DNA bases, can you give us distances in nm?

Duncan Graham replied: The signal increases most when going from an individual nanoparticle to a dimer and then increases again as more nanoparticles are added. We published a study where we examined this effect and found the following: "As the number of particles present in the aggregate increases, the relative surface enhanced resonance Raman Spectroscopy (SERRS) activity also increases, and the relative SERRS activities of single particles/dimers/trimers/aggregates of 4–9 particles/aggregates of 10–20 are estimated to be 1/12.4/15.6/23.2/43".[1] At this stage we do not have the suspension state distance in nm between the nanoparticles but this is something we are working on.

1 I. Khan, D. Cunningham, R. E. Littleford, D. Graham, W. E. Smith and D. W. McComb, *Anal. Chem.*, 2006, **78**, 224–230.

Abdu Subaihi commented: You mentioned 3 different orientations of your target, did you compare how the target was orientated and interaction with the analyte and the enhancement effect? Head to head, tail to tail, *etc.*?

Duncan Graham replied: We only did that in a limited way in this study as we have already published a comparison of the different orientations in a previous paper.[1]

1 F. McKenzie and D. Graham, *Chem. Commun.*, 2009, 5757–5759.

Robbin Vernooij queried: The silver nanoparticles were initially suspended in sodium hydroxide solutions, did you experience any issues with nanoparticle precipitation or aggregation prior to the hybridisation reactions when going to neutral and/or pH buffered solutions?

Duncan Graham responded: We resuspended the functionalised nanoparticles in the buffer and this takes the pH back to neutrality. As such we do not see any issues with stability.

Alois Bonifacio commented: Could you please comment on the stability, in terms of shelf-life, of the nanoparticles after DNA conjugation?

Duncan Graham replied: After DNA conjugation the nanoparticles are stable in buffers up to 0.7 M salt, and under normal buffer conditions, *e.g.* 0.1 M salt, will remain stable for a few weeks.

Royston Goodacre asked: For multiplexing you need good quantification. Do all dyes fall within the same linear range? If you had lower linearity with some (*i.e.*, more sensitivity), would you redesign the assay to target a microorganism that might be at a lower infective dose than other pathogens?

Duncan Graham responded: Not all dyes have the same Raman cross section and as such we use different amounts to give a similar SERS response per nanoparticle. We have thought about using stronger scatterers for lower infective doses but we haven't done this yet.

Kathleen Gough addressed Duncan Graham and Royston Goodacre: You have been talking about clinical applications of SERS. The complex nature of fungal infections and their detection in serum from a patient has led you to a rational design of targets and multiplexed probes, employing DNA functionalized nanoparticles, to ensure specificity. Another area where fungal infections cause massive health and economic impact is the infection of agricultural crops: *fusarium spp.* on grains, rice blast fungus *etc.* Would it be reasonable to consider the development of similar test methods that could be applied in the field, if not on the harvester equipment, then perhaps in grain storage centers? The approach that you have developed might work perfectly for this application, because they know what they are looking for. For example, *fusarium* produces a mycotoxin, deoxynivalenol (DON), also known as vomitoxin. Could your approach have significant economic value for the global agricultural industry?

Royston Goodacre responded: The detection of many different mycotoxins in grain processing/storage is very important and multiplexed detection of the different fungi responsible within these facilities would be highly beneficial for food safety/security.

Duncan Graham added: Very good point and this could be used in a high throughput centralised testing facility with samples logged and brought in for testing. The approach at this point isn't ready for field deployment but in future this could be a very interesting avenue to explore. Commenting on the economic value to the global agricultural industry is beyond my knowledge I'm afraid!

Matthew J. Baker opened a general discussion of the paper by Zachary D. Schultz: Can you explain the process involved in the background subtraction and peak intensities in the background and sample spectra?

Zachary D. Schultz responded: We do not do anything to the background or intensities. The dynamic surface enhanced Raman scattering (DSERS) algorithm is performed on the raw spectra acquired from the experiment. Similarly, the background in the capillary zone electrophoresis SERS (CZE-SERS) experiments contains information and is useful for analysis (see the reply to the next comment from Prof. Dluhy). There is one figure where we used background correction (Fig. 3A, DOI: 10.1039/C5FD00155B). Here we used a spectrum of the background for subtraction to compare to the DSERS result, where the DSERS algorithm removes the background peaks.

Alois Bonifacio remarked: Can the DSERS method be used for a quantitative approach as well?

Zachary D. Schultz responded: We initially were hopeful to use DSERS to quantify without separations, but at low concentrations the signal is less reproducible (see Fig. 4, DOI: 10.1039/C5FD00155B). We believe this arises from the low number of molecules giving rise to the detected signal, which produces signal variation common in single molecule detection. Quantitative analysis has been more straightforward in the SERS coupled to a separation.

Karen Faulds said: Do you think you are detecting single molecules, and is it possible to do so using this approach?

Zachary D. Schultz replied: We believe we are either detecting single or very low numbers of molecules at low concentrations. We have additional unpublished work where we have applied the statistical analysis proposed by Etchegoin and colleagues to our data.[1] Etchegoin *et al.* showed that single molecule detection is associated with a change in the frequency of detection rather than intensity. In unpublished work with rhodamine 6G, we see a transition from intensity changes to frequency changes between concentrations of 100 nM and 1 μM. This suggests that our improved detection limits are associated with the improved detection of single molecules in flow. It is important to note that the adsorption coefficient for the analyte appears to be an important factor that alters the relevant concentrations for different analytes. We look forward to verifying this claim soon.

1 J. E. Bohn, E. C. Le Ru and P. G. Etchegoin, *J. Phys. Chem. C*, 2010, **114**, 7330–7335.

Richard Dluhy said: The analysis of the CZE-SERS spectrum using the Pearson correlation coefficient requires the use of a reference spectrum. Can you use the same reference spectrum for multiple analytes?

Zachary D. Schultz replied: Essentially you are looking for a spectroscopic match between the spectra obtained during the CZE separation and your compound of interest. The "reference" spectrum needs to correspond to the analyte of interest. What we are doing is a reverse database look-up. In techniques like gas-chromatography with mass spectrometry detection, you try to match a spectrum from your separation against a database to find a match. Ultimately that would be the ideal case in CZE-SERS, however, there is no database at the moment. Instead, you can compare the spectra from the separation to a known compound's spectrum to see if it was detected. In Fig. 6A (DOI: 10.1039/C5FD00155B), we see that the highest correlation is observed at about 1000 s and comparing the spectra at this point shows good agreement. In my presentation I showed data that is included in Fig. 1, which is the correlation between the background spectrum and the spectra collected in the CZE run. What we see is that the background correlates well with the data, except when strong Raman signals are observed. In contrast to Fig. 6A (copied onto Fig. 1), there is increased correlation when the analyte migrates into the detection region. The electropherogram in Fig. 1 (copied from Fig. 5A) shows peaks that match well against the dips in the background correlation spectrum, illustrating the utility of the correlation plots to identify analytes as they are detected by SERS migrating off the column.

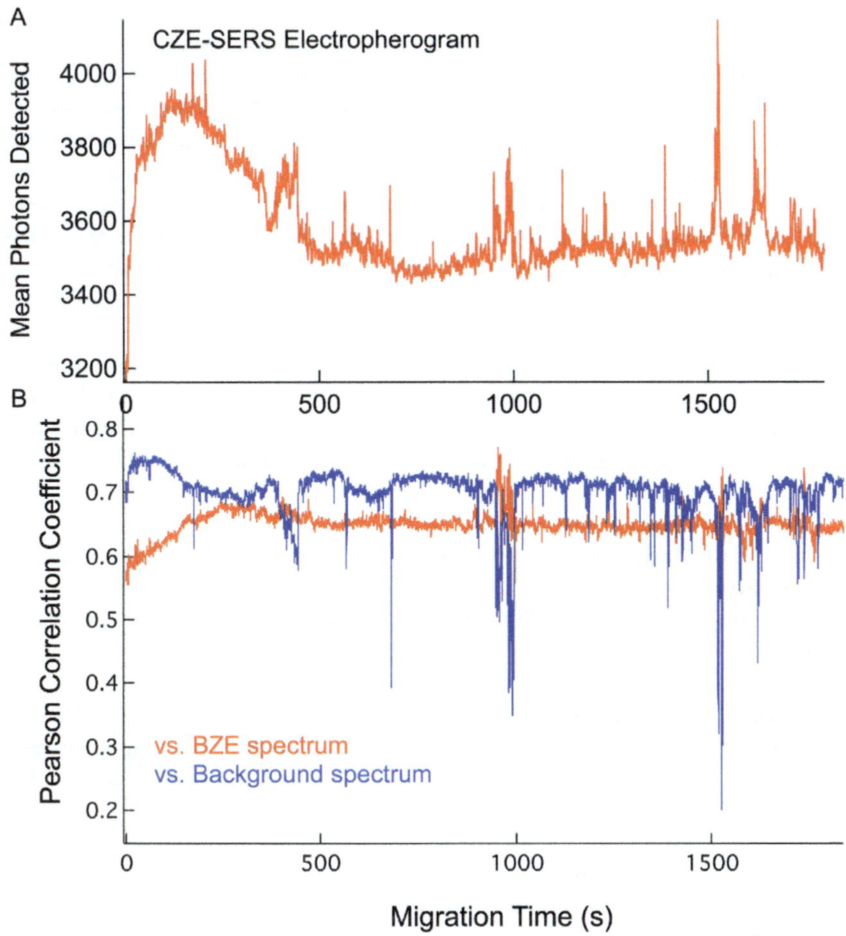

Fig. 1 The SERS electropherogram of urine spiked with BZE (A) is compared with the plot of the correlation coefficient (B) produced from the data set compared to a background spectrum (blue) and the spectrum of a specific analyte (BZE, red). The change in correlation coefficient with time can be used to identify compounds in the electropherogram.

Royston Goodacre asked: Mass spectrometry is a very good detector but in general not very quantitative, unless coupled with other techniques like isotopic labelling. As you seem to have an almost on–off switch with your sheath flow SERS, why don't you spike in internal standards and use isotopic ratios (*e.g.* from ^{13}C or D containing vibrations) to achieve quantification?

Zachary D. Schultz replied: This is an excellent idea, which we are exploring. The use of internal standards appears to be a promising approach for quantitative SERS.

Nick Stone commented: How does your approach compare with the other techniques available? Will it have an impact in the real world? Are your substrates single use? How long do they take to "foul" up?

Zachary D. Schultz responded: I believe our detector fits a niche where it will have real utility. For example, the analysis of small molecules by mass spectrometry is challenging for isobaric compounds such as structural isomers. The SERS spectrum of such molecules is often easy to differentiate. Since we are detecting in solution, we also avoid challenges associated with ionization. This complementary information is advantageous and not really available by other techniques. To have an impact in the real world, we need to translate this into a commercial device. We are working with Snowy Range Instruments to assess the technical specifications necessary for a commercial product. If we can commercialize this approach, I believe it will have impact. With regards to the second question, we have used the same substrate for up to a week in simple flow experiments and up to 3–4 days in capillary electrophoresis experiments. Applying a potential to the surface does cause the substrate to degrade more rapidly. Without a potential, as long as the substrate is rinsed well after use and kept in solution, it appears to remain stable and reusable. Good substrates are important though.

Royston Goodacre asked: Can you do the analysis in-line like a diode array is used in chromatography? If you are only interacting with the surface and then the analytes are coming off unchanged you could then do mass spectrometry afterwards?

Zachary D. Schultz replied: We believe so and are working toward this. There is one challenge, the sheath flow does dilute the sample which may complicate mass spectrometry afterwards. However, we think this is ultimately where this approach is going.

Abdu Subaihi remarked: You have dissolved the metabolite in methanol – does this affect binding to the surface? Have you tried dissolving it in water?

Zachary D. Schultz responded: The metabolite standard is shipped to us at 1 mg mL^{-1} in methanol. We then dilute this in urine to the concentration of 1 μg mL^{-1} (or less) in urine and buffer. The methanol is not likely to affect anything as it is diluted significantly. Because the reagent (benzoylecgonine) is a controlled substance, we do not have access to the pure compound, though it likely does not matter. As far as methanol inhibiting adsorption to the surface, we recently published results using sheath-flow liquid chromatography SERS.[1] The mobile phase was 5% methanol and we were still able to detect the species in our sample. We don't think methanol affects surface binding significantly.

1 A. Nguyen and Z. D. Schultz, *Analyst*, 2016, DOI: 10.1039/C6AN00155F.

Michael Hermes opened a general discussion of the paper by Valter Sergo: How is the therapeutic window discussed in the introduction (DOI: 10.1039/C5FD00173K) related to the concentrations investigated in this study? Is the SERS approach capable of covering the full concentration window expectable in blood?

Valter Sergo replied: In this study, we have tried to cover the maximum dynamic range possible. Apparently SERS is capable of covering the full concentration range expectable in blood, however a bigger cohort of samples/patients is needed to finally assess the issue.

Maria Paula Marques queried: Have you ever tried your approach with inorganic drugs, such as cisplatin or cisplatin-like agents?

Valter Sergo responded: Not yet, it would be interesting to try.

Royston Goodacre asked: You mentioned that methotrexate is cleared from the body by detoxification (metabolism) in the liver and then excretion (urination). Given that these methotrexate metabolites will have similar chemistry to the drug, can you discriminate between drugs and the xenometabolites in serum?

Valter Sergo answered: We have not measured the xenometabolites of methotrexate yet. However we have been able to distinguish methotrexate from folic acid, the structurally closest compound naturally present in the body (this has not been reported in the paper). For another drug, Irinotecan CPT-11, we have unpublished preliminary data proving that it is possible to distinguish it from its main metabolite SN-38.

Holly Fleming remarked: How difficult do you think it will be to implement technology like this into clinics when there are already similar techniques being used?

Valter Sergo responded: HPLC-MS is precise but time consuming, expensive and requires highly skilled personnel. Immunoassays are faster, but still expensive, not very accurate and the response has a narrow linear range. Surface plasmon resonance (SPR) is a serious competitor for SERS, however it provides only an indirect measurement of methotrexate concentration, being based on an enzymatic approach with limitations similar to those of immunoassays. The ultimate implementation in hospitals will be likely decided based on economical considerations.

Michael Hermes addressed Valter Sergo and Alois Bonifacio: If the chemical selectivity of the nanoparticles for the drug is increased by attaching a specific protein, does this make the advantage of having spectroscopic selectivity of the Raman scattering obsolete and the use of a fluorescent marker more suitable?

Alois Bonifacio replied: As a comment to the question raised by Michael Hermes about the competition between fluorescence-based quantitative methods (*e.g.* immunoassays) and SERS-based methods using a chemically modified metal surface, using recognition systems such as small peptides or molecularly imprinted polymers, I would like to point out that fluorescence detection is based on one single band (often only on one wavelength), whereas SERS spectra are more structured in many bands, which is a distinctive advantage when using a multivariate approach for building a predictive model, especially in the presence of complex biofluids such as serum.

Royston Goodacre asked: How close are you to measuring methotrexate in real samples? Do you think you would be able to get the same accuracy say if the subject had just eaten gravy compared to someone who had just had a chicken tikka masala?

Valter Sergo replied: This question can be answered at two levels: 1) intra- and inter-individual variations (due to diet, age, sex, co-administered drugs *etc.*) are not likely to affect significantly the SERS signal of methotrexate on our substrates, since, as shown in the ESI in our paper (DOI: 10.1039/C5FD00173K), the SERS bands due to serum itself are very weak, when compared with the bands of methotrexate. Different substrates might yield different results. 2) Intra- and inter-individual variations may occur because of competition for albumin of other compounds with respect to methotrexate, but, in our opinion, this might affect the limit of quantification, but should not affect heavily the accuracy of the model.

Nick Stone questioned: So for chemotherapy effects, presumably the drug has to be where the tumor/cancerous cells are; if you wanted to monitor local drug kinetics in solid tumours could you look at the amount of drug in the tumour by exploring tissue fluid or local lymphatic fluid?

Valter Sergo responded: As far as we understand, the approach entailed by the question would require a local acquisition of tissue/fluid, in close proximity of the cancerous tissues (for all practical purposes, a biopsy). It is a difficult question to answer, since we do not know i) how feasible it is to obtain a pure sample (lymphatic fluid or tissue fluid) in reasonable quantities from a patient (or animal), and ii) how the local drug kinetics in a solid tumour relate to the surrounding biofluid, considering the complex chemical and physiological equilibria involved.

Richard Dluhy said: The partial least-squares regression (PLSR) plots of the known MTX concentrations in diluted serum shown in Fig. 2B contain a large standard deviation (DOI: 10.1039/C5FD00173K). Can alternate spectral feature selection methods potentially decrease the variability of these measurements in serum?

Valter Sergo replied: Maybe yes. However, the stability selection is not aimed at minimizing the standard error of prediction, but to make the model more robust.

Abdu Subaihi commented: The drugs have been spiked to water and serum and the error prediction in serum is bigger compared to water. Do you think serum protein has an effect and increases the variations to the spectra?

Valter Sergo responded: The complexity of the serum matrix (hence, not only proteins, but also small molecules) is very likely to contribute to the spectral variability.

Nick Stone opened a general discussion of the paper by Colin J. Campbell: Do you have a handle on the redox potential and pH within tumours in the body? Do the ranges of sensitivity with this technique match what would be necessary to study tumours *in vivo* either in humans or animal models?

Colin J. Campbell answered: Making measurements in real tumours *in vivo* is something that we have never done. One would expect the range of potentials within the body to be determined by NAD(P)H at one end and oxygen at the other, so spanning from approximately −350 mV *vs.* NHE to approx +800 mV *vs.* NHE. We have probes that measure the very hypoxic end of this scale and far into the oxidative side (approx −400 mV to +150 mV *vs.* NHE) so in principle we could measure the pathologically relevant potentials (detailed by Jiang *et al.*[1] and Thomson *et al.*[2]).

1 J. Jiang, C. Auchinvole, K. Fisher and C. J. Campbell, *Nanoscale*, 2014, **6**, 12104–12110.
2 P. I. T. Thomson, V. L. Camus, Y. Hu and C. J. Campbell, *Anal. Chem.*, 2015, **87**, 4719–4725.

Alois Bonifacio said: The usual statistics tool such as the calculation of the mean (as an approximation of the "real value") and the standard deviation (as a measure of the data variability) are valid under the assumption that the data are following a normal distribution. However, in most of the boxplots in your paper (DOI: 10.1039/C5FD00216H), and especially in those in Fig. 9, the mean and the median differ, and the interquartiles are asymmetric, suggesting that the data are not normally distributed. Did you check if your SERS data are normally distributed (*e.g.* with a Shapiro–Wilk test)? And in case they are not, how does this non-normality impact on the overall results of your paper?

Colin J. Campbell responded: Honestly, we didn't check if our data are normally distributed, but it could be worth our while to look into it. When the data from biological systems is not normally distributed it sometimes tells you something interesting about the regulation of the processes that you are measuring, for example that cells die beyond a particular potential.

Duncan Graham asked: There are some new fluorescent dyes which appear to preferentially localise in areas of hypoxia. I am not sure what this mechanism is or why it happens. Do you see any effect like this with your nanoparticles?

Colin J. Campbell replied: That is not something that we have looked for to date. When you stain tissues for hypoxia using (for example) Hypoxyprobe, you see staining in areas with the highest nitroreductase activity but this is because the reporter has been chemically changed by the enzyme. We can similarly measure hypoxic cellular environments with our probes because their structure is changed by a redox reaction.[1]

1 J. Jiang, C. Auchinvole, K. Fisher and C. J. Campbell, *Nanoscale*, 2014, **6**, 12104–12110.

Petra Hellwig commented: In your study, quinoid redox-active reporter molecules are introduced in cells in order to obtain information on the overall redox potential. Do the particles themselves or the reporter molecules influence the pH or the redox potential within the cell?

Colin J. Campbell answered: That is something that we specifically sought to measure using established measures of oxidative stress,[1] and we saw no evidence that our nanosensors change the redox potential within the cell.

1 C. A. R. Auchinvole, P. Richardson, C. McGuinnes, V. Mallikarjun, K. Donaldson, H. McNab and C. J. Campbell, *ACS Nano*, 2012, **6**, 888–896.

Royston Goodacre asked: Can you comment on the biocompatibility of the reporter molecule? If you have other molecules in the cell that have a similar structure to the reporter molecule, do you get any exchange of these going on/off the nanoparticle surface inside cells?

Colin J. Campbell responded: We have investigated the stability of our reporters on the surface of nanoparticles inside cells and we see no evidence of their loss into the cellular milieu. We believe that the reporter molecules are biocompatible because we have carried out measures of toxicity, viability and oxidative stress and see no evidence of detrimental effects.[1,2]

1 C. A. R. Auchinvole, P. Richardson, C. McGuinnes, V. Mallikarjun, K. Donaldson, H. McNab and C. J. Campbell, *ACS Nano*, 2012, **6**, 888–896.
2 M. A. Ochsenkuhn, P. R. Jess, H. Stoquert, K. Dholakia and C. J. Campbell, *ACS Nano*, 2009, **3**, 3613–3621.

Karen Faulds commented: You are using gold nanoshells that have been shown to give strong localised heating when interrogated with a laser wavelength that coincides with the extinction band of the nanoshells. What do you think is the localised effect of the heating within a cell? Do you observe damage to the cell or degradation of the reporter molecule?

Colin J. Campbell answered: It is true that you have to be very careful with the laser power – too much and you will fry the cells. However, when you look at how much laser power is used when these types of particles are used for photoablation therapy, we are orders of magnitude lower.

Matthew J. Baker asked: Do you know the process of nanoparticle uptake within the cells and when you collect spectra to ensure that the nanoparticle uptake process does not affect the redox status of the cells?

Colin J. Campbell responded: We have assayed the redox status of cells with and without our nanosensors,[1] and found that our nanosensors do not change the redox status of the cells. The process by which the cells take up nanosensors is difficult to definitively confirm, but we have carried out a study that shows that the process is most efficient in the absence of tissue culture media, and is not detrimentally affected by ATP depletion, clathrin depletion or cholesterol depletion.[2]

1 C. A. R. Auchinvole, P. Richardson, C. McGuinnes,V. Mallikarjun, K. Donaldson, H. McNab and C. J. Campbell, *ACS Nano*, 2012, **6**, 888–896.
2 M. A. Ochsenkuhn, P. R. Jess, H. Stoquert, K. Dholakia and C. J. Campbell, *ACS Nano*, 2009, **3**, 3613–3621.

Royston Goodacre asked: Have you looked at your cells under electron microscopy (EM) to see if your nanoparticle reporter is actually bound by a membrane when inside the cytoplasm?

Colin J. Campbell replied: We have carried out several EM studies to investigate intracellular localisation.[1,2] We found that localisation was dependent on the uptake mechanism and the experimental conditions you use play a big role (*e.g.* whether you have serum in your tissue culture media).[1] We also saw a bit of variability between different cell types.

1 M. A. Ochsenkuhn, P. R. Jess, H. Stoquert, K. Dholakia and C. J. Campbell, *ACS Nano*, 2009, **3**, 3613–3621.
2 C. A. R. Auchinvole, P. Richardson, C. McGuinnes,V. Mallikarjun, K. Donaldson, H. McNab and C. J. Campbell, *ACS Nano*, 2012, **6**, 888–896.

Stefano Fornasaro queried: Please comment on the redox potential measurement and the viability of the cells. It seems paradoxical that ZnO NPs act as natural antioxidants and trigger an antioxidant response. Was this behaviour concentration dependent? Why was this not shown in the paper (DOI: 10.1039/C5FD00216H)?

Colin J. Campbell answered: The oxidative change in the redox potential corresponds to a loss of viability in the cells and this is in line with the known relative toxicity of silver and zinc oxide (the cells treated with silver have a lower viability than those treated with zinc oxide). We didn't measure redox potential changes as a function of concentration of either zinc oxide or silver, we made our measurements at 0.5 mg mL^{-1} of either silver or zinc oxide. I would suggest that the cells survive the treatment with zinc oxide because they are able to launch an antioxidant response (shown in the mRNA data) in a way that they are unable to do when treated with silver.

Michael Hermes remarked: Which effect limits the accuracy of the pH value determined? Is this limitation of stochastic nature or more likely to be related to spectroscopic effects?

Colin J. Campbell replied: I think that the limits of accuracy are determined by the sensitivity of the instrument. The signals from single particles in single cells can be low so the lower the noise the better the measurement will be.

Royston Goodacre commented: As Faraday Discussion meetings on this topic occur every 10 years or so, how do each of you think your specific field will have advanced within this period?

Duncan Graham responded: I think the main thing that will occur in the next 10 years will be an increase in the use of SERS for us in a clinical diagnostic setting. The research base is strong and the first examples of successful translation are emerging. I hope this will accelerate in this time period!

Zachary D. Schultz replied: In 10 years, we hope that sheath flow SERS detection is a commercial product that is available for everyone. We hope that we will have moved from the challenges of detection to focusing on the challenges of the biomedical problems.

Valter Sergo responded: The application of SERS to Therapeutic Drug Monitoring (TDM) will be viable provided that reliable, repeatable substrates will be commercially available, similar to what happens for electrochemical glucose sensing. The use of SERS of biofluids for mass screening of certain types of cancer is possibly closer than TDM to a practical application in clinics.

Colin J. Campbell answered: I would like to think that our field will have advanced such that we are making much faster measurements and achieving organellar resolution in pH and redox potentials. More broadly I hope that we can shed some light on the complementary roles that the redox potential and pH play in determining the phenotype (at the cellular level and beyond).

Royston Goodacre resumed the general discussion of the paper by Duncan Graham: Is there already any evidence of clinical use of your SERS system for diagnostics or screening?

Duncan Graham responded: We have a CE-marked diagnostic for fungal infections using SERS and some of our older science and is available from Renishaw Diagnostics Ltd. This is now being evaluated in hospitals in Europe. The research in this discussion paper (DOI: 10.1039/C5FD00167F) is potentially next generation and is aimed at being single tube without separation steps.

Lauren Jamieson asked: Your group has done a lot of very good work multiplexing multiple SERS reporters. What do you think the limit is to the number of reporters that you could realistically multiplex and detect in one assay?

Duncan Graham replied: We have published 6 in a study led by my colleague Karen Faulds and in collaboration with Roy Goodacre from Manchester.[1] Within Renishaw Diagnostics Ltd a 10-plex has been measured but in clinical analysis the question really is what multiplexing makes sense? To this end targets can be grouped according to therapy so although identification of a unique species may be possible it can be easier to detect two or three similar species that use the same treatment.

1 K. Faulds, R. Jarvis, W. E. Smith, D. Graham and R. Goodacre, *Analyst*, 2008, **11**, 1505–1512.

Bayden Wood opened a general discussion of the paper by Ganesh Dhruvananda Sockalingum: I was interested to see the effect of the anticoagulant EDTA on the serum spectra. Does this mean you cannot use EDTA (or Acid Citrate Dextrose (ACD)) as the anticoagulant when detecting infectious agents or cancer markers in serum? Can you combine data from Li-heparin and EDTA and achieve the same type of sensitivity and specificity?

Ganesh Dhruvananda Sockalingum answered: As serum is obtained from blood samples collected without anticoagulant, identification of markers in the serum is not hampered by the use of Li-heparin and EDTA. This is one of the advantages of using serum instead of plasma for identifying biomarkers.

If plasma is the biofluid of choice, the choice of the anticoagulant is an important issue for spectroscopic analysis. Plasma collected using Li-heparin tubes appears as the most appropriate. For EDTA, we have shown that additional peaks appear in the FTIR spectrum and that these can be eliminated after dialysis (see Fig. 4, DOI: 10.1039/C5FD00184F). However, this is time-consuming and adds an additional step in a clinical workflow. We have not used ACD as the anticoagulant since it is a liquid anticoagulant (compared to Li-heparin and EDTA which are spray-coated) and induces a dilution of the plasma.

Petra Hellwig commented: In your paper you present different aspects of the use of FTIR spectroscopy as an analytical high-throughput tool for the study of diseases. Could you give an overview on the diseases have you been looking at, and further diseases that you feel can be accessed with FTIR high throughput monitoring?

Ganesh Dhruvananda Sockalingum responded: We have been looking at various diseases such as cancer, fibrosis, sepsis, metabolic diseases (diabetes, galactosemia *etc.*). In general, all disorders could be analysed as long as markers or signaling molecules are present (a few examples are: nosocomial infections, malaria, ocular diseases, neurodegenerative diseases, cardiovascular diseases *etc.*).

Bayden Wood remarked: Others have shown the effects of drying on serum, which induces differential crystallisation (coffee ring effect) and differences in water content. How are you controlling the drying to remove all the water, and is this a confounding variable in the model prediction?

Ganesh Dhruvananda Sockalingum answered: We have analysed the impact of biofluid drying modalities on spectral quality and reproducibility (see part 5 of the Results section of the paper, DOI: 10.1039/C5FD00184F). This experimental step must be standardized to avoid spectral variability due to the presence of water. When measuring the spectra the software (OPUS ver6.5) calculates the signal-: water ratio to validate the spectral quality. The signals are taken in the ranges 1700–1600 cm^{-1} (S1) and 1200–960 cm^{-1} (S2) and the water content in the range 1847–1837 cm^{-1} (W). We also monitor the water content from the PCA loadings (see supplementary Fig. ESI10 in the paper).

Nick Stone addressed Ganesh Dhruvananda Sockalingum: I am trying to understand conceptually what impact the coffee ring might have. Having done a fair bit of Raman and IR looking at coffee rings, we get a large deposition around the outside of the ring which often results in saturation of the amide peaks. Do you think the impact of having a local saturation in the outer coffee ring will impact on the ensemble spectrum you are measuring? Do you know the thickness of the deposited layer? Can you put the plates under a microscope to see if there is local saturation?

Ganesh Dhruvananda Sockalingum replied: When using the HTS-FTIR mode a single spectrum is measured in the macro mode from a large dried drop. The coffee ring is present microscopically. Local saturations can be observed by IR microscopy but not in the macro mode. Once diluted, serum gave spectra that are highly reproducible in the macro mode.

We have performed measurements using high resolution microscopy which show a gradient of 0–50 μm from the centre to the periphery of a 0.5 μL drop diluted to 1/3.

When the silicon plate is analysed under an FTIR microscope, there are indeed regions of local saturation, especially at the drop periphery. These heterogeneities are not visible in the macro mode with the HTS as a global spectrum is acquired. However, it is worth noting that FTIR microscopy does not use the same detector and optics as the HTS-FTIR instrument.

Wolfgang Petrich remarked: Thanks for pointing out the importance of pre-analytics. It is detrimental to what you do. When we started doing mid-IR analysis of body fluids some years ago we looked at the spectral impact of anticoagulants such as EDTA and what actually struck me is that in clinics, the phlebotomists are supposed to draw the blood and gently shake or roll the filled *e.g.* EDTA tube for some time. What I saw in different clinics, however, was that this is not always done. In addition, there is also variations of EDTA loaded in the unfilled vials due to production tolerances. To cut a long story short this to me meant that using plasma will add quite a bit of variability merely due to what we might call pre-analytics, such that we ended up focusing on serum.

Ganesh Dhruvananda Sockalingum responded: We agree with this comment and we came to the same conclusion that spectroscopy of serum is more straightforward unless downstream tests are required for specific diseases where coagulating proteins are necessary for biochemical and hemostasis assays (see Baker *et al.*[1]).

1 M. J. Baker, S. R. Hussain, L. Lovergne, V. Untereiner, C. Hughes, R. A. Lukaszewski, G. Thiéfin and G. D. Sockalingum, *Chem. Soc. Rev.*, 2016, **45**, 1803–1818.

Hugh J. Byrne remarked: The "coffee ring" is both physically and chemically inhomogeneous, but much is known about it, since Leo Vroman, in the late 70s early 80s, extensively studied the adsorption of proteins from human serum, and documented the precise sequence of deposition, which depends on their relative mobility and affinity for the surface, giving his name to the so-called "Vroman effect".[1,2]

1 F. Bonnier, F. Petitjean, M. J. Baker and H. J. Byrne, *J. Biophotonics*, 2014, 7, 167–179.
2 L. Vroman, A. L. Adams, G. C. Fischer and P. C. Munoz, *Blood*, 1980, 55, 156–159.

Ganesh Dhruvananda Sockalingum responded: The coffee ring effect is indeed an issue to be considered when performing microspectroscopy. We have previously reported this effect using FTIR imaging.[1] The present paper deals with the development of HT-FTIR spectroscopy where a single spectrum of a 3 mm dried drop is recorded in the macro mode with very good reproducibility.

1 L. Lovergne, G. Clemens, V. Untereiner, R. A. Lukaszweski, G. D. Sockalingum and M. J. Baker, *Anal. Methods*, 2015, 7, 7140–7149.

Gianfelice Cinque commented: Signal saturation is related to detector dynamic range; in FTIR microscopy we found that IR absorbance is linear between 0.04 and 1.8.

Max Diem stated: This is a comment about "detector non-linearity" when observing high absorbance values in FTIR, in particular in the "coffee ring" effect observed for dried films.

The original calculations for "photometric accuracy" were carried out for dispersive instruments, where it can be shown that the magnitude of the signal, and the increased noise due to higher absorbance, lead to an optimal S/N ratio at about 0.6 OD units.

This is not necessarily the case in FTIR, where a background spectrum is measured over the entire source/detector wavenumber range. The photometric accuracy in this background spectrum is determined by the non-linear response the detector may give at the centerburst. The non-linearity of the detector is subsequently smeared over all wavenumbers during fast Fourier transform (FFT). When a sample is inserted, the entire interferogram is reduced in intensity, but the absorbance at one wavenumber, say the amide I peak, will not cause the detector to be "more non-linear", but rather, just increase the noise at the wavenumber of the strongest absorption.

Klaus Gerwert added: If the transmission under the strongest band, which is usually the amide I band and water at around 1660 cm^{-1}, is still 10% then there is no problem with non-linearity.

Royston Goodacre addressed Ganesh Dhruvananda Sockalingum: I like the systematic nature of the study. In other fields phenotypic/biochemical methods have been used to analyse biofluids and have reported epidemiological changes related to age, gender, body mass index, smoking, blood pressure, *etc.* How do you plan to adjust for these confounders in any disease-related studies (especially for case-control studies) when you measure a spectral pattern rather than identify specific molecules?

Ganesh Dhruvananda Sockalingum responded: We agree on the importance of adjusting multivariate analysis for confounding variables such as age and gender.

In the biomedical field, a number of diagnostic and prognostic algorithms based on serum biochemical measurements include also age and gender.

Similarly, chemometric analysis aimed at classifying serum spectral data should test the impact of these variables on the diagnostic performance through multi-block methods such as those described by Hassani *et al.*[1]

Regarding case-control studies, controls should be matched with cases for age and sex but also with any other potential confounding variable according to the specific objective of the study. In most spectroscopic studies, the selection of the control group is not optimal. This issue was not in the scope of this paper, where we focused on the pre-analytical phase, but we have addressed this concern in a recent review.[2]

1 S. Hassani, H. Martens, E. M. Qannari, M. Hanafi, G. I. Borge and A. Kohler, *Chemom. Intell. Lab. Syst.*, 2010, **104**, 140–153.
2 M. J. Baker, S. R. Hussain, L. Lovergne, V. Untereiner, C. Hughes, R. A. Lukaszewski, G. Thiéfin and G. D. Sockalingum, *Chem. Soc. Rev.*, 2016, **45**, 1803–1818.

Bayden Wood asked: A difficulty that arose in a lot of the earlier work done on prion detection in serum using FTIR was the ability to distinguish between primary markers of the prion virus and secondary inflammation markers. Do you have a way to test this in order to discriminate cancer markers from the inflammation response? Have you combined your sepsis data with your cancer data to investigate this?

Ganesh Dhruvananda Sockalingum answered: No we have not done this but it is a good point to consider when using the method to diagnose cancer. This is the next step, the analytical phase, while the present work is dedicated to the pre-analytical step.

This comment however raises the important issue of confounding factors. The risk of identifying false positive biomarkers should be limited by a proper selection of the control group (optimal matching process). As an example, in the case of liver cancer, the risk is to identify markers in relation to underlying cirrhosis rather than with cancer (most liver cancers develop on underlying cirrhosis). To avoid this, the control group should be composed of cirrhotic patients without cancer and not of healthy individuals .[1] In addition, the results of exploratory studies must be confirmed by validation studies performed in independent large cohorts and tested in "real life" clinical settings (see Baker *et al.*[2]).

1 I. Taleb, G. Thiéfin, C. Gobinet, V. Untereiner, B. Bernard-Chabert, A. Heurgué, C. Truntzer, P. Hillon, M. Manfait, P. Ducoroy and G. D. Sockalingum, *Analyst*, 2013, **138**, 4006–4014.
2 M. J. Baker, S. R. Hussain, L. Lovergne, V. Untereiner, C. Hughes, R. A. Lukaszewski, G. Thiéfin and G. D. Sockalingum, *Chem. Soc. Rev.*, 2016, **45**, 1803–1818.

Parvez Haris resumed the general discussion of the paper by Ganesh Dhruvananda Sockalingum by communicating: There is great potential for the analysis of biofluids using FTIR and Raman spectroscopy for clinical diagnosis. This study is an attempt towards standardisation of procedures for sample handling and preparation prior to spectral analysis using infrared spectroscopy. However, some important issues have not been mentioned that can easily influence spectral data. These include issues such as the effect of diet, age, ethnicity and gender *etc.* Furthermore, how the samples were collected could also influence the spectral data. For example, a first morning urine sample will differ from a 24 h urine sample. Likewise, a fasting blood sample will differ from a non-fasting blood sample. These are fundamental pre-analytical issues that need to be considered before spectra are acquired and should be mentioned in an article that focuses on standardisation. I recommend Fig. 1 in your manuscript (DOI: 10.1039/C5FD00184F) to be modified to include a step regarding harmonisation/standardisation of the sample collection procedures to avoid confounding factors resulting from some of the factors mentioned above.

Ganesh Dhruvananda Sockalingum communicated in reply: Indeed, this study focuses on standardization, not of the analytical step but of the pre-analytical step. This step does not take into account patient characteristics. This is why these issues are not discussed here. Standardisation at the inter- and intra-analytical steps is indeed an important issue as they are also sources of errors, but the pre-analytical step represents nearly 70% of errors. However, we completely agree that factors related to patients should be carefully controlled as they can influence spectral data and lead to false positive or false negative results. This issue is not specific to spectroscopic methods. Inappropriate selection of case patients and control subjects has been widely reported as a major pitfall in other high throughput technologies such as proteomics, lipidomics and genomics. When comparing groups that are not matched for example by age, gender and physical conditions such as hormonal status, data analysis may be biased and differences between groups may be linked to confounding factors. Our paper focused specifically on the pre-analytical issues regarding the procedures for spectral acquisition. We have reviewed the pre-analytical issues related to patients in a recent paper.[1] Fig. 1 cannot be modified at this stage since it refers to a workflow of the pre-analytical step related to spectroscopic procedures and not to patient selection.

Parvez Haris resumed the general discussion of the paper by Ganesh Dhruvananda Sockalingum by communicating: It is important to have a good knowledge about the source of the samples and how they were collected before spectral analysis. Indeed, one could argue that this information is even more important for vibrational spectroscopic measurements since diverse chemicals can give rise to peaks in the spectrum and complicate the analysis. In this context, the experimental section should be more informative and provide sufficient details regarding the samples analysed. Were the serum and plasma samples collected under identical conditions or not? Were they fasting or non-fasting blood samples? How many samples were analysed and from how many people were these samples collected from? It would be useful to have some demographic information regarding the samples.

Ganesh Dhruvananda Sockalingum communicated in reply: This paper is focused on the standardisation of pre-analytical steps of sample preparation, and looking at considerations for clinical applications of biofluid spectroscopy. The focus of the paper is looking at the impact of volume, dilution, freeze-thaw, inter-operator reproducibility, inter-plate reproducibility and the impact of drying. This is not a clinical diagnostic study. Serum and plasma samples were collected under the same conditions and then the process differed based upon standard clinical preparations. In order to achieve this objective all samples were collected and treated identically (see the experimental section of the manuscript, DOI: 10.1039/C5FD00184F). Six blood samples were collected to obtain serum and plasma from 3 patients and 3 bile samples were obtained from 3 different patients. Patient profiles were not considered in this study as the aim of this paper is not on pre-analytical issues related to patient recruitment for a clinical diagnosis. All were fasting blood samples.

Max Diem opened a general discussion of the paper by Christopher Phillips: I have a number of questions and comments.

1. It is appropriate to refer to the 1236 cm^{-1} region as the phosphodiester region, since both DNA and RNA contain this chemical moiety. However, as was pointed out in the lectures by Prof. Gough (DOI: 10.1039/C5FD00168D) and myself (DOI: 10.1039/C6FD00023A), this spectral region also contains strong vibrations of collagen, a protein found in connective tissue such as the stroma. Thus, a broad-band measurement of this spectral region is not a good indicator of the nucleic acid content of a sample.

2. It is also a mistake to utilize the phosphodiester group as a surrogate for the nuclear size (and therefore, for the nucleus : cytoplasm ratio) since RNA occurs at high concentrations outside the nucleus (see Notingher *et al.*, DOI: 10.1039/C5FD00172B).

3. The fact that the humidity in the instrument was not controlled deeply disturbs me. Water vapour has a gap in its vib-rotational spectrum at 1600 cm^{-1}, and a number of strong absorptions between 1650 and 1670 cm^{-1} (see Fig. 2). Their contribution will be superimposed on the protein absorption in a broad-band measurement, and vary with the humidity in the laboratory. Without control of the humidity, the intensity ratios can easily vary by 10 to 20%. This is well known in the FTIR community, where all IR microscopes are purged with dry air.

Christopher Phillips replied: Regarding comments 1 and 2, actually we have looked at this issue.[1] We looked at various sample areas with widely different concentrations of cell nuclei, and, as can be seen in Fig. 5 of that paper, we were able to show that the majority of the signal in this band correlates well with the amount of nuclear material.

In response to comment 3, the image acquisition protocol was carefully designed and tested to be immune to *e.g.* atmospheric absorption in the machine as well as spectral and spatial brightness fluctuations in the illumination. Details are reported elsewhere.[1,2]

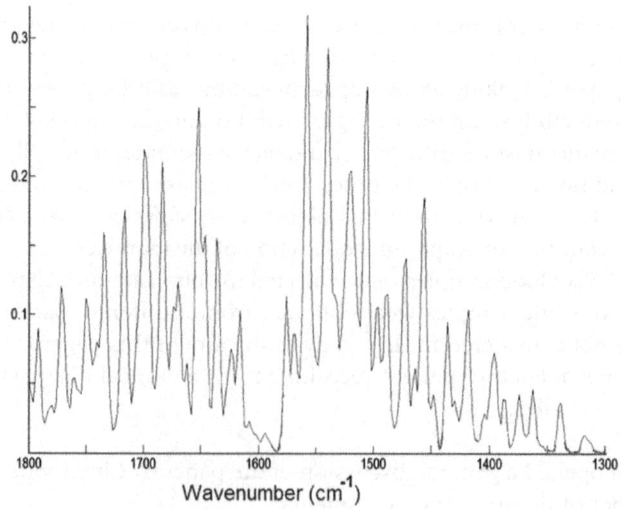

Fig. 2 Vib-rotational spectrum of water vapour.

1 H. Amrania, G. Antonacci, C.-H. Chan, L. Drummond, W. R. Otto, N. A. Wright and C. Phillips, *Opt. Express*, 2012, **20**, 7290–7299.
2 C. C. Phillips and H. Amrania, *GB Pat.*, GB1507314.1, 2015.

Nick Stone commented: Can you help us understand what metrics are used to provide the prediction of disease from the images you obtain? Is it based on the pathologists interpreting the "stained" image obtained? Does this actually help them get a better result? If the peak ratio is provided by the system of the region selected by the pathologist, why can't you provide a prediction of the sample pathology without a pathologist? Is it difficult in your images to find the region of interest (ROI), or is that not the point?

Christopher Phillips responded: Please see my response to Gianfelice Cinque's question about the rationale for using these wavelengths. Yes, the pathologist chooses the ROI, and Digistain is used to generate a "Digistain Index" (DI) value for the ROI, which we are finding has a prognostic value that exceeds that of the current H+E staining protocol.

Please see the answers to the question asked by James Livermore in the session titled Clinical Spectroscopy about the prospects for fully automating this process. We find very little appetite for this among the clinical profession; to their ears the very suggestion sounds hubristic.

Catherine Kendall asked: In your paper in section 1.1.3 (DOI: 10.1039/C5FD00150A) you state that we should be presenting spectroscopic data in simple visual ways, as this is how clinicians currently interpret images to inform their clinical decision making, for example in radiology and pathology. Clinicians do like to visualise the information, however are you clear that this is the way forward in terms of having to have experts undertaking subjective measurements, with all the associated problems we are aware of? Surely spectroscopic data can provide "added value" with objective measurement of tissue biochemistry and some quantification of risk and overcome some of the current issues?

Christopher Phillips replied: We see our method as a way of adding objectivity to the process. Yes the clinicians select the ROI by eye (as they have to anyway in the standard H+E protocol), but after that the Digistain measurement is an objective one, and it's statistical correlation, *e.g.* with patient outcome, is an objective fact that we believe can be established quantitatively. This last point is where the "added value" lies.

Josep Sulé-Suso stated: The data regarding survival was obtained from 80 patients. I think 80 is a small number. You need to do a bigger study to convince clinicians. No doubt you have already considered this.

Christopher Phillips answered: Indeed we have. The initial pilot study was intended to be of 200 cases, but we were confounded by patchy record keeping on the part of the clinicians and in the end only ~80 samples were usable. We are currently embarking on a much larger (n ~2000) study.

Peter Gardner commented: In order to simplify your system you have reduced it to just the key basic components making it fast and presumably cheap. Having locked it down, you must have to make compromises in terms of the data coming out. In essence this is an extreme case of discrete frequency imaging. We already know that the information content isn't high compared with measuring the full IR spectrum. Can you really get much more from the data compared with just looking at the visible H&E image? Many would argue that it is the spectroscopic content that allows us to add biochemical information, does your system give you anything extra compared to IR work? Regarding your trial, I would be interested in whether you have encountered any unexpected confounding factors. Early trials on this were confounded by things like differences between where the biopsy was taken, *e.g.* in two different hospitals, the type of biopsy, *etc.* We had no idea that these subtle changes would be confounding factors. Have you taken these things into account?

Christopher Phillips responded: We have indeed considered the points you have raised. Since tissue banks in the UK at least have their fixing protocols locked down and governed by the Royal College of Pathologists, we are confident that a multi-institutional study would produce similar results.

These 80 patients were pooled from 1 institution. This is the norm, for trials of this kind. In the next phase of this study we will stratify by institution as dictated by the ICH E9 guidelines.

As a general point, all these spectroscopic imaging techniques generate very large amounts of data, the majority of which will be discarded before being presented to a clinician. We chose to do this data reduction "up front", because that way the measurement is can be made robust and reproducible, and the process is simple enough to be understandable, and therefore usable, by the clinicians. We argue that "less is more" in this regard.

Gianfelice Cinque remarked: Practically in the imaging system you take 4 wavelengths, 2 for setting the background and 2 to measure the ratio of the phosphate group band with respect to the amide I band. Could you explain how that ratio is capable of discriminating between cancer and non-cancer tissue, *i.e.*, on the basis of which scientific information?

Christopher Phillips replied: The contrast mechanism here arises because the cell cycle is disrupted in tumourous tissue, with the result that it contains a different (usually greater) fraction of DNA and ribosomal RNA than does healthy tissue. This fact is widely understood and accepted by pathologists, and is reflected by the fact that many tissue biopsy scoring systems include measures of nuclear pleomorphism and mitotic activity. One of the strengths of Digistain is that it can measure this chemical ratio quantitatively and reproducibly.

Alois Bonifacio queried: Breast cancer is a molecularly heterogeneous disease, and it can be classified into several types and subtypes, Luminal A, Luminal B, *etc.* This subtyping is known to be related to prognosis: different subtypes are correlated with better or worse prognosis. Did you, or do you plan to, correlate your IR data collected from tissues with the subtype for each patient?

Christopher Phillips answered: Yes we have created a logistic regression model formed of several parameters including tumour size, ER status and HER2 status. This produces a model with good sensitivity and specificity (AUC = 0.86).

Max Diem asked: 1. Which of the two common breast cancers are in your study, ductal or lobular adenocarcinoma?

2. In order to predict disease outcome with any kind of acceptable confidence interval, one needs datasets much larger than the 80 samples reported in this study. What happened to the other 120 samples that were originally in the study?

Christopher Phillips responded: 1. In the study, ~90% of the tumors were invasive ductal carcinomas (IDCs).

2. Roughly 200 tumors were imaged in the pilot study, but, unfortunately, the follow up medical records were incomplete on many of them, so in the end only ~80 could be used for the statistical analysis. Nevertheless the results are encouraging, the small sample size notwithstanding.

Samir El-Mashtoly asked: I really like your approach, performing imaging at 4 wavenumber regions. We use the same approach for imaging of cells and tissues using non-linear coherent anti-Stokes Raman spectroscopy (CARS) and it saves a lot of time. We usually focus on the CH stretching region, which can reflect most of the cellular and tissue components. Have you tried looking at this region?

Christopher Phillips replied: No not yet. We focused on the phosphate-to-amide ratio because it is a known biomarker for the disease.

Zachary D. Schultz opened a general discussion of the paper by Kathleen M. Gough: Can you comment on how far scattering scanning near-field optical microscopy (s-SNOM) has come *vs.* tip-enhanced Raman spectroscopy (TERS)? Can you generate hyperspectral images, or are images single frequency measurements? What about reproducibility and resolution?

Kathleen Gough answered: The growth in the near-field IR methodologies has been very rapid, as evidenced by the proliferation and diversity of near-field methods that have been demonstrated in the last few years. The designation "scattering-type SNOM" was suggested in 2001,[1] a few years after the strategy was proposed and the vibrational contrast was demonstrated with a CO_2 laser.[2] Since that time, broad band spectral acquisition and single frequency imaging *via* s-SNOM have been explored by numerous labs. Other strategies have been designed, including AFM–IR, pioneered by Dazzi and co-workers.[3] Their original method relies on AFM detection of photothermal expansion; advances were recently reviewed,[4] illustrating how far it has come in the space of a decade. With AFM–IR, as you may know, following irradiation with light tuned to a vibrational absorption wavelength, the photothermal expansion of the sample is detected by the cantilevered AFM tip response, and converted into an absorption spectrum. Yet another approach is to heat the AFM tip, so that it becomes both the thermal IR source and the scattering probe.[5] The s-SNOM approach has been implemented at the Advanced Light Source (ALS, Berkeley),[6] where we have obtained most of our data. Other approaches have recently been implemented at other synchrotrons,[7,8] and two commercial companies, based on the design principles of Keilmann and of Dazzi, are now producing benchtop instruments with multiple modalities. Nanoscale IR modalities are valuable to many researchers, as they provide chemical information to complement the many topographical and physical nanoscale methods. I think these rapid developments are a testament to the success of these new methods and to the excellent match between nanoscale spectra and those from far field. This list is representative, but certainly not comprehensive, and is illustrative of very rapid growth indeed, possibly more rapid than TERS, with a very promising future.

With regard to your second question, the answer is yes, hyperspectral images can be obtained. The experiment design depends on the technique (s-SNOM, AFM–IR, *etc.*) and the available light sources. Broad band spectra can be obtained at a point with a broad band laser or a synchrotron source. The area sampled is presently automated for a line scan at the ALS. To my knowledge, imaging is also feasible with the neaspec instrument, using their broad band source. In both cases, single points are collected and the image created afterwards. If the source is a Quantum Cascade Laser (QCL), then it is necessary to obtain data at multiple different wavelengths. There, the strategy would be to tune the laser frequency over the desired frequency range and take a near-field image at each wavelength. Since the tuning range of QCLs is only a few hundred cm^{-1} at present, and since time would be a consideration, the number of frequencies sampled would still be a restriction; the resulting image would be more like discrete frequency or sparse frequency, rather than full hyperspectral imaging. My guess is that this is an area where improvements are being planned already by all interested players.

Scattering SNOM is very close to TERS in terms of spatial resolution and the acquisition of full spectral information at every point. Single molecular detection may be achievable with s-SNOM, with appropriate experimental enhancements. The tip condition is critical for the acquisition of good data in all tip-based methods. Provided the tip has not been damaged or contaminated, the reproducibility of the spectra is fine, in our experience. Both techniques suffer from drift issues which limit the reproducibility. If the stage position is highly controlled, drift is not an issue. For example, in the images displayed in Fig. 7 of our paper (DOI: 10.1039/C5FD00168D), acquired by Dr Max Eisele on a neaspec system, the drift and reposition between the acquisition at the resonant wavenumber (1655 cm^{-1}) and the background contrast non-absorbing wavenumber (1850 cm^{-1}) was about 3 pixels, or 15 nm. This was already on the order of the estimated spatial resolution and was easily addressed in post-processing of the data.

1 R. Hillenbrand, B. Knoll and F. Keilmann, *J. Microsc.*, 2001, **202**, 77–83.
2 B. Knoll and F. Keilmann, *Nature*, 1999, **399**, 134–137.
3 A. Dazzi, S. Goumri-Said and L. Salomon, *Opt. Commun.*, 2004, **235**, 351–360.
4 A. Dazzi, C. B. Prater, Q. Hu, D. B. Chase, J. F. Rabolt and C. Marcott, *Appl. Spectrosc.*, 2012, **66**, 1365–1384.
5 A. C. Jones and M. B. Raschke, *Nano Lett.*, 2012, **12**, 1475–1481.
6 H. A. Bechtel, E. A. Muller, R. L. Olmon, M. C. Martina and M. B. Raschke, *Proc. Natl. Acad. Sci. U. S. A.*, 2014, **111**, 7191–7196.
7 P. M. Donaldson, C. S. Kelley, M. D. Frogley, J. Filik, K. Wehbe and G. Cinque, *Opt. Express*, 2016, **24**, 1852–1864.
8 B. Pollard, F. C. B. Maia, M. B. Raschke and R. O. Freitas, *Nano Lett.*, 2016, **16**, 55–61.

Peter Gardner said: First of all I would like to congratulate you on such excellent data. It is some of the nicest tip-based IR data I've seen. I think that routine tip-based IR is really competitive compared with SERS (in most cases it blows TERS out of the water). You have used three or four different tip-based IR systems and so are really well placed to give an objective opinion of these methods. Can you give us an overview on this exciting and dynamic field?

Kathleen Gough replied: I have indeed had the opportunity to get data from three different systems, and managed to examine the identical samples, but the limited experience does not make for a fair comparison, nor would I endorse one method over another. We have spent about 4 weeks, spread over 2 years, collecting data at the ALS, and we have experienced the growth and developments from early commissioning in April 2014 to its recent status, February 2016. There are still issues with drift, but these are being worked on. The "nicest tip-based IR data" to which you refer was acquired, as we acknowledged, by Dr Max Eisele at the Neaspec headquarters in Germany. While we were provided with the recorded data and processed it ourselves in-house, we did not get to work with the system hands on. We have had many useful conversations with the scientists there, and the data collection certainly appeared to be a routine process, highly reproducible, with very stable positioning. Finally, in February 2016, I was fortunate enough to be able to visit the Anasys headquarters in California, where they produce the system developed by Dazzi. They had already very kindly acquired some nice spectra for us using the AFM–IR method, and we tried a few experiments using their version of the s-SNOM method, again with very helpful discussions. Unfortunately the brevity of my visit meant that we were not able to evaluate it under optimal conditions, so, as I say, a comparison would be unfair. That said, we have noted some very interesting differences in band intensities between the AFM–IR spectra and that from s-SNOM (unpublished data), which we currently ascribe to the very different physical phenomena being detected, and the different volumes probed. First, both these nanoscale IR techniques are indirect measurements of a vibrational spectrum at the nanoscale. In s-SNOM we are measuring light scattered by the sample and that light, as described in the manuscript (DOI: 10.1039/C5FD00168D), is processed to recover the absorption spectrum. This is very sound Physics by the way, well established, see for example the earlier work of Keilmann and of Hillenbrand over the past 15 years and recent papers detailing the physical basis, including refinements related to size and thickness.[1–3] In the AFM–IR approaches, one measures the thermal expansion of the sample under the probe tip, which is affected by the expansion of all the illuminated material that surrounds the probe (see the review by Dazzi *et al.*[4]). In s-SNOM, the sample volume is truly nanoscale, as small as 20 nm in all dimensions, as only the near-surface is probed; this can be controlled with some variations in operating conditions. In AFM–IR, the volume is often much larger, at least a hundred nm or more, due to thermal diffusion within non-isolated objects.

1 A. A. Govyadinov, I. Amenabar, F. Huth, P. S. Carney and R. Hillenbrand, *J. Phys. Chem. Lett.*, 2013, **4**, 1526–1531.
2 S. Mastel, A. A. Govyadinov, T. V. A. G. de Oliveira, I. Amenabar and R. Hillenbrand, *Appl. Phys. Lett.*, 2015, **106**, 023113.
3 E. A. Muller, B. Pollard and M. B. Raschke, *J. Phys. Chem. Lett.*, 2015, **6**, 1275–1284.

4 A. Dazzi, C. B. Prater, Q. Hu, D. B. Chase, J. F. Rabolt and C. Marcott, *Appl. Spectrosc.*, 2012, **66**, 1365–1384.

Gianfelice Cinque commented: The synchrotron-based scanning SNOM method was pioneered 2 years ago at the ALS. It works very well giving a clear near field signal in the 10s nm resolution scale (based on the lightning rod principle). This is a surface technique, *i.e.*, for thin samples, with a major limit that the signal is not a direct absorbance. At Diamond we did the world's first Synchrotron PhotoThermal nanoIR work published at the start of this year.[1] The method is not limited to nano-thin samples, it also provides bulk information based on measuring the mechanical expansion *via* AFM tip in resonance: most importantly the signal gives a true absorption spectrum. The spatial resolution is physically dictated by thermal diffusivity: experimentally the ~100nm range was achieved on a polymeric material, working at a <100kHz light chopping frequency. For higher resolution a faster measurement is needed.

The key point is that the two methods are complementary since both give nanometric resolution but different vibrational information since they are based on different physical mechanisms. However, only photothermal nanoIR (AFM IR) gives both a direct and a linear absorption spectrum.

1 P. M. Donaldson, C. S. Kelley, M. D. Frogley, J. Filik, K. Wehbe and G. Cinque, *Opt. Express*, 2016, **24**, 1852–1864.

Kathleen Gough responded: As explained in my response to the previous question from Prof. Gardner, since s-SNOM and photothermal methods are based on the detection of different physical phenomena, differences in spectra are only to be expected. In fact, some of these are well understood, through modelling and comparison with experimental results. Additionally, the volume probed will be different, and the strength of the response from a heterogeneous material within that volume may also differ. A thorough understanding of the physics behind each method is important when choosing a methodology, and when interpreting the data collected. Spatial resolution is greater for s-SNOM, with voxels on the order of 20–30 nm, but in some cases sub-micron information on the order of 100 nm or more spatial resolution might be preferred. For each technique, the chemical and physical properties of the sample present other factors for consideration, as do the tapping amplitude, tip, substrate material, detector, *etc.* There is still much to be learned in this rapidly expanding field.

Stan Remiszewski returned to the general discussion of the paper by Christopher Phillips: Regarding the Quantum Cascade Laser (QCL) work in feature selection for biological samples and analysis: An "educated guess" forming the basis of the work as stated in your paper (DOI: 10.1039/C5FD00150A) is naturally biased by prior knowledge or experience with the data and or the technology. Although a likely starting point for discovery, it may bring false results as it may be construed as a form of overfitting when not converted into a hypothesis independently validated by unbiased true-true test data and not a presumptive data set. How are your claims protected from bias and overfitting and what evidence is there in your study design and data management, either planned or already available, to refute claims of overfitting and interpretive bias?

Christopher Phillips responded: Our imaging protocol generates a single parameter, the so-called "Digistain Index", and the way that this parameter is derived, is clearly defined in the standard operating protocol (SOP). By the standards of vibrational spectroscopy it is extremely reproducible. True there was an element of choice in formulating it this way, and it may be sub-optimal for some clinical applications, but we cannot see how this approach could be seen as one of "over fitting". The way the DI is used in clinical trails (double blind, anonymised *etc.*) is carefully designed to deny the possibility of any bias taking place, as any suspicion of this would invalidate the process.

Lauren Jamieson returned to the general discussion of the paper by Kathleen M. Gough: This is very nice and interesting work with important application to what can be a very debilitating condition. If you wanted to look at human tendons what would you need to do to get access to such samples? Do you foresee that this information could help improve therapies and reverse tendon damage?

Kathleen Gough replied: We have been actively collecting human sartorius tendons for several months now, with the aim of acquiring several tendons from people in each decade of life from approximately 20 to 60 years of age. Very shortly, we will begin experimental work aimed at studying how age affects fibril and molecular-level structural disruption, and how these relate to fascicle-scale mechanics. We also plan to assess how the accumulation of non-enzymatic advanced glycation end-product crosslinks affect tendon structure/function through the comparison of tissue from diabetic and non-diabetic donors.

In response to the second part of the question, designing an intelligent solution often requires fundamental understanding of the underlying problem as a prerequisite. There is still remarkably little known about the structural changes that occur at a molecular level both leading to and as a result of tendinopathy and ligament sprains and tears, and how these changes interact with biological processes, particularly for healing. While we are still in the very early stages of understanding these ailments at their tissues' fundamental structural levels, our motivation for doing so is exactly that: to be able to aid in the eventual development of novel therapies and injury prevention strategies.

Ganesh Dhruvananda Sockalingum remarked: Have you planned to combine polarised FTIR data with other biophotonic techniques such as polarised Second Harmonic Generation (SHG), which is sensitive to only collagen and not to other extracellular matrix proteins like elastin? Polarised SHG can give interesting information on collagen fiber orientation, molecular order *etc.*

Kathleen Gough responded: SHG information is complementary to our micron-scale imaging data in the sense that SHG is sensitive to the packing quality of the molecules within a fibril; however, the spatial resolution is typically much poorer, being on the order of 300 nm at best for intact tissues. See for example some recent papers on SHG microscopy of collagen fibrillary structure and the considerations for experimental measurements.[1,2] One group has reported the diameter of isolated collagen fibrils at 30 nm spatial resolution, using correlative SHG electron microscopy.[3] This method would not work for intact tendon sections but the paper provides an excellent overview of the background for the SHG of collagen fibrils and the experimental challenges. My co-author, Laurent Kreplak, is currently testing the sensitivity of SHG to mechanical damage along single fibrils in collaboration with Francois Legare from INRS.

1 X. Chen, O. Nadiarynkh, S. Plotnikov and P. J. Campagnola, *Nat. Protoc.*, 2012, 7, 654–668.
2 C.-A. Couture, S. Bancelin, J. Van der Kolk, K. Popov, M. Rivard, K. Legare, G. Martel, H. Richard, C. Brown, S. Laverty, L. Ramunno and F. Legare, *Biophys. J.*, 2015, **109**, 2501–2510.
3 S. Bancelin, C. Aime, I. Gusachenko, L. Kowalczuk, G. Latour, T. Coradin and M.-C. Schanne-Klein, *Nat. Commun.*, 2014, 5, 4920.

Faris Sinjab communicated: I have a few short questions regarding s-SNOM IR microscopy with QCLs. I was wondering if there are any additional requirements for the probe, other than being sufficiently sharp? Can standard silicon AFM tips be used with this technique, or at least modified in a relatively simple way (*e.g.* roughly coated with gold for an enhanced lightning-rod effect)? Also, is the typical lifetime of these probes noticeably different for s-SNOM measurements compared with AFM measurements? Finally, could this technique potentially be used for studying active biological samples (*e.g.* live cells) in aqueous media?

Christopher Phillips communicated in reply: As the s-SNOM signal directly relies on the lightning rod effect, it needs metallic tips (usually platinum or gold-plated). If they are blunted it reduces the overall signal recovered, and s-SNOM tends to be more susceptible to tip degradation than AFM measurements. That said, careful operation of the s-SNOM will allow a probe lifetime similar to that of AFM. s-SNOM could potentially be used for such live samples, although there would be a number of substantial technical challenges, *e.g.* strong absorption of the IR light by an aqueous medium, and the need for a high scan speed to image moving samples.

Kathleen Gough communicated in reply: For s-SNOM, AFM tips must be coated with a metal, typically gold. Glass tips can be used where the method involves tip heating or heat response, but the signal is still poor. See, for example, experiments with the tip material by Donaldson *et al.*[1] In principle, commercial probes that are already gold-coated should work fine; this is what we have used at the ALS. Regarding lifetime, as long as you operate the probe in the tapping mode and the sample is not rough, it can be used for tens of measurements without a problem. In our experience, a "sticky" sample, such as the carbohydrate-rich fungal wall, is more likely to contaminate the tip within a day, whereas for the collagen fibril experiments that we recently conducted at the ALS (February 2016), we only used one tip over a period of 4–5 days, collecting >100 spectra and dozens of AFM images. However, in comparison with typical AFM experiments, the metal-coated tips are in general much more easily damaged: the coating flakes off or becomes contaminated, and the lifetime of several weeks that one gets for AFM tips is simply not possible. As for bioactive samples, the AFM technique is compatible with hydrated samples,[2] so in principle slow biological processes could be observed. Unfortunately, the usual caveats must be recognized: water is strongly absorbing in the infrared at the amide I and II bands, and it will evaporate, thereby changing the sample and filling the air with water vapor lines.

1 P. M. Donaldson, C. S. Kelley, M. D. Frogley, J. Filik, K. Wehbe and G. Cinque, *Opt. Express*, 2016, **24**, 1852–1864.
2 S. J. Baldwin, A. S. Quigley, C. Clegg and L. Kreplak, *Biophys. J.*, 2014, **107**, 1794–1801.

Gianfelice Cinque added: In our work at Diamond on AFM IR,[1] Au-coated tips were used, even when made in glass.

1 P. M. Donaldson, C. S. Kelley, M. D. Frogley, J. Filik, K. Wehbe and G. Cinque, *Opt. Express*, 2016, **24**, 1852–1864.
1 1 M. J. Baker, S. R. Hussain, L. Lovergne, V. Untereiner, C. Hughes, R. A. Lukaszewski, G. Thiéfin and G. D. Sockalingum, *Chem. Soc. Rev.*, 2016, **45**, 1803–1818.

Sergei Kazarian returned to the general discussion of the paper by Christopher Phillips by communicating: The first part of the presented work (DOI: 10.1039/C5FD00150A) is similar to your previous work,[1] which contains more information on this system. In this work only univariate analysis was used, but perhaps dimensional reduction (PCA, clustering *etc.*) can beneficially be used. Were any quality tests performed (assessing effects of tissue thickness or water content)? The latter (hydration) is known to affect the amide I band.

As to the second modality, s-SNOM, how do the actual spectra appear? Are there examples of spectral assessment of different components within the cell?

1 H. Amrania, G. Antonacci, C.-H. Chan, L. Drummond, W. R. Otto, N. A. Wright and C. Phillips, *Opt. Express*, 2012, **20**, 7290–7299.

Christopher Phillips communicated in reply: The Digistain imaging protocol was designed to be insensitive to the tissue thickness (see my response to the first question from Prof. Diem for details). In the clinical trials we adopt an empirical approach. The tissues are dried according to the SOP, but any variations due to residual water content would propagate through the analysis like any other randomising effect. In the end, the stratification of patients by risk using the Digistain Index alone was sufficiently compelling (log rank test $P = 0.02$) to essentially discount these quality issues. We have also constructed a multivariate model through logistical regression. This work is due to be published soon.

Regarding the s-SNOM, full spectra have not yet been collected, as we are still developing a laser based system with sufficient wavelength agility.

Faraday Discussions

ROYAL SOCIETY OF CHEMISTRY

PAPER

Translating vibrational spectroscopy into clinical applications – vision or reality?

Wolfgang Petrich

Received 2nd May 2016, Accepted 10th May 2016
DOI: 10.1039/c6fd00108d

The Faraday Discussion meeting "Advanced Vibrational Spectroscopy for Biomedical Applications" provided an excellent opportunity to share and discuss recent research and applications on a highly interdisciplinary level. Spectral pathology, single cell analysis, data handling, clinical spectroscopy, and the spectral analysis of biofluids were among the topics covered during the meeting. The focus on discussion rather than "merely" presentation was highly appreciated and fruitful discussions evolved around the interpretation of the amide-bands, optical resolution, the role of diffraction and data analysis procedure, to name a few. The meeting made clear that the spectroscopy of molecular vibrations in biomolecules has evolved from a purely academic research tool to a technology used in clinical practice in some cases. In this sense, biomedical vibrational spectroscopy has reached a pivotal point at which questions like diagnostic value, therapeutic consequence and financial viability are gaining more and more importance.

Two centuries ago, Michael Faraday was considered one of the most excellent analytical chemists. Among his achievements are the discovery of benzene, the synthesis of compounds made of carbon and chlorine, and the elucidating experiments on gold colloids. He was also honoured for his contribution to physics, most notably for his work on electricity and magnetism. The discoveries of electromagnetic induction and diamagnetism may serve as examples. He triggered a controversy debate around what he called the "lines of force", whereby his ideas and concepts became very important for both the physical insights into electromagnetism and the technical applications thereof. The lines of force provided a way to visualize the findings without the need to condense his discovery of electromagnetic induction into mathematical terms. In this sense, he had the spirit to think in mathematical concepts such that James Clark Maxwell later wrote that the "*…way in which Faraday made use of his idea of lines of force in co-ordinating the phenomena of magneto-electric induction shews him to have been in reality a mathematician of a very high order – one from which the mathematicians of the future may derive valuable and fertile methods*".[1]

During his career Faraday was always interested in exploiting his knowledge for practical purposes, be it for seafaring purposes, environmental protection or

Kirchhoff-Institute for Physics, Heidelberg University, INF 227, 69120 Heidelberg, Germany. E-mail: wolfgang.petrich@kip.uni-heidelberg.de

mining. Last but not least, he engaged in philosophical discussions as well as education, with the famous Christmas Lectures at the Royal Institution in London serving as an excellent example for the latter.

Having said this it should not be surprising that I appreciate this Faraday Discussions meeting. Indeed, we can learn from Faraday's highly interdisciplinary skills as well as from his drive to solving real-world problems.

Starting with the chemistry of biomedical vibrational spectroscopy I enjoyed the discussion around the identification of various proteins from the amide-I and amide-II band, which extended to a reflection of the general role of band assignments in this field. While it is my perception that the opinions are still somewhat controversial, I believe that the underlying question is simply a question of conceivable numbers of proteins of very similar spectral features *versus* special cases of pronounced deviations from "the usual protein spectrum". Collagen may serve as a good example for the latter since it has some distinguished spectral signatures and is the body's most abundant structural protein, *i.e.* it is present in high concentrations. For most other proteins, however, in my point of view, it will be very hard to identify, differentiate among and/or quantify these proteins solely on their spectral basis since their spectra resemble each other to a large extent. The multiplicity of the proteins' similar spectral peaks is altogether hidden within the prominent protein bands. Speaking in simple terms, I am afraid that it will not be possible to fit, say, 50 Gaussians, to two peaks in a meaningful way, but I am looking forward to following the further discussion.

As far as another of Faraday's disciplines – namely physics – is concerned, it is my impression that two topics were repeatedly addressed in this Faraday Discussions meeting. The first refers to the term "resolution", which is well defined in physics as the distance between two identical points that can be observed as two separate points ("resolved") when using the system under investigation. The distance was calculated by Ernst Abbe based on his observations and systematic investigations, and it is proportional to the wavelength divided by the numerical aperture. Nowadays, Abbe's diffraction limit can also readily be derived from the very first principles of the quantum nature of light using Heisenberg's uncertainty principle. While minor differences exist in the proportionality constant between the lateral resolution and the ratio of wavelength to numerical aperture,† such an optical resolution limit holds also true for indistinguishable particles in any Raman or infrared far-field microscope and, even more so, it refers to the best case scenario. In reality, aberrations, density of sensors *etc.* can severely deteriorate this optimum case. In turn, any improvement in *e.g.* sensor density is, strictly speaking, not helping to break the optical resolution limit *per se*.‡ Hence, I believe that this limit needs to be very clear when the term "resolution" is being used.

† The proportionality constant may be 1 (distinguishing adjacent lines under illumination along the optical axis according to Abbe) or 0.5 (adjacent lines under tilted illumination, following Abbe) or 0.61 (diffraction pattern of self-radiating points, see Rayleigh) or 0.51 (full width at half maximum of point-spread function).

‡ See also E. Abbe, "Denn es ist offenbar ebenso nutzlos, den physischen Bedingungen der Abbildung in weitem Umfang genügt zu haben, als die erreichbare Vergrösserung zu verwerthen erlaubt, wie es nutzlos ist, die Vergrösserungskraft des Mikroskops höher zu steigern, als die physische Capacität der Objective nöthig macht." (ref. 2).

The situation changes, however, if a measurement is performed outside the validity of Abbe's diffraction limit such as using near-field optics or localizing distinguishable particles. The combination of microscopy with spectroscopy, *i.e.* hyperspectral imaging, provides means to distinguish among different molecular species by virtue of their spectra such that, from a fundamental standpoint, it will be possible to localize two spectrally different particles even if they are closer together than the diffraction limit. It follows that there may be situations in hyperspectral imaging in which it could be appropriate to provide detection tools beyond Abbe's resolution limit dictated by the objective lens.

Already in 1873, Abbe also elaborated on the meeting's second topic in physics, namely diffraction. He stressed the point that a microscope will actually always measure two types of images at the same time, an absorption image and a diffraction image. In this sense, diffraction patterns should not be called "artefacts" – on the contrary, they are part of the imaging procedure from the very basics of microscopy. Even more so, the two "types of images" are closely interrelated by virtue of the Kramers–Kronig relation.

Mathematics obviously constitutes another important discipline in Faraday's as well as this meeting's interdisciplinary endeavour. Some decades ago, spectrometers became easier to use and computational skills started to enable the application of rather complex mathematics to "easy-to-obtain" spectra. However, (hyper-)spectral data is prone to overfitting due to the multivariate nature of the problem and the chronic lack of sufficiently many samples. Overfitting may nicely be illustrated on a relatively simple set of data,[3] namely mid-infrared FT-IR spectra of dried serum samples and the task to determine the concentration of glucose in each sample. A multivariate algorithm was optimized within a training set and subsequently applied to an independent validation set for the first case of correct assignments between the sample number and glucose concentration (as determined by clinical chemistry) and for the second case in which the assignments were randomized on purpose (*i.e.* "nonsense data"). It is no surprise that, within the training set, one could achieve fantastic accuracies for the concentration estimates based on FT-IR spectroscopy by simply enlarging the number of "fit" parameters§ regardless of whether the data assignment was correct or nonsense. The situation changed as soon as the trained algorithm was applied to the independent validation set whereby acceptable accuracies were obtained for the first case and extremely poor prediction accuracies were revealed in the second case. The necessity for independent validation prevails throughout other multivariate methods such as neural networks, support vector machines, *etc.* A brief look into analysis results from us and others indicates that the number of samples should exceed the number of adjustable parameters (latent variables, input neurons, *etc.*) by a factor of ~10 in order to obtain dependable information. I believe that, for example, the (in sum) more than 1000 gastric patients examined by means of Raman spectroscopy in Zhiwei Huang's research (DOI: 10.1039/C5FD00151J) provide an excellent impression of the number of patients we need to enrol in such spectroscopic investigations and multivariate data analysis procedures in order to enable acceptance by the medical community.

§ "Latent variables" in the case of partial least square regression.

If biomedical vibrational spectroscopy shall be utilized by medical doctors it is a prerequisite that we use the language of medicine to report spectroscopic results. Throughout this conference I was delighted to see that spectroscopic and mathematical achievements were often presented in terms of sensitivity, specificity, ROC, AUC, *etc.* Furthermore, the quest for medical gold standards was mentioned repeatedly throughout this meeting up to the point at which the replacement of histopathologists by hyperspectral imaging was discussed. While different standpoints may be taken in this question (and I personally think that our technologies should help the daily work of a histopathologist, maybe one day possibly even shift roles and responsibilities throughout the diagnostic process, but biomedical vibrational spectroscopy will not and should not "replace" the histopathologist) what appears much more important to me is the time scale: it is my impression that the path to application is still very long. Some of the contributions to the Faraday Discussions meeting state that the work presented is "*...paving the road...*" and, while I truly respect the value of these contributions, it must be allowed to ask whether we want to keep paving roads or whether this is really the time to see infrared and Raman spectroscopy be used in clinics in day-to-day life. If the majority of people in this audience were medical doctors, I would agree that some roads to application are in their final stage, but, obviously, a lot of things remain to be done. My "credo" is: let's stop paving roads and let us now deliver our promises to the clinics at the end of those roads!

In delivering our promises I believe that we need to be very precise in how far our methods are improving the present diagnostic capabilities or whether it is just another technology. I have, for instance, provided the above example of measuring the concentration of glucose in serum – but I do not believe that any hospital would switch from an established, well documented, FDA-approved and cheap reagent-based laboratory testing for glucose to vibrational spectroscopy. Measuring the concentration of glucose in this context can thus only serve as an academic example. From a health care standpoint, we need to answer the questions:

- What is the unmet need?
- What precisely is the diagnostic value?
- What is the therapeutic consequence?
- Will this application be financially viable?

In my point of view it is the detailed answers to these questions that point the way to the most attractive applications.

Having started with *Faraday*, let me end the closing remarks with the *Discussion* part of the meeting. I truly appreciate the valuable, open-minded, controversial and yet constructive and respectful discussions during this meeting. This point was nicely illustrated by Sergei Kazarian's way to reply to an objection during the discussions: "*Yes, you are right, if this were what we are doing – but we are not*". Last but not least I am thankful to the organizers, sponsors and the many "helping hands" for enabling such a great opportunity. So, will biomedical vibrational spectroscopy in the clinics remain an everlasting dream or is it becoming reality? To me, the best way to close this meeting is to simply cite Hugh Barr, one of our yet-to-be customers from the medical field: "*I would be enormously surprised, if this doesn't take off.*"

References

1 J. C. Maxwell, Cambridge University Press, 1890.
2 E. Abbe, *Archiv für mikroskopische Anatomie*, 1873, **9**, 467.
3 W. Petrich, in *Biomedical Vibrational Spectroscopy*, ed. P. Lasch and J. Kneipp, John Wiley & Sons, 2008, pp. 315–332.

Poster titles

Combining field effect scanning electron microscopy, deep UV fluorescence, Raman, classical and synchrotron radiation Fourier transform infrared spectroscopy in the study of crystal-containing kidney biopsies and breast samples, **D. Bazin**, *UPMC, France*

A 3D phantom spectral study for surface enhanced spatially offset Raman spectroscopy (SESORS) for Tissue Diagnostics, **L. Clark, N. Stone and F. Palombo**, *University of Exeter, UK*

A study of variability due to hormonal effects in cervical smear samples using Raman spectroscopy, **D. Traynor, P. Kearney and I. Ramos**, *Focas Research Institute, Dublin Institute of Technology, Republic of Ireland*

Application of bio-spectroscopy towards diagnosis of neurodegenerative pathology, **M. Paraskevaidi, K. I. Essa, D. M. A. Mann, P. L. Martin-Hirsch, D. Allsop and F. L. Martin**, *Lancaster University, UK*

Application of Raman microscopy for the diagnosis of the chronic lymphocytic leukemia (CLL), **M. Féré, C. Gobinet, A. Beljebbar, V. Untereiner, L. H. Liu, T. Happillon, M. Manfait, E. Cornet, X. Troussard, B. Chatelain, J. Klossa and O. Piot**, *URCA Champagne-Ardenne, France*

A new approach for malaria diagnosis using attenuated total reflection Fourier transform infrared (ATR-FTIR) spectroscopy, **D. Perez-Guaita, P. Heraud, A. Khoshmanesh, P. Jearanaikoon, M. Wongwattanakul, P. Chatchawal, T. Patcharaporn, D. McNaughton and B. R. Wood**, *Monash University, Australia*

Cell death induced in tumoral epithelial cells: SR-IRMS analysis, **S. Sabbatini, E. Giorgini, C. Rubini, R. Rocchetti, C. Conti, E. Mitri and L. Vaccari**, *Università Politecnica delle Marche, Italy*

Cellular response, distribution and metabolism of tyrosine kinase inhibitors in cancer cells by Raman spectral imaging, **S. El-Mashtoly, H. Yosef, W. Alsaidi, M. Hammoud, T. Lechtonen and K. Gerwert**, *Ruhr-University Bochum, Germany*

Characterising an osteogenic signature in primary cells using Raman spectroscopy, **S. Marfy-Smith, R. Emery, A. Pitsilledes, C. Clarkin and S. Mahajan**, *University of Southampton, UK*

Chemically-activatable alkyne tag for Raman imaging, **S. Yamaguchi, M. Ura, T. Matsushita and A. Okamoto**, *The University of Tokyo, Japan*

Confocal Raman micro-spectroscopy from multiple *ad hoc* positioned sample-points measured simultaneously, **F. Sinjab, K. Kong and I. Notingher**, *University of Nottingham, UK*

Detection of glycan-protein interactions *via* SERS, **R. Dluhy, X. Li , Z. S. Chinoy, G.-J. Boons and S. Martin**, *University of Alabama at Birmingham, USA*

Developments in infrared spectroscopy of colorectal pathology, **E. Upchurch, C. Kendall, R. Griggs, J. Wood, G. Lloyd, M. Isabelle, N. Shepherd, T. Cook, J. Nallala and N. Stone**, *GHNHSFT, UK*

Effects on human mammary epithelial cells of the exposition to x-ray radiation: a single cells study by visible micro-Raman spectroscopy, **I. Delfino, G. Perna, M. Lasalvia, V. Capozzi, L. Manti, C. Camerlingo and M. Lepore**, *Dipartimento di Scienze Ecologiche e Biologiche - Università della Tuscia, Italy*

Gold nanoparticles as a substrate in bio-analytical near-IR surface enhanced Raman spectroscopy, **H. Butler, S. Fogarty and F. Martin**, *Lancaster University, UK*

Imaging drug localisation with in skin SRS microscopy, **A. Rae and N. A. Belsey**, *National Physical Laboratory, UK*

Impact of dietary antioxidants on chemotherapy response probed by vibrational microspectroscopy in metastatic human breast cancer, **A. Batista de Carvalho, M. Pilling, P. Gardner, J. Doherty, G. Cinque, K. Wehbe, C. Kelley, L. Batista de Carvalho and M. P. Marques**, *Molecular Physical-Chemistry R&D Unit, University of Coimbra, Portugal*

Improved protocols for pre-processing Raman spectra from formalin fixed paraffin preserved tissue sections, **O. Ibrahim, A. Maguire, A. D. Meade, S. Flint, H. J. Byrne and F. M. Lyng**, *Focas Research Institute, Dublin Institute of Technology, Republic of Ireland*

Infrared spectroscopy for the biomolecular assessment of vulval intraepithelial neoplasia and lichen sclerosus found in association with squamous cell carcinoma, **J. Frost, L. Ludeman, K. Hillaby, R. Gornall, G. Lloyd, C. Kendall, A. C. Shore and N. Stone**, *Biophotonics Research Group, Gloucestershire Hospitals NHS Foundation Trust, UK*

A new wave of imaging? Intracellular drug imaging by Raman microscopy, **W. Tipping, M. Lee, A. Serrels, V. Brunton and A. Hulme**, *University of Edinburgh, UK*

Label-free surface-enhanced Raman spectroscopy of biofluids: diagnostic applications in oncology, **A. Bonifacio, S. Cervo, R. Spizzo and V. Sergo**, *University of Trieste, Italy*

When spectral analysis reveals hidden features: mid-infrared FEWS spectroscopy for medical early diagnosis, **M. Le Corvec, O. Loreal, S. Lardjane, V. Monbet, H. Tariel and O. Sire**, *Université de Bretagne Sud, France*

Optimisation of deparaffinizing and baking adherence techniques for breast tissue on calcium fluoride slides for FTIR & Raman imaging, **A. Meade, N. Jackson, N. Chaudhary, H. Lambkin and A. Malkin**, *Dublin Institute of Technology, Republic of Ireland*

Photon migration study by spatially offset Raman spectroscopy for sub-surface probing in various bone types, **K. Sowoidnich, J. H. Churchwell, K. Buckley, A. E. Goodship, A. W. Parker and P. Matousek**, *STFC Rutherford Appleton Laboratory, UK*

Providing a robust Raman spectroscopy-based cancer diagnostic platform to integrate into the clinic, **K. Lau, I. M. Bell, M. Isabelle, G. R. Lloyd, O. Old, J. Dorney, A. Lewis, M. Rodriguez-Justo, C. Kendall, N. Stone, G. Thomas and D. Reece**, *Renishaw plc, UK*

Raman spectroscopy for the detection and diagnosis of prostate cancer using peripheral blood lymphocytes, **D. Medipally, A. Maguire, A. Meade, F. Lyng, J. Bryant, V. Untereiner, G. Sockalingum, J. Armstrong and M. Dunne**, *Focas Research Institute, Dublin Institute of Technology, Republic of Ireland*

Raman mapping for pathology classification in Barrett's oesophagus, **M. Isabelle, O. Old, G. Lloyd, J. Dorney, A. Lewis, G. Thomas, N. Shepherd, H. Barr, I. Bell, N. Stone and C. Kendall**, *Biophotonics Research Group, Gloucestershire Hospitals NHS Foundation Trust, UK*

Raman microspectroscopic study of exfoliated oral cells, **I. Behl, O. Ibrahim, A. Malkin, H. J. Byrne and F. M. Lyng**, *DIT Centre for Radiation and Environmental Science, Focas Research Institute, Dublin Institute of Technology, School of Physics, Dublin Institute of Technology, Republic of Ireland*

Raman spectroscopy for imaging articular cartilage, **L. Messager, M. Fields, P. F. McMillan and J. Dudhia**, *The Royal Veterinary College, UK*

Re-evaluating the Mahalanobis distance measure: application to biomedical spectroscopy, **G. Lloyd and R. Brereton**, *Gloucestershire Hospitals NHS Foundation Trust, UK*

Spectroscopic ellisometry characterization of nano-structured gold films for LSPR biosensing, **A. Al-Rubaye, A. Nabok and A. Tsargorodska**, *Sheffield Hallam University, UK*

Study of Alzheimer's diseased tissue by Infrared and Raman microspectroscopies, **P. Hellwig, A. Schirer, C. Klein, L. Meyer, C. Patte-Mensah, E. Hartée, S. Lecompte and A. G. Mensah-Nyagan**, *CNRS- University of Strasbourg - UMR 7140, France*

Targeted SERS nanosensors in live 3D cell cultures as a new tool for drug discovery, **L. Jamieson, V. Camus, P. Bagnaninchi, K. Fisher, D. Harrison and C. Campbell**, *University of Edinburgh, UK*

The effect of *Plectranthus barbatus* and *P. zuluensis* infusions on protein secondary structure investigated using FTIR spectroscopy, **E. Brito, M. L. M. Serralheiro, P. I. Haris and R. Pacheco**, *Faculty of Sciences University of Lisbon, Portugal*

The use of a Raman needle probe as a histopathological tool for the diagnosis of lymphomas and lesions in lymph nodes, **L. Fullwood, I. I. Petterson, A. Dudgeon, G. Lloyd, O. Old, C. deCates, S. Hopcroft, C. Hall, H. Barr, C. Kendall, J. Day and N. Stone**, *University of Exeter, UK*

Ex vivo pH sensing *via* fibre optic based surface-enhanced Raman scattering (SERS), **S. McAughtrie, L. Mati, M. G. Tanner, H. Fleming, S. Seth, C. K. I. Williams and C. J. Campbell**, *University of Edinburgh, UK*

Vibrational spectroscopic studies of photoactivatable diazido Pt(IV) anticancer complexes, **R. Vernooij, T. Joshi, D. Perez-Guaita, B. Wood, E. Izgorodina, B. Graham, P. Sadler and L. Spiccia**, *Monash University, Australia*

The chemical shape of life: laser spectroscopy studies of gas phase conformation in biomolecules and their clusters, **I. A. Lobo, P. A. Robertson, D. J. D. Wilson and E. G. Robertson**, *La Trobe University, Australia*

Water correction method of live cell imaging using quantum cascade laser microscope, **Z. Zhang, A. Henderson, M. Pilling and P. Gardner**, *University of Manchester, UK*

Topographical characterization of human bladder cancer extracellular matrix by IR microprobe, **G. Bellisola, R. Lucianò, M. Nebuloni, G. Cinque, K. Wehbe and M. Alfano**, *Azienda Ospedaliera Universitaria Integrata Verona, Italy*

A comparison of 785nm and 830nm laser wavelengths for optimum clinical measurements of tissue samples, **J. Dorney, M. Isabelle, G. Lloyd, O. Old, H. Barr, N. Shepherd, C. Kendall, A. Lewis, G. Thomas, K. Lau, I. Bell and N. Stone**, *University of Exeter, UK*

Using Raman spectroscopy to monitor and modulate nano-particle temperature at depth, **B. Gardner, N. Stone and P. Matousek**, *University of Exeter, UK*

Amyloid nanoscale chemical analysis by AFM-IR: structural basis of proteins aggregation and toxicity, **F. S. Ruggeri, G. Dietler and T. Knowles**, *University of Cambridge, UK*

Spectal differentiation of breast cancer cell lines in 2D and 3D culture by infrared imaging, **M. Smolina and E. Goormaghtigh**, *Université Libre de Bruxelles (ULB), Belgium*

Quantitative detection of codeine in human plasma using surface enhanced Raman scattering *via* adaptation of the isotopic labelling (IL) principle, **A. Subaihi**, *Manchester Institute of Biotechnology, UK*

Raman spectroscopy and Raman imaging of human breast cancerous and noncancerous tissues and epithelial cancer MCF7, MDA-MB-231 and epithelial normal MCF10A cell lines, **B. Brozek-Pluska, J. Surmacki, M. Kopec, J. Musial, R. Kordek and H. Abramczyk**, *Lodz University of Technology, Poland*

Algorithm for the removal of the glass signal from Raman cytology spectra, **L. T. Kerr and Bryan M. Hennelly**, *Biomedical Engineering Research group, Maynooth University, Ireland*

Characterisation of human melanoma tissues by IR spectral micro-imaging, **D. Sebiskveradze, C. Gobinet, N. Cardot-Leccia, M. Manfait, P. Jeannesson, V. Vuiblet and O. Piot**, *MEDyC CNRS UMR 7369, France*

Label-free optical analusis of live lung cancer cells – methods to understand oxidative stress in chemotherapy response, **J. M. Surmacki, B. J. Woodhams and S. E. Bohndiek**, *Lodz University of Technology, Poland*

Fiber spectroscopy method to select for tumor margin detection, **V. Artyushenko, V. Belikova, U. Zabarylo, F. Schulte, I. Usenov, V. Galyanin and A. Bogomolov**, *Art Photonics GmbH, Germany*

The Skinner Prize for the best poster was jointly awarded to Jonathan Frost of Biophotonics Research Unit, UK, for his poster on Infrared spectroscopy for the biomolecular assessment of vulval intraepithelial neoplasia and lichen sclerosus found in association with squamous cell carcinoma and Lauren Jamieson of University of Edinburgh, UK, for her poster on Targeted SERS nanosensors in live 3D cell cultures as a new tool for drug discovery.

List of participants

Mr Ali Al-Rubaye, *Sheffield Hallam University*, *United Kingdom*
Dr Hemmel Amrania, *Imperial College London*, *United Kingdom*
Dr Alexander Apolonskiy, *Ludwig Maximilian University of Munich*, *Germany*
Dr Viacheslav Artyushenko, *Art Photonics GmbH*, *Germany*
Dr Matthew J. Baker, *University of Strathclyde*, *United Kingdom*
Professor Hugh Barr, *Biophotonics Research Unit, Gloucestershire Hospitals NHS Foundation Trust*, *United Kingdom*
Dr Ana Batista de Carvalho, *Molecular Physical-Chemistry R&D Unit, University of Coimbra*, *Portugal*
Dr Dominique Bazin, *UPMC*, *France*
Miss Isha Behl, *Dublin Institute of Technology*, *Ireland*
Dr Giuseppe Bellisola, *Azienda Ospedaliera Universitaria Integrata Verona*, *Italy*
Dr Natalie Belsey, *National Physical Laboratory*, *United Kingdom*
Professor Rohit Bhargava, *University of Illinois at Urbana-Champaign*, *USA*
Dr Alois Bonifacio, *University of Trieste*, *Italy*
Miss Elsa Brito, *Faculty of Sciences, University of Lisbon*, *Portugal*
Miss Julie Brooks, *University of Manchester*, *United Kingdom*
Miss Holly Butler, *Lancaster University*, *United Kingdom*
Professor Hugh J. Byrne, *Dublin Institute of Technology*, *Ireland*
Dr Colin J. Campbell, *University of Edinburgh*, *United Kingdom*
Professor Eleanor Campbell, *University of Edinburgh*, *United Kingdom*
Miss Neha Chaudhary, *Dublin Institute of Technology*, *Ireland*
Dr John Churchwell, *University College London*, *United Kingdom*
Dr Gianfelice Cinque, *Diamond Light Source*, *United Kingdom*
Miss Louise Clark, *University of Exeter*, *United Kingdom*
Mrs Joëlle De Meutter, *Université Libre de Bruxelles*, *Belgium*
Dr Ines Delfino, *Dipartimento di Scienze Ecologiche e Biologiche, Università della Tuscia*, *Italy*
Professor Max Diem, *Northeastern University*, *USA*
Professor Richard Dluhy, *University of Alabama at Birmingham*, *USA*
Mr James Doherty, *Manchester Institute of Biotechnology*, *United Kingdom*
Dr Jennifer Dorney, *University of Exeter*, *United Kingdom*
Dr Max Eisele, *Neaspec GmbH*, *Germany*
Dr Samir El-Mashtoly, *Ruhr-University Bochum*, *Germany*
Dr Karen Faulds, *University of Strathclyde*, *United Kingdom*
Mr Michael Féré, *URCA Champagne-Ardenne*, *France*
Dr Laura Fisher, *Royal Society of Chemistry*, *United Kingdom*
Ms Holly Fleming, *University of Edinburgh*, *United Kingdom*
Dr Stefano Fornasaro, *Department of Engineering and Architecture, University of Trieste*, *Italy*
Dr Jonathan Frost, *Biophotonics Research Unit, Gloucestershire Hospitals NHS Foundation Trust*, *United Kingdom*
Miss Leanne Fullwood, *University of Exeter*, *United Kingdom*

Dr Benjamin Gardner, *University of Exeter*, *United Kingdom*
Professor Peter Gardner, *University of Manchester*, *United Kingdom*
Professor Klaus Gerwert, *Ruhr-University Bochum*, *Germany*
Mr Adrian Ghita, *University of Exeter*, *United Kingdom*
Mr Matthias Godejohann, *MG Optical Solutions GmbH*, *Germany*
Professor Royston Goodacre, *University of Manchester*, *United Kingdom*
Dr Erik Goormaghtigh, *Université Libre de Bruxelles*, *Belgium*
Professor Kathleen Gough, *University of Manitoba*, *Canada*
Professor Duncan Graham, *University of Strathclyde*, *United Kingdom*
Dr Frederik Großerüschkamp, *Ruhr-University Bochum*, *Germany*
Mr Ravi Guntupalli, *Princeton Instruments*, *USA*
Dr Parvez Haris, *De Montfort University*, *United Kingdom*
Professor Petra Hellwig, *CNRS, University of Strasbourg, UMR 7140*, *France*
Dr Alex Henderson, *University of Manchester*, *United Kingdom*
Mr Michael Hermes, *University of Exeter*, *United Kingdom*
Dr Katherine Hollywood, *University of Manchester*, *United Kingdom*
Professor Zhiwei Huang, *National University of Singapore*, *Singapore*
Dr Caryn Hughes, *University of Manchester/University of Strathclyde*, *United Kingdom*
Miss Ola Ibrahim, *Focas Research Institute, Dublin Institute of Technology*, *Ireland*
Dr Shayz Ikram, *LOT-QuantumDesign*, *United Kingdom*
Dr Martin Isabelle, *Biophotonics Research Unit, Gloucestershire Hospitals NHS Foundation Trust*, *United Kingdom*
Miss Naomi Jackson, *Focas Research Institute, Dublin Institute of Technology*, *Ireland*
Dr David James, *Thermo Fisher Scientific*, *United Kingdom*
Miss Lauren Jamieson, *University of Edinburgh*, *United Kingdom*
Professor Sergei Kazarian, *Imperial College London*, *United Kingdom*
Dr Catherine Kendall, *Gloucestershire Hospitals NHS Foundation Trust*, *United Kingdom*
Ms Laura Kerr, *Maynooth University*, *Ireland*
Professor Achim Kohler, *Norwegian University of Life Sciences*, *Norway*
Dr Niels Kröger-Lui, *Kirchhoff Institute for Physics*, *Germany*
Dr Peter Lasch, *Robert Koch-Institute*, *Germany*
Dr Katherine Lau, *Renishaw plc*, *United Kingdom*
Miss Maena Le Corvec, *Université de Bretagne Sud*, *France*
Dr Guillaume Lepert, *Imperial College London*, *United Kingdom*
Mr Yuandi Li, *Royal Society of Chemistry*, *United Kingdom*
Dr James Livermore, *University of Oxford*, *United Kingdom*
Dr Manuel Llansola-Portoles, *CEA Saclay*, *France*
Dr Gavin Lloyd, *Gloucestershire Hospitals NHS Foundation Trust*, *United Kingdom*
Mrs Isabella A Lobo, *La Trobe University*, *Australia*
Miss Kirsty Lunnon, *Royal Society of Chemistry*, *United Kingdom*
Professor Fiona Lyng, *Dublin Institute of Technology*, *Ireland*
Dr Kerstin Mader, *Sheffield Hallam University, Materials and Engineering Research Institute, Sheffield*, *United Kingdom*
Mrs Stephanie Marfy-Smith, *University of Southampton*, *United Kingdom*
Professor Maria Paula Marques, *University of Coimbra*, *Portugal*
Professor Pavel Matousek, *Rutherford Appleton Laboratory*, *United Kingdom*
Dr Sarah McAughtrie, *University of Edinburgh*, *United Kingdom*
Dr Aidan Meade, *Dublin Institute of Technology*, *Ireland*
Mr Dinesh Kumar Reddy Medipally, *Focas Research Institute, Dublin Institute of Technology*, *Ireland*

Miss Jane Meharry, *Royal Society of Chemistry*, *United Kingdom*
Dr Lea Messager, *The Royal Veterinary College*, *United Kingdom*
Mr Clive Moss, *Specac Ltd*, *United Kingdom*
Ms Fiona Nalden, *Royal Society of Chemistry*, *United Kingdom*
Professor Ioan Notingher, *University of Nottingham*, *United Kingdom*
Ms Drishya Rajan Parachalil, *Focas Research Institute, Dublin Institute of Technology*, *Ireland*
Miss Maria Paraskevaidi, *Lancaster University*, *United Kingdom*
Dr Vibhuti Patel, *Royal Society of Chemistry*, *United Kingdom*
Dr Carl Paterson, *Imperial College*, *United Kingdom*
Dr David Perez-Guaita, *Monash University*, *Australia*
Professor Wolfgang Petrich, *Kirchhoff-Institute for Physics*, *Germany*
Professor Christopher Phillips, *Imperial College*, *United Kingdom*
Miss Claire Pickering, *University of Strathclyde*, *United Kingdom*
Dr Michael Pilling, *University of Manchester*, *United Kingdom*
Professor Luca Quaroni, *Institute of Nuclear Physics PAN*, *Poland*
Mr Ali Rae, *National Physical Laboratory*, *United Kingdom*
Miss Elisabeth Ratcliffe, *Royal Society of Chemistry*, *United Kingdom*
Mr Stan Remiszewski, *CIRECA Theranostics LLC*, *USA*
Dr Massimiliano Rocchia, *Thermo Fisher Scientific*, *Italy*
Mrs Sarah Rogers, *Royal Society of Chemistry*, *United Kingdom*
Dr Francesco Simone Ruggeri, *University of Cambridge*, *United Kingdom*
Dr Simona Sabbatini, *Università Politecnica delle Marche*, *Italy*
Professor Christopher Sammon, *Sheffield Hallam University*, *United Kingdom*
Mr Ángel Sánchez-Illana, *Neonatal Research Group, IIS La Fe*, *Spain*
Dr Zachary Schultz, *University of Notre Dame*, *USA*
Mr David Sebiskveradze, *MEDyC CNRS UMR 7369*, *France*
Professor Valter Sergo, *University of Trieste*, *Italy*
Dr André Filipe Couto Cardoso da Silva, *Faculdade de Ciências da Universidade de Lisboa*, *Portugal*
Dr Faris Sinjab, *University of Nottingham*, *United Kingdom*
Professor Ganesh Dhruvananda Sockalingum, *University of Reims Champagne-Ardenne*, *France*
Dr Kay Sowoidnich, *Central Laser Facility, Research Complex at Harwell, STFC Rutherford Appleton Laboratory*, *United Kingdom*
Miss Katie Spalding, *University of Strathclyde*, *United Kingdom*
Miss Katie Springham, *Royal Society of Chemistry*, *United Kingdom*
Professor Nick Stone, *University of Exeter*, *United Kingdom*
Mr Abdu Subaihi, *University of Manchester? United Kingdom*
Professor Josep Sulé-Suso, *Keele University*, *United Kingdom*
Dr Jakub Surmacki, *Department of Physics, University of Cambridge, Cavendish Laboratory*, *United Kingdom*
Dr Phillipa Timmins, *Princeton Instruments*, *United Kingdom*
Mr William Tipping, *The University of Edinburgh*, *United Kingdom*
Mr Damien Traynor, *Focas Research Institute, Dublin Institute of Technology*, *Ireland*
Miss Emma Upchurch, *Biophotonics Research Unit, Gloucestershire Hospitals NHS Foundation Trust*, *United Kingdom*
Miss Sarah Vanden-Hehir, *The University of Edinburgh*, *United Kingdom*
Mr Robbin Vernooij, *Monash University*, *Australia*
Professor Peter Weightman, *University of Liverpool*, *United Kingdom*
Mr Ludger Weisser, *Anasys Instruments*, *USA*
Professor Bayden Wood, *Monash University*, *Australia*

Mr Ben Woodhams, *University of Cambridge*, *United Kingdom*
Dr Satoshi Yamaguchi, *The University of Tokyo*, *Japan*
Mr Zhe Zhang, *University of Manchester*, *United Kingdom*
Dr Mihaela Zigman, *Max-Planck-Institut für Quantenoptik*, *Germany*

